Methods and applications of crystal structure prediction

University of Cambridge, UK

11–13 July 2018

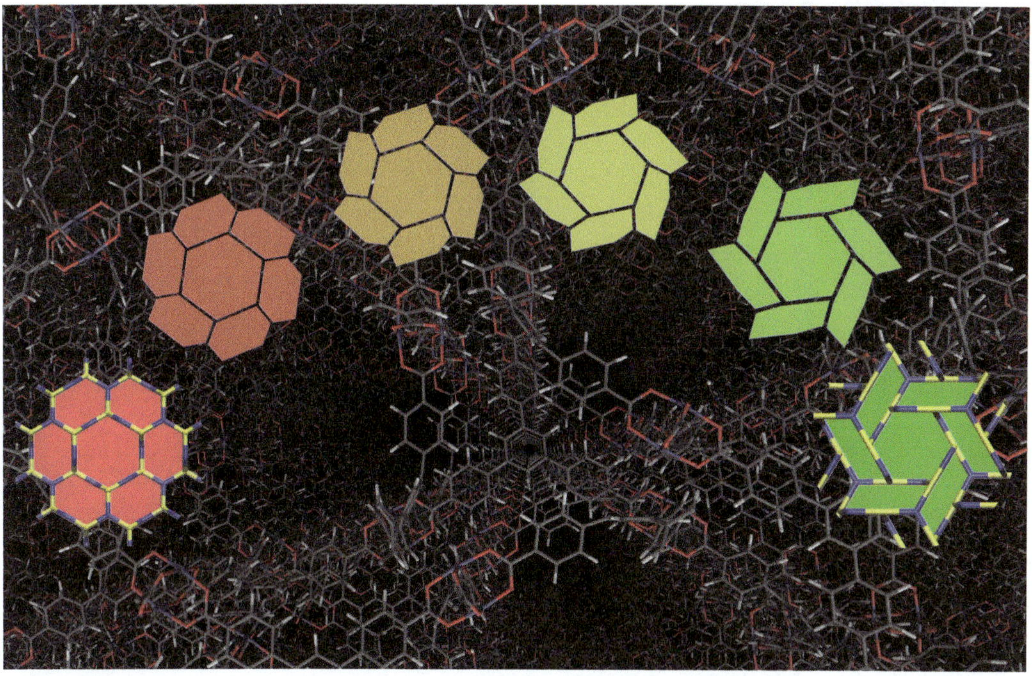

FARADAY DISCUSSIONS

Volume 211, 2018

ROYAL SOCIETY OF CHEMISTRY

The Faraday Division of the Royal Society of Chemistry, previously the Faraday Society, was founded in 1903 to promote the study of sciences lying between chemistry, physics and biology.

Faraday Discussions are unique international discussion meetings that focus on rapidly developing areas of chemistry and its interfaces with other scientific disciplines.

Faraday Discussions (Print ISSN 1359-6640, Electronic ISSN 1364-5498) is published 7 times a year by the Royal Society of Chemistry, Thomas Graham House, Science Park, Milton Road, Cambridge, UK CB4 0WF.

Volume 211 ISBN 13: 978-1-78801-170-9

2018 annual subscription price: print+electronic £1103 US $1940; electronic only £1050, US $1848.
Customers in Canada will be subject to a surcharge to cover GST. Customers in the EU subscribing to the electronic version only will be charged VAT.

All orders, with cheques made payable to the Royal Society of Chemistry, should be sent to the Royal Society of Chemistry Order Department, Royal Society of Chemistry, Thomas Graham House, Science Park, Milton Road, Cambridge, CB4 0WF, UK
Tel +44 (0)1223 432398; E-mail **orders@rsc.org**

If you take an institutional subscription to any Royal Society of Chemistry journal you are entitled to free, site-wide web access to that journal. You can arrange access via Internet Protocol (IP) address at **www.rsc.org/ip**

Customers should make payments by cheque in sterling payable on a UK clearing bank or in US dollars payable on a US clearing bank.

Printed in the UK

MIX
Paper from responsible sources
FSC www.fsc.org FSC® C013604

Methods and applications of crystal structure prediction

Faraday Discussions

www.rsc.org/faraday_d

A General Discussion on Methods and applications of crystal structure prediction was held at the University of Cambridge, UK on the 11th , 12th and 13th of July 2018.

RSC Publishing is a not-for-profit publisher and a division of the Royal Society of Chemistry. Any surplus made is used to support charitable activities aimed at advancing the chemical sciences. Full details are available from www.rsc.org

CONTENTS

ISSN 1359-6640; ISBN 978-1-78801-170-9

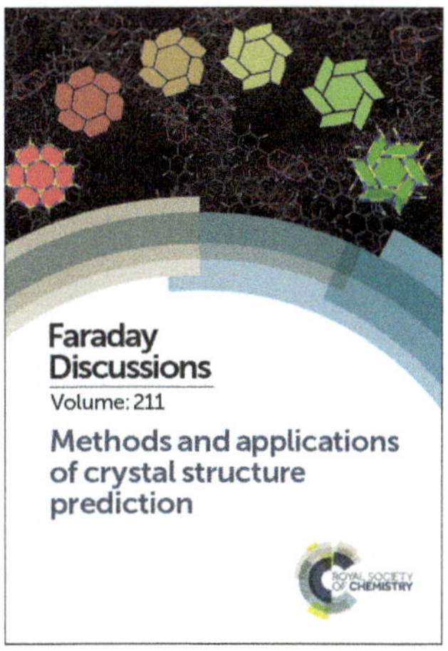

Cover
Front cover image courtesy of Prof. Dr Rochus Schmid.

An embedded topology is transformed by TopoFF to serve the needs of the building blocks, resulting in the MOF shown in the background.

See Rochus Schmid and Julian Keupp, *Faraday Discuss.*, 2018, **211**, 79–101.

INTRODUCTORY LECTURE

PAPERS AND DISCUSSIONS

CONCLUDING REMARKS

ADDITIONAL INFORMATION

Faraday Discussions

PAPER

Is zeroth order crystal structure prediction (CSP_0) coming to maturity? What should we aim for in an ideal crystal structure prediction code?

Sarah L. Price

Received 13th June 2018, Accepted 2nd July 2018
DOI: 10.1039/c8fd00121a

Crystal structure prediction based on searching for the global minimum in the lattice energy (CSP_0) is growing in use for guiding the discovery of new materials, for example, new functional materials, new phases of interest to planetary scientists and new polymorphs relevant to pharmaceutical development. This Faraday Discussion can assess the progress of CSP_0 over the range of types of materials to which CSP is currently and could be applied, which depends on our ability to model the variety of interatomic forces in crystals. The basic hypothesis, that the outcome of crystallisation is determined by thermodynamics, needs examining by considering methods of modelling relative thermodynamic stability not only as a function of pressure and temperature, but also of size, solvent and the presence of heterogeneous templates or impurities (CSP_thd). Given that many important materials persist, and indeed may be formed, when they are not the most thermodynamically stable structure, we need to define what would be required of an ideal CSP code (CSP_aim).

Introduction

Intellectual curiosity as to whether we can predict crystal structures predates the computer era, with Kitaigorodsky's mechanical structure seeker fitting the "projections" (bumps) formed by the atoms of one molecule into the hollows of another so that the molecules dovetailed into a close packed structure, or various radius ratio rules for simple inorganic crystals, that also are based on the principle of close packing. Maddox's famous quote[1] in 1981, "One of the continuing scandals in the physical sciences is that it remains in general impossible to predict the structure of even the simplest crystalline solids from a knowledge of their chemical composition", can be seen as reflecting the common assumption at the time, that a given molecule or ionic composition always crystallised in the same crystal structure. The term Crystal Structure Prediction (CSP) originates

Department of Chemistry, University College London, 20 Gordon Street, London WC1H 0AJ, UK. E-mail: s.l.price@ucl.ac.uk

from the days when they were seeking to predict the crystal structure, which was that of the first crystal that could be grown to a size and quality suitable for a crystallographic determination. The crystallography was usually done to prove that the correct molecule had been synthesised. As McCrone said in 1965 "In spite of the fact that different polymorphs of the same compound are, in general, as different in structure and properties as the crystals of two different compounds, most chemists are almost completely unaware of the nature of polymorphism and the potential usefulness of knowledge of this phenomenon in research".[2]

Computers make it possible to test whether we really understand what determines how a molecule will crystallise, by programming a theory and applying it to test the "predictions" against experiment. There is a big distinction between a theory or program that seeks to predict the crystal structure and one which seeks to predict all polymorphs. This distinction needs to be clear in this discussion. Most CSP programs are based on the theory that the crystal structure is the most thermodynamically stable structure, and, at least initially, assume that the relative thermodynamic stability can be approximated by the lattice energy, the energy of the static lattice relative to infinitely separated molecules in their lowest energy conformation (or relative to all electrons and nuclei at infinite separation).

I would like to define CSP_0, the zeroth order model, as the attempt to predict the most likely (thermodynamically stable) crystal structure as the most stable in lattice energy. That some of the competitive local minima in the lattice energy correspond to polymorphs is fortunate for the practical interest in CSP.

We can then define CSP_thd as the full implementation of the assumption that realistically estimates the relative thermodynamic stability of the crystal structures, and so will predict the most thermodynamically stable crystal structure under given conditions of temperature and pressure and any other thermodynamic variables that are relevant. The pharmaceutical industry would really welcome such a code, but currently the calculation of the phase diagram of crystalline methanol represents a major step forward in this regard.[3] The situation is more advanced for other materials, with a revision of the phase diagram of $CaCO_3$ over half the pressure range within the earth's mantle,[4] leading to the discovery of a new phase with implications for carbon storage in the deep mantle.[5] Differences in the approximations required for estimating the appropriate relative thermodynamic stability for different material types are currently important.

If we calculate the free energy minima by calculating the true thermodynamic stabilities of the CSP_0 structures, we expect that some lattice energy minima will have merged into the same free energy minimum. This structure could have a higher symmetry on average than any lattice energy minimum, if it is a dynamically disordered structure. The structures that are within the likely energy range of possible polymorphism constitute the crystal energy landscape, the set of thermodynamically plausible structures. Thus CSP_thd methods will undoubtedly improve the prediction of polymorphism at practically relevant temperatures, but how do we define which local minima on the crystal energy landscape are possible, practically relevant polymorphs?

One practical justification for developing CSP methods is in the design of new materials, to guide synthetic work. The main benefit of such studies is to avoid the synthesis of molecules or materials that will not readily crystallise in a structure which has the desired properties. This requires the calculation of the property of interest from the computer-generated structures, with sufficient realism to convince the experimental group that the synthetic effort is worth making. If more than the thermodynamically stable structure is of interest, which will generally be the case if metastable polymorphs (or variants in chiral composition) are likely to complicate product manufacture, then the property needs to be calculated for all the structures on the crystal energy landscape, giving crystal-structure–property maps. CSP_0 may be adequate for excluding an organic molecule from the synthetic program, but CSP_thd is desirable to test whether the desired novel organic material is the most thermodynamically stable at ambient. Other materials can survive over a much greater range of temperatures and pressures, and for these CSP_thd can be essential, with pressure being particularly widely used for planetary science.

The second practical justification for developing CSP is polymorphism. The importance of polymorphism for the quality control of industrial materials, which is particularly acute for the pharmaceutical industry, has inspired the development and commercialisation of CSP. However, intellectually, it messes up the ability to test the theory for predicting crystal structures, as it is unclear what crystal structures have to be predicted. Conclusions from the Cambridge Crystallographic Data Centre's blind tests of organic crystal structure prediction[6] have always been qualified by the possibility of unreported polymorphs. Over the lifetime of CSP, the question has changed from which systems are polymorphic to which are not. Systems, such as aspirin, that used to be quoted as examples of monomorphic systems now have polymorphs in the Cambridge Structural Database (CSD).[7]

It is now fifteen years since a group of UK scientists convinced Research Councils UK that a computational method of crystal structure prediction would be a "Basic Technology" for many areas of science. It is gratifying that one of our industrial contributions raises the question as to whether CSP is changing from basic science to applied technology (DOI: 10.1039/c8fd00033f). I would like to start this discussion meeting by asking a few questions about the state of CSP_0 and CSP_thd and considering what we would expect of a genuine crystal structure prediction code (CSP_aim) in terms of polymorph prediction. Here we must recognise that over-predicting polymorphs is not helpful, though it is more useful than having no computational guidance about the completeness of the experimental polymorph screen. Should we be aiming for a code that only produces the polymorphs that can be experimentally found, along with sufficiently reliable predicted properties to ensure that they could be found or safely dismissed as irrelevant?

The other dimension to this discussion is the extension of CSP methods to the widest range of materials. The ability to predict the crystal structures of benzene will not guarantee that all pharmaceuticals can be predicted by the same methods, any more than a CSP method that works for NaCl will work for zeolites. We now are developing CSP for many new forms of materials such as MOFs (metal organic frameworks), organometallics, and anything that the synthetic materials chemist can develop, including interface structures. This discussion should exchange ideas between the CSP communities working on different types of materials. The differences arise from the nature of the building blocks being used, and different interatomic interactions being balanced. A lot of the challenges in organic CSP for flexible molecules come from trying to balance the different interactions within the molecules, such as flexible torsions, with the variety of possible intermolecular interactions. When the molecules are linked by a great diversity of atoms, such as metals, the distinctions between inter- and intramolecular, ionic and covalent, oxidation and spin states may become less clear-cut and so cause inaccuracy for a given type of lattice energy modelling. As the coverage of the periodic table increases, the possibility of chemical substitutions in the lattice increases, as different ions can be exchanged with less effect on structures and relative energies than changing organic functional groups. When CSP is applied to geological materials there is a problem of defining the composition. The different classes of materials have different plausible starting assumptions in CSP, such as (approximately) rigid building blocks, but the complexity and difficulty comes as we combine different types of interactions of different strength and directionality. Hence "variety of interatomic interactions" is a major aspect of the material complexity axis on a two dimensional diagrammatic scheme on which to assess our progress as shown in Fig. 1.

Dirac's famous quote, "The underlying physical laws necessary for the mathematical theory of a large part of physics and the whole of chemistry are thus completely known", is essentially saying that we have the physical laws for a universal CSP code. Developments in computer power are allowing us to make huge strides in applying CSP_0, however we are all limited by resources in testing the theoretically more correct CSP_thd. How long will it be necessary to make material and application-dependent short cuts? Against the huge ambition of a universal CSP_aim code, we can plot our progress on Fig. 1. At each point, we have to tension reliability for all materials in that class against McCrone's famous

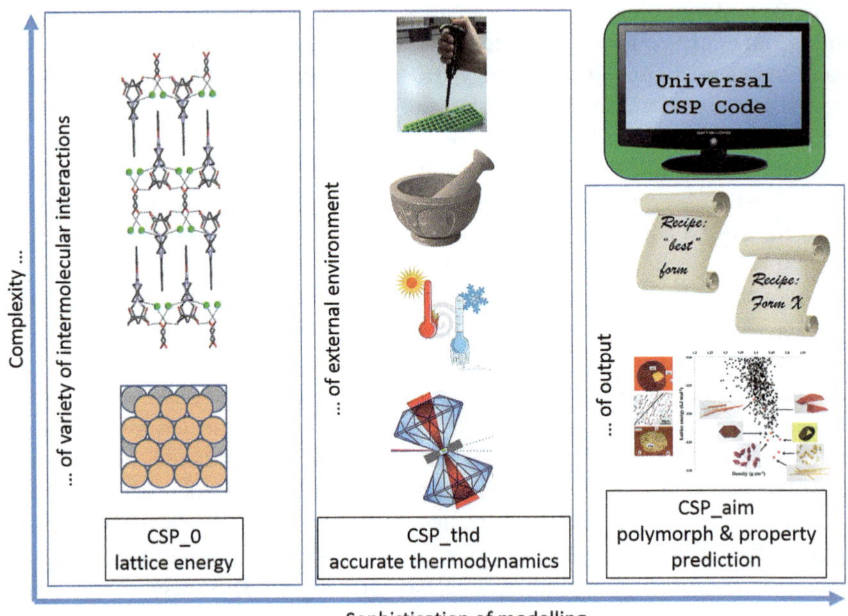

Fig. 1 Schematic of the two directions of progress required for a universal crystal structure prediction code, which predicts only all the experimentally observable polymorphs of a system, and sufficient properties to enable them to be found.

quote "every compound has different polymorphic forms, and that, in general, the number of forms known for a given compound is proportional to the time and money spent in research on that compound".[2]

CSP_0, finding the structure that has the lowest lattice energy

Accurate determination of the relative lattice energies

Most theoretical chemists will see an analogy between Fig. 1 and the diagram for systematically improving approximate wavefunctions as a function of basis set size (one axis) and electron correlation (the other axis) to reach true solution to the Schrödinger equation for a molecule or solid. Density functional methods have lost the advantages of the variation theorem, but there is still a hierarchy of methods[8–10] and the methods widely used in CSP are on the lower rungs. The hierarchy of what method of calculating the lattice energy is good enough for CSP is very system dependent. To what extent is the quantum mechanical method good enough for the type of material that the final CSP_0 energies can be relied upon as being accurate enough? For inorganic systems, are the possible variants in oxidation states, spin states and even relativistic effects being adequately represented? For some materials, theory shows that there is no hope of a worthwhile CSP study without evaluating the lattice energy by using an electronic structure theory accuracy calculation on all competitive structures (Ψ_{crys}).

Electronic structure theory methods suffer from the intimate link between the dispersion energy and electron correlation. Hence a high level of theory is required to incorporate dispersion realistically *ab initio*. Modelling the dispersion (DOI: 10.1039/c8fd00066b) is very important for organics, particularly if you are contrasting crystal structures which have only dispersion between layers with those that have stronger forces, such as hydrogen bonding in all three directions. Structures such as molecular ionic cocrystals (DOI: 10.1039/c8fd00036k) have many competing forces. This can lead to complex systems *e.g.* a pharmaceutical salt which exhibits many solid forms, including a series of alcohol solvates, where the alcohol interdigitates between the ionic and molecular layers.[11] At one level this is a simple series of salt solvates, but there are subtle differences that can give rise to disorder in the alcohol hydrocarbon tail layer (causing problems in crystal growth and characterisation) and disorder in Cl^- ions only detectable with the most accurate crystallography. In addition, this study that started from a CSP on the pharmaceutical salt, was further complicated by the late appearance of a highly metastable polymorph.[11] This illustrates the issue of the level of atomic detail and energetic accuracy at which we want to predict crystal structures, in systems where there can be considerable stabilisation by solvents in a disordered form (DOI: 10.1039/c8fd00031j), or multiple similar ions can occupy the same position. The level of atomic detail required is intimately linked with the quality of the lattice energy model needed to accurately balance all the different types of intermolecular interactions present.

We now know that calculating the lattice energy by electronic modelling (Ψ_{crys}) is very expensive, if it will be applied to a sufficient number of structures. We have different approaches to this challenge. One innovative approach is to use data-driven learning of the potential energy surface. There are applications of this approach to the CSP of medium and large-sized boron clusters (DOI: 10.1039/c8fd00055g) and to phosphorus (DOI: 10.1039/c8fd00034d) starting our conference.

The more traditional approach is to have a hierarchical scheme, increasing in accuracy as it is possible to focus on the more plausible structures. This needs care but CSP_0 is a valuable test for the developers of different lattice energy codes, from the variety of lattice energy models that can be used, to the vital details that improve efficiency and accuracy, such as use of symmetry, optimisation methods, lattice summations and convergence criteria. We should be careful. Tests made on the X23 set of molecular crystal lattice energies, such as the recent one using 4 codes and 18 functionals[12] are valuable, but this is a data set of very small, virtually rigid molecules. For pharmaceuticals or other molecules with significant flexibility and considerable variation in the intramolecular dispersion and packing density, then the reliability of different methods could be different (DOI: 10.1039/c8fd00010g). Balancing the effects of flexibility/conformational change with the other forces is a problem for a wide range of materials, from pharmaceuticals, to novel inorganic materials such as ultra-flexible boron oxide frameworks (DOI: 10.1039/c8fd00052b) but the forces differ according to the types of atoms and structures involved.

The demands of the search mean that many methods will use atomistic modelling somewhere in the hierarchy, with an appropriate force-field. The accuracies of force-fields for different applications vary according to the crystals, but they are widely used for ionic and molecular systems. The use of molecules or fixed ionic groups makes a lot of sense for restricting the search to the molecule or structural type of interest. For molecular systems, it seems that a molecule-specific electrostatic model is usually essential. For flexible molecules, this requires wavefunction calculations on the molecule covering the possible range of conformations in the crystal (Ψ_{mol}), which also provides the conformational energy penalty. Atomistic modelling of molecular crystals can then be performed with the addition of a set of empirically fitted repulsion–dispersion intermolecular atom–atom potentials. We are now revisiting the early work of Williams in empirically parameterising such atom–atom potentials for organics, with Day's group recently showing that such parameterizations combined with distributed multipoles can rival popular DFT-D methods in accuracy.[13] A new scheme is being proposed for empirical repulsion–dispersion potentials for use with a specific level of Ψ_{mol} (DOI: 10.1039/c8fd00064f). Empirically fitted potentials have the advantage of absorbing some of the approximations into the parametrisation, but that can also be a disadvantage in preventing extrapolation to other regions, *e.g.* predicting a high pressure phase of pyridine required a non-empirical anisotropic atom–atom intermolecular potential.[14] We also risk double counting thermal effects when these potentials are used in Molecular Dynamics (MD) approximations to CSP_thd. On the other hand, empirically fitted potentials, or other CSP schemes that are based on experimental crystal structures, could be more effective than CSP_0 from implicitly absorbing non-thermodynamic effects, *i.e.* partially moving towards CSP_thd.

Any system of CSP_0 is reliant on relative lattice energies, and relies on cancelation of errors between different crystal structures of the same system. This accounts for the huge successes that have been achieved by CSP_0, when the absolute lattice energies may be very inaccurate and the Ψ_{crys} or Ψ_{mol} calculations are so far from even state-of-the-art approximate solutions to the real wavefunction. Hence the need to be aware of the theoretical basis of the different approximations used as we increase the variety of types of interatomic interactions in our crystals.

Search coverage in CSP_0 structure generation

Choices are made in defining the search space to be covered. We can use sophisticated methods to try to ensure that the search is complete within a defined range, but a lack of computer and human time often prevents this limit of confidence being reached. Refining our methods to make them more efficient can potentially extend the scope of CSP studies, but to what extent are these refinements universal or material dependent, for example using genetic algorithms for molecular crystals (ref. 15, DOI: 10.1039/c8fd00067k) or zeolites (DOI: 10.1039/c8fd00035b)? Recognised patterns in the appropriate class of materials may be incorporated into the structure generation methodology very explicitly for the class of materials, from oxidation states (DOI: 10.1039/c8fd00032h), to properties at given site-symmetries in zeolites (DOI: 10.1039/c8fd00040a), optimised blueprints for MOFs (DOI: 10.1039/c8fd00051d), or probe structure for the inorganic compositional space (DOI: 10.1039/c8fd00045j). Ensuring that this provides an efficient encapsulation of the possible chemistry without restricting the ability to propose exciting new materials, particularly those that bridge traditional classes, is the real challenge in improving the complexity of structures being created first in the computer by CSP (Fig. 1). We need to ensure that assumptions made for computational convenience do not become outdated. For example, when CSP_0 was first instigated for organic molecules, the assumption of $Z' = 1$ (one molecule in the asymmetric unit cell) was more reasonable than it is now, as more $Z' > 1$ structures are determined from better crystallographic methods and $Z' > 1$ is common for metastable polymorphs.[16] Disorder plays a different role in different types of materials, but it is notable that one prediction made at the CCDC's 50th party[17] was that in another 50 years, all organic crystal structures would be disordered. Does using data on known crystal structures (such as the Cambridge Structural Database)[18,19] risk bias from sociological influences on research and reporting or historic limitations in analytical capabilities?

We need to be careful in defining our search and the expectations of users of the CSP codes. There is a history of claims being made that CSP has shown that all the polymorphs of a molecule are already known, which is what industry would love the codes to be able to do, but the earliest commercial CSP code had problems when claiming this for paracetamol, when many experimental polymorphism experts were aware of the literature evidence of another form. Subsequent CSP work has identified models for form III of paracetamol,[20] and there are now claims of more polymorphs formed under pressure[21] requiring a structure determination. More recently, CSP led to the experimental finding of the room temperature stable form of creatine[22] whereas an earlier CSP had helped determine the first structure and claimed that there would be no polymorphs.[23]

Currently, the most effective search strategy and the type of lattice energy evaluation that is good enough is very dependent on the type of material being studied. The shortcuts allowed by choices of composition and stoichiometry also differ – different elements can substitute into the same inorganic crystal structure, but even the best functional group swap favoured by crystal engineers, methyl for chloro in organic molecules, leads to a change of crystal structure unless there are no close or strong (*e.g.* hydrogen bonding) interactions involved at the substitution site.[24] The importance of polymorphism appears to differ between materials, though this may reflect the range of crystallisation/synthesis methods that can be applied, or the motivation for experimental screening for polymorphs. However, the ease of computation means that CSP_0 is applied to the widest range of materials. Whilst generally focussing on the global minimum in the lattice energy, we need to look with care at the energy gap to alternative structure(s) to determine whether it is within the likely accuracy of the method, *i.e.* a reasonable approximation to CSP_thd. For organic crystals, this is sometimes the case, but this monomorphism does require a uniquely good packing defining the structure in all three dimensions, which will rarely be true of all members of a class of materials that adopts low symmetry crystal structures.

CSP_thd; defining the crystal energy landscape

CSP_thd is calculating the thermodynamics accurately, so that you are genuinely predicting the most thermodynamically stable crystal structure at a given set of thermodynamic conditions. If thermodynamics was the only determinant of crystal structures, then the global minimum in CSP_thd at ambient would be the only structure observed at ambient conditions. Since this is often not the case, with strictly metastable polymorphs being sufficiently long-lived to appear to be stable, we interpret the crystal energy landscape, the set of all thermodynamically plausible structures, and expect that this would show all the polymorphs that would be stable under those thermodynamic conditions.

How this relates to the assumption that CSP_0 should be adequate, as a starting point, can be seen by considering the thermodynamics of chiral separation by crystallisation, *i.e.* the ability to predict when you would get a mixture of enantiopure crystals from a racemic solution. Using the sublimation thermodynamic cycle, where you separate your molecules by subliming the crystal and then solvate them to form the solution, we can predict the solubility ratio of the racemic crystal (RS) and the enantiopure crystal (S):

$$\alpha = \frac{x_{\text{RS}}}{x_{\text{S}}} = \exp\left(-\frac{\Delta_{\text{RS-S}}\Delta G_{\text{solv}}}{RT}\right) \approx \exp\left(-\frac{\Delta_{\text{RS-S}}\Delta G_{\text{subl}}}{RT}\right) \approx \exp\left(-\frac{\Delta_{\text{RS-S}}\Delta H_{\text{subl}}}{RT}\right) \approx \exp\left(\frac{\Delta_{\text{RS-S}}E_{\text{latt}}}{RT}\right)$$

Since the solutions are composed of the same molecule (apart from 50% of the RS solution being the mirror image molecule) the energy of solvation should be the same, assuming the solutions are ideal or non-ideal to the same extent, so the solubility difference is determined by the crystal thermodynamics. Assuming the thermal entropy terms are the same, the free energy of sublimation differences reduce to the enthalpy of sublimation differences, and that, ignoring heat capacity and zero-point energy differences, leads to the lattice energy determining the solubility ratio. Thus CSP_0 covering both chiral and enantiopure space groups provides the most stable structure for each to determine whether chiral separation is possible. If this ratio is turned into the eutectic excess, then the thermodynamic prediction of the crystal that is formed is shown to be very sensitive to the approximated energy difference, as shown in Fig. 2.

The steepness of the curve in Fig. 2 shows that if the lattice energy difference is large, you can confidently predict which crystal will form by CSP_0, but if it is small, the outcome is horribly dependent on the energy difference and hence the cancellation of errors in all the other energy terms. Our recent comparison of measured heat capacities at low and ambient temperatures for three pairs of enantiopure and racemic crystals of diverse organic molecules, and other measured thermodynamic quantities,[25] shows that the approximations listed above are not good enough for any of these molecules. The contributions differ in their sensitivity to the molecular and structural differences, particularly the extent to which the molecular vibrational frequencies are unchanged on crystallisation. Differences in the hydrogen bonding motif and consequent frequency shifts in the IR spectra between the two crystals, can lead to a temperature dependence of the heat capacity difference around ambient. The performance of our attempts to predict thermal corrections using the harmonic approximation,[25] using rigid molecule Ψ_{mol} or Ψ_{crys} methods similar to those used for larger systems (DOI: 10.1039/c8fd00010g), shows the challenge of predicting free energies or relative solubility at a useful accuracy.

These arguments also apply to polymorphs and can be extended. What are the free energy difference implications of thermal expansion properties (DOI: 10.1039/c8fd00048d), which can differ so much in their anisotropy between polymorphs or systems? Correcting experimental quantities back to give an "experimental" lattice energy, as done in benchmarking set X23, will be limited by

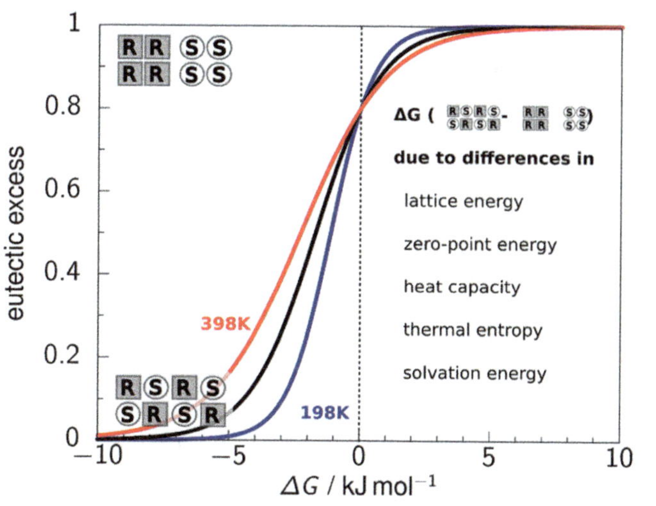

Fig. 2 The proportion of enantiopure and racemic crystal structures as a function of their energy difference $\Delta G = \Delta G(\text{RS}) - \Delta G(\text{S})$, plotted for three temperatures. The asymmetry comes from the definition of one mole of the racemic crystal corresponding to 1/2 a mole of each enantiomer, as used often in experimental work[25] but not in CSP where the reference state is a mole independent of chirality.[26] The spread of energies that changes the crystallisation outcome is not dependent on the reference state.

the accuracy of the calculated[27] or experimental[28] thermal corrections (which are nevertheless a big recent improvement on the $2RT$ correction that initially justified the use of CSP_0 for chiral resolution). The cancellation of the small thermodynamic terms between different structures of the same compound will be very dependent on the structural differences.

Perhaps a more significant implication of Fig. 2 is a limit on the energy differences between polymorphs obtained by solution crystallisation, which is comparable to the estimated range of polymorphic lattice energy differences of observed polymorphs,[29] and the cutoffs often used for progressing CSP_0 to more demanding calculations. This can mean that polymorphs obtained by other methods, for example by desolvating solvates[11] or the solid state synthesis of the molecule, could be much more metastable than polymorphs obtained by solution crystallisation. The concept of a crystal energy landscape requires a cut-off of thermodynamic plausibility. If this is dependent on the methods of crystallisation that can be applied, then it could be a context-determined parameter.

There are other thermodynamic contributions that should be included, given that real crystals are not perfect and infinite, such as disorder. Configurational disorder is an energy term which has been estimated to account for why the low temperature phase of caffeine is statically disordered.[30] Going from CSP generating potential disorder components to realistic modelling of the thermodynamics of disorder, let alone estimating disorder that is not thermodynamic but frozen in during crystallisation, is important for pharmaceuticals (DOI: 10.1039/c8fd00072g), and yet order–disorder phase transitions are challenging to experimentally characterise or model (DOI: 10.1039/c8fd00042e). Some forms of disorder, such as polytypism, imply that coverage of the search space will never be complete.

The size of a crystallite and the presence of solvent have recently been shown to be important in thermodynamic stability by the demonstration of a reversible cross over in relative stability of the polymorphs of a cocrystal and an aromatic disulphide compound in liquid-assisted grinding experiments.[31] These were rationalised by considering solvent-dependent surface energies relative to the bulk lattice energy. The effect that size can have on the relative stability of nuclei of different forms comes out clearly from consideration of the nano-cluster modelling of inorganic oxides (DOI: 10.1039/c8fd00060c).

The development of accurate calculations of the relative thermodynamic stability for CSP_thd will be a great step forward for the theoretical modelling of important properties, such as the solubility, morphology and stability range for pharmaceuticals. Doing this accurately for all properties related to thermodynamics for all known phases, let alone the entire crystal energy landscape, provides a significant aim for theory and computer modelling. However, a major benefit of the move from CSP_0 to CSP_thd is to eliminate structures that are artefacts of the use of a static lattice in CSP_0, *i.e.* are not minima in the free energy crystal landscape. The number of lattice energy minima that are eliminated by some form of molecular dynamics, from just a short shake-up[32] to a metadynamics treatment,[33–35] is very dependent on the molecule and structures concerned. A realistic CSP_thd should produce the more symmetrical structures that arise from dynamic disorder averaging over a variety of lattice energy minima (such polymorphs do not correspond to a lattice energy minimum and so are not produced by CSP_0). Experimental identification of such plastic crystals may be aided by simulations, allowing insight into the energetic and mechanistic forces that drive order–disorder phase transitions in plastic crystals (DOI: 10.1039/c8fd00042e). There may be experimentally observed intermediate phases that we may feel are beyond the scope of CSP, for example, CSP gave the low temperature ordered phase III of cyclopentane, and MD the hexagonal plastic phase I, but the size of the MD cell probably contributed to the inability of the dynamical simulations to get good agreement with the experimental powder pattern of the intermediate plastic phase II.[36] The development of sophisticated MD based methods to explore the free energy surface and study phase transitions, such as meta-shooting (DOI: 10.1039/c8fd00053k) will improve our understanding of the effects of the potential energy surface of CSP_thd.

CSP_aim

What do we consider that a genuine crystal structure prediction code (CSP_aim) should do?

Most people would want a CSP code to predict all practically important crystal structures that could be found for a defined system, and those involved in material synthesis or solid form screening would also want a recipe for how to find them. Full CSP_thd would be adequate for systems where you expect to form the most stable structure at a given set of thermodynamic conditions, so experimentally you should just move to the relevant thermodynamic conditions.

CSP is undoubtedly leading to an increase in the number of polymorphs whose structures are now known. The use of CSP has helped to solve the structures of extremely fibrous polymorphs of coumarin grown from the melt,[80] new needle forms of resorcinol[37] (another example of a new form being found despite published predictions that there were no more forms), and another form of aspirin[38] from the melt. The system ROY, with its various colours, is one where it has long been known that there are further uncharacterised forms[39] that exist at ambient, but CSP produces too many candidates.[40] The use of CSP to increase the number of polymorphs that have their structures determined, and the challenges involved will come up in discussion of the 8th solved structure (DOI: 10.1039/c8fd00039e). Can CSP also, by proposing new experiments or encouraging more detailed analysis, increase the number of polymorphs that are detected (DOI: 10.1039/c8fd00069g), as well as aiding their characterisation?

Disappearing polymorphs, where there is an inability to maintain control of the crystallisation of an apparently stable form after a more thermodynamically stable form appears, have been a major justification for developing CSP because of their practical importance to the pharmaceutical industry.[41] CSP_0 must generate any structure that is significantly more stable than the known forms, as well as the known polymorphs, with CSP_thd helping confirm the relative thermodynamic stability at ambient. However, the scientific rationalisation of disappearing polymorphs would conclude that you ought to be able to reproduce any polymorph that has ever been observed, using identical crystallisation conditions, even if this requires a fresh laboratory and student[41] to avoid seeding. Having identical crystallisation conditions may be practically unachievable, as a 50 year old sample of the disappearing form II of progesterone contained a cocktail of impurities not found in the modern material.[42] How does the issue of elusive or disappearing polymorphs, and the effects of impurities on the polymorph formed, impact the computational prediction of metastable polymorphs?

This is illustrated by the new polymorph (γ) of succinic acid that was found in an attempt to purify a synthesiser-made peptide from the resultant soup of impurities, by cocrystallisation with succinic acid (Fig. 3).[43] We were challenged to predict the structure. It was readily generated as a metastable polymorph by CSP. Periodic DFT-D lattice energy calculations had γ slightly more stable than the other known polymorphs for most dispersion corrections, though using the PBE + TS model, harmonic phonon estimates of the free energy differences made the β form more stable at room temperature. This closeness in energy is consistent with the γ form being found concomitantly with β in the failed cocrystallisation experiment, and illustrates the improvement of CSP_thd on CSP_0. This case is relevant to defining CSP_aim because the γ form has not been crystallised again, despite a large series of experiments testing the different impurities.[43]

The serendipitously-found γ crystal is a conformational polymorph of succinic acid, with the conformation that NMR and MD simulations in water show as being dominant in solution.[43] MD on the γ form does not give a solid state transformation, but reproduces the structure with a small change in the monoclinic angle, which, if forced by metadynamics, leads to extensive defect formation. That is the best explanation we can find to the problem of reproducing the new polymorph: the γ form can be rather susceptible to defect formation that may lead to a ready transformation to the β form. Perhaps γ succinic acid will be found again, and it is in the CSD as a challenge to anyone who can design a method of finding it. The high-temperature α phase of succinic acid was once considered elusive at normal temperatures, but has been found as a contaminant after grinding[44,45] and in liquid- (but not air-) segmented flow crystallisation[46] or by spray-drying from water.[47] This suggests that the α polymorph is observed when it forms first and there has not been sufficient time for the solvent mediated transformation to β. Gradually we are beginning to see the effects of kinetics and the role of impurities in catalysing or inhibiting polymorphic transformations. The serendipitous observation of γ succinic acid raises the question of what is wanted from CSP_aim: a recipe for producing the γ form and predicting its stability as a function of crystal size? If it had not been observed, would its apparent feasibility in the computational modelling have appeared to be an over-prediction of polymorphs? An early CSP study had noted that the planar conformation of succinic acid was less stable than that in the γ form, but had used the dominance of the planar conformation in the many crystal structures of

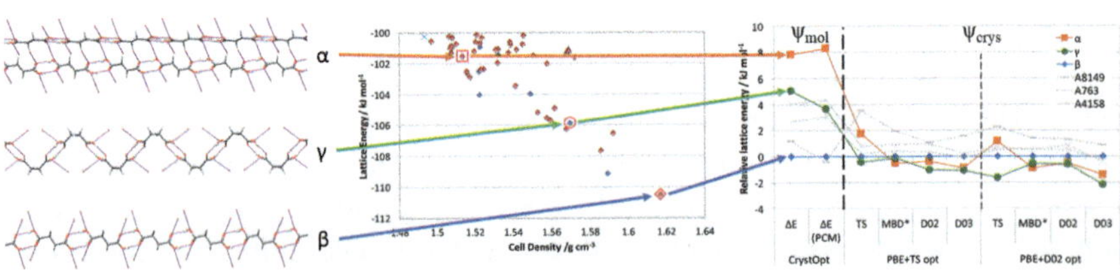

Fig. 3 The three polymorphs of succinic acid, showing the prediction of the novel γ form by CSP_0 using a Ψ_{mol} approach, and the relative lattice energies by a variety of Ψ_{crys} methods. Despite its relative thermodynamic stability, the γ polymorph has only been crystallised once in a failed cocrystallisation experiment.[43]

succinic acid, and a very limited search in the alternative conformation, to dismiss the possibility of any conformational polymorphs.[48]

Thus, if we are to achieve CSP_aim, we need to work with a knowledge of the variables in crystallisation/synthesis conditions that can lead to the formation of different structures. This will be very much a function of the types of materials and the range of interatomic interactions involved, *i.e.* going up the vertical axis of Fig. 1 and will essentially encapsulate the whole field of crystal engineering into our CSP theories and eventually codes. The possible synthetic methods in every field are evolving rapidly, and CSP plays the crucial role of predicting structures that should be achievable, and there may be structure dependent methods that can help find them. For example, CSP has inspired an isomorphous heterogeneous solution seeding experiment to produce a CSP predicted cocrystal which could not be made in four expert labs without the seeds initiating the first cocrystallisation.[49] Metastable pharmaceutical polymorphs have been successfully targeted by sublimation onto an isomorphous template crystal.[50] Will we ever have the control to be able to make a crystal by placing each atom in the position corresponding to the CSP-generated desired crystal structure? This is more possible for inorganic materials than pharmaceuticals. The pharmaceutical industrial scientists recognise that there is no such thing as a standard solid form screen; not only can the basic workflow vary between companies,[11,29] but also between molecules from the same drug discovery program, *e.g.* if one is able to obtain an amorphous starting material for one, and not for the other[51] then there may be a difference in the extent to which the starting point of the crystallisation has lost the memory of the input material structure.

This sensitivity of crystal structure outcome to exact material synthesis conditions probably extends to all types of materials, which is why CSP is most effective when there is very close collaboration between the people doing the calculations and those in the laboratory. In the aim for a code that can predict all the crystal structures that can be found, but not those that could never be made, we will have to consider relative nucleation and growth rates. Nucleation has been the focus of another recent Faraday Discussion.[52] There is an increasing realisation of the variety of mechanisms for crystallisation by particle attachment, and how this can vary in synthetic, biogenic and geological environments.[53] The aspect of whether crystallisation occurs by monomer attachment, let alone according to the models of classical nucleation theory, is up for debate and observation. Atomic force microscopy (AFM) showed dense hillocks forming at ledges of the dominant surface of olanzapine in water.[54] This two-step nucleation process, with a dense disordered solution forming, can be rationalised as the suspected growth unit (a dimer) has a variety of attachment sites on the ledge that are more stable than docking in the crystallographic site on the surface. The growth unit will not often dock directly into the most stable crystallographic ledge site when it can get caught in so many ledge, or solute–solute, and solute–water complexes. The AFM also showed that the dense blobs on the surface could grow into orientated crystals of the thermodynamically stable olanzapine dihydrate D, whereas dissolution of olanzapine in water usually resulted in the metastable dihydrate B. We can computationally rationalise the better registry (in the correct orientation) of a small nanocluster of the stable dihydrate D on the specific olanzapine anhydrate surface than the metastable dihydrate B, representing the support the surface gives to nucleating one form over another.[54]

As we consider interfaces and growth rates, the problem of distinguishing thermodynamics from kinetics gets worse. Morphology predictions based on minimising the surface energy give unrealistically spherical structures (consistent with classical nucleation theory), which led to the attachment energy model, which is strictly only appropriate for vapour grown crystals if all surfaces are below their roughening temperature. The attachment energy model is the cheap model for morphology–structure–energy landscapes, with relative growth volume rates being a possible means of eliminating unlikely polymorphs.[55] Predicting the morphology in different solvents uses models based on kinetic rates of attachment, dissolution, loss of solvent shell *etc.* derived from MD.[56] If we need to be reliant on MD to predict nucleation, growth and transformation rates, we are starting to enter the realm of multi-scale approximations to solving the appropriate nuclear and electronic time-dependent Schrödinger equation for the real system. We have lost the simplification of distinguishing between thermodynamics and kinetics.

The question of the extent to which CSP_aim will need to consider kinetic effects arising from differences in the experimental growth conditions is very problem dependent. The organic electronic community are obviously interested in substrate-induced polymorphs.[57] Are these cases where full thermodynamic treatment including the interfacial effects would predict these structures? Concomitant polymorphs raise further questions[37,58,59] as to whether we could ever devise a computational method that would allow the confident prediction of the conditions needed to crystallise phase pure samples. Extensive crystallisation work could not obtain phase pure samples of the metastable forms II and III of olanzapine. After the structure of form II was solved from a serendipitous single crystal and CSP suggested a structure for form III, the degree of similarity, (differing only in the stacking of layers), made it rather improbable that the crystallisation could ever be controlled to crystallise the forms separately.[60]

The question as to how changing crystallisation or synthesis conditions can affect which structures are formed, and how readily they transform to the most thermodynamically stable structure at the specified temperature and pressure, brings us back to the original assumption behind CSP. Does the thermodynamically most stable form have to be obtainable? It is the central tenet of CSP methods, and yet when so many molecules are difficult to crystallise at all, the first structure formed is quite unlikely to be the most stable by Ostwald's rule.[61] (Indeed, if CSP could guarantee that a pharmaceutical would never crystallise, so that the amorphous form could be confidently developed, that would be a valuable application.) It is quite conceivable that the most stable form could be so kinetically hindered that it would not form. The monohydrate of 4-aminoquinaldine[62] is an important example, as CSP predicted that a more thermodynamically stable structure should exist. It was later found by careful experimentation using either hydrothermal conditions or an impurity, but the metastable polymorph is kinetically favoured in both nucleation and growth. There are examples where analysis of the CSP_0 most stable structure, in contrast with the observed forms, can show that its formation seems rather unlikely on either crystal engineering or statistical likelihood.[63] It is noteworthy that despite extensive polymorph screening, a significant number of drugs are being marketed on the false assumption (as shown by CSP_0) that they are in the most stable thermodynamic form (DOI: 10.1039/c8fd00069g). Predicting polymorphs that could never be found also has major industrial implications (DOI: 10.1039/c8fd00033f). If CSP_0 is now an applied technology for the pharmaceutical industry (DOI: 10.1039/c8fd00033f), we must have a clear concept of the limitations of the technology, which involves determining how it differs from CSP_aim for this class of material.

Practicality: CSP_aim and energy–structure–function maps

The field of CSP as an aid to the discovery of new materials is becoming a reality across many types of materials. For example, in the field of the 18-valence-electron ABX family, consideration of the unreported family members led to the first synthesis of 15 compounds, including HfIrAs, a topological semi-metal of interest in quantum electronics, ZrNiPb, a small-gap semiconductor with a large Seebeck coefficient suitable for thermoelectric applications, and ZrIrSb, a rare example of a transparent p-type conductor with a high conductivity of holes.[64] This approach has also led to the design and discovery of the transparent conductor TaIrGe.[65] The use of computation for the accelerated discovery of new crystal structures (rather than just changing elements within known structures) in the complex inorganic Y–Sr–Ca–Ga–O phase field has led to new structures with ordered variants $Sr_2Ca_3Ga_6O_{14}$ and $SrCa_2Ga_2O_6$.[66] CSP on varying stoichiometries has also impacted energy materials research, with Li_7Ge_3 being predicted[67] and then found in research on germanium anodes in lithium batteries.[68]

The progress in the CSP-led design and realisation of the lowest density porous organic cage crystal, with its gas storage and selectivity properties[69] is a particular landmark, as a case when an organic molecule was synthesised following CSP prediction. However, it is noteworthy that the synthesis was targeting a highly metastable structure on the energy–structure–function map. The most thermodynamically stable form (according to the CSP_0) was not reported. This aspect of solvents stabilising porous structures (DOI: 10.1039/c8fd00031j) further illustrates how crystallisation conditions affect the observed structure. Zeolites, another important porous material (DOI: 10.1039/c8fd00035b, DOI: 10.1039/c8fd00040a), can also be more specifically templated,[70] though the organic molecules that direct the structure are burnt off afterwards.

Organic functional materials were already expanding in scope at the start of CSP, when the emphasis was on the most stable possible structure, with density being the key property for energetic materials, or non-centrosymmetric packing for non-linear optically active materials. Nowadays, energy–structure–function maps can be made for advanced properties, both in terms of technological applications and the challenge of calculating the property such as charge carrier mobility in organic semiconductors.[71] The differences between an organic crystal or film and the traditional silicon semi-conductors illustrates how spanning different types of materials can combine other properties. If a material is being produced industrially, there are a myriad of properties that are relevant to its crystallisation, storage, use, and the quality control procedures that need to be in place. Our CSP studies are just the start of the multi-scale modelling that is involved in the digital design agenda, requiring information on the behaviour of the material on a range of length and time scales, as the pharmaceutical materials science tetrahedron[72] relates the structure, properties, performance and processing of a drug. Hence the role of CSP in polymorph screening, including the design of experiments to find new forms, but the number of published examples contrasting the number of polymorphs found by an industrial screening process with those found by CSP is limited.[73] This iterative work where CSP may inspire experiments that find more forms, *versus* the problem of over-prediction, really illustrates the need to define what is required from CSP. Pressure is becoming increasingly important in the discovery of CSP "predicted" forms for pharmaceuticals,[14,74,75] whereas CSP at pressure has long been established as a route to finding new phases of interest to planetary scientists, including changes in bonding, such as ionic ammonia.[76,77]

Hence, CSP is playing a huge role in our ability to make new crystalline solids for many purposes. The combination of CSP and calculating (non-thermodynamic) properties, whether for targeting new functional materials or for structural characterisation, such as various diffraction[78] or solid state NMR experiments,[79] is really contributing to science. The question though is how much this relies on human scientific understanding interpreting the results of the CSP_0 (or CSP_thd) for experimental collaborators, or whether we can have a black-box computer code.

Conclusion

CSP_0 searches for the lowest lattice energy structures are becoming an established technique for a wide range of materials research, with an increasing diversity of types of interatomic forces within the materials. Hence the area of coverage of different materials in this discussion of the CSP_0 column in Fig. 1 is very wide. However, the most stable structure in CSP_0 is not always observed, and certainly is often not the only structure that can be found. This sometimes reflects the limitations of the thermodynamic modelling accuracy (*i.e.* really needing CSP_thd) but also the basic assumption that thermodynamics determines crystallisation. If the energy gaps are sufficiently large, then CSP_0 is good for determining whether a material is sufficiently likely to crystallise with a desired property that it is worth some experimental work.

However, when CSP_0 produces many structures that are close in energy, then eliminating those that are artefacts of approximating the free energy is more challenging. This is not only a challenge in obtaining the free energy of the perfect infinite crystal at the practically relevant temperatures and pressures. There is also correctly modelling the thermodynamics related to size, solvent, and the presence of specific surfaces or other molecules in solution. How realistic are thoughts of a CSP_thd code, spanning a wide diversity of interatomic interactions?

A CSP_aim code would reliably output the crystal structures that could be formed, and no more. It seems likely that calculating the true relative thermodynamic stability will sometimes incorporate information relevant to the experiments to find them, but much more understanding is needed of the mechanisms for nucleation catalysts and growth inhibitors.

CSP is maturing in the sense that an increasing number of experimental groups, including in the pharmaceutical industry, are taking it seriously enough to invest in CSP. It has been a long haul to get to this state. Over-prediction could lose this impetus. We need to be realistic about how much the codes can deliver, and the extent to which it is the experience of the scientist in interpreting the results in conjunction with the experimentalists that makes CSP practically useful. Hopefully this discussion will establish the current state of the art.

Conflicts of interest

There are no conflicts to declare.

Acknowledgements

There are too many people that have contributed to forming the views in this lecture, through discussions or through collaborative work, to be acknowledged but I am very grateful to them. Some are authors of the papers listed on http://www.cposs.org.uk where we have been developing CSP in collaboration with experimental or other computational groups, and this list extends far beyond those funded through GR/S24114, EPSRC EP/F03573X/1, EP/K039229/1 E.U. Horizon 2020 736899 and Eli Lilly.

Notes and references

1 J. Maddox, Crystals from First Principles, *Nature*, 1988, **335**(6187), 201.

2 W. C. McCrone, Polymorphism, in *Physics and Chemistry of the Organic Solid State*, ed. D. Fox, M. M. Labes and A. Weissberger, Wiley Interscience, New York, 1965, Vol. II, pp. 725–767.

3 C. Cervinka and G. J. O. Beran, Ab initio prediction of the polymorph phase diagram for crystalline methanol, *Chem. Sci.*, 2018, 4622–4629.

4 C. J. Pickard and R. J. Needs, Structures and stability of calcium and magnesium carbonates at mantle pressures, *Phys. Rev. B: Condens. Matter Mater. Phys.*, 2015, **91**(10), 104101.

5 D. Smith, K. V. Lawler, M. Martinez-Canales, A. W. Daykin, Z. Fussell, G. A. Smith, C. Childs, J. S. Smith, C. J. Pickard and A. Salamat, Postaragonite phases of CaCO3 at lower mantle pressures, *Phys. Rev. Mater.*, 2018, **2**(1), 013605.

6 A. M. Reilly, R. I. Cooper, C. S. Adjiman, S. Bhattacharya, A. D. Boese, J. G. Brandenburg, P. J. Bygrave, R. Bylsma, J. E. Campbell, R. Car, D. H. Case, R. Chadha, J. C. Cole, K. Cosburn, H. M. Cuppen, F. Curtis, G. M. Day, R. A. DiStasio Jr, A. Dzyabchenko, B. P. van Eijck, D. M. Elking, J. A. van den Ende, J. C. Facelli, M. B. Ferraro, L. Fusti-Molnar, C.-A. Gatsiou, T. S. Gee, R. de Gelder, L. M. Ghiringhelli, H. Goto, S. Grimme, R. Guo, D. W. M. Hofmann, J. Hoja, R. K. Hylton, L. Iuzzolino, W. Jankiewicz, D. T. de Jong, J. Kendrick, N. J. J. de Klerk, H.-Y. Ko, L. N. Kuleshova, X. Li, S. Lohani, F. J. J. Leusen, A. M. Lund, J. Lv, Y. Ma, N. Marom, A. E. Masunov, P. McCabe, D. P. McMahon, H. Meekes, M. P. Metz, A. J. Misquitta, S. Mohamed, B. Monserrat, R. J. Needs, M. A. Neumann, J. Nyman, S. Obata, H. Oberhofer, A. R. Oganov, A. M. Orendt, G. I. Pagola, C. C. Pantelides, C. J. Pickard, R. Podeszwa, L. S. Price, S. L. Price, A. Pulido, M. G. Read, K. Reuter, E. Schneider, C. Schober, G. P. Shields, P. Singh, I. J. Sugden, K. Szalewicz, C. R. Taylor, A. Tkatchenko, M. E. Tuckerman, F. Vacarro, M. Vasileiadis, A. Vazquez-Mayagoitia, L. Vogt, Y. Wang, R. E. Watson, G. A. de Wijs, J. Yang, Q. Zhu

and C. R. Groom, Report on the sixth blind test of organic crystal structure prediction methods, *Acta Crystallogr., Sect. B: Struct. Sci., Cryst. Eng. Mater.*, 2016, **72**(4), 439–459.

7 P. Vishweshwar, J. A. McMahon, M. Oliveira, M. L. Peterson and M. J. Zaworotko, The Predictable Elusive Form II of Aspirin, *J. Am. Chem. Soc.*, 2005, **127**(48), 16802–16803.

8 J. Klimes, D. R. Bowler and A. Michaelides, van der Waals density functionals applied to solids, *Phys. Rev. B: Condens. Matter Mater. Phys.*, 2011, **83**(19), 195131.

9 L. Goerigk and S. Grimme, A thorough benchmark of density functional methods for general main group thermochemistry, kinetics, and noncovalent interactions, *Phys. Chem. Chem. Phys.*, 2011, **13**(14), 6670–6688.

10 Y. Zhao and D. G. Truhlar, Benchmark databases for nonbonded interactions and their use to test density functional theory, *J. Chem. Theory Comput.*, 2005, **1**(3), 415–432.

11 D. E. Braun, S. R. Lingireddy, M. D. Beidelschies, R. Guo, P. Muller, S. L. Price and S. M. Reutzel-Edens, Unraveling Complexity in the Solid Form Screening of a Pharmaceutical Salt: Why so Many Forms? Why so Few?, *Cryst. Growth Des.*, 2017, **17**(10), 5349–5365.

12 L. M. LeBlanc, A. Otero-de-la-Roza and E. R. Johnson, Composite and Low-Cost Approaches for Molecular Crystal Structure Prediction, *J. Chem. Theory Comput.*, 2018, **14**(4), 2265–2276.

13 J. Nyman, O. S. Pundyke and G. M. Day, Accurate force fields and methods for modelling organic molecular crystals at finite temperatures, *Phys. Chem. Chem. Phys.*, 2016, **18**(23), 15828–15837.

14 A. A. Aina, A. J. Misquitta and S. L. Price, From dimers to the solid-state: Distributed intermolecular force-fields for pyridine, *J. Chem. Phys.*, 2017, **147**(16), 161722.

15 F. Curtis, X. Y. Li, T. Rose, A. Vazquez-Mayagoitia, S. Bhattacharya, L. M. Ghiringhelli and N. Marom, GAtor: A First-Principles Genetic Algorithm for Molecular Crystal Structure Prediction, *J. Chem. Theory Comput.*, 2018, **14**(4), 2246–2264.

16 K. M. Steed and J. W. Steed, Packing Problems: High Z′ Crystal Structures and Their Relationship to Cocrystals, Inclusion Compounds, and Polymorphism, *Chem. Rev.*, 2015, **115**(8), 2895–2933.

17 C. R. Groom and F. H. Allen, The Cambridge Structural Database in Retrospect and Prospect, *Angew. Chem., Int. Ed.*, 2014, **53**(3), 662–671.

18 C. R. Groom, I. J. Bruno, M. P. Lightfoot and S. C. Ward, The Cambridge Structural Database, *Acta Crystallogr., Sect. B: Struct. Sci., Cryst. Eng. Mater.*, 2016, **72**, 171–179.

19 J. C. Cole, C. R. Groom, M. G. Read, I. Giangreco, P. McCabe, A. M. Reilly and G. P. Shields, Generation of crystal structures using known crystal structures as analogues, *Acta Crystallogr., Sect. B: Struct. Sci., Cryst. Eng. Mater.*, 2016, **72**, 530–541.

20 M. A. Perrin, M. A. Neumann, H. Elmaleh and L. Zaske, Crystal structure determination of the elusive paracetamol form III, *Chem. Commun.*, 2009, **22**, 3181–3183.

21 S. J. Smith, M. M. Bishop, J. M. Montgomery, T. P. Hamilton and Y. K. Vohra, Polymorphism in Paracetamol: Evidence of Additional Forms IV and V at High Pressure, *J. Phys. Chem. A*, 2014, **118**(31), 6068–6077.

22 D. E. Braun, M. Orlova and U. J. Griesser, Creatine: Polymorphs Predicted and Found, *Cryst. Growth Des.*, 2014, **14**(10), 4895–4900.

23 M. D. King, T. N. Blanton, S. T. Misture and T. M. Korter, Prediction of the Unknown Crystal Structure of Creatine Using Fully Quantum Mechanical Methods, *Cryst. Growth Des.*, 2011, **11**(12), 5733–5740.

24 M. K. Corpinot, R. Guo, D. A. Tocher, A. B. M. Buanz, S. Gaisford, S. L. Price and D. K. Bucar, Are Oxygen and Sulfur Atoms Structurally Equivalent in Organic Crystals?, *Cryst. Growth Des.*, 2017, **17**(2), 827–833.

25 H. K. Buchholz, R. K. Hylton, J. G. Brandenburg, A. Seidel-Morgenstern, H. Lorenz, M. Stein and S. L. Price, Thermochemistry of Racemic and Enantiopure Organic Crystals for Predicting Enantiomer Separation, *Cryst. Growth Des.*, 2017, **17**(9), 4676–4686.

26 A. Otero-de-la-Roza, B. H. Cao, I. K. Price, J. E. Hein and E. R. Johnson, Predicting the Relative Solubilities of Racemic and Enantiopure Crystals by Density-Functional Theory, *Angew. Chem., Int. Ed.*, 2014, **53**(30), 7879–7882.

27 A. M. Reilly and A. Tkatchenko, Understanding the role of vibrations, exact exchange, and many-body van der Waals interactions in the cohesive properties of molecular crystals, *J. Chem. Phys.*, 2013, **139**(2), 024705.

28 J. Yang, W. Hu, D. Usvyat, D. Matthews, M. Schutz and H. Chan, Ab initio determination of the lattice energy in crystalline benzene to sub-kilojoule per mole accuracy, *Science*, 2014, **345**(6197), 640–643.

29 A. J. Cruz-Cabeza, S. M. Reutzel-Edens and J. Bernstein, Facts and fictions about polymorphism, *Chem. Soc. Rev.*, 2015, **44**, 8619–8635.

30 M. Habgood, Form II Caffeine: A Case Study for Confirming and Predicting Disorder in Organic Crystals, *Cryst. Growth Des.*, 2011, **11**(8), 3600–3608.

31 A. M. Belenguer, G. I. Lampronti, A. J. Cruz-Cabeza, C. A. Hunter and J. K. M. Sanders, Solvation and surface effects on polymorph stabilities at the nanoscale, *Chem. Sci.*, 2016, **7**(11), 6617–6627.

32 W. T. M. Mooij, B. P. van Eijck, S. L. Price, P. Verwer and J. Kroon, Crystal structure predictions for acetic acid, *J. Comput. Chem.*, 1998, **19**(4), 459–474.

33 P. Raiteri, R. Martonak and M. Parrinello, Exploring Polymorphism: The Case of Benzene, *Angew. Chem., Int. Ed.*, 2005, **44**, 3769–3773.

34 T. Zykova-Timan, P. Raiteri and M. Parrinello, Investigating the Polymorphism in PR179: A Combined Crystal Structure Prediction and Metadynamics Study, *J. Phys. Chem. B*, 2008, **112**(42), 13231–13237.

35 P. G. Karamertzanis, P. Raiteri, M. Parrinello, M. Leslie and S. L. Price, The Thermal Stability of Lattice Energy Minima of 5-Fluorouracil: Metadynamics as an Aid to Polymorph Prediction, *J. Phys. Chem. B*, 2008, **112**(14), 4298–4308.

36 A. Torrisi, C. K. Leech, K. Shankland, W. I. F. David, R. M. Ibberson, J. Benet-Buchholz, R. Boese, M. Leslie, C. R. A. Catlow and S. L. Price, The solid phases of cyclopentane: a combined experimental and simulation study, *J. Phys. Chem. B*, 2008, **112**(12), 3746–3758.

37 Q. Zhu, A. G. Shtukenberg, D. J. Carter, T. Q. Yu, J. X. Yang, M. Chen, P. Raiteri, A. R. Oganov, B. Pokroy, I. Polishchuk, P. J. Bygrave, G. M. Day, A. L. Rohl, M. E. Tuckerman and B. Kahr, Resorcinol Crystallization from the Melt: A New Ambient Phase and New "Riddles", *J. Am. Chem. Soc.*, 2016, **138**(14), 4881–4889.

38 A. G. Shtukenberg, C. H. T. Hu, Q. Zhu, M. U. Schmidt, W. Q. Xu, M. Tan and B. Kahr, The Third Ambient Aspirin Polymorph, *Cryst. Growth Des.*, 2017, **17**(6), 3562–3566.

39 L. Yu, Polymorphism in Molecular Solids: An Extraordinary System of Red, Orange, and Yellow Crystals, *Acc. Chem. Res.*, 2010, **43**(9), 1257–1266.

40 M. Vasileiadis, A. V. Kazantsev, P. G. Karamertzanis, C. S. Adjiman and C. C. Pantelides, The polymorphs of ROY: application of a systematic crystal structure prediction technique, *Acta Crystallogr., Sect. B: Struct. Sci.*, 2012, **68**, 677–685.

41 D. K. Bucar, R. W. Lancaster and J. Bernstein, Disappearing Polymorphs Revisited, *Angew. Chem., Int. Ed.*, 2015, **54**(24), 6972–6993.

42 R. W. Lancaster, L. D. Harris and D. Pearson, Fifty-year old samples of progesterone demonstrate the complex role of synthetic impurities in stabilizing a metastable polymorph, *CrystEngComm*, 2011, **13**(6), 1775–1777.

43 P. Lucaioli, E. Nauha, I. Gimondi, L. S. Price, R. Guo, L. Iuzzolino, I. Singh, M. Salvalaglio, S. L. Price and N. Blagden, Serendipitous isolation of a disappearing conformational polymorph of succinic acid challenges computational polymorph prediction, *CrystEngComm*, 2018, **20**, 3971–3977.

44 V. Chikhalia, R. T. Forbes, R. A. Storey and M. Ticehurst, The effect of crystal morphology and mill type on milling induced crystal disorder, *Eur. J. Pharm. Sci.*, 2006, **27**(1), 19–26.

45 A. V. Trask, N. Shan, W. D. S. Motherwell, W. Jones, S. H. Feng, R. B. H. Tan and K. J. Carpenter, Selective polymorph transformation via solvent-drop grinding, *Chem. Commun.*, 2005, **7**, 880–882.

46 K. Robertson, P. B. Flandrin, A. R. Klapwijk and C. C. Wilson, Design and Evaluation of a Mesoscale Segmented Flow Reactor (KRAIC), *Cryst. Growth Des.*, 2016, **16**(8), 4759–4764.

47 K. M. Carver and R. C. Snyder, Unexpected Polymorphism and Unique Particle Morphologies from Monodisperse Droplet Evaporation, *Ind. Eng. Chem. Res.*, 2012, **51**(48), 15720–15728.

48 P. G. Karamertzanis, A. V. Kazantsev, N. Issa, G. W. A. Welch, C. S. Adjiman, C. C. Pantelides and S. L. Price, Can the Formation of Pharmaceutical Cocrystals Be Computationally Predicted? 2. Crystal Structure Prediction, *J. Chem. Theory Comput.*, 2009, **5**(5), 1432–1448.

49 D. K. Bucar, G. M. Day, I. Halasz, G. G. Z. Zhang, J. R. G. Sander, D. G. Reid, L. R. MacGillivray, M. J. Duer and W. Jones, The curious case of (caffeine) (benzoic acid): how heteronuclear seeding allowed the formation of an elusive cocrystal, *Chem. Sci.*, 2013, **4**(12), 4417–4425.

50 V. K. Srirambhatla, R. Guo, S. L. Price and A. J. Florence, Isomorphous template induced crystallisation: a robust method for the targeted crystallisation of computationally predicted metastable polymorphs, *Chem. Commun.*, 2016, **52**(46), 7384–7386.

51 D. E. Braun, J. A. McMahon, L. H. Koztecki, S. L. Price and S. M. Reutzel-Edens, Contrasting Polymorphism of Related Small Molecule Drugs Correlated and Guided by the Computed Crystal Energy Landscape, *Cryst. Growth Des.*, 2014, **14**(4), 2056–2072.

52 In *Nucleation – A transition state to the directed assembly of materials: Faraday Discussion*, 2015.

53 J. J. De Yoreo, P. U. P. A. Gilbert, N. A. J. M. Sommerdijk, R. L. Penn, S. Whitelam, D. Joester, H. Z. Zhang, J. D. Rimer, A. Navrotsky, J. F. Banfield, A. F. Wallace, F. M. Michel, F. C. Meldrum, H. Colfen and

P. M. Dove, Crystallization by particle attachment in synthetic, biogenic, and geologic environments, *Science*, 2015, **349**(6247), aaa6760.

54 M. Warzecha, R. Guo, R. M. Bhardwaj, S. M. Reutzel-Edens, S. L. Price, D. A. Lamprou and A. J. Florence, Direct Observation of Templated Two-Step Nucleation Mechanism during Olanzapine Hydrate Formation, *Cryst. Growth Des.*, 2017, **17**(12), 6382–6393.

55 T. Beyer, G. M. Day and S. L. Price, The prediction, morphology, and mechanical properties of the polymorphs of paracetamol, *J. Am. Chem. Soc.*, 2001, **123**(21), 5086–5094.

56 Y. Y. Sun, C. J. Tilbury, S. M. Reutzel-Edens, R. M. Bhardwaj, J. J. Li and M. F. Doherty, Modeling Olanzapine Solution Growth Morphologies, *Cryst. Growth Des.*, 2018, **18**(2), 905–911.

57 A. O. F. Jones, B. Chattopadhyay, Y. H. Geerts and R. Resel, Substrate-Induced and Thin-Film Phases: Polymorphism of Organic Materials on Surfaces, *Adv. Funct. Mater.*, 2016, **26**(14), 2233–2255.

58 J. Bernstein, R. J. Davey and J. O. Henck, Concomitant Polymorphs, *Angew. Chem., Int. Ed.*, 1999, **38**(23), 3440–3461.

59 R. I. Goldstein, R. Guo, C. Hughes, D. P. Maurer, T. R. Newhouse, T. J. Sisto, R. R. Conry, S. L. Price and D. M. Thamattoor, Concomitant conformational dimorphism in 1,2-bis(9-anthryl)acetylene, *CrystEngComm*, 2015, **17**(26), 4877–4882.

60 R. M. Bhardwaj, L. S. Price, S. L. Price, S. M. Reutzel-Edens, G. J. Miller, I. D. H. Oswald, B. Johnston and A. J. Florence, Exploring the Experimental and Computed Crystal Energy Landscape of Olanzapine, *Cryst. Growth Des.*, 2013, **13**(4), 1602–1617.

61 T. Threlfall, Structural and thermodynamic explanations of Ostwald's rule, *Org. Process Res. Dev.*, 2003, **7**(6), 1017–1027.

62 D. E. Braun, H. Oberacher, K. Arnhard, M. Orlova and U. J. Griesser, 4-Aminoquinaldine monohydrate polymorphism: prediction and impurity aided discovery of a difficult to access stable form, *CrystEngComm*, 2016, **18**(22), 4053–4067.

63 R. K. Hylton, G. J. Tizzard, T. L. Threlfall, A. L. Ellis, S. J. Coles, C. C. Seaton, E. Schulze, H. Lorenz, A. Seidel-Morgenstern, M. Stein and S. L. Price, Are the Crystal Structures of Enantiopure and Racemic Mandelic Acids Determined by Kinetics or Thermodynamics?, *J. Am. Chem. Soc.*, 2015, **137**(34), 11095–11104.

64 R. Gautier, X. Zhang, L. Hu, L. Yu, Y. Lin, T. O. L. Sunde, D. Chon, K. R. Poeppelmeier and A. Zunger, Prediction and accelerated laboratory discovery of previously unknown 18-electron ABX compounds, *Nat. Chem.*, 2015, **7**, 308.

65 F. Yan, X. Zhang, Y. G. Yu, L. Yu, A. Nagaraja, T. O. Mason and A. Zunger, Design and discovery of a novel half-Heusler transparent hole conductor made of all-metallic heavy elements, *Nat. Commun.*, 2015, **6**, 7308.

66 C. Collins, M. S. Dyer, M. J. Pitcher, G. F. S. Whitehead, M. Zanella, P. Mandal, J. B. Claridge, G. R. Darling and M. J. Rosseinsky, Accelerated discovery of two crystal structure types in a complex inorganic phase field, *Nature*, 2017, **546**, 280.

67 A. J. Morris, C. P. Grey and C. J. Pickard, Thermodynamically stable lithium silicides and germanides from density functional theory calculations, *Phys. Rev. B: Condens. Matter Mater. Phys.*, 2014, **90**(5), 054111.

68 H. Jung, P. K. Allan, Y. Y. Hu, O. J. Borkiewicz, X. L. Wang, W. Q. Han, L. S. Du, C. J. Pickard, P. J. Chupas, K. W. Chapman, A. J. Morris and C. P. Grey, Elucidation of the Local and Long-Range Structural Changes that Occur in Germanium Anodes in Lithium-Ion Batteries, *Chem. Mater.*, 2015, **27**(3), 1031–1041.

69 A. Pulido, L. J. Chen, T. Kaczorowski, D. Holden, M. A. Little, S. Y. Chong, B. J. Slater, D. P. McMahon, B. Bonillo, C. J. Stackhouse, A. Stephenson, C. M. Kane, R. Clowes, T. Hasell, A. I. Cooper and G. M. Day, Functional materials discovery using energy-structure-function maps, *Nature*, 2017, **543**(7647), 657–664.

70 D. W. Lewis, D. J. Willock, C. R. A. Catlow, J. M. Thomas and G. J. Hutchings, De novo design of structure-directing agents for the synthesis of microporous solids, *Nature*, 1996, **382**(6592), 604–606.

71 B. Rice, L. M. LeBlanc, A. Otero-de-la-Roza, M. J. Fuchter, E. R. Johnson, J. Nelson and K. E. Jelfs, A computational exploration of the crystal energy and charge-carrier mobility landscapes of the chiral [6] helicene molecule, *Nanoscale*, 2018, **10**(4), 1865–1876.

72 C. C. Sun, Material Science Tetrahedron-A Useful Tool for Pharmaceutical Research and Development, *J. Pharm. Sci.*, 2009, **98**, 1671–1687.

73 S. L. Price and S. M. Reutzel-Edens, The potential of computed crystal energy landscapes to aid solid-form development, *Drug Discovery Today*, 2016, **21**(6), 912–923.

74 M. A. Neumann, J. V. de Streek, F. P. A. Fabbiani, P. Hidber and O. Grassmann, Combined crystal structure prediction and high-pressure crystallization in rational pharmaceutical polymorph screening, *Nat. Commun.*, 2015, **6**, 7793.

75 I. D. H. Oswald, I. Chataigner, S. Elphick, F. P. A. Fabbiani, A. R. Lennie, J. Maddaluno, W. G. Marshall, T. J. Prior, C. R. Pulham and R. I. Smith, Putting pressure on elusive polymorphs and solvates, *CrystEngComm*, 2009, **11**(2), 359–366.

76 C. J. Pickard and R. J. Needs, Highly compressed ammonia forms an ionic crystal, *Nat. Mater.*, 2008, **7**(10), 775–779.

77 S. Ninet, F. Datchi, P. Dumas, M. Mezouar, G. Garbarino, A. Mafety, C. J. Pickard, R. J. Needs and A. M. Saitta, Experimental and theoretical evidence for an ionic crystal of ammonia at high pressure, *Phys. Rev. B: Condens. Matter Mater. Phys.*, 2014, **89**(17), 174103.

78 S. Habermehl, P. Morschel, P. Eisenbrandt, S. M. Hammer and M. U. Schmidt, Structure determination from powder data without prior indexing, using a similarity measure based on cross-correlation functions, *Acta Crystallogr., Sect. B: Struct. Sci., Cryst. Eng. Mater.*, 2014, **70**, 347–359.

79 M. Zilka, D. V. Dudenko, C. E. Hughes, P. A. Williams, S. Sturniolo, W. T. Franks, C. J. Pickard, J. R. Yates, K. D. M. Harris and S. P. Brown, Ab initio random structure searching of organic molecular solids: assessment and validation against experimental data, *Phys. Chem. Chem. Phys.*, 2017, **19**(38), 25949–25960.

80 A. G. Shtukenberg, Q. Zhu, D. J. Carter, L. Vogt, J. Hoja, E. Schneider, H. Song, B. Pokroy, I. Polishchuk, A. Tkatchenko, A. R. Oganov, A. L. Rohl, M. E. Tuckerman and B. Kahr, *Chem. Sci.*, 2017, **8**(7), 4926–4940.

Faraday Discussions

PAPER

Accelerating CALYPSO structure prediction by data-driven learning of a potential energy surface†

Qunchao Tong,[ab] Lantian Xue,[ab] Jian Lv, *[abc] Yanchao Wang[ab] and Yanming Ma *[abd]

Received 4th March 2018, Accepted 5th April 2018
DOI: 10.1039/c8fd00055g

Ab initio structure prediction methods have been nowadays widely used as powerful tools for structure searches and materials discovery. However, they are generally restricted to small systems owing to the heavy computational cost of the underlying density functional theory (DFT) calculations in structure optimizations. In this work, by combining a state-of-art machine learning (ML) potential with our in-house developed CALYPSO structure prediction method, we developed two acceleration schemes for the structure prediction of large systems, in which a ML potential is pre-constructed to fully replace DFT calculations or trained in an on-the-fly manner from scratch during the structure searches. The developed schemes have been applied to medium- and large-sized boron clusters, both of which are challenging cases for either the construction of ML potentials or extensive structure searches. Experimental structures of B_{36} and B_{40} clusters can be readily reproduced, and the putative global minimum structure for the B_{84} cluster is proposed, where the computational cost is substantially reduced by ~1–2 orders of magnitude if compared with full DFT-based structure searches. Our results demonstrate a viable route for structure prediction in large systems *via* the combination of state-of-art structure prediction methods and ML techniques.

1. Introduction

Theoretical structure prediction methods are nowadays in common use, playing an increasingly important role in computational material discovery, as they can provide putative ground-state structures of materials and can be used to design

[a]*State Key Laboratory of Superhard Materials, College of Physics, Jilin University, Changchun 130012, China. E-mail: mym@calypso.cn*

[b]*Innovation Center of Computational Physics Methods and Software, College of Physics, Jilin University, Changchun 130012, China*

[c]*College of Materials Science and Engineering, Jilin University, Changchun 130012, China. E-mail: lvjian@calypso.cn*

[d]*International Center of Future Science, Jilin University, Changchun 130012, China*

† Electronic supplementary information (ESI) available. See DOI: 10.1039/c8fd00055g

structures with desired functionality to aid experimental synthesis.[1] These methods generally involve the exploration of the potential energy surfaces (PESs) of materials through various structure sampling techniques and optimization algorithms.[2–12] We have developed a swarm-intelligence based CALYPSO structure prediction method,[13,14] which has been successfully applied to a variety of material systems.[15,16] Despite widespread successes, structure prediction methods are inherently limited by the underlying computational workhorse of energy evaluation based on quantum mechanical methods, *e.g.*, density functional theory (DFT), which is prohibitively expensive for large systems.

Machine learning (ML) as a data-driven method for making predictions, decisions or classifications has now become a pervasive and deeply entrenched component of modern science and technology. It is now starting to enter the heart of computational physics, chemistry and material sciences in a manifold way.[17] Particularly, with plenty of quantum mechanical data available, various ML schemes have been established, enabling accurate predictions of a rich variety of materials' properties including atomization energies,[18–22] formation energies[23–25] and electronic properties.[20,26–28] Moreover, ML techniques have also been used to reconstruct the potential energy surfaces of materials based on data obtained from first-principles calculations.[29] A series of ML potentials have been developed by utilizing techniques such as artificial neural networks,[30–33] Gaussian process regression,[34–37] support vector machines[38] kernel ridge regression,[39,40] *etc.* They show accuracy and transferability comparable to those of quantum-mechanical simulations but require less computational effort by many orders of magnitude. This makes computational simulations accessible to large systems over long timescales, such as those applicable to phenomena such as phase transitions[41] and crystallization.[42]

Given the virtue in the cost-accuracy tradeoff of ML potentials, they are a promising substitute for DFT for accelerating global structure searches. However, for a ML potential to be useful for structure predictions, it is essential to take into account structures lying in different regions of the PES during the training process. This is in sharp contrast to most previous works where structures close to local minima of the PES are considered. The utilization of ML potentials to accelerate structure predictions is now in its early stages but has already generated several encouraging results.[40,43,44]

In this work, we first demonstrate the applicability of a current state-of-art ML potential, the Gaussian approximation potential (GAP),[34] in the reconstruction of the PES for global structure searches. Then two efficient schemes have been developed for accelerating structure predictions by combining our in-house developed CALYPSO method with GAP. Boron (B) clusters were used as testing systems. Our developed schemes can readily reproduce experimental ground-state structures for B_{36} and B_{40} clusters, as well as propose a putative global stable structure for a large-sized B cluster containing 84 atoms, where the computational cost is significantly reduced by $\sim$1–2 orders of magnitude compared with full DFT-based structure predictions. The generalization of the current schemes to crystal structure predictions is straightforward by taking into account the stress tensor during the GAP training process.

2. Method and applications

2.1 The applicability of a state-of-art ML potential in the reconstruction of a PES for structure prediction

There are two central components of any ML potential: one is the descriptor that represents a structure numerically in a unique and unambiguous way; the other is a ML method that reconstructs the PES using the designed descriptor as input. Currently, a number of descriptors are available for ML models, among which atom centered symmetry functions (ACSF),[45] the bispectrum of neighbor density and the smooth overlap of atomic positions[46] have been demonstrated to be particularly suitable for fitting PESs. As for the learning method, high-dimensional neural networks[47] and Gaussian process regression[34] show the most promising solutions to this problem. In principle, one can choose from the above descriptors and ML methods arbitrarily to construct a ML potential. Here we employ ACSF as the descriptor, while the Gaussian process regression technique as implemented in the GAP is used for the ML method.

The ACSF descriptor[45] is a set of radial and angular functions which describe the coordination environment of an atom depending on the positions of the neighboring atoms up to a cutoff radius. An arbitrary number (typically 50–100) of symmetry function values for an atom i can be obtained by adjusting parameters in the ACSF, which can be used to represent the environment of atom i in the structure and as input vectors ($\boldsymbol{d}_i$) for the ML model (the detailed formulation of the ACSF is given in the ESI†).

Within the GAP framework,[34,35] the total energy of a system is described as a sum of atomic energies,

$$E = \sum_{i=1}^{N} \varepsilon_i(\boldsymbol{d}_i), \tag{1}$$

where ε_i is the i-th atomic energy. Atomic energies are interpolated in the ACSF descriptor space through Gaussian process regression, which is a kernel-based ML technique. The central results of GAP are predictions of atomic energies (ε_*) and the corresponding variances (σ_*) described with the following two formulas,

$$\varepsilon_*(\boldsymbol{d}_*) = \boldsymbol{k}_*^T \boldsymbol{C}^{-1} \boldsymbol{t}, \tag{2}$$

$$\sigma_*(\boldsymbol{d}_*) = \kappa - \boldsymbol{k}_*^T \boldsymbol{C}^{-1} \boldsymbol{k}_*, \tag{3}$$

where $\boldsymbol{t}$ is the vector of the reference data, *i.e.*, the DFT total energies and atomic forces, $\boldsymbol{k}_*$ is the covariance vector of the functional values $\boldsymbol{k}_* \equiv \langle \varepsilon_* \boldsymbol{t} \rangle$, $\boldsymbol{C}$ is the covariance matrix defined as $\boldsymbol{C} \equiv \langle \boldsymbol{t}\boldsymbol{t}^T \rangle$, and κ is the covariance between the predicted atomic energy and itself $\kappa \equiv \langle \varepsilon_* \varepsilon_* \rangle$. In practice, a sparsification procedure is introduced to eliminate similar atomic environments to improve the numerical stability and reduce the computational cost.

In the current work, we further define the variance of the predicted total energy for a structure u as follows,

$$\Sigma_u = \frac{1}{N} \sum_{i \in u} \sigma_i(\boldsymbol{d}_i), \tag{4}$$

where N is the number of atoms in structure u. It can be used as a metric for estimating the reliability of the predicted total energy, which is of vital importance for our accelerated structure prediction scheme with on-the-fly training of the GAP.

Below we illustrate the performance of a current GAP model in the reconstruction of a PES using B clusters as testing systems. The complex energy landscapes of B clusters originate from the electron-deficient nature of B, leading to the occurrence a number of intriguing structures with distinct geometries (*e.g.*, planar,[48] tubular,[49] cage-like,[50–52] bi-layered[53] and core–shell structures[54,55]).[56] This poses formidable challenges for ML potentials to reconstruct the PES.

We start with generating two reference datasets at the DFT level as training and testing sets, respectively (details for DFT calculations can be found in "Computational Details" in the ESI†). The training set contains 12 053 data points (structures and the corresponding total energies as well as the forces on each atom) corresponding to 357 646 atomic environments. The testing set contains 3571 data points with 102 438 atomic environments. The data points are obtained by local structure optimizations of a number of B_n clusters that are generated randomly by CALYPSO in the size range of $n = 12$–42. In CALYPSO, the symmetry-constraint technique for random structure generation ensures a diverse sampling of the PES, while the structure characterization technique can avoid the inclusion of similar structures in the datasets. Thus, both the training and testing datasets are expected to cover a wide range of the PES (geometrically and energetically) and contain various atomic environments.

We trained a series of GAPs by gradually enlarging the size of the training data set from 100 to 12 000. An ACSF descriptor with 89 function values is used (the detailed parameters for the ACSF descriptor are listed in Table S1 in the ESI†). Fig. 1(a) and (b) show the evolution of the root mean square errors (RMSEs) for the energy and force for these GAPs tested on the testing set, respectively. Each data point was obtained by averaging over the RMSEs of ten GAPs which are constructed with training data randomly extracted from the training set (except for the data point at a training size of 12 000 where only one GAP can be obtained). It is clearly seen that the mean values of the RMSEs of both the energy and force

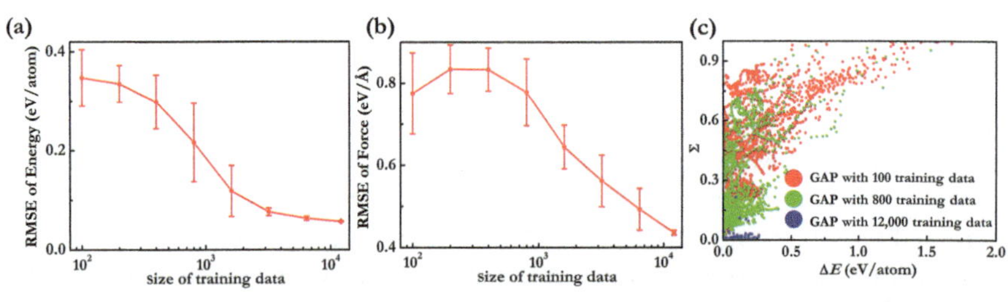

Fig. 1 Evolution of RMSEs for the GAP predicted energy (a) and force (b) for the testing set as a function of training set size. Each data point was obtained by averaging over the RMSEs of ten GAPs with training data randomly extracted from the training set, except for the data point at a training set size of 12 000 where only one GAP can be obtained. The error bar denotes the standard deviation. (c) Correlations between the energy error (ΔE, energy difference between GAP and DFT) and predicted variance (Σ) for structures in the testing set. Data points for GAPs using 100, 800 and 12 000 training data are denoted in red, green and blue, respectively.

significantly decrease as the size of training data increases, indicating a gradual improvement of the potentials. The standard deviation of the RMSEs is also reduced as the training size increases. The GAP with 12 000 training data gives the best performance, and will be used in the following tests. Comparisons between its predictions and DFT results for energy and force are shown in Fig. S1(a) and (b) in the ESI,† respectively. The data points are mainly distributed along the diagonal line corresponding to a perfect fit, leading to small RMSEs of the energy at 53 meV per atom and the force at 430 meV Å^{-1}, which are comparable to those of GAP results for silicon clusters.[46] We expect the RMSEs of the energy and force can be further reduced by increasing the training set size and number of function values of the ACSF descriptor but at the expense of larger computational cost during the training process.

To examine whether the defined variance (Σ) in eqn (4) can be used as a metric for estimating the reliability of the predicted total energy, we plotted the correlation between the energy error (ΔE, energy difference between GAP and DFT) and predicted variance (Σ) for structures in the testing set. As expected from the above results, the distribution of the data points gradually shrinks toward zero as the training set size increases from 100, to 800, to 12 000, indicating gradual improvement of the GAPs. Notably, the up-triangle shape of the plotted distribution implies that large Σ corresponds to a large uncertainty of ΔE. Thus Σ indeed can be used as a good indicator for the reliability of the predicted total energies.

To evaluate the transferability of the GAP, we performed local optimizations using DFT and GAP for 257 structures randomly generated by CALYPSO. The distributions of the interatomic distances and energies of the optimized structures are shown in Fig. 2(a) and (b). The distributions of the interatomic distances for optimized structures using the different methods are quite similar, with the first peak centered at ~1.65 Å. The energy distribution for the GAP-optimized structures shifts slightly toward higher-energy regions compared with the DFT results, but the overall difference (88 meV per atom) is close to the RMSE of the GAP (53 meV per atom). Visual inspection of optimized structures confirms that the two procedures generally give similar configurations. Thus, the above results indeed suggest high transferability of the current GAP model.

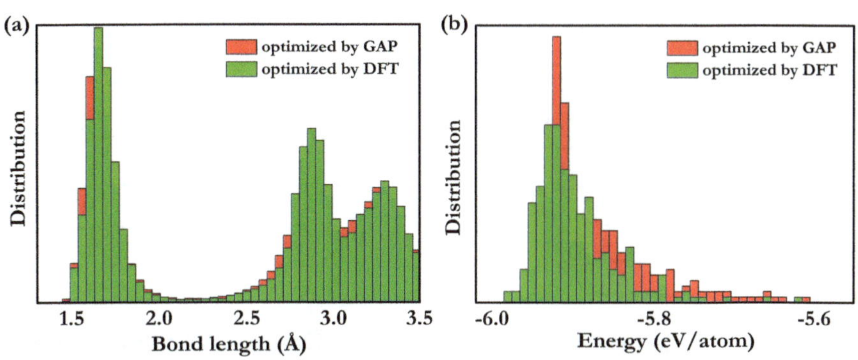

Fig. 2 The distributions of interatomic distances (a) and energies (b) for 257 random structures of B clusters optimized using GAP and DFT, respectively.

2.2 Acceleration schemes for CALYPSO structure prediction

Encouraged by the above results, we have developed two acceleration schemes for structure predictions. This is done on the top of our CALYPSO method,[13,14] which can explore the PES and locate stable/metastable structures intelligently by using the chemical composition alone for a material. Its validity and efficiency have been manifested by its successful applications in a variety of material systems at different dimensions,[15,16] such as 0D nanoclusters,[51,57] 2D layered materials[58,59] and surface reconstruction,[60] as well as 3D crystals.[61,62] The key features of this methodology are a swarm-intelligence based global optimization strategy and several particularly devised structure handling techniques. For more detailed theory on the CALYPSO method we refer the readers to ref. 13, 14, 63–65 but give a description of its main procedure below.

Structure prediction through CALYPSO comprises mainly four steps as depicted in the flow chart on the left panel of Fig. 3. First, the initial structures are randomly generated with the constraint of symmetry to allow a diverse sampling of the PES. Then, bond characterization matrix[14] or coordination characterization function[66] techniques will be used to characterize the new structures and examine their distances (similarities) to all the previous ones. Structures with a distance less than a threshold will be eliminated. After a user specified number of structures (a population or generation) have been generated, local structure optimizations based on DFT are performed to eliminate the noise of the energy surface and drive the systems to the local minima. Eventually, swarm-intelligence algorithms (*e.g.*, particle swarm optimization or symmetric artificial bee swarm) are applied to produce new structures for the next generation. This process continues iteratively until the maximum step number is reached.

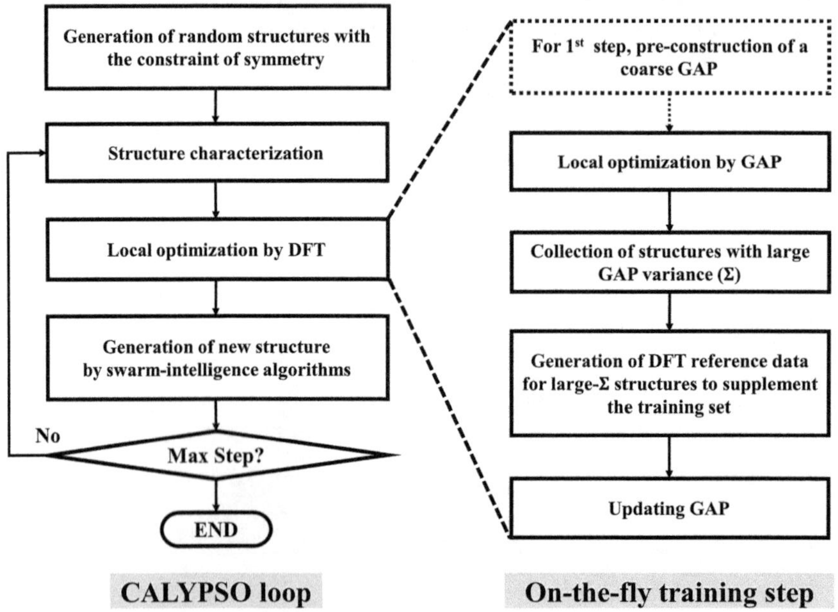

Fig. 3 Flow chart of CALYPSO structure prediction method (left panel) and acceleration scheme with on-the-fly training of GAP (right panel).

In the first acceleration scheme, we replace DFT calculations with GAP-based ones in local structure optimizations. This scheme has been tested on predictions of the structures of B_{36} and B_{40} clusters; it is very challenging for structure searches to reproduce the ground-state structures of these systems, with exotic quasi-planar and cage-like geometries, respectively.[52,67] The best GAP obtained above is adopted in the tests. The experimental quasi-planar B_{36} and cage-like B_{40} structures[52,67] are purposely removed from the training set during the construction of the GAP. The evolution of the energy distributions during the CALYPSO structure searches is shown in Fig. 4(a) and (b). It is remarkable to find that the two experimental structures can be readily reproduced during the searches. Although the two structures are not predicted as ground-state structures, they are lie in the low-energy regions below an energy threshold (defined as the predicted lowest energy in the structure search plus the RMSE of the GAP), and subsequent DFT-based local optimizations for refining the structures below the energy threshold (shaded areas in Fig. 4(a) and (b)) will reveal the two structures as the ground states. For each test case, the structure search lasts for 50 generations with each generation containing 140 structures. About 900 and 300 DFT-based local optimizations were performed in cases of B_{36} and B_{40}, respectively, for training the GAPs and refining the low-lying structures. Other local optimizations were performed at the GAP level, whose computational costs are negligible. For DFT-based structure predictions, 7000 local optimizations at the DFT level will be performed in each case. Therefore, the acceleration scheme reduces the computational cost by ~1 order of magnitude compared with a full DFT-based structure search in the test case.

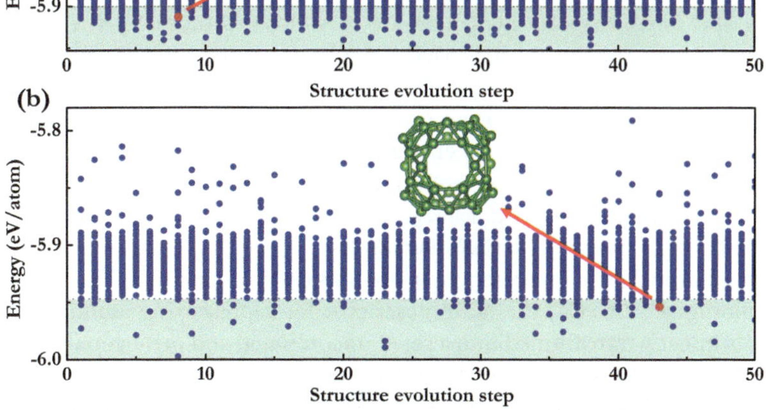

Fig. 4 Evolution of energy distributions during CALYPSO structure search for B_{36} (a) and B_{40} (b) clusters, where local optimizations are conducted by GAP.

The first acceleration scheme is appropriate for structure predictions in which a well-trained GAP for the system under investigation exists. But in most cases we carry out structure searches for a system from scratch without a well-trained GAP. Below we present an alternative scheme that can accelerate the structure search, while constructing the GAP in an on-the-fly manner.

In the second acceleration scheme, we replace the step of DFT-based local optimizations by a process of on-the-fly training of the GAP which is depicted on the right panel of Fig. 3. At the first step of the CALYPSO structure search, an arbitrary GAP can be provided. All subsequent local optimizations are performed at the GAP level with final configurations subjected to DFT single-point calculations. All structures during local optimizations are monitored, and the ones with a predicted variance Σ larger than a threshold are gradually accumulated. When a certain number of such structures are collected, DFT calculations are performed for these structures with the data being supplemented to the training set, and the GAP is updated according to the new training set. This process ensures a gradual improvement of the GAP by constructing the training set in an information-efficient way (*i.e.*, avoiding the inclusion of similar structures in the training set). Moreover, as the structure search proceeds, progressively fewer training processes and DFT calculations are needed.

We here illustrate the performance of this scheme through a structure search of a large 84-atom B cluster whose structural elucidation is still elusive though both planar and core–shell structures are suggested to be the ground state.[68,69] In the first step of the CALYPSO search, a coarse GAP is trained from data obtained by DFT-based local optimization of a random structure of B_{84} cluster. In the subsequent steps, all the structures are locally optimized at the GAP level, but the energies of the final configurations are evaluated at the DFT level. Structures with a predicted variance Σ larger than 0.03 are gradually collected, and DFT calculations are performed when 300 large-variance structures are obtained, whose data are supplemented to the training set to improve the GAP.

To evaluate the evolution of the GAP during the structure search, we constructed a testing set containing 3500 random structures of B_{84} clusters. Fig. 5(a) shows the evolution of the RMSEs of the GAP predicted energy and force for the testing set. At the first five GAP training steps, the RMSEs decrease rapidly. After that, the RMSEs gradually converge to $\sim$65 meV per atom and 440 meV Å^{-1} for energy and force, respectively, which are very close to those for the best GAP obtained above. This maybe approaches the best performance of a current GAP model for the description of a PES for a B cluster. Since B is a very challenging system, we expect much better performance by GAP for other simpler systems, such as metals. As the training set enlarges and the GAP improves, more structure evolution steps are needed to collect a sufficient number of large-variance structures (Fig. 5(a)). So progressively fewer DFT calculations are needed, and the frequency of training the GAP decreases.

We performed principal component analysis (PCA) to examine the evolution of the training set. In PCA, the ACSF descriptor for representing atomic environments is linearly transformed into a set of uncorrelated and orthogonal variables, known as principal components (PCs). The majority of the information contained within the ACSF descriptor can be captured by a few such PCs. We found that the first two PCs (PC1 and PC2) capture 98% of the information of the ACSF descriptor. Training data at the 1st, 5th, 7th and 11th training step are projected

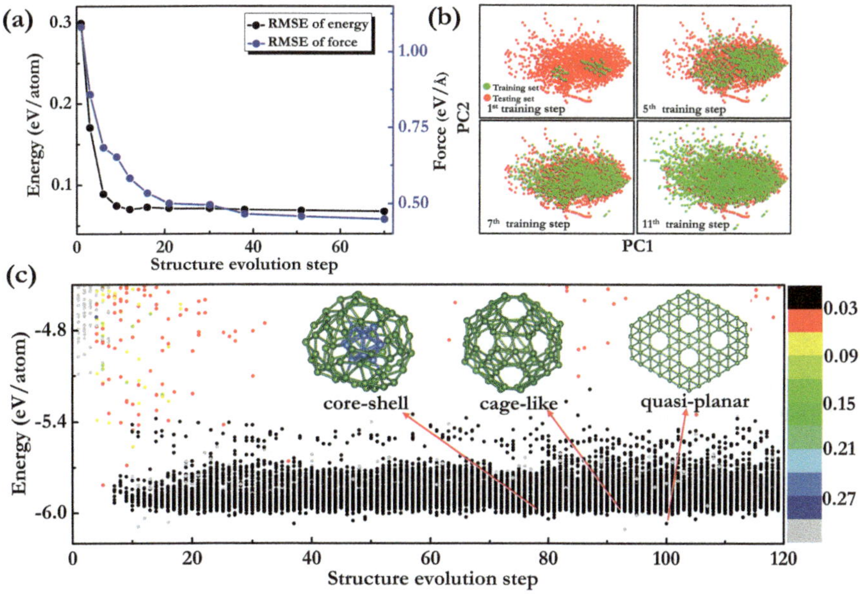

Fig. 5 (a) Evolution of the RMSEs of the GAP predicted energy and force for the testing set during the CALYPSO structure search for a B_{84} cluster, (b) PCA analysis of the training set and testing set at the 1st, 5th, 7th, and 11th GAP training steps, and (c) the evolution of the energy distributions during the CALYPSO structure search. Data points are colored according to the value of the predicted variance Σ. Lowest-energy B_{84} structures for core–shell, quasi-planar and cage-like configurations are shown in the insets.

onto the 2-dimensional space of the first two PCs as shown in Fig. 5(b) and compared with the testing data. It is clearly seen that the training set rapidly covers the testing set, indicating the gradual completeness of the training set.

The evolution of the energy distribution during the structure search is shown in Fig. 5(c) with data points colored according to the predicted variance Σ. It can be seen that the predicted variances Σ of the structures overall decrease during the structure search as a consequence of enlarging the training set and improving the GAP. The newly developed scheme allows a fast sampling of about ~17 000 local minimum structures, while only $\sim 2 \times 10^4$ DFT single-point calculations are performed during the structure search. A full DFT-based structure search in such a case needs ~17 000 DFT-based local optimizations in which about 1.7×10^6 DFT single-point calculations should be performed (supposing each local optimization contains 100 single-point calculations). Thus, the acceleration scheme reduces the computational cost by ~2 orders of magnitude compared with a DFT-based structure search in this test case.

The lowest-energy isomers for core–shell, quasi-planar and cage-like configurations in the current structure search are shown in the inset of Fig. 5(c) (more low-lying isomers are given in Fig. S2 in the ESI†). The lowest-energy structure of the quasi-planar configuration with four hexagonal holes is the same as that in a recent theoretical study.[68] New structures for core–shell and cage-like configurations are predicted. The energies of these low-lying structures calculated at different levels of theory are listed in Table 1. At the PBE level, the quasi-planar

Table 1 Relative energies of the predicted core–shell, quasi-planar and cage-like structures at PBE/PW, PBE0/6-31G(d) and TPSSh/6-31G(d) levels of theory. Energies are relative to the core–shell structure. Computational details can be found in the ESI

meV per atom	Core–shell	Quasi-planar	Cage-like
PBE/PW	0.0	−12.1	−1.29
PBE0/6-31G(d)	0.0	27.5	29.0
TPSSh/6-31G(d)	0.0	15.2	28.1

structure possesses the lowest energy. But at the PBE0 and TPSSh levels of theory, our newly predicted core–shell structure is lower in energy than the quasi-planar structure by 27.5 and 15.2 meV per atom, respectively. Recent benchmark calculations indicate that the TPSSh functional can give results consistent with those of the CCSD(T) high level theory for large-sized B clusters.[53] Thus, the core–shell structure is likely to be the ground state for the B_{84} cluster. In fact, our structure search calculations revealed a large number of low-lying structures with irregular core–shell configurations. This indicates a glasslike PES of B_{84} dominated by core–shell configurations.[70] Thus, the core–shell structure is preferred for B_{84} from both energetic and kinetic viewpoints.

3. Conclusions

In summary, using B clusters as testing systems, we have demonstrated the predictive capability of a well-trained GAP, which is qualified to be used to replace DFT during a global structure search. Then, two acceleration schemes for the structure prediction of large systems have been developed by combining GAP with our CALYPSO structure prediction method. One can choose between the two schemes depending on whether a well-trained GAP is available. Testing on medium- and large-sized B clusters demonstrated the efficiency and reliability of the current acceleration schemes. Moreover, a putative global minimum structure of the B_{84} cluster is proposed. We stress that although we use isolated clusters as testing systems in this work, the generalization of the proposed schemes to periodic systems (*e.g.* crystals) is straightforward by taking into account the stress tensor during the GAP training process. Note that there is a cutoff applied to the ACSF descriptors, which leads to strict localization of the atomic energies in eqn (1). Thus, long-range interactions outside the cutoff are not included in the GAP. This approximation puts a limit on the accuracy of the GAP but is expected to only have a marginal effect for most systems in the context of crystal structure prediction. We used a GAP and CALYPSO method in the current work, but the proposed schemes are general for current ML potentials and structure prediction methods. Thus, the current results represent a significant step toward the structure prediction of large systems.

Conflicts of interest

There are no conflicts to declare.

Acknowledgements

The authors acknowledge funding support from the National Natural Science Foundation of China (Grant Nos. 11534003, 11604117, 11404128 and 61775081), the Science Challenge Project, No. TZ2016001, the National Key Research and Development Program of China (Grant No. 2016YFB0201200), and the Program for JLU Science and Technology Innovative Research Team. Part of the calculations was performed in the high performance computing center of Jilin University and Tianhe2-JK in the Beijing Computational Science Research Center.

References

1 Y. Wang and Y. Ma, *J. Chem. Phys.*, 2014, **140**, 40901.
2 S. Kirkpatrick, C. D. Gelatt and M. P. Vecchi, *Science*, 1983, **220**, 671.
3 S. M. Woodley, P. D. Battle, J. D. Gale and C. R. A. Catlow, *Phys. Chem. Chem. Phys.*, 1999, **1**, 2535.
4 R. Martoňák, A. Laio and M. Parrinello, *Phys. Rev. Lett.*, 2003, **90**, 75503.
5 S. Goedecker, *J. Chem. Phys.*, 2004, **120**, 9911.
6 D. J. Wales and J. P. K. Doye, *J. Phys. Chem. A*, 1997, **101**, 5111.
7 A. R. Oganov and C. W. Glass, *J. Chem. Phys.*, 2006, **124**, 244704.
8 D. C. Lonie and E. Zurek, *Comput. Phys. Commun.*, 2011, **182**, 372.
9 C. J. Pickard and R. J. Needs, *J. Phys.: Condens. Matter*, 2011, **23**, 53201.
10 S. Q. Wu, M. Ji, C. Z. Wang, M. C. Nguyen, X. Zhao, K. Umemoto, R. M. Wentzcovitch and K. M. Ho, *J. Phys.: Condens. Matter*, 2014, **26**, 35402.
11 C. Shang, X.-J. Zhang and Z.-P. Liu, *Phys. Chem. Chem. Phys.*, 2014, **16**, 17845.
12 Y.-Y. Zhang, W. Gao, S. Chen, H. Xiang and X.-G. Gong, *Comput. Mater. Sci.*, 2015, **98**, 51.
13 Y. Wang, J. Lv, L. Zhu and Y. Ma, *Phys. Rev. B*, 2010, **82**, 94116.
14 Y. Wang, J. Lv, L. Zhu and Y. Ma, *Comput. Phys. Commun.*, 2012, **183**, 2063.
15 H. Wang, Y. Wang, J. Lv, Q. Li, L. Zhang and Y. Ma, *Comput. Mater. Sci.*, 2016, **112**, 406.
16 Y. Wang, J. Lv, L. Zhu, S. Lu, K. Yin, Q. Li, H. Wang, L. Zhang and Y. Ma, *J. Phys.: Condens. Matter*, 2015, **27**, 203203.
17 A. von Lilienfeld, *Angew. Chem., Int. Ed.*, 2018, **57**, 4164.
18 K. Hansen, F. Biegler, R. Ramakrishnan, W. Pronobis, O. A. Von Lilienfeld, K. R. Müller and A. Tkatchenko, *J. Phys. Chem. Lett.*, 2015, **6**, 2326.
19 M. Rupp, A. Tkatchenko, K.-R. Müller, V. Lilienfeld and O. Anatole, *Phys. Rev. Lett.*, 2012, **108**, 58301.
20 B. Huang and O. A. von Lilienfeld, *J. Chem. Phys.*, 2016, **145**, 161102.
21 K. T. Schütt, F. Arbabzadah, S. Chmiela, K. R. Müller and A. Tkatchenko, *Nat. Commun.*, 2017, **8**, 13890.
22 J. Sun, J. Wu, T. Song, L. Hu, K. Shan and G. Chen, *J. Phys. Chem. A*, 2014, **118**, 9120.
23 F. A. Faber, A. Lindmaa, O. A. von Lilienfeld and R. Armiento, *Phys. Rev. Lett.*, 2016, **117**, 135502.
24 F. Faber, A. Lindmaa, O. A. Von Lilienfeld and R. Armiento, *Int. J. Quantum Chem.*, 2015, **115**, 1094.

25 B. Meredig, A. Agrawal, S. Kirklin, J. E. Saal, J. W. Doak, A. Thompson, K. Zhang, A. Choudhary and C. Wolverton, *Phys. Rev. B*, 2014, **89**, 94104.

26 G. Montavon, M. Rupp, V. Gobre, A. Vazquez-Mayagoitia, K. Hansen, A. Tkatchenko, K.-R. Müller and O. Anatole von Lilienfeld, *New J. Phys.*, 2013, **15**, 95003.

27 K. T. Schütt, H. Glawe, F. Brockherde, A. Sanna, K. R. Müller and E. K. U. Gross, *Phys. Rev. B*, 2014, **89**, 205118.

28 F. A. Faber, L. Hutchison, B. Huang, J. Gilmer, S. S. Schoenholz, G. E. Dahl, O. Vinyals, S. Kearnes, P. F. Riley and O. A. Von Lilienfeld, *J. Chem. Theory Comput.*, 2017, **13**, 5255.

29 J. Behler, *J. Chem. Phys.*, 2016, **145**, 170901.

30 J. Behler and M. Parrinello, *Phys. Rev. Lett.*, 2007, **98**, 146401.

31 J. Behler, *J. Phys.: Condens. Matter*, 2014, **26**, 183001.

32 J. Behler, *Int. J. Quantum Chem.*, 2015, **115**, 1032.

33 P. E. Dolgirev, I. A. Kruglov and A. R. Oganov, *AIP Adv.*, 2016, **6**, 85318.

34 A. P. Bartók, M. C. Payne, R. Kondor and G. Csányi, *Phys. Rev. Lett.*, 2010, **104**, 136403.

35 A. P. Bartók and G. Csányi, *Int. J. Quantum Chem.*, 2015, **115**, 1051.

36 V. L. Deringer and G. Csányi, *Phys. Rev. B*, 2017, **95**, 94203.

37 Z. Li, J. R. Kermode and A. De Vita, *Phys. Rev. Lett.*, 2015, **114**, 96405.

38 R. M. Balabin and E. I. Lomakina, *Phys. Chem. Chem. Phys.*, 2011, **13**, 11710.

39 V. Botu, R. Batra, J. Chapman and R. Ramprasad, *J. Phys. Chem. C*, 2017, **121**, 511.

40 T. L. Jacobsen, M. S. Jørgensen and B. Hammer, *Phys. Rev. Lett.*, 2018, **120**, 26102.

41 R. Z. Khaliullin, H. Eshet, T. D. Kühne, J. Behler and M. Parrinello, *Nat. Mater.*, 2011, **10**, 693.

42 S. Gabardi, E. Baldi, E. Bosoni, D. Campi, S. Caravati, G. C. Sosso, J. Behler and M. Bernasconi, *J. Phys. Chem. C*, 2017, **121**, 23827.

43 V. L. Deringer, G. Csányi and D. M. Proserpio, *ChemPhysChem*, 2017, **18**, 873.

44 V. L. Deringer, C. J. Pickard and G. Csányi, 2017, arXiv:1710.10475.

45 J. Behler, *J. Chem. Phys.*, 2011, **134**, 74106.

46 A. P. Bartók, R. Kondor and G. Csányi, *Phys. Rev. B*, 2013, **87**, 184115.

47 J. Behler, *Angew. Chem., Int. Ed.*, 2017, **56**, 12828.

48 H. J. Zhai, B. Kiran, J. Li and L. S. Wang, *Nat. Mater.*, 2003, **2**, 827.

49 B. Kiran, S. Bulusu, H.-J. Zhai, S. Yoo, X. C. Zeng and L.-S. Wang, *Proc. Natl. Acad. Sci. U. S. A.*, 2005, **102**, 961.

50 J. Zhao, X. Huang, R. Shi, H. Liu, Y. Su and R. B. King, *Nanoscale*, 2015, **7**, 15086.

51 J. Lv, Y. Wang, L. Zhu and Y. Ma, *Nanoscale*, 2014, **6**, 11692.

52 H.-J. Zhai, Y.-F. Zhao, W.-L. Li, Q. Chen, H. Bai, H.-S. Hu, Z. A. Piazza, W.-J. Tian, H.-G. Lu, Y.-B. Wu, Y.-W. Mu, G.-F. Wei, Z.-P. Liu, J. Li, S.-D. Li and L.-S. Wang, *Nat. Chem.*, 2014, **6**, 727.

53 L. Sai, X. Wu, N. Gao, J. Zhao and R. B. King, *Nanoscale*, 2017, **9**, 13905.

54 F. Li, P. Jin, D. Jiang, L. Wang, S. B. Zhang, J. Zhao and Z. Chen, *J. Chem. Phys.*, 2012, **136**, 74302.

55 J. Zhao, L. Wang, F. Li and Z. Chen, *J. Phys. Chem. A*, 2010, **114**, 9969.

56 W.-L. Li, X. Chen, T. Jian, T.-T. Chen, J. Li and L.-S. Wang, *Nat. Rev. Chem.*, 2017, **1**, 71.

57 J. Lv, Y. Wang, L. Zhang, H. Lin, J. Zhao and Y. Ma, *Nanoscale*, 2015, **7**, 10482.
58 Y. Li, Y. Liao and Z. Chen, *Angew. Chem., Int. Ed.*, 2014, **53**, 7248.
59 Y. Wang, F. Li, Y. Li and Z. Chen, *Nat. Commun.*, 2016, 7, 11488.
60 S. Lu, Y. Wang, H. Liu, M. Miao and Y. Ma, *Nat. Commun.*, 2014, **5**, 3666.
61 J. Lv, Y. Wang, L. Zhu and Y. Ma, *Phys. Rev. Lett.*, 2011, **106**, 15503.
62 L. Zhu, H. Liu, C. J. Pickard, G. Zou and Y. Ma, *Nat. Chem.*, 2014, **6**, 644.
63 Y. Wang, M. Miao, J. Lv, L. Zhu, K. Yin, H. Liu and Y. Ma, *J. Chem. Phys.*, 2012, **137**, 224108.
64 X. Zhang, Y. Wang, J. Lv, C. Zhu, Q. Li, M. Zhang, Q. Li and Y. Ma, *J. Chem. Phys.*, 2013, **138**, 114101.
65 J. Lv, Y. Wang, L. Zhu and Y. Ma, *J. Chem. Phys.*, 2012, **137**, 84104.
66 C. Su, J. Lv, Q. Li, H. Wang, L. Zhang, Y. Wang and Y. Ma, *J. Phys.: Condens. Matter*, 2017, **29**, 165901.
67 Z. A. Piazza, H.-S. Hu, W.-L. Li, Y.-F. Zhao, J. Li and L.-S. Wang, *Nat. Commun.*, 2014, **5**, 3113.
68 A. B. Rahane and V. Kumar, *Nanoscale*, 2015, **7**, 4055.
69 B. Shang, L. F. Yuan, X. C. Zeng and J. Yang, *J. Phys. Chem. A*, 2010, **114**, 2245.
70 S. De, A. Willand, M. Amsler, P. Pochet, L. Genovese and S. Goedecker, *Phys. Rev. Lett.*, 2011, **106**, 225502.

Faraday Discussions

PAPER

Data-driven learning and prediction of inorganic crystal structures

Volker L. Deringer, *[ab] Davide M. Proserpio, [cd] Gábor Csányi[a] and Chris J. Pickard[ef]

Received 16th February 2018, Accepted 6th March 2018
DOI: 10.1039/c8fd00034d

Crystal structure prediction algorithms, including *ab initio* random structure searching (AIRSS), are intrinsically limited by the huge computational cost of the underlying quantum-mechanical methods. We have recently shown that a novel class of machine learning (ML) based interatomic potentials can provide a way out: by performing a high-dimensional fit to the *ab initio* energy landscape, these potentials reach comparable accuracy but are orders of magnitude faster. In this paper, we develop our approach, dubbed Gaussian approximation potential-based random structure searching (GAP-RSS), towards a more general tool for exploring configuration spaces and predicting structures. We present a GAP-RSS interatomic potential model for elemental phosphorus, which identifies and correctly "learns" the orthorhombic black phosphorus (A17) structure without prior knowledge of any crystalline allotropes. Using the tubular structure of fibrous phosphorus as an example, we then discuss the limits of free searching, and discuss a possible way forward that combines a recently proposed fragment analysis with GAP-RSS. Examples of possible tubular (1D) and extended (3D) hypothetical allotropes of phosphorus as found by GAP-RSS are discussed. We believe that in the future, ML potentials could become versatile and routine computational tools for materials discovery and design.

Introduction

Exploring and cataloguing new crystal structures is one of the principal tasks in chemistry. Decades of careful experimental work are collected in the Inorganic Crystal Structure Database (ICSD)[1] and the Cambridge Structural Database (CSD),[2] which have been impressive examples of "big data" since long before the

[a]*Department of Engineering, University of Cambridge, Cambridge CB2 1PZ, UK. E-mail: vld24@cam.ac.uk*
[b]*Department of Chemistry, University of Cambridge, Cambridge CB2 1EW, UK*
[c]*Dipartimento di Chimica, Università degli Studi di Milano, Milano, Italy*
[d]*Samara Center for Theoretical Materials Science (SCTMS), Samara State Technical University, Samara 443100, Russia*
[e]*Department of Materials Science and Metallurgy, University of Cambridge, Cambridge CB3 0FS, UK*
[f]*Advanced Institute for Materials Research, Tohoku University, 2-1-1 Katahira, Aoba, Sendai, 980-8577, Japan*

phrase became fashionable. While structural space has traditionally been explored by synthesis, more recent work has shown that *ab initio* structure searching can play an important and complementary role in this regard. Computational tools, including genetic algorithms,[3–5] particle swarm optimisation,[6,7] or *ab initio* random structure searching (AIRSS),[8,9] can predict structures that are (sometimes) far from what chemical intuition would suggest. Many of these predictions have subsequently been validated by experiments[10–14] or have given rise to databases of their own.[15] We only mention in passing the prediction of organic molecular crystal structures, which has likewise seen fundamental breakthroughs.[16–19]

Despite their predictive power, these structure searching methods are normally driven by quantum-mechanical density-functional theory (DFT) computations, and therefore they are limited to systems with relatively small unit cells. The allotropes of elemental phosphorus,[20–22] which are the topic of the present paper, directly illustrate the problem at hand. The thermodynamically stable "black" (orthorhombic) form, as well as the high-pressure As-type (rhombohedral) allotrope, exhibit simple crystal structures with only one symmetry-independent atom each (Table 1). Such systems are easily amenable to *ab initio* crystal structure prediction, and various, especially layered, hypothetical allotropes have been proposed in recent years.[23–32] On the other hand, "violet" (Hittorf's) phosphorus[33] has a notoriously complex structure that was solved more than 100 years after the initial synthesis[34] and contains no fewer than 21 symmetry-independent atoms in the unit cell. Systems of this size have hitherto been out of reach for *ab initio* crystal structure prediction.

It has recently been suggested that machine learning (ML) based interatomic potentials could help with this long-standing issue.[35–38] Such potentials are fitted to reference databases of DFT energies and forces and, once generated, they allow one to perform atomistic simulations with close to DFT quality but at a computational cost that is orders of magnitude lower.[39] Indeed, we have recently shown

Table 1 Experimentally known crystal structures of phosphorus.[20] Z is the number of atoms in the conventional unit cell; Z' gives the number of symmetry-inequivalent atoms in the cell, and is therefore a measure of structural complexity. The molecular ("white") and amorphous ("red") forms are omitted from this study for simplicity[b]

	Space group	Z	Z'	Source	Ref.
Black P	*Cmca*	8	1	High-pressure synthesis	43
As-type P[a]	$R\bar{3}m$	2	1	From black P ($p > 5$ GPa)	44
Simple cubic P	$Pm\bar{3}m$	1	1	From black P ($p > 11$ GPa)	44
Hittorf's P	*P*2/*c*	84	21	Slow cooling from Pb melt	34
Fibrous P	$P\bar{1}$	42	21	Resublimation with I catalyst	45

[a] Occasionally referred to as "blue P" in recent literature,[46] especially in the monolayer form.
[b] To explore whether our GAP would be able to find molecular ("white") P, we carried out a set of preliminary searches at low density (1.0 g cm^{-3}). The resulting structures did include distorted P_4 units, but also other small fragments, and we expect that the GAP will need to have "seen" these in iterative training to distinguish them more clearly. We also expect that fully capturing the intricate structural details of white P, including its packing variants,[74] will require additional reference data; this is the topic of ongoing work. Regarding amorphous forms, our searches are restricted to relatively small periodic unit cells which cannot fully represent the amorphous allotrope(s).

that an ML-based Gaussian approximation potential (GAP), initially developed for amorphous carbon,[40] can be used to identify crystalline allotropes.[36] This has added around 150 entries to the Samara Carbon Allotrope Database (SACADA, ref. 15). Very recently, the use of ML to speed up global searching and crystal structure optimization itself has been reported,[41,42] but we focus here on the use of explicit interatomic potentials ("force fields") for this purpose. What is more, in recent work,[37] we have argued that RSS can be used to construct the interatomic potential from scratch, exploring and fitting a complex potential energy surface (PES) at the same time. This points towards a more general "data-driven" strategy for atomistic materials modelling.

In this Discussion paper, we aim to take new steps in this direction and further develop our emerging method (which we call "GAP-RSS"). After briefly summarising its components, we present results for elemental phosphorus, generating and applying the first GAP-RSS potential for this material. Our protocol "discovers" the crystal structure of black P during the iterations and, by construction, adds it to the reference database without prior knowledge of any existing allotropes. We then use this potential for some exercises in crystal structure prediction: we show that a fibrous allotrope of P (Table 1) appears to be prohibitively hard to find in free searches, and we outline the use of an alternative approach, showing exemplary predicted structures in 1D and 3D. We also discuss open questions and expected future directions.

Methodology

The protocol for GAP-RSS potential fitting and structure searching consists of three components: single-point DFT computations, GAP fitting to an updated reference database, and structural optimisation using GAP. We give details of these in sequence, and an overview is provided in Fig. 1.

DFT computations

These provide the input data for GAP fitting: initially, single-point DFT computations are done for randomised atomic configurations, later, for intermediates or local minima of GAP-RSS searches. We obtain these data using standard DFT procedures, with high numerical accuracy to minimise noise in the input for fitting. In this work, we used the PBEsol functional,[47] which has been validated for black P before,[48] and on-the-fly ultra-soft pseudopotentials as implemented in CASTEP.[49] Reciprocal space was sampled on dense *k*-point meshes (maximum spacing 0.02 $Å^{-1}$). The cut-off energy for plane-wave expansions was 600 eV, and an extrapolation scheme was used to counteract finite-basis errors.[50] In Fig. 1, all parts that involve single-point DFT computations (*viz.* the construction of the reference database) are enclosed by dashed lines.

GAP fitting

With reference data available, an interatomic potential is fitted to these using GAP. The framework was introduced in 2010 (ref. 51) and has since been used to generate potentials for diverse molecular and solid-state materials.[40,52–54] A high-dimensional fit to reference energy and force data is performed based on structural similarity or kernel functions, comparing atomic environments one by one. The initial choice for

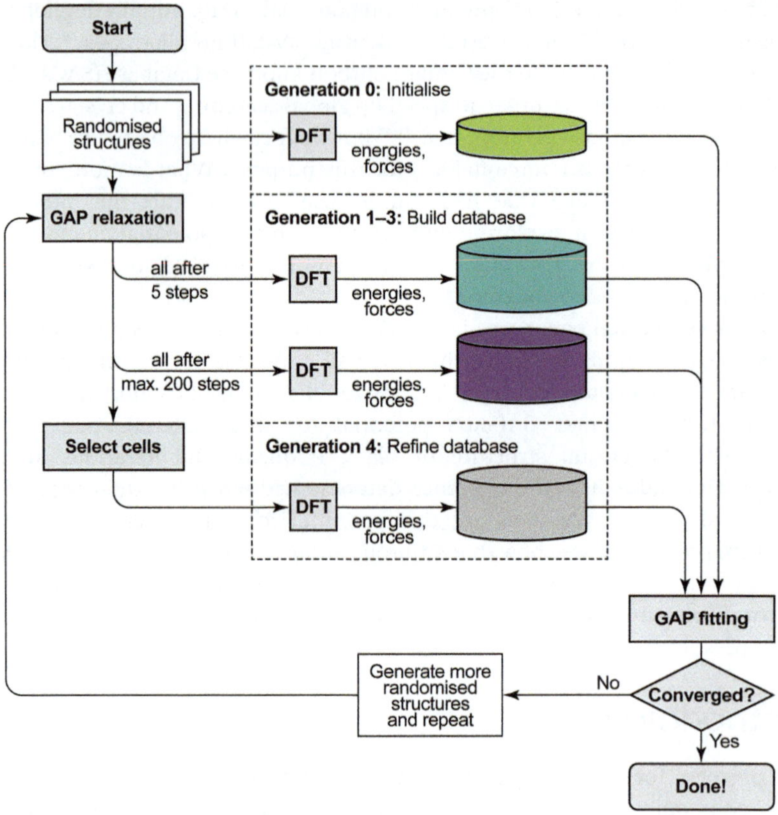

Fig. 1 Overview of the GAP-RSS protocol as introduced in ref. 37 and extended here by a selection step. We start from a set of randomised seed structures for a given chemical composition (in generations **0–3**, 250 each; in generation **4**, 5000). The reference database of energies and forces is then built by single-point DFT computations: first, for unrelaxed structures (generation **0**), later, at various stages of structural relaxation (**1–4**), performed by fitting interim GAPs to the evolving database. Initially, we feed all structures back into the training (up to generation **3**); later, we only add selected structures. As a criterion to select cells, here we use coordination numbers, demanding that all atoms in a candidate structure must have three nearest neighbours, and discarding any structures which do not comply. The GAP-RSS iterations are deemed "converged" when the resulting potential shows satisfactory performance. Here, we do so after generation **4**, as that version of the potential has "found" black P from scratch (see Fig. 3 and 4).

these have been many-body descriptors, most importantly the Smooth Overlap of Atomic Positions (SOAP), which includes all neighbours of an atom up to a cut-off radius.[55] To improve the stability of the fit, similar to our previous work on amorphous materials modelling[40] and structure searching,[36,37] we combine the many-body SOAP expansion with non-parametric two-body ("2b") and three-body ("3b") terms that encode interatomic distances and bond angles, respectively. The 2b and SOAP descriptors have radial cut-offs of 5.0 Å, whereas that for the 3b term is 2.6 Å (to include only "true" bond angles involving nearest-neighbour contacts). The final GAP uses 10 sparse points (that is, fitting coefficients) for the 2b term, 100 for the 3b term, and 3000 for SOAP. The convergence and smoothness parameters for the SOAP expansion are $n_{\max} = l_{\max} = 8$ and $\sigma_{\text{at.}} = 0.75$ Å, respectively, which were found to be suitable for GAP-RSS in ref. 37. For a more detailed walk-through of the underlying ML framework, we refer the reader to ref. 56.

GAP-RSS relaxation

This is the heart of the technique: structural space is explored by optimising random structures, akin to the well-established AIRSS technique, but now using GAP. In AIRSS (and consequently in GAP-RSS), it is important to make a sensible choice of randomised initial structures. For example, a reasonable minimum interatomic distance ("hard-sphere" criterion) is imposed. Furthermore, exploiting space-group symmetry can significantly reduce the number of attempts required. Here we search with 2–16 atoms in the unit cell, allowing for either 1, 2, 4, or 8 symmetry operations to be present. To some extent, this penalises rhombohedral space groups and their subgroups, but our data below show that the rhombohedral A7 structure is found by our protocol nonetheless. We choose the initial densities to be distributed around 2.5 g cm^{-3}, in between the black and red forms of phosphorus, and we constrain the P–P distances in the initial structures to at least 1.8 Å. An independent set of randomised input structures is generated for initialisation and for each new generation of the potential.

Results and discussion

Exploring the potential energy surface of elemental phosphorus

To showcase the principle of GAP-RSS, in this paper we explore the potential energy surface (PES) of elemental phosphorus. This is summarised in Fig. 2, where we show the energy–volume data for the reference database that is built

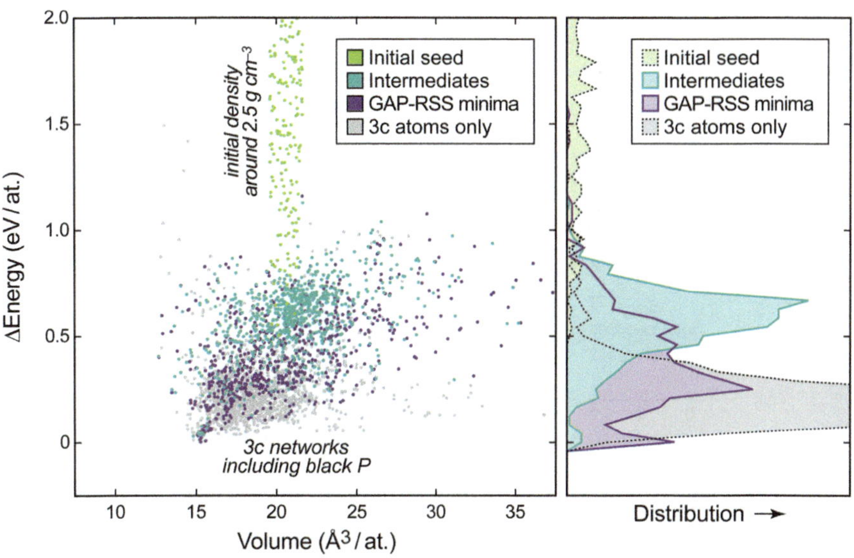

Fig. 2 Left: energy–volume plot of single-point DFT data during the generation of a GAP-RSS model for phosphorus, given relative to the most stable structure. The different stages of building the database (*cf.* Fig. 1) are highlighted in different colours. Right: binned distribution for energies in these datasets, drawn on the same vertical axis.

during construction of the potential. We start with randomised input structures, the data for which are added to the database without relaxation (light green points in Fig. 2; generation **0**). We then extend the database by three rounds (generations **1–3**), where snapshots of both intermediate (teal) and relaxed (purple) structures are added; finally, in generation **4**, we perform a larger search and keep only the unique structures with all-threefold coordinated atoms ("3c", determined using a 2.4 Å bond-length cut-off; light grey points). After this, we deem the database "converged" for the purpose of the present study, by the (subjectively chosen) criterion that the GAP has "learned" black P; see below. Future work will be concerned with less heuristic criteria for convergence. The distributions shown on the right-hand side of Fig. 2 illustrate how the different parts of the database, and thus of the PES, comprise progressively lower-lying structures.

We will show in the following that, concomitant with increasing exploration of the phase space, the GAP-RSS interatomic potential becomes more accurate, ultimately describing crystal structures without prior knowledge. We stress that in most of the previous literature on fitting ML potentials, reference databases are constructed by including known crystal structures (and distorted variants thereof).[39,40,53]

To assess the quality of the potential, we use it to compute energies for independent test sets of structures not included in the training, for which the DFT energies are known. The results are shown in Fig. 3a. The description of the higher-energy regions ("RSS intermediates", open symbols) appears to converge quite early on, but it carries a residual root-mean-square prediction error of the order of 0.05–0.10 eV per at. that does not improve with further iterations. The same is observed, and is even more pronounced, for a test set of randomised seed structures, where the error is practically constant at ≈0.10 eV per at. for all the GAP generations (not shown in Fig. 3a for clarity). This is intuitively understandable: the PES is smooth, and since the high-energy structures are so diverse,

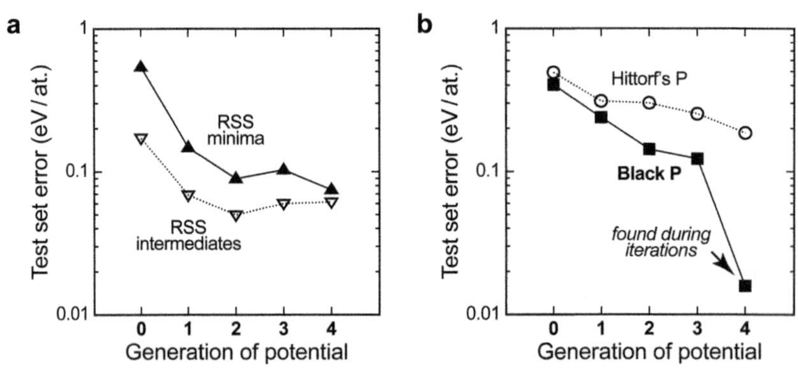

Fig. 3 Quality of the evolving interatomic potential, assessed by computing the energies for independent test sets (configurations outside the database) with different generations of the GAP, and reporting the root-mean-square error against the DFT data. (a) Errors for sets of structures from an independent GAP-RSS run without symmetry, taken after 5 CG steps ("intermediates") and 200 CG steps ("minima"); (b) the same for sets of distorted unit cells of known allotropes, *viz.* Hittorf's and black P. In generation **4**, our GAP has "seen" black P and therefore the predicted error for this allotrope falls close to zero (see the arrow).

only a handful of them are required for acceptable sampling. A similar observation was made for liquid carbon in our earlier work: the high-temperature liquid contains many different environments and is thus straightforward to "learn" from a single *ab initio* MD trajectory. In contrast, the amorphous regions of the PES required several rounds of GAP-driven MD and iterative refinement of the potential.[40]

For the purposes of GAP-RSS, these findings are encouraging: the early steps of relaxation appear to be easily "learned" by the potential, and even a notable degree of residual error can be tolerated if the potential is successful in navigating the high-energy region of the PES (say, the green data points in Fig. 2). In contrast, a growing database is needed for the relaxed minima. Indeed, testing for a set of local RSS minima outside the database initially leads to a significant prediction error above 0.5 eV per at. (filled symbols in Fig. 3a), but this gradually improves and falls below 0.1 eV per at. with increasing quality of the potential.

The latter accuracy turns out to be acceptable for exploratory searches, but it is still worse than what one would expect from an ML potential for the stable allotropes. Our long-term vision for GAP-RSS is therefore to find stable minima (such as black or orthorhombic P) during the iterations and automatically include them in the reference database. We expect this to improve the accuracy substantially, which is supported by the following result. Fig. 3b shows test set errors as before, but now for ensembles of distorted unit cells of two experimentally known allotropes. The orthorhombic structure of black P is a particularly interesting test, as it competes with other, more highly symmetric structures that are very close in energy on the PES.[22] Our potential, during generation **4**, did succeed in finding black P, and hence includes this structure type in the final reference database. This is reflected in Fig. 3b: the energy errors for black P get progressively lower, but only in generation **4** do they drop to very close to zero.

Likewise, the computed energy–volume scans for black P (Fig. 4a) progressively improve during the GAP-RSS iterations, but the structure needs to have been "seen" and included in the fitting (in generation **4**) to achieve an accurate result. The same is true for computed structural properties (Fig. 4b): for this test, we fully relaxed the black P structure with each version of our potential, as well as with DFT. In generation **4**, all three GAP-computed lattice parameters come very close to the DFT benchmark.

"Learning" high-pressure structures

Two important phosphorus allotropes, As-type and simple-cubic, are obtained from the black form at high pressure (Table 1).[44] With this in mind, we decided to explore what effect external pressure would have on the result of GAP-RSS for this element; doing so has already been suggested and proven to be beneficial for exploring the PES of elemental boron.[37] To sample higher-pressure structures, we apply hydrostatic external pressure during the GAP-RSS relaxation runs, with a randomised value in each run. The values are drawn from an exponential distribution with the probability density P of finding a given pressure p,[37]

$$P(p) = \lambda \exp(-\lambda p) \times p_0$$

where λ is the rate parameter (we chose $\lambda = 5$, a narrower distribution, to sample more around the low-pressure region but still include some high-pressure points),

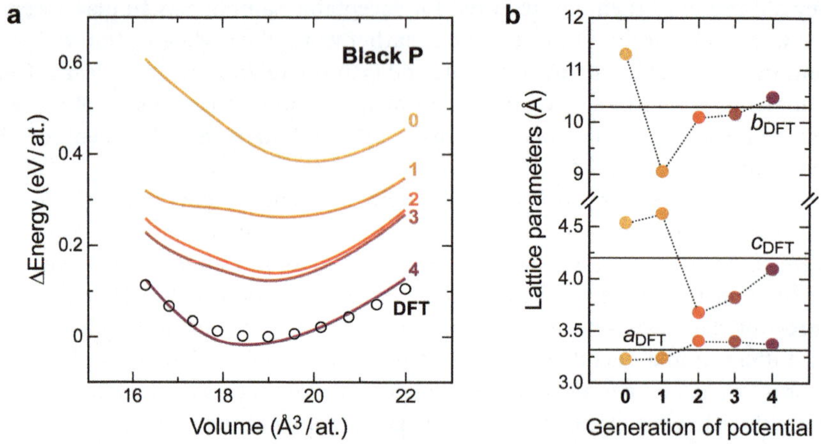

Fig. 4 Quality of the evolving interatomic potential, assessed by computing properties of black P. (a) Energy–volume scans for the orthorhombic unit cell (scaling lattice vectors and atomic coordinates without further relaxation), comparing progressive GAP generations (lines) to DFT reference data (circles). (b) Lattice parameters, obtained by fully relaxing the crystal structure of black P with each generation of the GAP. DFT results, obtained at the same computational settings as for the reference database, are given as horizontal lines.

and p_0 is an arbitrarily chosen reference. In this work, we tested the settings of $p_0 = 10$ GPa and $p_0 = 100$ GPa, respectively. For both, we generated new sets of potentials through iterative GAP-RSS fitting, starting with the same ensembles of seed structures as used in the pressure-free searches reported above ($p_0 = 0$ GPa).

The results are easily rationalised by looking at the evolving energy–volume scans again, but now for the high-pressure forms (Fig. 5). For reference, our pressure-free search ($p_0 = 0$ GPa) already yields an acceptable description of both allotropes in generation **4** (purple lines), but it fails to reach quantitative accuracy, especially for the simple cubic form. Searching with moderate external pressure ($p_0 = 10$ GPa) visibly improves the description of both high-pressure forms: the GAP and DFT data are now in better agreement. Note that due to the nature of the exponential distribution, most of the pressure values drawn are significantly smaller than p_0, and thus correspond to typical experimental conditions. When using external pressures that are an order of magnitude higher ($p_0 = 100$ GPa), the description of the As-type form deteriorates visibly, whereas the simple cubic form is already "learned" very easily in generation **2** (bottom right panel in Fig. 5). All this is intuitively understood, as the experimentally observed sequence of pressure-induced phase transitions is black → As-type → simple cubic, with transition pressures around 5 GPa and 11 GPa, respectively.[44,57]

Fibrous phosphorus

"Fibrous" P, described by Ruck *et al.* in 2005,[45] is structurally reminiscent of Hittorf's P: both allotropes exhibit characteristic tubes built from covalently bonded cage motifs, reminiscent of organophosphorus compounds.[58,59] The difference between the two allotropes is in the arrangement of the tubes, which is simpler (all tubes parallel) in fibrous P. Both allotropes contain 21 independent

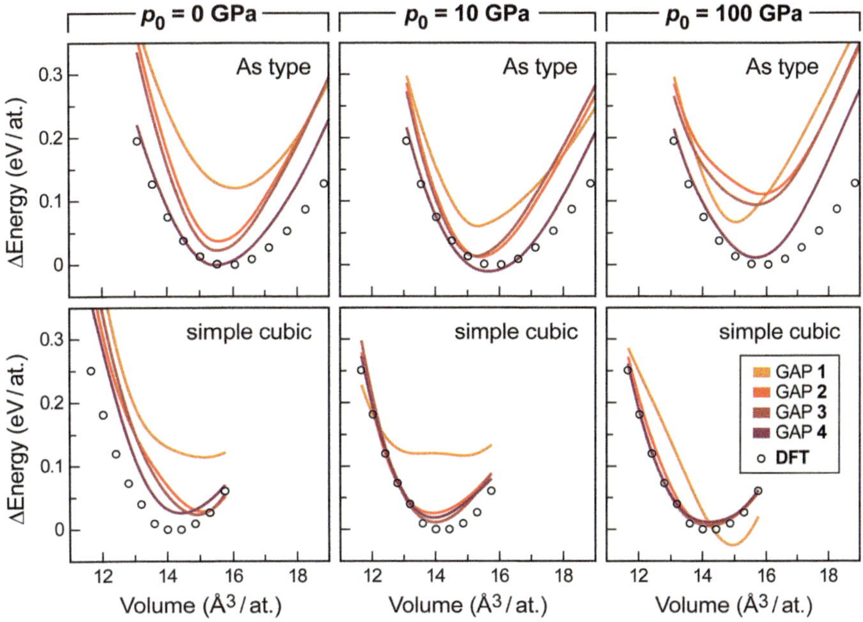

Fig. 5 Energy–volume scans similar to those in Fig. 4a, but now assessing the high-pressure forms, As-type (top) and simple cubic P (bottom). Data were obtained with three separate sets of GAP-RSS iterations that utilise, from left to right: no external pressure (*i.e.* the main dataset), 10 GPa, or 100 GPa of reference pressure p_0, as defined in the text. In all cases, the respective lowest DFT energy point is set as zero. The data for generation **0** are outside the range of the energy axis.

atomic positions (Table 1), which are repeated by space-group symmetry two and four times, respectively.

We performed 100 000 attempts to find the structure of fibrous P using GAP-RSS, seeding with 21 independent, randomised atomic positions in space group $P\bar{1}$. However, this search was unsuccessful, and none of the attempts led to the known structure after minimisation. We show three of the resulting structures, as mere examples, in Fig. 6. This reflects the more general problem that in global structure determination, the searching task becomes exponentially more complex with system size (and each independent atom adds three structural degrees of freedom), as discussed by Stillinger[60] and in ref. 9. While the GAP still provides an approximation of the DFT potential energy surface, and therefore some care must be taken when transferring conclusions from GAP-RSS to (DFT-based) AIRSS, we take the present findings as clear evidence that the structure of fibrous P (and, by extension, Hittorf's P) is too complex to find in free random searches.

Merging modular decomposition and GAP-RSS: towards fully data-driven crystal structure prediction

Inevitably, a fully unconstrained random search for more and more complex structures will fail at some point, as discussed above. Very recent work has shown that molecular network analysis can be used to great effect in this regard.[61]

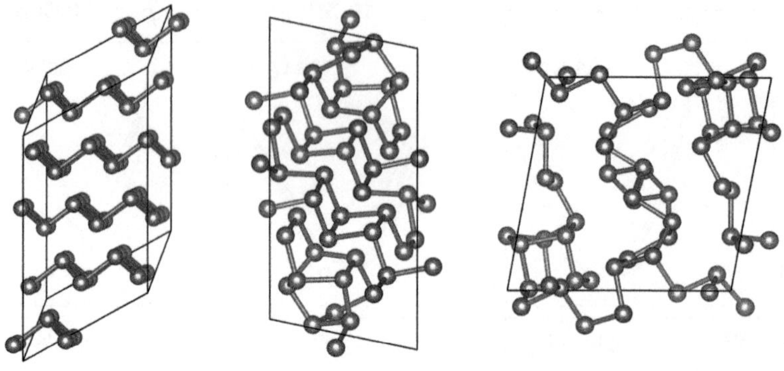

Fig. 6 Unsuccessful attempts to find fibrous P in free searches. Exemplarily, we show three low-energy, DFT-relaxed structures resulting from 100 000 GAP-RSS attempts (21 symmetry-independent atoms in space group $P\bar{1}$). One of these structures is a stacking variant of As-type phosphorus (left); the two other are low-symmetry configurations that are clearly different from the tubular structure of fibrous (and Hittorf's) P.

Inspired by Pauling's iconic rule of parsimony, stating that the number of constituents in a crystal structure tends to be small,[62] it was proposed in ref. 61 to decompose a crystal structure into building blocks such as to minimise the (mathematically quantifiable) information content. Here we combine this with GAP-RSS searching, and argue that both together provide a useful way forward for the prediction of complex crystal structures.

Fig. 7a summarises the procedure and illustrates how the structure of fibrous P is decomposed into two fragments using the approach of ref. 61. While this is a purely automated, data-driven procedure, its outcome is often in line with chemical intuition.[61] Indeed, in the original publication on fibrous P, the structure was described as a sequence of alternating [P_8] and [P_9] cages, interspersed with P_2 dumbbells.[45] Similar fragments have been identified by Thurn and Krebs in the crystal structure of Hittorf's P;[34] they have been discussed in Baudler's seminal work on the nomenclature of phosphorus cages,[58] and have been the topic of comprehensive theoretical analyses both in the gas phase and in the bulk.[22,59,64] Without chemical knowledge, but with an algorithm to find the most simple and representative building units, the network analysis likewise "cut" the chains into fragments.[61] In these, the well-known [P_9] cage was combined with P_2 dumbbells on either side, forming a [P_{13}] unit ("Fragment A"), whereas the [P_8] cage was recovered directly ("Fragment B"). We believe that alternative ways of decomposition, *e.g.* isolating the [P_9] cage, may also be useful as starting points for structure searches—as will combinations of different fragments. We also note the advantage of a chemically "agnostic" approach: while there is an extensive body of literature on the building blocks of P allotropes,[22,58,59] future work might be concerned with new materials where the local atomic structure is *a priori* unknown.

Even with an exemplary and simplified approach that uses only one fragment or the other (Fig. 7a), our GAP-RSS searches readily identified several hypothetical phosphorus allotropes with different dimensionalities. We filtered the output with more stringent criteria than in the preceding GAP-RSS generations, enforcing a minimum "non-bonded" (beyond-nearest-neighbour) distance of 2.9 Å,

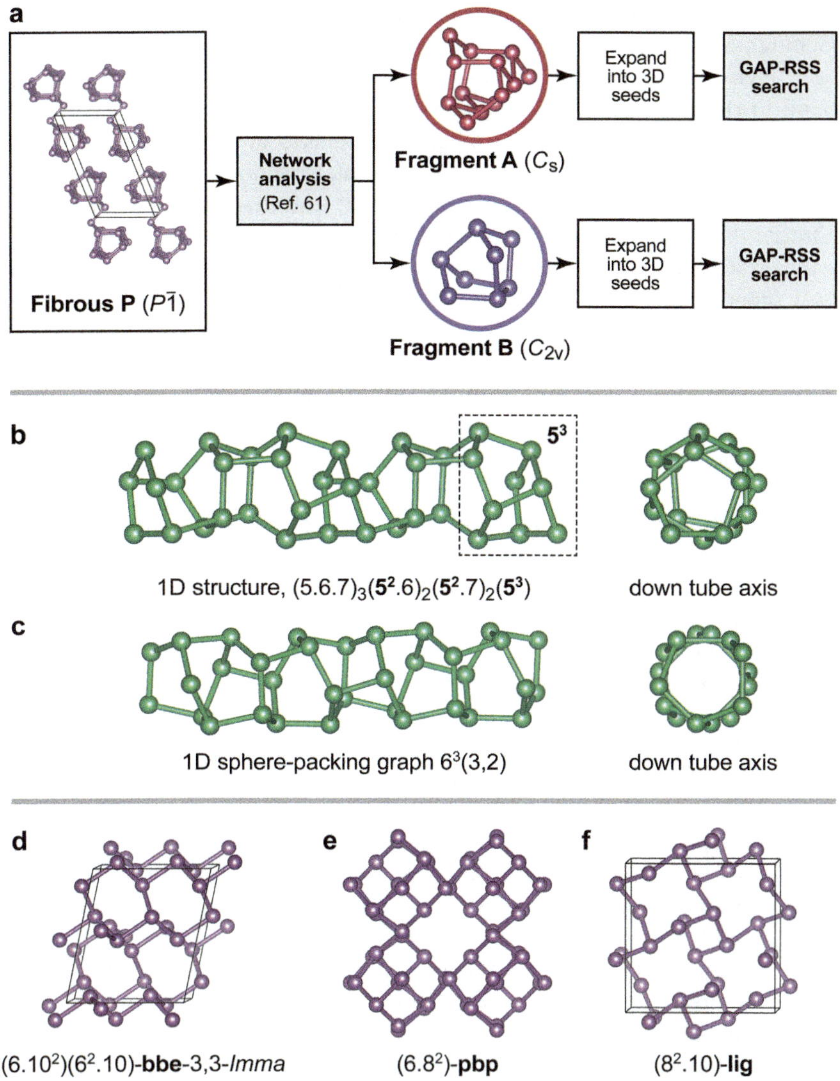

Fig. 7 (a) Overview of the fragment-based approach to structure analysis and structure searching introduced in ref. 61 that we combine here with GAP-RSS. Two fragments have previously been identified in the structure of fibrous P (see text),[61] and here we use both separately to generate seed structures for GAP-RSS structure searches. (b) Example of a 1D-periodic structure identified by this procedure. The tube "inherits" some of the structural information from Fragment B, namely, fused five-membered rings (dashed box). Full topology information is given in the text. (c) The same for a 1D structure composed of six-membered rings only. (d–f) Examples of 3D-periodic structures from the same search, labelled according to their network topologies.[63]

removing any structures that contain three- or four-membered rings (which are expected to be under significant strain), and post-relaxing selected structures using dispersion-corrected DFT.[65–67] We explored the effect of feeding structural data back into the GAP as before, but now for the more ordered structures coming from fragment-based searches, and generated two additional generations of the potential. An intermediate generation was used to find the structures shown in Fig. 7 (generation **5a**); another, final one additionally contains all the minima found in that search (**5b**). We observed that this reduces the energy error for distorted unit cells of Hittorf's P (*cf.* Fig. 3b) from 0.19 eV per at. (generation **4**) to 0.16 eV per at. (**5a**), and further to 0.13 eV per at. (**5b**); note that in none of these generations has our GAP-RSS potential "seen" the actual crystal structure of Hittorf's P. A full account of this, including the results of much more diverse searches, will be published elsewhere, and for the sake of brevity we present only a few salient examples here.

As expected, several 1D tubular motifs were found in our GAP-RSS searches, packed in different ways to form extended structures. One of these tubes, shown in Fig. 7b, retains some features of the [P_8] cage from which the initial seed is constructed: namely, fused five-membered rings, of which there are four in the [P_8] cage (topology symbol [5^4]), and up to three in the chain structure (dashed box in Fig. 7b). Another structure found by GAP-RSS, shown in Fig. 7c, breaks up the initial seed fragment, forming a tube that consists of only six-membered rings. This corresponds to a sphere-packing graph for a hexagonal net (6^3) with coincidence vectors $(m, n) = (3, 2)$.[68] In this, it is also equivalent to one of the smallest theoretically possible carbon nanotubes.

The same protocol also identified fully three-dimensional structures. One of them is a rare binodal net, **bbe**-3,3-*Imma* (Fig. 7d), which has occasionally been observed in MOFs.[69] We find another structure with a uninodal **pbp** net containing six- and eight-membered rings (Fig. 7e) that had originally been proposed for a hypothetical carbon structure dubbed "6.8^2P" (ref. 70) and also investigated as a hypothetical allotrope of P.[27] We finally find the uninodal **lig** net (Fig. 7f), which corresponds to the anion network in the Zintl phase LiGe.[71] Such a structure has not been proposed for P thus far, but it is in conceptual agreement with the Zintl–Klemm–Busmann framework (in which the group-14 element is viewed as "Ge^-", and thus takes the structural signature of a group-15 element due to its excess electron).[72] The computed dispersion-corrected DFT energies[65–67] for the structures shown in Fig. 7d–f are $\approx$+2 kJ mol^{-1} (**bbe**-3,3-*Imma*) and $\approx$+8 kJ mol^{-1} (**pbp** and **lig**) above that of black P, placing them well within the experimentally and computationally derived stability range of known allotropes.[22]

In ongoing work, beyond the scope of this paper, we are performing much larger-scale searches, including attempts to find "fibrous" P and possible related structures using GAP-RSS, and trying to understand the observed preference for the experimentally known form. It seems especially interesting to study such tubular forms since some of these structural units can be chemically extracted from phosphorus-rich compounds.[73]

Conclusions

Machine learning-based interatomic potentials can speed up random structure searching by orders of magnitude. Therefore, they seem well poised to become useful tools for crystal structure prediction and materials discovery. We expect them not to replace DFT-driven searching, but to provide a valuable complement, especially for very large and complex structures that are outside the reach of DFT. In this paper, we have discussed and further developed our recently introduced technique, dubbed Gaussian approximation potential-driven random structure searching (GAP-RSS), which combines ideas from the fields of potential fitting and structure prediction. We believe that this technique will be of interest not only for finding new structures, but also for the automated generation of fast, flexible, and accurate interatomic potentials for diverse materials.

Conflicts of interest

There are no conflicts to declare.

Acknowledgements

We thank Dr Noam Bernstein for ongoing valuable discussions. V. L. D. gratefully acknowledges a Feodor Lynen fellowship from the Alexander von Humboldt Foundation, a Leverhulme Early Career Fellowship, and support from the Isaac Newton Trust. C. J. P. is supported by the Royal Society through a Royal Society Wolfson Research Merit award. This work used the ARCHER UK National Supercomputing Service (http://www.archer.ac.uk) *via* EPSRC grant EP/P022596/1. Data access statement: original data supporting this publication are available at https://doi.org/10.17863/CAM.25572.

Notes and references

1 A. Belkly, M. Helderman, V. L. Karen and P. Ulkch, *Acta Crystallogr., Sect. B: Struct. Sci.*, 2002, **58**, 364–369.
2 C. R. Groom, I. J. Bruno, M. P. Lightfoot and S. C. Ward, *Acta Crystallogr., Sect. B: Struct. Sci.*, 2016, **72**, 171–179.
3 A. R. Oganov and C. W. Glass, *J. Chem. Phys.*, 2006, **124**, 244704.
4 A. R. Oganov, A. O. Lyakhov and M. Valle, *Acc. Chem. Res.*, 2011, **44**, 227–237.
5 D. C. Lonie and E. Zurek, *Comput. Phys. Commun.*, 2011, **182**, 372–387.
6 Y. Wang, J. Lv, L. Zhu and Y. Ma, *Phys. Rev. B: Condens. Matter Mater. Phys.*, 2010, **82**, 094116.
7 Y. Wang, J. Lv, L. Zhu and Y. Ma, *Comput. Phys. Commun.*, 2012, **183**, 2063–2070.
8 C. J. Pickard and R. J. Needs, *Phys. Rev. Lett.*, 2006, **97**, 045504.
9 C. J. Pickard and R. J. Needs, *J. Phys.: Condens. Matter*, 2011, **23**, 053201.
10 Y. Ma, M. Eremets, A. R. Oganov, Y. Xie, I. Trojan, S. Medvedev, A. O. Lyakhov, M. Valle and V. Prakapenka, *Nature*, 2009, **458**, 182–185.
11 M. Marqués, M. I. McMahon, E. Gregoryanz, M. Hanfland, C. L. Guillaume, C. J. Pickard, G. J. Ackland and R. J. Nelmes, *Phys. Rev. Lett.*, 2011, **106**, 095502.
12 H. Wang, J. S. Tse, K. Tanaka, T. Iitaka and Y. Ma, *Proc. Natl. Acad. Sci. U. S. A.*, 2012, **109**, 6463–6466.
13 W. Zhang, A. R. Oganov, A. F. Goncharov, Q. Zhu, S. E. Boulfelfel, A. O. Lyakhov, E. Stavrou, M. Somayazulu, V. B. Prakapenka and Z. Konôpková, *Science*, 2013, **342**, 1502–1505.
14 M. Jansen, *Adv. Mater.*, 2015, **27**, 3229–3242.
15 R. Hoffmann, A. A. Kabanov, A. A. Golov and D. M. Proserpio, *Angew. Chem., Int. Ed.*, 2016, **55**, 10962–10976.
16 S. L. Price, *Acc. Chem. Res.*, 2009, **42**, 117–126.

17 A. M. Reilly, *et al.*, *Acta Crystallogr., Sect. B: Struct. Sci.*, 2016, **72**, 439–459.

18 A. Pulido, L. Chen, T. Kaczorowski, D. Holden, M. A. Little, S. Y. Chong, B. J. Slater, D. P. McMahon, B. Bonillo, C. J. Stackhouse, A. Stephenson, C. M. Kane, R. Clowes, T. Hasell, A. I. Cooper and G. M. Day, *Nature*, 2017, **543**, 657–664.

19 F. Musil, S. De, J. Yang, J. E. Campbell, G. M. Day and M. Ceriotti, *Chem. Sci.*, 2018, **9**, 1289–1300.

20 J. Donohue, *The Structures of the Elements*, John Wiley & Sons, New York, 1974.

21 A. Pfitzner, *Angew. Chem., Int. Ed.*, 2006, **45**, 699–700.

22 F. Bachhuber, J. von Appen, R. Dronskowski, P. Schmidt, T. Nilges, A. Pfitzner and R. Weihrich, *Angew. Chem., Int. Ed.*, 2014, **53**, 11629–11633.

23 A. J. Karttunen, M. Linnolahti and T. A. Pakkanen, *Chem.–Eur. J.*, 2007, **13**, 5232–5237.

24 M. Wu, H. Fu, L. Zhou, K. Yao and X. C. Zeng, *Nano Lett.*, 2015, **15**, 3557–3562.

25 T. Zhao, C. Y. He, S. Y. Ma, K. W. Zhang, X. Y. Peng, G. F. Xie and J. X. Zhong, *J. Phys.: Condens. Matter*, 2015, **27**, 265301.

26 J.-R. Feng and G.-C. Wang, *RSC Adv.*, 2016, **6**, 22277–22284.

27 J. Liu, S. Zhang, Y. Guo and Q. Wang, *Sci. Rep.*, 2016, **6**, 37528.

28 D. Liu, J. Guan, J. Jiang and D. Tománek, *Nano Lett.*, 2016, **16**, 7865–7869.

29 Z. Zhuo, X. Wu and J. Yang, *J. Am. Chem. Soc.*, 2016, **138**, 7091–7098.

30 W. H. Han, S. Kim, I.-H. Lee and K. J. Chang, *J. Phys. Chem. Lett.*, 2017, **8**, 4627–4632.

31 H. Wang, X. Li, Z. Liu and J. Yang, *Phys. Chem. Chem. Phys.*, 2017, **19**, 2402–2408.

32 G. Sansone, L. Maschio and A. J. Karttunen, *Chem.–Eur. J.*, 2017, **23**, 15884–15888.

33 W. Hittorf, *Ann. Phys. Chem.*, 1865, **202**, 193–228.

34 H. Thurn and H. Krebs, *Acta Crystallogr., Sect. B: Struct. Sci.*, 1969, **25**, 125–135.

35 P. E. Dolgirev, I. A. Kruglov and A. R. Oganov, *AIP Adv.*, 2016, **6**, 085318.

36 V. L. Deringer, G. Csányi and D. M. Proserpio, *ChemPhysChem*, 2017, **18**, 873–877.

37 V. L. Deringer, C. J. Pickard and G. Csányi, *Phys. Rev. Lett.*, 2018, **120**, 156001.

38 Q. Tong, L. Xue, J. Lv, Y. Wang and Y. Ma, *Faraday Discuss.*, 2018, DOI: 10.1039/c8fd00055g.

39 J. Behler, *J. Chem. Phys.*, 2016, **145**, 170901.

40 V. L. Deringer and G. Csányi, *Phys. Rev. B: Condens. Matter Mater. Phys.*, 2017, **95**, 094203.

41 T. L. Jacobsen, M. S. Jørgensen and B. Hammer, *Phys. Rev. Lett.*, 2018, **120**, 026102.

42 T. Yamashita, N. Sato, H. Kino, T. Miyake, K. Tsuda and T. Oguchi, *Phys. Rev. Mater.*, 2018, **2**, 013803.

43 A. Brown and S. Rundqvist, *Acta Crystallogr.*, 1965, **19**, 684–685.

44 J. C. Jamieson, *Science*, 1963, **139**, 1291–1292.

45 M. Ruck, D. Hoppe, B. Wahl, P. Simon, Y. Wang and G. Seifert, *Angew. Chem., Int. Ed.*, 2005, **44**, 7616–7619.

46 Z. Zhu and D. Tománek, *Phys. Rev. Lett.*, 2014, **112**, 176802.

47 J. P. Perdew, A. Ruzsinszky, G. I. Csonka, O. A. Vydrov, G. E. Scuseria, L. A. Constantin, X. Zhou and K. Burke, *Phys. Rev. Lett.*, 2008, **100**, 136406.

48 A. S. Rodin, A. Carvalho and A. H. Castro Neto, *Phys. Rev. Lett.*, 2014, **112**, 176801.

49 S. J. Clark, M. D. Segall, C. J. Pickard, P. J. Hasnip, M. J. Probert, K. Refson and M. C. Payne, *Z. Kristallogr.*, 2005, **220**, 567–570.

50 G. P. Francis and M. C. Payne, *J. Phys.: Condens. Matter*, 1990, **2**, 4395–4404.

51 A. P. Bartók, M. C. Payne, R. Kondor and G. Csányi, *Phys. Rev. Lett.*, 2010, **104**, 136403.

52 A. P. Bartók, M. J. Gillan, F. R. Manby and G. Csányi, *Phys. Rev. B: Condens. Matter Mater. Phys.*, 2013, **88**, 054104.

53 W. J. Szlachta, A. P. Bartók and G. Csányi, *Phys. Rev. B: Condens. Matter Mater. Phys.*, 2014, **90**, 104108.

54 A. P. Bartók, S. De, C. Poelking, N. Bernstein, J. R. Kermode, G. Csányi and M. Ceriotti, *Sci. Adv.*, 2017, **3**, e1701816.

55 A. P. Bartók, R. Kondor and G. Csányi, *Phys. Rev. B: Condens. Matter Mater. Phys.*, 2013, **87**, 184115.

56 A. P. Bartók and G. Csányi, *Int. J. Quantum Chem.*, 2015, **115**, 1051–1057.

57 P. W. Bridgman, *Proc. Am. Acad. Arts Sci.*, 1948, **76**, 55–70.

58 M. Baudler, *Angew. Chem., Int. Ed. Engl.*, 1982, **21**, 492–512.

59 S. Böcker and M. Häser, *Z. Anorg. Allg. Chem.*, 1995, **621**, 258–286.

60 F. H. Stillinger, *Phys. Rev. E: Stat., Nonlinear, Soft Matter Phys.*, 1999, **59**, 48–51.

61 S. E. Ahnert, W. P. Grant and C. J. Pickard, *npj Comput. Mater.*, 2017, **3**, 35.

62 L. Pauling, *J. Am. Chem. Soc.*, 1929, **51**, 1010–1026.

63 M. O'Keeffe, M. A. Peskov, S. J. Ramsden and O. M. Yaghi, *Acc. Chem. Res.*, 2008, **41**, 1782–1789.

64 F. Bachhuber, J. von Appen, R. Dronskowski, P. Schmidt, P. Nilges, A. Pfitzner and R. Weihrich, *Z. Kristallogr.*, 2015, **230**, 107–115.

65 J. P. Perdew, K. Burke and M. Ernzerhof, *Phys. Rev. Lett.*, 1996, **77**, 3865–3868.

66 A. Tkatchenko and M. Scheffler, *Phys. Rev. Lett.*, 2009, **102**, 073005.

67 E. R. McNellis, J. Meyer and K. Reuter, *Phys. Rev. B: Condens. Matter Mater. Phys.*, 2009, **80**, 205414.

68 E. Koch and W. Fischer, *Z. Kristallogr.*, 1978, **148**, 107–152.

69 V. A. Blatov and D. M. Proserpio, *Acta Crystallogr., Sect. A: Found. Crystallogr.*, 2009, **65**, 202–212.

70 M. O'Keeffe, G. B. Adams and O. F. Sankey, *Phys. Rev. Lett.*, 1992, **68**, 2325–2328.

71 E. Menges, V. Hopf, H. Schäfer and A. Weiss, *Z. Naturforsch. B Chem. Sci.*, 1969, **24**, 1351–1352.

72 R. Nesper, *Z. Anorg. Allg. Chem.*, 2014, **640**, 2639–2648.

73 A. Pfitzner, M. F. Bräu, J. Zweck, G. Brunklaus and H. Eckert, *Angew. Chem., Int. Ed.*, 2004, **43**, 4228–4231.

74 A. Simon, H. Borrmann and J. Horakh, *Chem. Ber.*, 1997, **130**, 1235–1240.

Faraday Discussions

PAPER

Evolutionary niching in the GAtor genetic algorithm for molecular crystal structure prediction†

Farren Curtis,[ab] Timothy Rose [a] and Noa Marom *[abc]

Received 19th March 2018, Accepted 3rd April 2018

DOI: 10.1039/c8fd00067k

The goal of molecular crystal structure prediction (CSP) is to find all the plausible polymorphs for a given molecule. This requires performing global optimization over a high-dimensional search space. Genetic algorithms (GAs) perform global optimization by starting from an initial population of structures and generating new candidate structures by breeding the fittest structures in the population. Typically, the fitness function is based on relative lattice energies, such that structures with lower energies have a higher probability of being selected for mating. GAs may be adapted to perform multi-modal optimization by using evolutionary niching methods that support the formation of several stable subpopulations and suppress the over-sampling of densely populated regions. Evolutionary niching is implemented in the GAtor molecular crystal structure prediction code by using techniques from machine learning to dynamically cluster the population into niches of structural similarity. A cluster-based fitness function is constructed such that structures in less populated clusters have a higher probability of being selected for breeding. Here, the effects of evolutionary niching are investigated for the crystal structure prediction of 1,3-dibromo-2-chloro-5-fluorobenzene. Using the cluster-based fitness function increases the success rate of generating the experimental structure and additional low-energy structures with similar packing motifs.

1 Introduction

Molecular crystals have numerous applications in pharmaceuticals, organic electronics, pigments, and explosives.[1–11] Hence, there has been increasing interest in reliable crystal structure prediction (CSP) methods. CSP methods computationally explore various crystal packing arrangements of a given organic

[a]Department of Materials Science and Engineering, Carnegie Mellon University, Pittsburgh, Pennsylvania 15213, USA. E-mail: nmarom@andrew.cmu.edu

[b]Department of Physics, Carnegie Mellon University, Pittsburgh, Pennsylvania 15213, USA

[c]Department of Chemistry, Carnegie Mellon University, Pittsburgh, Pennsylvania 15213, USA

† Electronic supplementary information (ESI) available. See DOI: 10.1039/c8fd00067k

compound, aiming to generate any experimentally determined crystal structure(s) in addition to any potential low-energy polymorphs that may be possible to synthesize. CSP methods are increasingly being used to complement experimental investigations of molecular crystal polymorphs.[12–16]

A robust CSP method must be able to produce the global minimum structure in addition to other structures that are close in energy, as molecular crystal polymorphs are typically within a few kJ mol^{-1}.[17–20] The global optimization method used needs to efficiently search an enormous configuration space with many local minima that are very close in energy. Genetic algorithms (GAs) are a versatile class of global optimization methods inspired by the evolutionary principle of survival of the fittest.[21–23] GAs are suitable for organic CSP because they can handle systems with complex multidimensional search spaces, including those with many extrema. A GA starts from an initial population of locally optimized structures and proceeds by repeatedly mating the fittest structures in the population through crossover and mutation operators until a convergence criterion is reached. Typically the GA fitness function is based on the relative energies of structures in the population. In this scenario, breeding operators drive the evolutionary search by exploiting the structural motifs associated with lower total energies. GAs have been used extensively for CSP of crystalline solids[24–36] and clusters.[21–23,37–46]

One drawback of GAs is that they may be prone to 'genetic drift', meaning that the algorithm over-samples certain regions of the potential energy surface associated with the fittest structures in the population, while under sampling other regions. The origin of genetic drift may be related to the nature of the potential energy surface, with some stable crystal packing motifs found in wide basins, whereas others are found in isolated funnels that are rarely sampled.[47] Systematic biases of the total energy method towards or against particular packing motifs may also contribute to genetic drift.[48] Another problem of GAs is that the composition of the initial population may bias the outcome of the search.[44]

To overcome these issues, GAs can be adapted to perform multi-modal optimization by incorporating evolutionary niching methods. These methods aim to increase diversity in the population and converge several promising solutions simultaneously. In order to identify niches, *i.e.* clusters comprised of structures with similar geometric features, unsupervised clustering techniques from machine learning may be employed to influence the GA search strategy as new data is accumulated.[45,47] Clustering can help identify the structural motifs that are over- or under-represented in the population. This information can be used to modify the fitness function and/or selection strategy.

Recently, we presented the GAtor genetic algorithm for molecular crystal structure prediction from first principles.[47] GAtor offers the option to perform evolutionary niching by dynamically clustering the population by structural similarity and then employing a cluster-based fitness function. In ref. 47, we demonstrated the effectiveness of evolutionary niching for generating the experimental structure of tricyano-1,4-dithiino[*c*]-isothiazole (Target XXII from the sixth CSP blind test[49] organized by the Cambridge Crystallographic Data Centre (CCDC)). Therein, clustering was performed with respect to a descriptor based on the lattice parameters. Here, we further investigate the effect of using cluster-based fitness functions based on different molecular crystal descriptors *versus* a traditional energy-based fitness function. As an example, we have chosen

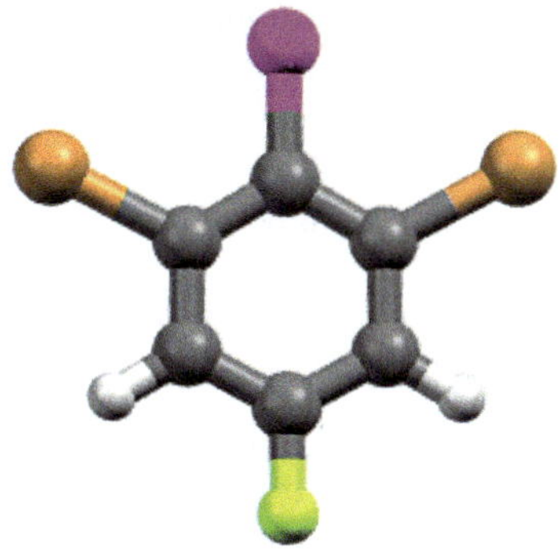

Fig. 1 The molecular diagram of 1,3-dibromo-2-chloro-5-fluorobenzene (Target XIII).[50] C, H, Br, Cl, and F atoms are colored in grey, white, brown, pink, and yellow, respectively.

1,3-dibromo-2-chloro-5-fluorobenzene (Target XIII from the fourth CCDC blind test[50]), illustrated in Fig. 1, which contains several halogen elements, namely Cl, F, and Br. The theoretical description of halogen bonds is challenging because it requires an accurate treatment of both electrostatic and dispersion interactions.[51–54] In ref. 47, the experimental structure of Target XIII was rarely generated when using a traditional energy-based fitness function, making this molecule a good test case for exploring the effect of incorporating evolutionary niching *via* a modified cluster-based fitness function. In the following, we show that the use of evolutionary niching provides uniform sampling of the potential energy surface by evolving several subpopulations simultaneously. This helps find the experimental structure of Target XIII and several additional low-energy structures, which are not otherwise generated in control runs that employ an energy-based fitness function.

2 Methodology

To study the effect of evolutionary niching, a set of GA runs was conducted starting from two initial pools, as described in Section 2.1. Section 2.2 describes the DFT settings used for the energy evaluations and local optimizations performed in GAtor. For each initial pool, the GA was run four times with the settings described in Section 2.3. One run was used as a control, which employed a traditional energy-based fitness function. The other three runs used evolutionary niching by employing a cluster-based fitness function with three different molecular crystal descriptors, as described in Section 2.4. The effect of using a cluster-based fitness function on the final population of structures produced was analyzed for each run and descriptor.

2.1 Initial population

Two initial pools consisting of 45 molecular crystal structures of Target XIII with four molecules per unit cell ($Z = 4$) were prepared using the random and diverse workflows of the molecular crystal structure generation package Genarris,[55] previously generated as reported in ref. 55. Briefly, a raw pool of 5000 structures was generated in all the space groups compatible with $Z = 4$. For the random initial pool, the final structures were randomly selected from the raw pool. For the diverse initial pool, fragment-based density functional theory (DFT) and clustering techniques from machine learning were used for the selection of the final structures. For a more detailed explanation of the preparation of the initial pools, see ref. 55. Both initial pools were locally optimized using the DFT settings presented in Section 2.2.

2.2 DFT settings

For energy evaluations and local structural optimizations within the GA, the generalized gradient approximation of Perdew–Burke–Ernzerhof (PBE)[56,57] is used with the pairwise Tkatchenko–Scheffler (TS) dispersion-correction[58], with *lower-level* numerical settings, which correspond to the tier 1 basis sets and light numerical settings of FHI-aims.[59] For all calculations, a $3 \times 3 \times 3$ *k*-point grid is used. During local optimization, the space group symmetry is allowed to vary. Atomic ZORA scalar relativity[59] settings are used for the heavier halogen elements. For post-processing, the top structures produced are re-relaxed and re-ranked using a $3 \times 3 \times 3$ *k*-point grid, PBE+TS, and *higher-level* numerical settings, which correspond to the tier 2 basis sets and tight numerical settings of FHI-aims.[59]

2.3 GA settings

Four different GA runs were performed starting from each initial pool using GAtor.[47] For all runs, the GA was terminated when the final population reached a total of 385 structures, including the initial pool. All runs used 50% crossover and 50% mutation, and tournament selection with a tournament size of 20 structures. For closeness checks, the minimum distance between two atoms of different molecules was set to 0.7 of the sum of their van der Waals radii. A relative energy cutoff of 0.75 eV per molecule was used for single-point energy (SPE) evaluations, such that structures generated with energies per molecule greater than the current global minimum plus 0.75 eV were rejected without performing local optimization.

For each initial pool, a control run was conducted using a traditional energy-based fitness function. In this fitness scheme, the total energy E_i of the ith structure in the population is evaluated using dispersion-inclusive DFT as detailed in Section 2.2. The energy-based fitness f_i of each structure is given by

$$f_i = \frac{\varepsilon_i}{\sum_i \varepsilon_i} \qquad 0 \leq f \leq 1 \tag{1}$$

$$\varepsilon_i = \frac{E_{\max} - E_i}{E_{\max} - E_{\min}} \tag{2}$$

where ε_i is the ith structure's relative energy, and $E_{\max}$ and $E_{\min}$ correspond to the structures with the dynamically updated highest and lowest total energies in the population, respectively.[43,44,47,60] Hence, in this fitness scheme, structures with lower relative energies have higher fitness values.

For each initial pool, three additional GA runs were conducted using a cluster-based fitness function. In this scheme, the population is dynamically clustered into niches of structural similarity using different molecular crystal descriptors (see Section 2.4). Then, a fitness sharing scheme is implemented such that the cluster-based fitness f_i^c of each structure is given by

$$f_i^c = \frac{\bar{f}_i}{\bar{f}_{max}} \quad 0 \le f^c \le 1 \tag{3}$$

$$\bar{f}_i = \frac{f_i}{m_i} \tag{4}$$

where m_i is each structure's niche count given by the number of structures in each individual's cluster and f_i is the ith structure's energy-based fitness. $\bar{f}_{max}$ is the maximum cluster-based fitness in the population. With this fitness scheme, structures in less populated clusters with under-represented structural motifs have higher fitness values, and hence a higher probability of being selected for mating. Penalizing the fitness of over-sampled clusters and steering the GA towards under-sampled clusters provides a more uniform sampling of the configuration space. If all the cluster sizes were equal, this fitness scheme would be equivalent to energy-based fitness.

2.4 Clustering

When using GAtor's cluster-based fitness function, every time a new structure is added to the common pool of structures, clustering is performed to group the common population into niches. As the GA evolves, the clusters are updated automatically, affecting each structure's cluster-based fitness value. For clustering, the affinity propagation (AP)[61] clustering algorithm is used, as implemented in scikit-learn.[62] AP detects the number of clusters in a data set as opposed to specifying it *a priori*. The input of AP is a pairwise similarity matrix between data points. All data points are initially considered as possible "exemplars", *i.e.* points that are representative members of each cluster. The algorithm iteratively refines the candidate exemplars until well-defined exemplars and corresponding clusters are identified. For a more detailed description of the AP algorithm, see ref. 61. AP clustering has been shown to successfully resolve small isolated clusters with distinct structural motifs.[55]

We used three different molecular crystal descriptors for clustering. The first descriptor, L, is given by[47]

$$\mathrm{L} = \frac{1}{\sqrt[3]{V}}(a, b, c) \tag{5}$$

where V is the unit cell volume and a, b, and c are the structure's lattice parameters after employing Niggli reduction[63–66] and unit cell standardization. All the unit cells are standardized such that $\vec{a}$ points along the $\hat{x}$ direction, $\vec{b}$ lies in the xy plane, and the convention $a \le b \le c$ is used. For the lattice parameter based descriptor, a negative squared Euclidean metric is used for the construction of the pairwise similarity matrix for AP clustering.

The second descriptor combines several radial distribution function (RDF) vectors of selected interatomic contacts. The radial distribution function for atom types A and B is given by

$$G_{\mathrm{AB}}(r) = \frac{\sum_{ij} \exp\left(-(r - r_{ij})^2\right)}{N_{\mathrm{A}}} \tag{6}$$

where r_{ij} is the distance between the ith and jth atoms of atom types A and B, respectively, and N_A is the number of A atoms. RDF vectors for atom types A and B are constructed by sampling $G_{AB}(r)$ at finite intervals in a user-defined range. Multiple RDF vectors of different atom type pairs are concatenated to form a combined RDF descriptor. For Target XIII, $G_{AB}(r)$ is sampled for interatomic Br···Br, H···Br, H···F, and H···Cl contacts in bins of 1 Å for 1 Å $\leq r \leq$ 8 Å. For the RDF descriptor, a negative squared Euclidean metric is used for the construction of the pairwise similarity matrix for AP clustering.

The third molecular crystal descriptor is the relative coordinate descriptor (RCD).[55] RCD was developed to capture the packing motif of molecular crystals by using the relative center of mass position $\vec{P}$ and orientation $\vec{Q}$ of N (16 by default) neighboring molecules with respect to a reference molecule. The relative orientation $\vec{Q}$ between two molecules is computed by taking the dot product of orthogonal reference axes centered on each molecule, constructed as described in ref. 55. The RCD descriptor for a given molecular crystal is given by

$$\mathrm{RCD} = \left[\left(\vec{P}^1, \vec{Q}^1\right), \ldots, \left(\vec{P}^N, \vec{Q}^N\right)\right]. \tag{7}$$

AP clustering allows the user to input custom similarity metrics that determine how similar two input vectors are to one another. When using the RCD descriptor with AP clustering in GAtor, the distance matrix between two RCD vectors is given by

$$D_{i,j} = \left(\frac{\left|\vec{P}_1^{\,i} - \vec{P}_2^{\,j}\right|^2}{\left|\vec{P}_1^{\,i}\right|\left|\vec{P}_2^{\,j}\right|}\right) + \frac{1}{3}\left(\left|\vec{Q}_1^{\,i} - \vec{Q}_2^{\,j}\right|^2\right). \tag{8}$$

The M (8 by default) smallest entries of D are selected, such that no two entries have the same value for i or j. The sum of these M entries serves as the input pairwise similarity for the AP clustering algorithm.

3 Results and discussion

GAtor was run four times for each initial pool. One of the four runs employed a traditional energy-based fitness function, while the other three runs used a cluster-based fitness function with the lattice parameter based descriptor L, the combined RDF descriptor, and the RCD descriptor. Panel (a) of Fig. 2 shows the average energy of the common population as a function of GA iteration for all eight runs, scaled with respect to the total energy of the global minimum structure. Here an iteration corresponds to when a single structure has been added to the common population. The average energy of the initial population corresponds to iteration 0. The common populations produced from the random (R) and diverse (D) energy-based control runs, shown in light and dark red, respectively, displayed the lowest average energies (approximately 9 kJ mol^{-1} per molecule) when the GA was terminated. The run that used the random initial pool and the cluster-based fitness function with the RDF descriptor, shown in light blue, had an average energy slightly above 9 kJ mol^{-1} per molecule when the GA was terminated. All the other runs that used a cluster-based fitness function showed

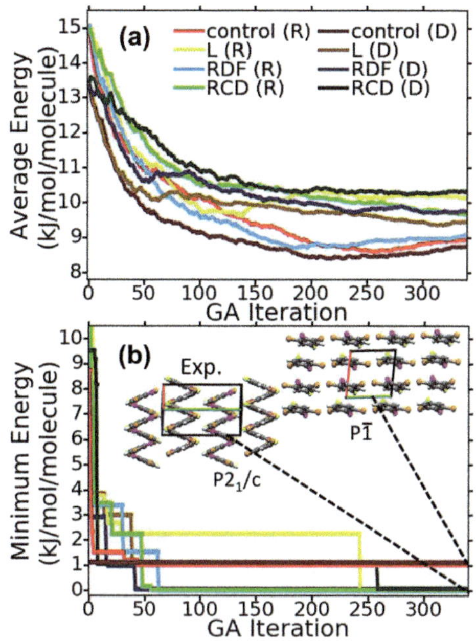

Fig. 2 (a) The average energy of the common population and (b) the global minimum structure produced as a function of GA iteration for different GA runs. For the displayed crystal structures, the $\vec{a}$, $\vec{b}$, and $\vec{c}$ crystallographic lattice vectors are colored in red, green, and blue, respectively.

average energies of 1–2 kJ mol^{-1} per molecule higher than the energy-based fitness runs for the last 200 iterations. This behavior may be attributed to the fact that the use of a cluster-based fitness function promotes the generation of structures with under-represented structural motifs, which may have higher energies. Panel (b) of Fig. 2 shows the minimum energy structure as a function of GA iteration. The experimental structure (corresponding to the PBE+TS global minimum energy) and the second-lowest energy structure are displayed. By the time the GA was terminated, five of the six runs that employed a cluster-based fitness function generated the experimental structure, the exception being the run that used the diverse initial pool and the lattice parameter based descriptor. Neither of the control runs generated the experimental structure, but both generated the second-lowest energy structure within the first 50 GA iterations. The layered packing motif of the second-lowest energy structure is prevalent in the low-energy structures generated by the control runs.[47]

Fig. 3 demonstrates the effect of evolutionary niching by visualizing the structures generated by each GA run in a 3D space consisting of each structure's unique, Niggli-reduced *a*, *b*, and *c* lattice parameters. All points are colored by their total PBE+TS energy (with lower-level numerical settings), relative to the global minimum. For the runs that generated the experimental structure, its location is indicated by a green 'X'. Panels (a) and (b) show the random and diverse initial populations, respectively. The diverse initial pool displays structures with a greater variety of lattice parameters than the random pool, as expected. Panels (c) and (d) depict the structures produced by the GA runs that used

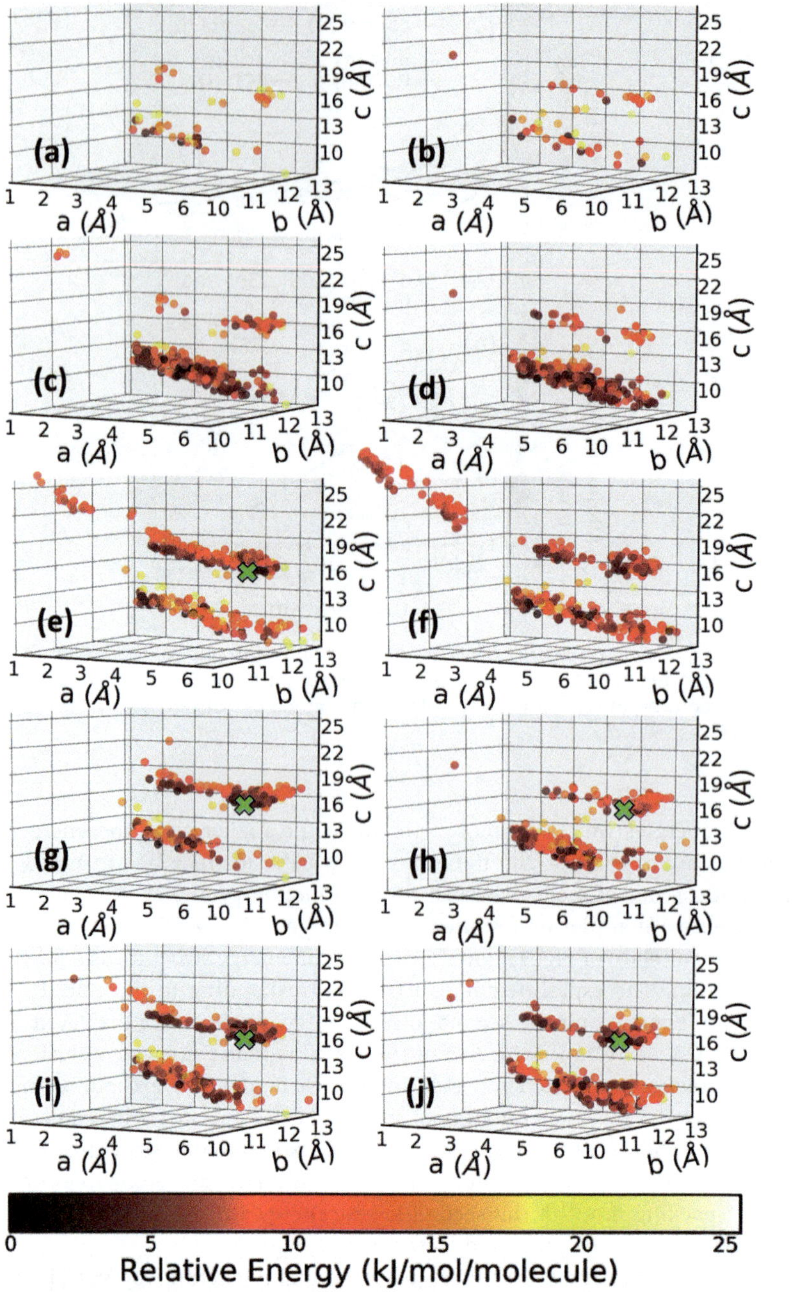

Fig. 3 Structures of Target XIII produced from the evolution of the random (left) and diverse (right) initial pools. Structures are represented in a 3D space consisting of each structure's Niggli-reduced lattice parameters *a*, *b*, and *c*, and colored according to their energy relative to the global minimum. The distributions of the initial population structures are shown in panels (a and b). The structures produced by the GA runs that used the energy-based fitness function are shown in (c and d). The structures produced by the GA runs that used cluster-based fitness functions with L, RDF, and RCD are shown in panels (e and f), (g and h), and (i and j), respectively. The location of the experimental structure, if generated in a given run, is indicated by a green 'X'.

the energy-based fitness function with the random and diverse initial pools, respectively. Both of these control runs exhaustively explored the low-lying region of structures that have *c*-parameters with a maximum value of approximately 13 Å, a region that contains many low-energy structures with layered packing motifs but does not contain the experimental structure. Structures representative of this region are shown in Fig. 4, panels (a) and (b). Although the control run that used the diverse initial pool had more low-energy structures with a variety of lattice parameters to start with, over time it thoroughly sampled the same regions of the PES as the control run that used the random initial pool. This is an example of

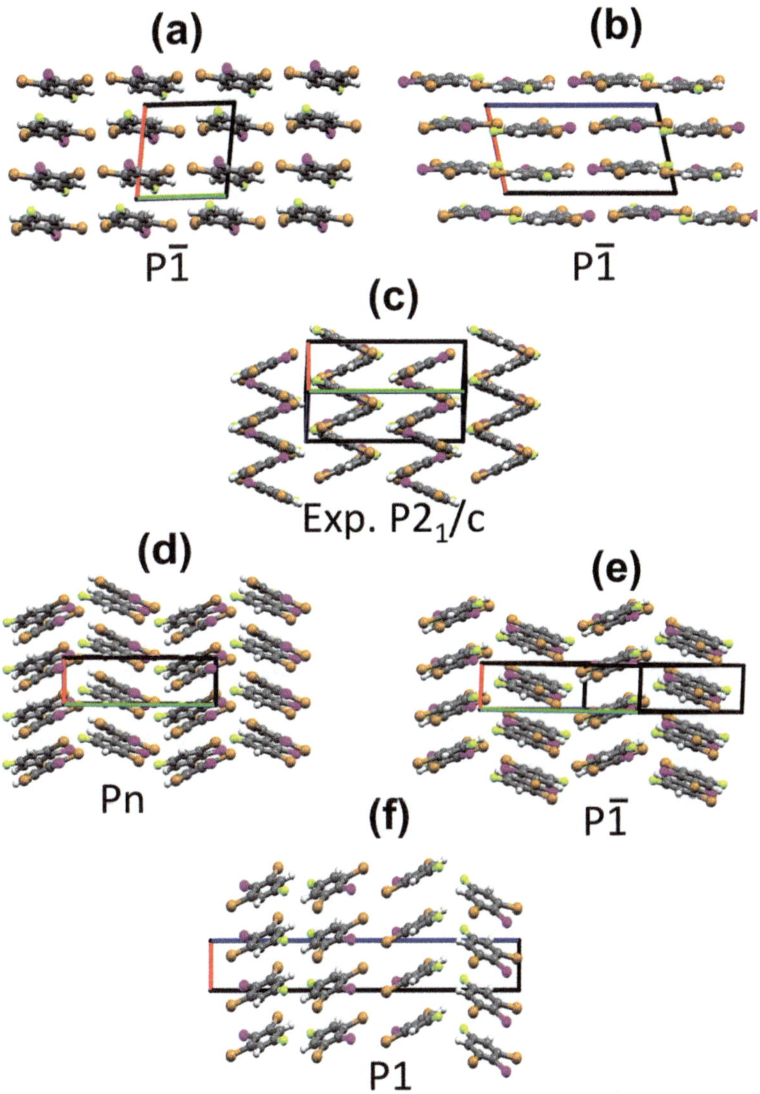

Fig. 4 Structures representative of (a and b) layered packing motifs (c) zig-zag packing motifs (d and e) herringbone packing motifs and (f) large *c*-parameters. The $\vec{a}$, $\vec{b}$, and $\vec{c}$ crystallographic lattice vectors are colored in red, green, and blue, respectively.

genetic drift, where the search is biased towards exhaustively exploring particular basins of the potential energy surface, which may or may not contain the experimental structure and/or other desirable low-energy crystal structures. Panels (e and f), (g and h), and (i and j) of Fig. 3 show the final populations of structures produced by the GA runs that used cluster-based fitness with the lattice parameter descriptor, the combined RDF descriptor, and the RCD descriptor, respectively. All of the runs that used a cluster-based fitness function explored the region containing the experimental structure more frequently than the control runs, with five out of the six runs successfully generating the experimental structure. The region containing the experimental structure contains numerous structures with zig-zag and herringbone packing motifs and elongated unit cells, examples of which are shown in Fig. 4, panels (c–e). The runs that used the cluster-based fitness function with the lattice parameter descriptor, shown in Fig. 3, panels (e and f), additionally explored a higher-energy region containing structures with large c lattice parameters greater than approximately 20 Å. A representative structure of this region is shown in panel (f) of Fig. 4. The exploration of structures with large c-parameters was particularly the case for the run that used the diverse initial pool with the lattice parameter descriptor, which did not generate the experimental structure. The lattice parameter descriptor, given by eqn (5), scales as the inverse of the cube root of the unit cell volume. Therefore, the cluster-based fitness function based on it occasionally promotes the selection of structures with larger unit cell volumes, which may have higher energies.

This demonstrates how evolutionary niching helps overcome initial pool biases and genetic drift by evolving several subpopulations simultaneously. GA runs employing cluster-based fitness more frequently sample regions of the potential energy surface that are not well represented in the initial pool, compared to control runs employing energy-based fitness. Furthermore, GA runs using cluster-based fitness more frequently sample different regions than those sampled preferentially by GA runs employing energy-based fitness. Some of these under-represented populations contain important low-energy structures, whereas others may not necessarily produce structures that could be viable polymorphs. Our findings for Target XIII are consistent with the findings reported for Target XXII in ref. 47. We now proceed to perform a more detailed analysis of the effect of the descriptor on evolutionary niching.

Fig. 5 presents an analysis of the final clusters generated by the GA runs that were started from the random initial pool and used the lattice parameter descriptor (panel a), the RDF descriptor (panel b), and the RCD descriptor (panel c). For each run that utilized evolutionary niching, AP was used to cluster the final population of structures with respect to the same descriptor used in the GA. The resulting distributions are shown in blue. AP was then used to predict the cluster labels of the structures generated in the energy-based control run, shown in red. For all the descriptors, a negative squared Euclidean metric is used for the construction of the pairwise similarity matrix for AP clustering. The distribution of initial pool structures contained in each cluster is shown in grey. The cluster assigned to the experimental structure is indicated by a green arrow. Additionally, the average and standard deviation of the relative energies of the structures in each cluster are plotted in orange. Histograms produced from the GA runs that were started from the diverse initial pool are provided in the ESI.†

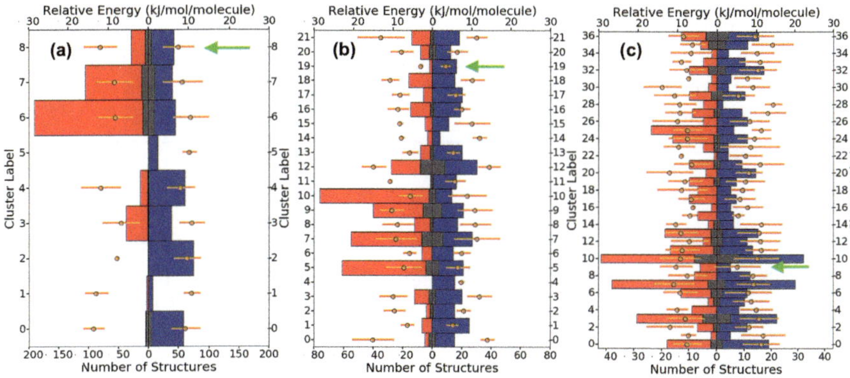

Fig. 5 The distribution of clusters produced by different GA runs initialized from the random initial pool. Histograms corresponding to runs that used evolutionary niching are shown in blue and histograms corresponding to the control run are shown in red. The structures are clustered with respect to (a) L, (b) RDF, and (c) RCD. In all the plots the average relative energy and standard deviation for each cluster are shown in orange. The cluster assigned to the experimental structure for each run is indicated by a green arrow.

For all three descriptors, the use of a cluster-based fitness function suppressed the sampling of clusters that were over-sampled in the control run (*e.g.* clusters 6–7 in panel (a), clusters 5 and 10 in panel (b), and clusters 7 and 10 in panel (c)). In addition, the GA runs that utilized evolutionary niching generally sampled more structures in clusters that were not represented in the initial pool than the control runs. For the runs that used the lattice parameter descriptor and the RDF descriptor, the clusters that contained the experimental structure (clusters 8 and 19, respectively) were more frequently sampled than in the control run. The runs based on L and RDF yield more uniform distributions than the run based on RCD. Additionally, the RCD descriptor produces more overall variance in the average energies of the clusters than L and RDF. The fact that fewer structures with similar energies are grouped together with RCD may be related to the weaker correlation of RCD with unit cell volume, demonstrated in ref. 55.

The three descriptors differ in the final number of clusters produced. The lattice parameter based run yields 9 final clusters, while the runs that used the RDF and RCD descriptors yield 22 and 37 clusters, respectively. By construction, the RDF and RCD descriptors capture the various packing motifs of Target XIII better than the lattice parameter descriptor. While elongated unit cells may correlate with zig-zag and herringbone motifs, and more box-like unit cells may correlate with layered motifs, there are more subtle differences between packing motifs that are not captured with the lattice parameter descriptor. Additionally, the lattice parameter descriptor promotes sampling of higher-energy structures with large *c*-parameters. For these reasons, the lattice parameter descriptor is probably too crude for the purposes of evolutionary niching.

The recommended best practice when using GAtor is to run the GA several times with different settings and collect all the resulting structures, then remove duplicates, and re-relax all the unique structures with *higher-level* numerical settings. The final structures may then be re-ranked using a series of increasingly accurate energy methods.[47] The primary focus of this work is to study the effect of

Table 1 The top 10 crystal structures of Target XIII as ranked using PBE+TS and *higher-level* numerical settings. In the last column, 'Control' indicates runs that used energy-based fitness, while L, RDF, and RCD indicate runs that used cluster-based fitness. The initial pool used in a given run is indicated in parentheses, with R indicating the random pool and D indicating the diverse pool

Rank	ΔE (kJ mol^{-1} per molecule)	Volume (Å^3)	a (Å)	b (Å)	c (Å)	α (°)	β (°)	γ (°)	Z	SG	Runs generated
1	0.00	783.6	4.0	13.6	14.5	90	87	90	4	$P2_1/c$	L (R), RDF (R,D), RCD (R,D)
2	0.57	403.5	7.0	8.0	8.2	67	74	78	2	$P\bar{1}$	Control (R,D), L (D), RDF (D)
3	0.84	784.7	4.0	13.3	14.9	90	96	90	4	$P2_1$	Control (D), RCD (R)
4	0.98	797.9	7.4	9.2	13.0	84	77	69	4	$P\bar{1}$	Control (R,D)
5	1.11	799.8	8.8	7.0	15.7	90	125	90	4	$P2_1/c$	Control (R,D), L (D), RDF (R), RCD (D,R)
6	1.15	788.0	4.0	13.4	15.8	110	90	90	4	Pn	L (D), RDF(R)
7	1.23	794.3	7.6	7.8	14.0	97	96	103	4	$P\bar{1}$	Control (R,D), L (D), RDF (R), RCD (R)
8	1.34	791.0	4.0	13.6	16.0	68	86	89	4	$P\bar{1}$	RDF (R), RCD (R)
9	1.43	798.2	7.2	9.3	12.7	83	89	71	4	$P\bar{1}$	Control (R,D), L (D)
10	1.46	795.2	7.9	9.3	10.9	94	96	92	4	$P\bar{1}$	Control (R,D), RDF (D)

evolutionary niching. Therefore, the best structures produced from the GA runs here are locally optimized using PBE+TS and *higher-level* numerical settings, and no further re-ranking is performed. Thermal contributions to the total energy, which may influence the relative stabilities of polymorphs,[67] are also not included in the present study.

Table 1 presents the top 10 lowest energy structures produced from all the runs combined, as computed using PBE+TS and *higher-level* numerical settings. In the last column, the specific runs that generated a given low-energy structure are indicated. The experimental structure, which has a zig-zag packing motif and elongated unit cell, is ranked as #1 and was only generated in GA runs that used clustering. The structures ranked #6 and #8, shown in panels (d and e) in Fig. 4, have herringbone packing motifs and were also only generated in runs that used clustering. These structures were not produced in our previous study of this molecule, in which evolutionary niching was not used.[47] Seven of the ten top structures (ranked #2–#5, #7, #9 and #10) were generated by the control runs using energy-based fitness. The majority of these structures display a layered packing motif. For example, the structures ranked #2 and #4 are shown in panels (a) and (b) of Fig. 4, respectively. Six of the seven top structures found in runs that used energy-based fitness were also generated by at least one GA run that used cluster-based fitness. This shows that evolutionary niching can be a valuable tool for generating novel structures that may be overlooked when using a traditional energy-based fitness function.

4 Concluding remarks

The effect of using evolutionary niching in the GAtor genetic algorithm was investigated for the crystal structure prediction of 1,3-dibromo-2-chloro-5-fluorobenzene (Target XIII from the fourth CCDC blind test[50]). Evolutionary niching was performed by using the affinity propagation machine learning algorithm to dynamically cluster the population based on structural similarity. Clustering was conducted with respect to three structural descriptors: the first was based on the lattice parameters (L); the second was based on the combined radial distribution functions (RDFs) of the Br···Br, H···Br, H···F, and H···Cl interatomic distances; the third was a relative coordinate descriptor (RCD), based on the relative positions and orientations of neighboring molecules. A cluster-based fitness function was then used to steer the GA towards under-sampled regions of the potential energy surface by penalizing the fitness of structures found in populous clusters, thus reducing their probability of being selected for mating.

The results of GA runs that used evolutionary niching based on L, RDF, and RCD were compared to control runs that employed a traditional energy-based fitness function. To examine the effect of the initial population, two sets of runs were launched from a random pool and from a diverse pool, generated by Genarris. Evolutionary niching promoted sampling in regions of the potential energy landscape that were not well represented in the initial population, as well as in regions rarely sampled by the control runs. We have thus demonstrated that evolutionary niching can help overcome initial pool biases and evolutionary drift in a genetic algorithm.

The experimental structure of Target XIII is characterized by a zig-zag packing motif, in contrast to most of the other low-energy structures that are characterized by layered packing arrangements. The region of the experimental structure was therefore rarely sampled by the control runs with the energy-based fitness function. Employing evolutionary niching based on all three descriptors significantly enhanced the likelihood of generating the experimental structure, as well as other low-energy structures with herringbone packing motifs.

Some differences were observed in the GA behavior when evolutionary niching was performed based on different descriptors. Using the L descriptor resulted in increased sampling of a region characterized by structures with very long *c*-parameters, particularly when the GA was started from the diverse initial pool. This was counterproductive in the case of Target XIII because this region did not contain any important low-energy structures. The L descriptor was additionally found to produce fewer clusters than the RDF and RCD descriptors. This suggests that the sensitivity of the L descriptor may be insufficient to resolve different packing motifs that do not necessarily correlate with the unit cell shape and volume.

In conclusion, the recommended best practice for molecular crystal structure prediction with GAtor is to run the GA several times with different settings. At least one of the runs should employ a cluster-based fitness function using the RDF or RCD descriptors. The best structures produced from all the GA runs should be combined for post-processing. This increases the likelihood of generating more low-energy structures overall than when running GAtor with the energy-based fitness function alone. Evolutionary niching using cluster-based fitness functions is a promising strategy for generating important structures located in different basins of the potential energy surface for CSP applications.

Conflicts of interest

There are no conflicts to declare.

Acknowledgements

Work at CMU was funded by the National Science Foundation (NSF) Division of Materials Research through grant DMR-1554428. An award of computer time was provided by the Innovative and Novel Computational Impact on Theory and Experiment (INCITE) program. This research used resources of the Argonne Leadership Computing Facility, which is a DOE Office of Science User Facility supported under Contract DE-AC02-06CH11357. Part of this research was performed while the authors were visiting the Institute for Pure and Applied Mathematics (IPAM), which is supported by the National Science Foundation.

References

1 J. Bernstein, *Polymorphism in Molecular Crystals*, Oxford University Press, Oxford, England, 2002, vol. 14.
2 G. M. Day, W. D. S. Motherwell and W. Jones, *Phys. Chem. Chem. Phys.*, 2007, **9**, 1693–1704.
3 A. M. Reilly and A. Tkatchenko, *Phys. Rev. Lett.*, 2014, **113**, 055701.
4 D. P. Elder, J. E. Patterson and R. Holm, *J. Pharm. Pharmacol.*, 2015, **67**, 757–772.
5 C. Reese and Z. Bao, *Mater. Today*, 2007, **10**, 20–27.
6 T. Hasegawa and J. Takeya, *Sci. Technol. Adv. Mater.*, 2009, **10**, 024314.
7 S. Bergantin and M. Moret, *Cryst. Growth Des.*, 2012, **12**, 6035–6041.
8 P. Cudazzo, M. Gatti and A. Rubio, *Phys. Rev. B: Condens. Matter Mater. Phys.*, 2012, **86**, 195307.
9 P. Cudazzo, F. Sottile, A. Rubio and M. Gatti, *J. Phys.: Condens. Matter*, 2015, **27**, 113204.
10 N. Panina, F. J. J. Leusen, F. F. B. J. Janssen, P. Verwer, H. Meekes, E. Vlieg and G. Deroover, *J. Appl. Crystallogr.*, 2007, **40**, 105–114.
11 M. Fitzgerald, M. G. Gardiner, D. Armitt, G. W. Dicinoski and C. Wall, *J. Phys. Chem. A*, 2015, **119**, 905–910.
12 M. A. Neumann, J. van de Streek, F. P. A. Fabbiani, P. Hidber and O. Grassmann, *Nat. Commun.*, 2015, **6**, 7793.
13 S. L. Price, D. E. Braun and S. M. Reutzel-Edens, *Chem. Commun.*, 2016, **52**, 7065–7077.
14 A. G. Shtukenberg, Q. Zhu, D. J. Carter, L. Vogt, J. Hoja, E. Schneider, H. Song, B. Pokroy, I. Polishchuk, A. Tkatchenko, A. R. Oganov, A. L. Rohl, M. E. Tuckermanejk and B. Kahral, *Chem. Sci.*, 2017, **8**, 4926–4940.
15 A. G. Shtukenberg, C. Hu, Q. Zhu, M. U. Schmidt, W. Xu, M. Tan and B. Kahr, *Cryst. Growth Des.*, 2017, **17**, 3562–3566.
16 B. Meredig and C. Wolverton, *Nat. Mater.*, 2013, **12**, 123–127.

17 N. Marom, R. A. DiStasio, V. Atalla, S. Levchenko, A. M. Reilly, J. R. Chelikowsky, L. Leiserowitz and A. Tkatchenko, *Angew. Chem., Int. Ed.*, 2013, **52**, 6629–6632.
18 A. J. Cruz-Cabeza, S. M. Reutzel-Edens and J. Bernstein, *Chem. Soc. Rev.*, 2015, **44**, 8619–8635.
19 G. J. Beran, *Angew. Chem., Int. Ed.*, 2015, **54**, 396–398.
20 G. J. Beran, *Chem. Rev.*, 2016, **116**, 5567–5613.
21 R. L. Johnston, *Dalton Trans.*, 2003, 4193–4207.
22 M. Sierka, *Prog. Surf. Sci.*, 2010, **85**, 398–434.
23 S. Heiles and R. L. Johnston, *Int. J. Quantum Chem.*, 2013, **113**, 2091–2109.
24 A. R. Oganov and C. W. Glass, *J. Chem. Phys.*, 2006, **124**, 244704.
25 C. W. Glass, A. R. Oganov and N. Hansen, *Comput. Phys. Commun.*, 2006, **175**, 713–720.
26 N. L. Abraham and M. I. J. Probert, *Phys. Rev. B: Condens. Matter Mater. Phys.*, 2006, **73**, 224104.
27 G. Trimarchi and A. Zunger, *Phys. Rev. B: Condens. Matter Mater. Phys.*, 2007, **75**, 104113.
28 S. Wu, K. Umemoto, M. Ji, C. Z. Wang, K. M. Ho and R. M. Wentzcovitch, *Phys. Rev. B: Condens. Matter Mater. Phys.*, 2011, **83**, 184102.
29 S. Woodley, P. Battle, J. Gale and C. A. Catlow, *Phys. Chem. Chem. Phys.*, 1999, **1**, 2535–2542.
30 G. Trimarchi and A. Zunger, *J. Phys.: Condens. Matter*, 2008, **20**, 295212.
31 D. C. Lonie and E. Zurek, *Comput. Phys. Commun.*, 2011, **182**, 372–387.
32 G. H. Jóhannesson, T. Bligaard, A. V. Ruban, H. L. Skriver, K. W. Jacobsen and J. K. Nørskov, *Phys. Rev. Lett.*, 2002, **88**, 255506.
33 Q. Zhu, A. R. Oganov, C. W. Glass and H. T. Stokes, *Acta Crystallogr., Sect. B: Struct. Sci.*, 2012, **68**, 215–226.
34 A. M. Lund, G. I. Pagola, A. M. Orendt, M. B. Ferraro and J. C. Facelli, *Chem. Phys. Lett.*, 2015, **626**, 20–24.
35 P. Avery, Z. Falls and E. Zurek, *Comput. Phys. Commun.*, 2017, **217**, 210–211.
36 Z. Falls, D. C. Lonie, P. Avery, A. Shamp and E. Zurek, *Comput. Phys. Commun.*, 2016, **199**, 178–179.
37 J. Morris, D. Deaven, K. Ho, C. Wang, B. Pan, J. Wacker and D. Turner, *IMA Vol. Math. Its Appl.*, 1999, **111**, 167–176.
38 A. N. Alexandrova and A. I. Boldyrev, *J. Chem. Theory Comput.*, 2005, **1**, 566–580.
39 J. M. C. Marques and F. B. Pereira, *Chem. Phys. Lett.*, 2010, **485**, 211–216.
40 B. Hartke, *J. Comput. Chem.*, 1999, **20**, 1752–1759.
41 C. R. A. Catlow, S. T. Bromley, S. Hamad, M. Mora-Fonz, A. A. Sokol and S. M. Woodley, *Phys. Chem. Chem. Phys.*, 2010, **12**, 786–811.
42 V. E. Bazterra, O. Oña, M. C. Caputo, M. B. Ferraro, P. Fuentealba and J. C. Facelli, *Phys. Rev. A: At., Mol., Opt. Phys.*, 2004, **69**, 053202.
43 S. Bhattacharya, S. V. Levchenko, L. M. Ghiringhelli and M. Scheffler, *Phys. Rev. Lett.*, 2013, **111**, 135501.
44 S. Bhattacharya, S. V. Levchenko, L. M. Ghiringhelli and M. Scheffler, *New J. Phys.*, 2014, **16**, 123016.
45 M. S. Jørgensen, M. N. Groves and B. Hammer, *J. Chem. Theory Comput.*, 2017, **13**, 1486–1493.
46 W. W. Tipton and R. G. Hennig, *J. Phys.: Condens. Matter*, 2013, **25**, 495401.

47 F. Curtis, X. Li, T. Rose, A. Vázquez-Mayagoitia, S. Bhattacharya, L. M. Ghiringhelli and N. Marom, *J. Chem. Theory Comput.*, 2018, **14**, 2246.
48 F. Curtis, X. Wang and N. Marom, *Acta Crystallogr., Sect. B: Struct. Sci., Cryst. Eng. Mater.*, 2016, **72**, 562–570.
49 A. M. Reilly, R. I. Cooper, C. S. Adjiman, S. Bhattacharya, A. D. Boese, J. G. Brandenburg, P. J. Bygrave, R. Bylsma, J. E. Campbell, R. Car, D. H. Case, R. Chadha, J. C. Cole, K. Cosburn, H. M. Cuppen, F. Curtis, G. M. Day, R. A. DiStasio Jr, A. Dzyabchenko, B. P. van Eijck, D. M. Elking, J. A. van den Ende, J. C. Facelli, M. B. Ferraro, L. Fusti-Molnar, C.-A. Gatsiou, T. S. Gee, R. de Gelder, L. M. Ghiringhelli, H. Goto, S. Grimme, R. Guo, D. W. M. Hofmann, J. Hoja, R. K. Hylton, L. Iuzzolino, W. Jankiewicz, D. T. de Jong, J. Kendrick, N. J. J. de Klerk, H.-Y. Ko, L. N. Kuleshova, X. Li, S. Lohani, F. J. J. Leusen, A. M. Lund, J. Lv, Y. Ma, N. Marom, A. E. Masunov, P. McCabe, D. P. McMahon, H. Meekes, M. P. Metz, A. J. Misquitta, S. Mohamed, B. Monserrat, R. J. Needs, M. A. Neumann, J. Nyman, S. Obata, H. Oberhofer, A. R. Oganov, A. M. Orendt, G. I. Pagola, C. C. Pantelides, C. J. Pickard, R. Podeszwa, L. S. Price, S. L. Price, A. Pulido, M. G. Read, K. Reuter, E. Schneider, C. Schober, G. P. Shields, P. Singh, I. J. Sugden, K. Szalewicz, C. R. Taylor, A. Tkatchenko, M. E. Tuckerman, F. Vacarro, M. Vasileiadis, A. Vazquez-Mayagoitia, L. Vogt, Y. Wang, R. E. Watson, G. A. de Wijs, J. Yang, Q. Zhu and C. R. Groom, *Acta Crystallogr., Sect. B: Struct. Sci.*, 2016, **72**, 439–459.
50 G. M. Day, T. G. Cooper, A. J. Cruz-Cabeza, K. E. Hejczyk, H. L. Ammon, S. X. M. Boerrigter, J. S. Tan, R. G. Della Valle, E. Venuti, J. Jose, S. R. Gadre, G. R. Desiraju, T. S. Thakur, B. P. van Eijck, J. C. Facelli, V. E. Bazterra, M. B. Ferraro, D. W. M. Hofmann, M. A. Neumann, F. J. J. Leusen, J. Kendrick, S. L. Price, A. J. Misquitta, P. G. Karamertzanis, G. W. A. Welch, H. A. Scheraga, Y. A. Arnautova, M. U. Schmidt, J. van de Streek, A. K. Wolf and B. Schweizer, *Acta Crystallogr., Sect. B: Struct. Sci.*, 2009, **65**, 107–125.
51 K. E. Riley and P. Hobza, *J. Chem. Theory Comput.*, 2008, **4**, 232–242.
52 G. Cavallo, P. Metrangolo, R. Milani, T. Pilati, A. Priimagi, G. Resnati and G. Terraneo, *Chem. Rev.*, 2016, **116**, 2478–2601.
53 S. Kozuch and J. M. L. Martin, *J. Chem. Theory Comput.*, 2013, **9**, 1918–1931.
54 A. Otero-de-la-Roza, E. R. Johnson and G. A. DiLabio, *J. Chem. Theory Comput.*, 2014, **10**, 5436–5447.
55 X. Li, F. Curtis, T. Rose, C. Schober, A. Vázquez-Mayagoitia and N. Marom, *J. Chem. Phys.*, 2018, **148**, 241701.
56 J. P. Perdew, K. Burke and M. Ernzerhof, *Phys. Rev. Lett.*, 1996, **77**, 3865–3868.
57 J. P. Perdew, K. Burke and M. Ernzerhof, *Phys. Rev. Lett.*, 1997, **78**, 1396.
58 A. Tkatchenko and M. Scheffler, *Phys. Rev. Lett.*, 2009, **102**, 073005.
59 V. Blum, R. Gehrke, F. Hanke, P. Havu, V. Havu, X. Ren, K. Reuter and M. Scheffler, *Comput. Phys. Commun.*, 2009, **180**, 2175–2196.
60 S. Bhattacharya, B. H. Sonin, C. J. Jumonville, L. M. Ghiringhelli and N. Marom, *Phys. Rev. B: Condens. Matter Mater. Phys.*, 2015, **91**, 241115.
61 B. J. Frey and D. Dueck, *Science*, 2007, **315**, 972–976.
62 F. Pedregosa, G. Varoquaux, A. Gramfort, V. Michel, B. Thirion, O. Grisel, M. Blondel, P. Prettenhofer, R. Weiss, V. Dubourg, J. Vanderplas, A. Passos, D. Cournapeau, M. Brucher, M. Perrot and E. Duchesnay, *J. Mach. Learn. Res.*, 2011, **12**, 2825–2830.

63 P. Niggli, *Krystallographische Und Strukturtheoretische Grundbegriffe*, Akademische Verlagsgesellschaft, Leipzig, Germany, 1928.

64 B. Gruber, *Acta Crystallogr., Sect. A: Cryst. Phys., Diffr., Theor. Gen. Crystallogr.*, 1973, **29**, 433–440.

65 I. Křivý and B. Gruber, *Acta Crystallogr., Sect. A: Cryst. Phys., Diffr., Theor. Gen. Crystallogr.*, 1976, **32**, 297–298.

66 R. W. Grosse-Kunstleve, N. K. Sauter and P. D. Adams, *Acta Crystallogr., Sect. A: Found. Crystallogr.*, 2004, **60**, 1–6.

67 J. Nyman and G. M. Day, *CrystEngComm*, 2015, **17**, 5154–5165.

Faraday Discussions

PAPER

TopoFF: MOF structure prediction using specifically optimized blueprints

Julian Keupp and Rochus Schmid *

Received 27th February 2018, Accepted 1st May 2018
DOI: 10.1039/c8fd00051d

For the structure prediction of MOFs and related crystalline framework materials we have proposed the Reversed Topological Approach (RTA), where the default embedding of a topology is used as a blueprint. The optimal rotational insertion of the building blocks (BBs) at the fixed vertex positions of the blueprint is performed by minimizing the target function of the average angle deviation (AAD). Here we extend this idea by pre-optimizing the maximum symmetry embedding of a topology in order to minimize the overall mean AAD for the given set of BBs. By this fast and essentially parameter-free topoFF method, the vertex positions and cell parameters of the blueprint are further optimized in order to fit the structural needs of the BBs, which speeds up the overall search for the most energetically favorable structure. In addition, different topologies can be ranked in a quantitative and intuitive way. The definition and implementation of topoFF is explained and its application for the RTA-based structure prediction of MOFs is demonstrated with a number of instructive examples.

1 Introduction

Crystalline polymeric materials like Metal Organic Frameworks (MOFs) or Covalent Organic Frameworks (COFs) are interesting materials due to their potential porosity and structural tunability: large accessible surface areas make these materials natural candidates for a broad range of adsorption applications.[1,2] Their well-defined cages and channels can feature specialized interaction sites that can lead to the selective adsorption of specific guest molecules, which is useful in separation processes.[3] Due to their modularity in terms of building blocks, catalytic functions can be introduced into the matrix.[4] A remarkable phenomenon is the flexibility in terms of their crystal lattice, which is apparent (to some degree) in many known MOFs: depending on external stimuli like temperature, mechanical pressure or exposure to guest molecules, the cell size and shape, and therefore the pore landscape, can change drastically while the network of covalent and coordination bonds remains intact.[5,6]

Computational Materials Chemistry Group, Lehrstuhl für Anorganische Chemie 2, Ruhr-Universität Bochum, Bochum, Germany. E-mail: rochus.schmid@rub.de; Web: www.rochusschmid.de

Beyond these attractive properties, it is the fact that similar networks can be achieved by systematic variation of the discrete molecular constituents of the framework, referred to as building blocks (BBs), while their mutual connection is maintained, which has attracted a large interest in MOFs and related framework materials in recent years. The scaling of pore space was observed early by Yaghi and coworkers[7] and the concept of (iso)reticular chemistry was developed.[8] The possibility to combine different building blocks (assuming that identical inorganic building blocks are formed under the chosen synthesis conditions) allows for rational design. In this context, the topological deconstruction of crystalline lattices to their underlying network topologies has been found to be very beneficial for the analysis and comparison of different systems.

The network topology of an atomistic periodic crystal structure can be obtained as depicted in Fig. 1 (from left to right): first the structure is discretized into building blocks, which are then collapsed into single vertex positions. Chemical bonds between building blocks define the connections of the vertices. As a result the still 3D-periodic embedding of the topology (the "blueprint") is obtained. The actual topology is a further abstraction from the blueprint: it is obtained by removing the position information from the blueprint's vertices, keeping only their (periodic) connectivity. Note that in each step a certain amount of information is abstracted and thus lost on purpose.

Since the chemical space of building blocks for MOFs is seemingly infinite, especially with respect to the organic linkers, an exhaustive experimental evaluation of a material's properties for a given application is not possible. Hence, chemical intuition and insights from computational modeling are needed to guide synthetic efforts. An important method is the large-scale screening of large numbers of structures with respect to a desired property like adsorption selectivity[9] or the working capacity to store gases.[10] Furthermore, structure–property relationships can be derived from screening data *via* machine learning.[10–14] Such computational screening relies on systematic concepts to generate new hypothetical MOF structures for a given set of building blocks. A strategy to assemble building blocks into periodic nets, later referred to as a "bottom-up" strategy,[15] was proposed by Wilmer *et al.*[16] A large database of more than 100 000 hypothetical MOFs was constructed, but only a limited number of different topologies were formed in this process. The alternative "top-down" approach starts with a blueprint of a given topology and generates the atomistic crystal structure by inserting and orienting the building blocks at the vertex positions in a reverse topological approach (RTA), as shown in Fig. 1. This was first demonstrated by

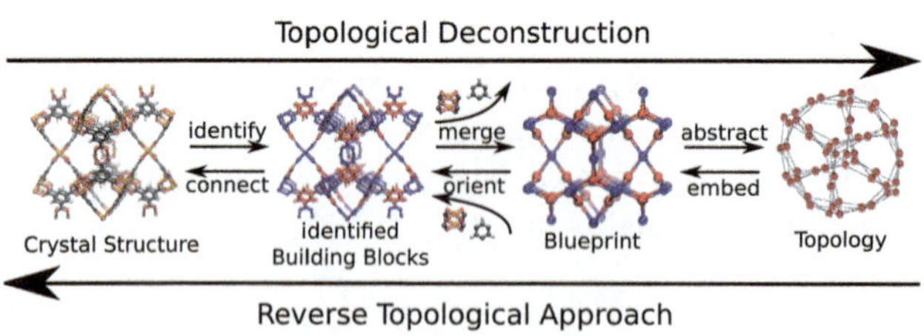

Fig. 1 Relationship between topology and crystal structure. Left to right: topological deconstruction. Right to left: reverse topological approach.

Baburin *et al.* (and later also by Lewis *et al.*) by constructing zeolitic imidazolate frameworks (ZIFs) from the blueprints defined by the corresponding zeolite structures.[17] It was pointed out by us, however, that for lower symmetry building blocks there is no unique solution for this insertion process and the energy screening of isoreticular isomers is necessary for complete structure prediction. This complete RTA, including a global search algorithm based on a first principles derived force field, has been applied for both MOFs and COFs[18–20] and is implemented as a complete workflow in our *weaver* code.

Meanwhile, a number of technically similar approaches to construct framework structures in a "top-down" fashion from topology blueprints have been proposed and implemented. Among these is the *AutograFS* code and method by Addicoat *et al.*[21] Haranczyk *et al.* implemented a version of it in their *Zeo++* code,[22] and recently the Snurr group published a similar methodology using their *ToBACCO* code.[15] Note, however, that to the best of our knowledge none of these algorithms include automated screening for potential isoreticular isomers. In contrast, our RTA approach, as implemented in *weaver*, solves the orientation problem by minimizing the target function of the average angle deviation (AAD) for arbitrary building blocks and vertex coordinations. The AAD essentially measures how well the building block fits on a given vertex position by an average deviation in degrees. Importantly, at the same time it provides the number of different insertion modes leading to isoreticular isomerism by Monte Carlo type screening.

This orientation optimization method for each building block, which is completely structure agnostic and needs no special code for particular topological blueprints (as for example the *AutograFS* method), is now the basis for the further extension of RTA proposed in this work. As shown in Fig. 1 the RTA starts from a topological blueprint, which is the scaled embedding of the topology itself, as provided by the RCSR database, for example.[23] By an isotropic scaling of this maximum symmetry barycentric embedding of the periodic graph with a scaling factor related to the size of the building blocks, reasonable vertex positions for the initial structure can be obtained in simpler cases. However, for more complex shapes, linkers of different length or asymmetric BBs, these blueprints do not fit their structural needs and the corresponding atomistic structures are far from reasonable. The fundamental idea of our topoFF approach proposed here is to extend the individual rotational optimization of the building blocks to an overall optimization of building block orientation and blueprint vertex positions (including the cell parameters of the unit cell) to arrive at an optimized blueprint. Since the optimized target function itself can be compared for different topologies, topoFF is not just useful for optimizing topological blueprints, but can serve in screening like a coarse-grained method.[24–26] Note that Boyd and Woo have also proposed a different approach to optimize topological blueprints.[27] However, this method does not include the handling of isoreticular isomerism at the same time.

In the first section, the original RTA method is briefly summarized, since it is the starting point for topoFF. Then the definition of the topoFF target function is given and the technical implementation is explained. In the remaining sections, topoFF is applied to a series of problems. In the first example the question is raised: which alternative topologies are feasible for the well known HKUST-1 (ref. 28), given its BBs? Due to the high symmetry of the BBs there is no isoreticular isomerism involved in screening. In the second example, the inorganic cluster

Zn_4O leads to isoreticular isomerism at the atomistic level. The topoFF pre-screening method is used to predict potential polymorphs of MOF-5 with a bent linker. In the third example it is shown how topoFF can be used to pre-screen for metric-breaking isoreticular isomers at the blueprint level in order to speed up the screening process.

2 Methods

2.1 The reverse topological approach

MOF structure prediction *via* the RTA, as it has been used in previous investigations,[19,20] starts from the choice of building blocks, namely the organic linker and the inorganic fragment. Note that under different synthesis conditions different inorganic clusters can in principle be formed, but in the RTA a well-defined structure and coordination mode must be assumed. Conceptually it consists of four steps (see Fig. 1): (i) embedding of the periodic graph, (ii) scaling of the embedding to serve as a blueprint, (iii) assigning and orienting the BBs and (iv) optimizing the atomistic structure using an appropriate theoretical model. In each step essential information has to be generated by fixing certain degrees of freedom (DoF) in order to proceed to the next step. Depending on the nature of the BBs, in each step it is possible that more than one value for a DoF is possible, leading to degeneracy of the solutions. It is very important to detect these ambiguities and consider them in the process in order not to miss relevant isomeric structures.

The RTA usually starts from the maximum symmetry embedding of the topology (with vertices with proper coordination), as obtained from the RCSR database,[23] serving as the raw blueprint. However, in principle it is also possible to start directly with the periodic graph of the topology, which then needs to be embedded in space (fixing the DoF of vertex positions in 3D space) in order to serve as a topological blueprint.[29] This raw blueprint usually features unit edge lengths such that the cell parameters have to be scaled by a factor corresponding approximately to the centroid distance of the connected BBs. Note that this is an isotropic scaling and works well if a single critical length exists, but becomes difficult for systems with two organic linkers of different lengths. With a properly scaled blueprint the BBs have to be assigned to their corresponding vertices. For two (or more) BBs with equal coordination numbers and symmetry-inequivalent vertices, this leads to ambiguous solutions, and in principle all possible choices have to be taken into account. Finally, the atomistic BB models have to be positioned with respect to the vertices and rotationally oriented. At this step, two types of information have to be derived for each BB–vertex combination: first, an assignment has to be made as to which connector of the BB connects to which neighboring vertex. Second, with a given assignment, the rigid BB model needs to be oriented. Therefore, the BB model is decomposed into its skeleton, as shown in Fig. 2. The skeleton is defined by a center $\boldsymbol{c}$ and (normalized) connector vectors $\boldsymbol{v}_i$ that point towards a connector position $\mathbf{r}_i$. The center position and the connecting atoms must be provided along with the atomistic BB model.

From the connector vectors, the matrix $\boldsymbol{V}$ is constructed and the matrix $\boldsymbol{B}$ is extracted by taking the vectors that point from a given vertex towards the neighboring vertices of the scaled blueprint. The assignment problem now

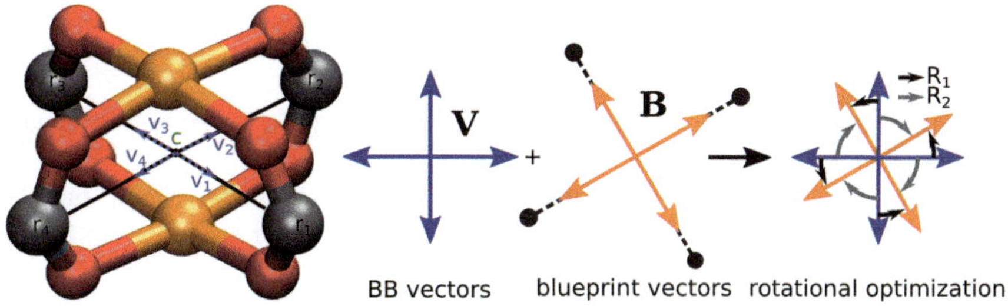

Fig. 2 The BB is decomposed into its skeleton and the resulting vectors are rotated onto the corresponding set of vectors defined by the blueprint's vertices.

corresponds to assigning which of the BB vectors $\boldsymbol{v}_i$ belong to which $\boldsymbol{b}_i$. Since the vectors are arranged as matrices as

$$\mathbf{V} = \begin{bmatrix} \boldsymbol{v}_1 \\ \boldsymbol{v}_2 \\ \cdots \\ \boldsymbol{v}_{n_c} \end{bmatrix} \in R^{n_c \times 3} \text{ and } \mathbf{B} = \begin{bmatrix} \boldsymbol{b}_1 \\ \boldsymbol{b}_2 \\ \cdots \\ \boldsymbol{b}_{n_c} \end{bmatrix} \in R^{n_c \times 3}$$

the assignment can be encoded as an index mapping

$$w(i) = j \forall \boldsymbol{b}_i \text{ assigned to } \boldsymbol{v}_j.$$

The total number of possible assignments is the number of permutations of a set with n_c entries, hence in total there are $n_c!$ possibilities. However, only a few of them are relevant, since some lead to very strained structures and some simply encode the same atomistic model in the end due to the BBs' symmetry. Therefore the assignment problem is solved by a Monte Carlo type random sampling of initial orientations of the BB vector matrix **V**, followed by performing the assignment so that the target function by which the subsequent rotational optimization is performed is as small as possible. By this approach a set of mappings w is obtained with many unreasonable assignments already excluded. The optimum orientation for each assignment is the one that superimposes the two sets of vectors as well as possible. To obtain it we perform a numerical optimization of an objective function f, constructed to be a continuous shape measure, by varying the rotation transformation matrix **R**.

$$\min_{\mathbf{R}} f(\mathbf{R} \,|\, \mathbf{V}, \mathbf{B}, \mathbf{w})$$

The definition of this target function is key to the RTA process but is also fundamental to the topoFF method proposed here. In principle, different target functions are possible, but in numerous tests we found the average angle deviation (AAD) to be the most practical and intuitive. The function f_{AAD}, used for the optimization, measures by how much the vectors $\boldsymbol{v}_i$ and $\mathbf{b}_i$ deviate on average, in degrees. It becomes zero for equal shapes and optimal orientations (perfect superposition). The target function is defined as

$$f_{\mathrm{AAD}}(\mathbf{R} \mid \mathbf{V}, \mathbf{B}, \mathbf{w}) = \frac{1}{n_{\mathrm{c}}} \sum_{i=1}^{n_{\mathrm{c}}} \arccos\langle \mathbf{b}_i, \mathbf{R}\boldsymbol{v}_{w(i)} \rangle, \tag{1}$$

where $\boldsymbol{v}_{w(i)}$ is the properly mapped blueprint vector using the assignment map $w(i)$ introduced above.

In practice, for the numerical optimization we do not use f_{AAD} itself, but the corresponding root mean square f_{RMSAD}, because it is better suited to optimization algorithms that assume a quadratic objective function, defined as

$$f_{\mathrm{RMSAD}}(\mathbf{R} \mid \mathbf{V}, \mathbf{B}, \mathbf{w}) = \sqrt{\frac{1}{n_{\mathrm{c}}} \sum_{i=1}^{n_{\mathrm{c}}} \arccos\langle \boldsymbol{b}_i, \mathbf{R}\boldsymbol{v}_{w(i)} \rangle^2}. \tag{2}$$

Due to the process of random sampling of initial orientations, a set of mappings and correspondingly a set of final optimized orientations is obtained, but only those with the lowest AADs are considered further (*e.g.* as denoted by the black and gray arrows in Fig. 2). For each orientation in this subset the atomistic BB model is translated to put $\boldsymbol{c}$ on the vertex position and rotated by the optimized rotation matrix. By comparing the root mean square deviation (RMSD) of the Cartesian coordinates (considering equal atom types), duplicates in the atomistic structure are removed from the subset. If in the end only one atomic configuration for the overall process remains, no ambiguity exists in the orientation for the given BB. This is the case if the symmetry of the BB is equal to or higher than the symmetry of the vertex in the blueprint. However, for a lower symmetry more than one (nonequivalent) orientation of the BB with equal AAD can result. With sufficient initial random sampling, all the potential orientations can be detected with a high probability.

As a result of orientational ambiguity, isoreticular isomers can be formed.[20,30] To illustrate this, a selection of BBs with and without isoreticular isomerism are shown in Fig. 3. The square planar copper paddlewheel BB (1) allows for eight

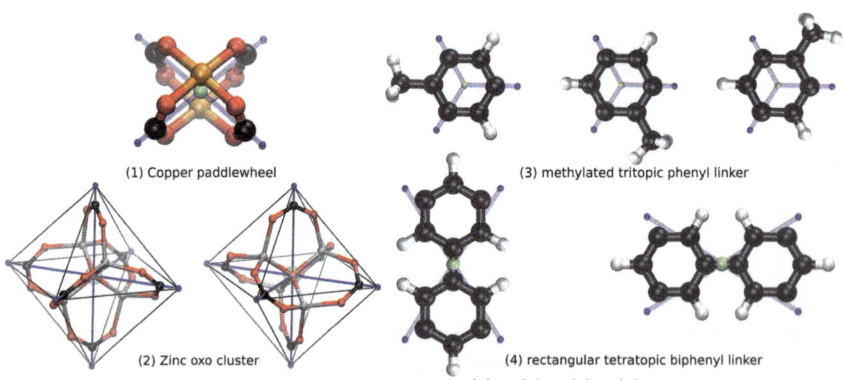

	(1)	(2)	(3)	(4)
number of assignments	24	720	6	24
minimum AAD assignments	8	24	6	8
distinct orientations	1	2	3	2
metric breaking	no	no	no	yes

Fig. 3 Examples of BBs that give rise to isoreticular isomerism. The distinct orientations are displayed.

assignments in **w**, which have an AAD of zero and all lead to an identical atomistic orientation in the end. In contrast, for the Zn_4O cluster (2), twelve zero AAD assignments in **w** lead to one of the two shown orientations (flipped tetrahedra in an octahedral site). For a methylated phenyl fragment (3) in a trigonal site, all the possible assignments in **w** have zero AAD and lead to three unique orientations (or six if the hydrogen atoms are included in the RMSD test). Note that for the biphenyl fragment (4), the 8 minimum AAD assignments lead to only two distinct orientations, but due to the rectangular shape of the BB, the connector atoms are not at the same positions in space, which we refer to as metric-breaking isomerism.

For RTA structure prediction it is necessary to evaluate the relative energies of all these isomeric forms, *e.g.* by an accurate force field like MOF-FF.[31] This means fully optimizing each form, starting from the above described structure building process with rigid BBs. However, if multiple orientations for one or more BBs have been found in the screening process, a large number of isoreticular isomers are possible. Each vertex–BB pair i that has been found to possess n_i orientations contributes to the overall number of $n_{\text{tot}} = \prod_i n_i$ possible supramolecular isomers. Suppose that 8 vertices with two possible orientations and 4 vertices with three possible orientations exist in the unit cell; in principle this amounts to $2^8 \cdot 3^4 = 20\,736$ isomeric structures to be screened. The complexity can be reduced by identifying identical structures from the space group symmetry of the blueprint, but an exhaustive search for the global minimum structure is numerically still out of reach.

However, the problem of the structure lends itself perfectly to a global search using a genetic optimization algorithm, because the orientations of all the BBs can be straightforwardly encoded as a gene: by labeling each orientation of a BB with an identifier (*e.g.* 0, 1, 2, …), each structure is uniquely defined by its genome of strings composed of such identifiers. For the above example of 8×2 and 4×3 orientations, a genome could be anything from 00000000-0000 to 11111111-2222. Due to the selection pressure that arises from creating new individuals preferentially by the crossover or mutation of existing individuals with low energy, convergence is reached much more quickly than with random sampling. Further details of this global search can be found in ref. 19 and 20, where this strategy was successfully employed.

2.2 TopoFF: optimizing the blueprint

The idea behind topoFF is the realization that modifying the vertex positions and cell parameters of the blueprint also affects the AAD for each BB–vertex pair, since this changes the matrix $\boldsymbol{B}$, which was previously fixed in the original RTA. Instead of just an isotropic scaling of the barycentric embedding, an optimization of the blueprint can save a lot of numerical effort in optimizing the atomistic models. This is especially true when the initial vertex coordination environment deviates noticeably from the BB structure. In order to incorporate the topoFF optimization into the RTA scheme we need to change its protocol slightly: in the second step, the simple scaling of the raw blueprint's unit cell is replaced by an optimization with topoFF. However, some steps that are performed in the original RTA at a later stage, have to be done earlier now during the topoFF optimization. Note that again we use the maximum symmetry barycentric embedding as a starting point for the optimization of the blueprint.

First of all, the assignment of the BBs to the blueprint vertices obviously has to be done in this step. Afterwards the local assignment problem (finding the mapping $w(i)$) is carried out. This is analogous to the third step of the original RTA, with one difference: rather than checking for isoreticular isomers on the level of the atomistic model of the BBs (by means of comparing the RMSD of the models' Cartesian coordinates after rotation), only the connector positions are used and compared w.r.t. their RMSD. As a consequence, only BBs that lead to metric-breaking isoreticular isomerism (see Fig. 3) lead to different solutions and need to be screened at the topoFF level of optimization in the same way as is done for the atomistic systems in the RTA. Depending on the total number of possible solutions, either an exhaustive search of all the possibilities or a genetic algorithm-based global search must be performed.

After this assignment, the target function of the topoFF optimization can be set up as a fictitious potential energy. One part is the characteristic edge lengths defined by the BBs' centroid distances as in the original scaling step in the regular RTA. However, here each pair of connected BBs can have an individual length. For simplicity, this part is described by a harmonic stretch potential with an equilibrium distance defined by the nature of the pair of BBs. The second part of this fictitious potential energy is the sum of the above introduced RMSAD values for each BB in its optimum orientation. The total topoFF energy for a given blueprint and a set of BB models is given by

$$U_{\text{topoFF}} = U_{\text{dist}} + U_{\text{angle}}$$

with

$$U_{\text{dist}}{}^{ij}(r_{ij}) = \frac{1}{2}k^{ij}\left(r_{ij} - r_0{}^{ij}\right)^2,$$

where k^{ij} is a force constant, $r_0{}^{ij}$ is the characteristic distance between the centroids of the BBs on vertices i and j, and r_{ij} is the vertex distance at the current configuration. The geometry contribution is computed by

$$\begin{aligned} U_{\text{angle}} &= \frac{1}{n_{\text{v}}}\sum_{i=1}^{n_{\text{v}}} c_i \min_R f_{\text{RMSAD}}{}^i(\mathbf{R}_i\,|\,\mathbf{B}_i, \mathbf{V}_i, \mathbf{w}_i) \\ &= \frac{1}{n_{\text{v}}}\sum_{i=1}^{n_{\text{v}}} c_i f_{\text{RMSAD}}{}^i\left(\mathbf{R}_i = \mathbf{R}_{\text{opt}}{}^i\,|\,\mathbf{B}_i, \mathbf{V}_i, \mathbf{w}_i\right), \end{aligned}$$

where n_{v} is the number of vertices, c_i is a weighting coefficient for vertex i and $\mathbf{R}_{\text{opt}}$ is the rotation matrix that minimizes the RMSAD for vertex i. The force constants k are the only free parameters of the method and can be used to define the relative importance of the bonded part with respect to the angle-dependent part, U_{geom}. We used a value of $c_i = 1$ kcal mol^{-1} deg^{-1} to convert the mean of the AAD to an energy unit.

Note that the matrices $\mathbf{V}_i$ still enter the geometry penalty function parametrically, but during a topoFF optimization of the vertex coordinates and cell parameters, $\mathbf{V}$ actually changes, which means that after every update of positions or cell parameters, the optimum rotation matrices $\mathbf{R}_{\text{opt}}$ of all the vertices have to be adjusted to the new blueprint geometry.

Formally, the topoFF optimization target function can be formulated as

$$\min_{\text{cell}}\min_{\mathbf{r}_{xyz}} U_{\text{topoFF}} = \min_{\text{cell}}\min_{\mathbf{r}_{xyz}}\left[\begin{array}{l}\sum_{ij\in\text{edges}} U_{\text{dist}}{}^{ij}\left(r_{ij}\left(\mathbf{r}_{xyz}\right)\right) \\ +\frac{1}{n_{\text{v}}}\sum_{i=1}^{n_{\text{v}}} c_i \min_{R} f_{\text{RMSAD}}{}^{i}\left(\mathbf{R}_i \,|\, \mathbf{B}_i, \mathbf{V}_i\left(\mathbf{r}_{xyz}\right), \mathbf{w}_i\right)\end{array}\right]$$

which is basically the topoFF potential energy U_{topoFF} optimized with respect to the cell parameters and the vertex positions $\mathbf{r}_{xyz}$. Note that because the vertex positions depend on $\mathbf{r}_{xyz}$, and $\mathbf{V}_i$ depends on $\mathbf{r}_{xyz}$, the optimization is a triply nested minimization. The optimum embedding of a topology for a given set of BBs is found at the global minimum of this fictitious potential energy. Both parts are defined as zero when the geometries match perfectly, and increase steadily with an increasing geometric difference. In order to find this optimum embedding, a lattice geometry optimization of the blueprint has to be performed, which is technically not different from any geometry optimization utilized for atomistic systems.

For the gradient-based optimization, the derivatives of U_{topoFF} with respect to the vertex positions $\boldsymbol{r}_i$ and the cell parameters are needed. Whereas the gradients for the distance-dependent terms are readily calculated for the harmonic potential, the angle-dependent part U_{geom} is more involved. These gradients with respect to the vertex positions are obtained *via* a finite difference method as

$$\frac{\partial U_{\text{angle}}}{\partial r_i(d)} \approx \frac{U_{\text{angle}}{}^{i}(r_i(d)+\Delta r) - U_{\text{angle}}{}^{i}(r_i(d)-\Delta r)}{2\Delta r} \forall d \in \{x, y, z\},\ i \in \ \text{vertices}.$$

Note that upon changing the position of one vertex, the AAD of not only this vertex, but also of all the neighboring vertices is changed, as is schematically shown in Fig. 4. This implies that the rotation matrix $\boldsymbol{R}_{\text{opt}}{}^{j}$ is reoptimized for all j adjacent to i during the evaluation of each $U_{\text{geom}}{}^{i}$ after a displacement of vertex i.

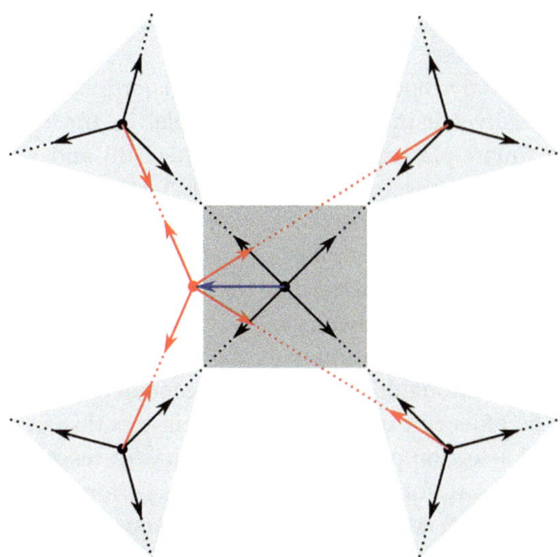

Fig. 4 By a displacement (blue arrow) of the central vertex, the AAD of this vertex and the neighboring vertices is modified.

The local change in energy $U_{\text{angle}}{}^{i}$ used for the finite difference method can then be calculated as

$$U_{\text{angle}}{}^{i} = f_{\text{RMSAD}}{}^{i} + \sum_{j \text{ adjacent to } i} f_{\text{RMSAD}}{}^{i}.$$

The derivative of the topoFF energy with respect to the cell parameters is also evaluated *via* a finite difference method. However, in this case it is necessary to update the optimum rotations $\mathbf{R}_{\text{opt}}{}^{i}$ of all the vertices, since the local coordination environment of all the vertices could potentially change.

Gradient-based optimizers typically find only neighboring local optima and there is no guarantee of identifying the optimal blueprint when starting from a single starting point (*e.g.* the scaled maximum symmetry barycentric embedding). Since the shape of the topoFF energy landscape is unknown beforehand, a gradient-free global optimization method not usually used for structural optimizations was also successfully tested in this context. The efficient Covariance Matrix Adaptation Evolutionary Strategy (CMA-ES)[32–34] is a stochastic evolutionary algorithm that allows local minima to be overcome and enables a global search of the target function without the need to compute gradients. We simply used fractional vertex coordinates and cell parameters as the parameters.

2.3 Choice of topologies

The RTA methodology aims to predict the optimal (one or many) structure(s) of a system defined by BBs and topology. For the complete structure prediction it is necessary to enumerate all the relevant topologies. Due to the arbitrary number of topologies in principle (including large unit cells with low space group symmetry), one needs to make a reasonable choice here. One can rely on concepts like reducing the set of high symmetry cases, *e.g.* edge transitive topologies, but the well known example of MOF-177 (ref. 35) in its non-edge transitive **qom** topology indicates that this could lead to overlooked topological isomers of particular systems.

In the first step, only topologies with proper coordinations of vertices and respective connectivity are included. For example, in the case of the typical HKUST-1 with a square-planar copper paddlewheel BB and a trigonal organic linker, only 3,4-hetero-connected nets, with the 3-coordinated (3c) vertex connected only to 4-coordinated (4c) vertices, and *vice versa*, are possible.

In order to further limit the number of possibilities, in the second step, the remaining topologies are ranked according to the extent to which they support the geometry of the given BBs by means of first computing the AADs of all the topologies in their maximum symmetry embedding. Only those topologies that already fit within a reasonable threshold are considered further. Clearly, the optimization of the blueprint by topoFF then optimizes this fit further.

This search and selection process of topologies with respect to coordination, connectivity and local AAD for particular shapes was performed using our recently launched online platform MOF+.[36] Note that the source of the topologies and embeddings for MOF+ is the RCSR database.[23] The aim of MOF+ is to facilitate the search for topologies in the context of RTA structure prediction and to connect MOF structures, BBs and topologies by a relational (SQL-based) database.

2.4 Implementation and computational details

All the computational methods presented in this work were implemented in the in-house developed force field code *pydlpoly*,[31] which is a fork of DL_Poly Classic[37] with a completely flexible *Python*-based scripting front-end. Apart from the rotational optimization, implemented in the F90 compiled language, all the other methods were implemented either in *Python* or *Python* interfaces employing the CMA-ES library, for example. For the implementation of topoFF, the *pydlpoly* harmonic bond terms for U_{dist} and the gradient-based cell optimizer were employed, whereas an additional module was programmed for the computation of U_{angle} and its derivatives. The finite difference scheme used displacement $\Delta r = 0.01$ Å for the vertex positions and $\Delta c = 0.1$ Å for the cell tensor components. As explained previously, the only free parameter in topoFF is the force constant in U_{dist}, which controls the relative importance of U_{dist} with respect to U_{angle} in this overall fictitious energy. After some testing we used the same value of $k^{ij} = 0.1$ mdyn per Å throughout all the applications. In the gradient-free optimizations with CMA-ES, standard settings of the CMA-ES library where used together with a population size of twenty-four individuals (matching the number of cores employed in the computations). Note that in this case a couple of thousand iterations were needed for convergence, which were, however, numerically efficient since no gradients needed to be computed. For the final optimization of the atomistic structures we used our first principle parameterized force field MOF-FF,[31] as in previous investigations.[19,20]

Pore size distributions, void fractions and surface areas were calculated by our in-house developed poreanalyzer code using a probe radius of 1.4 Å. Its methodology is described in detail in ref. 25.

3 Applications of topoFF

3.1 Hypothetical phases of HKUST-1 and related systems

One of the best investigated MOFs is HKUST-1 (Hong Kong University of Science and Technology 1).[28] It is synthesized from 1,3,5-benzenetricarboxylic acid (btc) and a copper(II) salt, forming a paddlewheel (PW) type BB that connects with the linker in the twisted boracite (**tbo**) topology. However, the introduction of even a small substitution at the btc results in a MOF with a different topology. A btc with a single methyl group leads to the formation of the **fmj** topology.[38] There are numerous other examples where linker functionalization leads to a different topology,[39–43] which suggests that every MOF can potentially form topological polymorphs.

The reasons for topological preferences have already been studied for the high symmetry edge transitive **tbo** and **pto** nets.[18,44] It was found that, even though the **pto** topology geometrically fits the BBs better than **tbo** does, the **pto** variant of HKUST-1 cannot avoid an energetically disfavored and strained structure. However, other less symmetric topologies might also be possible in principle and could potentially be synthesized by a specific synthetic protocol or by introducing a different functionalization. To get an idea of the possible hypothetical structures, we screened for possible HKUST-1 polymorphs by considering all the topologies available in the RCSR database that feature proper 3c–4c connectivity.

First, all the topologies were optimized using the topoFF method, as described in the previous section, with respect to the BBs of HKUST-1 in order to obtain a ranking based on the mean AAD (Fig. 5). Note that in the following, for clarity we discuss U_{angle}/c_i in units of degrees.

Since the BBs have high symmetry themselves, the maximum symmetry embedded topologies that served as the initial blueprints are in most cases already close to the topoFF optimized blueprints, as can be seen by comparing the mean AADs in Fig. 5 before and after optimization. Three topologies, namely **pto**, **iab** and **fjh**, were found to have a mean deviation of zero, which means these fit perfectly to the BBs. In contrast, the **tbo** topology, observed for HKUST-1, has a mean AAD of 4.2°. Interestingly, for the **sln** topology the initial and optimized mean AAD values are very different: in the initial embedding the 4c vertices show a tetrahedral coordination, but by modifying the blueprint a substantially better fit to the square planar paddlewheel can be achieved.

In the second step, atomistic models were constructed from these optimized blueprints and were optimized with MOF-FF to obtain relative energies with respect to the minimum energy structure, normalized to one unit formula Z = PW_3btc_4.

The **tbo** model (and its siblings **yao** and **yac**) has the lowest energy structure. The energetic differences for the low mean AAD topologies arise from atomistic details that cannot be captured in the coarse-grained topoFF pre-screening: due a twist in the Cu_2O_2C–phenyl plane the **pto**, **iab** and **fjh** topologies possess a torsional energetic penalty of about 25 kcal mol^{-1} Z in comparison to **tbo** (see also the corresponding discussion in ref. 18 and 44). For the system in the **fmj** topology only two in every three fragments show this twist, and hence its relative torsional penalty is about 17 kcal mol^{-1} Z. Still, **pto**, **iab** and **fjh** are more stable than **fmj** since there are other stabilizing interactions. Apart from this, the relative energies correlate reasonably well with the geometric mean AADs obtained from the topoFF optimization and illustrate its usefulness in order to pre-rank the topologies (Fig. 6).

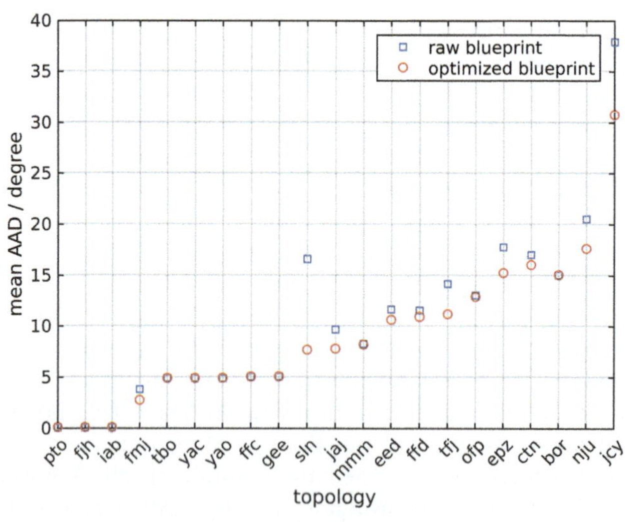

Fig. 5 Mean AADs of the trial 3c–4c topologies before and after optimization with topoFF.

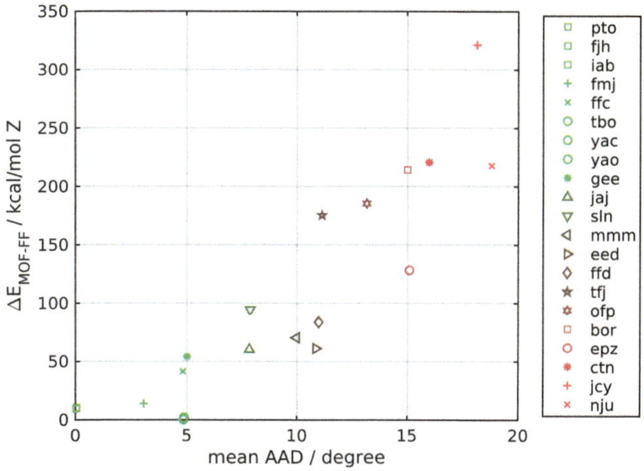

Fig. 6 Mean AADs of the topoFF optimized blueprints *vs.* ΔE(MOF-FF) of the atomistic models (Z = PW_3btc_4).

As discussed in previous work,[18] the 1,3,5-benzenetribenzoic acid (btb) linker features the same trigonal geometry, but due to steric repulsions between the hydrogen atoms of adjacent phenyl systems, a non-planar structure results. In addition, rotation around the biphenyl bond is more facile then around the conjugated phenyl carboxylate bond, which allows for more flexibility. Note that the interwoven MOF-14 with **pto** topology is found experimentally for this linker.[45] Again, atomistic systems were constructed from the optimized blueprints with this linker and optimized using MOF-FF. As observed previously,[18] due to the non-planar linker conformation, **tbo** is not the most stable structure anymore. The most energetically preferred topologies are **pto**, **iab** and **fjh**, which also show the lowest topoFF energies and mean AADs. Interestingly, the relative energy of the **sln** system is even lower than that of **tbo** and **fmj**. Note that **sln** is most stabilized by topoFF, converting the tetrahedron into a more or less square planar shaped vertex. At the atomistic level the stabilization with respect to the other topologies goes even further and during its optimization it folds to create a framework with a 'double wall' structure and hexagonal pores (see Fig. 7). This leads to an increase in dispersive interactions and is also reflected by the low void volume of 66%, which is smaller than that of about 75% for the other low energy systems. This hypothetical MOF polymorph of MOF-14 is structurally very interesting since it features one-dimensional pore channels, with the open metal sites of all the adjacent PWs pointing towards the center of the channels. However, this structure becomes feasible only with the more flexible btb linker. Without topoFF pre-optimization, the **sln** topology would probably not even be considered for the construction of a PW-based MOF because of the tetrahedral 4c vertex.

In Fig. 8 the correlation between MOF-FF relative energies and mean AADs is shown for the btb linker. Due to the higher flexibility of btb, the range of strain energies is smaller than in the case of btc and the correlation with the mean AAD also becomes better. The reason is clearly that the energetic penalty for a twist in the linker, which is ignored in topoFF, is smaller for btb. The experimentally

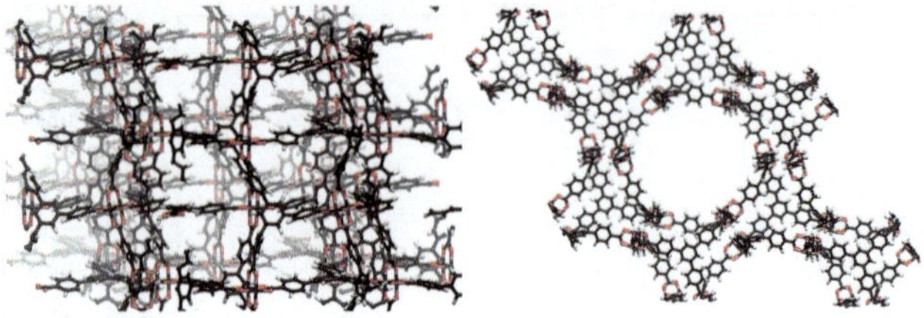

Fig. 7 MOF-FF optimized structure of the hypothetical PW-based MOF with a btb linker in the **sln** topology.

observed **pto** topology is correctly predicted by MOF-FF, however, here we are comparing non-interpenetrated systems without the inclusion of solvent. Despite the necessary deformation for the interweaving in MOF-14, the interpenetration of the framework leads to additional stabilization.

3.2 A MOF-5 type system with a bent linker?

Another well known MOF is MOF-5, or IRMOF-1,[46] which is formed from a zinc oxo cluster and 1,4-benzenedicarboxylic acid (terephthalic acid). The MOF that is formed can be deconstructed into the primitive cubic (**pcu**) topology featuring $Zn_4O(COO)_6$ and *para*-benzenedicarboxylic acid (*p*-bdc) building units. Due to the fact that the skeleton of the inorganic BB possesses octahedral O_h symmetry, whereas its atomistic model has tetrahedral T_d symmetry, there is intrinsic orientational ambiguity that leads to isoreticular isomerism in the MOF structures (see also Fig. 3).[30] From a topological point of view, MOFs with ditopic BBs are special, since two-fold coordinated BBs are not represented as vertices, but as edges. In the RTA scheme, for such linkers a blueprint has to be generated by inserting an additional vertex in the center of each edge. By default these are then linear linkers with angles of 180°. However, for a non-linear BB, *e.g.* 1,3-

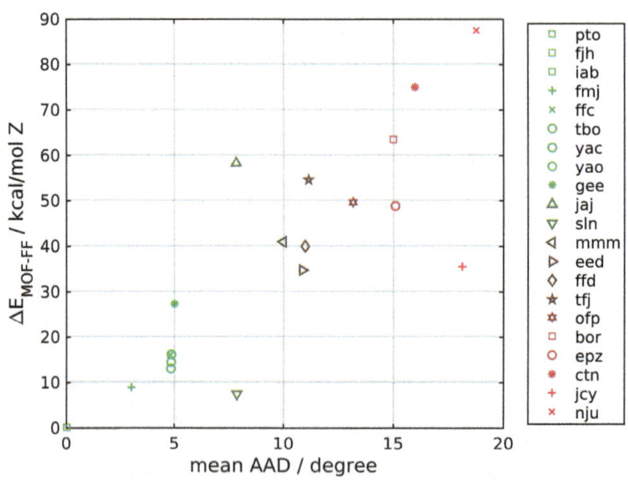

Fig. 8 Mean AADs after topoFF optimization *vs.* ΔE(MOF-FF) of the atomistic systems (Z = PW_3btb_4).

benzenedicarboxylic acid (*m*-bdc), any blueprint has to be altered drastically in order to match its 120° angle.

A priori it is not clear which topologies can be embedded in a way to meet the need for the 120° angle in *m*-bdc. In principle any topology that features only 6c–6c connections is feasible. Herein we restricted our study to consider only small-transitivity topologies with a maximum number of symmetry distinct vertices $p \leq 2$ and edges $q \leq 2$, in order to reduce the number of topologies from about 300 to 30, according to the minimum transitivity principle.[47,48] For each topology, 2c vertices were added to the blueprint at the edge centers and topoFF optimization was conducted in various supercells, ranging from $1 \times 1 \times 1$ to $2 \times 2 \times 2$, with both the gradient-based optimization and CMA-ES optimization algorithms. For each topology the minimum mean AAD was extracted from the data and the corresponding blueprint was used for further investigation of this topology. In almost all cases the CMA-ES topoFF energy optimizer managed to find blueprints that are lower in energy, because the gradient-based optimizers often got stuck in local minima. Supercell screening is necessary because it encodes for periodicity: **pcu** contains a single vertex in its $1 \times 1 \times 1$ cell. The MOF-5 structure, however, requires an alternating orientation of the zinc oxo clusters in its minimum energy structure, which can only be realized in a $2 \times 2 \times 2$ supercell.

The data shown in Fig. 9 demonstrates that by using a non-linear ditopic linker, the number of topologies that can be embedded with a small mean AAD is relatively large. Almost any topology considered here can be embedded with a mean AAD of less than ten degrees. The initial configuration of the octahedral sites does not matter, as mirrored by the fact that the low-ranking average AAD topologies do not preferentially have small octahedral AAD vertices in the initial blueprints (with the exception of **pcu**). Fig. 10 shows two examples of the optimized blueprints.

In order to evaluate the relative stabilities of the actual atomistic MOFs, RTA runs were carried out for some of the low mean AAD blueprints and **nia**, as an

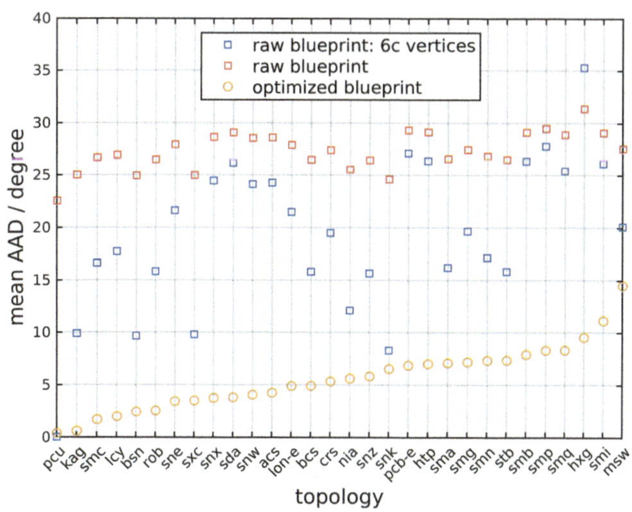

Fig. 9 Mean AADs of the trial topologies before and after optimization with topoFF.

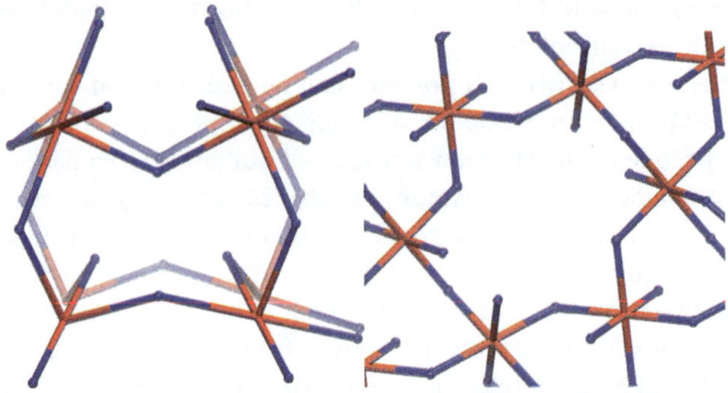

Fig. 10 TopoFF optimized blueprints of **pcu** (left) and **kag** (right).

example of a blueprint with a mean AAD of more than five. Due to the Zn_4O cluster the number of isoreticular isomers for each case is $2^{n_{6c}}$, where n_{6c} is the number of six-coordinated vertices in the supercell. These were evaluated exhaustively for a total of up to 512 structures. For larger systems, the established genetic algorithm search was employed. In Fig. 11, the range of MOF-FF optimized energies (normalized to one formula unit $Z = Zn_4O(m\text{-bdc})_3$) for each of the investigated topologies is shown together with the respective accessible surface area.

Solely based on the MOF-FF total energy computed in the gas phase, the systems in the **rob** topology are the most stable. However, MOFs are synthesized in a solvent, which can stabilize open pores. By assuming that this stabilization effect is roughly proportional to the accessible surface area or correlates with the void fraction, one can rationalize the results of Fig. 11: the low energy structures of each topology tend to be more dense, but since the density is inversely proportional to the void fraction, the high density structures are those that are stabilized the least by the solvent.

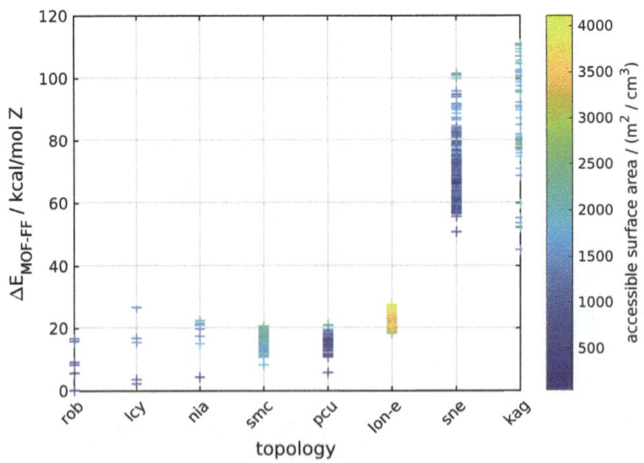

Fig. 11 Relative energies of *m*-bdc Zn_4O models. The color encodes the accessible surface area.

From a visual inspection of the optimized structures (see Fig. 12), it is evident that the more dense models gain dispersion energy, sometimes at the great expense of the bond, angle and torsion terms, and that the high energy structures are not necessarily irrelevant if they possess higher void fractions or larger surface areas. For the ultimate prediction of the most probable candidate, the inclusion of the solvation free energy would be necessary.[49]

It can be concluded that none of the topologies are a perfect match for this hypothetical MOF. The same kind of strained geometry is present to some degree in all of the structures, which is presumably the reason why no isophthalate Zn_4O MOF has yet been reported. Instead, a structure with different inorganic structural motifs has been reported, where the zinc(II) ions form tetrahedral structures that connect to the carboxylic oxygens of four different isophthalates.[50] As has already been pointed out, the RTA method is bound to predefined BBs. On the other hand, this investigation shows that the topological space for non-linear linkers is drastically increased. For such a problem, a two-step process with pre-optimization of a blueprint by topoFF with the robust CMA-ES method and an RTA screening of atomistic models based on these optimized blueprints is necessary to keep the effort tractable.

The structural difficulties we encountered with the isophthalate linker might be overcome by isoreticular chemistry, namely by extending the linker to achieve more conformational flexibility. In fact, for the benzene-1,3-dibenzoate (*m*-BDB) linker, a MOF featuring a Zn_4O BB was reported that possesses **pcu** topology.[51]

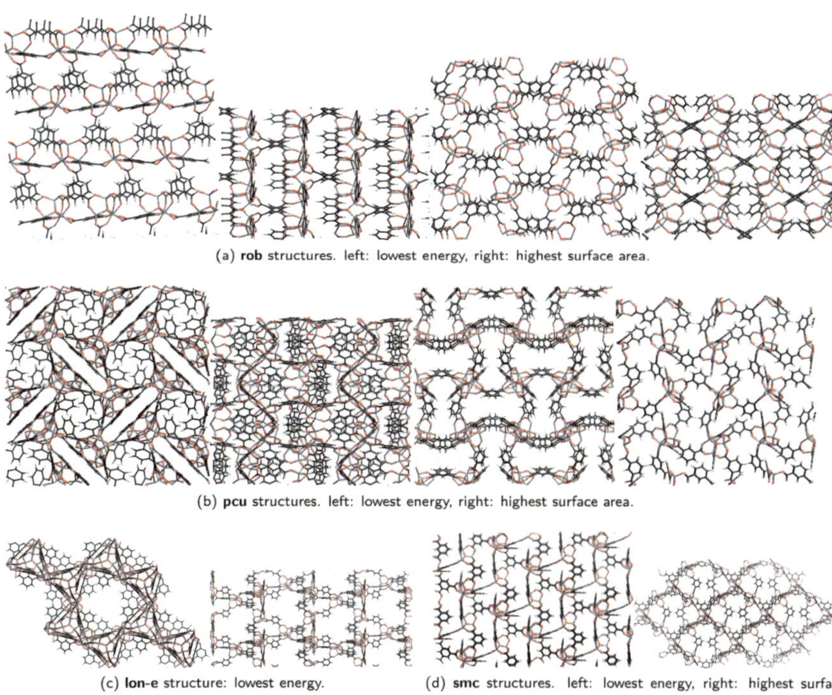

(a) **rob** structures. left: lowest energy, right: highest surface area.

(b) **pcu** structures. left: lowest energy, right: highest surface area.

(c) **lon-e** structure: lowest energy.

(d) **smc** structures. left: lowest energy, right: highest surface area.

Fig. 12 Selection of MOF-FF optimized structures of various topologies (two views along different crystallographic axes for (a) to (c)).

3.3 Metric-breaking building blocks in topoFF

In the previous examples, no metric-breaking BBs were involved and isoreticular isomerism appeared only at the atomistic level. For the final example we have deliberately selected a metric-breaking BB with a rectangular geometry, namely the tetratopic linker 5,5′-(1,4-phenylenedi-2,1-ethinyl)bis(1,3-benzenedicarboxylic acid) (bdpe), together with the usual square planar copper paddlewheel BB. In its relaxed structure the organic linker has a large aspect ratio of about four, which is why it does not fit well into the maximum symmetry embedded blueprints of most topologies. The metric-breaking isoreticular isomerism means that orientation screening needs to be performed at the blueprint level before the topoFF optimization.

Rather than performing a full structure prediction, we restricted our study to investigate only the topology that was observed experimentally,[52] in order to demonstrate the method. The MOF is reported to have the **stx** topology, which is a 4c–3c net and is a "sibling" topology of the parent 4c–4c topology **ssb. stx** can be derived from **ssb** by splitting one of the 4c vertices into two adjacent 3c vertices. Thus, in the topological deconstruction the organic linker is not reduced to one vertex, but is represented by two 3c vertices. This representation provides more insight and is thus the preferred choice for the topological analysis. However, for structure prediction by RTA it is preferential to use the minimum vertex topology **ssb**, where the linker is treated as a single vertex. Orientational screening is then used to predict the correct optimal structure. Note that **stx** is only one of many topologies that can be derived from **ssb.** In the RCSR database there are currently four topologies derived from **ssb** in this way (**stw, stu, stj** and **stx**), but in principle there are many more lower symmetry possibilities. Even the choice of which 4c vertex is split is ambiguous in the **ssb** case, with its two symmetry distinct 4c vertices. Without space group symmetry (*P*1), the **ssb** blueprint has 16 vertices in its unit cell, with eight of them labeled as 0 and 1. The rectangular BB can be put on either the 0 or 1 vertex (the choice of 0 is referred to as **ssb** 0, and **ssb** 1 for the 1 vertex), and it can be inserted in two orientations on each vertex, resulting in $2 \cdot 2^8 = 512$ unique raw blueprints, which is sufficient for an exhaustive evaluation with RTA and topoFF. The blueprints were optimized with topoFF, using both CMA-ES and gradient-based optimizers, which in this case gave the same results. The atomistic models were constructed and optimized from the optimized blueprints, and the relative total energies were normalized to one formula unit Z = CuPW(bdpe) (Fig. 13).

In contrast to the previously discussed examples, here the topoFF total energy is not correlated with the relative MOF-FF energies of the atomistic models, and it cannot be used to pre-rank the optimized blueprints. All the models provided in Fig. 14 show visible buckling of the rectangular BB, which means that its structure deviates from the initial planar BB model used for topoFF. Yet the topoFF method works under the assumption that the BBs are rigid, which is clearly a limitation of the method. However, the optimizations of the blueprints are beneficial for the overall workflow because the initial structures for the atomistic optimization are still much closer to the later optimum than the structures from the scaled raw blueprint would be.

The resulting structures were further analyzed with respect to their pore space and topology. The topological deconstruction is performed as depicted in Fig. 1,

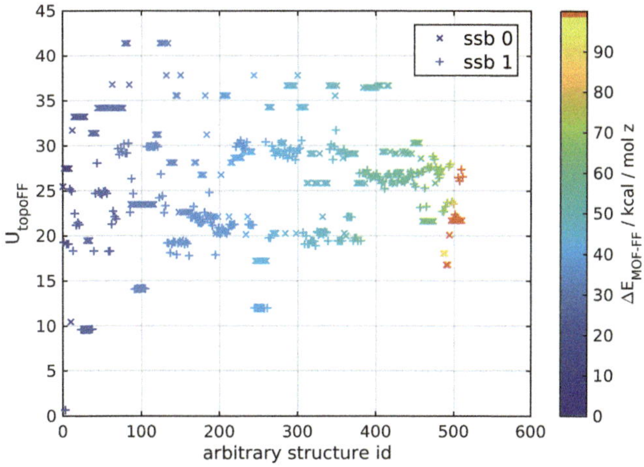

Fig. 13 Total MOF-FF energy $\Delta E_{\text{MOF-FF}}$ rank *vs.* topoFF geometry energy. The color map corresponds to the relative energy. Structures belonging to the highest bin (red) are above 100 kcal mol^{-1} Z. The relative energies of **ssb** 0 and **ssb** 1 each start at zero.

by identifying fragments, merging them onto vertices and analyzing the vertex connector graph by computing the coordination sequences and vertex symbols, which are then compared to the known topologies. The methodology is described in ref. 53. Note that for this analysis we deconstructed not into the minimum vertex topology, but into the 3c–4c topologies. As expected, only four of the 512 evaluated structures have a topology known in the database. These are the high symmetry (vertex transitivity $p = 2$) topologies **stw**, **stu**, **stj** and **stx**. However, structures with similar energies were found with a larger number of topologically distinct vertices. The results are summarized in Table 1 for the energetically preferred systems, which are all shown in Fig. 14.

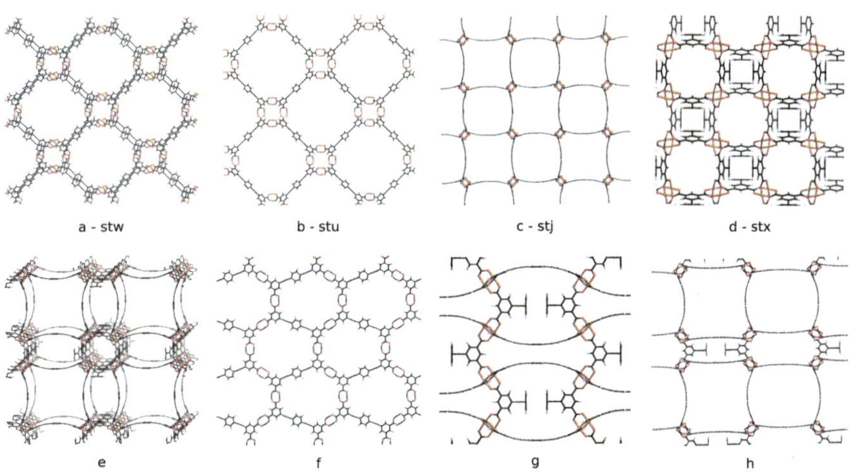

Fig. 14 Selection of low energy structures after optimization.

Table 1 Collection of data obtained from the characterization of systems (a)–(h)

ID	ssb vertex	$\Delta E_{\text{MOF-FF}}$ [kcal mol^{-1} Z]	Void fraction	Surface area [m^2 cm^{-3}]	Number of vertices (p)
a – **stw**	0	31.8	0.35	1542	2
b – **stu**	0	38.3	0.51	3784	2
c – **stj**	1	9.9	0.63	4917	2
d – **stx**	1	0.0	0.26	4206	2
e	1	10.2	0.55	4776	3
f	0	33.1	0.26	1775	3
g	1	11.7	0.25	3695	3
h	1	21.7	0.51	4285	5

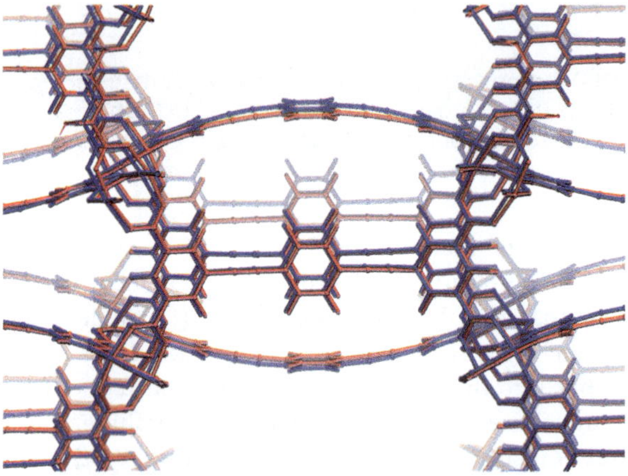

Fig. 15 Overlay of the experimental structure and the global energetic minimum (d). Blue: predicted model, red: experimentally derived model.

It turns out that it is more beneficial for the stability to place the rectangular BB onto the **ssb** vertex labeled as 1, whose minimum energy structure (d) is lower in energy by 31.8 kcal mol^{-1} Z and has a higher surface area, and thus potentially has larger solvent stabilization effects. The **ssb** 0 structures (with the exception of (b)) fold themselves to create π-stacking domains, at the expense of a distorted paddlewheel. It is interesting to note that the number of unique vertices correlates with the rank. Structures with a greater number of vertices p are a mixture of the structural motifs of the low transitivity models, as can be seen for example in (e), which features the barrel motif of (d), or in (h), which features the structure motifs of (c) and (g). The global energy optimum (d) also corresponds to the experimentally derived structure.[52] Their overlay is plotted in Fig. 15.

4 Conclusion

The topoFF approach is a rapid and essentially parameter-free method to optimize blueprints for the reverse topological approach, based on the shape of the building blocks used in the framework constructions. A fictitious energy expression is employed as a target function for the optimization of vertex positions and cell parameters, which is based on the average angle deviation, used previously for the local optimization of the rotational positions of the building blocks. In topoFF, the overall mean AAD, together with a harmonic vertex–vertex distance potential, is minimized with either a gradient-based or a gradient-free evolutionary strategy, which helps to find the global minimum. The integration of the topoFF methodology into the RTA was demonstrated and its advantages over the conventional RTA approach were shown. Due to the optimization of the blueprint, the atomistic structure optimization can become significantly faster due to a more reasonable initial structure, especially when the BB models used possess lower symmetry. Note that for computationally more demanding quantum mechanic methods, this can serve as an efficient way to generate starting structures.

The application examples demonstrate that the topoFF fictitious energy or its main component, namely the mean AAD, can serve as a tool to rank potential topologies with respect to how well they are suited to form a framework for a given set of building blocks. The interesting example of the hypothetical copper paddlewheel MOF with a 1,3,5-benzenetribenzoic acid linker in the **sln** topology (a polymorph of MOF-14) shows that certain topologies might seem implausible from their original embedding (tetrahedral vertex), but can adapt. However, the prediction of the best topological isomer by topoFF is restricted by its coarse-grained nature, which is not able to capture the detailed conformational energy landscapes of more complex building blocks. The gain in computational efficiency upon the incorporation of topoFF into the overall structure prediction becomes evident for the case of a bent linker variant of MOF-5. A bent ditopic linker for 6c–6c networks leads to an enormously wide range of possible structures, which would be impossible to screen purely on the atomistic level, but screening becomes possible with topoFF pre-optimization and selection.

We are currently working to incorporate the RTA method into our MOF+ web platform so that it is accessible for every user. The inclusion of the topoFF method offers a very robust pre-optimization technique for blueprints. In addition, this approach will allow for the exploration of more complex (multilinker) framework compounds by computational structure prediction.

Conflicts of interest

There are no conflicts of interest to declare.

Acknowledgements

The authors would like to thank Johannes P. Dürholt for discussions and help with the topological deconstruction. The Deutsche Forschungsgemeinschaft (DFG) is thanked for financial support within PP-1928 and FOR-2433 (grants SCHM1389/9-1 and SCHM1389/10-1). Further financial support from the Cluster of Excellence RESOLV (EXC 1069) funded by the DFG is acknowledged.

References

1 L. J. Murray, M. Dincă and J. R. Long, *Chem. Soc. Rev.*, 2009, **38**, 1294–1314.
2 S. Ma and H.-C. Zhou, *Chem. Commun.*, 2010, **46**, 44–53.
3 J.-R. Li, R. J. Kuppler and H.-C. Zhou, *Chem. Soc. Rev.*, 2009, **38**, 1477–1504.

4 W. Kleist, F. Jutz, M. Maciejewski and A. Baiker, *Eur. J. Inorg. Chem.*, 2009, **2009**, 3552–3561.
5 A. Schneemann, V. Bon, I. Schwedler, I. Senkovska, S. Kaskel and R. A. Fischer, *Chem. Soc. Rev.*, 2014, **43**, 6062–6096.
6 L. Sarkisov, R. L. Martin, M. Haranczyk and B. Smit, *J. Am. Chem. Soc.*, 2014, **136**, 2228–2231.
7 M. Eddaoudi, J. Kim, N. Rosi, D. Vodak, J. Wachter, M. O'Keeffe and O. M. Yaghi, *Science*, 2002, **295**, 469–472.
8 O. M. Yaghi, M. O'Keeffe, N. W. Ockwig, H. K. Chae, M. Eddaoudi and J. Kim, *Nature*, 2003, **423**, 705–714.
9 S. Li, Y. G. Chung and R. Q. Snurr, *Langmuir*, 2016, **32**, 10368–10376.
10 M. Z. Aghaji, M. Fernandez, P. G. Boyd, T. D. Daff and T. K. Woo, *Eur. J. Inorg. Chem.*, 2016, **2016**, 4505–4511.
11 M. Pardakhti, E. Moharreri, D. Wanik, S. L. Suib and R. Srivastava, *ACS Comb. Sci.*, 2017, **19**, 640–645.
12 J. D. Evans and F.-X. Coudert, *Chem. Mater.*, 2017, **29**, 7833–7839.
13 M. Fernandez, P. G. Boyd, T. D. Daff, M. Z. Aghaji and T. K. Woo, *J. Phys. Chem. Lett.*, 2014, **5**, 3056–3060.
14 G. Borboudakis, T. Stergiannakos, M. Frysali, E. Klontzas, I. Tsamardinos and G. E. Froudakis, *npj Comput. Mater.*, 2017, **3**, 40.
15 Y. J. Colón and R. Q. Snurr, *Chem. Soc. Rev.*, 2014, **43**, 5735–5749.
16 C. E. Wilmer, M. Leaf, C. Y. Lee, O. K. Farha, B. G. Hauser, J. T. Hupp and R. Q. Snurr, *Nat. Chem.*, 2012, **4**, 83–89.
17 (*a*) I. A. Baburin, S. Leoni and G. Seifert, *J. Phys. Chem. B*, 2008, **112**, 9437–9443; (*b*) D. W. Lewis, A. R. Ruiz-Salvador, A. Gomez, L. M. Rodriguez-Albelo, F.-X. Coudert, B. Slater, A. K. Cheetham and C. Mellot-Draznieks, *CrystEngComm*, 2009, **11**, 2272–2276.
18 S. Amirjalayer, M. Tafipolsky and R. Schmid, *J. Phys. Chem. C*, 2011, **115**, 15133–15139.
19 S. Bureekaew and R. Schmid, *CrystEngComm*, 2013, **15**, 1551–1562.
20 S. Bureekaew, V. Balwani, S. Amirjalayer and R. Schmid, *CrystEngComm*, 2014, **17**, 344–352.
21 M. A. Addicoat, D. E. Coupry and T. Heine, *J. Phys. Chem. A*, 2014, **118**, 9607–9614.
22 R. L. Martin and M. Haranczyk, *Cryst. Growth Des.*, 2014, **14**, 2431–2440.
23 M. O'Keeffe, M. A. Peskov, S. J. Ramsden and O. M. Yaghi, *Acc. Chem. Res.*, 2008, **41**, 1782–1789.
24 J. P. Dürholt, R. Galvelis and R. Schmid, *Dalton Trans.*, 2016, **45**, 4370–4379.
25 J. P. Dürholt, J. Keupp and R. Schmid, *Eur. J. Inorg. Chem.*, 2016, **2016**, 4517–4523.
26 R. Semino, J. P. Dürholt, R. Schmid and G. Maurin, *J. Phys. Chem. C*, 2017, **121**, 21491–21496.
27 P. G. Boyd and T. K. Woo, *CrystEngComm*, 2016, **18**, 3777–3792.
28 S. S.-Y. Chui, S. M.-F. Lo, J. P. H. Charmant, A. G. Orpen and I. D. Williams, *Science*, 1999, **283**, 1148–1150.
29 J.-G. Eon, *Acta Crystallogr., Sect. A: Found. Crystallogr.*, 2011, **67**, 68–86.
30 S. Amirjalayer and R. Schmid, *J. Phys. Chem. C*, 2008, **112**, 14980–14987.
31 S. Bureekaew, S. Amirjalayer, M. Tafipolsky, C. Spickermann, T. K. Roy and R. Schmid, *Phys. Status Solidi B*, 2013, **250**, 1128–1141.

32 N. Hansen and A. Ostermeier, *Evol. Comput.*, 2001, **9**, 159.
33 C. Igel, N. Hansen and S. Roth, *Evol. Comput.*, 2007, **15**, 1.
34 N. Hansen, *Workshop Proceedings of the GECCO Genetic and Evolutionary Computation Conference*, 2009, p. 2389.
35 H. K. Chae, D. Y. Siberio-Pérez, J. Kim, Y. Go, M. Eddaoudi, A. J. Matzger, M. O'Keeffe and O. M. Yaghi, *Nature*, 2004, **427**, 523–527.
36 R. Schmid, J. Keupp and J. P. Dürholt, MOFplus Project Webpage, https://www.mofplus.org/.
37 W. Smith and T. Forester, *J. Mol. Graphics*, 1996, **14**, 136–141.
38 Y. Cai, A. R. Kulkarni, Y.-G. Huang, D. S. Sholl and K. S. Walton, *Cryst. Growth Des.*, 2014, **14**, 6122–6128.
39 S. Helten, B. Sahoo, P. Müller, D. Janßen-Müller, N. Klein, R. Grünker, V. Bon, F. Glorius, S. Kaskel and I. Senkovska, *Microporous Mesoporous Mater.*, 2015, **216**, 42–50.
40 P. Müller, R. Grünker, V. Bon, M. Pfeffermann, I. Senkovska, M. S. Weiss, X. Feng and S. Kaskel, *CrystEngComm*, 2016, **18**, 8164–8171.
41 Y.-B. Zhang, H. Furukawa, N. Ko, W. Nie, H. J. Park, S. Okajima, K. E. Cordova, H. Deng, J. Kim and O. M. Yaghi, *J. Am. Chem. Soc.*, 2015, **137**, 2641–2650.
42 A. Schneemann, R. Rudolf, S. Henke, Y. Takahashi, H. Banh, I. Hante, C. Schneider, S.-i. Noro and R. A. Fischer, *Dalton Trans.*, 2017, **46**, 8198–8203.
43 H. Furukawa, Y. B. Go, N. Ko, Y. K. Park, F. J. Uribe-Romo, J. Kim, M. O'Keeffe and O. M. Yaghi, *Inorg. Chem.*, 2011, **50**, 9147–9152.
44 P. Müller, R. Grünker, V. Bon, M. Pfeffermann, I. Senkovska, M. S. Weiss, X. Feng and S. Kaskel, *CrystEngComm*, 2016, **18**, 8164–8171.
45 B. Chen, M. Eddaoudi, S. T. Hyde, M. O'Keeffe and O. M. Yaghi, *Science*, 2001, **291**, 1021–1023.
46 H. Li, M. Eddaoudi, M. O'Keeffe and O. M. Yaghi, *Nature*, 1999, **402**, 276–279.
47 L. Pauling, *J. Am. Chem. Soc.*, 1929, **51**, 1010–1026.
48 M. Li, D. Li, M. O'Keeffe and O. M. Yaghi, *Chem. Rev.*, 2014, **114**, 1343–1370.
49 J. A. Gee and D. S. Sholl, *J. Phys. Chem. C*, 2013, **117**, 20636–20642.
50 J. Kulesza, B. Silva Barros, S. Alves Júnior, C. A. Fernandes de Oliveira, D. M. de Araújo Melo and J. Chojnacki, *Mater. Chem. Phys.*, 2014, **143**, 1522–1527.
51 H. He, Y. Song, F. Sun, Z. Bian, L. Gao and G. Zhu, *J. Mater. Chem. A*, 2015, **3**, 16598–16603.
52 D. Frahm, F. Hoffmann and M. Fröba, *Cryst. Growth Des.*, 2014, **14**, 1719–1725.
53 E. V. Alexandrov, V. A. Blatov, A. V. Kochetkov and D. M. Proserpio, *CrystEngComm*, 2011, **13**, 3947–3958.

Faraday Discussions

PAPER

Zeolite structure determination using genetic algorithms and geometry optimisation†

Xuehua Liu,[a] Soledad Valero, [a] Estefanía Argente[a] and German Sastre *[b]

Received 17th February 2018, Accepted 13th April 2018

DOI: 10.1039/c8fd00035b

The recently presented software zeoGAsolver is discussed, which is based on genetic algorithms, with domain-dependent crossover and selection operators that maintain the size of the population in successive iterations while improving the average fitness. Using the density, cell parameters, and symmetry (or candidate symmetries) of a zeolite sample whose resolution can not be achieved by analysis of the XRD (X-ray diffraction) data, the software attempts to locate the coordinates of the T-atoms of the zeolite unit cell employing a function of 'fitness' (F), which is defined through the different contributions to the 'penalties' (P) as $F = 1/(1 + P)$. While testing the software to find known zeolites such as LTA (zeolite A), AEI (SSZ-39), ITW (ITQ-12) and others, the algorithm has found not only most of the target zeolites but also seven new hypothetical zeolites whose feasibility is confirmed by energetic and structural criteria.

1. Introduction

The design of new solid state materials is strongly based on our knowledge of the general rules which dictate order at the short and long range in matter. In zeolites, with both covalent (short range) and ionic (long range) factors playing an important role, the rigidity of the tetrahedral primary building units is in stark contrast with the flexibility of the relative orientations of such tetrahedra. This flexibility is the reason for the large number of existing zeolites, and can be numerically studied through the analysis of T–T–T (T = Si, Al) angles.[1]

In addition to the database of zeolites compiled by the IZA,[2] hypothetical zeolites (either feasible or not feasible) have also been proposed. In spite of exhaustive investigations since the first databases of hypothetical zeolite structures were presented,[3–8] we still wonder about the rules that nature dictates which

[a]*Departamento de Sistemas Informaticos y Computacion, Universitat Politècnica de València, Calle Camino de Vera s/n, 46022 Valencia, Spain. E-mail: gsastre@itq.upv.es*

[b]*Instituto de Tecnologia Quimica U.P.V.-C.S.I.C, Universitat Politècnica de València, Avenida Los Naranjos s/n, 46022 Valencia, Spain*

† Electronic supplementary information (ESI) available. See DOI: 10.1039/c8fd00035b

include some and omit others from the reduced set of *feasible zeolites*, those that have been or might be synthesised. Criteria of zeolite feasibility, presented below, can be combined with hypothetical zeolites in order to generate new zeolites using computer algorithms, which is the main aim of the present study.

Román-Román and Zicovich-Wilson,[9] similarly to earlier ideas by O'Keeffe and Hyde,[10] pointed to the crucial role played by the long-range forces that operate in zeolites and contribute significantly to their stabilities, and also explain the relative instability of large voids. Large micropores are however possible, provided that they are sustained by thick pillars. In chemical language, large micropores need the proximity of high density regions.

This concept is based on the old and most successful idea by Brunner and Meier[11] that very small rings are needed when very large rings are present. Almost 30 years later, no zeolites have been found to contradict the postulated lower threshold of density *versus* 'average size of smallest ring'. Less dense structures than those predicted by Brunner and Meier might however be feasible if a large number of small rings is present. This is the argument employed by Zwijnenburg and Bell,[12] where several hypothetical zeolites containing a large percentage of 4-rings are proposed, giving low densities and stable energetics, even below 11.8 T-atoms per $Å^3$, the smallest density predicted by Brunner and Meier for zeolites whose average size of smallest ring is the 4-ring.

A different criterion has been defined in terms of structural factors extracted from zeolites optimised with a force field.[13] For silica zeolites, eight mathematical equations involving O···O, T···T, and T–O distances are proposed in the so called LID criteria (local interatomic distances). Unlike the energetic criterion which is highly dependent on the methodology, this is a more general definition although numerically the boundaries can always be subject to controversy. In general, these and other definitions work well for the most stable structures, and so in that region they can safely determine feasibility. However, in the less stable region there is not any algorithm that can ensure that a structure is not feasible. The existence of RWY zeolite, utterly amongst the most energetic (and less feasible!)[14] is proof that algorithms work confidently only in one direction: meeting the criteria means 'feasible', but not meeting the criteria does not necessarily mean 'unfeasible'.

During the past two decades, two different types of global optimization techniques have been developed: (i) techniques based on different versions of the Monte Carlo method, such as simulated annealing[15–17] and parallel tempering;[18] and (ii) techniques based on evolution theory, such as genetic algorithms[19,20] and differential evolution.[21]

ZEFSA II[17] is based on Monte Carlo methods and aims to generate the most feasible zeolite structures in real space under a set of restraints on the framework densities, bonding geometries, and powder X-ray diffraction intensities. FraGen,[18] also based on Monte Carlo techniques, generates framework structures according to one or more user-defined density maps. These two programs have also been used for *a priori* zeolite structure prediction, working without any experimental input, thus being able to generate millions of hypothetical zeolite frameworks. SCIBS[22] aims to enumerate all possible four-connected networks within each space group given the number of unique T-atoms. GRINSP[23] aims to generate all possible framework structures constructed exclusively from corner-sharing polyhedra. EZs[24] is based on genetic algorithms and simulates the pore, cage and even small void space structures in zeolites, by imposing exclusion zones in the crystalline space which atoms are forbidden to enter. AASBU (automated assembly of secondary building blocks)[25–27] is based on the accommodation in a given unit cell of differing modes of inter-SBU connection, allowing for corner-, edge-, and face-sharing modes, with broad flexibility in the definition of the linkage points. Finally, P-GHAZ[28] improves a previous work by Baumes and coworkers[29] using evolutionary algorithms combined with general purpose computation on graphic processing units, and aims to find hypothetical zeolite structures close to a thermodynamic feasibility criterion.

Our zeoGAsolver software, based on genetic algorithms, has been presented recently[30,31] and will not be explained in detail. The initial aim is to help with synthesised structures which can not be determined with algorithms for the treatment of the XRD (X-ray Diffraction) spectrum. zeoGAsolver tries to find the atomic coordinates of the unit cell using as input the cell parameters (fixed during the search), some candidate space groups, and the material density. Using a domain-dependent initial generation of possible zeolite individuals and dealing with the material density as a variable, the software has been used to generate hypothetical zeolites belonging to several implemented space groups while the final aim to implement all 230 space groups is under development. Furthermore, only zeolite structures with up to four inequivalent T-sites have been successfully obtained by the current version of the software. Hence it has been possible to obtain success in a number of tests which aimed to determine known zeolites such as LTA, NPT, BIK, AFO, AHT, AEI, ATS, CDO, RTH, STF, and ITW. In these tests, only RTH and STF were not found by the software. While running these tests, other new zeolites have been obtained by a combination of the zeoGAsolver plus new functionality based on geometry optimisation that we describe in this study. Ultimately, the aim of zeoGAsolver is to determine the atomic positions of synthesised zeolites where only the cell parameters (or occasionally also a hint of the space group) can be obtained by XRD and any other methods.

2. Methods

2.1. zeoGAsolver

The software works using the symmetry of the crystal and it needs only to find the location of as many T-atoms as are present in the asymmetric unit. The program employs a 'fitness' (F) function specifically defined and successfully tested in previous work,[30,31] including the different contributions to the 'penalties' (P), as $F = 1/(1 + P)$, with:

$$P = w_{SG} \times P_{SG} + w_D \times P_D + w_{TC} \times P_{TC} + w_{GC} \times P_{GC} + w_A \times P_A + w_{AA} \times P_{AA} + w_{TP} \times P_{TP} + w_M \times P_M$$

where each penalty is multiplied by its corresponding weight (w), used to normalise the results of each penalty. The penalties include: space group (SG); density (D); tetra coordination, either local (TC) or global (GC); angles, either individual (A) or averaged (AA); tetrahedron planarity (TP); and molecularity (M).

Each chromosome or individual for the genetic algorithm is composed of the T-atoms of the asymmetric unit of the zeolite. In zeoGAsolver, an initial population of individuals for the genetic algorithm is randomly generated taking into account: (i) the minimum and maximum number of possible T atoms of the asymmetric unit; (ii) the coordinate constraints specified in the space group to which the zeolite under study belongs; (iii) the minimum and maximum number of possible T atoms in the unit cell; and (iv) the Wyckoff positions specified in the space group. Then, an iterative process starts using a genetic algorithm containing hybrid domain-dependent operators (that combine the effects of crossover and mutation) and selection operators that maintain the size of the population in successive iterations while improving the average fitness. Successful execution is characterised by the generation of a set of T-atoms (CIF file) with F very close to unity.

2.2. Geometry optimisation

A modified version of the zeoTsites[32–34] software has been adapted by including the feature of adding oxygen atoms to the T-only cells, taking into account the cell boundaries. A Linux script is applied to the result that performs a geometry optimisation of the unit cells using four different force fields and also performs a topological analysis to test if the resulting zeolites belong to the IZA database. The script calculates the zeolite feasibility with the LID[13] criteria and also the energies with respect to quartz.

2.3. Force fields employed

Four force fields have been selected, taking into account different styles and parameterisation strategies. The force field by Bushuev and Sastre[35] (BS), developed for silicates and extended to alumino-silicates and silico-aluminophosphates,[36,37] employs Lennard-Jones functions for the Si···O and O···O interactions, avoiding the wrong behaviour at close distances of the Buckingham function. The force field employs partial charges ($Q_{Si} = 2.1$) and three body harmonic terms for O–Si–O (with eq. at 109.47°) and Si–O–Si (with eq. at 142°) bond angles. This force field is particularly accurate for calculating the relative energies of zeolites. Two core–shell (for oxygens) force fields with formal charges ($Q_{Si} = 4.0$) were used, one developed by Sanders, Leslie, and Catlow[38] (SLC), and a derivative by Sastre and Corma[39] (SC). They both employ Buckingham functions for the Si···O and O···O interactions as well as a harmonic function for the O–Si–O angles. SLC does not include a bonding term for the Si–O–Si angles whilst SC includes a screened Vessal anharmonic function. Most parameters remain similar in SC with respect to SLC, and although the O···O Buckingham parameters are markedly different, above 2 Å the interaction is controlled by electrostatics, with both terms behaving similarly. The difference in behaviour between these two core–shell force fields is in the Si–O–Si term, present in SC and absent in SLC. The fourth force field employed, by Vessal, Leslie, and Catlow[40] (VLC), was parameterised for amorphous silica, with formal charges (like SLC and SC) and an anharmonic term for the O–Si–O angles, and using a rigid ion model (like BS).

These force fields were parameterised using mainly structural properties, widely available for tectosilicates and zeolites,[41,42] and have been largely tested giving good accuracy.[43,44]

3. Results and discussion

3.1. Candidate zeolites found by zeoGAsolver

From the candidate zeolites found by the software, those containing 3-rings were not considered, following the strategy by Baumes and coworkers,[28,29] since they are rarely found as all-silica or alumino-silicate compositions, which are of interest for applications in catalysis and separation.[45] zeoGAsolver has found 28 new zeolites whose smallest ring is the 4-ring, corresponding to the space group numbers 12, 63, and 221, and with 1–4 unequivalent T-sites in the asymmetric unit.

3.2. Geometry optimisations of the candidate zeolites

A table in the ESI† shows details about the volume and cell parameters of all the optimised candidate zeolite structures. The geometry optimisations have been performed using the GULP software with 25 cycles using the BFGS optimiser plus a subsequent 10 cycles using the RFO optimiser.[46,47] The "conp" (constant pressure) option has been used instead of "conv" (constant volume) since one of the features of this procedure is to allow new structures to be obtained by expanding or shrinking the original unit cell, hence introducing an additional possibility to relocate atoms in positions of minimum energy using more degrees of freedom, leading to a larger chance of obtaining a stable structure. An analysis of the volume variation (between the T-only CIF file given by zeoGAsolver and the optimised geometry) is shown in Table 1, with averages over the zeolite structures (avg. zeo) and over the force fields (avg. ff).

Throughout the zeolite structures (avg. zeo), variations of up to 30% indicate how flexible the optimisation procedure is, which allows it to find a zeolite significantly different from the initial candidate. These are the cases in which the force field is contributing actively to finding a new zeolite. In spite of the large fitness, very close to unity, of the original (T-only) structure, the introduction of oxygens, and the geometry optimisation using the force field lead to a more realistic description of the system. An analysis throughout the force fields (avg. ff) reveals that, on average, the two core–shell force fields (SC and SLC) lead to similar values of volume variation (10% and 9%), with BS giving a slightly larger volume difference (11%) and VLC giving the lowest (7%) average difference between the optimised and the initial volumes. It is more important to analyse the number of failed optimisations ("X") with each force field, with BS performing poorly (15 failures out of 28 structures), the core–shell force fields (SC and SLC) showing intermediate behaviour (6 failures), and VLC performing extremely well (1 failure). The success of VLC must be ascribed to it being parameterised for amorphous silica and it performs particularly well in regions far from the zeolite minima. The bad performance of BS is probably due to the presence of two three-body harmonic bonding terms (O–Si–O and Si–O–Si) instead of only one (O–Si–O) as in the SLC and VLC force fields. This introduces too much order in the requirements for optimisation and it fails if the initial structure is far from the minimum. The SC force field, also with O–Si–O and Si–O–Si terms, performs clearly better than BS, by having an anharmonic function to describe the flexible Si–O–Si angle, while the harmonic function in BS is less realistic far from the equilibrium.

Table 1 Volume variation (from initial zeoGAsolver to the optimised geometry) of 28 candidate zeolites. "X" means that the zeolite structure optimisation did not result in a tetracoordinated structure

	Force-fields				
Zeolite codename	BS	SC	SLC	VLC	Avg. zeo (%)
0303_64567	X	28	X	33	**30**
0310_48918	X	X	X	4	**4**
0902_181848	X	10	X	17	**14**
0_30894	X	5	X	3	**4**
0_34894	−10	−6	−7	6	**7**
0_44627	−12	−7	−11	−1	**8**
1_11481	−3	0	−3	3	**2**
3_1365021	9	14	11	21	**14**
3_27767	−3	3	1	9	**4**
3_31105	X	−18	−20	−5	**14**
4_15364	−5	−7	−11	6	**7**
4_76033	−8	X	−7	4	**6**
4_98439	16	−11	−12	−1	**10**
5_27095	24	X	−1	9	**11**
5_51886	X	10	2	15	**9**
6_18831	X	22	18	29	**23**
6_45037	X	−13	−8	−4	**8**
7_31194	X	−23	−28	−9	**20**
7_5188	X	−9	X	7	**8**
8_111416	X	−9	−7	0	**5**
8_33283	−8	−5	−9	2	**6**
8_46824	13	18	14	X	**15**
9_13403	22	13	6	17	**14**
9_25035	X	6	2	14	**7**
9_28607	X	X	−9	−9	**9**
9_3540	X	X	4	11	**8**
9_3746	X	X	X	−8	**8**
9_42398	−5	3	−1	8	**4**
Avg. ff (%)	**11**	**10**	**9**	7	

3.3. Structural and topological analysis of the optimised candidate zeolites

The resulting connectivities and vertex symbols of the candidate zeolites have been compared with those in the IZA database, with 8_46824 giving the same result as CHA, and 9_42398 giving the same result as SOD. In the case of 9_42398, the small cell variation upon optimisation suggests that it is found by zeoGAsolver, whilst in the case of 8_46824, the cell variations are 13–18% depending on the force field, which suggests a significant role for the geometry optimisation process in finding this zeolite. Most of the structures in Table 2 show the same number of inequivalent T-sites before and after geometry optimisation. Five structures show a reduction in the number of inequivalent T-sites (3_31105, 8_111416, 8_46824, 9_3540 and 9_42398), while one structure shows an increase in that number (0310_48918).

The symmetry has also been analysed after geometry optimisation, although this is a non-trivial task. The phonopy software[48] has been used with an increasingly large tolerance, from 10^{-4} to 0.02, until a cell different to P1 is

Table 2 Number of inequivalent T-sites and symmetry number of space group of 28 candidate zeolites obtained by zeoGAsolver and the corresponding values after geometry optimisation. The number of T-sites after geometry optimisation has been found by calculating their connectivity sequences. Asterisks indicate structures with a changing number of inequivalent T-sites upon geometry optimisation

	T-sites		Symmetry	
Codename	Origin.	Optimi.	Origin.	Optimi.
0303_64567	4	4	63	62
0310_48918	4	18*	63	2
0902_181848	4	4	63	63
0_30894	3	3	12	14
0_34894	3	3	12	2
0_44627	3	3	12	1
1_11481	2	2	63	63
3_1365021	2	2	221	221
3_27767	3	3	12	4
3_31105	4	2*	12	2
4_15364	3	3	12	2
4_76033	3	3	12	2
4_98439	4	4	63	51
5_27095	3	3	12	1
5_51886	3	3	12	14
6_18831	2	2	221	221
6_45037	3	3	12	14
7_31194	3	3	12	2
7_5188	3	3	12	1
8_111416	5	4*	12	2
8_33283	3	3	12	12
8_46824	3	1*	12	166
9_13403	2	2	63	63
9_25035	3	3	12	14
9_28607	3	3	12	1
9_3540	3	2*	12	1
9_3746	3	3	12	2
9_42398	3	1*	12	148

obtained (Table 2). The resulting unit cells have in turn been modified by moving the atomic positions according to the maximum tolerance, trying to achieve an overlap between symmetry related atoms. Increasingly larger tolerances lead to higher symmetries but in 7 cases cells with $P\bar{1}$ symmetry could not be upgraded to a higher symmetry. This far from optimum procedure does not allow us to ensure that the symmetry obtained is the largest possible in the corresponding topology. Those cases where the number of inequivalent T-sites after geometry optimisation is lower than in the original cells (3_31105, 8_111416, and 9_3540) should give higher symmetry in the optimised structure than in the original cell, and this is not the case. The correct trend is observed for 8_46824 and 9_42398, where a reduction in the number of inequivalent T-sites leads to an increase in the symmetry.

Table 2 does not indicate which force field(s) succeeded in finding the highest space group for each candidate zeolite. The respective number of successes was 10, 12, 9 and 16 for BS, SC, SLC and VLC respectively.

3.4. Feasibility of the candidate zeolites

Two criteria for feasibility have been employed to assess the feasibility of the candidate zeolites, the energy and the structural LID criteria based on the study by Li. *et al.*[13] described earlier. The results are indicated in Table 3. The energies of α-quartz calculated with each force field are also indicated for the sake of comparison. A justification of feasibility in terms of energetic criteria does not seem possible since there is not a clear trend that can be considered general for all the force fields tested. In order to be applied to a large number of structures, optimisation procedures tend to be simplified, as in this case with a maximum number of 35 geometry optimisations, which results in some structures still being far from the minimum. It could then be suggested that any structure whose geometry optimisation does not fail should be considered as feasible regardless of its energy. The second criterion (LID), based on the analysis of local interatomic distances explained above, requires that the optimisation be carried out with the SLC force field and in order to apply this to all candidate structures, those whose SLC optimisation failed have been reoptimised using a better starting geometry obtained from a successful optimisation with any

Table 3 Relative energies per SiO_2 unit (kJ mol^{-1}) of the structures optimised with the BS, SC, SLC and VLC force fields, and feasibility according to the LID criteria.[13] Feasible structures are indicated as "yes", and those not feasible are indicated by the number of conditions failed (out of a total of 8). "X" means that the zeolite structure optimisation did not result in a tetracoordinated structure

Codename	*E*(BS)	*E*(SC)	*E*(SLC)	*E*(VLC)	LID
0303_64567	X	6	X	2	1
0310_48918	X	X	X	30	3
0902_181848	X	32	X	69	2
0_30894	X	50	X	84	2
0_34894	37	27	31	69	1
0_44627	12	10	16	17	Yes
1_11481	21	8	13	0	Yes
3_1365021	23	19	25	49	Yes
3_27767	26	33	26	45	Yes
3_31105	X	43	45	84	4
4_15364	41	27	30	69	1
4_76033	43	X	37	49	4
4_98439	268	19	21	49	Yes
5_27095	202	X	25	45	2
5_51886	X	26	30	60	Yes
6_18831	X	37	43	67	1
6_45037	X	36	56	62	4
7_31194	X	30	36	85	3
7_5188	X	51	X	101	3
8_111416	X	95	81	116	6
8_33283	43	30	34	68	2
9_13403	144	33	36	54	3
9_25035	X	25	30	60	Yes
9_28607	X	X	103	79	4
9_3540	X	X	36	43	3
9_3746	X	X	X	80	2
α-Quartz	0	0	0	2	

other force field. This whole process gives 7 feasible zeolites, and 4 zeolites where only one LID condition fails.

The 7 candidate zeolites assessed as feasible (0_44627, 1_11481, 3_1365021, 3_27767, 4_98439, 5_51886, 9_25035) are shown in Fig. 1. Regarding the channel systems, 3_27767 contains medium sized cavities containing 12-rings, with each cavity connected by six 8-rings, giving a 3-D channel system. 4_98439 contains a unidimensional straight 10-ring channel connected perpendicularly by short 8-ring channels, giving a 2-D channel system. 5_51886 contains large cavities containing 12-rings connected by 8-rings giving a straight channel system, connected perpendicularly by a small pore channel of elliptic shape. 9_25035 contains an elliptic medium pore channel made by 12-rings not oriented perpendicularly to the channel. This medium pore straight channel is connected

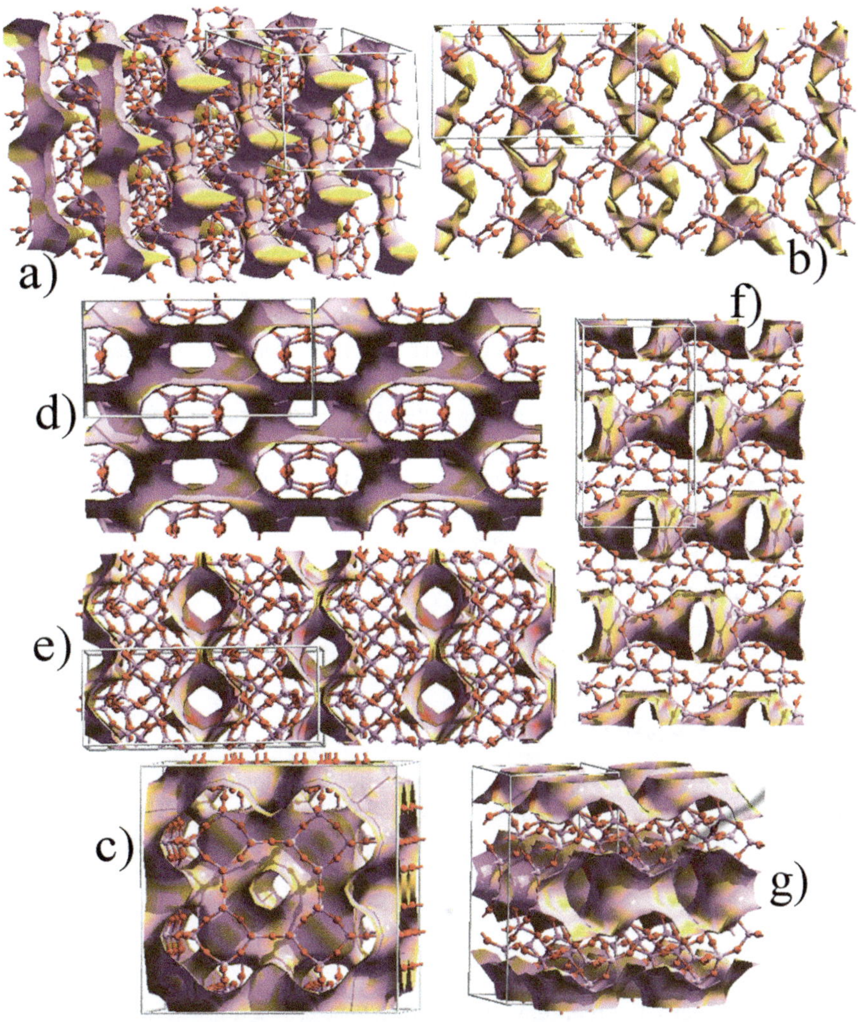

Fig. 1 Seven feasible hypothetical zeolites. (a) 0_44627; (b) 1_11481; (c) 3_1365021; (d) 3_27767; (e) 4_98439; (f) 5_51886; (g) 9_25035.

perpendicularly through an 8-ring channel, giving a 2-D channel system. 0_44627 contains a 1-D channel system made by sinusoidal 8-rings, 1_11481 is a clathrasil, and 3_1365021 displays 3-D small pore (8-ring) channels with a very large cube-shaped cavity containing as many as 18 intersections.

4. Conclusions

A new methodology to perform geometry optimisation on candidate zeolites obtained by a genetic algorithm is proposed. Since the fitness function employed by the genetic algorithm (zeoGAsolver) does not refer to oxygen but only T-atoms, it is perfectly possible, albeit not probable, that structures found with fitness = 1 do not finally correspond to real zeolites due to the absence of evaluation of the effects due to the presence of oxygen atoms. With the aim of not only testing if the candidate zeolite is stable, but also with the idea of allowing changes in the atomic positions until a stable zeolite is found, an optimisation procedure has been added as new software to be executed after zeoGAsolver. The software not only introduces the oxygen atoms and makes four parallel unconstrained geometry optimisations using as many force fields, but also it is script-integrated with algorithms to perform this automatically for an arbitrarily large number of output files obtained from zeoGAsolver. The four force fields have been deliberately chosen to be different to each other in order to widen the landscape of zeolite structures that can be found. The VLC force field, although not the best among them by the usual definition, has proved to be the best performing here by giving 27 zeolites out of 28 candidate structures.

As a result of testing zeoGAsolver trying to find 11 known zeolites (LTA, NPT, BIK, AFO, AHT, AEI, ATS, CDO, RTH, STF and ITW), 28 candidate zeolites containing 4-rings as the smallest have been found after the automated geometry optimisation using force fields. Using criteria based on structure and energetics, a short list of seven feasible hypothetical zeolites is proposed. The present algorithms are being updated in order to be able to tackle more complex zeolites (containing more than four inequivalent T-sites) in reasonable CPU time.

Conflicts of interest

There are no conflicts of interest.

Acknowledgements

G. S. thanks the ASIC computational centre and Paco Rosich of UPV for making their computational resources available. G. S. thanks the Spanish government for the provision of the Severo Ochoa (SEV 2016-0683), CTQ2015-70126-R and MAT2015-71842-P projects.

Notes and references

1 X. Liu, S. Valero, E. Argente, V. Botti and G. Sastre, The importance of T···T···T angles in the feasibility of zeolites, *Z. Kristallogr. - Cryst. Mater.*, 2015, **230**, 291–299.

2 C. Baerlocher, L. B. McCusker and D. H. Olson, *Atlas of Zeolite Framework Types*, Elsevier, 6th revised edn, 2007, (176 structures); the web version [www.iza-structure.org] currently contains 232 structures.

3 M. M. J. Treacy, K. H. Randall, S. Rao, J. A. Perry and D. J. Chadi, Enumeration of Periodic Tetrahedral Frameworks, *Z. Kristallogr.*, 1997, **212**, 768–791.

4 O. Delgado-Friedrichs and D. H. Huson, Tiling space by platonic solids, *Discrete Comput. Geom.*, 1999, **21**, 299–315.

5 O. Delgado-Friedrichs, A. W. M. Dress, D. H. Huson, J. Klinowski and A. L. Mackay, Systematic enumeration of crystalline networks, *Nature*, 1999, **400**, 644–647.

6 M. M. J. Treacy, I. Rivin, E. Balkovsky, K. H. Randall and M. D. Foster, Enumeration of periodic tetrahedral frameworks. Polynodal graphs, *Microporous Mesoporous Mater.*, 2004, **74**, 121–132.

7 M. D. Foster and M. M. J. Treacy, Database of Hypothetical Zeolite Structures, http://www.hypotheticalzeolites.net/NEWDATABASE/SILVER_UNIQ/query.php.

8 D. J. Earl and M. W. Deem, Toward a Database of Hypothetical Zeolite Structures, *Ind. Eng. Chem. Res.*, 2006, **45**, 5449–5454.

9 E. I. Román-Román and C. M. Zicovich-Wilson, The role of long-range van der Waals forces in the relative stability of SiO_2 zeolites, *Chem. Phys. Lett.*, 2015, **619**, 109–114.

10 M. O'Keeffe and B. G. Hyde, The role of nonbonded forces in crystals, *Structure and Bonding in Crystals*, Academic Press, New York, 1981, vol. 1, ch. 10, pp. 227–254.

11 G. O. Brunner and W. M. Meier, Framework density distribution of zeolite-type tetrahedral nets, *Nature*, 1989, **337**, 146–147.

12 M. A. Zwijnenburg and R. G. Bell, Absence of Limitations on the Framework Density and Pore Size of High-Silica Zeolites, *Chem. Mater.*, 2008, **20**, 3008–3014.

13 Y. Li, J. Yu and R. Xu, Criteria for Zeolite Frameworks Realizable for Target Synthesis, *Angew. Chem., Int. Ed.*, 2013, **52**, 1673–1677.

14 D. Majda, F. A. Almeida Paz, O. Delgado Friedrichs, M. D. Foster, A. Simperler, R. G. Bell and J. Klinowski, Hypothetical Zeolitic Frameworks: In Search of Potential Heterogeneous Catalysts, *J. Phys. Chem. C*, 2008, **112**, 1040–1047.

15 M. W. Deem and J. M. Newsam, Framework crystal structure solution by simulated annealing: test application to known zeolite structures, *J. Am. Chem. Soc.*, 1992, **114**, 7189–7198.

16 S. Pagola and P. W. Stephens, PSSP, a computer program for the crystal structure solution of molecular materials from X-ray powder diffraction data, *Appl. Crystallogr.*, 2010, **43**, 370–376.

17 M. Falcioni and M. W. Deem, A biased Monte Carlo scheme for zeolite structure solution, *J. Chem. Phys.*, 1999, **110**, 1754–1766.

18 Y. Li, J. Yu and R. Xu, FraGen: a computer program for real-space structure solution of ex-tended inorganic frameworks, *Appl. Crystallogr.*, 2012, **45**, 855–861.

19 B. M. Kariuki, H. Serrano-González, R. L. Johnston and K. D. Harris, The application of a genetic algorithm for solving crystal structures from powder diffraction data, *Chem. Phys. Lett.*, 1997, **280**, 189–195.

20 S. Woodley, P. Battle, J. D. Gale and C. R. A. Catlow, The prediction of inorganic crystal structures using a genetic algorithm and energy minimisation, *Phys. Chem. Chem. Phys.*, 1999, **1**, 2535–2542.

21 M. Tremayne, C. C. Seaton and C. Glidewell, Structures of three substituted arenesulfonamides from X-ray powder diffraction data using the differential evolution technique, *Acta Crystallogr., Sect. B: Struct. Sci.*, 2002, **58**, 823–834.

22 M. M. J. Treacy, K. Randall, S. Rao, J. Perry and D. Chadi, Enumeration of periodic tetrahedral frameworks, *Z. Kristallogr.*, 1997, **212**, 768–791.

23 A. Le Bail, Inorganic structure prediction with GRINSP, *J. Appl. Crystallogr.*, 2005, **38**, 389–395.

24 S. M. Woodley, C. R. A. Catlow, P. D. Battle and J. D. Gale, The prediction of inorganic crystal framework structures using excluded regions within a genetic algorithm approach, *Chem. Commun.*, 2004, **1**, 22–23.

25 C. Mellot Draznieks, J. M. Newsam, A. M. Gorman, C. M. Freeman and G. Férey, De novo prediction of inorganic structures developed through automated assembly of secondary building units (AASBU method), *Angew. Chem., Int. Ed.*, 2000, **39**, 2270–2275.

26 C. Mellot-Draznieks, S. Girard, G. Férey, J. C. Schön, Z. Cancarevic and M. Jansen, Computational design and prediction of interesting not-yet-synthesized structures of inorganic materials by using building unit concepts, *Chem.–Eur. J.*, 2002, **8**, 4102–4113.

27 C. Mellot-Draznieks, S. Girard and G. Férey, Novel inorganic frameworks constructed from double-four-ring (D4R) units: Computational design, structures, and lattice energies of silicate, aluminophosphate, and gallophosphate candidates, *J. Am. Chem. Soc.*, 2002, **124**, 15326–15335.

28 O. Abdelkafi, L. Idoumghar, J. Lepagnot, J.-L. Paillaud, I. Deroche, L. Baumes and P. Collet, Using a novel parallel genetic hybrid algorithm to generate and determine new zeolite frameworks, *Comput. Chem. Eng.*, 2017, **98**, 50–60.

29 L. A. Baumes, F. Kruger, S. Jimenez, P. Collet and A. Corma, Boosting theoretical zeolitic framework generation for the determination of new materials structures using GPU programming, *Phys. Chem. Chem. Phys.*, 2011, **13**, 4674–4678.

30 X. Liu, S. Valero, E. Argente and G. Sastre; Determining zeolite structures with a domain-dependent genetic algorithm; *Iberian Conference on Information Systems and Technologies, CISTI*, 2017, DOI: 10.23919/CISTI.2017.7976059.

31 X. Liu, E. Argente, S. Valero and G. Sastre, Applying Genetic Algorithms in Chemical Engineering for Determining Zeolite Structures, *Advances in Intelligent Systems and Computing*, 2018, **649**, 34–43, DOI: 10.1007/978-3-319-67180-2_4.

32 G. Sastre and J. D. Gale, ZeoTsites: a code for topological and crystallographic tetrahedral sites analysis in zeolites and zeotypes, *Microporous Mesoporous Mater.*, 2001, **43**, 27–40.

33 G. Sastre and A. Corma, Topological Descriptor for Oxygens in Zeolites. Analysis of Ring Counting in Tetracoordinated Nets, *J. Phys. Chem. B*, 2009, **113**, 6398–6405.

34 D. Bermudez and G. Sastre, Calculation of pore diameters in zeolites, *Theor. Chem. Acc.*, 2017, **136**, 116.

35 Y. G. Bushuev and G. Sastre, Feasibility of Pure Silica Zeolites, *J. Phys. Chem. C*, 2010, **114**, 19157–19168.

36 G. Sastre, Computational study of diffusion of propane in small pore acidic zeotypes AFX and AEI, *Catal. Today*, 2014, **226**, 25–36.

37 A. Ghysels, S. L. C. Moors, K. Hemelsoet, K. De Wispelaere, M. Waroquier, G. Sastre and V. Van Speybroeck, Shape-Selective Diffusion of Olefins in 8-Ring Solid Acid Microporous Zeolites, *J. Phys. Chem. C*, 2015, **119**, 23721–23734.

38 M. J. Sanders, M. Leslie and C. R. A. Catlow, Interatomic potentials for SiO_2, *J. Chem. Soc., Chem. Commun.*, 1984, 1271–1273.

39 G. Sastre and A. Corma, Rings and strain in pure silica zeolites, *J. Phys. Chem. B*, 2006, **110**, 17949–17959.

40 B. Vessal, M. Leslie and C. R. A. Catlow, Molecular Dynamics Simulation of Silica Glass, *Mol. Simul.*, 1989, **3**, 123–136.

41 M. O'Keeffe and B. G. Hyde, On Si–O–Si configurations in silicates, *Acta Crystallogr., Sect. B: Struct. Crystallogr. Cryst. Chem.*, 1978, **34**, 27–32.

42 D. S. Wragg, R. E. Morris and A. W. Burton, Pure Silica Zeolite-type Frameworks: A Structural Analysis, *Chem. Mater.*, 2008, **20**, 1561–1570.

43 J.-R. Hill, C. M. Freemanand L. Subramanian, Use of force fields in materials modeling, *Rev. Comput. Chem.*, ed. K. B. Lipkowitz and D. B. Boyd, Wiley-VCH, New York, 2000, vol. 16, pp. 141–216.

44 A. F. Combariza, D. A. Gomez and G. Sastre, Simulating the properties of small pore silica zeolites using interatomic potentials, *Chem. Soc. Rev.*, 2013, **42**, 114–127.

45 B. C. Knott, C. T. Nimlos, D. J. Robichaud, M. R. Nimlos, S. Kim and R. Gounder, Consideration of the Aluminum Distribution in Zeolites in Theoretical and Experimental Catalysis Research, *ACS Catal.*, 2018, **8**, 770–784.

46 J. D. Gale, GULP: a computer program for the symmetry adapted simulation of solids, *J. Chem. Soc., Faraday Trans.*, 1997, **93**, 629–637.

47 J. D. Gale and A. L. Rohl, The General Utility Lattice Program (GULP), *Mol. Simul.*, 2003, **29**, 291–341.

48 A. Togo and I. Tanaka, First principles phonon calculations in materials science, *Scr. Mater.*, 2015, **108**, 1–5.

Faraday Discussions

PAPER

The Flexible Unit Structure Engine (FUSE) for probe structure-based composition prediction

C. Collins, G. R. Darling and M. J. Rosseinsky *

Received 20th February 2018, Accepted 25th April 2018
DOI: 10.1039/c8fd00045j

Computationally led materials discovery requires efficient methods to generate either exact or approximate crystal structures that span the composition range of a chosen phase space. Here we present a new tool, the Flexible Unit Structure Engine (FUSE), for the generation of approximate 'probe structures' to predict regions of composition space where compounds can be experimentally realised. We then test FUSE by applying it to 42 compositions in the Y^{3+}–Sr^{2+}–Ti^{4+}–O^{2-} phase field. FUSE correctly identifies all of the target compounds in the regions of stability and identifies the exact crystal structure for 8 out of the 10 compositions.

1 Introduction

The computational search for new materials at its simplest involves a method for choosing an initial structure to test with a particular choice of elements, and a methodology for evolving this structure[1–3] or choosing an alternative from a suitable database.[4,5] There have been notable successes from such approaches[6,7] to identifying missing elemental compositions or new high-pressure phases of materials. Much effort in inorganic materials chemistry has focused on the discovery of new materials with complex compositions or on the prediction of new compositions that can adopt previously known crystal structures – in either case, the problem becomes very complex once we consider systems containing more than three elements due to the number of possible phase fields and the size of the compositional phase space.

In materials discovery our aim is to experimentally discover materials, guided by computational composition prediction. We do not seek to predict the properties of the materials computationally *a priori*, instead constructing phase fields from combinations of elements or compounds based on an understanding of the likely physical and chemical properties of the constituents, for example combining elements which are known to produce high levels of ionic conductivity. We then use well-established substitution and materials processing

Department of Chemistry, University of Liverpool, Liverpool, UK. E-mail: rossein@liverpool.ac.uk

techniques to alter or enhance the desired properties of materials as they are experimentally realised.

While compositional complexity has the prospect of yielding multi-functional materials and offers more possibilities to tune functional behaviour, it has the drawback of greatly increasing the computational effort required to explore the range of structures and compositions. The initial choice of elements to investigate is of course determined by chemical knowledge and understanding. For example, Ga-containing oxides have been known to produce high oxide ion conductivity;[8,9] it would therefore be desirable to seek new Ga-containing oxides.

Recently we developed a methodology for addressing the complex materials discovery problem, which has been successfully applied to realise two new materials of previously unknown composition and structure. To make the search through complex structure space tractable, we build crystal unit cells from fragments (we refer to these as modules) found in materials of simpler composition containing the chosen elements. These modules are then combined in a chemically sensible fashion (*e.g.* with the stacking rules of the cubic perovskite structure type) to construct a unit cell containing ions with plausible coordination environments. Relaxation to the local potential energy minimum determines the energy ranking of a structure. The modules are then permuted to generate new structures to be energy ranked. This Extended Module Materials Assembly (EMMA) method[10] was employed to aid in the determination of the structure of a large layered perovskite, $Y_{2.24}Ba_{2.28}Ca_{3.48}Fe_{7.44}Cu_{0.56}O_{21}$, where the modular approach worked well to impose chemically sensible co-ordination chemistry on all of the atoms.

We extended EMMA with the addition of a Monte Carlo basin hopping algorithm to the permutation of modules (MC-EMMA) to obtain a structure searching tool that is sufficiently flexible and fast to permit exploration of the composition space in addition to the structure at given compositions.[11] Building crystals by EMMA imposes restrictions on the possible outcomes. In particular, the modules are of fixed composition and they cannot be updated on-the-fly. They also have the same fixed periodicity perpendicular to the stacking direction, although during relaxation the unit cell shape and volume and all of the ionic positions can vary, resulting in crystal structures very different to the initial MC-EMMA constructions.

The best structures generated by MC-EMMA at any given composition in a completely unexplored phase space are typically approximations to the true global energy minimum structure, which we refer to as probe structures. If the probe structures contain physically realistic local environments for the constituent ions, they should then be close enough in energy to a local minimum in comparison to the energies of all the known competing phases (*i.e.* the convex hull – the energy surface derived from stable compositions) to indicate the possibility of obtaining a new compound with that composition. We can then initiate an experimental synthesis programme targeting the indicated composition, with a successful outcome being the experimental identification of a previously unknown phase.

We chose to explore combinations of Y^{3+}, Sr^{2+}, Ca^{2+}, Ga^{3+} and O^{2-} to search for possible new materials. These elements have different charge states, ionic radii and bonding characteristics, giving a range of coordination geometries. Although there are related gallate materials with functional optical and transport properties, there were no known materials in this chosen phase-field. With MC-EMMA, we generated probe structures based upon large layered cubic perovskite or melilite structure types. Coarse energy ranking with force-fields and refinement with Density Functional Theory indicated a target region of composition space, and detailed experimental synthesis and structural characterisation led to the identification of two new crystal structures of previously unknown composition. One material is related to the melilite structure and the other is a $4 \times 4 \times 4$ supercell of perovskite. We subsequently modified the perovskite structure to generate a new down conversion phosphor host.

In the present study we introduce a new approach to generating/representing candidate structures that operates below the module level of MC-EMMA, choosing building blocks and the unit cell size and shape within the modules before assembling the full structure with chemically sensible ionic environments. We illustrate this method, the Flexible Unit Structure Engine (FUSE), by applying it to the generation of probe structures to identify the compositions of the stable phases in the experimentally explored Y^{3+}–Sr^{2+}–Ti^{4+}–O^{2-} phase field.[12–21] Although we have used FUSE to generate probe structures and implemented structure search by a generalization of the MC basin hopping steps similar to those used in MC-EMMA, we anticipate that the basic construction approach could also be implemented in other structure evolution approaches, such as genetic algorithms or particle swarm optimisation, as used in other inorganic crystal structure prediction methods.[22–25]

2 Building the unit cell with the Flexible Unit Structure Engine (Fig. 1)

FUSE has been implemented in Python (2.7/3.6 or higher) and is dependent on the atomic simulation environment (ASE)[26] and its subsequent dependencies. We generate an initial random seed structure for a given composition in the phase space under consideration in a series of steps outlined in detail below. First we choose randomly from five approximate lattice types, *e.g.* cubic or trigonal (Table 1), but without imposing internal symmetry, there is no triclinic approximation since we restrict two of the three unit cell angles to 90°. This step determines the number of units (sub-modules) and the number of layers (modules) that these should be spread across. A particular sub-module set is then generated by assigning the in-plane coordination of the cations. This set is split into equal-sized groups and spread across the modules that are used to assemble the 3D crystal structure, with stacking rules appropriate to the chosen lattice type, as implemented previously in EMMA. Before moving to geometry optimisation, we carry out a final error check to reduce the number of cations that are in physically unreasonable environments. Should a structure fail at this point, a fresh structure is generated. The geometry optimisation can be performed with an external Density Functional Theory (DFT) or force-field code.

2.1 The generation of sub-modules (Fig. 1a)

The sub-modules are composed of one cation position accompanied by 0–3 nearest-neighbour anions, as shown in Fig. 1a. This defines eight basic motifs for the sub-modules based upon a single unit of $AX_{3-\delta}$, where A is a cation species, X

Table 1 Initial choices of lattice, with the dimensions (in integer units of sub-modules) of the unit cells and cell angles; z is restricted to being even for tetragonal and orthorhombic by cubic stacking and the requirement that each adjacent layer must have a different translation to assemble the structure. As we only vary the angle γ in our approximations, monoclinic unit cells are created in a non-standard setting (for the standard setting $\beta \neq 90°$)

Approximate lattice	Dimensions (x, y, z = # sub-modules along a, b, c)	γ (°)
Cubic	$x = y, z = 2x$	90
Tetragonal	$x = y, z =$ even	90
Trigonal	$x = y, z$	120
Orthorhombic	x, y, z (z = even)	90
Monoclinic	x, y, z	120

is an anion species and $\delta = 0, 1, 2$ or 3. For $\delta = 1$ or 2, there are multiple non-equivalent ways in which anions can be arranged, giving one motif containing three anions, three containing two anions, three containing one anion and one motif with zero anions (Fig. 1a). We select randomly from this set, subject to the constraint that the total numbers of each ion type give the correct stoichiometry. Within the computer code, each sub-module is stored as three descriptors: the first labels the cation species, the second stores the motif label, and the final descriptor labels the anion species accompanying the cation in the sub-module. When all of the sub-modules have been generated they are stored in three lists, each containing all the descriptors of one type (*e.g.* one list contains all of the cation labels).

We describe the integer number of sub-modules in a given crystallographic direction using the symbols x, y and z, referring to the a, b and c crystallographic axes, respectively. The chosen lattice type also affects the total number of sub-modules, *e.g.* for a cubic lattice if we choose to have 2 sub-modules in the x-plane (Table 1), then we should have $2x^3 = 16$ sub-modules in total. During optimisation, the size of the unit cell, *i.e.* the number of modules, can grow, so it is not necessary for the initial structures to be very large. In this study we have imposed an arbitrary upper limit of 50 atoms in total in the unit cells of the initial structures. In these small unit cells it might prove impossible to fit the correct numbers of modules to satisfy stoichiometric requirements into the cubic lattice choice, for example. If FUSE is unsuccessful in populating a lattice of a particular type, it will try a different lattice type until a unit cell can be generated. The monoclinic lattice type is the default which can always be populated – in this lattice it will always be possible to at least assemble the sub-modules into a 1D chain.

2.2 The formation of modules from sub-modules (Fig. 1b)

The order in which the sub-modules are stored is randomised and the set is divided into z groups, where z is the number of modules determined by the choice of lattice. $x \times y$ sub-modules are selected and stored in separate descriptor arrays for each module. The positions are converted from fractional coordinates to Å scaling by:

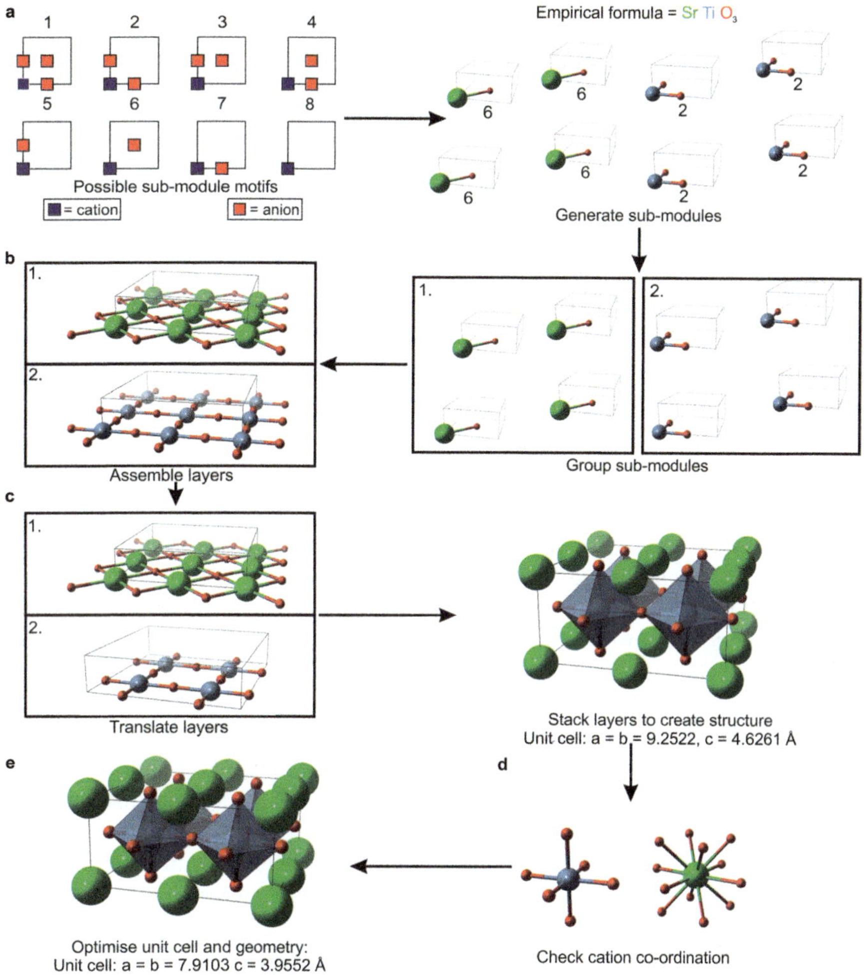

Fig. 1 Example of how FUSE generates a structure, with an example cell containing 4 formula units of $SrTiO_3$. (a) The motifs used to generate sub-modules labelled 1–8. FUSE chooses to target a tetragonal unit cell where $a = b = 2$ sub-modules and $c = 2$ sub-modules; the eight required sub-modules are then selected, with four of type 6 each containing SrO and four of type 2 each containing TiO_2. (b) The unit cell indicates 2 modules; therefore the sub-modules are organised into two groups, which are then assembled into 2D modules. (c) The required translations are applied and the two modules are stacked to produce a full 3D structure. (d) All of the cation co-ordination environments are checked against tabulated data. (e) An external program is called to optimise the geometry and obtain the total energy.

$$I_p = 2(R_{cat} + R_{an})$$

where R_{cat} is the average cation radius and R_{an} is the average anion radius. This gives lattice parameters for the module: $a = xl_p \times b = yl_p \times c = 0.5l_p$ where l_p is constant for all the sub-modules in the system, with the angles determined by the chosen lattice type. To construct the module, the positions for a sub-module are translated according to the sub-module's position within the module. For example, for the second module in the a direction, all of the positions will be translated by l_p along a. The whole process is then repeated for each of the modules within the structure.

2.3 Stacking the modules to produce the 3D structure (Fig. 1c)

To avoid stacking modules with like-charged ions packed on top of each other, each module is assigned a translation to be applied to all of its constituent ions. The translations are selected such that each adjacent module has a different translation instruction. If the selected lattice has $\gamma = 90°$ the translations are based upon cubic packing, such that they alternate between $[0l_p, 0l_p, 0l_p]$ and $[0.5l_p, 0.5l_p, 0.0l_p]$. When $\gamma = 120°$ the translations are based on a close packed structure, with possible translations being $[0l_p, 0l_p, 0l_p]$, $[1/3l_p, 2/3l_p, 0l_p]$ or $[2/3l_p, 1/3l_p, 0l_p]$. Applying the translations to the module list completes the generation of an initial structure.

2.4 Error checking the co-ordination chemistry (Fig. 1d)

With a fully assembled structure, it is possible to determine the number of nearest-neighbour anions for each cation, giving an initial co-ordination number. As each sub-module is of the same size, nearest-neighbour cations and anions are separated by less than l_p. To determine the initial co-ordination of a cation, we need to count the number of anions within $0.72l_p$. If the calculated co-ordination number of a cation differs by more than one from the value listed in the Shannon radii tables,[27,28] it is counted as an 'error'. If the fraction of these co-ordination errors exceeds the threshold b_{tol}, then this structure is rejected and a new one is generated until the fraction of errors is acceptable. For the systems investigated in this study, we optimised b_{tol} to maintain a high percentage of structures converging successfully in the geometry optimisation, while minimising the time taken to generate a structure with the number of errors below the threshold. We found that a b_{tol} value of 0.25 resulted in typically 95% of structures converging, with minimal impact on the time taken to generate structures.

2.5 Geometry optimisation

Before relaxing the assembled structure it is important to apply a random displacement to each ion to prevent the optimisation routine failing to relax ions out of artificially high symmetry sites. We use the ASE 'rattle' function, which selects random displacements from a normal distribution of a specified standard deviation (in this study we have arbitrarily used 0.025 Å). Having completed the initial structure generation, we generate input files for the chosen geometry optimisation. We have implemented this for the DFT code VASP[29] and the force-field based GULP[30] code.

When the external code has completed its geometry optimisation calculation(s), the co-ordinates and energy of the relaxed structure are recorded by FUSE. During structure evolution, we generate lattices of different sizes containing different numbers of ions; therefore we need to use the energy per atom for ranking structures. If geometry optimisation fails (*e.g.* due to a physically unreasonable starting geometry, or because the structure cannot relax within the allotted run time), then the energy per atom is artificially set to a very high value to prevent any subsequent search routines from selecting it as a structure to evolve.

3 Using FUSE to search for a probe structure at a given composition

Structure modifications are performed on the unassembled module representation, *e.g.* we swap ions between modules by simply swapping the appropriate labels within the arrays describing the modules, and then repeat the module stacking and error checking steps above. In principle, the basic FUSE construction could be paired with many existing search algorithms. In this study we implement a Monte Carlo basin-hopping algorithm, building on our earlier study with MC-EMMA[11] (Fig. 2).

We begin the search for an optimum structure with an initialisation stage; we build structures until we are able to successfully perform geometry optimisation on 1000 structures. The lowest energy structure is then selected for evolution. The initialisation step ensures that the search starts from a physically reasonable structure. Since the number of atoms in the initial structures is limited (to a maximum of 50 in this study), the initialisation has a small computational cost compared to the geometry optimisation of the larger evolving structures. For the systems studied here, we limited the size of structures in the Monte Carlo search to no more than 150 atoms.

In each step of basin-hopping the current evolving structure (CES) can be altered by six different permutations:

3.1 Swap atoms within the structure

We first select whether to swap cations or anions. If there is only one anion or one cation species, FUSE selects to swap the other, *i.e.* if the only anion in a system is O^{2-}, FUSE will only swap cations. When swapping cations, for example, we simply select two cations of different types from the cation arrays and swap their locations before the modules are assembled into a crystal. Both intra-module and inter-module swapping are allowed. This is a key improvement on our original MC-EMMA basin-hopping, in which the module compositions are fixed. For anions we currently have to keep both the anion type and the motif type together

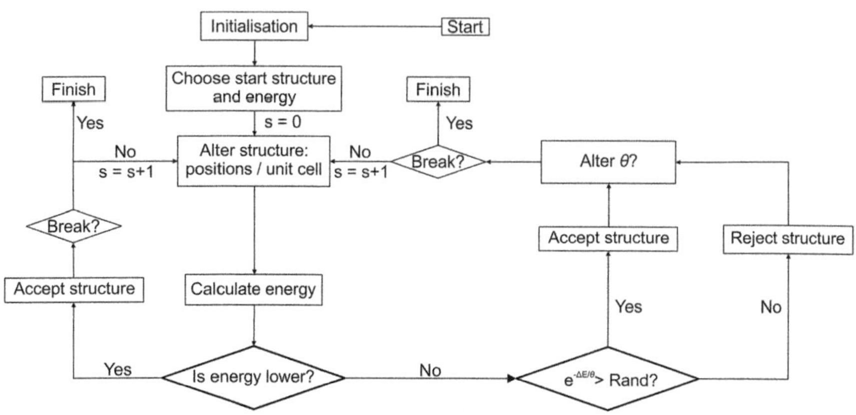

Fig. 2 Flow diagram of the Monte Carlo search used by FUSE. Rand indicates a randomly generated number between 0 and 1.

in order to preserve the overall composition of the unit cell. Note that if only one cation and one anion species are present (*e.g.* for SrO) then this permutation type cannot be used as no swaps will be possible.

3.2 Alter the stacking sequence

Two modules within the sequence are switched before the structure is assembled.

3.3 Grow the lattice by doubling the structure

If the current number of atoms in the unit cell is less than or equal to half of the permitted maximum, FUSE doubles the structure along a randomly chosen crystallographic axis.

3.4 Alter the lattice by changing γ

The lattice angle, γ, is swapped from 90 to 120° or *vice versa*. This is performed prior to assembling the modules into a structure since the lattice angle determines which translation vectors are used in stacking.

3.5 Generate a new random structure of similar volume

A new random structure is generated from the sub-module motifs, as detailed above, but with the maximum number of permitted atoms equal to that for the current structure.

3.6 Generate a new random structure

A new random structure is generated with a random number of sub-modules. This differs from the generation of the initial structures in that we do not constrain the unit cell to small values, but allow a choice up to the maximum of number of atoms permitted in structure evolution.

During the MC search, we use the simpler permutations more frequently. For the results presented here we use permutation 1 53% of the time. Permutations 2 and 3 each occur 21% of the time and the remaining permutations occur much less frequently, only ~1% of the time (permutations are rounded to the nearest whole percent).

Once a new trial structure has been generated, geometry optimisation is performed to obtain its energy. The new trial structure replaces the CES if its energy is lower than that of the CES, or randomly if the usual Monte Carlo condition is fulfilled, *i.e.* if $e^{-\Delta E/\theta} > x$, where x is a random number between 0 and 1, ΔE is the energy difference between the CES and the trial structure, and θ is the MC temperature parameter. Higher values of θ give higher probabilities of accepting an uphill energy step. In this study we use a value of $\theta = 0.02$ eV unless the basin-hopping gets trapped in a particular minimum.

The MC loop contains a break condition, at which point the search is considered converged. This occurs when r_{max} structures have been rejected since finding the structure with the lowest known energy. If the number of structures rejected since the last structure acceptance, r, is greater than $0.1 \times r_{max}$ the MC temperature parameter is gradually increased. θ is increased by a random number (<0.001) for every ten steps that r is above $0.1 \times r_{max}$. While the system is 'heating up', FUSE only uses permutation types 5 and 6 (in a 2 : 1 ratio), *i.e.* the entire

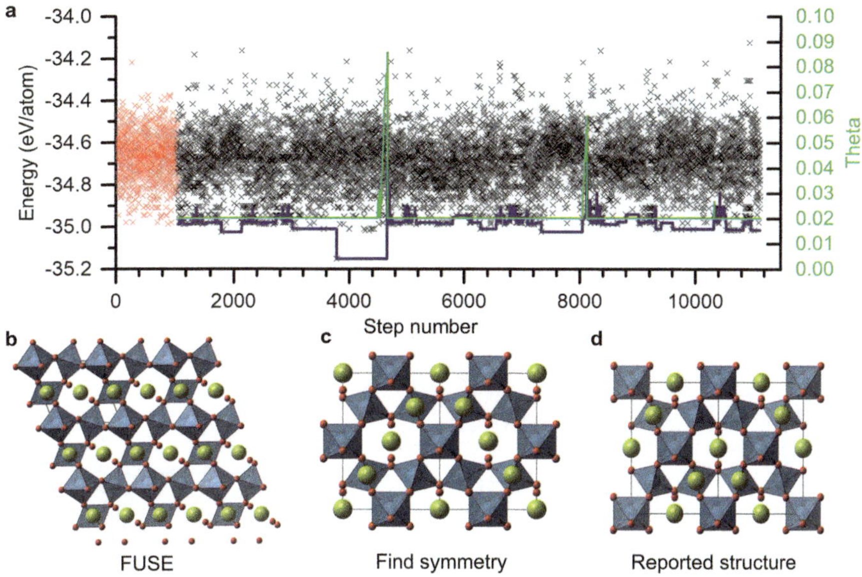

Fig. 3 (a) Example FUSE run for $Y_2Ti_2O_7$. The red markers indicate the 1062 structures sampled in the initialisation loop; black markers indicate structures sampled in the MC search routine; the blue line indicates the energy of the current evolving structure that FUSE is carrying forward in the MC search; the green line indicates the value of the MC temperature parameter (right axis) during the MC search. (b) The raw structure obtained in MC step 2715. (c) The structure obtained after passing the structure in (b) through FINDSYM. (d) The experimental crystal structure for $Y_2Ti_2O_7$; the FUSE structure differs from the ICSD structure by an origin shift. Atoms are coloured as follows: Y (yellow), Ti (blue) and O (red).

crystal structure is replaced. For the example presented in this study, these values were found to be effective in helping the MC search escape into a new basin (as can be seen in Fig. 3a). When the system escapes into a new basin, θ and the permutation frequencies return to the original values.

During probe structure generation, we performed three independent searches at each composition, taking the lowest energy structure from the three runs as the probe structure for that composition.

4 Testing the methodology: the Y^{3+}–Sr^{2+}–Ti^{4+}–O^{2-} phase field

We used FUSE to explore the Y^{3+}–Sr^{2+}–Ti^{4+}–O^{2} phase field. Using the Inorganic Crystal Structure Database (ICSD)[31] we identified 10 compounds with well-defined structures including rutile, rock-salt, perovskite, pyrochlore and Ruddlesden–Popper phases. This diverse range provided a strong test of our ability to identify energy minima corresponding to different ionic environments using the simple construction encoded in FUSE to generate approximate probe structures.

4.1 Geometry optimisation

For the calculations presented within this study, we used the General Utility Lattice Program (GULP)[30] for geometry optimisation. Buckingham potentials were

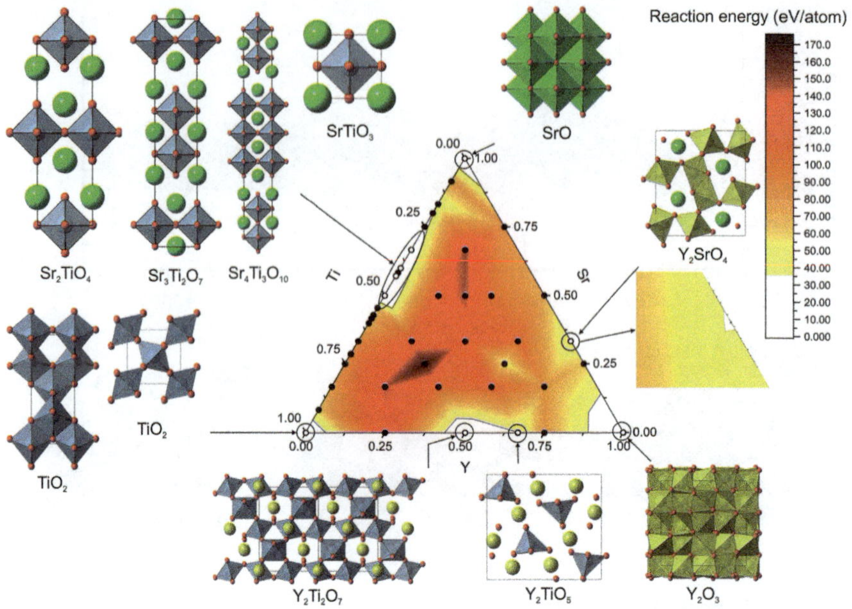

Fig. 4 FUSE-based probe structure calculations on the Y–Sr–Ti–O phase field. Points on the phase field indicate where we generated probe structures; open circles indicate compositions at which compounds are reported in the ISCD, with their crystal structures shown outside the phase field; regions highlighted in white indicate regions where the energy is less than 35 meV per atom above the convex hull; inset on the right shows shallow minima found around Y_2SrO_4. Atoms are coloured as follows: Y (yellow), Ti (blue), Sr (green) and O (red).

used for the short-range component of the potential, with a cut-off radius of 12 Å. For all structures, the unit cell parameters and atomic positions were optimised until the norm of the gradient was lower than 0.001 (a.u.).

Parameters for the O^{2-}–O^{2-} and Sr^{2+}–O^{2-} interactions were obtained from the literature.[32,33] The parameters for Ti^{4+}–O^{2-} and Y^{3+}–O^{2-} were based upon literature values,[22,34] but the *A* parameters were adjusted so that the lowest energy experimental structures were obtained as the energetic ground states for TiO_2 and Y_2O_3 in comparison to other polymorphs (for TiO_2) and random structures generated using FUSE. The final potential parameters are presented in Table 2.

5 Results

5.1 The ground-state structure for a single composition: $Y_2Ti_2O_7$

As an example of a FUSE calculation from within the Y–Sr–Ti–O phase field, we focus on the composition $Y_2Ti_2O_7$. Experimentally, $Y_2Ti_2O_7$ crystallises in the pyrochlore structure,[16] containing a mixture of octahedrally co-ordinated Ti^{4+} and eight co-ordinate Y^{3+} ions in the space group $Fd\bar{3}m$. The full unit cell contains 88 ions, with 22 in the primitive cell, with four unique sites (Fig. 3d). Obtaining the ground-state structure in the absence of symmetry therefore presents a substantial challenge.

Fig. 3a shows the energies of all the structures generated and tested for this composition. The energy of the CES is shown by the blue line. After the initiation

Table 2 Buckingham potential parameters used within this study

Interaction	A (eV)	ρ (Å)	C (eV Å^{-6})
O^{2-}–O^{2-}	1388.77	0.36262	175
Y^{3+}–O^{2-}	23 000	0.24203	0
Sr^{2+}–O^{2-}	1952.39	0.33685	19.22
Ti^{4+}–O^{2-}	4590.7279	0.261	0

stage the CES cycled up and down in energy before dropping into a deep minimum at MC step 2715. This minimum actually corresponds to the experimental crystal structure and is the lowest energy structure obtained for this composition. Since the energy could not go lower, the MC routine was stuck in this basin until the MC temperature (also shown) was ramped up and the full structure replacement permutations, 5 and 6, were employed. Several other basins were explored, with temperature jumps required to exit two of them. The structure evolution ended after ~10 000 MC steps.

The structure in the lowest energy basin was found to be a small, low symmetry unit cell (Fig. 2b) containing 22 atoms. This was then passed through the FINDSYM program,[35] which returned the correct space group and structure, with four unique sites (Fig. 3c), although with an origin shift relative to the reported crystal structure (Fig. 3d).

5.2 Exploring the composition space (Fig. 4)

Across the phase field we sampled 42 different compositions from the binary tie lines and within the ternary space. For each composition we performed three runs, taking the lowest energy structure as the probe structure. The control parameters for FUSE were determined by finding a set for which the correct ground-state structure could be reliably obtained for the $Y_2Ti_2O_7$ composition. The initialisation loop was run until 1000 structures converged during geometry optimisation or until 1500 structures had been tried. The error tolerance parameter was set to 0.25, the default θ to 0.02, and the run termination parameter r_{max} was set to 7000 for each composition. The weightings given to the MC moves were not optimised and therefore we hypothesise that given optimisation, the MC search routine would converge on the lowest energy structure more quickly.

Using the results from the FUSE calculations, we assembled the convex hull using pymatgen.[36] Accounting for the approximate nature of the probe structures, energies lying slightly above the convex hull could still identify islands of stability. In the previous phase diagrams explored using probe structures, we found experimentally realisable materials lying ~35 meV per atom (ref. 11) above the convex hull and so we used this as a cutoff value to identify stable compositions. All of the known stable compositions were found to lie within this cutoff, *i.e.* the probe structures correctly identified the compositions of all the known phases. The highest energy above the convex hull was found to be 34 meV per atom for Y_2TiO_5.[15] Plotting the compositional phase space, we observed that only these known compositions were found under the 35 meV per atom threshold. Of these, FUSE obtained the exact ICSD crystal structures for: TiO_2 (anatase[18] and rutile[17]), SrO,[20] Sr_2TiO_4,[21] $Sr_3Ti_2O_7$,[12] $Sr_4Ti_3O_{10}$,[12] $SrTiO_3$,[19] $Y_2Ti_2O_7$ (ref. 16) and Y_2TiO_5.[15]

6 Comparing probe structures with ICSD structures (Fig. 5)

For SrO, TiO_2, Sr_2TiO_4, $SrTiO_3$ and $Y_2Ti_2O_7$, the lowest energy structures obtained in FUSE matched the crystal structures in the ICSD database (Fig. 5a–e) when we found symmetry within the probe structure using FINDSYM, although some structures contained an origin shift *versus* the published structure. For TiO_2 FUSE

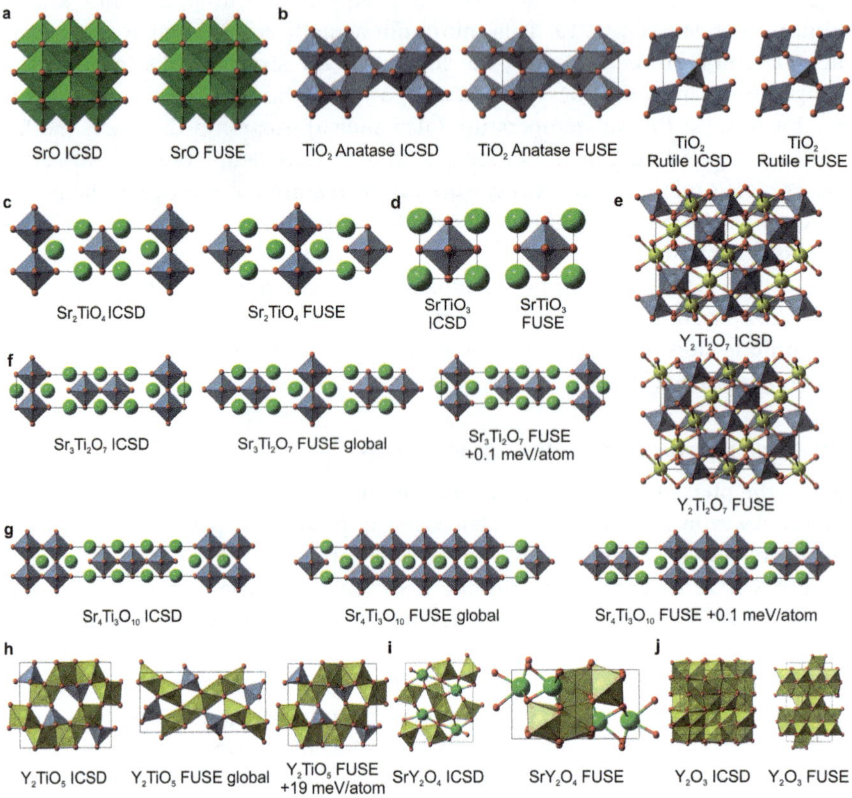

Fig. 5 Probe structures for each of the 10 ICSD compounds studied with FUSE in the Y–Sr–Ti–O phase field. (a) SrO in the rock-salt structure. (b) For TiO_2 FUSE correctly identifies the rutile polymorph as the ground-state structure (right) and is also able to locate the anatase polymorph (left). (c) The Sr_2TiO_4 Ruddlesden–Popper $n = 1$ phase. (d) $SrTiO_3$ perovskite. (e) $Y_2Ti_2O_7$ pyrochlore structure; FUSE identifies the correct structure although with the origin shifted such that a Y ion is at the origin rather than Ti. (f) The $Sr_3Ti_2O_7$ Ruddlesden–Popper $n = 2$ phase; for this composition, there are several similar stacking sequences within 0.2 meV per atom around the minimum energy; the experimental structure is identified by FUSE at +0.1 meV per atom from the minimum energy. (g) The $Sr_4Ti_3O_{10}$ Ruddlesden–Popper $n = 3$ phase; as with (f), there exist several stacking sequences with similar energies, and the experimental structure is within 0.1 meV per atom of the minimum. (h) The Y_2TiO_5 structure; by force-field calculations, FUSE found the experimental structure at +19 meV per atom above the minimum. (i–j) For SrY_2O_4 (ref. 14) and Y_2O_3 (ref. 13) FUSE was unable to locate the experimental structure, although the cation co-ordinations in the probe structures created are similar to those in the experimental structures. Atoms are coloured as follows: Y (yellow), Ti (blue), Sr (green) and O (red).

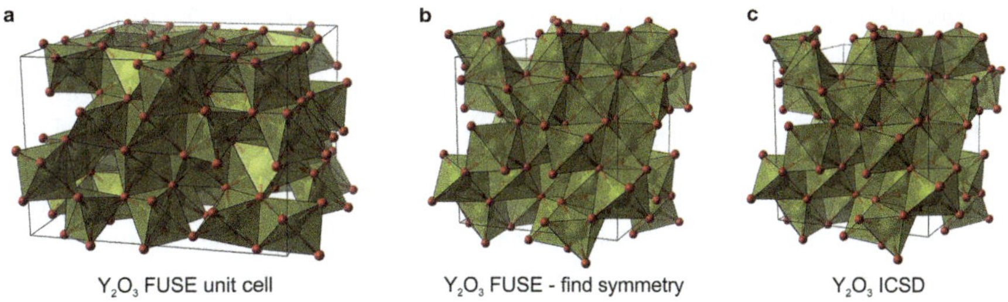

Fig. 6 (a) The unit cell of Y_2O_3 assembled by hand within the FUSE description (after geometry relaxation). (b) The FUSE unit cell of Y_2O_3 after being run through FINDSYM, reducing the cell to the experimental $Ia\bar{3}$ cell. (c) The ICSD unit cell for Y_2O_3. Atoms are coloured as follows: Y (yellow) and O (red).

identified the exact crystal structure of both the anatase and rutile polymorphs, with rutile calculated to be the ground-state structure.

For the compositions $Sr_3Ti_2O_7$ and $Sr_4Ti_3O_{10}$ (Fig. 5f and g), which both form Ruddlesden–Popper phases, there are multiple stacking sequences which differ by less than 0.2 meV per atom. With the force-field used within this study, the experimental crystal structures were found to be 0.1 meV per atom above the lowest energy structure obtained in FUSE.

For Y_2TiO_5, the probe structure obtained was found to be 19 meV per atom lower in energy than the experimental structure (Fig. 5h), although the experimental structure was obtained during the FUSE runs, but not identified as the global minimum. This highlights the need to recalculate the energies of the probe structures with more accurate methods (DFT), as we have done previously.[11]

The compositions SrY_2O_4 and Y_2O_3 were correctly identified by the probe structures as lying in regions of stability, however the present MC search was unable to obtain the exact ICSD crystal structures (Fig. 5i and j). The FUSE probe structures for SrY_2O_4 and Y_2O_3 were found to be +9 and +40 meV per atom above the ICSD structures. Although we do not set out to fully predict structures, but rather the regions of stable composition, we believe that FUSE will correctly obtain the structures of these two outliers once adjustments are made to the search/construction routine. To demonstrate that FUSE can assemble a model that will relax to the correct structure for Y_2O_3, we used the FUSE rules to build it by hand. Building a cell for Y_2O_3 from the mixed anion–cation sub-modules used in the present version of FUSE requires stacking along the [110] axis of the experimental cell. This yields a unit cell containing 160 ions, a greater number than was allowed in the MC search in this study. However, assembling this structure by hand, geometry optimisation (Fig. 6a) followed by symmetry searching did yield the ICSD structure (Fig. 6b and c).

7 Conclusions

In this study, we have presented a new method, FUSE, for probe structure-based inorganic materials discovery, based upon the construction of unit cells from randomly generated modules of varying shape, size and composition. FUSE can generate a wide range of physically plausible structures, with only modest restrictions applied to the type of crystal structure being created. The method was tested by exploring the known Y^{3+}–Sr^{2+}–Ti^{4+}–O^{2-} phase field, and successfully identified all of the compositions where structures are known to form as lying in low energy regions of the phase diagram. Additionally, in the search we found the exact crystal structure of 8 out of the 10 compositions. FUSE provides a new tool specifically designed for the efficient generation of probe structures across the compositional phase space. It has the potential to accelerate the computational prediction of regions of the phase space that can be profitably explored to experimentally realise new materials.

8 Data availability

The data used within this study is available from the authors on request.

Conflicts of interest

There are no conflicts of interest to declare.

Acknowledgements

We thank the EPSRC for funding under EP/N004884 and the University of Liverpool for access to local high-performance computing resources. We would like to thank Dr F. McBride (University of Liverpool) for useful discussions when developing the FUSE concept and Dr J. B. Claridge (University of Liverpool) for useful discussions when deciding upon the test system for FUSE. All structural images were created using VESTA.[37]

References

1 I. V. Pentin, J. C. Schon and M. Jansen, *Phys. Chem. Chem. Phys.*, 2010, **12**, 8491–8499.
2 J. C. Schon, K. Doll and M. Jansen, *Phys. Status Solidi B*, 2010, **247**, 23–39.
3 S. M. Woodley and R. Catlow, *Nat. Mater.*, 2008, **7**, 937–946.
4 G. Hautier, C. Fischer, V. Ehrlacher, A. Jain and G. Ceder, *Inorg. Chem.*, 2011, **50**, 656–663.
5 J. C. Schön, *Z. Anorg. Allg. Chem.*, 2014, **640**, 2717–2726.
6 H. Chen, G. Hautier and G. Ceder, *J. Am. Chem. Soc.*, 2012, **134**, 19619–19627.
7 R. Gautier, X. Zhang, L. Hu, L. Yu, Y. Lin, T. O. L. Sunde, D. Chon, K. R. Poeppelmeier and A. Zunger, *Nat. Chem.*, 2015, **7**, 308–316.
8 T. Ishihara, H. Matsuda and Y. Takita, *J. Am. Chem. Soc.*, 1994, **116**, 3801–3803.
9 X. Kuang, M. A. Green, H. Niu, P. Zajdel, C. Dickinson, J. B. Claridge, L. Jantsky and M. J. Rosseinsky, *Nat. Mater.*, 2008, 7, 498–504.
10 M. S. Dyer, C. Collins, D. Hodgeman, P. A. Chater, A. Demont, S. Romani, R. Sayers, M. F. Thomas, J. B. Claridge, G. R. Darling and M. J. Rosseinsky, *Science*, 2013, **340**, 847–852.
11 C. Collins, M. S. Dyer, M. J. Pitcher, G. F. S. Whitehead, M. Zanella, P. Mandal, J. B. Claridge, G. R. Darling and M. J. Rosseinsky, *Nature*, 2017, **546**, 280–284.
12 S. N. Ruddlesden and P. Popper, *Acta Crystallogr.*, 1958, **11**, 54–55.
13 M. G. Paton and E. N. Maslen, *Acta Crystallogr.*, 1965, **19**, 307–310.

14 H. Müller-Buschbaum, *Z. Anorg. Allg. Chem.*, 1968, **358**, 138–146.

15 W. G. Mumme and A. D. Wadsley, *Acta Crystallogr., Sect. B: Struct. Crystallogr. Cryst. Chem.*, 1968, **24**, 1327–1333.

16 W. J. Becker and G. Will, *Z. Kristallogr. - New Cryst. Struct.*, 1970, **131**, 278–288.

17 W. H. Baur and A. A. Khan, *Acta Crystallogr., Sect. B: Struct. Crystallogr. Cryst. Chem.*, 1971, **27**, 2133–2139.

18 M. Horn, C. F. Schwerdtfeger and E. P. Meagher, *Z. Kristallogr. - New Cryst. Struct.*, 1972, **136**, 273–281.

19 R. J. Nelmes, G. M. Meyer and J. Hutton, *Ferroelectrics*, 1978, **21**, 461–462.

20 J. Bashir, R. T. A. Khan, N. M. Butt and G. Heger, *Powder Diffr.*, 2002, **17**, 222–224.

21 K. Kawamura, M. Yashima, K. Fujii, K. Omoto, K. Hibino, S. Yamada, J. R. Hester, M. Avdeev, P. Miao, S. Torii and T. Kamiyama, *Inorg. Chem.*, 2015, **54**, 3896–3904.

22 S. M. Woodley, P. D. Battle, J. D. Gale and C. R. A. Catlow, *Phys. Chem. Chem. Phys.*, 1999, **1**, 2535–2542.

23 Y. Wang, J. Lv, L. Zhu, S. Lu, K. Yin, Q. Li, H. Wang, L. Zhang and Y. Ma, *J. Phys.: Condens. Matter*, 2015, **27**, 203203.

24 A. O. Lyakhov, A. R. Oganov, H. T. Stokes and Q. Zhu, *Comput. Phys. Commun.*, 2013, **184**, 1172–1182.

25 J. P. Chris and R. J. Needs, *J. Phys.: Condens. Matter*, 2011, **23**, 053201.

26 L. Ask Hjorth, M. Jens Jørgen, B. Jakob, E. C. Ivano, C. Rune, D. Marcin, F. Jesper, N. G. Michael, H. Bjørk, H. Cory, D. H. Eric, C. J. Paul, J. Peter Bjerre, K. James, R. K. John, K. Esben Leonhard, K. Joseph, K. Kristen, L. Steen, M. Jón Bergmann, M. Tristan, O. Thomas, P. Lars, P. Andrew, R. Carsten, S. Jakob, S. Ole, S. Mikkel, S. T. Kristian, V. Tejs, V. Lasse, W. Michael, Z. Zhenhua and W. J. Karsten, *J. Phys.: Condens. Matter*, 2017, **29**, 273002.

27 Database of Ionic Radii, http://abulafia.mt.ic.ac.uk/shannon/ptable.php, accessed October, 2016.

28 R. D. Shannon, *Acta Crystallogr., Sect. A: Cryst. Phys., Diffr., Theor. Gen. Crystallogr.*, 1976, **32**, 751–767.

29 G. Kresse and J. Furthmuller, *Phys. Rev. B: Condens. Matter Mater. Phys.*, 1996, **54**, 11169–11186.

30 J. D. Gale and A. L. Rohl, *Mol. Simul.*, 2003, **29**, 291–341.

31 The Inorganic Crystal Structure Database, FIZ Karlsruhe, www2.fiz-karlsruhe.de/icsd_home.html, accessed February 2017.

32 B. W. H. Vanbeest, G. J. Kramer and R. A. Vansanten, *Phys. Rev. Lett.*, 1990, **64**, 1955–1958.

33 S. M. Woodley, http://www.ucl.ac.uk/klmc/Potentials/, accessed August 2016.

34 R. C. Baetzold, *Phys. Rev. B: Condens. Matter Mater. Phys.*, 1988, **38**, 11304–11312.

35 H. T. Stokes and D. M. Hatch, *J. Appl. Crystallogr.*, 2005, **38**, 237–238.

36 S. P. Ong, W. D. Richards, A. Jain, G. Hautier, M. Kocher, S. Cholia, D. Gunter, V. L. Chevrier, K. A. Persson and G. Ceder, *Comput. Mater. Sci.*, 2013, **68**, 314–319.

37 K. Momma and F. Izumi, *J. Appl. Crystallogr.*, 2011, **44**, 1272–1276.

Faraday Discussions

DISCUSSIONS

Structure searching methods: general discussion

Matthew Addicoat, Claire S. Adjiman, Mihails Arhangelskis, Gregory J. O. Beran, Jan Gerit Brandenburg, Doris E. Braun, Virginia Burger, Asbjoern Burow, Christopher Collins, Andrew Cooper, Graeme M. Day, Volker L. Deringer, Matthew S Dyer, Alan Hare, Kim E. Jelfs, Julian Keupp, Stefanos Konstantinopoulos, Yi Li, Yanming Ma, Noa Marom, David McKay, Caroline Mellot-Draznieks, Sharmarke Mohamed, Marcus Neumann, Sten Nilsson Lill, Jonas Nyman, Artem R. Oganov, Sarah L. Price, Susan Reutzel-Edens, Michael Ruggiero, German Sastre, Rochus Schmid, Julia Schmidt, J. Christian Schön, Peter Spackman, Seiji Tsuzuki, Scott M. Woodley, Shiyue Yang and Qiang Zhu

DOI: 10.1039/c8fd90030b

Graeme M. Day opened the discussion of the introductory lecture by Sarah L. Price: You have nicely outlined a vision for a crystal structure prediction code and what its capabilities should be. In this, you have not said much about what analysis methods this code should have for understanding and interrogating the structure sets that are generated. The crystal structure prediction community, at least in the molecular organic field, has mainly borrowed methods from the analysis of experimental crystal structures. These can be useful, but do we need more sophisticated approaches to get the most information from computed sets of structures? I am interested in hearing your thoughts on what analysis methods need to be part of a general crystal structure prediction code.

Sarah L. Price answered: This is a difficult question, as it is still a matter of debate how different two structures have to be to count as polymorphs when working with experimental structures, with many of the difficulties relating to polytypes or disorder, where CSP analysis needs further development. Most of the methods used to compare experimental structures are appropriate for CSP generated structures, assuming that duplicates which only differ in the cell choice have already been eliminated. We certainly need to be able to cluster structures into families so that members of the same family could be observed as disorder components of experimental structures or be deemed to be so readily able to transform into the most stable structures of the family somewhere along the nucleation and growth pathway that they could not be observed, or are closely related structures which may appear in the same sample or interconvert. At the moment this is mainly done by experienced human judgement, using tools such as the size of the coordination sphere that can be overlapped, powder pattern similarity, hydrogen bonding motifs, XPAC, *etc.* For complex flexible molecules, deciding how to classify the low energy structures into families is fraught with difficulties, but the families of CSP structures may help interpret the experimental data since the ability of the molecules to rearrange into a more stable structure can depend on the size and quality of the crystallites. The analysis tools we require will be system dependent and closely linked to the prevalence of practically important metastable polymorphs and the accuracy to which their structures and properties are required.

J. Christian Schön remarked: Regarding the issue of what we should aim for, I would suggest looking at an often neglected issue: the time scale – equilibration + observation + escape – on which the system equilibrates, is observed, and (for metastable compounds/structures) leaves the region in its respective configuration space (*e.g.* by transformation into a different structure), respectively. If we talk about a relevant crystal structure/polymorph, we must keep in mind the time scale on which we are interested in the structure. People often confuse the stability of a compound with thermodynamic stability – the fact that the potential energy of a structure is low does not mean that it exhibits high stability (*e.g.* there might be a simple transformation route with a low energetic barrier to the global minimum). On the other hand, for a given observational time scale, many structures are kinetically stable even if they are rather high in potential energy. An important step would thus be to classify the structures by their lifetimes (escape times). We already have the computational technology at hand to analyse the kinetic stability of crystalline structures, *e.g.* based on the threshold algorithm[1,2] or transition path sampling.[3,4] A second very important goal should be the development of theoretical methodologies and algorithms that allow us to optimally reach target modifications/structures with a high probability, thus aiding and guiding the experimental synthesis.[5] An important tool could be the application of finite-time thermodynamics concepts to chemical processes.[6,7] Of course, this needs to be done on top of standard crystal structure prediction, and typically requires an order of magnitude more computational effort for the analysis of the barrier structure of the configurational space of the chemical system, and several orders of magnitude more for the analysis and optimization of synthesis routes. Nevertheless, one would envision a multi-level code package that provides such a range of computational capabilities – *e.g.* the G42+ code already does contain such algorithms for both structure prediction and the analysis of barrier structures[8] – even though the predictive modelling of the synthesis route is now only at a stage comparable to that of crystal structure prediction twenty-five years ago.

1 J. C. Schön, *Ber. Bunsenges. Phys. Chem.*, 1996, **100**, 1388–1391.
2 J. C. Schön, H. Putz and M. Jansen, *J. Phys.: Condens. Matter*, 1996, **8**, 143–156.
3 C. Dellago, P. Bolhuis, F. S. Csaijka and D. Chandler, *J. Chem. Phys.*, 1998, **108**, 1964–1977.
4 P. G. Bolhuis, C. Dellago and D. Chandler, *Faraday Discuss.*, 1998, **110**, 421–436.
5 J. C. Schön, *Adv. Chem. Phys.*, 2015, **157**, 125–134.
6 B. Andresen, *Angew. Chem., Int. Ed.*, 2011, **50**, 2690–2704.
7 J. C. Schön, *Z. Anorg. Allg. Chem.*, 2009, **635**, 1794–1806.
8 J. C. Schön, *Proc. Appl. Ceram.*, 2015, **9**, 157–168.

Sarah L. Price answered: Thank you for clearly emphasising the point about lifetimes of metastable structures. I hope that these approaches will prove suitable for extension to pharmaceutical polymorphs.

David McKay remarked: On CSP-AIM, how does/should the role of the conversation between CSP people and experimental people fit in? Would this no longer be necessary by writing CSP software that can understand and interpret experimental data?

Sarah L. Price answered: A CSP_aim code should have encapsulated this vital conversation, so that the code can be used without distinction between computational and experimental scientists. A clear output of the software should be the structures of the possible polymorphs of the system, with the corresponding powder X-ray patterns, solid state NMR spectra, phonon (infrared, Raman, terahertz) spectra *etc.* to allow the polymorphs to be identified even when observed in a poorly crystalline mixed-phase sample. The code should also be capable of providing a suggestion of the experiment most likely to find these polymorphs (*i.e.* specifying pressure, temperature, solvent system or template for sublimation, seeds, *etc.*). There would also be estimates of the relative thermodynamic and kinetic stability, and other properties of interest in the application. We should aim for software that allows seamless comparisons of the experimental data and corresponding data for computer-generated target polymorphs. However, it is essential that there is always a clear distinction between the structures that have been realised in the laboratory, and those which have yet to be verified experimentally.

Alan Hare opened the discussion of the paper by Yanming Ma:† Because the quasi-planar B_{36} species is hexagonal, this can be formulated as $B_{6\Sigma_{i=1}^{n} i}$, where in this instance, $n = 3$; *i.e.* 6 times the sum of the first n natural numbers. Can it be shown that n is always a local energy minimum? Will $B_{3n(n+1)}$ always be the lowest quasi-planar energy state? Should we expect to see B_{60} too, for instance ($n = 4$)? The structure has a hole in the centre, $i = 0$ (presumably the triangular ring would be under too much strain); now, $B_{6\Sigma_{i=0}^{1} i}$ gives the visual impression of a nucleus. Does this suggest that if the B_{84} cluster were to be either planar or cage-like it would require multiple nuclei to coalesce? Perhaps this would explain the lower energy of the core-shell. When you first get a glass-like structure – at B_{84}, say – is that "game over" for the traditional crystal? Or is it still possible for the structure to revert to a periodic one at some higher number (such as $n = 5$, B_{90})?

Yanming Ma communicated in response: It cannot be said that n corresponds to the local energy minimum for a planar configuration. For technical reasons, experimental data are only available for the charged clusters. Up to now, all the anionic B_n clusters with $n < 38$ have been experimentally confirmed to be planar, and B_{40} is cage-like. The structures for larger neutral B_n clusters with $n > 40$ are still elusive, with planar, core-shell, bilayered and cage-like configurations

† The paper by Yanming Ma was not presented at the meeting. Questions were submitted to the author after the meeting.

competing with each other. Whether B_{60} exhibits a planar or core-shell structure should be further checked.

Hexagonal vacancies are crucial for the stability of B clusters. It is reasonable to expect the existence of multiple vacancies in the B_{84} cluster. However the vacancies should be separate from each other (as seen in the alpha-boron sheet[1]). It is generally accepted that B_n clusters with $n > 68$ have core-shell structures, which can be seen as embryos of bulk boron.[2]

1 H. Tang and S. Ismail-Beigi, *Phys. Rev. Lett.*, 2007, **99**, 115501.
2 J. Zhao, L. Wang, F. Li and Z. Chen, *J. Phys. Chem. A*, 2010, **114**, 9969–9972.

J. Christian Schön asked: Concerning the "flat" structures shown in Fig. 4a and 5c, are they perfectly two-dimensional? Is there another layer of atoms directly behind them, so that they are actually multi-2D-layered structures?

Yanming Ma communicated in reply: No, they are not perfectly two-dimensional, since the plane is curved, but they are single-layered structures without additional layers behind them.

J. Christian Schön commented: Considering the first equation, *i.e.* where the energy is written as the sum over "atomic" energies, where these individual energies take some aspects of the local neighborhood of the atoms into account; this procedure is reminiscent of work done in the phase diagram theory community on intermetallic phases, led by Sanchez and DeFontaine since the early 1980s,[1,2] the so-called cluster expansion method. This method takes many (locally optimised) lattice structures, computes their energy (*e.g.* on *ab initio* level), and then derives expressions for the energy as the sum over binary, ternary, *etc.* local clusters of atoms surrounding each atom in the structure. These parameters can then be used to very quickly compute the energy of very large arrangements of atoms in a lattice (originally) or in a more general solid structure. There might be some valuable insights for the approach discussed here.

1 J. M. Sanchez and D. DeFontaine, *Phys. Rev. B*, 1980, **21**, 216–228.
2 J. M. Sanchez, *Phys. Rev. B*, 2010, **81**, 224202.

Yanming Ma communicated in response: Agreed! As far as I know, the cluster expansion method is particularly useful for alloys or systems where the atomic environment is known to some extent. I think the ML method may be more versatile.

Sten Nilsson Lill asked: On page 8, you discuss an 84-atom boron cluster and mention that there are two candidates for the ground state. What are the spectroscopic fingerprints of these and how do the predicted spectroscopic data compare to the experimental data?

Yanming Ma communicated in reply: To the best of our knowledge, there is no experimental spectroscopic data. In theory, we can calculate the photoelectron spectrum, which will aid future experiments to determination the ground-state structure.

J. Christian Schön remarked: It is not quite clear to me whether the experimentally observed structures correspond to the global minima of the potential energy.

Yanming Ma communicated in response: Any local minima on the potential energy surface correspond to mechanically stable configurations, where the gradient vanishes and hence so do all the forces. The experimentally observed structures generally correspond to the global minimum according to the principle of minimum energy, which is essentially a restatement of the second law of thermodynamics.

Virginia Burger commented: In your paper, it says that "[ML potentials] show accuracy and transferability comparable to those of quantum-mechanical simulations…". I understand that they have similar accuracy, but it is not clear to me that they would have similar transferability, since I thought ML potentials tended to perform much more poorly when extrapolated than on data similar to their training set. I think this sentence would be more useful and clearer if it were supported by citations with examples for accuracy and transferability.

Yanming Ma communicated in reply: Thanks for the good reminder. In the current work, both the training and testing sets consist of random structures generated by CALYPSO and are expected to cover a wide range of the PES. This to some extent demonstrates the high transferability of the ML potential. However, we agree that the current expression is not very accurate, since the transferability of ML potentials generally depends on the quality of the training set.

Volker L. Deringer asked: You mention in the Conclusions the possible extension of your potential to the bulk phases of boron. Have you tested the current version, say, on crystalline α-B_{12}? How much information about the bulk can already be "learned" from clusters?

Yanming Ma communicated in response: Currently, we have not tested the potential for bulk boron, since the stress tensor is not included during the training process, but we would like to perform a test according to your comments and provide you with the results as soon as possible.

Julia Schmidt communicated: Regarding the sentence in your paper: "We expect the RMSEs of the energy and force can be further reduced by increasing the training set size and number of function values of the ACSF descriptor but at the expense of larger computational cost during the training process."; the fact that the RMSE will be reduced if you train the model with more training data, hence with more outliers, makes sense. You currently have a reasonable training : testing ratio of 78% : 22% (12053 : 3571), but if I follow correctly the suggestion above would decrease the test set proportion. Why would increasing the training set size only be beneficial and not lead to overfitting?

Yanming Ma communicated in reply: Since the training and testing sets consist of random structures generated by CALYPSO, they are independent of each other and should cover a wide range of the PES. Therefore, the RMSE of the testing set is a reasonable indicator for the quality of the potential, regardless of the training : testing ratio. We expect the training set not to fully cover the PES. Thus, increasing the number of function values of the ACSF descriptor will lead to a more complete description of the structures, while increasing the training size will include more outliers. This will improve the quality of the potential. However, both the parameters of the ACSF descriptor and the newly included structures should be chosen carefully to avoid overfitting.

J. Christian Schön opened the discussion of the paper by Volker L. Deringer: How do the machine learning potentials perform compared with the refined empirical potentials people have been developing and using over a long time – in particular with respect to accuracy, speed, robustness and generalisability?

Volker L. Deringer replied: These are two conceptually different approaches, so let me start by saying clearly that machine learning (ML) potentials do not aim to replace empirical ones. As a rule of thumb, ML potentials are significantly more computationally demanding than empirical potentials, but are also significantly more accurate—reaching "chemical accuracy" in the kJ mol^{-1} region in many cases. In turn, empirical potentials are normally very robust, because they have physically and chemically motivated forms (so they will correctly describe the repulsion between two atoms that come very close, something that ML models need to "pick up" from the input data). Indeed, robustness is one of the major challenges for making good ML potentials, and it is precisely what we are addressing in this work: the initial atomic configurations are largely randomised, and the fact that our potential can perform progressively better relaxations of these configurations attests to its robustness and flexibility. This is particularly important for systems (such as phosphorus and boron) for which no good general empirical potentials exist.

Michael Ruggiero asked: Given that your machine learning methods typically generate an interatomic potential, is it possible to determine vibrational dynamics? Can they be taken a step further, to determine properties such as thermal displacement or thermodynamic parameters (*i.e.* due to the dependence on the vibrational partition function)?

Volker L. Deringer answered: Anything that is based on energies or forces is accessible. If the forces are accurate, any other properties derived from them will be, too. For example, people have computed vibrational (phonon) modes using ML-based interatomic potentials with very good accuracy; once these are there, you could go and obtain vibrational contributions to the free energy, move on to the quasi-harmonic approximation, and so on. I think these are very interesting research directions for the future.

Jan Gerit Brandenburg remarked: The discussion of the machine learning potentials leads to the comparison of Gaussian kernel regression and artificial neural networks. While comprehensive comparisons are out of reach for this discussion and one might refer to review articles (*e.g.* ref. 1), the computational speed has been mentioned. Independent of the employed descriptor, feed-forward neural networks do not need any matrix operation once they have been optimised. Thus, they can potentially be evaluated extremely quickly on graphic processing units. This holds for increasingly large reference datasets as the evaluation depends on the network geometry and not on the reference set. However, this might not be a major issue for the relatively small datasets used in potential energy interpolations, and the fitting process itself or the smoothness of the resulting potential may be more important. So the best suited machine learning model will strongly depend on the type of application.

1 J. Behler, *J. Chem. Phys.*, 2016, **145**, 170901.

Volker L. Deringer responded: Without going too deep into aspects of algorithms and implementations, which should be addressed by my co-author Gábor Csányi, I do note that we use a standard sparsification procedure in practice: the fitting algorithm selects a number (say, a few thousand) of the most representative atomic configurations, and there are regression parameters only for these. The cost of a GAP simulation, therefore, scales linearly with the number of "sparse points", not with the size of the reference database.

After the meeting, Gábor Csányi (a co-author of the paper under discussion, DOI: 10.1039/c8fd00034d) communicated in writing: There is a widely shared misconception that kernel-based methods scale badly with system size. This is due to the pedagogical way these methods are typically introduced in textbooks: a square kernel matrix is formed between every pair of input data points, and is then inverted to obtain the coefficients of the fit, leading to $O(N^3)$ scaling. In practice, one never does this for large data sets. Instead, a rectangular matrix is formed between the N data points and a much smaller set of M data points (typically picked from the N set using various algorithms—this is the "sparsification" Volker L. Deringer refers to). This leads to an $M \times N$ matrix, for which the operations needed to determine the fit are $O(N)$. After the coefficients are calculated, the prediction of properties is $O(M)$, and this does not scale with the database size at all, just like for neural network models. Formally, an equivalence can be drawn between kernel regression and 1-hidden-layer feed-forward neural networks, and in this case M corresponds to the number of neurons in the hidden layer.

In practice, for the large and accurate models published to date, the speed of predicting a force is within the same order of magnitude as our GAPs and Behler's neural network models. Note that in the GAP case, a significant fraction of time is taken up by computing the representation, before the sum over M is taken, so significant optimisations are still possible in picking better (more quickly converging) basis functions. Typical prediction speeds are 10–100 milliseconds/atom/core, which parallelises extremely well due to the relatively large time spent on computing the force of a single atom. Therefore, applications with >50 000 atoms are entirely feasible and are currently being undertaken.

Qiang Zhu enquired: In order to a make a good force field, how many structures do you need? Are there any stop criteria? Can this type of force field be used for MD simulations with large cells?

Volker L. Deringer replied: A typical value is a couple of thousand reference structures—after which the predicted properties converge to within what one is aiming for. The energy and force error is one possible stopping criterion, but experience has told us that this question is much more intricate. One can fit potentials with a reasonably low numerical error that still fail to capture important physical properties; and one can have potentials with a notable residual error (let's say, for the case of carbon, $\approx$ 1 eV Å^{-1} on the forces for highly disordered liquid and amorphous structures) but which perform exceptionally well for structural and mechanical properties (especially when, as in the liquid, the absolute forces are high anyway).[1] These machine-learned force fields can be, and have been, used for large systems: for example, we recently simulated the deposition of amorphous carbon films using system sizes of > 10 000 atoms,[2] and others have modelled GeTe nanowires to a similar extent,[3] approaching the real physical sizes of devices—and all at close-to-DFT accuracy!

1 V. L. Deringer and G. Csányi, *Phys. Rev. B*, 2017, **95**, 094203.
2 M. A. Caro *et al.*, *Phys. Rev. Lett.*, 2018, **120**, 166101.
3 S. Gabardi *et al.*, *J. Phys. Chem. C*, 2017, **121**, 23827–23838.

Qiang Zhu remarked: If you did not sample enough configurations (just a couple of thousand), it seems that this would be inadequate to reconstruct the entire potential energy surface. You might get good estimations of the local minima, but the transition state and liquid-like states are unlikely to be described very well.

Volker L. Deringer responded: This would be true if there were no high-energy structures in the reference database, but the use of RSS here works to our distinct advantage. Perhaps surprisingly, a few thousand configurations are indeed enough to describe the system, including the relaxation trajectories that lead to the local minima. Other databases and potentials have been built for carbon and silicon with comparable numbers of reference structures, and both potentials capture liquid, amorphous, and other higher-energy regions reliably, including excellent agreement with diverse experimental probes for amorphous carbon[1] and amorphous silicon.[2] Interestingly, the liquid structures are often comparably easy to fit using an ML algorithm: they are so highly diffusive that just a handful of snapshots gives you lots of inequivalent atomic environments, and each of these simulation cells contains $3N$ force components, all of which contain valuable information about the potential-energy surface (PES). We also take distinct advantage of the fact that the PES is smooth, and therefore a limited number of sampling points are sufficient to describe it.

1 M. A. Caro *et al.*, *Phys. Rev. Lett.*, 2018, **120**, 166101.
2 V. L. Deringer *et al.*, *J. Phys. Chem. Lett.*, 2018, **9**, 2879–2885.

Matthew S Dyer queried: From a practical point of view, it is very simple to use interatomic potentials by reading a small number of parameters from a published table. In practice, how would a researcher use the GAP potential that you have generated for phosphorus?

Volker L. Deringer replied: They can (and we would encourage them to!) download the potential parameter files, which are provided in XML format.[1] The GAP code, which directly reads these parameter files, is free to download and use for any non-commercial research from the same web address, and my colleagues have interfaced it to the QUIP, ASE, and LAMMPS simulation software. In that, it is arguably much easier and error-proof to use than coding up even a simple interatomic potential model by oneself!

1 www.libatoms.org

Mihails Arhangelskis asked: My question concerns the type of DFT calculations used for fitting the interatomic potentials. Is there any advantage in performing DFT geometry optimizations? Presumably geometry optimization would provide more information in the form of atom coordinates and atomic forces around the local minimum of the crystal structure. Is there any advantage to doing that, or are single-point calculations sufficient for deriving the interatomic potentials?

Volker L. Deringer answered: This is an interesting alternative strategy, which indeed several authors have followed recently.[1,2] We do not do relaxations with DFT, and this a conscious decision—we want to make a potential with the lowest necessary computational effort in terms of reference data, in turn sacrificing some of the advantages (in the quality of early-generation structures) that DFT relaxation would have. Of course, we do perform relaxations too: with the GAP directly, which is progressively improved, as evidenced by the fact that we do find the structure of black P in the fourth generation.

1 S. Hajinazar *et al.*, *Phys. Rev. B*, 2017, **95**, 014114.
2 E. L. Kolsbjerg *et al.*, *Phys. Rev. B*, 2018, **97**, 195424.

Artem R. Oganov remarked: For pure elements, this approach works well. For compounds, how many parameters do you need in the force field? What would you do about long-range interactions?

Volker L. Deringer responded: For compounds, I think the key step would be fragment-based approaches (as partly seen in our attempts to "find" fibrous phosphorus); trying to find silica polymorphs (or worse, complex silicate compounds) by seeding with isolated atoms would likely waste a lot of resources; initialising the search with [SiO_4] building units should narrow down the search space to a more manageable size. Regarding long-range interactions, see my answer to the previous question—there are existing routes, and it would indeed be extremely interesting to couple these with GAP-RSS and other ML-driven structure-searching techniques.

Virginia Burger enquired: I noticed in the paper that you deemed the database converged once you had "learned" black phosphorus. Were there other indicators that the space had been sufficiently sampled that might help in making more a general convergence criterion for molecules in the future?

Volker L. Deringer replied: This is a very interesting point, to which I think the community hasn't found a definitive answer yet. A more general stopping criterion might be the energies of entries in the database (we've done this for boron) or the energy errors (as we've exemplified for Hittorf's phosphorus), but these would still make it necessary to define a numerical convergence criterion, and would hence still be somewhat specific to the system. Another more general approach could mirror what is being done in *ab initio* RSS: identifying minimum structures, and deeming the potential converged only once it finds all the minimum structures repeatedly.

Artem R. Oganov asked: You mentioned that for you it takes 2000 structure relaxations to get alpha-boron with 12 atoms/cell. Can you get gamma-boron with 28 atoms/cell and beta-boron with > 100 atoms/cell, and how long does this take you?

Volker L. Deringer answered: We have not tried this yet, but it would be a very interesting next step. Getting the β-rhombohedral boron structure with its ~106 atoms per unit cell out of a free search would certainly be a major accomplishment.

Christopher Collins remarked: When you are fitting your force fields, you get a large number of parameters. Do you have any upper limits on how many parameters your fitting is permitted to use? Do you use any methods to check to see if you are overfitting your training data?

Volker L. Deringer responded: There are two different types of parameters that one needs to control when fitting GAP models. There are the hyper-parameters, including the expected noise in the input data and the smoothness of the SOAP kernel (that is, how "fuzzy" the comparison between two atomic environments is being made). These need to be controlled by hand but are limited in number, typically fewer than ten, and the sensible ranges are small. There is also the number of regression parameters, normally a couple thousand for SOAP, which can be converged. Generally, overfitting is checked by taking parts of the training database and evaluating the error for those—or by having a separate test set, as was the case in this paper.

Gregory J. O. Beran queried: Given that your machine learning model is based on local geometric descriptors (SOAP), could you comment more on how pressure is included in the model? Is it another input descriptor?

Volker L. Deringer replied: The (DFT-computed) virial stresses enter the fit, alongside the energies and forces. However, as descriptors we use solely the atomic environments—which are, of course, influenced by the applied pressure. For example, if we fit a GAP-RSS model for phosphorus using applied external pressure during the structural relaxations, it will have a much easier time picking up the six-fold coordinated environments of a simple cubic lattice (see Fig. 5 in the paper).

Gregory J. O. Beran asked: How well does the machine learning model reproduce the DFT enthalpy *versus* pressure curves?

Volker L. Deringer answered: Once the DFT energy–volume curve can be reproduced with confidence, this curve can be directly translated into an enthalpy–pressure plot of the same confidence level. Fig. 5 suggests that data for the high-pressure forms (rhombohedral and simple cubic) can be fitted in principle, but this could of course be improved by including more distorted cells explicitly for this purpose. This is the topic of ongoing work.

Jan Gerit Brandenburg communicated: The authors cite ref. 1 for using the PBEsol density functional approximation (DFA) in their data generation. As the GAP potential (and actually the whole structure search) builds upon this, more careful validation might be worthwhile. Specifically, it is not clear whether the semi-local DFA is capable of describing the subtle balance of covalent interactions of phosphorus layers and the van der Waals (vdW) interactions between them. To this aim, Table 1 summarizes low-temperature diffraction data for black phosphorus that has been corrected for thermal and zero-point effects, as well as very recent high-level calculations of exfoliation energies. This is compared to results from PBEsol as well as a few other DFAs. While PBE is unsatisfying, PBEsol has errors of 1.9 Å/atom and 40 meV/atom for the unit cell volume and exfoliation energy, respectively, and is reasonably well suited for describing the electronic structure of black phosphorus. However, it should be noted that inclusion of vdW interactions in the DFA framework substantially improves the agreement with the reference data.

1 A. S. Rodin, A. Carvalho and A. H. Castro Neto, *Phys. Rev. Lett.*, 2014, **112**, 176801.
2 A. Brown and S. Rundqvist, *Acta Crystallogr. B*, 1965, **19**, 684–685.
3 L. Cartz, S. R. Srinivasa, R. J. Riedner, J. D. Jorgensen and T. G. Worlton, *J. Chem. Phys.*, 1979, **71**, 1718–1721.
4 L. Shulenburger, A. D. Baczewski, Z. Zhu, J. Guan and D. Tománek, *Nano Lett.*, 2015, **15**, 8170–8175.
5 M. Schütz, L. Maschio, A. J. Karttunen and D. Usvyat, *J. Phys. Chem. Lett.*, 2017, **8**, 1290–1294.
6 G. Sansone, L. Maschio, D. Usvyat, M. Schütz and A. Karttunen, *J. Phys. Chem. Lett.*, 2016, **7**, 131–136.

Table 1 Structural data and exfoliation energy of black P from experiment and high-level methodologies compared to density functional approximations

Reference data	E_{exf} [meV/atom]	V [Å^3/atom]	a [Å]	b [Å]	c [Å]
Powder diffr.[2]	—	19.0	3.31	10.48	4.38
Neutron diffr.[3]	—	19.0	3.31	10.47	4.37
Diffr. [QHA corrected][a]		**18.8**	**3.31**	**10.42**	**4.36**
Quantum Monte Carlo[4]	−80	—			
Ring coupled cluster[5]	**−92**	—			
Local MP2[6]		18.6	3.27	10.42	4.36
Density functional approximations					
PBE	−10	21.4	3.31	11.33	4.57
PBEsol	−52	17.9	3.32	10.29	4.20
PBE-MBD	−78	19.0	3.31	10.52	4.36
PBE-D3[b]	−74	19.8	3.31	10.74	4.46
PBE0-D3[b]	−85	19.4	3.29	10.62	4.44

[a] Thermal and zero-point effects extracted from diffraction data employing the quasi-harmonic approximation at the PBE0-D3(0,ATM) level.[7] [b] London dispersion scheme used in the zero-damping scheme including Axilrod–Teller–Muto-type three-body terms, data from ref. 7.

7 G. Sansone, A. J. Karttunen, D. Usvyat, M. Schütz, J. G. Brandenburg and L. Maschio, *Chem. Commun.*, 2018, **54**, 9793–9796.

Volker L. Deringer communicated in reply: I recall the dispersion interactions in phosphorus allotropes being a generally challenging issue (see also ref. 1); in this light, the performance of state-of-the-art dispersion correction schemes is really encouraging. Our current discussion paper is a proof-of-concept study for GAP-RSS, where one of the milestones has been "discovering" the structure of black phosphorus at all (and at this stage I'm personally very happy with how PBEsol performs); ongoing work beyond this will indeed deal with the inclusion of dispersion interactions in a more general phosphorus potential.

1 F. Bachhuber *et al.*, *Angew. Chem., Int. Ed.*, 2014, **53**, 11629–11633.

Marcus Neumann enquired: Where do you see the advantages of the random search method? It is surely easy to implement but is generally considered less efficient than methods such as genetic algorithms or Monte Carlo parallel tempering.

Volker L. Deringer responded: One distinct advantage is precisely the broad selection of initial structures for generating the potential ("broad" here being subject to reasonable constraints): we need an appropriate coverage of the potential-energy surface to start with, and we do want to include high-energy structures. That being said, the general ideas from our paper can be transferred to other structure-searching methods—in fact, the preceding paper coupled GAP fitting to particle-swarm optimisation, and Artem Oganov and Alex Shapeev have combined a novel class of so-called "moment tensor potentials" with evolutionary searching. All these methods have given highly encouraging results so far. I believe the implementation itself is less of an issue: as one example, GAP models can be directly loaded into the popular LAMMPS package, and the latter in turn has been interfaced to many structure-prediction codes.

Peter Spackman opened the discussion of the paper by Noa Marom: Did you try using different clustering algorithms? How sensitive were the results you found to the choice of clustering algorithm and parameters?

Noa Marom responded: In ref. 55 we compared the performance of affinity propagation to *k*-means clustering.[1] In general, clustering algorithms that automatically determine the number of clusters (as opposed to specifying them *a priori*) are most suitable for evolutionary niching because the clusters dynamically evolve with the growing GA population.[2] In addition, affinity propagation as implemented in sci-kit learn enables the use of custom distance metrics.

1 X. Li, F. S. Curtis, T. Rose, C. Schober, A. Vázquez-Mayagoitia, K. Reuter, H. Oberhofer and N. Marom, *J. Chem. Phys.*, 2018, **148**, 241701.
2 M. S. Jørgensen, M. N. Groves and B. Hammer, *J. Chem. Theory Comput.*, 2017, **13**, 1486–1493.

Shiyue Yang asked: The cluster-based fitness is defined by $\bar{f}_{\max}$ as the denominator, and according to the previous definition, this term seems to be 0 all the time. Is it my misunderstanding or is something wrong with the definition?

Noa Marom answered: This is a misunderstanding of the notation. $\bar{f}_{max}$ is the maximum cluster-based fitness in the population, which is never zero. It does not correspond to the maximal energy in the population. A clarification of the definition of $\bar{f}_{max}$ has been added to the paper (DOI: 10.1039/c8fd00067k).

Seiji Tsuzuki enquired: I would like to know the details of the new structure generation procedure in the optimization process using the genetic algorithm. How do you generate new structures with the genetic algorithm? What kind of structure parameters do you change to get new structures?

Noa Marom replied: The crossover and mutation schemes available in GAtor are described in detail in Sections 3.4 and 3.5 of ref. 47.[1]

1 F. Curtis, X. Li, T. Rose, A. Vázquez-Mayagoitia, S. Bhattacharya, L. M. Ghiringhelli and N. Marom, *J. Chem. Theory Comput.*, 2018, **14**, 2246–2264.

Virginia Burger remarked: I really liked your niching strategy. I am wondering if you tried tuning the fitness function (equation 4), so that either energy or cluster size would have more weight? For example, since the L descriptor is proportional to the inverse cube root of the unit cell volume, perhaps something like cubing f_i in equation 4 would help reduce the occasional promotion of large unit cell volume structures that is mentioned on page 10 of your paper (DOI: 10.1039/c8fd00067k). Also, I think that the definitions would be clearer if $\bar{f}_{max}$ were defined in mathematical notation. I had trouble understanding whether $\bar{f}_{max} = \max_i(\bar{f}_i)$ while reading the paper.

Noa Marom responded: GAtor is an open source code. Its modular structure enables the user to implement and test various fitness functions as they see fit for their own purposes. A clarification has been added to the paper that $\bar{f}_{max}$ is the maximum cluster-based fitness in the population.

Asbjoern Burow asked: In how far is the construction of your fitness functions motivated by structural and physical constraints from experimental observations and conditions, if the (free) energy alone is not employed? Can you point out the driving principle behind your fitness constructions?

Noa Marom replied: The driving principle behind the construction of fitness functions is the objective of the search. As explained in the paper, for the purpose of evolutionary niching, the fitness function is constructed to promote diverse sampling. For the purpose of property-based optimization, fitness functions may be constructed based on target electronic properties, as we have done in ref. 60.[1] To date, we have not used information from experiments to tailor fitness functions, although this can in principle be done.

1 S. Bhattacharya, B. H. Sonin, C. J. Jumonville, L. M. Ghiringhelli and N. Marom, *Phys. Rev. B: Condens. Matter Mater. Phys.*, 2015, **91**, 241115.

Artem R. Oganov enquired: Maybe you have brought more complexity into the algorithm than is necessary. We have been developing evolutionary algorithms for organic and inorganic compounds for many years.[1–4] For us, genetic drift hasn't been a major problem. It is sufficient to use energy itself as fitness and, as for similar structures, simply kill duplicate structures and among (nearly) identical structures keep one, the structure with the lowest energy.

1 A. R. Oganov and C. W. Glass, *J. Chem. Phys.*, 2006, **124**, 244704.
2 A. R. Oganov, A. O. Lyakhov and M. Valle, *Acc. Chem. Res.*, 2011, **44**, 227–237.
3 A. O. Lyakhov, A. R. Oganov, H. T. Stokes and Q.Zhu, *Comp. Phys. Comm.* 2013, **184**, 1172–1182.
4 Z. Allahyari, A. R. Oganov, Coevolutionary search for optimal materials in the space of all possible compounds, 2018, arXiv:1807.00854.

Noa Marom replied: As explained in detail in Section 3.6.2 of ref. 47,[1] GAtor performs similarity checks twice, before and after relaxation, and duplicate structures are eliminated. As explained here and in ref. 47,[1] evolutionary drift may be the result of the structure of the potential energy landscape with some wide basins that are more likely to be sampled and some narrow funnels that are rarely sampled. The choice of energy method (DFT functional or force field) may also affect the structure of the potential energy landscape and different energy methods may favor particular packing motifs, as we have demonstrated in ref. 47 and 48.[2]

1 F. Curtis, X. Li, T. Rose, A. Vázquez-Mayagoitia, S. Bhattacharya, L. M. Ghiringhelli and N. Marom, *J. Chem. Theory Comput.*, 2018, **14**, 2246–2264.
2 F. Curtis, X. Wang and N. Marom, *Acta Crystallogr., Sect. B: Struct. Sci., Cryst. Eng. Mater.*, 2016, **72**, 562–570.

Artem R. Oganov asked: Why do you need to divide fitness by the number of structures in the cluster of structures?

Noa Marom responded: As explained in the paper: "With this fitness scheme, structures in less populated clusters with under-represented structural motifs have higher fitness values, and hence a higher probability of being selected for mating. Penalizing the fitness of over-sampled clusters and steering the GA towards under-sampled clusters provides a more uniform sampling of the configuration space. If all the cluster sizes were equal, this fitness scheme would be equivalent to energy-based fitness."

J. Christian Schön addressed Noa Marom and Artem R. Oganov: I got the impression that you found many basins containing many local minima separated by very small energy barriers. In that case, even at low temperatures, the system can easily switch between these "structures" on the observational time scale. When looking for free energies, a description including configurational entropy would then be more appropriate than just computing the (harmonic) phonon contribution to the free energy. Do the experiments produce data that would agree with an average over such nearby local minima? Did you try to explore the set of closely related minima, or did you just consider one "representative" minimum?

Noa Marom answered: The GA does not provide any information on the energy barriers between minima. Therefore, we cannot determine whether the system is likely to switch between structures at finite temperature.

Artem R. Oganov noted: This is not what we usually find in our studies of inorganic crystals (but let us remember that Noa Marom's work is on organic substances). What we usually find is well-defined global minima of the static energy (or enthalpy) at $T = 0$ K, and these persist when you include dynamics. Of course, there are cases similar to what you describe, *e.g.* plastic phases of molecular crystals, where molecules can freely rotate.

Qiang Zhu enquired: Can you describe the performances of different descriptors? Would it be useful to develop some universal descriptor by using them as the primary descriptors?

Noa Marom responded: The performance of the lattice parameter-based descriptor (L), the radial distribution function (RDF), and the relative coordinate descriptor (RCD) is discussed in detail in the paper. For this system both RDF and RCD perform well, although RCD may perform better for systems that do not have distinct intermolecular interactions like the halogen bonds of Target XIII (for example, molecules comprising only C and H atoms). The L descriptor may be unable to resolve the differences between packing motifs that do not correlate strongly with the unit cell size and shape. GAtor is an open source code and the user may implement and test various descriptors as they see fit for their own purposes.

Virginia Burger asked: As a follow-up question, why were you not satisfied with the L descriptor? It appears to have the most diverse sampling, and it looks like it came close to finding the experimental structure.

Noa Marom replied: As explained in the conclusion of the paper: "Some differences were observed in the GA behavior when evolutionary niching was performed based on different descriptors. Using the L descriptor resulted in increased sampling of a region characterized by structures with very long *c*-parameters, particularly when the GA was started from the diverse initial pool. This was counterproductive in the case of Target XIII because this region did not contain any important low-energy structures. The L descriptor was additionally found to produce fewer clusters than the RDF and RCD descriptors. This suggests that the sensitivity of the L descriptor may be insufficient to resolve different packing motifs that do not necessarily correlate with the unit cell shape and volume."

Graeme M. Day queried: As a general question about genetic algorithms (GAs) for crystal structure prediction, I'm interested in your thoughts about how high up the energy landscape a GA can provide a complete sampling of possible crystal structures. I ask because we are sometimes interested in structures that are in high-energy regions, outside of the typical energy range of polymorphism (7–10 kJ mol^{-1}).

I can see how a GA can be directed to structures with specific properties by including the predicted property in the fitness function. However, this requires that we can quantify a property of interest, which is not always the case; we might not know what property the unknown high-energy structure might have, but still want to predict its existence. This might be the case in a pharmaceutical materials context. So I wanted to know whether we can ensure that we explore the landscape thoroughly, and up to what energy a GA gives all or most of the observable structures when the fitness function is defined by the calculated energy, perhaps along with structural diversity such as in the paper you presented.

Noa Marom responded: Genetic algorithms are designed to perform optimization by definition. Furthermore, GAs are designed to converge to the regions of interest with respect to the target property, rather than thoroughly explore other regions of the landscape. To promote broader exploration, one may run the GA several times with different fitness functions. For example, in ref. 60 we tailored GAs to search for structures with low energy, high electron affinity, and low ionization potential.[1] We found that the optimization for target electronic properties rather than energy minimization promoted the exploration of higher energy regions (see Fig. 1 therein). In addition to energy and electronic properties, the fitness function may target diversity. If one is completely unable to define target quantities to optimize then perhaps a different approach (*e.g.* random search, Monte Carlo, or MD) would be more suitable than a GA.

1 S. Bhattacharya, B. H. Sonin, C. J. Jumonville, L. M. Ghiringhelli and N. Marom, *Phys. Rev. B: Condens. Matter Mater. Phys.*, 2015, **91**, 241115.

Graeme M. Day added: As an example, we might be interested in knowing all the polymorphs that are possible because the occurrence of an unanticipated polymorph with associated changes in properties can have important consequences for a pharmaceutical molecule. The property that changes could be solubility, crystal habit, stability or a number of other properties. We don't know which of these might change, but a change in any of them is important and needs to be anticipated. So we cannot use one particular property to include in our fitness function. Also, these structures might be quite high in energy, such as structures resulting from dehydration of a hydrate crystal structure, which might be much higher in energy than the usual few kJ mol^{-1} related to polymorphism. Therefore, from a pharmaceutical polymorphism perspective, we need to know what range of lattice energy we are fully exploring with the method we choose for crystal structure prediction.

Marcus Neumann asked: It can be quite misleading to test and optimize the performance of a global search algorithm against a single compound. What prevented you from applying the same approach to a larger number of compounds?

Noa Marom answered: As explained in the paper, Target XIII was chosen as a case study for evolutionary niching because "In ref. 47, the experimental structure of Target XIII was rarely generated when using a traditional energy-based fitness function". Evolutionary niching using the L descriptor was previously applied to Target XXII in ref. 47 and proved beneficial.[1] The evolutionary niching algorithm is by no means optimized for a particular system. Other than the choice of atom pairs for the RDF, which is system specific, the method is completely general. We have emphasized, both here and in ref. 47,[1] that because it is impossible to know *a priori* which GA settings will perform best for a particular system, the recommended best practice is to run GAtor several times with different settings and collect the structures found in all runs, rather than running it with the same settings for many iterations. Furthermore, we recommend using evolutionary niching in at least one run. Due to the short deadline for submitting papers to the *Faraday Discussion* we were unable to run tests for more than one target. We will certainly test GAtor on additional targets in the future.

1 F. Curtis, X. Li, T. Rose, A. Vázquez-Mayagoitia, S. Bhattacharya, L. M. Ghiringhelli and N. Marom, *J. Chem. Theory Comput.*, 2018, **14**, 2246–2264.

Susan Reutzel-Edens enquired: As a follow-on, how can you tell *a priori* that regions or clusters of structures in the 3D representations outside of where a known experimental structure is found are not important?

Noa Marom replied: It is impossible to tell *a priori* which regions of the PES will be important. This is exactly the reason for using evolutionary niching to promote the exploration of more diverse regions.

Jonas Nyman remarked: You sound very confident in GAtor finding all the energy minima. You could compare all your predicted structures to those submitted in the blind test to see if your search is really complete. You should get all their structures and then some. Has such a comparison been done?

Noa Marom responded: In general, it is impossible to know whether all the low-lying energy minima have been found in any CSP attempt. Hence, our recommended best practice is to run GAtor several times with different settings and collect the structures found in all runs. If the same low-energy structures consistently appear, it is reasonable to assume that the search was exhaustive (at least within the range of lattice parameters explored). Prior to the sixth blind test, submissions contained only three structures and CIF files of the predicted structures are not available, which makes the comparison difficult for older blind test targets, such as Target XIII. Our experience so far indicates that GAtor's ability to spontaneously generate structures with $Z' > 1$ yields additional low-energy minima, as discussed in ref. 47.[1] For example, for Target II GAtor found a low-energy structure with $Z' = 2$ and a scaffold packing motif, which had not previously been reported by others. Several low-energy structures with $Z' = 2$ were also found for Target XIII.

1 F. Curtis, X. Li, T. Rose, A. Vázquez-Mayagoitia, S. Bhattacharya, L. M. Ghiringhelli and N. Marom, *J. Chem. Theory Comput.*, 2018, **14**, 2246.

Sharmarke Mohamed opened a general discussion of the papers by Yanming Ma, Volker L. Deringer and Noa Marom: We know from various crystal structure prediction (CSP) studies that if the structure generation step does not employ physically realistic intermolecular potentials for sampling the range of forces between the atoms/molecules/ions in question, it will often not be possible to locate the experimental structure in subsequent structure optimization steps, regardless of how accurate the energy model may be. Clearly there are some promising data on the value of using GAPs for accelerating the CSP process. I am curious about the views of both Yanming Ma (paper 11894, DOI: 10.1039/

c8fd00055g) and Volker L. Deringer (paper 19763, DOI: 10.1039/c8fd00034d) on how easy it would be to extend machine learning potentials/algorithms to organic molecular systems beyond the boron (paper 11894, DOI: 10.1039/c8fd00055g) and phosphorus (paper 19763, DOI: 10.1039/c8fd00034d) clusters studied by the authors. My concern is that the combination of the cut-off implicit in the Atom Centered Symmetry Function (ACSF) descriptors, combined with the poor description of long-range effects in DFT methods, may make it challenging to locate global minima corresponding to experimental crystal structures for hydrogen-bonded organic molecular systems (*e.g.* pharmaceuticals) where such long-range interactions are integral to the crystal packing.

Volker L. Deringer replied: The correct description of long-range interactions is indeed an extremely important goal—both regarding dispersion and electrostatic interactions. Jörg Behler and his co-workers showed how one can generate a separate machine learning (neural network-type) model for the long-range electrostatic interactions and add this to the short-range potential,[1] and later made and applied potentials for water that included van der Waals interactions.[2] There is indeed a recent machine-learned potential for organic molecules, called "ANI-1".[3] Note that even in the absence of an explicit treatment of long-range electrostatics, hydrogen bonding may often be adequately described using ML models with their typical cut-off radii of around 5–6 Å.

1 N. Artrith *et al.*, *Phys. Rev. B*, 2011, **83**, 153101.
2 T. Morawietz *et al.*, *Proc. Natl. Acad. Sci. USA*, 2016, **113**, 8368–8373.
3 J. S. Smith *et al.*, *Chem. Sci.*, 2017, **8**, 3192–3203.

Yanming Ma replied: I agree. In the case of organic molecular crystals, it is even challenging for DFT to give a very accurate energy order for different crystal packing, and some sophisticated force fields are superior to the ML potentials at the current stage. However, extending the current ML potential to molecular systems is not impossible. Many recent works have shown promising solutions to this problem, *e.g.* expressing the total energy as a sum of the short-range energy, the electrostatic energy and the dispersion correction energy.[1]

1 T. Morawietz and J. Behler, *J. Phys. Chem. A*, 2013, **117**, 7356–7366.

Scott M. Woodley addressed Volker L. Deringer and Yanming Ma: When I started my research in the field of crystal structure prediction (back in the late 1990s), I was told that it was too computationally expensive to search the energy landscape for the global minimum atomic configuration when the total energy was defined using interatomic potentials (IP) and included an Ewald Summation for the long-range terms. With the increased computer resources that are available today, many groups are now searching the "orders of magnitude more expensive to compute", DFT (or similar electronic structure-based) energy landscapes for low-energy local minima. However, there are still many groups, like mine, who typically first target finding the atomic configurations of the IP local minima on the IP energy landscape, and use these as the initial configurations for electronic structure calculations (including the refinement of the atomic configuration using a standard local optimisation method). My group specialises in developing IP for oxides, for example, where the latter approach is much more efficient. Are there IP models of sufficient quality for boron and phosphorus, and, if so, how efficient is your approach compared to first targeting IP local minima?

Volker L. Deringer responded: There are no general IP models for elemental boron and phosphorus, which both adopt extremely diverse structures despite being "only" elemental solids. (In fact, there is a single report on an IP model for B that came out just a few years ago, and has been fitted only for the α-rhombohedral form.[1]) This makes these systems particularly suitable targets for ML-driven exploration and potential fitting at the same time—this is why we chose them for GAP-RSS.

1 P. Pokatashkin *et al.*, *Model. Simul. Mater. Sci. Eng.*, 2015, **23**, 045014.

Yanming Ma responded: Thanks very much for the professional comments. I think both IP and ML potentials have their own advantages. However, to the best of our knowledge, there is no IP model available for isolated boron clusters.

Seiji Tsuzuki asked: I have a question on the fitting of the potential energy surface using machine learning. I would like to know the number of samples necessary to find the global energy minimum. I suspect that quite a large number of samples are needed to find the global energy minimum. I think that the structure of one of the samples must be close to the global energy minimum, if you want to find the global minimum by the fitting of the potential energy surface. I suspect that the number of samples necessary for finding the global minimum is close to the number of local potential minima.

Volker L. Deringer answered: The number of required samples certainly depends on the system, and on the shape and properties of its potential-energy surface (PES). As a rule of thumb, the reference database for our phosphorus GAP-RSS model contains on the order of 3000 DFT single-point computations; this includes structures that have been relaxed into local minima, but also intermediates and the initial randomised cells.

Michael Ruggiero addressed Volker L. Deringer and Yanming Ma: Machine learning is clearly becoming very popular. It is encouraging to see the success of ML in using learned potentials to predict structures, but it seems that armed with the knowledge of the entire potential energy surface there is so much more information available. Can we take this data a step further to predict things like dynamics? From this could we generate accurate thermodynamics? What about temperature dependent structures, anisotropic displacement parameters, phase transitions, *etc.*? Is this a limit of the method, or a limit of the development of tools for analyzing the data? Is there a "hard line" where ML will stop being effective?

Volker L. Deringer responded: You have stated it already: once the potential energy surface (PES) is known, one can describe all the static and dynamical properties within the Born–Oppenheimer approximation, including the description of thermodynamics. This holds irrespective of whether the description of this PES comes directly from a quantum-mechanical computation or from an (accurate) ML fit. This is what can make ML-based interatomic potentials such powerful tools! Indeed, the search for crystal structures is just one small (but growing) application of these simulation methods—we started GAP-based structure searching around two years ago,[1] but ML-based interatomic potentials have been around much longer. The first application of high-dimensional neural network potentials (in 2008!) dealt with the high-pressure phase transitions in Si;[2] we have discussed the usefulness of ML-based potentials for computing thermal properties in a recent overview paper.[3] Again, anything that you can extract from an atomistic simulation is principally in reach, and that is also where the limit is—we can go up to tens of thousands of atoms with ML potentials, and perhaps to millions in a few years, but even this will correspond to simulation cells sized a few tens of nanometres, not more.

1 V. L. Deringer *et al.*, *ChemPhysChem*, 2017, **18**, 873–877.
2 J. Behler *et al.*, *Phys. Rev. Lett.*, 2008, **100**, 185501.
3 G. C. Sosso *et al.*, *Mol. Simul.*, 2018, **44**, 866–880.

Yanming Ma responded: The accuracy and transferability of ML potentials are closely related to the training set. With a reasonable choice of reference dataset, I think it is promising to use ML potentials to deal with these problems. However, the training set should be constructed carefully, because the potential may become invalid during extrapolation.

Michael Ruggiero asked: Given the known structures of polymorphs, how far away are machine learning methods from predicting the relative thermodynamic stabilities of crystalline polymorphs?

Volker L. Deringer answered: I believe that what is keeping us furthest away is not the ML-*versus*-DFT error but the DFT-*versus*-reality error. In so far as ML potentials give accurate ground-state energies and accurate vibrational free energies, there is no reason not to use them for this purpose.

Scott M. Woodley enquired: Sample points on the *ab initio* energy landscape are evaluated in all three papers presented, but, if I remember correctly, only in Noa Marom's paper did I read that additional checks are first made in order to determine whether to evaluate the energy of a newly generated atomic configuration. Surely, particularly for the random sampling approach, you also make initial checks before deciding whether to evaluate a structure?

Volker L. Deringer replied: For the initial generations, we do not check at all, but instead evaluate all the configurations; we do this on purpose, because we want to sample both low- and (reasonably) high-energy regions. Later on, once the reference database has grown to a reasonable size, we do introduce such checks: we count coordination numbers, with the chemical knowledge in mind that P normally forms three bonds (8–*N* rule). We are currently looking into more general, less element-specific criteria, which I think are very much needed.

Yanming Ma replied: For the CALYPSO method, once a new structure is generated, the bond characterization matrix (BCM) is calculated and applied to examine the similarity/distance of this structure compared with all the previous ones.[1] Similar structures with BCM distances below a certain threshold will be eliminated without energy evaluation by DFT or ML potentials.

1 J. Lv, Y. Wang, L. Zhu and Y. Ma, *J. Chem. Phys.*, 2012, **137**, 084104.

Scott M. Woodley asked: Following on from my previous question, does the AIRSS software really perform local optimisations on *random* structures during its search of the energy landscape? If I understand correctly, the efficiency of electronic structure codes is much better near local minima, whereas outside local energy basins, or a sufficient distance from the local minimum, the convergence of the electronic structure can be problematic and it is likely that, during a search, more computer resources are spent evaluating regions of the landscape that you are not interested in if you do not avoid assessing such points using an electronic structure code.

Volker L. Deringer answered: There are reasonable boundary conditions on making the structures, as stated in our paper. Accordingly, the computed energies span a few eV per atom, but do not completely go through the roof. We've seen that happen in early computational experiments with more loose criteria, where we did still converge the electronic-structure calculations but ended up including regions much higher in energy; this definitely didn't help! Generating structures so unrealistic (atoms so close to one another) that the electronic-structure code doesn't converge at all would be another issue; if our GAP-RSS iterations had led to such structures (which they didn't), something would have surely been seriously wrong.

Graeme M. Day addressed Volker L. Deringer and Noa Marom: What do you think is a good measure of the performance of a crystal structure search method? In the most common applications of crystal structure prediction, such as pharmaceutical polymorph screening, we are interested in the entire landscape of possible structures. A common comparison between methods is how quickly they reach the global energy minimum structure. However, the importance of knowing about experimentally realisable higher energy structures gives you a situation where how quickly we get to global minimum is a poor measure of performance. Crystal structure prediction is not simply a global optimisation problem, but relies on global exploration.

Volker L. Deringer replied: This is an extremely interesting point because it can be linked to the synthesis of (possible) metastable compounds and to the description of amorphous phases. We believe that a broad, stochastic approach such as RSS does help with exactly that—in any case, the aim of our technique is to generate a general interatomic potential for a given chemical composition, not just for its minimum structures.

Noa Marom replied: The measure of performance depends on the objective of the search: If the objective is to find low-energy structures, we would define success as finding all the low-energy structures that could be potential polymorphs, not just the global minimum structure. As explained in Section 3.7 of ref.

47 in our paper:[1] "Because there is no unique way of converging a genetic algorithm, the user specifies simple conditions for when the code should terminate. One option is choosing to terminate the algorithm if a certain number of the best structures in the common population have not changed in a user-defined amount of iterations (*e.g.*, if the top 20 structures have not changed in 50 iterations of the GA). This tracks whether all low-energy structures have been located in a reasonable number of iterations." The GA fitness function can be tailored to optimize for properties other than the total energy, as we did for example in ref. 60.[2] In this case, success would be defined as finding structures with the target properties and/or identifying structure–property relations that provide insight on how to design structures with the target properties. If the objective of the search is to promote diverse sampling through the exploration of high-energy regions, one may construct a GA fitness function that weighs diversity more heavily than energy. In this case, a convergence criterion could be when the algorithm has detected the same number of clusters for a long time, meaning no new packing motifs are being generated so the space has been, theoretically, fully sampled.

1 F. Curtis, X. Li, T. Rose, A. Vázquez-Mayagoitia, S. Bhattacharya, L. M. Ghiringhelli and N. Marom, *J. Chem. Theory Comput.*, 2018, **14**, 2246–2264.
2 S. Bhattacharya, B. H. Sonin, C. J. Jumonville, L. M. Ghiringhelli and N. Marom, *Phys. Rev. B: Condens. Matter Mater. Phys.*, 2015, **91**, 241115.

Artem R. Oganov remarked: Our inorganic blind test[1] only tested search methods; all teams used identical ranking methods (force fields or DFT functionals and pseudopotentials, with the same force fields or DFT codes), and to prevent leakage of information, avoided experimentally studied systems – all this is unlike the CCDC test for organic crystals. It's about time we repeat this test; we need more teams, more methods. We need to see how different global optimization strategies compare.

A. R. Oganov, J. C. Schön, M. Jansen, S. M. Woodley, W. W. Tipton and R. G. Hennig, First blind test of inorganic crystal structure prediction, In *Modern Methods of Crystal Structure Prediction,* ed. A. R. Oganov, Berlin, Wiley-VCH, 2010, pp. 223–231.

Graeme M. Day responded: I agree that this would be good to repeat. If such a challenge were set up as a test of global optimization strategies, it would be good to see the range of target structures be quite broad, perhaps even including an organic molecular system. Of course, for organic molecular systems, a method's effectiveness for finding the global minimum in lattice energy is only one measure of a method. We know that observed polymorphs often do not correspond to the global minimum and we need to ensure that a good method locates these alternative structures as well.

Marcus Neumann addressed Artem R. Oganov and Graeme M. Day: CSP needs to solve two problems: the structure generation problem and the energy ranking problem. In his blind test on inorganic CSP organized several years ago, Artem Oganov focused on the first challenge only by imposing on every participant the use of the same energy calculation method. A constraint of this type may be an interesting option for the next blind test on organic CSP. Especially if all the submitted crystal structures are energy optimized, it would be easy to combine all the submitted structures for each compound into a single crystal energy landscape against which the completeness of each method can then be assessed.

Such a completeness estimate should take into account that:

– Many local minima on the potential energy hypersurface have small attractive basins and are closed to other, deeper local minima. Such minima are not found in every search and are also not relevant for practical CSP. The completeness assessment should only be based on structures that are found by at least two participants.

– CSP methods tend to be biased towards the more stable local minima, which is desirable in terms of performance. So the completeness estimate should be limited to the low-energy region of the potential energy landscape.

Graeme M. Day replied: I think that this is a good suggestion for a benchmarking exercise of search methods, since it isolates one aspect of the crystal structure prediction problem. However, I am not sure that such an exercise needs to be performed as a blind test. The analysis of these results would be to judge the completeness of the search by comparing all of the low-energy structures generated by different methods. Therefore, I do not see the advantage of this being performed on a blind test molecule (*i.e.* a molecule whose observed crystal structure is not known). The restriction to blind test conditions can really limit the choice of target molecule, which seems unnecessary. Perhaps an exercise of this format could be organised independently of the blind tests.

Artem R. Oganov replied: I completely agree. I hope that the community will continue the inorganic blind tests, and that such rigorous tests of search and – separately – ranking methods will also become part of the organic blind tests.

Sten Nilsson Lill said: We discussed CCDC blind tests for crystal structure prediction and it was suggested that maybe the blind tests need a new focus, not only who finds the energetically most stable structures. Marcus Neumann suggested combining all the results and seeing who has the best coverage of all the suggested structures. Is it time to also have a property blind test? For example regarding crystalline solubility, *i.e.* including a full thermodynamical cycle of one molecule going from the crystalline phase to the solution phase, or material properties (elasticity, hardness, slip planes, stickiness, *etc.*), or some other key property?

Graeme M. Day answered: My opinion is that the blind tests of crystal structure prediction should test methods for crystal structure prediction and leave property prediction as a separate test. To combine structure and property prediction in one blind test would add quite a lot of complexity in terms of analysis. Is it better for a method to predict the wrong structure, but get close to the property, or predict the correct structure, while being farther away on the property prediction? I think that we would end up judging the separate parts of the prediction on their own, in the way that we currently try to analyse the crystal structure search and ranking problems separately. In view of this, it is probably more effective to have separate blind challenges for structure and property prediction. There have been similar tests for solubility prediction.[1]

1 A. Llinàs, R. C. Glen and J. M. Goodman, *J. Chem. Inf. Model.*, 2008, **48**, 1289–1303.

Qiang Zhu addressed Volker L. Deringer and Noa Marom: What are the best methods for developing descriptors for crystals? We know that the current descriptors take into account the local chemical environment. Is it possible to develop a universal descriptor?

Volker L. Deringer responded: Developing descriptors is indeed a very active research topic—one needs mathematical ways to "encode" a crystal structure; this needn't be in an atom-centred picture, but in practice often is. In the case of interatomic potential fitting, this normally means describing all the neighbours of an atom up to a given cut-off radius ("many-body descriptor"). There is no one best descriptor, but there are several highly successful approaches, including atom-centred symmetry functions (ACSF)[1] and the Smooth Overlap of Atomic Positions (SOAP).[2,3] Both ACSF and SOAP are universal in the sense that they can be applied to any conceivable chemical structure, which to the algorithm is just a list of coordinates and atomic numbers within the cut-off sphere.

Two more comments on that are in order: First, in our current paper, we add non-parametric two-body (2b) and three-body (3b) descriptors to make the fit more robust—which is particularly important for GAP-RSS. The final model is a (properly scaled) linear combination of 2b, 3b, and SOAP terms, following ideas we introduced earlier.[4] Second, when it goes beyond the fitting of interatomic potentials (which is just one of the many current applications of ML methods), there are other directions: as one example, it was recently suggested to represent crystal structures by (simulated) diffraction images, thereby creating two-dimensional "fingerprints".[5]

1 J. Behler, *J. Chem. Phys.*, 2011, **134**, 074106.
2 A. P. Bartók *et al.*, *Phys. Rev. B*, 2013, **87**, 184115.
3 A. P. Bartók *et al.*, *Sci. Adv.*, 2017, **3**, e1701816.
4 V. L. Deringer and G. Csányi, *Phys. Rev. B*, 2017, **95**, 094203.
5 A. Ziletti *et al.*, *Nat. Commun.*, 2018, **9**, 2775.

Noa Marom responded: Many researchers are working on developing descriptors/representations using different strategies. Time will tell whether a universal descriptor will be developed. In analogy, decades of attempts to develop a universal DFT functional have led to a proliferation of different functionals and researchers typically pick a functional that works well for their problem.

Scott M. Woodley commented: Artem Oganov earlier mentioned the equivalent inorganic crystal structure blind test that he organised,[1] where an evolutionary and simulated annealing Monte Carlo (MC) algorithm proved much better than locally optimising random structures. In the slide presented at this meeting (see Fig. 1), I showed the output from the simulated annealing approach for one of these test systems, $BaMgAl_4Si_4O_{16}$. It captures two key features of the local landscape of ten independent walkers as they progress across the energy landscape. Each walker (simulation run) starts from a different random point on the energy landscape (corresponding to a random atomic configuration) and a series of small Monte Carlo (MC) steps (each MC step corresponds to small random

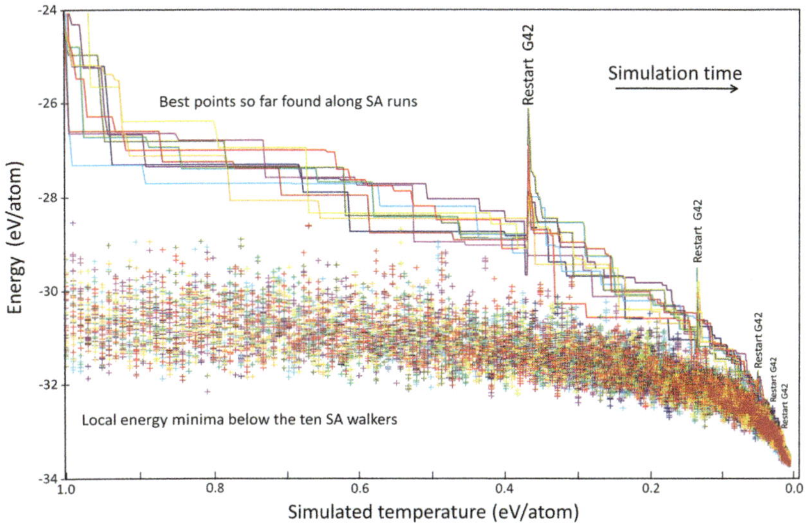

Fig. 1 The trajectories of the best energy found (lines) during ten simulated annealing Monte Carlo runs (each with a unique colour) for a unit cell containing 26 atoms. Each run starts (restarts) with a different random (the last accepted) configuration. For a number of configurations along each trajectory the local minima are reported (crosses). This figure is adapted from 'From Ergodicity to Extended Phase Diagrams', Scott M. Woodley and Alexey A. Sokol, *Angew. Chem., Int. Ed.*, 2012, **51**, 2–5.

atomic and cell displacements and, typically a change in energy) are attempted by each walker. The Metropolis Criterion is employed to decide whether or not to accept each MC step – the greater the step up in energy the less likely the move will be accepted (downhill MC steps are always accepted). The probabilities are also dependent upon the simulated temperature, which is initially set to a high value to decrease the initial rejection rate of uphill MC steps and ensure the walker is not initially stuck within a local energy basin. At each temperature 8000 MC steps are attempted (single-point energy evaluations). The temperature parameter of the Metropolis Criterion is slowly reduced so that the process of annealing is simulated; hence, there is an initial investment of computational resources that the user commits to. The first key feature that is collected by the G42 software[2,3] used to run these simulations (shown in Fig. 1 as a line) is the best energy found during the particular simulation. Given that the G42 software needed to be restarted several times for each walker, due to imposed computer resource time constraints, the first point upon a restart is the actual energy of where the walker was when the simulation was momentarily interrupted. The resulting increase in the lines shown mark temperatures at which the runs were restarted and also give a feeling of how far above the best line the walker typically moves. As temperature is decreased, the walkers gradually move down in energy. The second feature captured by G42 is displayed as crosses in the figure; these are the values of the local energy minima that are accessible to the walker (if only allowed to continue downhill) just before the temperature is dropped. Typically, the number of accessible local minima decreases with temperature. At high temperatures, the crosses represent the energy of the typical configurations that would be quickly found using a random sampling technique coupled with a local optimiser. It can be seen that at a temperature of 0.8 eV/atom there is a very low energy point, however, none of the crosses found at high temperature match the lower energies found from all ten simulated annealing runs. There is an initial computational effort, or cost, required for the simulated annealing runs as it is anticipated that the target structures will be found towards the end of the annealing process. The investment pays off for complex energy landscapes, as the high degree of complexity prevents the likely success of random searching; for less complex systems, the crosses at higher temperatures will be as low as those found at low temperatures.

1 Artem R. Oganov, J. Christian Schön, Martin Jansen, Scott M. Woodley, William W. Tipton and Richard G. Hennig, 'First blind test of inorganic crystal structure prediction methods', in *Modern Methods of Crystal Prediction*, Wiley VCH, 2010, pp. 223–231.
2 J. C. Schön and M. Jansen, *Angew. Chem.*, 1996, **108**, 1358–1377.
3 J. C. Schön and M. Jansen, *Angew. Chem., Int. Ed. Engl.*, 1996, **35**, 1286–1304.

J. Christian Schön enquired: Regarding the ability of random structure search to be competitive with global optimization algorithms such as genetic algorithms or simulated annealing, Artem Oganov organized a structure prediction competition between GA, SimAnn and RSS about ten years ago.[1] Very complicated multinary ionic chemical systems (26, 52 and 60 atoms per variable cell) were searched, with all three codes employing the same empirical potential. RSS was the clear loser, with GA being ahead in two of the problems and SimAnn winning the third search. The conclusion is similar to the experience many people in the combinatorial optimization community have had for decades, *i.e.* random starting point quench searches are not competitive with more refined global optimization searches once the systems become more complex. So, the question is why did you employ the RSS method for your global searches – perhaps you used chemical intuition to select the starting points (although then the name RSS would not really be appropriate!)?

1 A. R. Oganov, J. C. Schön, M. Jansen, S. M. Woodley, W. W. Tipton and R. G. Hennig, in *Modern Methods of Crystal Structure Prediction*, ed. A. R. Oganov, Wiley-VCh, Weinheim, 2011, pp. 223–231.

Volker L. Deringer responded: We use RSS as a tool for exploring structure space in this paper; I do not intend to compare different structure-searching methods side by side in this discussion. As said above, our paper is not so much about specifically coupling GAP with RSS, and more generally about coupling machine learning interatomic potentials with structure searching.

Noa Marom asked: Are the initial structures generated by Chris Pickard's algorithm truly random or are physical constraints applied?

Volker L. Deringer answered: There are physical constraints where they are reasonable. For example, we keep atoms from getting too close to one another by using a "hard-sphere" constraint. Another constraint is given by the density, which we choose to be intermediate between that of red and black phosphorus, and which in the random structures is distributed around that value (this can be seen on the left-hand side of Fig. 2 in the paper, DOI: 10.1039/c8fd00034d).

Matthew S Dyer opened the discussion of the paper by Julian Keupp: Could this approach be used for other material types, such as covalent organic frameworks or inorganic framework materials? If so, how would this be done in practice?

Julian Keupp and **Rochus Schmid** replied: The reverse topological approach (RTA) – including the blueprint pre-optimization by topoFF – works in principle for any crystalline polymer. We have already used it to predict new hypothetical COFs.[1] In practice, the specific building blocks (organic, inorganic or hybrid) and their connecting atoms need to be defined and the algorithm can start.

1 S. Bureekaew and R. Schmid, *CrystEngComm*, 2013, **15**, 1551–1562.

Marcus Neumann asked: How many distinct morphologies have been observed for MOFs so far?

Julian Keupp answered: Many known MOFs are observed only in a single morphology. Few examples exist where multiple polymorphs emerge due to different synthetic conditions; one case is MOF-505 and NOT-100 (where the precursors are the same, but the observed PXRD differs slightly). This has been investigated by us using the RTA.[1] This is an example of isotopological polymorphism. I do not know of a MOF featuring topological polymorphs given the exact same precursors, but often slight modifications in the used linker can lead to different topologies.[2–4]

1 S. Bureekaew, V. Balwani, S. Amirjalayer and R. Schmid, Isoreticular isomerism in 4,4-connected paddle-wheel metal–organic frameworks: structural prediction by the reverse topological approach, *CrystEngComm*, 2014, **17**, 344–352.
2 Y.-B. Zhang, H. Furukawa, N. Ko, W. Nie, H. J. Park, S. Okajima, K. E. Cordova, H. Deng, J. E. Kim and O. M. Yaghi, Introduction of Functionality, Selection of Topology, and Enhancement of Gas Adsorption in Multivariate Metal–Organic Framework-177, *J. Am. Chem. Soc.*, 2015, **137**, 2641–2650.
3 A. Schneemann, R. Rudolf, S. Henke, Y. Takahashi, H. Banh, I. Hante, C. Schneider, S. Noro and R. A. Fischer, Linker functionalisation triggers an alternative 3D-topology for Zn-isophthalate-4,4′-bipyridine frameworks, *Dalton Trans.*, 2017, **46**, 8198–8203.
4 Y. Cai, A. R. Kulkarni, Y.-G. Huang, D. S. Sholl and K. S. Walton, Control of Metal–Organic Framework Crystal Topology by Ligand Functionalization: Functionalized HKUST-1 Derivatives, *Cryst. Growth Des.*, 2014, **14**, 6122–6128.

Asbjoern Burow enquired: As far as I understood, topologies (of high symmetry) are required as input to predict MOF crystal structure candidates with your approach. Do you see any potential to generalize the rules to less symmetric topologies of H-bonded molecular systems? Is there a way to reverse the procedure to predict topologies? Do you know of any published approaches which attempt this?

Matthew Addicoat replied: I'm pleased to say that AuToGraFS[1] + UFF4MOF[2,3] does allow you to form hydrogen-bonded frameworks just as easily as standard covalently-bound frameworks. We published the methodology last year,[4] but essentially specifying the hydrogen bond as a "real" bond with a minimal bond order straightforwardly yields the framework you expect. We've tried both forming the hydrogen bond as the framework forms and embedding the hydrogen bond in the linker, and both give excellent agreement with the reference crystal structure.

1 M. A. Addicoat, D. E. Coupry and T. Heine, *J. Phys. Chem. A*, 2014, **118**, 9607–9614.
2 M. A. Addicoat, N. Vankova, I. F. Akter and T. Heine, *J. Chem. Theory Comput.*, 2014, **10**, 880–891.
3 D. E. Coupry, M. A. Addicoat and T. Heine, *J. Chem. Theory Comput.*, 2016, **12**, 5215–5225.
4 D. E. Coupry, M. A. Addicoat and T. Heine, *J. Chem. Phys.*, 2017, **147**, 161705.

Julian Keupp replied: Our approach is not restricted to topologies (blueprints) of high symmetry. Any blueprint can in principle be used to generate structures. However, we are currently focusing on topologies from the RCSR (and EPINET) database. The reverse of our Reverse Topological Approach (the Topological Deconstruction) is a well established way to deconstruct experimental crystal structures into their underlying topologies, as implemented *e.g.* in the TOPOS code and others. Note that the corresponding TOPOS Database contains a large number of topologies, partly with large unit cells and low symmetry. These could be used in principle, but we have not tried this yet. Furthermore (please consider the corresponding documentation on the RCSR website), many of the topologies have not been observed experimentally but derived from other nets. Another approach to derive new topologies is the EPINET project, where 2D hyperbolic tilings are used as a source to generate topologies.

Andrew Cooper said: The following is a draft version of Professor Cooper's question. Please answer as best you can, and we will contact you with the final version. You will then be able to update your answer if necessary. Some work by Eddaoudi shows MOFs with 12-coordinate linkers and linkers with even higher coordination numbers are now possible, but exotic topologies are not represented in known databases. Have you tried to push your method to that level of complexity?

Julian Keupp and **Rochus Schmid** responded: We need a topology as a starting point for the reverse topological approach. Thus, structures with yet unknown topologies cannot be predicted. However, building blocks with a large number of connections and atoms are not a problem for our method.

Andrew Cooper asked: The following is a draft version of Professor Cooper's question. Please answer as best you can, and we will contact you with the final version. You will then be able to update your answer if necessary. Do big systems have too many degrees of freedom?

Julian Keupp and **Rochus Schmid** answered: Of course, the numerical effort scales with the size of the system and the degrees of freedom, but there is no principal limit for our method in general.

Matthew Addicoat remarked: As we both know there are a number of frameworks that are not well or not completely described by a topology – things like MOF-74/CPO-27 and others that have 1D infinite connectors. How do you deal with those? Do you have a way to decide where to cut the infinite part?

Julian Keupp replied: In the case of MOF-74, there exist different possibilities to use a topology to construct the atomistic model; the topologies *eta*[1] and *etb*[2] both describe one particular realization of the three-fold screw symmetry and can be used with small inorganic cuts of the periodic inorganic structure motif. There also exists the parent net where this ambiguity is not resolved, which is the 5-connected *bnn* topology.[3] By using this topology one could in principle obtain one unit of the inorganic building block, which would in this case have to be periodic itself. For the moment we can only work with the former two topologies, because at the moment we cannot handle periodicity in the building blocks. For the cutting procedure we do not have a clean and off-the-shelf recipe. However, some heuristics we use are:

- Cut as few bonds as possible
- Try to distribute the number of atoms more or less equally
- Try to keep fragments together (*e.g.* don't cut a phenyl in half, …)

These are of course not carved in stone and depending on the application of the current BB, other things need to be considered. It is definitely one of the steps where human input is required.

1 www.mofplus.org/nets/net/eta
2 www.mofplus.org/nets/net/etb
3 www.mofplus.org/nets/net/bnn

J. Christian Schön enquired: Your network topology approach reminds one of the work by Klee,[1,2] a crystallographer from the 1980s, and also, of course, of the work by Friedrichs,[3] LeBail,[4] and Winkler[5] in the 1990s and early 2000s. How does your method compare to their work? A big issue for Klee (and also for Friedrichs, I seem to recall) was the embedding of the network in the three-dimensional space – is this an issue for you? Quite generally, I think the work of the above authors might be helpful when constructing new types of network structures.

1 S. J. Chung, Th. Hahn and W. E. Klee, *Acta Crystallogr. A*, 1984, **40**, 42–50.
2 W. E. Klee, M. Bader and G. Thimm, *Z. Kristallogr.*, 1997, **212**, 553–558.
3 O. D. Friedrichs, A. W. M. Dress, D. H. Huson, J. Klinowski and A. L. Mackay, *Nature*, 1999, **400**, 644–647.
4 A. Le Bail, *J. Appl. Cryst.*, 2005, **38**, 389–395.
5 B. Winkler, C. J. Pickard, V. Milman and G. Thimm, *Chem. Phys. Lett.*, 2001, **337**, 36–42.

Julian Keupp and **Rochus Schmid** replied: In the above works of Klee and also partially in Delgado-Friedrichs' work, their main focus is the generation of topologies in the first place. Herein we use the default embedding (note that we use the words embedding and blueprint synonymously) as obtained from a database (the RCSR) as a starting point and further optimize the cell and vertex positions based on the geometrical needs of the building blocks used. At least in the MOF field, for the embedding there now exists an established methodology to get maximum symmetry embeddings,[1] which restrains the embedding to have edge lengths of 1 and to have a high symmetry space group. This embedding can be computed by Systre (where Delgado-Friedrichs is the main author). Systre is the program used to preprocess topological data for the RCSR database.

The other authors also use topologies as their foundation to generate crystal structures, but for different kinds of systems, and therefore in a different way. Whereas we use larger building blocks that already consist of chemical entities, in other areas smaller units are put onto the vertices of the topology, which can even be single atoms.

1 O. Delgado-Friedrichs, M. D. Foster, M. O'Keeffe, D. M. Proserpio, M. M. J. Treacy, O. M. Yaghi, What do we know about three-periodic nets?, *J. Solid State Chem.*, 2005, **178**, 2533–2554.

J. Christian Schön asked: It looks to me that you could draw a straight line (with some scatter) through the correlation plot in Fig. 6 (mean AAD *vs.* DE(MOF-FF)). Is this just an accident or is there some real physics behind the "correlation"?

Julian Keupp and **Rochus Schmid** answered: That is indeed the case and we say in the paper that "the relative energies correlate reasonably well with the geometric mean AADs obtained from the topoFF optimization and illustrate its usefulness in order to pre-rank the topologies". The reason here is that the simpler topoFF optimization at the blueprint level corresponds well to the actual atomistic potential energy surface, since the energetics are mainly determined by the deformation of the building blocks, which corresponds well to the summed up average angle deviation (AAD) in topoFF. Only if, for example, torsional deformations (not considered in topoFF), start to dominate the energetics, does the topoFF optimization not anymore correlate well with the final atomistic relative energetics.

Caroline Mellot-Draznieks asked: Since your construction method uses a decoration strategy starting from dimensionless networks such as those enumerated by O'Keeffe, would it be appropriate for the construction or prediction of mixed-linker MOFs possessing, for example, linkers of different sizes and connectivities?

Julian Keupp and **Rochus Schmid** responded: Yes. This is exactly what we have in mind for the next steps. For example, considering the mixed-linker systems from *e.g.* the Shane Telfer group, we could use topoFF to quickly assess which kind of linkers with different lengths or shapes might give rise to stable networks.

J. Christian Schön queried: How complete are the sets of networks that you generate? Are there (network-based) structures that you cannot get to by twisting, *etc.*?

Julian Keupp and **Rochus Schmid** replied: We do not generate the networks but use existing databases (like RCSR) as a source, which might not be complete. What we do is use existing networks/topologies to generate "all" potential atomistic structures from these. Note that for molecular fragments – in contrast to zeolites – multiple polymorphs for the same network are possible. Note also that for decreasing symmetry, of course infinite different networks/topologies are possible in principle.

Marcus Neumann opened the discussion of the paper by German Sastre: You presented a diagram showing a linear increase with time of known zeolite structures. Are these predicted or experimental zeolite structures? How does the increase of distinct predicted structures compare to the increase of known experimental structures?

German Sastre responded: This corresponds to experimental zeolites; the pace is rather constant, about 5–10 new zeolites every year. There are typically millions of predicted zeolites, and in particular several millions are found in two such databases by Treacy-Foster and Deem, respectively. However, how many of these predicted zeolites are feasible? Not all of them, since most of the computer generated zeolites satisfy the criteria of connectivity, as well as some geometrical restraints, which are not enough to guarantee feasibility. Since the conditions of feasibility are not so easy to define, and energy alone is not enough, it becomes difficult to assess how many zeolites are feasible. This is a long-standing and interesting question within the community, to which efforts are being increasingly addressed. A good guide is the known correlation between energy and density, a straight line with a negative slope established by Akporiaye and Price in a remarkable paper in 1989,[1] but this is only for the pure silica (SiO_2) composition. Even for pure silica one wonders how far to the correlation line a zeolite may be in order to disqualify as feasible. But, more importantly, zeolites, initially defined as alumino-silicates, soon expanded into materials where tetrahedral atoms also include P, Ge, B, Mg, Zn, Ga, Be and Ti; and even N and S were found to replace oxygen. With these, and possibly more atoms that can be present, although not of course in any combination, it becomes beyond the current scope to foresee the number of feasible zeolites. Nature is, once more, breathtaking.

1 Akporiaye and Price, Systematic enumeration of zeolite frameworks, *Zeolites*, 1989, **9**, 23–32.

Sarah L. Price said: Your paper concludes that there are 7 new zeolite structures that appear feasible. Zeolites are made by crystallising around a template molecule – have you looked at possible templates for these new structures? Are any experiments planned to make your predicted new zeolites?

German Sastre answered: This is ongoing work, where we are using a newly developed software to search for appropriate templates from a database in order to find one suitable for a zeolite of our choice.

Yi Li asked: Why do you make a structure with *P*1 symmetry at the beginning and determine the symmetry again in your GA approach?

German Sastre replied: This approach applies to structures that do not qualify as "zeolites" by having a large, but not sufficiently large, fitness. We then try to apply our specific "geometry optimiser", which makes a *P*1 structure, locates oxygen atoms around the Si, and makes a lattice energy minimisation with specifically selected force fields. And, very importantly, this optimisation does allow the volume to change, so that the optimisation is driven from the "non-zeolite" to the "zeolite" realm through the force field without volume restrictions. Then, in a certainly significant number of cases the optimisation is able to generate a new zeolite whose symmetry should be determined since it can be different from the original.

Marcus Neumann said: You are using a force field for the energy optimization of zeolites. Making a link with earlier presentations, I wonder if a machine learning force field has already been generated for zeolites, and how such a force field would compare to your force field in terms of accuracy and CPU time requirements for energy calculations after parameterization.

German Sastre responded: Machine learning approaches have been used to calculate properties or generate zeolites, relying on geometric features related to local geometry, structure, and porosity. This is opposed to force field approaches which calculate properties explicitly after geometry optimisation, whilst machine learning is focused instead on finding correlations between – in this case – structures and properties. Hence I do not know of any approach which employs machine learning techniques to parameterise a force field for zeolites.

Volker L. Deringer commented: Following up on the previous question – I don't know of currently available "machine-learned" force fields specifically for zeolites; the long-range electrostatic interactions are a challenging issue. Note that there are interesting studies on artificial neural network potentials for bulk metal oxides [*e.g.* ref. 1], including fitting neural networks specifically for the long-range part.

1 N. Artrith, T. Morawietz and J. Behler, *Phys. Rev. B*, 2011, **83**, 153101.

Stefanos Konstantinopoulos asked: A question about force fields: I would like to ask why you use only non-bonded interactions (van der Waals and electrostatics) to generate zeolite structures. Is there any hindrance to not using bonded interactions?

German Sastre replied: For this particular approach we start with a structure that fails to fully qualify as a zeolite and therefore we need to use force fields capable of as much distortion as possible. "Non-bonding" force fields provide a wider landscape of geometry optimisation. Further, our initial structure will, in many cases, not contain proper bond connectivity, and "bonding" force fields typically preserve the connectivity throughout the geometry optimisation procedure, unlike "non-bonding" force fields.

David McKay commented: The speaker said bonds, angles and dihedral angles are not included in the force field. I support the speaker in this. By including bonds in the force field, structures found after geometry optimisation are limited to the connectivity of the guess structures, thus limiting the search space explored. In my experience, by allowing the structure to change more during minimisation, fewer structures would later be discarded as invalid and so a more diverse structural landscape would be probed.

Jan Gerit Brandenburg said: The research group of Jörg Behler has initial work on MOFs, *i.e.* it is indeed possible to generate machine learning potentials for these kind of structures. The corresponding manuscript is in preparation.[1]

1 Marco Eckhoff and Jörg Behler, Development of a Neural Network Potential for Metal–Organic Frameworks.

German Sastre responded: It is different for MOFs than for zeolites. In MOFs it makes more sense to apply machine learning techniques to parameterise force fields, since MOFs do have much wider chemical variety. Instead, in zeolites specific force fields have been developed for silica, aluminosilicate, silicoaluminophosphate, and silicoaluminogermanate. These four compositions cover a large percentage of the zeolites synthesised and the performance of the existing force fields is sufficiently accurate. Machine learning approaches are more appropriate when we do not know how to link structures and properties. When chemical knowledge is available, an explicit force field is usually more accurate than a machine learning force field. In the near future it is possible that the computer will move towards the human approach and the current trend may become reversed.

J. Christian Schön enquired: Concerning the force fields you are using, would it make sense to use/add shell terms to create more bond-like behaviour? Furthermore, regarding the earlier comment/question about symmetries, you might want to employ the KPLOT-code by Rudolf Hundt;[1] the KPLOT code is freely available on the webpage of CRYSTALIMPACT, where a number of algorithms are incorporated that search for symmetries in three-dimensional structures,[2] determine space groups,[3] and compare periodic three-dimensional structures.[4]

1 R. Hundt, *KPLOT Manual – a program for plotting and analysing crystal structures*, ed. J. C. Schön and D. Zagorac, Technicum Scientific Publishing, Stuttgart, 2015.
2 R. Hundt, J. C. Schön, A. Hannemann and M. Jansen, *J. Appl. Cryst.*, 1999, **32**, 413–416.
3 A. Hannemann, R. Hundt, J. C. Schön and M. Jansen, *J. Appl. Cryst.*, 1998, **31**, 922–928.
4 R. Hundt, J. C. Schön and M. Jansen, *J. Appl. Cryst.*, 2006, **39**, 6–16.

German Sastre replied: We have used two core-shell force fields (SLC and SC) but this is only for the oxygen atoms and does not imply bonds between atoms. Using bonds between atoms, either through a core-shell or any other function, would imply a connectivity fixed from the initial structure. Given that the initial structure is not satisfactory, and that is why we want to see if a force field is able to drive it towards a proper zeolite, we need to use force fields capable of as much distortion as possible, and hence "non-bonding" force fields provide a wider landscape of geometry optimisation.

Concerning the algorithms to find symmetry, so far we have used FINDSYM by Harold Stokes,[1] but this does not allow incorporation to analyse big data; and also SPGLIB by Atsushi Togo.[2] None of the approaches used so far are sufficiently robust to work consistently on a high-throughput set where a large diversity of situations can be found. Hence, your suggestion of KPLOT is highly welcome. Thank you.

1 stokes.byu.edu/iso/findsym.php
2 atztogo.github.io/spglib

Alan Hare communicated: Concerning the small voids we see: can you say why there might exist an exclusion zone forbidding an atom to enter? Years ago, attempts were made – principally by Daniel Kivelson, RIP[1] – to determine void size by the constraint it would impose on the rotation of a nitroxide radical (through its consequential effect on the ESR line-broadening).[2] I am now wondering if an impurity species might nucleate the zeolite, perhaps later vacating it to leave the void.

1 http://www.chem.ucla.edu/news/Kivelson_Obit.html
2 *J. Phys. Chem. B*, 1997, **101**, 8631–8634.

German Sastre communicated in reply: Kivelson's approach gave new insights into how molecules move in liquids, described as translations (in which the molecules move from place to place) and tumbling motions. Templates in zeolites tend to fill the empty voids and this results in significant motion restraints of the template with respect to the liquid phase, not only in translation but also in rotation. The change of the rotational modes of occluded templates in zeolites has recently been explored in our group[1] and sometimes can be used to assess the balance between electrostatic and van der Waals template–zeolite interations. Templates do fill the voids during the synthesis and are only evacuated through calcination, but water molecules may have space in the micropore to partially "solvate" the template and may leave the void under mild conditions. All of this (template + water) contributes to the exclusion zone, but in zeolites most of the exclusion zone (micropore voids) is permanently filled by the organic template.

1 T. Lemishko, J. Simancas, M. Hernández-Rodríguez, M. Jiménez-Ruiz, G. Sastre and F. Rey, *Phys. Chem. Chem. Phys.*, 2016, **18**, 17244–17252.

Doris E. Braun opened the discussion of the paper by Christopher Collins: With the FUSE-based probe structure calculations you have shown that you can generate structures with different compositions. Is it possible that different polymorphs are generated within a region of the same composition?

Christopher Collins responded: For all of the given probe structure generation runs with FUSE we do retain all of the structures visited (for this work we generated 1 253 298 structures across 42 compositions). So yes, many different polymorphs are generated for a given composition; as for assessing which ones are stable, we would still suffer from all the same issues discussed in the section on crystal structure evaluation.

Volker L. Deringer enquired: I'm more generally curious about higher-lying, metastable inorganic polymorphs. There are several others, say for TiO_2, with potentially useful applications (*e.g.* ref. 1) – and at the same time, experimental solid-state chemists are making remarkable progress in developing synthesis routes towards such metastable polymorphs. Aside from the two ICSD entries you show, and brookite, did you observe any other TiO_2 structures? Do you generally see a perspective for using your methodology in that direction?

1 J. Buckeridge *et al.*, *Chem. Mater.*, 2015, **27**, 3844–3851.

Christopher Collins answered: We did generate a large number of TiO_2 structures (46 889 that converged), although we did not investigate which ones are synthesisable. This is because our search was set up to find low-energy minima fast and then stop, since our overall objective was to efficiently compute realistic formation energies for the given compositions, not to precisely solve crystal structure(s) and polymorphs. In theory, yes, one could use a method such as FUSE to search for possible polymorphs; to do so we would need to run searches for much longer, although we would still suffer from the same issues as all of the speakers in the discussions on comparing the stability and energy ranking of polymorphs. Currently we have no plans to expand into this area.

Michael Ruggiero asked: Your method uses individual building blocks to generate crystal structures. How difficult would it be to introduce defects into the models? Can the method be used to find probable vacancy sites, or to determine favourable locations for impurity deposition? Can the method be used for solid solutions?

Christopher Collins replied: In principle, yes. Although not mentioned in the paper, as it's a feature we're still working on (for cation vacancies), it is possible to have vacant sites on both the anion and cation lattices. On the anion lattice, this is implemented in the standard sub-module definition, for example one could imagine constructing models of $SrTiO_3$-δ where we would still expect to construct perovskite structures, albeit with oxygens missing from the Ti coordination sphere. For cation vacancies, it is possible to specify a "vacancy" as a species in the initial formula unit; then every time we assemble a sub-module where the cation is labelled as vacancy the site is skipped, thereby creating vacancies in the cation lattice. It is then possible for FUSE to vary the number of vacancies on the lattice on-the-fly.

In practice, I'm not sure that using FUSE would be the most efficient way to study defect chemistry, given that there are a number of groups/software dedicated to looking at the defect chemistry of solids, where the structure you're starting with is pre-defined (examples being SimDope[1] or ChemDASH (currently unpublished)). With FUSE you would be relying on it to find the crystal structure of interest and locate defects at the same time; I suspect that this would be rather inefficient. Although if you are considering systems where the crystal structure is not known, one could use FUSE first to try to determine the base stoichiometric crystal structure, then follow this up with codes like SimDope to study the potential defect chemistry.

As for broader solid solutions, yes, we can definitely use FUSE for this. Essentially this would be using FUSE as we have in this paper, but in a lower dimensional search space. FUSE would be very well suited for considering solid solutions where the end members are different structures – since there's no way of knowing *a priori* at which point a structure should change between the structure type of each end member, or even if an entirely new structure is formed somewhere in between! An example of this is in our results, if you consider the area of the phase diagram between $(SrTiO_3)_{1-x}$ and $(SrO)_x$, where each end member is a different structure type, with 3 different, stable structures in between, the Ruddlesden–Popper $n = 1$, 2 and 3 phases.

1 L. Enciso-Maldonado, M. S. Dyer, M. D. Jones, M. Li, J. L. Payne, M. J. Pitcher, M. K. Omir, J. B. Claridge, F. Blanc and M. J. Rosseinsky, Computational identification and experimental realisation of lithium vacancy introduction into the olivine $LiMgPO_4$, *Chem. Mater.*, 2015, **27**, 2074–2091 (available from Matthew Dyer, University of Liverpool).

Artem R. Oganov queried: Key to the efficiency of your method is that you can feed into the input favourable atomic contacts: *e.g.* cations are surrounded by anions and *vice versa*. How useful would this be for intermetallics, for example, where every type of atom can be next to any other type of atom?

Christopher Collins responded: The discretisation used in the current version of FUSE is entirely aimed at constructing realistic models for ionic solids, and so it would not be well suited to intermetallics. If we were to look at an intermetallic system (or any other system where anything could neighbour anything else), we would have to design new sub-module motifs and modify the assembly rules to reflect the different chemistry.

Artem R. Oganov asked: When you put three or four atoms together, how do you choose the angles? Does it matter?

Christopher Collins answered: When the sub-modules are initially constructed, the angles on the sub-module are all 90° and with the grid-like placement of ions, the O–M–O angles will all be either 90° or 45°. When the full structure is assembled the unit cell angles are then set according to the lattice type that FUSE is using, which of course will then distort these angles. As to whether or not it matters, we don't know, as this is the only way we've tried assembling the sub-modules/the coordinations of the atoms. My best guess is that it doesn't matter as they will be refined during the geometry optimisation step.

Artem R. Oganov enquired: What do you do with stoichiometries? Your graph shows that you found many stable stoichiometries, but how do you sample the chemical space?

Christopher Collins replied: In most cases, we will design our search pattern based upon two main factors: how many compositions is it reasonable to search with available computing resources and how long is it reasonable to spend doing calculations before going into the lab? For this work, as it was a test case to see how FUSE performs, we initially sampled the $SrO–TiO_2$ line quite densely (20 compositions), and once this worked we then expanded out to include Y_2O_3. We decided that we had time for around 20 additional compositions. So we distributed 20 points evenly across the phase space, and included the two compositions which were not on this evenly distributed grid. This resulted in a total of 42 compositions being sampled in the phase field.

J. Christian Schön remarked: Concerning the stacking procedure in FUSE: It looks like you are stacking layers that are electrostatically neutral. But when stacking 3D-blocks like in the AASBU method,[1,2] one can run into the problem that either the blocks or the resulting structures are not charge neutral. Do you ever run into this problem, or how do you avoid that?

1 C. Mellot-Draznieks, J. M. Newsam, A. M. Gorman, C. M. Freeman and G. Ferey, *Angew. Chem., Int. Ed.*, 2000, **39**, 2270–2275.

2 C. Mellot-Draznieks, S. Girard, G. Ferey, J. C. Schön, Z. Cancarevic and M. Jansen, *Chem. Eur. J.*, 2002, **8**, 4103–4113.

Christopher Collins responded: The modules that FUSE generates are not required to be electrostatically neutral, it just happens that in the worked example in the paper they were. For example, there would be nothing stopping FUSE generating SrO_2 modules and having non-charge neutral modules does not cause any issues as long as the full structure is charge neutral overall.

When generating structures at random, there can be an issue with structures not being charge neutral. In practice, this is avoided by only choosing the lattice size/shape to correspond to *n* formula units. The code will then create two lists, one of cations and one of anions for *n* formula units, which are then distributed randomly across the sub-modules. The sub-module motifs used are constrained such that the total number of anions used is equal to the number required for *n* formula units.

J. Christian Schön asked: What local environments can you actually deal with? For example, if a cation is surrounded by six oxygen atoms, then the cation plus its first coordination sphere by anions is a charged building block, you would need to merge adjacent complex anions to form a real structure. Alternatively, you would have to assign only *e.g.* three of the oxygen atoms to the cation, which makes for strange building blocks. Does this lead to restrictions on the type of building blocks you can employ?

Christopher Collins answered: In order to avoid having to merge complexes to form a full 3D structure, as you suggest we work with the sub-modules where only 0–3 anions are assigned to a given cation, which may seem like strange building units to use (I seem to recall that avoiding this very problem was the origin of using the sub-module description used by FUSE). This is (in part) why we have the step to "error check" all of the cation coordination numbers once the full 3D structure is assembled. This also means that the number of building block motifs are restricted, with FUSE generating single atom thick modules. We restricted the sub-modules to be planar, and based upon a grid (as in Fig. 1a) this results in only 8 possible motifs.

Error checking of the full 3D structure avoids the sub-modules being assembled in a non-realistic fashion (*e.g.* with all of the anions clustering in one part of the unit cell). However, we currently only check the number of anions in the

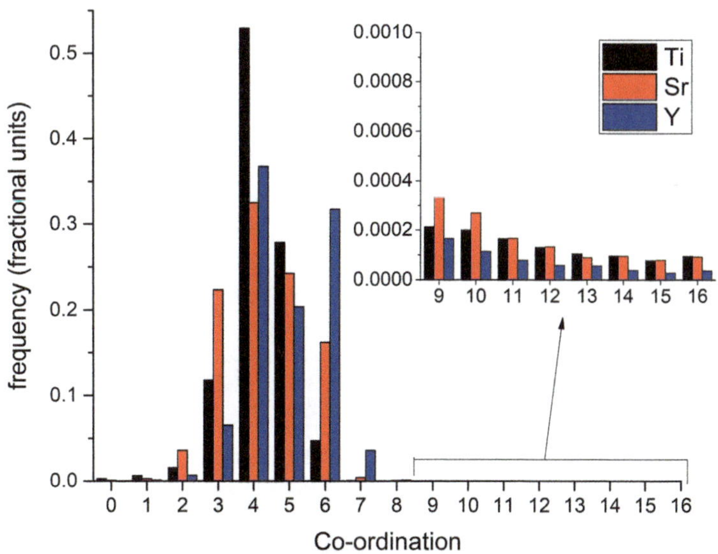

Fig. 2 Fractional occurrence of the coordination numbers between 0–16.

immediate coordination sphere and do not place any further restrictions on how they are geometrically arranged beyond that imposed by the use of the sub-module motifs.

This method of restricting the cation coordinations doesn't appear to result in the types of coordination environments in the resulting geometry optimised structures being too restrictive. Since the meeting, I have had the time to analyse the coordination environments of 924 403 of the structures generated in this work, containing 1.445×10^7 environments. We define our coordination environments by the cation environment's bond valence sum. We define a pair of ions as bonded when the ion pair contributes over 5% of the formal charge to the cation's total bond valence sum, as we have done previously.[1] I have included a bar chart, showing the fractional occurrence of the coordination numbers between 0–16 (see Fig. 2). While there are large peaks for each of the elements, all of the coordinations are non-zero (note, though, that some of these structures will be ones which have ended up in non-physical structures).

1 C. Collins, M. S. Dyer, M. J. Pitcher, G. F. S. Whitehead, M. Zanella, P. Mandal, J. B. Claridge, G. R. Darling and M. J. Rosseinsky, Accelerated Discovery of Two Crystal Structure Types in a Complex Inorganic Phase Field, *Nature*, 2017, **546**, 280–284 (10.1038/nature22374).

David McKay queried: On the phase diagram, you said that the white areas, which appear to be below 35 eV/atom, represent stable structures. Why this energy cut-off?

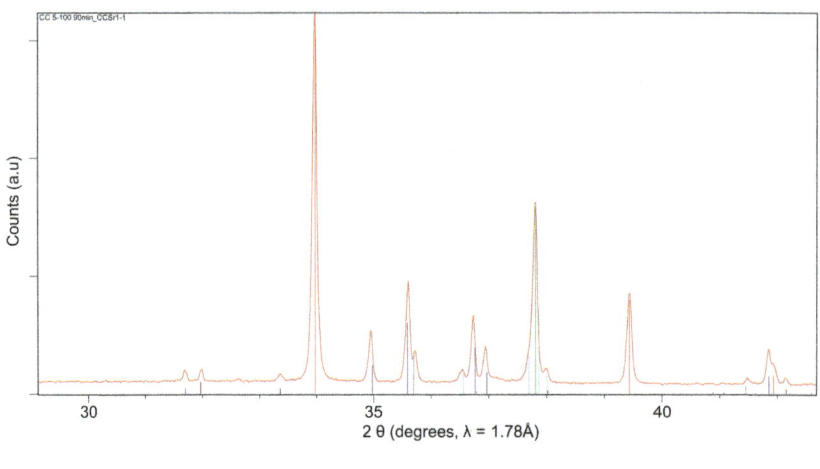

Pattern List:

Compound Name	Chemical Formula
Yttrium Oxide	Y_2O_3
Strontium Yttrium Oxide	SrY_2O_4
Strontium Titanium Oxide	$SrTiO_3$
Yttrium Titanium Oxide	$Y_2Ti_2O_7$
Rutile	TiO_2
Strontium Titanium Oxide	Sr_2TiO_4

Fig. 3 Indexed powder diffraction pattern for sample of Y_2TiSrO_6.

Christopher Collins replied: As stated in the paper, this is the threshold we have found previously for locating experimentally realisable compounds. It is quite likely that this is system dependent.

David McKay asked: How does the result change if this energy cut-off is increased?

Christopher Collins responded: If we start to raise the threshold that we consider, the next composition to come under the cut-off is Y_2TiSrO_6 at 48 meV/atom, which is not known to experimentally form a compound. When we tried this in the lab, we obtained a multi-phase sample containing Y_2O_3, TiO_2, SrY_2O_4, $SrTiO_3$, $Y_2Ti_2O_7$ and Sr_2TiO_4. This confirms that for this phase diagram at least, a cut-off of 35 meV/atom is a reasonable value; however, it is entirely possible that this may vary between phase diagrams. For reference, I have included an indexed powder diffraction pattern (see Fig. 3).

In practice, in the lab we will start synthesis at the lowest energy composition(s) and work up in energy until we obtain samples which only contain mixtures of known phases. So if the phase diagram in the paper were being explored for the first time we would have started with $SrTiO_3$ and finished with Y_2SrTiO_6, experimentally obtaining all of the compounds in the phase diagram.

Sarah L. Price remarked: In your paper, you had only found a structure for Y_2O_3 that was 40 meV above the ICSD crystal structure. In your presentation you showed that you had since adapted your procedure so that it now finds the correct structure. Please can you expand on what you needed to change so that this structure was found by your method? Does the updated procedure also find the correct structure for SrY_2O_4?

Christopher Collins replied: For reference, I've included the slide from my presentation about Y_2O_3 (see Fig. 4). With the method of constructing unit cells in

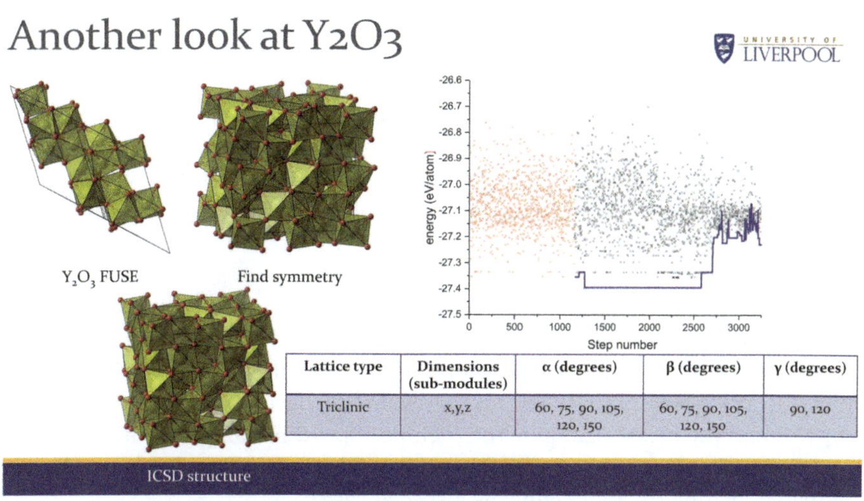

Fig. 4 Slide presented at the meeting about Y_2O_3 and the introduction of triclinic lattices.

the paper, it was not possible to construct the correct structure for Y_2O_3 as this would have required 160 ions, 10 above our limit. It was not possible to construct a 40 ion primitive unit cell, so we learnt that we needed to increase the flexibility of the possible unit cell shapes we could create.

The main change we made (we also made a few bug fixes/minor changes to the search routine) was to add in a triclinic lattice approximation. For the triclinic lattice approximation, we allow the *a*, *b* and *c* directions to be of any length, then the α and β angles are randomly chosen from possible values of 60, 75, 90, 105, 120 and 150 degrees. The γ angle is still chosen to be either 90 or 120 degrees. We retained this restriction, because the choice of γ governs what rules are in use to stack modules along *c*, and if we deviated from these values, it is not so easy to automate the stacking.

These changes allowed FUSE to create a 40 ion cell, which was then optimised in GULP into a primitive structure for Y_2O_3. Running the structure through FINDSYM yields the body-centred *Ia*-3 structure. As of yet we have still been unable to obtain the experimental crystal structure of SrY_2O_4 and are currently investigating why. Currently our best result has yielded a structure within 5 meV/atom of the ICSD structure (*vs.* 9 quoted in the paper), and our hypothesis is that there are a lot of low-energy minima in the system (within 10 meV/atom of the ICSD structure), so it is taking far longer to locate the ICSD structure than at the other compositions within our test phase diagram.

Sarah L. Price asked: Is this a matter of symmetry that's using a triclinic primitive cell to expand to a larger structure with higher symmetry?

Christopher Collins answered: The use of symmetry for tidying up structures after we have completed our search using FUSE is important for arriving at experimental structures. In most cases in our compositional phase diagram, the "raw" structure from FUSE is an odd shape/size in *P*1 symmetry, which FINDSYM then refines to a higher symmetry. Two notable examples of this are for Y_2O_3 and $Y_2Ti_2O_7$: in each case we find a smaller, primitive unit cell, from which FINDSYM then finds the correct, higher symmetry cells. It is also possible, however, to make a larger supercell, which is then reduced to a smaller, higher symmetry cell; this often happens with $SrTiO_3$, for example.

It should be noted, however, that we do not impose any internal symmetry on the structures as they are constructed; if anything, we go out of our way to assemble asymmetric structures by "rattling" all of the atoms off their starting positions, which can then relax into a higher symmetry structure if this is favourable.

Alan Hare communicated: In growing the lattice – besides doubling the structure along a random axis – would there also be value in creating a supercell other than randomly, from as many units as are needed to retain the shape of the unit cell (so for example in the cubic approximation having dimensions 3×, 5×, *et seq.*)? Could you please elaborate the triclinic approximation (as mentioned prior to the presentation)? What internal symmetry rules are imposed, and can they be tested by treating the monoclinic as a special case?

Christopher Collins communicated in reply: In practice there is no reason why we could not allow the unit cell to grow isotropically; as you suggest, it's just something that we have not tried. As for how much value there is to trying this, I simply do not know; I suspect it is something we would have to try and find out! For the triclinic lattice approximation, we allow the a, b and c directions to be of any length, then the α and β angles are randomly chosen from the possible values of 60, 75, 90, 105, 120 and 150 degrees, and the γ angle is still chosen to be either 90 or 120 degrees; this restriction was left in place, as the choice of γ governs what rules are in use to stack modules along c.

As for internal symmetry, we do not impose any, so all of the lattice approximations are purely geometric. It was created this way because if we impose internal symmetry (for example by imposing the lowest symmetry space group for each lattice type), then I suspect this would make the construction too restrictive and make it difficult for FUSE to discover crystal structure types with disorder present, for example perovskites with cation disorder.

Kim E. Jelfs opened a general discussion of the papers by Julian Keupp, German Sastre and Christopher Collins: Do you have any intention of calculating the properties of the structures you have found? Could you bias your search towards structures with interesting properties? Or investigate how the energy relates to the material's properties?

Christopher Collins answered: At the present time, we are not looking to make property predictions for structures we have found, especially for oxides (our main area of interest), the properties can be very sensitive to the structure. As our research is focused on synthetic chemistry, our focus is to use calculations to lead us to new materials. Once we have obtained them and experimentally verified/solved their crystal structures, we can compute/measure properties. From this point we could then use calculations to predict how to best modify them in order to improve their functionality. In the future, however, as we become more confident in the accuracy of our structural predictions, we may well revise this view! This is just our philosophy; there is no reason why one could not combine a method like FUSE with property prediction, especially if the properties are cheap to compute and one is confident enough in the structural predictions; one could very easily bias a search towards specific properties.

Claire S. Adjiman asked: To what extent can one use the desired properties of the MOF, rather than the desired topology, to guide the search?

Julian Keupp answered: Our method (reverse topological approach, RTA, with or without the blueprint optimization with topoFF) aims at a complete generation of all the possible structures by enumerating topologies. The MOF-FF force field energy is used for ranking the stability of the structures. At this point, properties like surface area can also be determined and a systematic search for certain key properties can be performed. However, at the level of structure generation it cannot be used to guide the search.

J. Christian Schön addressed Julian Keupp and German Sastre: Can your methods generate interpenetrating networks?

Julian Keupp and **Rochus Schmid** responded: Yes, of course. We can simply interpenetrate (duplicate and shift) the blueprint and start from there, or use the more general approach reported by Baburin.[1] The difficulty is ranking the energies. An interpenetrated system is usually more stable (due to dispersive interactions) than the non-interpenetrated system in the gas phase. However, to predict whether interpenetration is formed, one would need to include solvent effects present in the solvothermal synthesis conditions. This is possible in principle but up to now we have just used simple ideas like a fixed stabilization energy per accessible surface area to approximate this effect.

1 I. A. Baburin, On the group-theoretical approach to the study of interpenetrating nets, *Acta Crystallogr. A*, 2016, **72**, 366–375.

German Sastre responded: So far our methods have not generated interpenetrated structures. Indeed, interpenetrated structures require that not all atoms in the unit cell are part of the same connectivity tree. Our algorithm is sufficiently general to allow this.

J. Christian Schön asked: Is there any chemical reason why you would not get interpenetration in zeolites? Some MOFs look like they surely have sufficiently large pores, and large pore zeolites should be feasible and kinetically stable once they have been generated.

German Sastre replied: Although catenated metal–organic frameworks are common, this is not the case for zeolites. The catenation requires, geometrically speaking, a large pore in comparison to the thickness of the framework pillars, and this is not usually found in zeolites. I would expect that zeolites containing rings of at least 16 tetrahedral atoms would be necessary, and catenation also seems to appear more frequently in structures with monodimensional pores. However, zeolites containing extra-large pores do also frequently contain smaller pores in perpendicular directions. From the – currently – 11 zeolites containing such large pores, only two of them (IFO and VFI[1,2]) have a monodimensional channel. These would be my suggestion to try to observe catenation in zeolites; however the structural arguments above explain why this has yet not been observed, to the best of my knowledge.

1 www.iza-structure.org/IZA-SC/framework.php?STC=IFO
2 www.iza-structure.org/IZA-SC/framework.php?STC=VFI

Marcus Neumann remarked: In organic CSP we struggle to get from the knowledge of the crystal energy landscape to the crystallization of new forms. Do you know of success stories for rational crystallization experiment design in inorganic CSP?

Julian Keupp and **Rochus Schmid** answered: In the context of our topoFF method within RTA and the prediction of "soft" framework materials, things are probably quite different from classic inorganic CSP of hard materials like oxides, *etc.* To some extent, the notion of reticular chemistry[1] essentially means to design new systems in a rational way, based on the knowledge of the shape of the building blocks.

1 O. M. Yaghi, M. O'Keeffe, N. W. Ockwig, H. K. Chae, M. Eddaoudi and J. Kim, Reticular synthesis and the design of new materials, *Nature*, 2003, **423**, 705–714.

Christopher Collins answered: I know that others have commented on this and highlighted examples where research groups have been able to identify exact crystal structures of new compounds which have subsequently been synthesised. For our group we deem a "success" to be calculations which lead us to the synthesis of new compounds. As stated in our paper, our aim is primarily to be able to predict new chemical compositions; we do not explicitly aim to predict the crystal structure. As we are looking to synthesise compounds with complex compositions, we do not believe it is currently pragmatic to attempt to perfectly predict structures, hence our use of probe structures.

Providing a compositional target for synthesis sufficiently accelerates the discovery process, since finding the correct compositional regions to study is often an extremely difficult step in large phase spaces. Crystallography of ionic solids is currently good enough that experimental structure determination after computational composition prediction is currently the most pragmatic solution. An example of our philosophy was our computational composition exploration of the Y^{3+}–Sr^{2+}–Ca^{2+}–Ga^{3+}–O^{2-} phase diagram,[1] where synthesis around a target composition resulted in the synthetic identification of two new compounds in under 40 samples, both with new crystal structures, neither of which we would expect to be able to predict *a priori* at the current time, given both their structural complexity and disorder.

1 C. Collins, M. S. Dyer, M. J. Pitcher, G. F. S. Whitehead, M. Zanella, P. Mandal, J. B. Claridge, G. R. Darling and M. J. Rosseinsky, Accelerated Discovery of Two Crystal Structure Types in a Complex Inorganic Phase Field, *Nature*, 2017, **546**, 280–284 (10.1038/nature22374).

German Sastre answered: The stabilities of $(ZnO)_x$ nanoclusters could be an example, although they are not crystals but rather 0-D materials. Zeolites are not an example since here the challenge is not to know the crystal structure of SiO_2, unlike organic drugs where an important challenge is identifying the stable crystalline structures of each compound. Boron oxides, with a paper in this *Faraday Discussion*, are a good example of CSP inorganic materials. The structures of perovskites could be another example. Tungsten borides are a recent and elegant example where the structures and stoichiometries that are stable at different *P*,*T* conditions have been elegantly predicted in Artem Oganov's group.

J. Christian Schön enquired: In the field of inorganic compounds, there are two areas where there have been quite a number of successful syntheses after(!) prediction. One is high-pressure structure prediction: see *e.g.* work by Oganov *et al.* on various high-pressure structures,[1] or work by Cancarevic *et al.* on alkali metal sulfides.[2–9] The other is the synthesis of metastable variants in systems like alkali halides,[10–12] where difficult synthesis routes like the low-temperature atom-beam-deposition method[13] were needed to access these modifications. Nevertheless, the synthesis methods in solid state chemistry are noticeably less developed than those in organic molecular chemistry regarding the targeted synthesis of metastable compounds, which helps to explain why for many years the polymorph richness of bulk inorganic compounds has been gravely underestimated.

1 A. R. Oganov, C. W. Glass and S. Ono, *Earth Planet. Sci. Lett.*, 2006, **241**, 95–103.
2 J. C. Schön, M. A. C. Wevers and M. Jansen, *J. Mater. Chem.*, 2001, **11**, 69–77.
3 J. C. Schön, Z. Cancarevic and M. Jansen, *J. Chem. Phys.*, 2004, **121**, 2289–2304.
4 Z. Cancarevic, J. C. Schön and M. Jansen, *Phys. Rev. B*, 2006, **73**, 224114.
5 A. Grzechnik, A. Vegas, K. Syassen, I. Loa, M. Hanfland and M. Jansen, *J. Solid State Chem.*, 2000, **154**, 603–611.
6 A. Vegas, A. Grzechnik, K. Syassen, I. Loa, M. Hanfland and M. Jansen, *Acta Crystallogr. B*, 2001, **57**, 151–156.
7 A. Vegas, A. Grzechnik, M. Hanfland, C. Mühle and M. Jansen, *Solid State Sci.*, 2002, **4**, 1077–1081.
8 D. Santamaria-Perez, A. Vegas, C. Mühle and M. Jansen, *Acta Crystallogr. B*, 2011, **67**, 109–115.
9 D. Santamaria-Perez, A. Vegas, C. Mühle and M. Jansen, *J. Chem. Phys.*, 2011, **135**, 054511.
10 Z. Cancarevic, J. C. Schön and M. Jansen, *Chem. Asian J.*, 2008, **3**, 561–572.
11 Y. Liebold-Ribeiro, D. Fischer and M. Jansen, *Angew. Chem.*, 2008, **47**, 4428–4431.
12 A. Bach, D. Fischer and M. Jansen, *Z. Anorg. Allg. Chem.*, 2009, **635**, 2406–2409.
13 D. Fischer and M. Jansen, *J. Am. Chem. Soc.*, 2002, **124**, 3488–3489.

Christopher Collins responded: Thank you for your comments regarding successes of crystal structure prediction and subsequent synthesis.

Marcus Neumann addressed Christopher Collins, German Sastre and Artem R. Oganov: This question is about the economics and social dynamics of CSP. A previous comment insinuated that there is not much economic/industrial value in inorganic CSP. In organic CSP, the prediction of late-appearing or disappearing polymorph events for pharmaceutical compounds provides a strong economic incentive. At this meeting, inorganic CSP is very strongly represented, which seems to contradict the supposed lesser economic value of inorganic CSP. Where do you see the economic high-value applications of inorganic CSP?

Christopher Collins replied: The value of inorganic CSP is in the discovery of new materials. In order to increase the performance of devices, we constantly need to develop new materials from which to construct them. Previously, new materials came from purely synthetic research programs, where a number of revolutionary materials have been discovered, such as the discovery of $LiCoO_2$ (subsequently doped with Mn and Ni), which enabled the wide-scale use of the Li-ion batteries we now take for granted in our everyday lives. Inorganic CSP has the potential to dramatically accelerate the pace at which we are able to discover new materials. So the economic value (from our perspective) is in driving this initial discovery. As for the application, it should be able to contribute to any area of inorganic chemistry where materials discovery is the bottleneck preventing technological progress.

Artem R. Oganov replied: Inorganic CSP is essential for computational materials discovery. This is the future of numerous technologies, where the lack of new materials is the bottleneck. Let me just mention the acute need in new thermoelectric, photovoltaic, thermoresistant, superconducting materials, and the relentless search for novel materials for Li-ion batteries, fuel cells, *etc.* The economic and societal impact is, of course, huge.

German Sastre replied: Inorganic CSP also has an important economic impact in the case of zeolites. They were introduced in the late sixties by Plank and Rosinski, working at Socony/Mobil, in fluid catalytic oil cracking, since they needed catalysts with controlled pores of molecular dimensions in order to obtain higher selectivity and less coking than the previously used amorphous silica alumina. However, knowing the crystal structure is not necessary in order to patent the use of zeolites as catalysts or selective adsorbents. Hence the many industrial applications of zeolites is not a reason to justify the economic impact of CSP in this family of inorganic materials. However, knowing the structure is key to linking microporosity with properties and ultimately with industrial applications. In the last two decades, computational techniques have had the goal of being able to predict the conditions for the synthesis of new zeolites with specific properties (selected by design). This has only been claimed in a few cases so far,[1] but it is likely that this could be achieved soon in a more systematic and effective way.

1 T. M. Davis, A. T. Liu, C. M. Lew, D. Xie, A. I. Benin, S. Elomari, S. I. Zones and M. W. Deem, *Chem. Mater.*, 2016, **28**, 708–711, DOI: 10.1021/acs.chemmater.5b04578.

Michael Ruggiero asked: How important is the role of the solvent for predicting the right structure, *i.e.* structures stabilised by trapped solvent? How likely is your technique to incorrectly predict the experimentally observed structures due to solvent effects or captured gases? Can this be overcome computationally?

German Sastre answered: This is a very good question. The role of the solvent is important but it has usually been neglected in computer simulations regarding the synthesis of zeolites. Usually the solvent is water, with the hydrothermal synthesis being the most employed method of synthesis, but there is also the ionothermal method, using ionic liquids, where the solvent is also the template at the same time. Focusing on water, which is by far the most commonly employed solvent, water molecules alone (without organic template molecules) can certainly stabilise small cavities, although most of these molecules will be evacuated after the usual calcination, whose aim is to remove the template in order to activate the material by making the pores accessible. Water can also be coordinated to template molecules in large cavities. Water can also be responsible for the presence of internal silanol groups forming an internal connectivity defect and clearly affecting the stability of the final material. Water is also responsible, when the synthesis is at pH > 7 and when using cationic templates, for the presence of OH^- anions that neutralise the template and become stabilised in – presumably – small cavities. Finally, the commonly employed stabilisation of the material by zeolite–template interactions should also contain another term, usually neglected, for the desolvation energy between the template and the solvent (water molecules). A few of these topics are discussed in a previous publication.[1]

1 Y. G. Bushuev and G. Sastre, *J. Phys. Chem. C*, 2009, **113**, 10877–10886, DOI: 10.1021/jp9013306.

Sten Nilsson Lill queried: How flexible are these MOFs and zeolites? Do they breathe? Has this been studied with any molecular dynamics?

Julian Keupp and **Rochus Schmid** responded: In general MOFs can breathe (*e.g.* MIL-53), and this has also been investigated extensively by MD simulations. If you are referring to the particular MOFs we constructed here with topoFF, we have not investigated that in the context of the current paper.

German Sastre responded: Breathing is a kind of long-amplitude vibration that can be attributed to MOFs, but not to zeolites, whose vibrations are typically narrower. This also implies that zeolites show much stronger thermodynamic stability and are better suited for those catalytic processes taking place at temperatures higher than, say, 300 °C. The flexibility of zeolites is, nevertheless, a very important process that often contributes to enhancing their properties. This has been explored since the pioneering work by Deem, Creighton and Newsam in 1992.[1] The flexibility of zeolites contributes to a certain framework deformation when molecules of similar micropore size diffuse, resulting in a minimisation of attractive van der Waals interactions, and enhancing the diffusivity. This is one of the foundations of the concept of enhanced diffusivity, introduced by Eric Derouane, which plays a crucial role in the choice of specific zeolites for selective adsorption processes.

1 M. W. Deem, J. M. Newsam and J. A. Creighton, Fluctuations in Zeolite Aperture Dimensions Simulated by Crystal Dynamics, *J. Am. Chem. Soc.*, 1992, **114**, 7198–7207.

Scott M. Woodley remarked: One of the problems with using interatomic potentials is that you may find a suitable set for describing the framework and another set for the template molecules, but not necessarily the set that also includes the interaction between the framework and the template. In your talk, you kindly referenced a number of my publications in *Phys. Chem. Chem. Phys.*[1–3] These papers report how I have adapted a genetic, or evolutionary approach for crystal structure prediction (ref. 20 in your paper[4]) and demonstrated how to predict (starting from an initial population of random structures) the framework of microporous structures where the dimensions of the volume within the unit cell that the template molecules occupy are assumed to be known (from experimental measurements) or targeted (you want to predict the framework that has these microporous voids or channels). One of the advantages of any of these approaches is that the interatomic potentials for the template molecules are not required. If a new framework is predicted and a feasible template molecule is then targeted, one could of course then use the code developed by Lewis *et al.*,[5] which uses concepts originally applied in drug design where molecules are "grown" computationally to fit a receptor site. In template design, organic cations are grown to match a target microporous structure. Their software was used to find a template for the microporous aluminophosphate DAF-5. The predicted template succeeded in synthesizing the material rapidly as a pure phase, whereas previous lengthy syntheses had produced impure materials. As a bonus, the predicted site of the template within the host material was verified using microcrystalline single-crystal diffraction techniques.[6]

1 S. M. Woodley, P. D. Battle, J. D. Gale and C. R. A. Catlow, *Phys. Chem. Chem. Phys.*, 2004, **6**, 1815–1822.
2 S. M. Woodley, *Phys. Chem. Chem. Phys.*, 2004, **6**, 1823–1829.
3 S. M. Woodley, *Phys. Chem. Chem. Phys.*, 2007, **9**, 1070–1077.

4 S. M. Woodley, P. D. Battle, J. D. Gale and C. R. A. Catlow, *Phys. Chem. Chem. Phys.*, 1999, **1**, 2535–2542.
5 D. W. Lewis, C. R. A. Catlow, J. M. Thomas, D. J. Willock and G. J. Hutchings, De novo design of structure-directing agents for the synthesis of microporous solids, *Nature*, 1996, **382**, 604–606.
6 G. Sankar *et al.*, Structure of templated microcrystalline DAF-5 ($Co_{0.28}Al_{0.72}PO_4C_{10}H_{20}N_2$) determined by synchrotron-based diffraction methods, *Chem. Commun.*, 1998, **1**, 117–118.

German Sastre replied: The force field necessary to describe the template–zeolite interactions can (and should) be tested by comparing to adsorption energies obtained experimentally. Since this force field is independent of the two other force fields that describe the zeolite framework and the template, it should be possible to select a good choice of three force fields (zeolite, template, and template–zeolite) that describe the system accurately. A good choice of force fields would improve the success of the strategy that you describe in the second part of your comment, where software such as that by Lewis and Willock can be employed in order to "build" the template inside the micropore. Regarding the first part of your comment, your approach based on genetic algorithms is certainly able to predict structures containing predefined voids (mimicking the space filled by the template) without the need for force fields. One advantage is that this is a faster algorithm in terms of computer time and, when big data is at play, as is the case, this is a crucial advantage.

Virginia Burger asked: Zeolites are often found in nature with contaminants and can take thousands to millions of years to crystallize in nature. In the lab, are contaminants/impurities considered when synthesizing or predicting zeolites?

Julian Keupp and **Rochus Schmid** responded: Within MOFs, similar to zeolites, defects and impurities can dominate the properties (the surface must also be considered as a defect). Theoretical simulations still mostly focus on the ideal system in periodic boundary conditions, but studies of defects[1] and domains[2] (as well as surfaces[3]) have started to appear.

1 J. P. Dürholt, J. Keupp and R. Schmid, The Impact of Mesopores on the Mechanical Stability of HKUST-1: A Multiscale Investigation, *Eur. J. Inorg. Chem.*, 2016, 4517–4523.
2 A. K. Cheetham, T. D. Bennett, F.-X. Coudert and A. L. Goodwin, Defects and disorder in metal–organic frameworks, *Dalton Trans.*, 2016, **45**, 4113–4126.
3 S. Amirjalayer, M. Tafipolsky and R. Schmid, Surface Termination of the Metal–Organic Framework HKUST-1: A Theoretical Investigation, *J. Phys. Chem. Lett.*, 2014, **5**, 3206–3210.

German Sastre responded: We could consider "contaminants" in two senses. In a broad sense, zeolites do allow a very large range of chemical compositions that depend on each particular structure. The zeolites initially defined as "aluminosilicates" are now also called "zeotypes" in order to express other chemical compositions, where tetrahedral atoms also include phosphorus, germanium, titanium, boron, magnesium, zinc, gallium and beryllium. Oxygen can also be replaced by sulfur and nitrogen. The presence of these atoms has an important effect that drives the synthesis towards different structures than when they are absent. In a more strict sense, it has been reported that several synthetic procedures cannot be reproduced when the source of silica or alumina is changed, depending on the manufacturer, or even depending on the batch. It is often not clear in which ways impurities drive the synthesis towards a particular zeolite structure. Finally, an active research topic is "seeding". The presence of certain small crystallites of a zeolite can somehow influence the synthesis towards the appearance of another zeolite, not necessarily the one that has been seeded. Crystallites can dissolve into specific building units that can then be the starting point for a new structure. Finding rules and correlations between the seeding and the final outcome is an active topic of research.

Sarah L. Price remarked: I believe that zeolites are very important industrial materials, and that there was patent litigation on whether a zeolite contained Al or not, whose financial implications were very significant relative to those of pharmaceutical polymorphs.

German Sastre replied: ZSM-5 zeolite was patented by researchers at Mobil Oil Co. in 1972[1] as an aluminosilicate zeolite with a range of Al contents and without explicit reference to a pure silica phase that was later patented by researchers at Union Carbide in 1977.[2] A short notice in *Nature* covered the news.[3] Rather than the final legal result, it is important to note how minimal details on the composition can have an enormous impact in the industrial arena. It is important to investigate new ways to synthesise relevant zeolites and circumvent patent protections. An added aspect is that the structure of ZSM-5 was not solved (or published) until 1978 by Kokotailo, Lawton, Olson and Meier,[4] hence the structure is not necessary to patent the material.

1 ExxonMobil Oil Corp., *US Pat.*, US3702886A, 1972.
2 Union Carbide Corp., *US Pat.*, US4061724A, 1977.
3 S. Budiansky, *Nature*, 1982, **300**, 309.
4 G. T. Kokotailo *et al.*, *Nature*, 1978, **272**, 437–438.

Conflicts of interest

Asbjoern Burow and Marcus Neumann work for a company that develops software for organic crystal structure prediction. There are no other conflicts to declare.

Faraday Discussions

PAPER

Identifying pragmatic quasi-harmonic electronic structure approaches for modeling molecular crystal thermal expansion†

Jessica L. McKinley and Gregory J. O. Beran *

Received 26th February 2018, Accepted 5th March 2018
DOI: 10.1039/c8fd00048d

Quasi-harmonic approaches provide an economical route to modeling the temperature dependence of molecular crystal structures and properties. Several studies have demonstrated good performance of these models, at least for rigid molecules, when using fragment-based approaches with correlated wavefunction techniques. Many others have found success employing dispersion-corrected density functional theory (DFT). Here, a hierarchy of models in which the energies, geometries, and phonons are computed either with correlated methods or DFT are examined to identify which combinations produce useful predictions for properties such as the molar volume, enthalpy, and entropy as a function of temperature. The results demonstrate that refining DFT geometries and phonons with single-point energies based on dispersion-corrected second-order Møller–Plesset perturbation theory can provide clear improvements in the molar volumes and enthalpies compared to those obtained from DFT alone. Predicted entropies, which are governed by vibrational contributions, benefit less clearly from the hybrid schemes. Using these hybrid techniques, the room-temperature thermochemistry of acetaminophen (paracetamol) is predicted to address the discrepancy between two experimental sublimation enthalpy measurements.

1 Introduction

Molecular crystals occur in many areas of chemistry, including pharmaceuticals, organic semi-conductor materials, foods, and explosives. Crystal structure and polymorphism, or the tendency for molecules to adopt multiple distinct crystalline packing motifs, can have substantial impacts on the macroscopic properties of these materials. The difficulty in identifying or engineering crystal forms experimentally has generated considerable interest in predicting crystal

Department of Chemistry, University of California, Riverside, California 92521, USA. E-mail: gregory.beran@ucr.edu

† Electronic supplementary information (ESI) available: Sample free energy curves, examination of the negative thermal expansion in ice, and additional sensitivity analysis. See DOI: 10.1039/c8fd00048d

structures and properties from first principles.[1–3] Substantial progress has been made toward successful crystal structure prediction, as evidenced by the successes in the last few Blind Tests of Crystal Structure Prediction.[4–6]

Most molecular crystal modeling focuses on the 0 K electronic (internal) energy rather than the finite-temperature free energy.[7] However, temperature can play an important role in many molecular crystal properties. Crystals typically undergo thermal expansion, often at a rate of 0.8–2.5% expansion per 100 K.[8] Even larger expansion occurs in weakly bound crystals. As the crystal expands, its mechanical properties change. Bulk moduli often decrease by 50% and shear moduli by 30–50% at temperatures approaching the melting point.[8]

The Gibbs free energy also varies with temperature, with room-temperature values typically differing by 1–4 kJ mol^{-1} from the 0 K electronic energy.[8,9] Thermal expansion typically softens the lattice vibrations, resulting in a larger Helmholtz vibrational entropy, while simultaneously decreasing the lattice energy. Early investigations suggested that entropic effects were seldom large enough to reverse polymorph stability.[10] Though a significant fraction of the finite-temperature contribution cancels when considering polymorph energy differences, a more recent survey noted that at least 20% of polymorph pairs are enantiotropically related and reverse their relative thermodynamic stabilities upon heating.[8] Many other properties, including vibrational[11] and nuclear magnetic resonance spectroscopic properties, also vary with the unit cell size.

Various strategies exist for incorporating finite temperature contributions into the model. Molecular dynamics techniques provide natural methods for treating these effects.[12–21] However, obtaining force fields capable of describing the subtle energetic balances that occur in molecular crystal problems is not always easy. Alternatively, one can apply statistical thermodynamic harmonic or quasi-harmonic corrections to static models.[11] Harmonic phonon contributions are increasingly being employed in crystal structure prediction,[6] and they capture a significant fraction of the vibrational contribution to the free energy.[8,22] Harmonic vibrational contributions impact the polymorph stability ordering in glycine,[23,24] oxalyl dihydrazide,[25] aspirin,[26] and modern drug targets,[27] for example.

However, to capture thermal expansion effects, one must address how the phonons change with crystal volume. The quasi-harmonic approximation provides a straightforward means for doing so and therefore for predicting how the Gibbs free energy depends on temperature and pressure. The quasi-harmonic approximation has long been used in force field studies.[28–33] Recent DFT work has examined how free energy impacts structure rankings in crystal structure prediction.[34,35] It has sought to reproduce the structural and thermochemical properties of crystalline ammonia,[11] urea,[36] and carbamazepine (using density functional tight binding theory).[37]

These DFT-based quasi-harmonic studies have been successful, and wavefunction-based techniques are often too computationally demanding for quasi-harmonic crystal studies. However, fragment approaches such as the hybrid many-body interaction (HMBI) model enable the application of high-level, correlated wavefunction techniques that would otherwise be computationally prohibitive to molecular crystal problems. HMBI combines a quantum mechanical (QM) treatment of the unit cell monomers and their short-range dimer interactions with a classical polarizable force field treatment of longer-range and many-body effects.[38–41] We have demonstrated that the quasi-harmonic HMBI[42] with large-basis second-order Møller–Plesset perturbation theory (MP2) and/or coupled clusters singles and doubles with perturbative triples (CCSD(T)) allows the prediction of small-molecule structures and properties in excellent agreement with experiments. For example, it predicts the thermal expansion of carbon dioxide phase I to within 2–3%, the sublimation enthalpy to within 1–2 kJ mol^{-1}, and the sublimation temperature to within 3 K.[9,43] By comparing and contrasting the excellent agreement between the theoretical and experimental structures and Raman spectra for several different crystalline phases of carbon dioxide, we argued that the long-accepted structure for phase III carbon dioxide was incorrect, and that phases III and VII were actually identical.[44] Furthermore, these same techniques enabled the successful prediction of crystalline methanol thermochemistry.[45] More significantly, they allow *ab initio* prediction of the polymorph phase diagram with accuracy corresponding to Gibbs free energy errors of ~0.5 kJ mol^{-1} or less.[46] Other fragment-based studies have similarly employed the quasi-harmonic approximation to study thermochemical properties in a much broader range of crystals.[47,48]

However, even with fragment techniques, studies employing correlated wavefunction methods are computationally demanding. Predicting the methanol phase diagram entirely with MP2 and CCSD(T) requires several hundred thousand central processing unit hours, for example. Performing the same level of calculations on a typical pharmaceutical crystal would be impractical. Is there a useful middle ground, which achieves most of the accuracy found above, albeit at a much lower computational cost?

A quasi-harmonic calculation involves three main ingredients: energies, crystal geometries, and phonon frequencies. This study investigates how accurately one must compute each of those factors to accurately predict how crystal structures vary with temperature. Can one, for example, replace a computationally expensive MP2 treatment of phonon frequencies with a less-demanding one from periodic DFT? What about the geometries? Or, taken from the opposite perspective, in which dispersion-corrected DFT models are the baseline: How much do the predicted finite-temperature structures or thermochemistries benefit from the additional computational effort invested in replacing some or all of these quasi-harmonic approximation components with results obtained from correlated wave function methods? Even inexpensive semi-empirical density functional tight binding models reproduce thermal expansion in carbamazepine fairly well,[37] for example.

Here, to improve our understanding of how the performance of correlated wavefunction and DFT methods compare and the potential benefits of combining them in different ways, we examine the thermal expansion in several small-molecule crystals, systematically replacing the MP2 (or better) treatment of the energies, structures, and/or phonons with dispersion-corrected DFT treatment. We find that while portions of the calculations can often be performed with periodic density functional theory, there are clear benefits to including contributions from wavefunction techniques. Based on the results from the small molecule systems, quasi-harmonic calculations are performed for the pharmaceutical acetaminophen (paracetamol) both to predict the thermal expansion and to investigate a discrepancy between two experimentally reported enthalpies of sublimation.

2 Theory

2.1 Quasi-harmonic approximation

The Gibbs free energy is required to model molecular crystals at finite temperatures and pressures. From statistical thermodynamics, the Gibbs free energy combines the electronic internal energy U_{el}, the Helmholtz vibrational free energy F_{vib}, and a pressure–volume (PV) contribution:

$$G(T,P) = U_{el} + F_{vib}(T) + PV \tag{1}$$

In crystals at ambient pressure, the PV term contributes negligibly.

The internal electronic energy U_{el} is computed here *via* the fragment-based hybrid many-body interaction model (HMBI). HMBI treats individual molecules in the unit cell and their short-range pairwise interactions quantum mechanically (QM), while longer-range and many-body interactions are typically approximated using a classical molecular mechanics (MM) polarizable force field.

$$U_{el}^{HMBI} = E_{1\text{-body}}^{QM} + E_{SR\ 2\text{-body}}^{QM} + E_{LR\ 2\text{-body}}^{MM} + E_{many\text{-body}}^{MM} \tag{2}$$

In some cases, the MM terms are computed from periodic Hartree–Fock (HF) instead, in which case HMBI is equivalent to the method of increments.[49–51]

The Helmholtz vibrational free energy is computed from standard harmonic oscillator vibrational partition functions as:

$$F_{vib}(T) = N_a \sum_i \left(\frac{\hbar\omega_i}{2} + k_b T \ln\left[1 - \exp\left(-\frac{\hbar\omega_i}{k_b T}\right)\right]\right) \tag{3}$$

where N_a is Avogadro's number, $\hbar$ is Planck's constant, k_b is the Boltzmann constant, and ω_i is the vibrational frequency of mode i. The first term corresponds to the zero-point vibrational contribution, while the second gives the thermal vibrational contribution. Here, phonons are evaluated only at the Γ point.

To circumvent the high computational cost of computing the phonons repeatedly while minimizing $G(T,P)$ for the given thermodynamic conditions, the vibrational contribution can be approximated *via* the quasi-harmonic approximation, which estimates how the phonon frequencies and F_{vib} vary with the unit cell volume. Mode-specific Grüneisen parameters γ_i are employed to estimate how each individual phonon's frequency varies with volume:

$$\gamma_i = -\left(\frac{\partial \ln \omega_i}{\partial \ln V}\right) \tag{4}$$

Integrating this equation yields

$$\omega_i = \omega_i^{ref}\left(\frac{V}{V^{ref}}\right)^{-\gamma_i} \tag{5}$$

The reference volumes and phonon frequencies are obtained for the crystal structures which minimize the electronic energy U_{el}, and the Grüneisen parameters are obtained *via* the finite difference of the phonon frequencies with respect to changes in the unit cell volume about the reference structure.

2.2 Hierarchy of approximations

The key question addressed in this study is how accurately one needs to compute each ingredient in the quasi-harmonic model: the energies, crystal geometries, and phonon frequencies. Previous work has demonstrated that evaluating all three ingredients with HMBI using large-basis MP2 often performs very well (subject to the well-known limitations of MP2 (ref. 51)), and even better results are obtained if the single-point energies are refined with a CCSD(T) treatment of the HMBI monomers and dimers. However, large-basis correlated wave function method calculations can be computationally expensive. Here, a systematic hierarchy of four different tiers of approximation is adopted, as summarized in Table 1.

In Tier 1, all calculations are performed at the "High" level of theory—optimizing the structures, computing the phonons, and in some cases refining the single-point energies with correlated wavefunction methods *via* HMBI. Single-point energy refinement might include using larger basis sets, employing CCSD(T) or MP2C instead of MP2, or calculating the many-body contributions from periodic HF instead of AMOEBA. AMOEBA models electrostatics with atom-centered multipole moments up to quadrupoles, polarization with self-consistently induced dipoles and a Thole damping model, and van der Waals *via* a buffered 14-7 potential.[52] Although Tier 1 performs well,[9,43–46] it is computationally demanding and quickly becomes impractical beyond several dozen atoms per unit cell (assuming $P2_1/c$ symmetry or similar).

Tier 2 replaces the single most computationally demanding portion of the calculation, the HMBI phonon frequency evaluation, with phonons at the "Low" level, periodic DFT. Because harmonic frequency calculations require that the system geometry be at a stationary point of the potential energy surface with respect to the atomic positions, Tier 2 structures optimized with HMBI then have their atomic positions relaxed with DFT while the lattice parameters are held fixed at their HMBI values. This ensures real vibrational frequencies while maintaining the (hopefully) more reliable HMBI unit cell. Because the unit cell is constrained, DFT relaxation typically alters the atomic positions only slightly.

A similar approach was previously used to predict the pressure-dependent Raman spectra for several phases of carbon dioxide in excellent agreement with experiments.[44] In that instance, the unit cell was first predicted at the quasi-harmonic HMBI MP2/CBS limit, and then the atomic positions were relaxed and the Raman spectrum predicted at the HMBI MP2/aug-cc-pVDZ level to reduce

Table 1 The four model tiers considered in this work. Tier 1 is the most computationally demanding, and higher tiers subsequently reduce the computational cost by replacing portions of calculations at the "High" HMBI level with faster ones at the "Low" level of theory, DFT. HMBI corresponds to employing MP2 or other correlated wavefunction methods for monomer and dimer treatment, plus AMOEBA or periodic HF for many-body contributions

Tier	Energies	Structures	Frequencies
1	HMBI	HMBI	HMBI
2	HMBI	HMBI	DFT
3	HMBI	DFT	DFT
4	DFT	DFT	DFT

the computational cost of the phonon evaluation. Constraining the cell at the dimensions predicted with the large basis set helped ensure that the key lattice phonon modes were reproduced correctly despite the small basis set.

Tier 3 seeks to reduce the computational costs further, performing all structure optimizations and phonon frequency evaluations with periodic DFT. Only single-point energies are computed from wave function methods with HMBI. Depending on the electronic structure method used for monomers and dimers in the HMBI single-point energies, the computational cost of Tier 3 can be only moderately higher than that of a pure DFT calculation (Tier 4).

Our recent studies of crystalline methanol[45,46] considered a complete basis set (CBS) CCSD(T) HMBI treatment on geometries from either MP2/CBS HMBI (Tier 1) or periodic PBE-D3 calculations (Tier 3). In the initial work,[45] it was unclear how much better Tier 1 was than Tier 3. For example, the sublimation enthalpy was more accurate when computed from MP2 structures, while the thermal expansion agreed more closely with experiments when using the DFT structures. Subsequently, we recognized that AMOEBA was overestimating the many-body contribution. Replacing that AMOEBA contribution with one computed from periodic HF in the final single-point energies significantly improved the quality of the predictions using the MP2 geometries, while those using the PBE-D3 geometries became slightly worse.[46] In the present work, the examination of several additional crystals will provide further insight into the performance of Tier 3 models.

Finally, Tier 4 performs all calculations with periodic DFT. Many quasi-harmonic studies in the literature already employ this approach. Here, Tier 4 provides a baseline for assessing what, if any, improvement one finds upon incorporating energies, structures, or phonons from correlated wavefunction techniques on top of DFT. Of course, many different density functionals and dispersion corrections exist, and the results will depend somewhat on the particular choice of model.

2.3 Computational details

To evaluate the performance of the different hybrid tiers described in Section 2.2, we return to the four small-molecule crystals whose thermal expansion was previously investigated with Tier 1:[9,43] carbon dioxide phase I, ice Ih, acetic acid (orthorhombic phase), and α imidazole (Fig. 1). The initial crystal structures were taken from a low-temperature experimental crystal structure of carbon dioxide,[53] a zero net dipole 16-molecule supercell of ice,[54] a 278 K single crystal X-ray diffraction structure of acetic acid (Cambridge Crystal Structure Database (CSD) reference code ACETAC01),[55] and a 103 K neutron scattering structure of imidazole (IMAZOL06).[56] To investigate the performance for larger species, results are also presented for acetaminophen form I (monoclinic form), starting from a 20 K neutron scattering structure (HXACAN13).[57] Experimental temperature-dependent volume data was found in the literature for carbon dioxide,[58] ice,[59] acetic acid,[55,60,115] and imidazole.[56,61,62,116] Enthalpy and entropy data for carbon dioxide,[63–66] ice,[67–69] acetic acid,[68,70,71] and imidazole[72–74] were taken or derived from experimental data, as described previously.[9]

All Tier 1 results here were taken from our earlier work in ref. 9. The results were obtained by minimizing the Gibbs free energy (eqn (1)) on the fly at each temperature with quasi-harmonic evaluation of the phonon contribution. In that

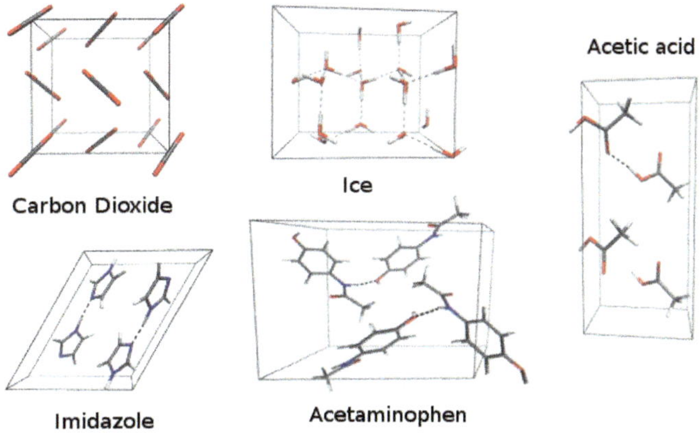

Fig. 1 The crystals modeled in this study.

work, the Grüneisen parameters were computed by isotropically expanding and contracting the lattice by 10 Å^3 in order to approximate the derivative in eqn (4) *via* the finite difference. Ice and carbon dioxide do in fact expand nearly isotropically. The other two crystals exhibit slightly more anisotropic expansion, so one might prefer an approach that estimates the Grüneisen parameters based on anisotropic volume changes. Nevertheless, agreement between the predicted and experimental volumes is fairly reasonable. Note that the PV term contributes negligibly at ambient pressure and was neglected in all cases. Also, slight irregularities occur in some of the Tier 1 thermal expansion results due to numerical convergence issues on the generally flat energy surfaces.

For the new Tier 2–4 results here, a slightly different approach is taken. First, energy *versus* volume curves $E(V)$ were mapped out as a function of volume, by minimizing the HMBI or DFT energy of the cell under positive (cell compression) or negative pressure (cell expansion). The resulting $E(V)$ curve was fitted to the Murnaghan equation of state,

$$E(V) = E_0 + \frac{B_0 V}{B'_0}\left[\frac{(V_0/V)^{B'_0}}{B'_0 - 1} + 1\right] - \frac{B_0 V_0}{B'_0 - 1} \tag{6}$$

where E_0, V_0, B_0, and B'_0 are the fit parameters. E_0 gives the electronic energy at the minimum, V_0 is the molar volume at the minimum energy, B_0 is the bulk modulus, and B'_0 is the dimensionless first derivative of the bulk modulus with respect to pressure.

Next, the reference phonon frequencies and Grüneisen parameters were computed. Instead of isotropically scaling the crystal volumes in the finite difference estimate of eqn (4), frequencies were computed on the positive and negative pressure structures surrounding the electronic energy minimum. This quasi-anisotropic estimate for the Grüneisen parameters should provide a better description for how the phonons vary with volume. Using eqn (3) and (5), the Helmholtz vibrational free energy F_{vib} was calculated as a function of volume for various temperatures and cubically splined. Care was taken to ensure that explicit data points extend outside the actual volume range to avoid spline artifacts in the region of interest. Summing the fitted $E(V)$ and splined F_{vib} contributions (neglecting the PV term at ambient pressure) gives the Gibbs free energy as a function of volume and temperature. Minimizing $G(V,T)$ with respect to volume provides the optimal unit cell structure at that temperature. Selected crystalline energy–volume curves, Helmholtz vibrational free energies, and Gibbs free energies are provided in ESI Fig. S1–S4.†

Enthalpy and entropy were computed from the same ingredients using standard statistical mechanical expressions (see ref. 43 for details). Similarly, standard ideal gas, rigid rotor, and harmonic oscillator partition function expressions were used to evaluate thermochemical contributions for the gas phase species when computing sublimation enthalpies and entropies.

For HMBI, density-fitted MP2 was carried out in the Dunning aug-cc-pVXZ basis sets (X = D, T, or Q, and abbreviated aXZ in some places) using Molpro v2012.1 (ref. 75). Because MP2 has well-known problems with van der Waals dispersion, we also refined some single-point energies with MP2C,[76,77] also using Molpro. MP2C replaces the uncoupled HF treatment of dispersion found in MP2 with an improved coupled Kohn–Sham treatment, and it performs very well for non-covalent interactions.[78] Because the MP2C dispersion correction is less sensitive to the basis set than MP2,[79] the correction was computed in the aug-cc-pVTZ basis set and then combined with MP2 in various basis sets (aug-cc-pVTZ or larger). Extrapolation to the complete basis set limit from the triple and quadruple zeta basis sets was performed in the usual fashion.[80,81] All dimer calculations in HMBI employed a counterpoise correction for basis set superposition error.

Most of the HMBI calculations used the AMOEBA force field for long-range and many-body contributions. These were evaluated using Tinker 7.1.[82] Existing force field parameters were used for water, acetic acid, and imidazole.[52] Poltype[83] was used to generate parameters for carbon dioxide[43] and acetaminophen. Only the intermolecular force field parameters are needed in the context of HMBI, since intramolecular contributions are treated quantum mechanically.

In select cases, single-point energies were refined with periodic Hartree–Fock and the pob-TZVP basis,[84] which is a variant of def2-TZVP adapted for periodic calculations, and were performed using CRYSTAL09.[85,86] Basis set superposition error is a potential problem in Gaussian basis set calculations on periodic crystals, but addressing it in the context of the many-body calculations here is challenging. This basis was chosen here because an earlier study found that many-body contributions in small molecular clusters computed in the pob-TZVP basis set without counterpoise correction compared well against those in large basis sets.[46] A 10^{-7} a.u. tolerance was used for the Coulomb overlap threshold, Coulomb penetration threshold, and exchange overlap threshold, and tolerances of 10^{-12} and 10^{-30} a.u. were used for the pseudo-overlaps (see ref. 86 for details). Monkhorst–Pack and Gilat shrinking factors of 8 were used for the four smaller crystals, and 4 for acetaminophen.

DFT calculations on carbon dioxide, ice, acetic acid, and imidazole were performed using the periodic boundary planewave/pseudopotential and projector augmented wave (PAW) approaches with an 80 Rydberg energy cutoff and a $5 \times 5 \times 5$ Monkhorst–Pack k-point grid. For the larger acetaminophen crystal, a lower plane wave energy cutoff of 60 Ry and a $1 \times 3 \times 3$ k-point grid were used. DFT energies and gradients were computed with Quantum Espresso v6.1 (ref. 87 and 88) and Γ-point frequencies were produced by Phonopy v1.11.2.[89] The BLYP and

B86bPBE PAW functionals for H, C, N and O were produced using A. Dal Corso's Atomic code v6.1. Most DFT results reported here were obtained using the B86bPBE functional[90,91] with the Becke-Johnson exchange-hole dipole method (XDM) dispersion correction,[92] which has performed well in earlier molecular crystal studies.[34,35,92,93] Selected results with other functionals, such as BLYP[94,95] and PBE[91] with the D2 (ref. 96) or XDM dispersion correction are also presented.

3 Results and discussion

The work here primarily focuses on phase I carbon dioxide, ice Ih, the orthorhombic phase of acetic acid, and the α form of imidazole. These crystals exhibit diverse non-covalent interactions, ranging from strongly cooperative hydrogen bonding to π systems with significant van der Waals dispersion interactions. Ice and acetic acid are dominated by hydrogen bond interactions. Ice has a familiar hexagonal pattern, with each molecule accepting two and donating two hydrogen bonds (Bernal–Fowler rules), while acetic acid molecules form one-dimensional chains. Carbon dioxide is bound by a mixture of electrostatics and dispersion, and the π-conjugated imidazole rings form one-dimensional hydrogen-bonded chains with strong dispersion interactions between them.

3.1 Molar volume and thermal expansion

To begin, we examine the performance of Tiers 1–4 on carbon dioxide, ice, acetic acid, and imidazole. We previously examined these crystals with MP2 and various basis sets (Tier 1).[9] With counterpoise-corrected dimer energies, smaller basis sets typically underestimate non-covalent attractions. For the crystals modeled with HMBI, this translates to the molar volumes being overestimated. In carbon dioxide or ice, for example, systematically increasing the basis set from aug-cc-pVDZ all the way to the complete basis set limit shrinks the molar volume, with the thermal expansion curves in the different basis sets being relatively parallel to one another.[9]

The present study focuses on Tier 1 results at the MP2/aug-cc-pVTZ level for a couple of reasons. First, Tier 1 calculations require the energies, geometries, and phonons all to be computed at the MP2 level, which is computationally expensive. The aug-cc-pVTZ basis set provides a practical compromise between basis-set completeness and computational efficiency, making Tier 1 calculations feasible for species like acetic acid and imidazole. Second, as shown in Fig. 2, this level of theory predicts volumes for carbon dioxide, ice, and acetic acid that are (fortuitously) in reasonably good agreement with experiments. MP2 performs worse for imidazole, since it exaggerates the van der Waals attractions between the molecules, which translates to the underestimated volume seen in Fig. 2. The Tier 1 results also generally reproduce the experimentally observed rates of thermal expansion.

Tier 2 replaces the MP2 phonons with those calculated from DFT (B86bPBE-XDM here). The MP2 phonon frequencies tend to differ moderately from the DFT ones, as shown for ice in Fig. 3. Most notably, MP2 predicts excessively large intramolecular frequencies in the O–H stretching region. In inelastic neutron scattering experiments,[97,98] these modes occur in the range of $\sim$3000–3500 cm^{-1}, while MP2 predicts them to be several hundred wave numbers higher in energy.[99]

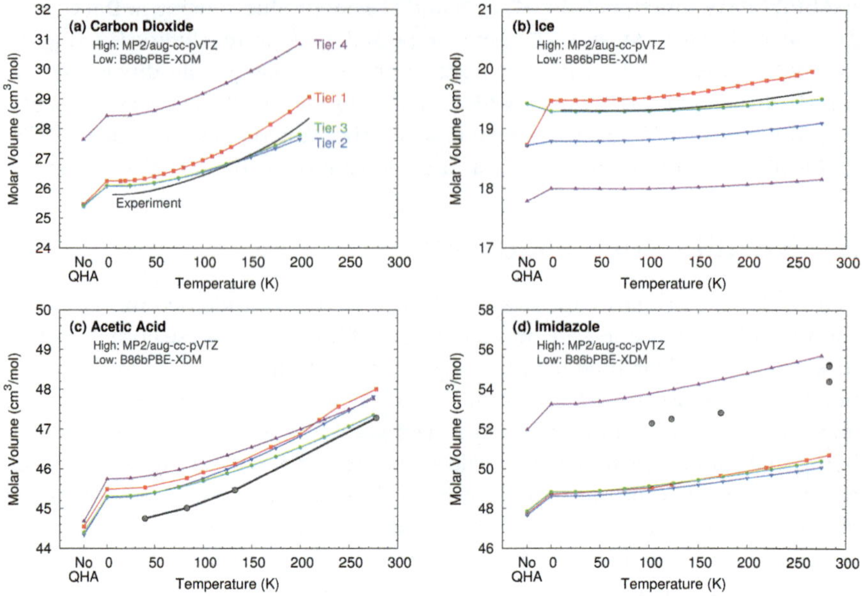

Fig. 2 Comparison of predicted molar volumes from Tier 1–4 approximations for (a) carbon dioxide, (b) ice, (c) acetic acid, and (d) imidazole using MP2/aug-cc-pVTZ + AMOEBA for the high-level calculations and B86bPBE-XDM for the low level calculations. The "No QHA" data refers to the electronic energy minimum, with no vibrational contribution.

DFT calculations provide a much better description of those high-frequency modes. On the other hand, the slightly lower 500–1100 cm^{-1} range of the MP2 librational modes is in better agreement with experiments than the DFT modes, which are shifted about 100 cm^{-1} higher. Both models predict generally similar frequencies for the HOH bending modes near 1600 cm^{-1} and pseudotranslational models below 400 cm^{-1}. See ref. 99 and 100 for more discussion of MP2 and DFT prediction of inelastic neutron scattering spectra in ice. MP2 and DFT phonon

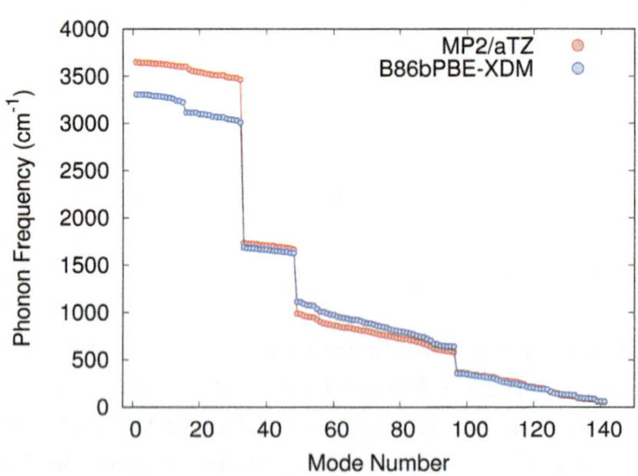

Fig. 3 Comparison of ice phonon frequencies predicted with MP2/aug-cc-pVTZ + AMOEBA or from periodic B86bPBE-XDM.

frequencies exhibited similar qualitative behaviors relative to experiments in crystalline methanol.[45] Errors in the high-frequency modes will impact the zero-point vibrational energy contribution, but they have a much smaller effect on thermal expansion. Differences in the low frequency modes will have a bigger impact on the thermal expansion.

When applied to quasi-harmonic thermal expansion, replacing the MP2 phonon frequencies with DFT ones in Tier 2 introduces only a modest decrease in the quality of the predicted molar volumes. In ice, for example, moving from Tier 1 to Tier 2 with B86bPBE-XDM phonons shrinks the molar volume by a few tenths of a cm^3 mol^{-1}. Increasing the MP2 basis set further would lead to further volume underestimation.[9] For carbon dioxide, the 0 K volumes are similar at Tiers 1 and 2, but the thermal expansion is underestimated at Tier 2.

The performance of Tier 3, which uses geometries and phonons computed with B86bPBE-XDM, is generally very similar to that of Tier 2. The largest molar volume differences occur for ice, where the two models differ by up to $\sim$1 cm^3 mol^{-1}. For acetic acid, they agree very closely at low temperature, but differ by $\sim$0.5 cm^3 mol^{-1} near room temperature. The differences between Tiers 2 and 3 are even smaller in the other two crystals.

However, Tier 3 ice exhibits the unusual feature that adding the zero-point vibrational energy contribution (*i.e.* from "No QHA" to 0 K in Fig. 2b) actually leads to a 0.1 cm^3 mol^{-1} lattice contraction, instead of the typical expansion. This does not occur for Tier 4, which uses the same phonons. As shown in ESI Fig. S2,† the quasi-harmonic B86bPBE-XDM vibrational free energy exhibits a shallow minimum at 18 cm^3 mol^{-1}. For Tier 4 B86bPBE-XDM, the electronic energy minimum occurs at a smaller volume, where F_{vib} has a negative slope, so adding the zero-point contribution leads to expansion. On the other hand, the Tier 3 MP2 electronic energy minimum occurs above 19 cm^3 mol^{-1}, in the regime where the B86bPBE-XDM vibrational free energy has a positive slope, and adding zero-point vibrational energy drives an initial contraction. At higher temperatures, the F_{vib} minima shift toward larger volumes, allowing expansion to occur in Tier 3. Interestingly, this minimum in F_{vib} does not occur for MP2 phonons (Tier 1). In Tier 2, the minimum occurs at larger volumes and does not cause the contraction seen in Tier 3. In other words, this behavior appears to be an unfortunate artifact of the Tier 3 combination of MP2 energies with DFT geometries and phonons. Nevertheless, the actual predicted molar volumes are reasonable.

Finally, employing pure B86bPBE-XDM (Tier 4) also predicts the molar volumes rather accurately. It significantly overestimates the molar volume in carbon dioxide, but it performs well for the other three crystals. Notably, B86bPBE-XDM reproduces the experimental molar volume in imidazole fairly well, where MP2 significantly overbinds the crystal and underestimates the volume due to its well-known problem describing van der Waals dispersion.[101] This MP2 problem will be addressed below.

Interestingly, the Tier 2 and 3 (and to a lesser extent Tier 4) results here tend to exhibit less thermal expansion than the Tier 1 results. One possible explanation could be that the former include only Γ-point phonons, while the Tier 1 results from ref. 9 include lattice dynamical phonon dispersion. We previously showed that including phonon dispersion increased the predicted rate of thermal expansion.[43] Supercell phonon calculations quickly become computationally demanding in DFT. With the fragment HMBI approach, on the other hand, they can be performed with very little additional cost over the Γ-point only calculation, since the unique additional contributions in the supercell are handled at the MM level. Other more subtle factors involving the interplay between the $E(V)$ curves and F_{vib} contributions likely contribute as well. In acetic acid, for instance, the Tier 2 model expands more rapidly with temperature than Tier 3, despite both neglecting phonon dispersion.

Ice Ih is unusual in that it exhibits negative thermal expansion, with the volume contracting by around 0.06% between 10 K and 70 K, before expanding. Most of the models studied here reproduce this trend qualitatively, though fewer are quantitatively correct. We previously found that Tier 1 MP2/CBS + AMOEBA predicts a slight contraction of only 0.1% and at a lower temperature of 40 K. Interestingly, a different MP2-based fragment study[102] also predicted the contraction in the aug-cc-pVDZ and aug-cc-pVTZ basis sets, though MP2 was found to exaggerate the contraction and the minimum was predicted at higher temperatures. The discrepancy between those results and our earlier ones might result from the use of counterpoise correction in our work, which leads to underbinding of the crystal, *versus* no counterpoise correction in ref. 102, which will lead to overbinding. Counterpoise correction will impact the balance between the key hydrogen bonding phonon modes and hydrogen stretching modes.

Here, we examine how the hybrid approaches behave. ESI Fig. S5† plots the relative molar volumes at low temperatures for several representative methods from Tiers 2–4. While Tier 2 MP2/aug-cc-pVTZ + AMOEBA demonstrates negligible contraction of the lattice, the Tier 3 and 4 results perform somewhat better. They correctly predict the location of the minimum near 60–70 K, but they underestimate the amount of contraction as only 0.02–0.03%.

Taken together, these results reiterate that dispersion-corrected density functionals like B86bPBE-XDM can predict thermal expansion well. Tier 1 MP2 molar volumes and thermal expansion often differ from the DFT results, but refining the single-point energies with MP2 already corrects for much of that difference. Such single-point energy refinements can be achieved with relatively modest additional computational cost. The further advantages of Tier 2 or Tier 1 for molar volumes are somewhat smaller, and will not be computationally worthwhile for predicting structural parameters in many applications.

Of course, these results represent only the B86bPBE-XDM functional. Additional insight is obtained by considering the behavior of Tiers 2–4 with several different density functionals and either the XDM or D2 dispersion correction. Fig. 4a plots the predicted thermal expansion with four different functionals: BLYP-D2, PBE-D2, PBE-XDM, and B86bPBE-XDM. The functionals overestimate the 0 K molar volume by ~10–15%, and the rate of thermal expansion differs considerably between the D2 and XDM dispersion corrections. Models corrected with D2 generally predict larger thermal expansion than those employing XDM. Because they derive their dispersion coefficients directly from electronic structure treatments, dispersion models like XDM, Tkatchenko–Scheffler (TS),[103] or many-body dispersion (MBD),[104–106] are generally expected to be superior to a purely empirical correction like D2.[107] The D3 or D4 models,[108,109] which adapt the dispersion coefficients based on the chemical environment, would provide better comparisons, but these are not currently implemented in Quantum Espresso.

Regardless, we focus not on the performance of specific density functionals and/or dispersion corrections, but rather on what happens when ingredients from

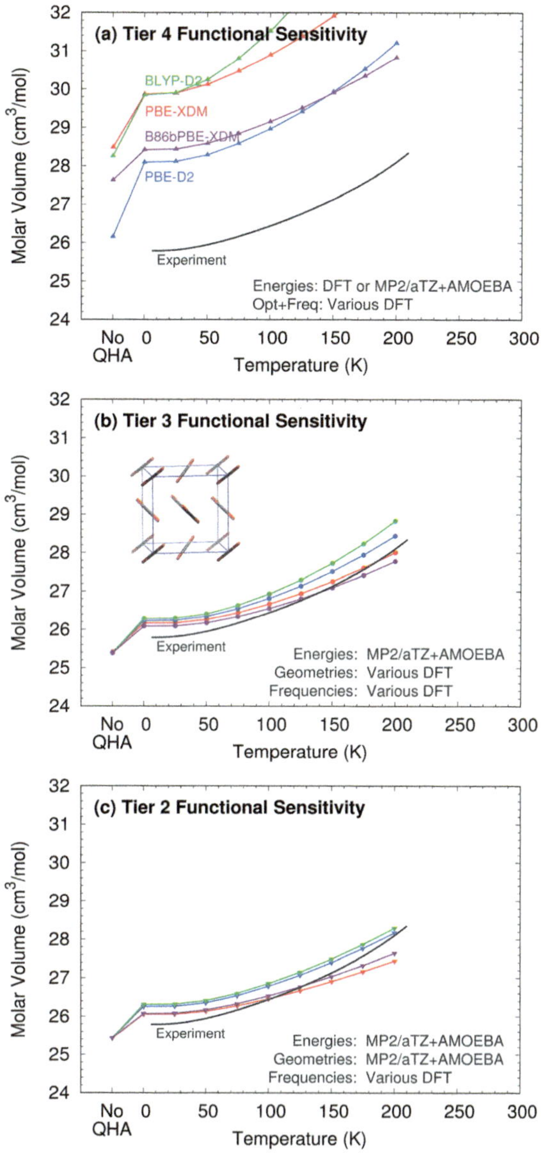

Fig. 4 (a) While different density functionals predict somewhat different molar volumes for carbon dioxide (Tier 4), the sensitivity of the predicted volumes decreases considerably when (b) MP2 single points are used on the DFT geometries (Tier 3), and (c) decreases further when only the DFT phonons are used (Tier 2).

those functionals are used in Tier 2 or 3 calculations. As shown in Fig. 4b, simply refining the calculations with MP2 single-point energies (Tier 3) dramatically reduces the sensitivity of the calculations to the density functional used to optimize the structures and compute the phonons. At Tier 2, where only the phonons are obtained from DFT, the sensitivity of the predicted molar volumes to the functional decreases further. Additional sensitivity data is provided in ESI Fig. S6 and S7.† These results suggest that although the DFT geometries defining the $E(V)$ curve clearly differ from each other and the MP2/aug-cc-pVTZ geometries, single-point energy refinement renormalizes much of the variation among them and gives $E(V)$ curves in generally good agreement with one another. In this context, it is also worth noting that density functional tight binding proved successful in modeling the thermal expansion in carbamazepine, and this could provide another route to low-cost phonons for Tiers 2 and 3.[37]

Given the good performance and relatively low computational cost of using HMBI-refined single-point energies with DFT optimizations and phonons (Tier 3), the next question is to see how reliably one can predict the molar volumes by further improving the quality of the HMBI single-point energies. These energies can be refined in three ways. First, one might increase the basis set size from aug-cc-pVTZ to the complete basis set limit *via* a triple-zeta/quadruple-zeta extrapolation. Counterpoise-corrected energies tend to underestimate the interactions in small basis sets, leading to overestimation of the molar volume.[9]

Second, one could replace the AMOEBA many-body treatment with one from periodic HF. Recent work found that AMOEBA overestimates many-body polarization associated with cooperative hydrogen bonding in crystalline methanol, and refining the energetics with periodic HF proved important for obtaining the correct phase diagram.[46] Similar over-polarization is found for AMOEBA treatments of ion–water interactions.[110] Strong polarization effects are present in several of the crystals here, so there could be appreciable benefits to periodic HF refinement here as well.

Third, MP2 exhibits well-known problems with the treatment of van der Waals dispersion.[101] Among the four crystals here, this is most apparent for imidazole, where MP2 substantially overbinds the lattice energy[38,39] and underestimates the molar volume (see Fig. 2d).[9] Ideally, one would correct this problem by replacing MP2 with CCSD(T). However, dispersion-corrected MP2C (ref. 76 and 77) provides a more pragmatic approach which corrects most of the MP2 error in these crystals at much lower computational cost.[79]

Fig. 5 examines this hierarchy of refinements for the four crystals. As seen in the earlier discussion of Fig. 2, switching from (1) pure B86bPBE-XDM (Tier 4) to (2) Tier 3 MP2/aug-cc-pVTZ + AMOEBA single-point energies leads to a substantial change in the predicted molar volumes, usually toward smaller unit cells (the slight volume increase in ice is the exception). (3) Replacing the AMOEBA many-body contribution with periodic HF increases the molar volume. For carbon dioxide, ice, and acetic acid, the effect is fairly small. For imidazole, on the other hand, the volume is increased by ~9%. As expected, (4) increasing the MP2 basis set toward the CBS limit induces a volume contraction of ~0.4 cm^3 mol^{-1} in ice and around 1 cm^3 mol^{-1} in the other three crystals. With the exception of imidazole, the MP2/CBS + periodic HF predicted volumes are slightly smaller than the experimental volumes. Surprisingly, for imidazole, the MP2/CBS + periodic HF volumes agree very well with experiments, despite overestimation of the lattice energy by MP2. In all four cases, (5) MP2C weakens the binding and shifts toward larger molar volumes. Unsurprisingly, the largest corrections are observed for carbon dioxide and imidazole, while hydrogen-bonded ice and acetic acid are less affected by the dispersion correction.

Overall, the combination of MP2C/CBS + periodic HF energies on B86bPBE-XDM geometries and phonons predicts the highest-temperature experimental

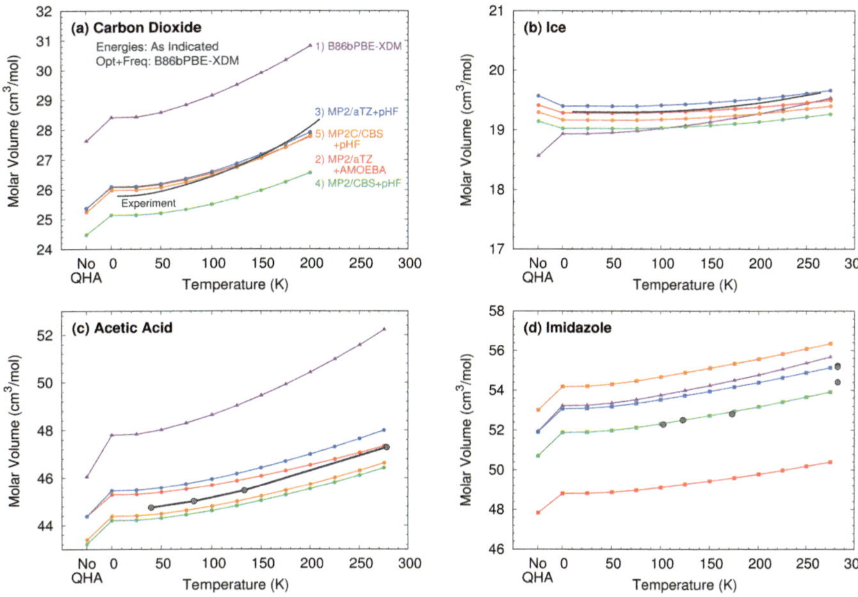

Fig. 5 Impact of single-point HMBI energy refinement on the DFT predictions for (a) carbon dioxide, (b) ice, (c) acetic acid, and (d) imidazole. Starting with (1) pure B86bPBE (Tier 4), we (2) first refine to Tier 3 with MP2/aug-cc-pVTZ + AMOEBA. Further Tier 3 refinements are made by (3) replacing the AMOEBA many-body treatment with periodic HF treatment, (4) extrapolating MP2 to the complete basis set limit, and (5) correcting MP2 with MP2C.

molar volumes of carbon dioxide, ice, and acetic acid to within 1–2%. For imidazole, the errors are 2% or 3.5%, depending on which high-temperature experimental structure one compares against. Finally, notice that the different possible Tier 3 single-point energy refinements considered here all predict similar thermal expansivity (curves (2)–(5) in Fig. 5 are largely parallel). The most significant variations are manifest as shifts in the 0 K molar volume. This reflects that while the minimum of the $E(V)$ curve shifts upon single-point energy refinement, changes to the potential energy well curvature are apparently minor (see ESI Fig. S8).†

3.2 Thermochemical data

A key reason for modeling thermal expansion in molecular crystals is to predict the temperature-dependence of other crystal properties. Accordingly, we consider the enthalpy and entropy of sublimation for these four crystals. Experimental values were taken and/or derived from the literature as described in ref. 9 and 43. Quantitative experimental uncertainties are not readily available for the thermochemical properties presented here. The temperature-dependent results were frequently derived using data from a variety of sources that did not always report uncertainties. Moreover, different studies do not always agree (*e.g.* the reported room-temperature sublimation enthalpies for imidazole differ by several kJ mol^{-1} (ref. 111)). Crystalline defects in the experimental solids can also play a role. Forming a point defect is typically an endothermic process, but it also leads to an increase in entropy. Quantifying the impact of the various factors here is difficult, but one should probably allow for uncertainties of up to a few kJ mol^{-1} in the sublimation enthalpies, for example.

Fig. 6 compares the sublimation enthalpies as predicted with several of the techniques described in the previous section against experimental results. Given the relatively small improvements in molar volume provided by Tier 2 over Tier 3, the discussion here focuses primarily on Tiers 3 and 4. Unless otherwise mentioned, all DFT results in this section employ B86bPBE-XDM.

As the temperature increases, the sublimation enthalpy $\Delta H_{sub} = H_{gas} - H_{crystal}$ typically first increases before reaching a maximum and then decreasing. Heating increases H_{gas} *via* destabilizing translational, rotational, and PV contributions. $H_{crystal}$ is derived from a balance between the thermal expansion and the vibrational contributions, but it is typically dominated by lattice energy destabilization due to thermal expansion. Little thermal expansion occurs in the crystal at low temperatures, and ΔH_{sub} increases primarily due to the gas contribution in that temperature regime. At higher temperatures, however, the larger thermal expansion in the crystal eventually leads to the increase in $H_{crystal}$ becoming more significant than the changes in H_{gas}, which produces the maximum and subsequent decrease in ΔH_{sub} seen for carbon dioxide, acetic acid, and imidazole. In ice, this maximum does not occur below the melting point due to the low thermal expansivity. See ref. 9 for more detailed discussion of the enthalpy behaviors.

Earlier Tier 1 work[9,43] demonstrated that changing the electronic structure method and basis set shifts the entire sublimation curve vertically but has a minor impact on the curvature. Larger basis sets in particular shift the enthalpy toward larger values. Obtaining the proper maximum and subsequent decrease of

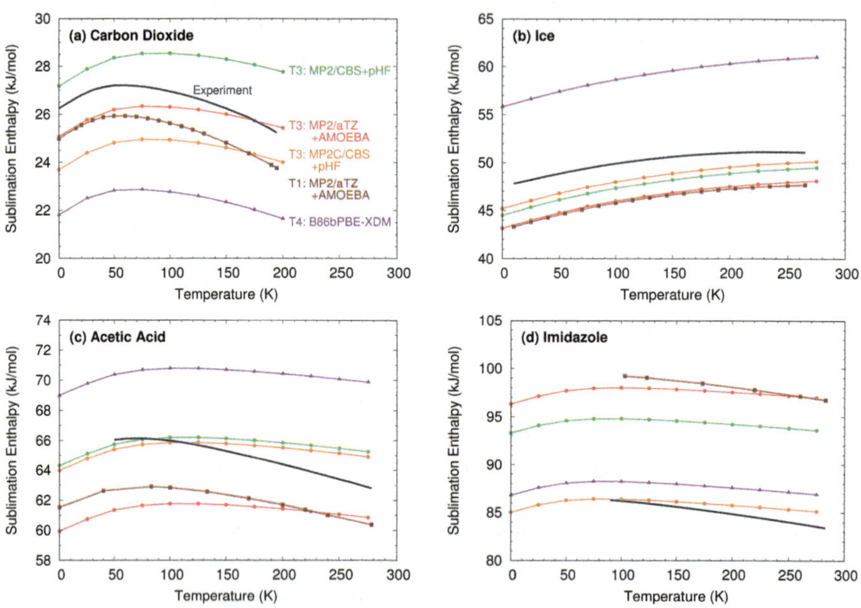

Fig. 6 Comparison of predicted sublimation enthalpies for (a) carbon dioxide, (b) ice, (c) acetic acid, and (d) imidazole using Tier 1 (T1) MP2/aug-cc-pVTZ + AMOEBA, Tier 4 (T4) B86bPBE-XDM, and Tier 3 (T3) with several different energy refinements.

the sublimation enthalpy at higher temperatures requires the thermal expansion to be captured correctly. A simple harmonic model that neglects thermal expansion agrees well with the quasi-harmonic result at low temperatures, but it overestimates the enthalpy at higher temperatures by about 1 kJ mol^{-1} in these crystals. Ice proves to be the exception: since it exhibits comparatively little thermal expansion, the error introduced in ΔH_{sub} by neglecting that expansion at 273 K is negligible.

In the present work, the Tier 2–3 results frequently underestimate the amount of thermal expansion compared to Tier 1. In Fig. 6, this underestimated thermal expansion is manifest in the Tier 3 MP2/aug-cc-pVTZ + AMOEBA sublimation enthalpy (red curve) being too large at higher temperatures, which is analogous to what was observed for the harmonic model.[9] Nevertheless, the Tier 3 results generally mimic the Tier 1 results to within 1–2 kJ mol^{-1} across the full temperature range.

Refining Tier 1 by increasing the basis set size to the CBS limit, applying the MP2C dispersion correction, and replacing AMOEBA with periodic HF does not resolve the issue of the erroneous slope in ΔH_{sub} at higher temperatures. Nevertheless, the MP2C/CBS + periodic HF results (orange curve) provide excellent agreement with experiments across the temperature range, with root-mean-square (rms) errors ranging from 0.9–2.1 kJ mol^{-1} for the four crystals (Table 2). Applying the dispersion correction in MP2C has minimal impact on hydrogen-bonded ice and acetic acid, but it alters the sublimation enthalpies in carbon dioxide and imidazole considerably (compare green *vs.* orange curves). In our earlier work, Tier 1 MP2/CBS and CCSD(T)/CBS both overestimated the sublimation enthalpy for carbon dioxide, with CCSD(T) predicting a slightly smaller ΔH_{sub}.[9,43] Here, it appears that MP2C over-corrects the MP2 result, with the net result of slightly increasing the rms error from 1.7 to 2.1 kJ mol^{-1}. On the other hand, MP2C performs very well for imidazole, reducing the rms error in ΔH_{sub} from 9.2 kJ mol^{-1} (MP2) to 0.8 kJ mol^{-1} (MP2C).

Compared to the best Tier 3 MP2C results, the errors for B86bPBE-XDM are 2–3 times larger for carbon dioxide and imidazole, where van der Waals dispersion contributes appreciably, and 5–6 times larger for acetic acid and ice, which are dominated by hydrogen bonding interactions (Table 2). Notably, B86bPBE-XDM performs very well for imidazole, which is problematic for MP2. Overall, the sublimation enthalpy data here bolsters the case for Tier 3 single-point energy refinement with MP2C/CBS + periodic HF on top of the DFT structures and phonons.

Table 2 Root-mean-square error (in kJ mol^{-1}) between the predicted and experimental sublimation enthalpies over the temperatures for which experimental data is available (see Fig. 6). The errors were computed by splining the data curves and taking the differences between them at 1 K intervals

Crystal	Temperature range	Tier 3 MP2/CBS + pHF	Tier 3 MP2C/CBS + pHF	Tier 4 B86bPBE-XDM
Carbon dioxide	0–195 K	1.7	2.1	4.2
Ice	10–265 K	2.5	1.9	8.9
Acetic acid	50–278 K	1.3	1.0	5.6
Imidazole	90–283 K	9.2	0.9	2.6

Next, consider the prediction of the sublimation entropy. Whereas $\Delta H_{\text{crystal}}$ is dominated by the lattice energy and is therefore sensitive to the final single-point energy, S_{crystal} is governed by the phonon frequencies. Tiers 2–4 all compute the phonons with DFT. The improved energies in Tier 2 or Tier 3 alter the predicted volume at a given temperature, which in turn impacts the phonons *via* their volume dependence (eqn (5)).

Like for the enthalpy, the temperature dependence of the sublimation entropy $\Delta S_{\text{sub}}(T) = S_{\text{gas}}(T) - S_{\text{crystal}}(T)$ arises from competition between the entropy of the gas phase and that of the crystal phase.[9] At low temperatures, $S_{\text{gas}}(T)$ rises more quickly with temperature. At higher temperatures, the situation reverses, with $S_{\text{crystal}}(T)$ increasing more quickly due to the phonon contribution. Together, these produce the overall concave shape of $\Delta S_{\text{sub}}(T)$. If thermal expansion is not accounted for, the phonon frequencies tend to be too large, and $S_{\text{crystal}}(T)$ will be too small. This causes overestimation of the entropy at high temperatures.

Here, the thermal expansion is frequently underestimated relative to both experimental and Tier 1 calculations. This translates to consistent overestimation of $\Delta S_{\text{sub}}(T)$ by up to 10–12% relative to experimental results at high temperatures (Fig. 7). In cases like ice and acetic acid, where the volume is not too sensitive to the specific method used for Tier 3 energy refinement, the variation in predicted sublimation entropies is fairly small across the different energy models. Larger variations are observed among the different Tier 3 models for imidazole and carbon dioxide, commensurate with the greater sensitivity of the molar volumes to the modeling approach in those crystals. Overall, in marked contrast to the molar volumes and sublimation enthalpies, Tier 3 energy refinement does not

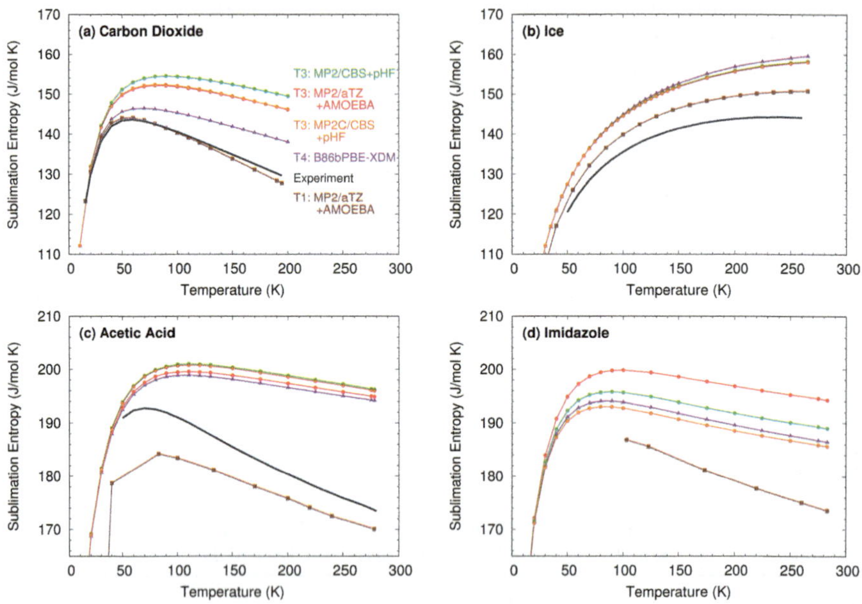

Fig. 7 Comparison of predicted sublimation entropies for (a) carbon dioxide, (b) ice, (c) acetic acid, and (d) imidazole using Tier 1 MP2/aug-cc-pVTZ + AMOEBA, Tier 4 B86bPBE-XDM, and Tier 3 with several different energy refinements. Experimental entropy data is unavailable for imidazole.

clearly improve the quality of the underlying DFT entropies. The same can be said for Tier 2 (not shown), since the Tier 2 and 3 sublimation entropies differ by less than 1%.

Calculating phase transition temperatures provides another means of assessing the predicted thermochemistry values, since the transition temperature can be sensitive to subtle changes in the Gibbs free energy.[9,43,46] Computing the Gibbs free energy accurately requires a balance between the enthalpic and entropic components, which may or may not be maintained in the hybrid approaches.

Experimentally, carbon dioxide sublimes at 194.7 K and atmospheric pressure.[64] Table 3 summarizes the sublimation temperatures predicted by several different methods. Large-basis Tier 1 MP2 and CCSD(T) perform excellently, predicting the sublimation temperature within 5 degrees Kelvin of the experimental value. In smaller basis sets, the sublimation temperature is underestimated. At the other extreme, B86bPBE-XDM underestimates the sublimation temperature by almost 40 K. Tier 3 refinement at the MP2 level improves the predicted sublimation temperature somewhat, though the Tier 3 temperatures are ~10 K lower than the corresponding Tier 1 temperatures. As noted earlier, MP2C seemingly over-corrects MP2, to the detriment of the predicted sublimation temperature. The ~30 K error in the MP2C/CBS + periodic HF sublimation temperature is only modestly better than that of B86bPBE-XDM.

3.3 Acetaminophen form I

To investigate how well these models perform in a pharmaceutically relevant species, Fig. 8 plots the thermal expansion data for acetaminophen. Twenty two experimental crystal structures of ambient pressure form I have been reported in the CSD. While most of the reported molar volumes exhibit reasonable agreement, a wide ~4 cm^3 mol^{-1} scatter exists in the reported room temperature values. Seven of the structures (connected by a line in Fig. 8) come from a single temperature-dependent neutron scattering study.[57] The discussion below focuses primarily on this consistent set of neutron scattering data.

Given the large size of acetaminophen ($C_8H_9NO_2$ and $Z = 4$ in form I), calculations were only performed at Tiers 3 and 4. Tiers 1 and 2 would require

Table 3 Comparison of experimental and predicted sublimation temperatures for carbon dioxide

Tier	Method	T_{sub} (K)
	Experiment[a]	194.7
Tier 1	MP2/aDZ + AMOEBA[b]	163.6
Tier 1	MP2/aTZ + AMOEBA[b]	185.3
Tier 1	MP2/aQZ + AMOEBA[b]	193.4
Tier 1	MP2/CBS + AMOEBA[b]	199.2
Tier 1	CCSD(T)/CBS + AMOEBA[b]	201.0
Tier 3	MP2/aTZ + AMOEBA	174.2
Tier 3	MP2/CBS + pHF	186.6
Tier 3	MP2C/CBS + pHF	165.7
Tier 4	B86bPBE-XDM	157.5

[a] Ref 64. [b] Ref 43.

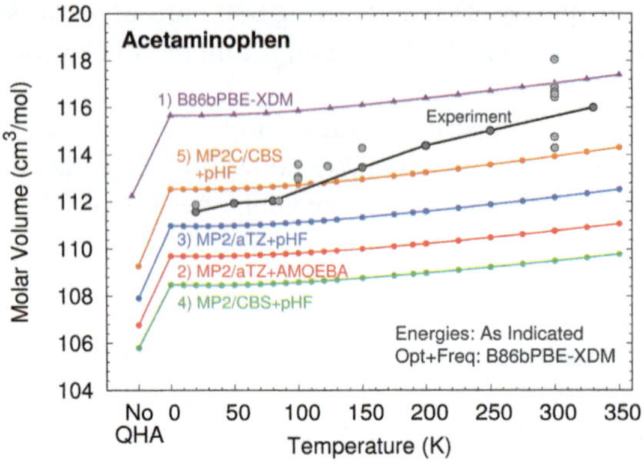

Fig. 8 Predicted acetaminophen molar volumes *versus* temperature. The connected dark gray points indicate data from a temperature-dependent neutron scattering study,[57] while light gray points represent other experimental data found in the CSD.

substantially more computational effort. With a predicted molar volume of 115.6 cm^3 mol^{-1} at 20 K, Tier 4 B86bPBE-XDM overestimates the experimental molar volume of 111.6 cm^3 mol^{-1} (HXACAN13 (ref. 57)) by 3.6%. It also significantly underestimates the thermal expansivity, with a B86bPBE-XDM predicted molar volume increase of only 1.3% *versus* 3.8% experimentally. The combination of an overestimated volume at low temperature and an underestimated thermal expansivity leads to much smaller errors in the molar volumes at high temperatures, *e.g.* only 1% error at 330 K (HXACAN19 (ref. 57)).

Switching to Tier 3 with MP2/aug-cc-pVTZ + AMOEBA leads to a considerable reduction in the predicted molar volume to 109.65 cm^3 mol^{-1}, which is in somewhat better agreement with experiments (1.7% error at 20 K). However, the underestimation of thermal expansivity persists, and the volume at 330 K is underestimated by 4.4%. Moving toward the CBS limit, replacing MP2 with MP2C and AMOEBA with periodic HF causes a shift back toward larger molar volumes, but has little impact on the thermal expansivity. In the end, the MP2C/CBS + periodic HF volumes are fairly close to the experimental values, especially at low temperatures (0.8% error at 20 K), but they underestimate the volume by 1.6% at 330 K due to the low thermal expansivity.

Next, we turn to the form I acetaminophen sublimation enthalpy and entropy. Perlovich and co-workers measured ΔH_{sub} (298 K) = 117.9 $\pm$ 0.7 kJ mol^{-1} from vapor pressure measurements,[112,113] while Picciochi *et al.* subsequently obtained 129.9 $\pm$ 1.4 kJ mol^{-1} from a combination of vapor pressure measurements and microcalorimetry experiments (and proposed a slight refinement of Perlovich's value to 118.9 $\pm$ 1.6 kJ mol^{-1}).[114] The $\sim$10 kJ mol^{-1} discrepancy between the experimental values is significantly larger than the reported experimental uncertainties of 1–2 kJ mol^{-1}. The quasi-harmonic calculations here can address this discrepancy.

Table 4 compares the predicted and experimental room-temperature sublimation enthalpies. Tier 4 B86bPBE-XDM predicts 131.0 kJ mol^{-1}, which lies within the stated uncertainty in the Picciochi *et al.* value of 129.9 $\pm$ 1.4 kJ mol^{-1}. MP2 predicts a sublimation enthalpy that is up to 23 kJ mol^{-1} higher, but

Table 4 Comparison of experimental and predicted 298 K sublimation enthalpies (kJ mol^{-1}) and entropies (J (mol K)$^{-1}$) for acetaminophen form I

Method	Source	ΔH_{sub} (298 K)	ΔS_{sub} (298 K)
Perlovich *et al.*	Ref. 112 and 113	117.9 ± 0.7	190 ± 2
Picciochi *et al.*	Ref. 114	129.9 ± 1.4	
B86bPBE-XDM	Tier 4	131.0	242.1
MP2/aTZ + AMOEBA	Tier 3	146.4	245.6
MP2/aTZ + pHF	Tier 3	147.4	244.8
MP2/CBS + pHF	Tier 3	154.1	246.3
MP2C/CBS + pHF	Tier 3	138.0	243.9

employing MP2C/CBS + periodic HF reduces this enthalpy to 138.0 kJ mol^{-1}. This is 8 kJ mol^{-1} larger than the Picciochi *et al.* experimental value. Some of this overestimation may stem from the underpredicted thermal expansion, as seen for acetic acid and imidazole earlier. In any case, both the MP2C and B86bPBE-XDM results support the larger 129.9 ± 1.4 experimental sublimation enthalpy instead of the smaller 117.9 ± 0.7 kJ mol^{-1} value.

Finally, Perlovich *et al.* reported a sublimation entropy of 190 ± 2 J (mol^{-1} K^{-1}) at 298 K. Tier 4 B86bPBE-XDM and various Tier 3 refinements all predict much larger values of 240–246 J (mol^{-1} K^{-1}). Overestimation of the sublimation entropy is expected based on the results for the other crystals described above and from how much the predictions underestimate the thermal expansion. Still, this ~25% disagreement between theory and experiment is significantly larger than the ~10–12% seen for the smaller crystals. On the other hand, it is not clear how reliable the reported experimental entropy value is, since it was derived from a sublimation enthalpy measurement[113] that is suspect in light of the Picciochi *et al.* experiments[114] and our calculations here.

4 Conclusions

The combination of fragment-based correlated wavefunction electronic structure techniques and quasi-harmonic approximations can be very successful in predicting thermal expansion and other properties, at least for rigid-molecule crystals. On the other hand, dispersion-corrected DFT models also often perform well at appreciably lower computational cost. By comparing a hierarchy of models ranging from pure DFT to pure MP2, we have demonstrated that introducing correlated wavefunction energies on top of DFT geometries and phonons can appreciably improve the predicted molar volumes and enthalpies. One must ensure, however, that the single-point energies are well-chosen, which in practice means employing large basis sets and accounting for any deficiencies in the models (*e.g.* by replacing MP2 with MP2C for the monomer and dimer contributions or replacing AMOEBA with HF for the many-body contributions).

In the small molecule crystals, the hybrid Tier 3 approach led to predicted molar volumes within a few percent and sublimation enthalpies within 1–2 kJ mol^{-1} of experimental values over broad temperature ranges. Though more computationally expensive than pure DFT approaches, such single-point energy refinements can be performed on crystals containing dozens of atoms per molecule, especially when crystal symmetry can be exploited. Using these techniques, we demonstrated that the predicted sublimation enthalpy for acetaminophen agrees better with the more recent experimental value of Picciochi *et al.* than the earlier reported value.

Issues remain, however. The hybrid approaches that combine DFT and wavefunction techniques do not appreciably improve the predicted entropies, since those are largely governed by phonon contributions. We also found that the hybrid calculations here generally underestimate the amount of thermal expansion. In the smaller crystals the impact was modest, but it was more significant in acetaminophen. Acetaminophen also exhibited considerably larger 8 kJ mol^{-1} errors in the predicted sublimation enthalpy than the smaller crystals. Accounting for phonon dispersion might help, but anharmonicity or other factors may also play an important role. More testing in larger, more flexible molecules is needed to assess the performance of the quasi-harmonic approach. Efficient and effective strategies for addressing anharmonicity would be especially valuable.

Moreover, though the performances of the hybrid approaches are quite good, they may not always be sufficient. The imbalances between enthalpy and entropy appreciably increased the error in the predicted carbon dioxide sublimation temperature, for example. More dramatically, in our recent study of the methanol phase diagram,[46] even a half kJ mol^{-1} error in the relative free energies between two polymorphs shifts the phase transition temperature by ~100 K. For such challenging properties, more reliable and computationally demanding Tier 1 approaches may be needed.

Conflicts of interest

There are no conflicts to declare.

Acknowledgements

Funding for this work from the National Science Foundation (CHE-1665212 to G. B. and Graduate Research Fellowship Grant 1326120 to J. M.) and supercomputer time from XSEDE (TG-CHE110064 to G. B. and TG-CHE170098 to J. M.) is gratefully acknowledged. We thank Dr Alberto Otero de la Roza for providing guidance on the B86bPBE-XDM calculations and Dr Ctirad Červinka for helpful discussions.

Notes and references

1 S. L. Price, *Chem. Soc. Rev.*, 2014, **43**, 2098–2111.

2 S. L. Price, D. E. Braun and S. M. Reutzel-Edens, *Chem. Commun.*, 2016, **52**, 7065–7077.

3 S. L. Price and S. M. Reutzel-Edens, *Drug Discov. Today*, 2016, **21**, 912–923.

4 G. M. Day, T. G. Cooper, A. J. Cruz-Cabeza, K. E. Hejczyk, H. L. Ammon, S. X. M. Boerrigter, J. S. Tan, R. G. Della Valle, E. Venuti, J. Jose, S. R. Gadre, G. R. Desiraju, T. S. Thakur, B. P. van Eijck, J. C. Facelli, V. E. Bazterra, M. B. Ferraro, D. W. M. Hofmann, M. A. Neumann, F. J. J. Leusen, J. Kendrick, S. L. Price, A. J. Misquitta, P. G. Karamertzanis,

G. W. A. Welch, H. A. Scheraga, Y. A. Arnautova, M. U. Schmidt, J. van de Streek, A. K. Wolf and B. Schweizer, *Acta Crystallogr. B*, 2009, **65**, 107–125.

5 D. A. Bardwell, C. S. Adjiman, Y. A. Arnautova, E. Bartashevich, S. X. M. Boerrigter, D. E. Braun, A. J. Cruz-Cabeza, G. M. Day, R. G. Della Valle, G. R. Desiraju, B. P. van Eijck, J. C. Facelli, M. B. Ferraro, D. Grillo, M. Habgood, D. W. M. Hofmann, F. Hofmann, K. V. J. Jose, P. G. Karamertzanis, A. V. Kazantsev, J. Kendrick, L. N. Kuleshova, F. J. J. Leusen, A. V. Maleev, A. J. Misquitta, S. Mohamed, R. J. Needs, M. A. Neumann, D. Nikylov, A. M. Orendt, R. Pal, C. C. Pantelides, C. J. Pickard, L. S. Price, S. L. Price, H. A. Scheraga, J. van de Streek, T. S. Thakur, S. Tiwari, E. Venuti and I. K. Zhitkov, *Acta Crystallogr. B*, 2011, **67**, 535–551.

6 A. M. Reilly, R. I. Cooper, C. S. Adjiman, S. Bhattacharya, A. D. Boese, J. G. Brandenburg, P. J. Bygrave, R. Bylsma, J. E. Campbell, R. Car, D. H. Case, R. Chadha, J. C. Cole, K. Cosburn, H. M. Cuppen, F. Curtis, G. M. Day, R. A. DiStasio Jr, A. Dzyabchenko, B. P. van Eijck, D. M. Elking, J. A. van den Ende, J. C. Facelli, M. B. Ferraro, L. Fusti-Molnar, C.-A. Gatsiou, T. S. Gee, R. de Gelder, L. M. Ghiringhelli, H. Goto, S. Grimme, R. Guo, D. W. M. Hofmann, J. Hoja, R. K. Hylton, L. Iuzzolino, W. Jankiewicz, D. T. de Jong, J. Kendrick, N. J. J. de Klerk, H.-Y. Ko, L. N. Kuleshova, X. Li, S. Lohani, F. J. J. Leusen, A. M. Lund, J. Lv, Y. Ma, N. Marom, A. E. Masunov, P. McCabe, D. P. McMahon, H. Meekes, M. P. Metz, A. J. Misquitta, S. Mohamed, B. Monserrat, R. J. Needs, M. A. Neumann, J. Nyman, S. Obata, H. Oberhofer, A. R. Oganov, A. M. Orendt, G. I. Pagola, C. C. Pantelides, C. J. Pickard, R. Podeszwa, L. S. Price, S. L. Price, A. Pulido, M. G. Read, K. Reuter, E. Schneider, C. Schober, G. P. Shields, P. Singh, I. J. Sugden, K. Szalewicz, C. R. Taylor, A. Tkatchenko, M. E. Tuckerman, F. Vacarro, M. Vasileiadis, A. Vazquez-Mayagoitia, L. Vogt, Y. Wang, R. E. Watson, G. A. de Wijs, J. Yang, Q. Zhu and C. R. Groom, *Acta Crystallogr. B*, 2016, **72**, 439–459.

7 S. L. Price, *Int. Rev. Phys. Chem.*, 2008, **27**, 541–568.

8 J. Nyman and G. M. Day, *Phys. Chem. Chem. Phys.*, 2016, **18**, 31132–31143.

9 Y. N. Heit and G. J. O. Beran, *Acta Crystallogr. B*, 2016, **72**, 514–529.

10 A. Gavezzotti and G. Filippini, *J. Am. Chem. Soc.*, 1995, **117**, 12299–12305.

11 J. Hoja, A. M. Reilly and A. Tkatchenko, *Wiley Interdiscip. Rev.: Comput. Mol. Sci.*, 2017, **7**, e1294.

12 P. Raiteri, R. Martoňák and M. Parrinello, *Angew. Chem., Int. Ed.*, 2005, **44**, 3769–3773.

13 P. G. Karamertzanis, P. Raiteri, M. Parrinello, M. Leslie and S. L. Price, *J. Phys. Chem. B*, 2008, **112**, 4298–4308.

14 T. Zykova-Timan, P. Raiteri and M. Parrinello, *J. Phys. Chem. B*, 2008, **112**, 13231–13237.

15 A. M. Reilly, S. Habershon, C. A. Morrison and D. W. H. Rankin, *J. Chem. Phys.*, 2010, **132**, 134511.

16 T.-Q. Yu and M. Tuckerman, *Phys. Rev. Lett.*, 2011, **107**, 015701.

17 E. Schneider, L. Vogt and M. E. Tuckerman, *Acta Crystallogr. B*, 2016, **72**, 542–550.

18 M. J. Schnieders, J. Baltrusaitis, Y. Shi, G. Chattree, L. Zheng, W. Yang and P. Ren, *J. Chem. Theory Comput.*, 2012, **8**, 1721–1736.

19 I. J. Nessler, J. M. Litman and M. J. Schnieders, *Phys. Chem. Chem. Phys.*, 2016, 36–38.

20 E. C. Dybeck, N. P. Schieber and M. R. Shirts, *J. Chem. Theory Comput.*, 2016, **12**, 3491–3505.

21 E. C. Dybeck, N. S. Abraham, N. P. Schieber and M. R. Shirts, *Cryst. Growth Des.*, 2017, **17**, 1775–1787.

22 J. Nyman and G. M. Day, *CrystEngComm*, 2015, **17**, 5154–5165.

23 S. A. Rivera, D. G. Allis and B. S. Hudson, *Cryst. Growth Des.*, 2008, **8**, 3905–3907.

24 N. Marom, R. A. DiStasio, V. Atalla, S. Levchenko, A. M. Reilly, J. R. Chelikowsky, L. Leiserowitz and A. Tkatchenko, *Angew. Chem., Int. Ed.*, 2013, **52**, 6629–6632.

25 S. Wen and G. J. O. Beran, *J. Chem. Theory Comput.*, 2012, **8**, 2698–2705.

26 A. M. Reilly and A. Tkatchenko, *Phys. Rev. Lett.*, 2014, **113**, 055701.

27 D. E. Braun, J. A. McMahon, L. H. Koztecki, S. L. Price and S. M. Reutzel-Edens, *Cryst. Growth Des.*, 2014, **14**, 2056–2072.

28 G. Filippini, C. M. Gramaccioli, M. Simonetta and G. B. Suffritti, *Chem. Phys. Lett.*, 1975, **35**, 17–20.

29 G. Filippini and C. M. Gramaccioli, *Acta Crystallogr., Sect. A: Found. Crystallogr.*, 1981, **37**, 335–342.

30 B. P. van Eijck, *J. Comput. Chem.*, 2001, **22**, 816–826.

31 G. M. Day and S. L. Price, *J. Am. Chem. Soc.*, 2003, **125**, 16434–16443.

32 G. M. Day, S. L. Price and M. Leslie, *J. Phys. Chem. B*, 2003, **107**, 10919–10933.

33 E. Sanz, C. Vega, J. Abascal and L. MacDowell, *Phys. Rev. Lett.*, 2004, **92**, 255701.

34 S. R. Whittleton, A. Otero-de-la Roza and E. R. Johnson, *J. Chem. Theory Comput.*, 2017, **13**, 441–450.

35 S. R. Whittleton, A. Otero-de-la Roza and E. R. Johnson, *J. Chem. Theory Comput.*, 2017, **13**, 5332–5342.

36 A. Erba, J. Maul and B. Civalleri, *Chem. Commun.*, 2016, **52**, 1820–1823.

37 J. G. Brandenburg, J. Potticary, H. A. Sparkes, S. L. Price and S. R. Hall, *J. Phys. Chem. Lett.*, 2017, **8**, 4319–4324.

38 G. J. O. Beran and K. Nanda, *J. Phys. Chem. Lett.*, 2010, **1**, 3480–3487.

39 S. Wen and G. J. O. Beran, *J. Chem. Theory Comput.*, 2011, **7**, 3733–3742.

40 K. Nanda and G. J. O. Beran, *J. Chem. Phys.*, 2012, **137**, 174106.

41 Y. Heit and G. J. O. Beran, *J. Comput. Chem.*, 2014, **35**, 2205–2214.

42 G. J. O. Beran, J. D. Hartman and Y. N. Heit, *Acc. Chem. Res.*, 2016, **49**, 2501–2508.

43 Y. N. Heit, K. D. Nanda and G. J. O. Beran, *Chem. Sci.*, 2016, **7**, 246–255.

44 W. Sontising, Y. N. Heit, J. L. McKinley and G. J. O. Beran, *Chem. Sci.*, 2017, **8**, 7374–7382.

45 C. Červinka and G. J. O. Beran, *Phys. Chem. Chem. Phys.*, 2017, **19**, 29940–29953.

46 C. Červinka and G. J. O. Beran, *Chem. Sci.*, 2018, **9**, 4622–4629.

47 C. Červinka, M. Fulem, R. P. Stoffel and R. Dronskowski, *J. Phys. Chem. A*, 2016, **120**, 2022–2034.

48 C. Červinka and M. Fulem, *J. Chem. Theory Comput.*, 2017, **13**, 2840–2850.

49 H. Stoll, *Chem. Phys. Lett.*, 1992, **191**, 548–552.

50 C. Müller and B. Paulus, *Phys. Chem. Chem. Phys.*, 2012, **14**, 7605–7614.

51 G. J. O. Beran, *Chem. Rev.*, 2016, **116**, 5567–5613.
52 P. Ren, C. Wu and J. W. Ponder, *J. Chem. Theory Comput.*, 2011, **7**, 3143–3161.
53 W. H. Keesom and J. W. L. Kohler, *Physica*, 1934, **1**, 655–658.
54 I. Morrison, J.-C. Li, S. Jenkins, S. S. Xantheas and M. C. Payne, *J. Phys. Chem. B*, 1997, **101**, 6146–6150.
55 I. Nahringbauer, *Acta Chem. Scand.*, 1970, **24**, 453.
56 R. K. McMullan, J. Epstein, J. R. Ruble and B. M. Craven, *Acta Crystallogr. B*, 1979, **35**, 688–691.
57 C. C. Wilson, *Z. Kristallogr.*, 2000, **215**, 693–701.
58 I. N. Krupskii, A. Prokhvatilov, A. I. Erenburg and A. S. Barylnik, *Fiz. Nizk. Temp.*, 1982, **8**, 533.
59 V. F. Petrenko and R. W. Whitworth, *Physics of Ice*, Oxford University Press, 1999.
60 P. Jönson, *Acta Crystallogr., Sect. B: Struct. Crystallogr. Cryst. Chem.*, 1971, **27**, 893–898.
61 B. M. Craven, R. K. McMullan, J. D. Bell and H. C. Freeman, *Acta Crystallogr. B*, 1977, **33**, 2585.
62 G. Will, *Z. Kristallogr.*, 1969, **129**, 211.
63 M. Azreg-Aïnou, *Monatsh. Chem.*, 2005, **136**, 2017–2027.
64 W. F. Giauque and C. J. Egan, *J. Chem. Phys.*, 1937, **5**, 45.
65 G. Herzberg, *Electronic Spectra and Electronic Structure of Polyatomic Molecules*, D. Van Nostrand Company, Inc., 1966.
66 T. Shimanouchi, *Tables of Molecular Vibrational Frequencies*, United States Department of Commerce, 1972, vol. 1.
67 R. Feistel and W. Wagner, *Geochim. Cosmochim. Acta*, 2007, **71**, 36–45.
68 NIST/TRC Web Thermo Tables (WTT), NIST Standard Reference Subscription Database 2—Lite Edition Version 2-2012-1-Lite.
69 *Handbook of Chemistry and Physics*, ed. W. M. Haynes, CRC Press, 96th edn, 2004.
70 S. Verevkin, *J. Chem. Eng. Data*, 2000, **45**, 953.
71 J. Martin and R. L. Andon, *J. Chem. Thermodyn.*, 1982, **14**, 679.
72 P. Jiménez and C. T. M. V. Roux, *J. Chem. Thermodyn.*, 1987, **19**, 985.
73 H. G. M. De Wit, C. G. De Kruif and J. C. Van Miltenburg, *J. Chem. Thermodyn.*, 1983, **15**, 891.
74 F. Billes, H. Endrédi and G. Jalsovszky, *J. Mol. Struct.*, 1999, **465**, 157.
75 H.-J. Werner, P. J. Knowles, G. Knizia, F. R. Manby, M. Schütz, P. Celani, T. Korona, R. Lindh, A. Mitrushenkov, G. Rauhut, K. R. Shamasundar, T. B. Adler, R. D. Amos, A. Bernhardsson, A. Berning, D. L. Cooper, M. J. O. Deegan, A. J. Dobbyn, F. Eckert, E. Goll, C. Hampel, A. Hesselmann, G. Hetzer, T. Hrenar, G. Jansen, C. Köppl, Y. Liu, A. W. Lloyd, R. A. Mata, A. J. May, S. J. McNicholas, W. Meyer, M. E. Mura, A. Nicklass, D. P. O'Neill, P. Palmieri, D. Peng, K. Pflüger, R. Pitzer, M. Reiher, T. Shiozaki, H. Stoll, A. J. Stone, R. Tarroni, T. Thorsteinsson and M. Wang, *MOLPRO, version.1, a package of ab initio programs*, 2012, see http://www.molpro.net75.
76 A. Hesselmann, *J. Chem. Phys.*, 2008, **128**, 144112.
77 M. Pitonak and A. Hesselmann, *J. Chem. Theory Comput.*, 2010, **6**, 168–178.
78 L. A. Burns, M. S. Marshall and C. D. Sherrill, *J. Chem. Phys.*, 2014, **141**, 234111.

79 Y. Huang, Y. Shao and G. J. O. Beran, *J. Chem. Phys.*, 2013, **138**, 224112.
80 A. Karton and J. M. L. Martin, *Theor. Chem. Acc.*, 2006, **115**, 330–333.
81 T. Helgaker, W. Klopper, H. Koch and J. Noga, *J. Chem. Phys.*, 1997, **106**, 9639–9646.
82 J. W. Ponder, *TINKER v7.1, 2015*, accessed September 21, 2016, http://dasher.wustl.edu/tinker/.
83 J. C. Wu, G. Chattree and P. Ren, *Theor. Chem. Acc.*, 2012, **131**, 1138.
84 M. F. Peintinger, D. V. Oliveira and T. Bredow, *J. Comput. Chem.*, 2013, **34**, 451–459.
85 R. Dovesi, R. Orlando, B. Civalleri, C. Roetti, V. R. Saunders and C. M. Zicovich-Wilson, *Z. Kristallogr.*, 2005, **220**, 571–573.
86 R. Dovesi, V. R. Saunders, C. Roetti, R. Orlando, C. M. Zicovich-Wilson, F. Pascale, B. Civalleri, K. Doll, N. M. Harrison, I. J. Bush, P. D'Arco and M. Llunell, *CRYSTAL09 User's Manual*, University of Torino, Torino, 2009, C. Science and A. Technologies.
87 P. Giannozzi, S. Baroni, N. Bonini, M. Calandra, R. Car, C. Cavazzoni, D. Ceresoli, G. L. Chiarotti, M. Cococcioni, I. Dabo, A. Dal Corso, S. de Gironcoli, S. Fabris, G. Fratesi, R. Gebauer, U. Gerstmann, C. Gougoussis, A. Kokalj, M. Lazzeri, L. Martin-Samos, N. Marzari, F. Mauri, R. Mazzarello, S. Paolini, A. Pasquarello, L. Paulatto, C. Sbraccia, S. Scandolo, G. Sclauzero, A. P. Seitsonen, A. Smogunov, P. Umari and R. M. Wentzcovitch, *J. Phys.: Condens. Matter*, 2009, **21**, 395502.
88 P. Giannozzi, O. Andreussi, T. Brumme, O. Bunau, M. Buongiorno Nardelli, M. Calandra, R. Car, C. Cavazzoni, D. Ceresoli, M. Cococcioni, N. Colonna, I. Carnimeo, A. Dal Corso, S. de Gironcoli, P. Delugas, R. A. DiStasio, A. Ferretti, A. Floris, G. Fratesi, G. Fugallo, R. Gebauer, U. Gerstmann, F. Giustino, T. Gorni, J. Jia, M. Kawamura, H.-Y. Ko, A. Kokalj, E. Küçükbenli, M. Lazzeri, M. Marsili, N. Marzari, F. Mauri, N. L. Nguyen, H.-V. Nguyen, A. Otero-de-la Roza, L. Paulatto, S. Poncé, D. Rocca, R. Sabatini, B. Santra, M. Schlipf, A. P. Seitsonen, A. Smogunov, I. Timrov, T. Thonhauser, P. Umari, N. Vast, X. Wu and S. Baroni, *J. Phys.: Condens. Matter*, 2017, **29**, 465901.
89 A. Togo and I. Tanaka, *Scr. Mater.*, 2015, **108**, 1–5.
90 A. D. Becke, *J. Chem. Phys.*, 1986, **85**, 7184–7187.
91 J. P. Perdew, K. Burke and M. Ernzerhof, *Phys. Rev. Lett.*, 1996, **77**, 3865.
92 A. Otero-de-la Roza and E. R. Johnson, *J. Chem. Phys.*, 2012, **136**, 174109.
93 A. Otero-De-La-Roza, B. H. Cao, I. K. Price, J. E. Hein and E. R. Johnson, *Angew. Chem., Int. Ed.*, 2014, **53**, 7879–7882.
94 A. D. Becke, *Phys. Rev. A*, 1988, **38**, 3098–3100.
95 C. Lee, W. Yang and R. G. Parr, *Phys. Rev. B*, 1988, **37**, 785–789.
96 S. Grimme, *J. Comput. Chem.*, 2006, **27**(15), 1787–1799.
97 J. Li and D. K. Ross, *Nature*, 1993, **365**, 327–329.
98 J. Li, *J. Chem. Phys.*, 1996, **105**, 6733–6755.
99 X. He, O. Sode, S. S. Xantheas and S. Hirata, *J. Chem. Phys.*, 2012, **137**, 204505.
100 P. Zhang, L. Tian, Z. P. Zhang, G. Shao and J. C. Li, *J. Chem. Phys.*, 2012, **137**, 044504.
101 K. E. Riley, M. Pitonak, P. Jurecka and P. Hobza, *Chem. Rev.*, 2010, **110**, 5023–5063.
102 M. A. Salim, S. Y. Willow and S. Hirata, *J. Chem. Phys.*, 2016, **144**, 204503.

103 A. Tkatchenko and M. Scheffler, *Phys. Rev. Lett.*, 2009, **102**, 073005.
104 R. A. DiStasio, O. A. von Lilienfeld and A. Tkatchenko, *Proc. Natl. Acad. Sci. U. S. A.*, 2012, **109**, 14791–14795.
105 A. Tkatchenko, R. A. DiStasio, R. Car and M. Scheffler, *Phys. Rev. Lett.*, 2012, **108**, 236402.
106 A. Ambrosetti, A. M. Reilly, R. A. DiStasio and A. Tkatchenko, *J. Chem. Phys.*, 2014, **140**, 18A508.
107 S. Grimme, A. Hansen, J. G. Brandenburg and C. Bannwarth, *Chem. Rev.*, 2016, **116**, 5105–5154.
108 S. Grimme, J. Antony, S. Ehrlich and H. Krieg, *J. Chem. Phys.*, 2010, **132**, 154104.
109 E. Caldeweyher, C. Bannwarth and S. Grimme, *J. Chem. Phys.*, 2017, **147**, 034112.
110 O. Demerdash, Y. Mao, T. Liu, M. Head-Gordon and T. Head-Gordon, *J. Chem. Phys.*, 2017, **147**, 161721.
111 W. Acree and J. S. Chickos, *J. Phys. Chem. Ref. Data*, 2010, **39**, 043101.
112 G. L. Perlovich, T. V. Volkova and A. Bauer-Brandl, *J. Therm. Anal. Calorim.*, 2007, **89**, 767–774.
113 G. L. Perlovich, T. V. Volkova and A. Bauer-Brandl, *J. Pharm. Sci.*, 2006, **95**, 2158–2169.
114 R. Picciochi, H. P. Diogo and M. E. Minas da Piedade, *J. Therm. Anal. Calorim.*, 2010, **100**, 391–401.
115 R. Boese, D. Bäser, R. Latz and A. Bäumen, *Acta Crystallogr., Sect. C: Cryst. Struct. Commun.*, 1999, **55**, 9900001.
116 J. R. Deschamps, J. M. Cook and Y. Teng, Personal communication, 2008.

Faraday Discussions

PAPER

The importance of configurational disorder in crystal structure prediction: the case of loratadine†

Grahame R. Woollam, *[a] Marcus A. Neumann,[b] Trixie Wagner[c] and Roger J. Davey [d]

Received 31st March 2018, Accepted 14th May 2018
DOI: 10.1039/c8fd00072g

Loratadine, an over-the-counter antihistamine medication, has two known monotropically related polymorphs, both of which feature disorder. A combined experimental and computational approach using variable temperature single crystal X-ray diffraction (VT-SCXRD) analysis and dispersion corrected density functional theory (DFT-D) reveals that the nature of the disorder in each form is markedly different and cannot be described by a simple isolated-site model with thermally populated conformations in either of the two cases. In Form I, the ethyl carbamate functionality adopts two different configurations, with adjacent moieties interacting along one-dimensional chains. The most stable arrangement features alternating configurations, but because of the low energetic cost of stacking faults, the domain sizes are short and an average crystal structure is observed experimentally. The configurational free energy of the disordered structure is lower than the energy of the two corresponding ordered crystal structures, but the energy decrease is dominated by the lower lattice energy of the alternating arrangement with a small entropic contribution. In Form II, the flexible cycloheptane bridge adopts two different configurations. The disorder is not an equilibrium property but is instead frozen-in during the crystallisation process. The configurational free energy of the disordered structure falls in between the lattice energies of the two corresponding ordered structures. The two ordered components of each disordered structure are all found in a crystal structure prediction (CSP) study with the GRACE programme. However, the experimentally observed stability relationship is only reproduced when the energy contribution of disorder is taken into account. The disordered model of Form I is found to be lower in energy than all the other predicted structures and there is no indication of a missing, thermodynamically more stable, form.

[a]*Novartis Pharma AG, Basel 4002, Switzerland. E-mail: grahame.woollam@novartis.com*
[b]*Avant-garde Materials Simulation, Alte Strasse 2, Merzhausen, D-79249, Germany*
[c]*Novartis Institutes for BioMedical Research, Basel 4002, Switzerland*
[d]*School of Chemical Engineering and Analytical Science, University of Manchester, M13 9PL, UK*

† Electronic supplementary information (ESI) available. CCDC 1835768–1835774. For ESI and crystallographic data in CIF or other electronic format see DOI: 10.1039/c8fd00072g

The case of loratadine demonstrates that experimentally observed disorder close to 50/50 does not necessarily correspond to a free energy decrease by $kT \ln 2$.

1 Introduction

Following the generation of a crystal structure landscape, it is usual to ascertain whether the putative structures obtained are sufficiently different that crystallisation conditions could be defined to enable their isolation as distinct polymorphs. Commonly, starting from the computed body of unique crystal structures, distinct families of structures are clustered according to packing similarity. An associated risk of the likelihood that a more stable form will appear is applied once the experimentally determined crystal structures are placed amongst the *de novo* generated crystal structures, assuming that the putative structures accurately represent the domain available to the system of interest.

To assess the prominence of disorder in organic molecular crystals, a Conquest search of the Cambridge Structural Database (CSD) revealed that 25% of the 935 981 structures contained a disordered component; a similar figure of 20% of 2646 structures was obtained from the Novartis internal structural database (excluding structures with unresolved errors and powder structures). Despite this relatively high occurrence, disordered structures have been avoided when selecting targets for the Cambridge Crystallographic Data Centre (CCDC) blind tests to date. However, one of the polymorphs of molecule XXIII from the sixth blind test was disordered and one of the polymorphs of molecule XXI from the fifth blind test contained disorder in the hydrogen bond network, highlighting the ubiquitous nature of disorder.

There are a small number of studies where modelling and rationalising structural disorder have previously been shown to be amenable to CSP, including two highly relevant examples using pharmaceutical compounds. Neumann *et al.*[1] studied the polymorphic compound dalcetrapib using DFT and high pressure crystallisation. Two out of three experimental polymorphs contained a disordered component and would have been described ineffectively without endeavours to determine the configurational contribution to the free energy by means of lattice energy calculations. Price[2] indicated that we are still at the stage of learning how to compare a crystal energy landscape with experimental structures of polymorphs, highlighting deviations between the computed structures when compared with the experimental structures obtained by crystallisation. A proportion of the low energy structures generated as discrete entities may in fact represent alternative arrangements modelled as disordered components by a crystallographer.

Braun *et al.*[3] recognised that further developments are required to enable the accurate calculation of the relative stabilities of crystal structures of active pharmaceutical ingredients (APIs), where conformational flexibility and functional group diversity yield a series of favourable structures. Hinting at, yet not explicitly highlighting, how configuration may play a role in stabilising structures, Braun *et al.* referred to single component systems without alternating conformations (*i.e.* modelled disorder). They noted how CSP methods have advanced to the point of complementing solid form screening efforts, offering a platform from which experimental observations can be rationalised. Thus, further work could be

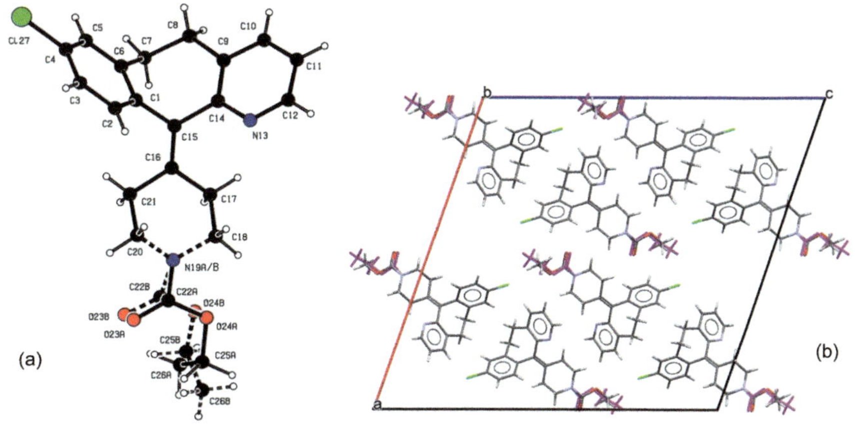

Fig. 1 Molecular and crystal structure of loratadine Form I. (a) Loratadine molecular structure and numbering scheme. (b) Projection down *b*; conformer 2 is coloured magenta from C22 to C26 in order to highlight the alternative "tail" orientation adopted by the minority component.

focused by demonstrating that a range of alternative structures are thermodynamically competitive.

Copley *et al.*[4] noted how the distinctive nature of the disorder in eniluracil accounted for the different structures acquired from powder X-ray diffraction (PXRD) data recorded for different samples. In retrospect, single crystal X-ray refinements of publishable quality may have been interpreted as polymorphism rather than disorder (eniluracil features interchangeable hydrogen bonded motifs). Here, the computed crystal energy landscape was used as an additional form of analysis, and was considered to be a valuable complement to X-ray diffraction and solid-state NMR when trying to understand and characterise disorder in organic solid state systems.

In this contribution we further address the theme of disorder, using loratadine as a case study; Fig. 1 shows its molecular structure. At the outset of this work the only available crystal structure was that of Form I (BEQGIN; with the structure determined at ambient temperature),[5] which was identified as exhibiting disorder. Here we describe the crystallisation and crystal structure of a second polymorph, Form II, and report the use of the GRACE programme for an *in silico* polymorphism assessment of loratadine and the incorporation of disorder into the computations.

2 Materials and methods

Loratadine free base (Form I) was purchased from ABCR (1 kg batch AB261650); 98% purity, toluene; Sigma Aldrich reagent grade ≥99%, *tert*-butyl methyl ether (TBME); Fluka 99%.

2.1 Preparation of Form I

Loratadine free base Form I crystallises from a variety of organic single and mixed solvent systems. Single crystals were prepared and analysed from acetonitrile, cumene, ethyl acetate, nitromethane, toluene and toluene/*tert*-butyl methyl ether.

2.2 Preparation of Form II

Form II was first recorded based on its PXRD pattern in a patent by Dibenedetto and Gala (Schering Corporation).[6] No other references to either its preparation or characterisation could be found in the open literature. The patent states "we have discovered specific solvents and experimental conditions which produce a distinctly different polymorph, Form II, of loratadine". It describes a complex process in which, briefly, loratadine was dissolved in hot toluene, with crystallisation taking place upon the addition of an antisolvent, TBME, at −3 to −10 °C. After stirring for about 1–6 hours at this temperature, the resulting crystals were of Form II.

Unfortunately, this written description omits many essential details, with no targeted concentrations, volumes, solvent compositions or cooling rates given. Eventually, with advice from Robert Wenslow of Crystal Pharmatech, the isolation of Form II was successfully achieved.[7] Briefly, 480 mg of loratadine Form I and 0.8 mL of toluene were loaded into a 5 mL glass vial and stirred magnetically at 50 °C for 90 minutes to produce a clear solution. This solution was cooled to 30 °C, 1 mL TBME was added and the solution was filtered through a Nylon membrane (with a pore size of 0.45 μm) into a clean vial. This clear solution was heated to 50 °C at 2 K min^{-1}, held at 50 °C for 16 hours, then cooled to 4 °C at 1 K min^{-1} and held at 4 °C for 60 minutes, whereupon Form II crystallised. The product was filtered under vacuum and dried at 40 °C for 3 hours and 50 °C for 1 hour. Using this method, Form II was successfully prepared on a scale of 0.5 g. While this recipe was successful as a consistent methodology for preparing Form II, it seemed essential to gain some scientific insight into the crystallisation processes occurring throughout the procedure. This was achieved by scaling the process up to ~100 g scale and observing the physical changes taking place.

This revealed that during cooling of the mixed loratadine/toluene/TBME solution to 4 °C, an amount of colourless liquid could be seen to condense and cascade down the crystallisation vessel walls into the cooling solution of loratadine. Assuming this to be TBME, it could offer a form of self-induced antisolvent crystallisation in combination with the cooling crystallisation. An independent experiment, in which TBME levels were measured chromatographically, confirmed that the 16 hour 50 °C hold period allows two important processes to occur: firstly, the loss of TBME results in an increase of the solution concentration, and secondly, the TBME in the head space condenses on cooling and enters the solution. It is this process which effectively provides a solvent drown out in addition to the cooling crystallisation process.

2.3 DSC measurements

Differential Scanning Calorimetry (DSC) data were recorded in order to measure the melting temperature and enthalpy of the loratadine free base polymorphs.

The analyses were performed using a TA Instruments Discovery DSC. Accurately weighed samples (0.2–1 mg) were placed into crimped aluminium pans. A reference was prepared using the same sample pan without any material added. The DSC thermogram was recorded as follows: the temperature of the apparatus was equilibrated at 20 °C, and heated to 300 °C at a heating rate of 10 K min^{-1} under a nitrogen flow of 50 mL min^{-1}. The temperature and enthalpy of the instrument were calibrated with indium (at least 99.9999% purity). Trios V

4.1.133073 was used to control the TA Discovery DSC instrument and collect and process the data.

2.4 Crystal structure determination

Single crystal X-ray diffraction data were collected at various temperatures (as specified) using a Bruker AXS SMART 6000 CCD detector on a three-circle platform goniometer with Cu K_α radiation (λ = 1.54178 Å) from a microsource generator equipped with multilayer mirrors. Data processing and global cell refinement were performed with Saint. A semi-empirical absorption correction (SADABS) was applied,[8] based on the intensities of symmetry-related reflections measured at different angular settings. All structures were solved by dual-space recycling methods and refined on F^2 with the SHELXTL suite of programmes.[9] Anisotropic displacement parameters were used for all non-hydrogen atoms; hydrogen atoms were calculated in idealised positions and refined using a riding model. For the refinement of disorder, all the bond lengths and angles in the minor occupancy orientation were restrained to be similar to those in the major occupancy orientation (SAME), and the displacement parameters of the corresponding atoms were restrained to be similar (SIMU). The 5 relevant structure solutions referred to in this study are given the reference codes GWO11a, b, c, d and e.

Additional single crystal diffraction data were collected at beamline PXII of the Swiss light source at 10 K with a Pilatus 6M detector on a single axis goniometer using a wavelength of λ = 0.711 Å. Data processing and global cell refinement were performed with XDS.[10] For full references see ref. 11. All structures were solved by dual-space recycling methods and refined on F^2 with the SHELXTL suite of programmes (using Mo K_α (λ = 0.71073 Å) as the wavelength). Anisotropic displacement parameters were used for all non-hydrogen atoms; hydrogen atoms were calculated in idealised positions and refined using a riding model. The two structure solutions obtained from these experiments are referred to as GWO29 and GWO30.

2.5 The Cambridge structural database (CSD) system software

The CSD and accompanying software packages provide a foundation for the analysis and interrogation of crystallographic and structural chemistry in organic crystal systems derived by experimental or *in silico* means. The database,[12] which contains in excess of 900 000 structures, facilitates the analysis of conformations, intermolecular interactions and crystal packing,[13] which are the cornerstones of crystal engineering (design), polymorphism (risk) and crystal structure prediction (rationalisation of the structures within the generated landscape). CSD System Software 2018 release CSD 5.39 (with November 17 and February 18 updates) was used. Searches were performed in Mercury CFC 3.10.1 and Conquest 1.21.

The following filters were methodically switched on; otherwise default settings were used: 3D coordinates determined, no errors, not polymeric, no powder structures and only organics, with medium geometric tolerances; 20%, distance and 20° angle. The crystal packing similarity tool was used to calculate the root-mean squared deviations in atomic coordinates with a 20 molecule comparison.

Mercury was used to generate images of the asymmetric units, packing motifs and non-bonded interactions.

2.6 Crystal structure prediction

The computer programme GRACE[14,15] was used to generate an accurate lattice energy landscape of loratadine. GRACE uses a dispersion-corrected density functional theory method (DFT-D) that combines calculations with the PBE functional in VASP[16,17] with an empirical dispersion correction.[14] Every crystal structure prediction starts by fitting a tailor-made force field (TMFF) to DFT-D reference data. The actual crystal structure prediction is a convergence-controlled three-step procedure, executed separately for one and two molecules per asymmetric unit. In the first step, a large number of crystal structures are generated with a Monte Carlo parallel tempering algorithm using the tailor-made force field. In the second step, some of these structures are subjected to a coarse lattice energy optimisation at the DFT-D level. In the final step, a small number of structures are subjected to a more stringent DFT-D lattice energy optimisation. DFT calculations use a plane wave cutoff energy of 520 eV and a *k*-point spacing of roughly 0.07 $Å^{-1}$. All lattice energy minimisations of the final step are converged to within at least 0.003 Å for atomic displacements, 0.00025 kcal mol^{-1} per atom for energy changes, 0.7 kcal mol^{-1} $Å^{-1}$ for atomic forces and 1.0 kbar for cell stress. In the second step the lattice energies are converged to within at least 0.02 Å for atomic displacements, 0.001 kcal mol^{-1} per atom for energy changes, 7.0 kcal mol^{-1} $Å^{-1}$ for atomic forces and 15.0 kbar for cell stress. The convergence criteria of the final step were applied when performing the lattice energy minimisations in the disorder models.

Molecules are considered fully flexible in the Monte Carlo parallel tempering crystal structure generation. A conformational analysis of all molecules is carried out prior to the crystal structure generation in order to flag wide-amplitude degrees of freedom. These wide-amplitude degrees of freedom are varied by the Monte Carlo parallel tempering algorithm alongside the molecular positions and rotations, as well as the unit cell parameters.

The following space groups were covered in the CSP of loratadine:

$Z' = 1$: $P1$, $P\bar{1}$, $P2_1$, $C2$, Pc, Cc, $P2/c$, $P2_1/c$, $C2/c$, $P2_12_12$, $P2_12_12_1$, $C222_1$, $Pca2_1$, $Pna2_1$, $Aba2$, $Fdd2$, $Iba2$, $Pcca$, $Pccn$, $Pbcn$, $Pbca$, $Fddd$, $P4_1$, $I4$, $I4_1$, $I\bar{4}$, $P4_2/n$, $I4_1/a$, $P4_12_12$, $I4_1cd$, $P\bar{4}2_1c$, $P3_1$, $R3$, $R\bar{3}$, $P3_121$, $R3c$, $P61$, $P6_122$.

$Z' = 2$: $P1$, $P\bar{1}$, $P2_1$, $C2$, Pc, Cc, $P2_1/c$, $C2/c$, $P2_12_12_1$, $Pca2_1$, $Pna2_1$.

2.6.1 Isolated-site disorder model. In molecular crystals, disorder and the associated free energy can often be described in terms of isolated sites with configurations that are thermally populated according to the Boltzmann distribution. For the sites to be isolated, disordered regions in space should be separated by ordered regions that are at least a few atoms thick. In practice it is convenient to express energies in terms of deviations from the energy of a reference configuration, E_{ref}. The reference configuration typically has the lowest energy of all the configurations and may correspond to the major component of a disordered experimental crystal structure or an ordered structure obtained by crystal structure prediction. The energies of other configurations are expressed in terms of energy differences, ΔE_i, relative to the reference configuration. The energy difference ΔE_0 associated with the reference configuration is zero by definition. To evaluate the energy differences ΔE_i, it is usual to start from an energy-minimised cell or a supercell with all molecules in the reference configuration. By exchanging one molecule with one in the alternative configuration and again minimising the lattice energy, the difference gives ΔE_i.

The partition function Z, the free energy F, the internal energy U and the thermal occupancies p_i can be evaluated according to the following formula:

$$Z = \sum_i \mathrm{e}^{-\frac{\Delta E_\mathrm{i}}{kT}} \tag{1}$$

$$F = E_\mathrm{ref} - kT \ln Z \tag{2}$$

$$U = E_\mathrm{ref} + \frac{1}{Z} \sum_i \Delta E_\mathrm{i} \mathrm{e}^{-\frac{\Delta E_\mathrm{i}}{kT}} \tag{3}$$

$$p_\mathrm{i} = \mathrm{e}^{-\frac{\Delta E_\mathrm{i}}{kT}} \Big/ Z \tag{4}$$

In our discussion we neglect the effect of hydrostatic pressure, and the internal energy U is a valid approximation for the enthalpy.

There is sometimes confusion with respect to how many molecules the energies E_ref, U and F refer to and it is important to clarify this issue. The asymmetric unit may contain several molecules and only one of the molecules may be disordered. The correct results are obtained when E_ref, U and F are evaluated per asymmetric unit. However, when the energy differences ΔE_i are determined by calculations on cells or supercells, the energy difference of the entire cell or supercell needs to be taken into account and not be divided by the number of molecules or asymmetric units per cell or supercell. When initially allowing for the procedure, this may appear counterintuitive, yet it should be considered that only one asymmetric unit is changed to another configuration, thus the obtained energy change is *de facto* an energy change per asymmetric unit. It is also important to note at this stage that the energies obtained in a crystal structure prediction study are typically specified per molecule. When performing calculations on disordered structures with more than one molecule per asymmetric unit, one first needs to determine E_ref by multiplying the energy per molecule by the number of molecules per asymmetric unit, then apply eqn (2) and/or (3), and finally divide by the number of molecules per asymmetric unit, in order that the configurational free energies obtained can be compared to the lattice energies of the fully ordered crystal structures from the crystal structure prediction study.

2.6.2 Symmetry-adapted ensemble theory disorder model. When disordered sites are close enough to interact strongly, the isolated site model is not applicable and in principle one needs to consider all the combinations of disordered configurations in a cell or supercell. However, many of these combinations are symmetry equivalent, and the computational effort can be greatly reduced by limiting lattice energy calculations to one representative per symmetry equivalent set and working out the correct multiplicities. This approach is called symmetry-adapted ensemble theory.

Eqn (2) remains valid but eqn (1), (3) and (4) need to be replaced by eqn (5), (6) and (7), respectively.

$$Z = \sum_i m_i e^{-\frac{\Delta E_i}{kT}} \tag{5}$$

$$U = E_{\text{ref}} + \frac{1}{Z}\sum_i \Delta E_i m_i e^{-\frac{\Delta E_i}{kT}} \tag{6}$$

$$p_A = \frac{1}{Z}\frac{1}{N}\sum_i n_A m_i e^{-\frac{\Delta E_i}{kT}} \tag{7}$$

The index i now covers all non-equivalent combinations of configurations with multiplicity m_i. To obtain the fraction of molecules with a configuration A (see eqn (7)), one needs to count the number of molecules with configuration A, n_A, for each combination of configurations. N is the total number of molecules.

It is again pertinent to ask how many molecules the energies E_{ref}, U and F refer to. Typically the number of molecules per cell or supercell would be the right choice, though with Forms I and II of loratadine an alternative selection allowed for computational time to be conserved. In both cases it was clear from geometrical calculations that the disordered sites interact along one-dimensional rods, with little interaction between the rods. Therefore, the energies refer to the numbers of molecules per rod, with two rods per supercell. The evaluation of the energy differences, ΔE_i, was again based on the energies of the entire supercell.

2.6.3 Frozen-in disorder. It may be that experimentally observed disorder is not in fact a thermodynamic equilibrium property and is instead frozen-in during the crystallisation process. The occupation factors, o_i, of the disordered configurations can be determined experimentally. The occupation factors can be used to determine the non-equilibrium internal energy and free energy of the system in the isolated site approximation:

$$U = E_{\text{ref}} + \sum_i o_i \Delta E_i \tag{8}$$

$$F = U + kT\sum_i o_i \ln o_i \tag{9}$$

3 Results

3.1 Crystal structures

Tables 1 and 2 provide crystallographic data associated with the structures of Form I and II as well as the temperature dependence of the Form I parameters. This is followed by a discussion of the key features and differences between the structures.

3.1.1 Form I. Loratadine Form I (GWO29, 10 K) crystallises in the monoclinic space group *C*2/*c*. The carbamate 'tail' is disordered, as seen in Fig. 1(a). In the structure solution this was modelled by refining two different conformations (starting at N19) with a ratio of 0.51 : 0.49. This result essentially mirrors the previous determination, BEQGIN,[18] performed at 300 K. Fig. 1(b) provides

Table 1 Crystallographic details of loratadine free base Form I and Form II

Polymorph	Form I	Form II
Identifier	GWO29	GWO30
Temperature (K/°C)	10/−263	10/−263
Empirical formula	$C_{22}H_{23}ClN_2O_2$	$C_{22}H_{23}ClN_2O_2$
Formula weight	382.87	382.87
Wavelength (Å)	0.71073	0.71073
Crystal system	Monoclinic	Monoclinic
Space group (no.)	$C2/c$ (15)	$C2/c$ (15)
Unit cell dimensions (Å, °)	a = 27.844(16)	a = 35.652(10)
	b = 4.867(3)	b = 5.206(2)
	c = 28.892(16)	c = 22.743(6)
	β = 108.98(3)	β = 117.418(14)
Volume (Å^3)	3703(4)	3747(2)
Z	8	8
D_{calc} (g cm^{-3})	1.374	1.357
μ (mm^{-1})	0.227	0.224
$F(000)$	1616	1616
Crystal size (mm^3)	0.46,·0.04,·0.02	0.18,·0.04,·0.02
Refl. measured	19 266	19 746
Independent	3194	3215
R_{int}	0.0454	0.0451
Θ range (°)	1.49–25.03	1.29–25.02
Completeness (%)	98.3	97.4
Refl. with $I > 2\sigma(I)$	3115	2867
Restraints	180	282
Parameters	293	265
R_1 ($I > 2\sigma(I)$)	0.0321	0.0466
R_1 (all data)	0.0324	0.0525
wR_2 ($I > 2\sigma(I)$)	0.0826	0.1134
wR_2 (all data)	0.0828	0.1246
GooF	1.055	1.088
Restrained GooF	1.031	1.115
Residual electron density (e$^-$ Å^{-3})	+0.29/−0.32	+0.41/−0.39

a packing diagram viewed down b, showing how the two conformers pack and how the alternate carbamate tail orientations are accommodated within the structure. These conformers are related by rotation around the torsion C22–O24–C25–C26, defined below as (Torsion 3) conformers 1 and 2. Structure determinations between 10 and 353 K indicated that this disorder could not be frozen-out. This is discussed in more detail later in this section.

Fig. 2 shows details of the molecular packing. Molecules are related by translation along the b axis through the utilisation of short C–H···N and C–H···O contacts, shown in Fig. 2(a). The two weak H-bonds in this b axis chain are between the piperidine axial proton and the pyridine nitrogen (2.713 Å), in addition to the carbamate ether oxygen and the methyl proton CH26 of the translated molecule (2.418 Å). Additionally, the pyridine nitrogen (N13) forms an intramolecular weak H-bond to the piperidine N···HC (CH17) (2.397 Å).

The existence of the two conformers in the structure results in different non-bonded interactions involving the carbamate moiety and, as such, the dimers in the (010) plane shown in Fig. 2(b). Conformers 1 and 2 are overlaid to depict the

Table 2 Crystallographic details of loratadine free base Form I at various collection temperatures

Polymorph	Form I					
Identifier	GWO29	GWO11d	GWO11a	GWO11b	GWO11e	GWO11c
Temperature (K/°C)	10/−263	100/−173	173/−100	273/0	298/25	353/80
Empirical formula	$C_{22}H_{23}ClN_2O_2$	$C_{22}H_{23}ClN_2O_2$	$C_{22}H_{23}ClN_2O_2$	$C_{22}H_{23}ClN_2O_2$	$C_{22}H_{23}ClN_2O_2$	$C_{22}H_{23}ClN_2O_2$
Formula weight	382.87	382.87	382.87	382.87	382.87	382.87
Wavelength (Å)	0.71073	1.54178	1.54178	1.54178	1.54178	1.54178
Crystal system	Monoclinic	Monoclinic	Monoclinic	Monoclinic	Monoclinic	Monoclinic
Space group (no.)	$C2/c$ (15)	$C2/c$ (15)	$C2/c$ (15)	$C2/c$ (15)	$C2/c$ (15)	$C2/c$ (15)
Unit cell dimensions (Å, °)	a = 27.844(16)	a = 27.906(4)	a = 27.997(4)	a = 28.201(6)	a = 28.301(6)	a = 28.502(15)
	b = 4.867(3)	b = 4.8860(10)	b = 4.918(1)	b = 4.972(1)	b = 4.998(1)	b = 5.051(4)
	c = 28.892(16)	c = 28.927(5)	c = 28.971(4)	c = 29.106(7)	c = 29.154(6)	c = 29.231(16)
	β = 108.98(3)	β = 109.016(5)	β = 108.987(1)	β = 109.077(10)	β = 109.217(6)	β = 109.51(3)
Volume (Å^3)	3703(4)	3728.9(11)	3772.0(11)	3857.0(15)	3894(8)	3966(4)
Z	8	8	8	8	8	8
D_{calc} (g cm^{-3})	1.374	1.364	1.348	1.319	1.306	1.282
μ (mm^{-1})	0.227	1.971	1.948	1.906	1.887	1.853
F(000)	1616	1616	1616	1616	1616	1616
Crystal size (mm^3)	0.46,·0.04,·0.02	0.55,·0.06,·0.03	0.55,·0.06,·0.03	0.55,·0.06,·0.03	0.55·0.06·0.03	0.55·0.06·0.03
Refl. measured	19 266	17 727	17 978	17 068	18 666	19 121
Independent	3194	3249	3284	3352	3398	3449
R_{int}	0.0454	0.0323	0.0309	0.0325	0.0346	0.0360
Θ range (°)	1.49–25.03	3.32–66.59	3.23–66.58	3.21–66.56	3.21–66.59	3.21–66.58
Completeness (%)	98.3	98.8	98.7	98.7	99.1	98.9
Refl. with $I > 2\sigma(I)$	3115	3045	3035	2881	2777	2434
Restraints	180	180	180	186	186	186
Parameters	293	292	292	292	292	292
R_1 ($I > 2\sigma(I)$)	0.0321	0.0356	0.0383	0.0458	0.0559	0.0798

Table 2 *(Contd.)*

Polymorph	Form I					
Identifier	GWO29	GWO11d	GWO11a	GWO11b	GWO11e	GWO11c
R_1 (all data)	0.0324	0.0383	0.0429	0.0541	0.0671	0.1010
wR_2 ($I > 2\sigma(I)$)	0.0826	0.1088	0.1129	0.1252	0.1587	0.2437
wR_2 (all data)	0.0828	0.1160	0.1297	0.1431	0.1753	0.2792
GooF	1.055	1.118	1.172	1.105	1.076	1.088
Restrained GooF	1.031	1.095	1.154	1.106	1.082	1.101
Residual electron density ($e^- Å^{-3}$)	+0.29/−0.32	+0.25/−0.38	+0.24/−0.42	+0.35/−0.44	+0.43/−0.40	+0.62/−0.35

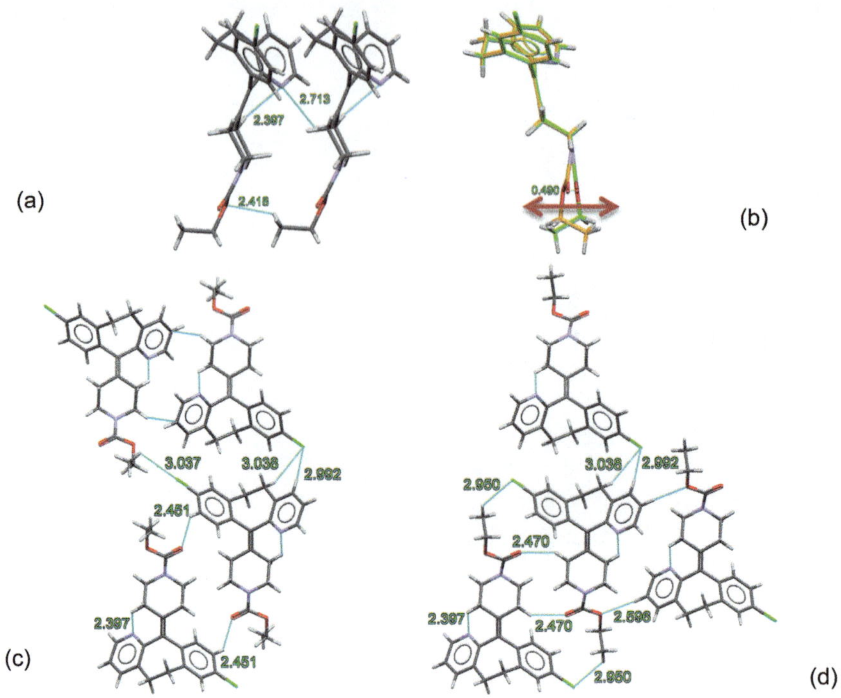

Fig. 2 Packing in loratadine Form I. (a) Molecules are related by translation along the *b* axis through utilisation of short C–H⋯N and CH⋯O contacts. (b) Overlay of conformers 1 (green) and 2 (orange), highlighting the deviation in the relative position of the carbonyl oxygen. (c) Conformer 1 short contacts and distances. (d) Conformer 2 short contacts and distances.

difference in the relative position of the carbonyl oxygen at a distance of 0.490 Å. As a result, the position of the carbamate tail in conformer 1 is closer to the translated molecule, thus a shorter contact distance results for the $R_2^2(22)$ centrosymmetric carbonyl⋯phenyl dimer (2.451 compared to 2.767 Å), as seen in Fig. 2(c). Meanwhile, conformer 2 has a shorter contact distance for the $R_2^2(12)$ carbonyl⋯piperidine dimer in the same plane (2.470 compared to 2.773 Å), as seen in Fig. 2(d).

Fig. 2(c) and (d) illustrate that both conformers have the same catemeric head-to-head chlorine⋯alicyclic cycloheptane CH8 and chlorine⋯aromatic pyridine CH10 interactions within the (010) plane, as depicted in Fig. 1(b).

3.1.2 Form II. Loratadine Form II (GWO30, 10 K) also crystallises in the monoclinic space group *C*2/*c*. In contrast to the carbamate tail disorder of Form I, in this form it is the cycloheptane CH_2–CH_2 bridge (C7–C8–C9) that is disordered. Conformational flexibility at this "head" is discussed further below. The single crystal structure was solved by modelling two different conformations in a ratio of 0.53 : 0.47, shown in Fig. 3 and defined below as (Torsion 1) conformers 1 and 2.

In contrast to Form I, the pyridine nitrogen no longer forms an intramolecular close contact to the piperidine, and the N⋯HC distance is increased to 2.8 Å. However, as with Form I, the molecules are related by translation along the *b* axis and the existence of dimers in the (010) plane is evident in Fig. 3(b). Fig. 4(a)

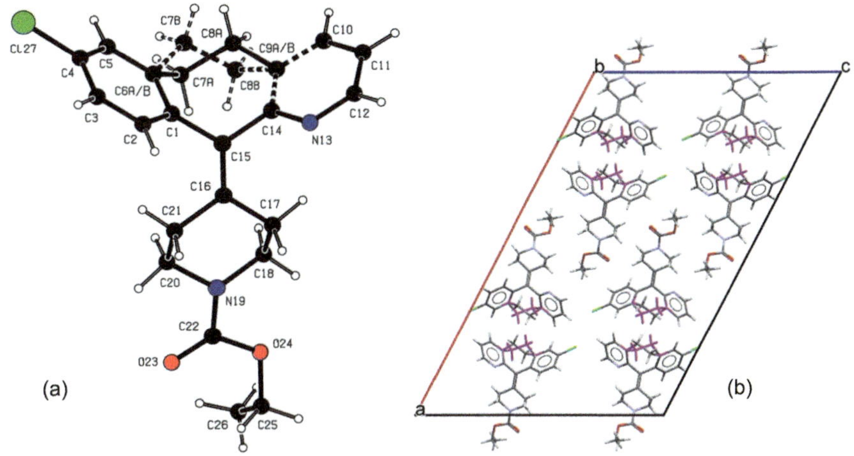

Fig. 3 Molecular and crystal structure of loratadine Form II. (a) Loratadine molecular structure and numbering scheme. (b) Projection down *b*; conformer 2 is coloured magenta from C6 to C9 in order to highlight the alternative "head" orientation adopted by the minority component.

shows the bifurcated interaction with the pyridine nitrogen utilising both a translated molecule and a rotated molecule along the *b* axis. In contrast to Form I, the contact is with the bottom of the translated piperidine ring (2.730 Å), owing to the conformation of the alicyclic system having the lower half of the piperidine ring (CH_2–N–CH_2) in the back position. The second interaction with pyridine is from the carbamate tail CH_2 (2.741 Å) of the rotated molecule (as depicted in

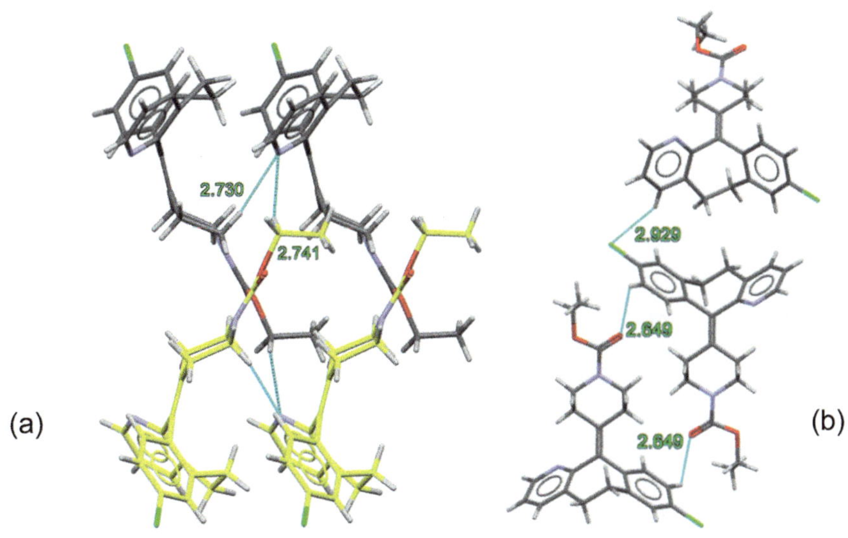

Fig. 4 Crystal packing in loratadine Form II. (a) Molecules are related by translation along the *b* axis through utilisation of short C–H⋯N contacts. (b) Conformer 1 short contacts and distance.

Fig. 4(a)). It is perhaps this pyridine···CH_2 carbamate interaction that prevents free rotation of the bond, and thus explains the absence of both back and forward conformations of the ethyl tail, as seen in Form I. Within the (010) plane the $R_2^2(22)$ centrosymmetric carbonyl dimer, seen in Fig. 4(b), involves the C=O and CH of the aromatic phenyl ring (inversion) (2.649 Å), however the $R_2^2(12)$ carbonyl···piperidine dimer seen in Form I conformer 2 is absent. The different components in the lattice have subtly different interactions to one another: conformer 1 has aliphatic cycloheptane CH7···π phenyl close contacts and conformer 2 has CH8···π pyridine interactions, specifically the weak H-bond between aliphatic CH8···N-pyridine (2.465 Å).

From the previous sections it is evident that Forms I and II differ in both conformational disorder and major intermolecular interactions. The use of the crystal packing similarity search in CCDC Mercury allows a wider perspective on these differences through a comparison of the overall packing of the molecules. This reveals that in overlaying 20 molecules from each structure, only 1 molecule shows a positional match, effectively indicating a total lack of similarity in the packings of Form I and Form II. It appears that this may be accounted for by the combined differences in conformation and weak H-bonding arrangements. Given that loratadine does not feature strong hydrogen bond donors and that, superficially, it would appear relatively rigid compared with many modern pharmaceutical molecules (*e.g.* ritonavir), this result is perhaps surprising.

3.1.3 Torsions. In this section the key torsions leading to the observed conformational variations in the two polymorphs are defined and discussed in

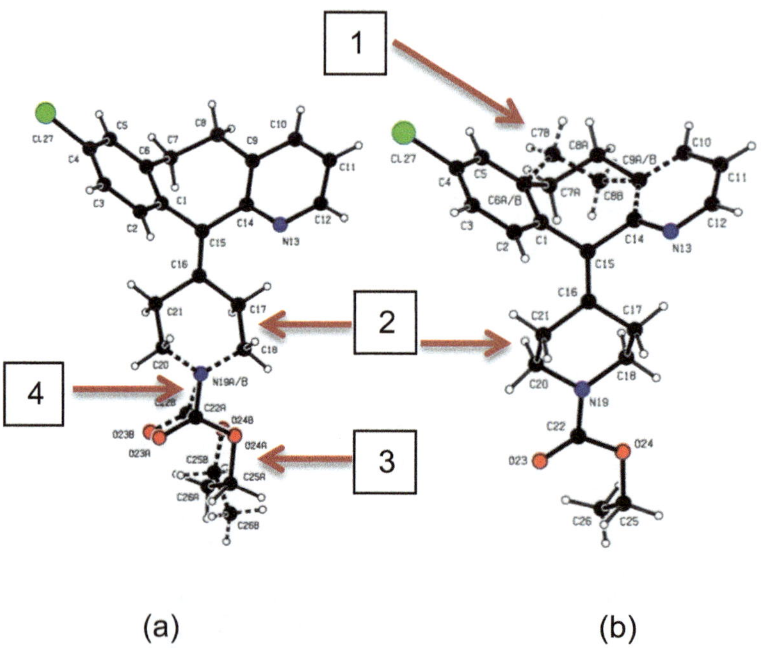

Fig. 5 Rotatable bonds in the asymmetric units of Forms I and II. (a) Form I and (b) Form II with numbering scheme marking and naming the torsions of interest, highlighting the alternative conformations in the carbamate tail and cycloheptane bridge, respectively.

the context of the previous crystal structures and from a wider structural perspective available through the CSD. Fig. 5 depicts the four torsions of interest that vary throughout the polymorphic pair. Torsion 1 refers to the cycloheptane methylene bridge (the feature of disorder in Form II), Torsion 2 to the piperidine ring (a key conformational difference between Forms I and II), Torsion 3 to the carbamate tail (the feature of disorder in Form I and thus the key torsion of interest) and Torsion 4 to the piperidine carbamate. Fig. 6(a) shows a molecular overlay of Form II conformers (Torsion 1), with cyan and magenta representing conformers 1 and 2, respectively. Fig. 6(b) gives a molecular overlay of Form I in green and Form II in blue to highlight the alternative piperidine chair conformations (Torsion 2) of the polymorphs. Fig. 6(c) shows a molecular overlay of Form I conformers (Torsion 3), with green and orange representing conformers 1 and 2, respectively.

Considering firstly Fig. 5, it is noted that loratadine has two formal rotatable bonds in the carbamate tail, indicated by the red arrows 3 and 4 in Fig. 5(a and b), giving rise to Torsion 3 C(22)–O(24)–C(25)–C(26) and Torsion 4 N(19)–C(22)–O(24)–C(25). It is this torsion which features in the conformational disorder observed in Form I. As far as the flexibility in loratadine's two ring systems is concerned, this gives rise to Torsions 1 and 2 in Fig. 5(a and b). Flipping of the cycloheptane ring can occur at C7 and C8, as seen in Fig. 5(b), and this is the origin of the conformational disorder in Form II (Fig. 3(a)), in which C7 and C8 effectively swap back and front positions. This flip can be examined formally *via* the torsion C(5)–C(6)–C(7)–C(8) (Torsion 1), which allows the monitoring of both C7 and C8. While conformational disorder in the cycloheptane group is a feature only of Form II, with both C7 in the back position (conformer 1) and C8 in the back position (conformer 2), the variation of conformers 1 and 2 is seen with a relatively equal distribution throughout the structural database. The second ring system to exhibit flexibility is the piperidine ring, Torsion 2 in Fig. 5(a and b),

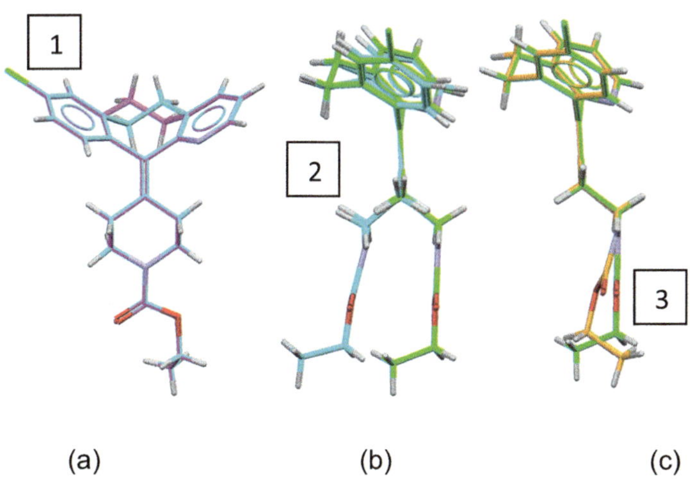

Fig. 6 Comparison of conformers in the polymorphs. (a) Torsion 1 variation highlighted with an overlay of Form II conformers 1 (cyan) and 2 (magenta). (b) Torsion 2 variation highlighted with an overlay of Form I (green) and Form II (blue). (c) Torsion 3 variation highlighted with an overlay of Form I conformers 1 (green) and 2 (orange).

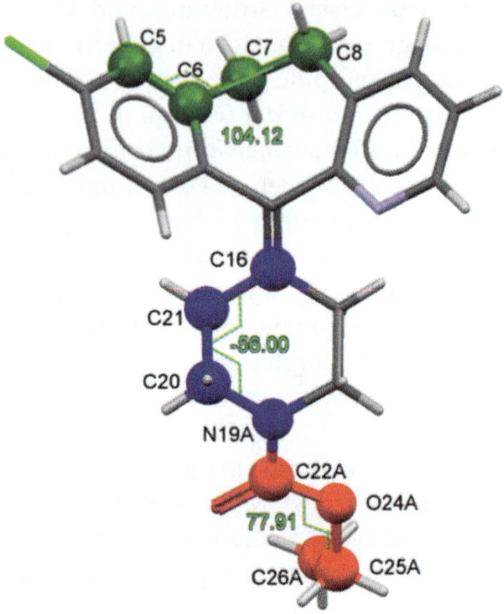

Fig. 7 The molecular structure of loratadine with colour coded torsions of interest (Form I conformer 1 torsions enumerated). Torsion 1: green; Torsion 2: blue; Torsion 3: red.

which, as shown in Fig. 6(b), has two well defined chair conformations involving the position of N19. Thus, Fig. 6(b) shows Form I, where the chair adopts an N19 forward position (in green), and Form II, where it adopts the backward position (in blue). The conformational change results in different inter- and intramolecular interactions and, ultimately, very different packing motifs between the polymorphic pair.

Further understanding of these conformational variations was gained by comparing the torsions found in loratadine with those in the CSD using Mogul. The four torsions studied were based on the previous discussion and are highlighted in Fig. 5, including Torsions 1 (green atoms), 2 (blue atoms) and 3 (red atoms), with the colours referring to Fig. 7. Torsion 4 has two well defined torsions, as reflected in the CSD search: ~179°, as observed in the polymorphs, and also ~−179° (maximum variance 8°), as found in the database. Attention is paid to Torsions 1 and 3, as they feature in the configurational disorder of Forms II and I, respectively, whereas Torsion 2 ultimately defines the major difference between Forms I and II.

3.1.3.1 Torsion 1. As Fig. 8 shows, the CSD population mirrors the dihedral angles found within the loratadine polymorphs, providing we accept that Torsion 1 and the related torsion C(7)–C(8)–C(9)–C(10) are equivalent and opposite, *i.e.* when C7 is back into the page, Torsion 1 is 104° and C(7)–C(8)–C(9)–C(10) is 179°; the angles are inverted when C8 is back into the page.

3.1.3.2 Torsion 3. According to the CSD search, Torsion 3 varies completely, with four major populations ranging from −179 to 179° (with −94 and 81° as alternative ranges), as seen in Fig. 9. On the left hand side of the distribution (±70° divide), Form I has C26 in the forward position (*i.e.* −94°, orange conformer 2). On the right hand side, C26 is in the backward position (*i.e.* 78°, green

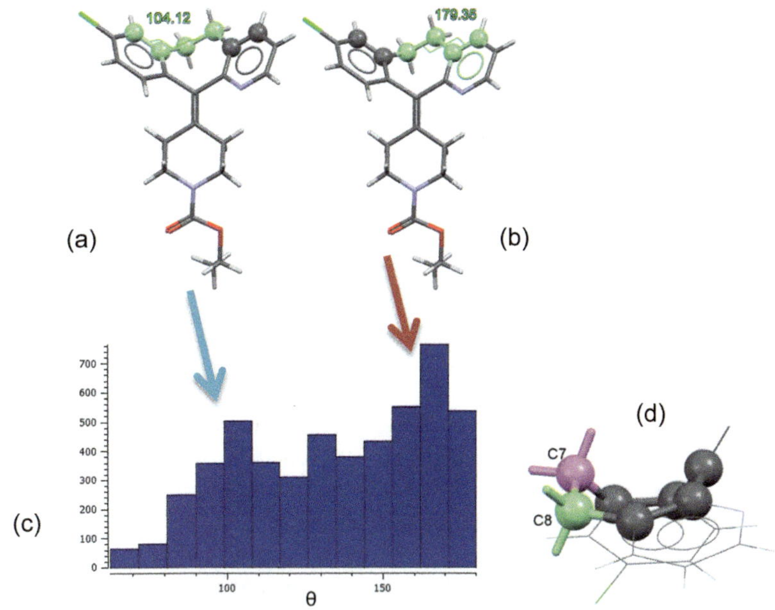

Fig. 8 The CSD results of Torsion 1. (a) Dihedral angle of Torsion 1 at 104°. (b) Dihedral angle of related torsion C(7)–C(8)–C(9)–C(10) at 179°. (c) Histogram of the distribution of dihedral angles for Torsion 1 (9560 hits), showing two populations. (d) Detail of the twisted boat conformation at the head of the cycloheptane group; a ring flip takes place to avoid the protons at C7 and C8 eclipsing.

conformer 1). It is noted that neither Form I nor Form II contains the ±180° orientation, *i.e.* a sideways facing ethyl group in relation to the face on the tricyclic ring system, despite a higher prevalence in the CSD and its occurrence in multicomponent forms of loratadine.

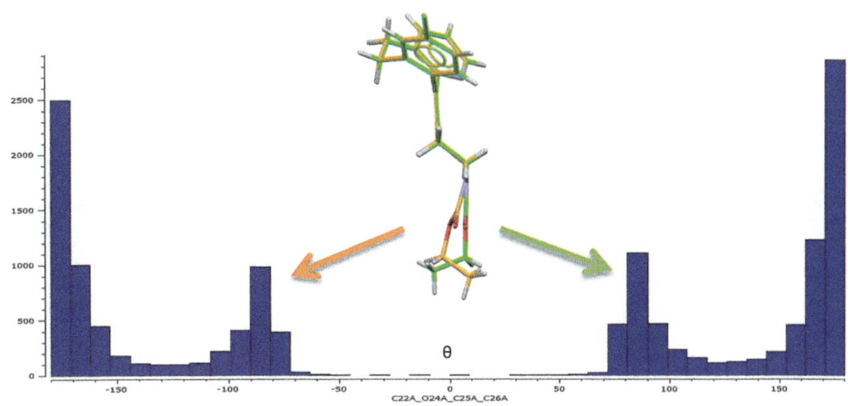

Fig. 9 The CSD results plotted as a histogram (14 141 hits returned) of the distribution of dihedral angles for Torsion 3, with the two conformations shown (major in green – conformer 1, 78° "back"; minor in orange – conformer 2, −94° "forward"), with alternative and more populous orientations at 180/−180°.

Table 3 The variation of conformer ratio, unit cell volume and *b* axis length with temperature

Temperature (K)	Ratio of conformers 1 : 2 (e.s.d.)	Volume (Å^3)	*b* axis (Å)
10	0.51 : 0.49 (0.005)	3703	4.867
100	0.52 : 0.48 (0.006)	3729	4.886
173	0.55 : 0.45 (0.006)	3772	4.918
273	0.55 : 0.45 (0.007)	3857	4.972
298	0.55 : 0.45 (0.008)	3894	4.998
353	0.52 : 0.48 (0.010)	3966	5.051

3.2 Variable temperature SCXRD

To assess whether the ratio of conformers found in Form I is a function of temperature, single crystal X-ray data were collected at six different temperatures: 10, 100, 173, 273, 298, and 353 K. Structure determinations between 10 and 353 K indicated that this disorder could not be frozen-out.

Table 3 shows how the ratio of Form I conformers 1 and 2 changed with increasing temperature; conformer 1 increased from 0.51 : 0.49 at 10 K through to 0.55 : 0.45 at 273 K, while at 353 K (just 54 K below the melting point of Form I) the increase in ratio was discontinuous. The estimated standard deviation (e.s.d.) is shown alongside the occupancy in Table 3 to provide an estimate of the error associated with the ratio of conformers found experimentally in the structures collected at various temperatures. In order to determine whether the variation of occupancy with temperature was significant, an independent refinement was performed by Richard I. Cooper (University of Oxford) using CRYSTALS (as opposed to SHELXTL); all occupancies were found to be equivalent within three times the estimated standard deviation. A plot of the experimentally determined occupation of conformer 2 in loratadine Form I as a function of SCXRD collection temperature, comparing both refinements, can be found in the ESI.† This illustrates the trend of the minor component occupancy decreasing with increasing temperature, which is evident from 10–273 K.

The unit cell expands by 7% and the *b* axis (the shortest axis, and also the needle axis where the ring systems stack) expands by 3.7% from 10 to 353 K.

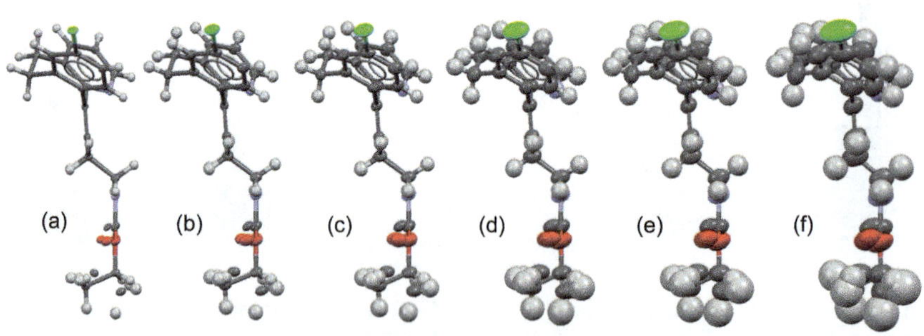

Fig. 10 The thermal ellipsoids (50% probability level) in Form I from left to right (a, b, c, d, e, and f) of structures collected at 10, 100, 173, 273, 298, and 353 K, respectively.

Table 4 Experimental melting onset and enthalpy of Forms I and II

Polymorph	Melting onset (°C)	Enthalpy of fusion (kcal mol^{-1})	Crystallisation exotherm (kcal mol^{-1})
Form I	134.13	6.83	—
Form II	118.79	5.09	0.33

Fig. 10 depicts the molecular structure of loratadine in Form I collected at the various temperatures. The molecule drawn with connected atoms is conformer 1; for conformer 2 only the thermal ellipsoids (50% probability level) of the atoms are shown. The temperature dependence of the thermal ellipsoids in the cycloheptane methylene bridge, the piperidine ring and throughout the whole carbamate function shown in Fig. 10 indicates a marked increase in molecular flexibility and motion beyond 273 K (Fig. 10(d–f)).

3.3 DSC data

Table 4 summarises the DSC results for loratadine Forms I and II. These data confirm that the forms are related monotropically, with Form I being the more stable.

3.4 Crystal structure prediction

The crystal energy landscape of loratadine with one and two molecules per asymmetric unit was determined using GRACE. Fig. 11 shows the 62 most stable predicted structures, grouped as Families 1–4 according to packing similarity. Tabulated lattice energies can be found in the ESI.† The predicted structures are named and referred to according to their rank number in the energy landscape.

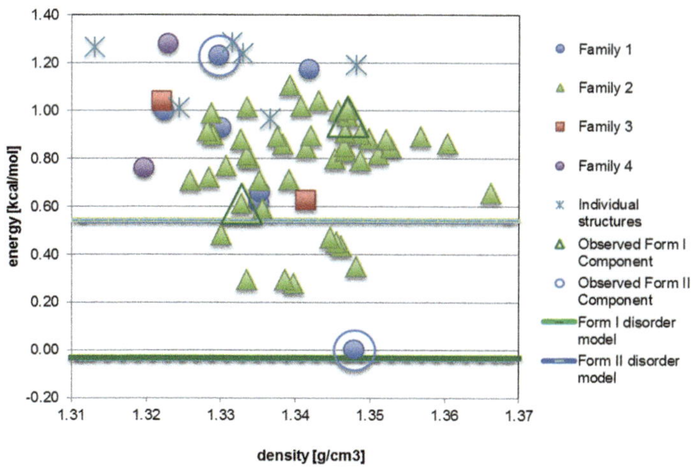

Fig. 11 Crystal energy landscape of loratadine, showing the four families, plus individual structures and the relative energies of the constructed Form I and II disorder models, to illustrate the comparative differences in accounting for disorder by stabilisation or destabilisation.

The two disordered configurations of Form I are found as ordered crystal structures with rank numbers 11 and 45, highlighted on the plot in Fig. 11 surrounded by open green triangles. Similarly, the two disordered configurations of Form II are found as ranks 1 and 58, highlighted on the plot in Fig. 11 surrounded by open blue circles. Fig. 11 also shows two horizontal lines that indicate the energies of the disorder models of Forms I and II, which will be discussed below. Structures with a high degree of similarity are grouped into families and represented by the same symbol, with the exception of the six individual structures which feature a single member (asterisks).

At first sight, the computed crystal energy landscape seems to be in rather stark disagreement with the experimentally observed crystal structures. The two ordered predicted structures matching the experimentally observed disorder in Form II (ranks 1 and 58) are separated by an energy difference of 1.2 kcal mol^{-1}, which is rather large compared to $kT = 0.59$ kcal mol^{-1}, such that one would expect hardly any disorder. For the two ordered structures associated with Form I (ranks 11 and 45), the energy difference is only 0.4 kcal mol^{-1}, such that disorder is to be expected. However, the stabilisation of rank 11 by the configurational free energy would be less than $kT \ln 2 \approx 0.41$ kcal mol^{-1}, and hence would not be

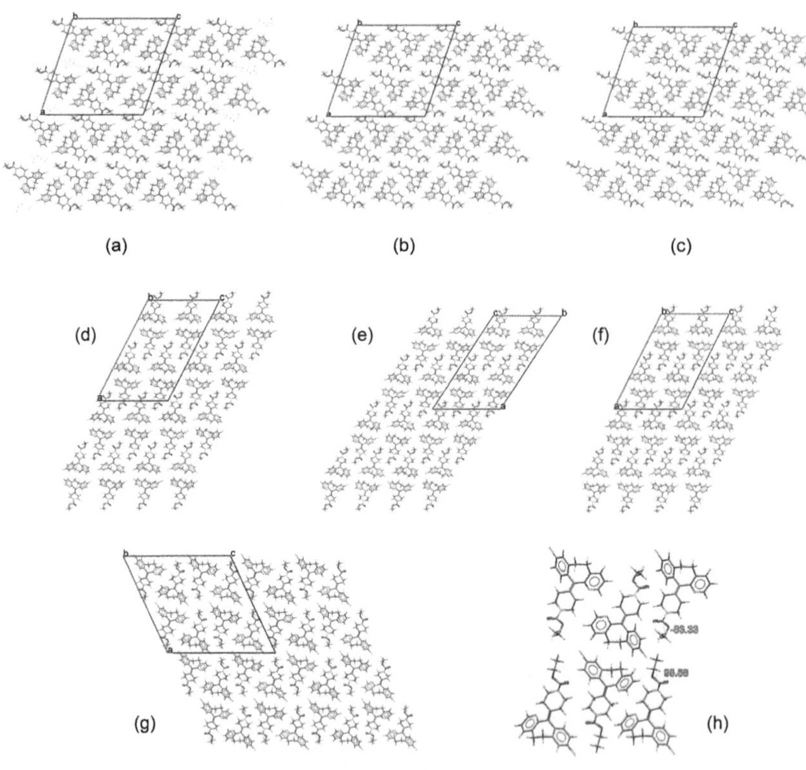

Fig. 12 Some comparisons of experimental and predicted structures. (a) Experimental Form I, (b) rank 11, (c) rank 45; 2 × 2 × 2 packing motifs viewed down *b*. (d) Experimental Form II, (e) rank 1, (f) rank 58; 2 × 2 × 2 packing motifs viewed down *b*. (g) Rank 2 packing (2 × 2 × 2) in the $Z' = 2$, $P2_1/c$ structure. (h) Rank 5, featuring both 83.33° and −95.56°; Torsion 3 conformer 1 and conformer 2 in the $Z' = 2$, $P\bar{1}$ structure.

enough to make the disordered Form I more stable than the predicted rank 1, which is associated with Form II. This contradicts the fact that Form II is experimentally observed to be less stable than Form I at all temperatures.

After more detailed analysis, we will show below that the computed crystal energy landscape is actually in rather good agreement with the experimental findings.

It is instructive to examine some of these predictions and compare them with the experimental structures set out in Fig. 12. Fig. 12(a) shows the experimental Form I, with conformer 1 corresponding to the structure in rank 11 (Family 2), shown in Fig. 12(b), and conformer 2 corresponding to that in rank 45 (Family 2), shown in Fig. 12(c). All structures are found in the space group *C*2/*c* and have the plate-like mosaic packing motifs of Form I. Fig. 12(d) shows the experimental Form II, in which conformer 1 corresponds to rank 1 (Family 1), as shown in Fig. 12(e), and conformer 2 corresponds to rank 58 (Family 1), shown in Fig. 12(f). Again, all belong to the space group *C*2/*c* and feature the rod-like mosaic packing of Form II. Fig. 12(g) indicates that conformational considerations alone are likely to be insufficient to discriminate between the families. This is a prediction (rank 2) with $Z' = 2$ and a Form II-like Torsion 2 $\sim$51° in both independent molecules in the structure, yet it packs in the plate-like mosaic of Family 2. With this in mind, the packing and potential growth units of the *de novo* structures and their similarity and relevance to the experimental structures are expected to be of primary importance. Finally, Fig. 12(h) shows the $Z' = 2$ structure rank 5 from Family 2, which features both Form I conformers 1 and 2.

3.4.1 Isolated-site disorder analysis. Visual analysis of Forms I and II, as seen in Fig. 13, shows that the disordered regions are relatively close and likely to interact with each other.

Nevertheless, it is interesting to apply the isolated-site model as a start. For both Forms I and II, the more stable of the two ordered predicted structures was selected as a starting point. The energy optimised structure was first converted from the C-centred unit cell to a reduced unit cell and then the reduced unit cell was doubled along its shortest axis, resulting in a model system containing 8 molecules. The configuration of one of these molecules was changed to the other

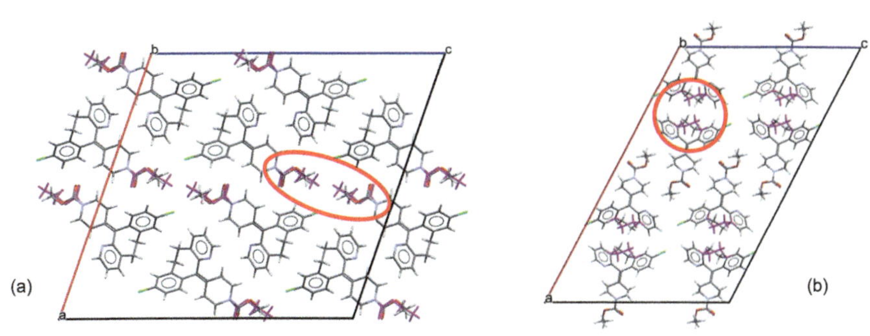

Fig. 13 Experimental Form I and II cells. (a) Form I projection down *b*; conformer 2 is coloured magenta from C22 to C26 in order to highlight the alternative "tail" orientation adopted by the minority component. (b) Form II projection down *b*; conformer 2 is coloured magenta from C6 to C9 in order to highlight the alternative "head" orientation adopted by the minority component.

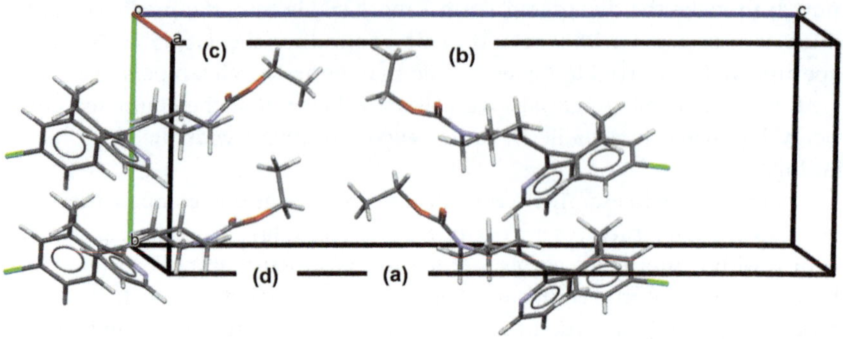

Fig. 14 1D rod of interacting molecules in 1 × 2 × 1 supercell of Form I. Molecules not belonging to the rod have been omitted for clarity.

(disordered) configuration, and the lattice energy of the model system was minimised.

For Form I, where 8 molecules initially have the rank 11 configuration and one molecule is changed to the rank 45 configuration, the lattice energy decreases by as much as 0.85 kcal mol^{-1}, indicating that rank 11 is actually not a valid 'ground state' for disorder analysis. Since the energy decreases, there must be another ordered arrangement with lower symmetry that was not generated by the crystal structure prediction procedure. This issue will be discussed further in the next section.

For Form II, where 8 molecules initially have the rank 1 configuration and one molecule is changed to the rank 58 configuration, the energy increases by 2.0 kcal mol^{-1}, resulting in a population of the less favourable configuration of only 3%; far less than the experimentally observed value of almost 50%. The low calculated population of the less favourable configuration indicates that the experimentally observed disorder in Form II may not be a thermodynamic equilibrium property, but may be frozen-in during the crystallisation process. However, this needs to be confirmed by symmetry-adapted ensemble theory disorder analysis.

3.4.2 Symmetry-adapted ensemble theory disorder analysis. To further investigate the disorder in Forms I and II, symmetry-adapted ensemble theory was used to study the interaction of the disordered sites along a one-dimensional rod. Fig. 14 shows one such rod in a 1 × 2 × 1 supercell with site labels. Molecules not belonging to the rod have been omitted for clarity. In total, there are 7 non-equivalent combinations of configurations. Table 5 shows the 7 cases together with their multiplicity, the number of molecules in the rank 45 configuration, and the energies after lattice energy minimisation compared to the starting cell in which all molecules have the rank 11 configuration.

The most stable combination of configurations by far was obtained when both sites a and c were changed from the rank 11 configuration to the rank 45 configuration, resulting in an energy decrease of almost 2 kcal mol^{-1}. This is the arrangement shown in Fig. 14. To a certain extent, Form I is a missed structure with two molecules per asymmetric unit. However, Form I is accurately observed to be disordered experimentally. The cost of a stacking fault within the rod is

Table 5 Form I supercell models with energies

Sites with rank 45 configuration	Multiplicity	Number of sites with rank 45 configuration	Relative energy (kcal mol^{-1})
—	1	0	0
a	4	1	−0.848
ab	2	2	−0.376
ac	2	2	−1.996
ad	2	2	0.468
abc	4	3	0.892
abcd	1	4	3.264

roughly the energy difference between the situation where both sites a and c are exchanged and the instance (with multiplicity 4) where only one site is exchanged, *i.e.* 1.15 kcal mol^{-1}. Such stacking faults are thermally populated at room temperature, hence destroying the long-range order. Yet, the alternating arrangement still reduces the lattice energy per molecule by 0.5 kcal mol^{-1}.

Using the values in Table 5 together with eqn (2), (5), (6) and (7), we compute an occupation factor for the rank 45 configuration of 0.446, which is in excellent agreement with the experimental value of 0.45. It is interesting to note (see Fig. 15) that the value of the occupation factor for the less stable configuration is actually predicted to decrease with increasing temperature. The predicted behaviour may be difficult to observe experimentally, because the change in the occupation factor is small and at low temperature the conversion to the equilibrium state can be slow.

For the average energy change and the configurational free energy change, compared to the lattice energy of rank 11, symmetry-adapted ensemble theory yields values of +0.405 kcal mol^{-1} and −0.645 kcal mol^{-1}. The symmetry-adapted ensemble theory disorder model of Form I is hence found to be −0.035 kcal mol^{-1} more stable than the predicted rank 1. The model is indicated by a horizontal green line in Fig. 11. Disordered Form I is correctly predicted to be more stable than any other ordered crystal structure.

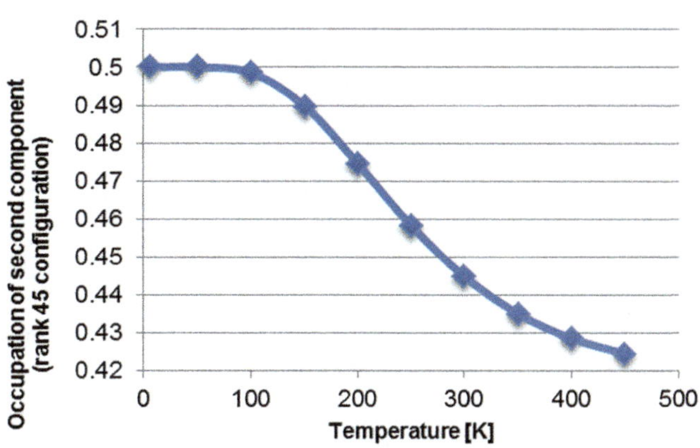

Fig. 15 Predicted occupation of the rank 45 configuration in Form I as a function of temperature.

In the case of Form II, symmetry-adapted ensemble theory confirms that the change of the rank 1 configuration to the rank 58 configuration is highly unfavourable and the high degree of disorder observed in Form II is not a thermodynamic equilibrium property. Energy values are provided in the ESI.†

3.4.3 Non-equilibrium free energy of Form II. Since both the isolated-site model and symmetry-adapted ensemble theory fail to explain the disorder observed in Form II as a thermodynamic equilibrium property, the disorder has to have been frozen-in during the crystallisation process.

Using an experimental occupation factor of 0.47 for the rank 58 conformation, and the energy difference between the rank 58 and the rank 1 conformations of the isolated site model in eqn (8) and (9), we obtain an internal energy of 0.94 kcal mol^{-1} and a free energy of 0.53 kcal mol^{-1} for the disordered, non-equilibrium Form II structure. The internal energy difference obtained for Forms II and I is 0.54 kcal mol^{-1}, which compares well to the experimentally measured crystallisation exotherm of 0.33 kcal mol^{-1}, considering the fact that we completely neglected the contribution of phonon dispersion. The calculated non-equilibrium free energy of Form II is indicated by the blue line in Fig. 11.

4 Energy calculations and crystallisation behaviour

The energy landscape and additional energy calculations can be used to rationalise the observed crystallisation behaviour of loratadine. From the energy landscape, we see that two structures have practically the same energy, the thermodynamically disordered Form I and rank 1, which is an ordered predicted structure that matches one of the experimentally observed disordered configurations in Form II.

Form I belongs to a highly populated family of structures. All members of this family have a short axis and essentially look like the structure(s) in Fig. 12(a–c) in the projection along the short axis, with some conformational variability around the common theme. In total, 45 out of the first 62 low-energy structures belong to this family. We call this Family 2. One of the authors has observed in numerous crystal structure prediction studies that highly populated families are dynamically favoured when it comes to crystallisation, and that within such a family it is typically the most stable predicted form that is experimentally obtained. The behaviour can be rationalised in terms of cross-seeding, resulting from the fact that since members of the family exhibit similar lattice parameters and similar surface arrangements, each member can act as a template for the crystallisation of any other member. The family is dynamically favoured because there are many nucleation starting points, and within the family the most stable form readily crystallises because it can start to grow on the surface of whichever other member starts to nucleate first. From the crystal energy landscape, we would expect Family 2 to be observed in most crystallisation experiments, which is actually the case.

The thermodynamically disordered Form I and the predicted ordered rank 1 structure are almost degenerate in terms of energy. This structure belongs to the structure Family 1, which is much less populated, with only 7 members out of the 62 structures. The members of this family also have a short axis, but look different in the projection along that short axis compared to Family 2 (see Fig. 12(d–f)). Family 1 is the second most populated family.

Form II is obtained by crystallisation at sub-ambient temperature, and it may be the frozen-in disorder that explains why, under very special conditions, it is no longer Form I that crystallises. In solution, both conformations of the cycloheptane bridge are populated. In the crystal, one of the two conformations is preferred, but the conversion rate may be slow if the crystal environment significantly increases the energy of the saddle point between the two conformations. At low temperature and sufficiently high growth rate, molecules will be incorporated into the crystal surface and overgrown before they have a chance to adopt the more favourable configuration. Under such growth conditions, the kinetically favoured crystal structure is actually not the thermodynamically most stable form, but that which best accommodates the disorder. Indeed, in rank 11 (corresponding to one of the disordered configurations of Form I), the energy cost of flipping the cycloheptane methylene bridge to the more unfavourable configuration is 2.5 kcal mol^{-1}, compared to just 2.0 kcal mol^{-1} for rank 1.

Overall, it is concluded that two ordered components of each disordered structure are found in a crystal structure prediction study using the GRACE programme. However, the experimentally observed stability relationship is only reproduced when the energy contribution of disorder is taken into account. The disordered model of Form I is found to be lower in energy than all the other predicted structures, while in the case of Form II the disorder appears to result from kinetic processes determined during crystallisation. The case of loratadine demonstrates that experimentally observed disorder close to 50/50 does not necessarily correspond to a free energy decrease by $kT \ln 2$.

From a wider perspective it would appear from these results that disorder can have an important effect on the prediction of relative stabilities in polymorphism investigations and crystal structure prediction. As such, it would be pragmatic to ensure that alternative configurations (beyond $Z' = 2$) are included as part of a risk assessment strategy in the use of CSP during development and discovery phases in the pharmaceutical industry to accurately represent experimental forms and their relative energies.

The symmetry-adapted ensemble method is effective for a range of systems of pharmaceutical interest. Here, it was used to distinguish whether disorder is thermodynamically favourable or frozen-in during the crystallisation process in the case of loratadine Forms I and II. The method has previously been used to assess the crystal structure landscapes of carbamazepine DMSO solvate[19] and the influence of disorder in the polymorphs of caffeine, indicating whether disorder may be thermodynamically feasible (Form II (β)).[20] Habgood contrasted the strikingly different systems of caffeine, which is polymorphic, with disorder playing an important role in stabilising Form II and isocaffeine, which was found to be monomorphic experimentally, with a large separation in the CSP energy landscape of the low energy structure corresponding to the experimental structure.

From a pharmaceutical development perspective, having accurately represented the relative stabilities of the experimentally observed dimorphic pair through the construction of disorder models from the putative ordered components found within the generated crystal structure landscape, there is no warning of a missing, thermodynamically more stable form of loratadine.

Conflicts of interest

M. A. N. is the founder, owner and director of the company Avant-garde Materials Simulation, which developed the GRACE programme for crystal structure prediction. The remaining authors declare no competing financial interests.

Acknowledgements

The authors thank Robert Wenslow of Crystal Pharmatech for the crystallisation procedure of loratadine Form II. We also thank Richard I. Cooper of the Department of Chemistry, University of Oxford, UK for the independent refinement of the loratadine Form I datasets collected at various temperatures.

References

1 M. A. Neumann, J. van de Streek, F. P. A. Fabbiani, P. Hidber and O. Grassmann, *Nat. Commun.*, 2015, **6**, 7793.
2 S. L. Price, *Acta Crystallogr., Sect. B: Struct. Sci., Cryst. Eng. Mater.*, 2013, **69**, 313–328.
3 D. E. Braun, J. A. McMahon, L. H. Koztecki, S. L. Price and S. M. Reutzel-Edens, *Cryst. Growth Des.*, 2014, **14**, 2056–2072.
4 R. C. B. Copley, S. A. Barnett, P. G. Karamertzanis, K. D. M. Harris, B. M. Kariuki, M. Xu, E. A. Nickels, R. W. Lancaster and S. L. Price, *Cryst. Growth Des.*, 2008, **8**, 3474–3481.
5 Search Results – Access Structures, https://www.ccdc.cam.ac.uk/structures-beta/Search?Ccdcid=beqgin, accessed January 26, 2017.
6 D. Gala and D. J. DiBenedetto, *US Pat.*, US 6,335,347 B1, 2002.
7 R. Wenslow, *Crystal Pharmatech*, 19 May 2015.
8 G. M. Sheldrick, *SADABS*, University of Göttingen, Göttingen, Germany, 1996.
9 G. M. Sheldrick, *SHELXTL*, Bruker AXS Inc., Madison, WI, USA, 2001.
10 W. Kabsch, *Acta Crystallogr., Sect. D: Biol. Crystallogr.*, 2010, **66**, 125–132.
11 XDSwiki, http://strucbio.biologie.uni-konstanz.de/xdswiki/index.php/Main_Page, accessed August 16, 2017.
12 F. H. Allen, *Acta Crystallogr., Sect. B: Struct. Sci.*, 2002, **58**, 380–388.
13 F. H. Allen and W. D. S. Motherwell, *Acta Crystallogr., Sect. B: Struct. Sci.*, 2002, **58**, 407–422.
14 M. A. Neumann and M.-A. Perrin, *J. Phys. Chem. B*, 2005, **109**, 15531–15541.
15 M. A. Neumann, *J. Phys. Chem. B*, 2008, **112**, 9810–9829.
16 J. Prywer, *J. Cryst. Growth*, 2004, **270**, 699–710.
17 R. Łazarski, A. M. Burow and M. Sierka, *J. Chem. Theory Comput.*, 2015, **11**, 3029–3041.
18 J. J. Kaminski, N. I. Carruthers, S.-C. Wong, T.-M. Chan, M. Motassim Billah, S. Tozzi and A. T. McPhail, *Bioorg. Med. Chem.*, 1999, **7**, 1413–1423.
19 A. J. Cruz-Cabeza, G. M. Day and W. Jones, *Phys. Chem. Chem. Phys.*, 2011, **13**, 12808–12816.
20 M. Habgood, *Cryst. Growth Des.*, 2011, **11**, 3600–3608.

Faraday Discussions

PAPER

Metashooting: a novel tool for free energy reconstruction from polymorphic phase transition mechanisms

Samuel Alexander Jobbins,[a] Salah Eddine Boulfelfel[b] and Stefano Leoni*[a]

Received 2nd March 2018, Accepted 13th March 2018

DOI: 10.1039/c8fd00053k

We introduce a novel scheme for the mechanistic investigation of solid–solid phase transitions, which we dub 'metashooting'. Combining transition path sampling, molecular dynamics and metadynamics, this scheme allows for both a complete mechanistic analysis and detailed mapping of the free energy surface. This is illustrated by performing metashooting calculations on the pressure-induced B4/B3 → B1 phase transition in ZnO. The resulting free energy map helps to clarify the role of intermediate configurations along this activated process and the competition between different mechanistic regimes with superior accuracy. We argue that metashooting can be efficiently applied to a broader class of activated processes.

Introduction

Molecular dynamics (MD) has become the method of choice to investigate complex chemical phenomena at the atomic level. Thanks to both efficiently written algorithms and high-performance computational infrastructures, the scope of numerical simulations has markedly widened. Nonetheless, the presence of inherent kinetic bottlenecks means that entire regions of configuration space may remain unexplored. If such states are separated by large energy barriers, even the most efficient MD will end up spending considerable time within long-lived, metastable states. Furthermore, depending on the height of those barriers, the event of interest may fully elude observation over the time scale of the simulation.[1–3]

Protein folding, nucleation, crystallisation, phase transitions and chemical reactions are all important phenomena which are characterised by activated steps. Their detailed mechanistic investigation would benefit considerably from precise mapping of the underlying free energy landscape. However, despite

[a]*School of Chemistry, Cardiff University, Cardiff, CF10 3AT, Wales, UK. E-mail: leonis@cf.ac.uk*

[b]*Georgia Institute of Technology, School of Chemical and Biomolecular Engineering, Atlanta, GA 30332-0100, USA*

improving numerical algorithms and increasing computational power, their systematic investigation remains extremely demanding and often completely impractical. Methods to explore multi-minima energy landscapes exist[4,5] and recently there has been a pronounced effort to develop novel approaches to overcome this time-scale problem.[3,6] Historically, the use of an external bias aimed at enhancing the frequency of rare events in MD simulations has been realised in methods such as umbrella sampling (US).[7] This is achieved by systematically adding a bias along well-chosen coordinates, in order to sample only small but overlapping regions from which a free energy can be reconstructed. Another bias-oriented method, metadynamics (metaD),[3] exploits the concept of collective variables (CVs), which span a lower dimensional coordinate space. Within the space of these CVs, a history-dependent bias is deposited, which rapidly drives the system away from its initial probable state. The definition of a good set of CVs is therefore key to the success of the method.[8] Within a CV space, standard MD would linger around a minimum close to the initial state corresponding to some value of the CV. Overcoming free energy barriers by metadynamics inevitably leads to broader exploration of the CV values by crossing over to different states within a finite computational time.

Metadynamics takes place within the space of one or a set of CVs, which must be known at the beginning of the calculation. A valid set of collective variables optimally covers the entire space of the conformations of interest. This clearly requires that CVs be engineered with a global free energy map in mind. However, this is evidently problematic, as often the underlying free energy landscape is the desired final result of the simulation and it often cannot be deduced by intuition. Additionally, the requirement for such 'global' CVs directly clashes with the effort to keep their dimensionality as low as possible.

Recently, a novel approach has been introduced, which allows for a "bespoke bias" based on a variational approach.[9,10] Within this approach, the CV problem is partially overcome by reinterpreting its meaning as a local descriptor of metastable states. Enhancing static fluctuations will amplify the probability of a crossover to another metastable state. Importantly, the bias can be deposited in such a way that partial filling can be achieved in all the basins of interest. While the success of the method still depends on an appropriate choice of CVs, this represents a novel general scheme for enhanced-sampling MD simulations.

To tackle the intrinsic disparity between the time scales of activated processes and molecular dynamics simulations, the transition path sampling method (TPS)[6,11] was introduced. TPS is based upon the idea that, in a complex system, crossing a barrier between (meta-)stable states may happen in a multitude of different ways. In such a scenario, the potential energy surface of a system is unlikely to be represented by saddle points alone. Instead, numerous points on the potential energy surface become relevant, some of which may still be stationary points, but many others not. The implementation of the method consists of collecting such an ensemble of reactive trajectories, without previous knowledge of the details of the process (that is, no *a priori* information or guess about the reaction coordinates is required – such information should naturally result from the simulation approach). Within a transition ensemble, the relevance of a path is weighted by its probability. TPS collects true dynamical trajectories and can be plugged on top of different molecular dynamics schemes, within different simulation ensembles.[2]

TPS belongs to the class of pathway-orientated methods and allows for an exquisite level of detail in mechanistic investigations.[12–18] Unlike metadynamics, TPS does not deposit any bias potential; similarly to metadynamics, it deploys a low-dimensional order parameter, which distinguishes between the basins of interest. TPS is designed to enhance statistics in intermediate regions between long-lived basins, however it is generally not a method of choice for free energy reconstruction as the main statistical weights are still contributed by (meta-) stable basins.

Here, we present a novel combination of TPS and metadynamics techniques, which we dub 'metashooting'. This novel approach allows for reconstruction of the underlying free energy of a process from TPS trajectories. This combination is based on the propagation of CVs from TPS to metaD. While TPS controls the rapidity of CV variation (inherent to true dynamical trajectories), metadynamics deposits a bias around regions of the CV that vary comparatively slowly. In doing so, TPS keeps the path crossing probability ratios (A $\rightarrow$ B to B $\rightarrow$ A) unchanged, thus preventing unilateral over-biasing. By rapidly varying the controlling CVs in TPS, partial basin filling can be achieved while maintaining the intermediate regions bias-free for most of the calculations. Only in its final phases do the details of the intermediate regions appear.

In the following we illustrate the details of our approach by performing a complete metashooting analysis of the solid–solid reconstructive phase transition between the B3 zincblende phase and the B1 high-pressure rocksalt phase of zinc oxide (ZnO). ZnO has proven to be extremely challenging over the years,[19–23] with several attempts to conclusively shed light on competitive mechanisms and intermediates. Saitta and Decremps first discussed the competitive nature of the 'tetragonal' and 'hexagonal' routes in the B4–B1 phase transition, and their work concluded that semiconductors involving d-electrons, such as ZnO, preferred the 'tetragonal' path.[21] However, other work seems to contradict this – notably the experimental work of Liu *et al.*, which utilised high resolution angular dispersive X-ray diffraction, which seemed to indicate that the 'hexagonal' pathway was the more favoured for ZnO at lower pressures.[24] Second harmonic generation experiments pinpointed the role of the pressure medium (hydrostatic *vs.* non-hydrostatic) in promoting the formation of centrosymmetric iH or acentric tetragonal iT, respectively.[22] In previous work,[15] TPS was used to produce a detailed mechanistic pathway for the B4–B1 transition, and it was shown that the mechanism proceeds *via* a rich series of transition states and intermediates.[12] Additionally, the two 5-coordinate intermediate motifs were visited and were shown to compete with one another in the mechanism. The analysis, based on dynamical trajectories, concluded that a hexagonal intermediate could indeed be visited, however it was not essential to the transformation. In fact, the reconstruction was found to propagate over the interfaces of local tetragonal motifs (Fig. 1).

It is therefore apparent that the phase transitions between different polymorphs of ZnO are non-trivial. Indeed, much debate still rages over their precise nature and work is still being published to attempt to unravel the minutiae of these elusive mechanisms, with a particular emphasis on the B4–B1 transformation. In the following, after a summary of the implementation details, the results of metashooting on a challenging solid–solid phase transition in this material between the B3 and B1 phases are discussed.

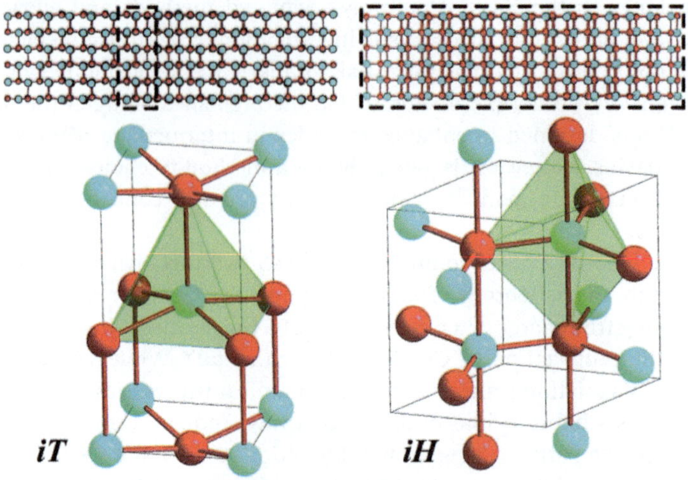

Fig. 1 Two 5-coordinate intermediates (tetragonal *iT* (*I*4*mm*, left) and hexagonal *iH* ($P6_3/mmc$, right)) were shown to compete against one another in the B4–B1 transformation. Whilst *iT* was only found at the interface between B4 and B1, the *iH* structure sometimes dominated the entire system under study. Despite this, the *iH* phase was shown not to be crucial to the transformation.[12]

Methods

Our complete analysis of the B3–B1 phase transition in ZnO proceeds over 2 steps: (i) TPS calculations, including the definition of a suitable order parameter and (ii) free energy mapping with metashooting, where TPS and metaD are combined.

Transition path sampling

The aim of TPS is to obtain the transition path ensemble,[6,11] and an efficient approach to this task involves the use of Monte Carlo techniques. A random walk is carried out in the space of the trajectories, whereby a new path $x^{(n)}(\tau)$ is generated from an old one, $x^{(o)}(\tau)$. The implementation of the random walk moves is based on the shooting algorithm,[2] whose general premise is to generate pathways by "shooting off" a new pathway from a randomly selected time slice $x_{t'}^{(o)}$ of the previous pathway. Modifications are introduced to $x_{t'}^{(o)}$, yielding $x_{t'}^{(n)}$. From this perturbed time slice at time t', two new path segments are generated forward to time τ and backward to time 0 in molecular dynamics runs within the chosen ensemble. A trajectory is rejected if the introduced modifications cause the new trajectory not to connect the basins of interest. Otherwise, it is accepted with a given probability.[2] The acceptance/rejection criteria thus depend only on the shooting point, and can easily be translated into an algorithm. The introduction of such a probability weighting scheme allows the system to cross intermediate activation barriers of the order of k_BT or lower. Perturbations are introduced as momenta modifications with strict conservation of the total linear and angular momenta.

Order parameter

Transition path sampling calculations require the initial and final configurations of the system to be strictly distinguishable, such that there is no overlap of their respective basins when projecting onto the order parameter.[25] A handy yet powerful order parameter used in previous path sampling calculations is the average coordination number of the system. The coordination sphere (CS) is a set of integers $n_1, n_2, \ldots n_i$, in which the ith term is the number of atoms in "shell" i that are bonded to atoms in "shell" $i-1$. Shell 0 consists of a single atom, and the number of atoms in the 1st shell n_1 is the conventional coordination number (CN). This quantity can be calculated for all atoms in a system, collated, and averaged over the whole structure. The 1st three shells of the coordination spheres of the three phases of zinc oxide are: zincblende (B3) [4,12,24], wurtzite (B4) [4,12,25] and rocksalt (B1) [6,18,38]. The 1st coordination numbers (1st CNs) of zincblende and rocksalt are 4 and 6, respectively. During TPS, the coordination spheres of each atom were calculated up to the 3rd shell, but only the 1st CN was utilised to trace the progress of the path sampling calculations. The 2nd and 3rd CNs were recorded simply to gain further insight into the underlying details of the transition pathways.

Using only the 1st coordination shell means that there is no discrimination between the B3 and B4 configurations. As the system starts from a metastable B3 structure (see below), changes in the final regime can be expected. Furthermore, using solely the 1st coordination number as the order parameter exerts absolutely no bias on the calculation. The variation in the order parameter arises simply as a result of the application of the transition pressure to the system. Hence, monitoring the 1st coordination sphere does not force the system to evolve in any particular way over the course of the path sampling iterations. In TPS, both the overall mechanism and the trajectory endpoints are rectified.[26] Therefore, the order parameter is chosen with minimal constraints in mind, in order to allow the process to express the most probable outcome. A higher-order CS would drive the system into a specific transformation regime within a biased TPS approach, in order to study specific, less probable transformation pathways.

Metashooting

Metashooting combines TPS and metadynamics to efficiently reconstruct the underlying free energy landscape from TPS trajectories. This way, computing time will not be expended by filling other regions of the configuration space that have no relevance to the transition.

The general procedure of the metashooting approach for a transition A → B is as follows:

(1) TPS step 1: Starting from a randomly chosen snapshot along the TPS trajectories, shoot off a new trajectory and time propagate.

(2) MetaD step 1: Once basin A is reached, deposit a bias using a metadynamics scheme (in this case, using the well-tempered approach), in order to locally partially fill the as-yet uncharacterised energy well. At this stage, a small number of metasteps are chosen, allowing the CV to vary about the characteristic value of the basin.

(3) TPS step 2: Generate another trajectory (as in step 1) and propagate into the now-biased basin A, to generate a true MD velocity distribution on the biased potential.

(4) TPS step 3: Similarly to TPS step 1, shoot off a new trajectory from the same snapshot as above and propagate, only this time backwards in time, to basin B.

(5) MetaD step 2: In basin B, apply the same metadynamics scheme as above for a number of steps, in order to partially fill this second basin of attraction.

(6) TPS step 4: Re-shoot a novel trajectory from a random snapshot of the previous trajectory. Propagate into the now-biased basin B.

Repeat the above iterations of metadynamics and TPS moves until the underlying free energy profile is fully converged.

The number of metadynamics steps for each metashooting move can be chosen such that only moderate variations around the current CV value are allowed, as ΔCV_{Max}. Alternatively, as a better means to control bias deposition without imbalance, the TPS path acceptance probability in TPS can be chosen to stay the same, *i.e.* at no point can crossing over the activation energy barrier become easier for either direction A → B or B → A. This is realised within an adaptive scheme of rescaling momenta modifications (see below).

Calculation details

Molecular dynamics. Born–Oppenheimer molecular dynamics simulations were performed using the *cp2k* package within the *NpT* ensemble at $T = 300$ K and $p = 9.8$ GPa. At this pressure, the enthalpies of B1 and B3 are equal. Constant pressure and temperature were ensured by the Martyna–Tobias–Klein[27] algorithm, allowing for anisotropic shape changes of the simulation box.[28] Interatomic forces were calculated from a Buckingham pair potential,[29] which has been previously validated.[15] Long-range electrostatic effects were accounted for using an Ewald summation. Time propagation was performed with a time-reversible predictor–corrector algorithm. An integration timestep of 0.2 fs was used. The simulation box contained 1200 Zn–O atom pairs.

Transition path sampling. The sampling starts from an initial trajectory connecting the limiting phases. A new trajectory is generated by randomly selecting a configuration from an existing trajectory and slightly modifying the atomic momenta. The modifications δp are applied to randomly chosen pairs of atoms (i,j) according to: $\vec{p}_{\mathrm{new}} = \vec{p}_{\mathrm{old}} + \delta p(\vec{r}_j - \vec{r}_i)/|\vec{r}_j - \vec{r}_i|$ and $\vec{p}_{\mathrm{new}} = \vec{p}_{\mathrm{old}} - \delta p(\vec{r}_j - \vec{r}_i)/|\vec{r}_j - \vec{r}_i|$, conserving both momentum and angular momentum. The resulting atomic momenta are rescaled by a factor of $\sqrt{E_{\mathrm{kin}}^{\mathrm{old}} \Big/ \left(\sum_{i=1} |\vec{p}_i|^2 \Big/ 2m_i\right)}$, in order to conserve the total kinetic energy. The propagation of the modified configuration in both directions of time $(-t, +t)$ generates a new trajectory. Repeating this step provides a set of trajectories, with successful examples representing the transition ensemble. TPS requires an initial guess trajectory, which we obtained from a topological/geometric approach of interpolation between the two limiting crystal structures (B1 and B3), as described elsewhere.[13,14,26] Momenta modifications were introduced using an adaptive scheme, which entailed rescaling δp to larger or smaller values, based on whether this δp led to a successful or a failed trajectory, respectively. The acceptance ratio for the production runs was generally between 40–60%.

Metadynamics. Metadynamics calculations were carried out using the plumed plug-in,[30] in which a history-dependent Gaussian bias is added to the underlying potential. The well-tempered[31] prescription of metadynamics was used, in which the heights of the Gaussian hills are rescaled at each step. The collective variables corresponded to the 1st and 3rd coordination spheres of the zinc oxide systems – analogous to the order parameter used in TPS. The coordination sphere in plumed is mapped onto a switching, differentiable function, $s(r) = \frac{1-\left(\frac{r-d_0}{r_0}\right)^n}{1-\left(\frac{r-d_0}{r_0}\right)^m}$. r is the Zn–O inter-atomic distance, while r_0, d_0, m and n are adjustable parameters. For the 1st and 3rd coordination spheres the parameters were set to (2.6, 0.1, 6, 12) and (5.3, 0.1, 6, 12), respectively. These parameters reflect a one-to-one relationship between the geometrically-calculated coordination numbers and the values based on the switching function, despite resulting in slightly exaggerated values for the 3rd coordination sphere of B4/B3 (28/29 instead of 24/25). Gaussian functions were deposited every 500 simulation steps. On average, 5 metasteps were calculated on each side of the trajectory during each iteration of the metashooting procedure. The height of the Gaussians was set at 1000 kJ mol^{-1}. The Gaussian widths were chosen in accordance with the variance of the CV in an unbiased simulation – as such, these were set to 0.2 and 1.0 for the 1st and 3rd coordination numbers, respectively. The well-tempered 'bias-factor' was set at 10 000, which is large enough to escape the minima and re-scale the deposited Gaussians within a meaningful time-frame.

Results

Mechanistic analysis

TPS was initiated from a B3–B1 trajectory regime, obtained from the modelling approach described and used elsewhere.[13,32] One of the advantages of initiating the process from a plausibly designed trajectory is that it takes fewer TPS iterations for the system to move away from the initial, unfavourable regime to a more probable one. The process of moving towards a probable regime is known as trajectory decorrelation. In fewer than 50 TPS iterations, the procedure steered the trajectory away from the collective motion encoded by the geometric model towards a regime characterised by nucleation events. The trajectories continued to evolve until no further major mechanistic changes were observed, which occurred after several hundred iterations of the path sampling procedure.

Fascinatingly, the final 4-coordinated product was not a pure zincblende (B3) structure, but a mixed B4/B3 system dominated by the wurtzite (B4) form. In all the successful trajectories, the final material consisted of only 10–20% B3, with the more stable hexagonal form forming the remainder of the structure. The order parameter based on the 1st coordination number permits both the wurtzite and zincblende configurations. Whilst the initial, unfavorable regime was B1 → B3, TPS rectified the mechanism into B1 → B4/B3, which contains predominantly hexagonal components.

The complexity of the system leads to several types of trajectories, with some degrees of (dis)similarity in the details of the reconstruction.

A representative B1 → B4/B3 trajectory is shown in Fig. 2. A nucleus of 4-connected ZnO (Fig. 2, 0.6 ps) forms within B1, followed by the formation of regions of 5-fold coordination at the interface with B1 (Fig. 2, 0.7 ps), which are subsequently outgrown by 4-fold coordinated regions (Fig. 2, 0.9–4.7 ps). Regions of 5-fold coordination (mostly *iH*) are found between the expanding 4-connected

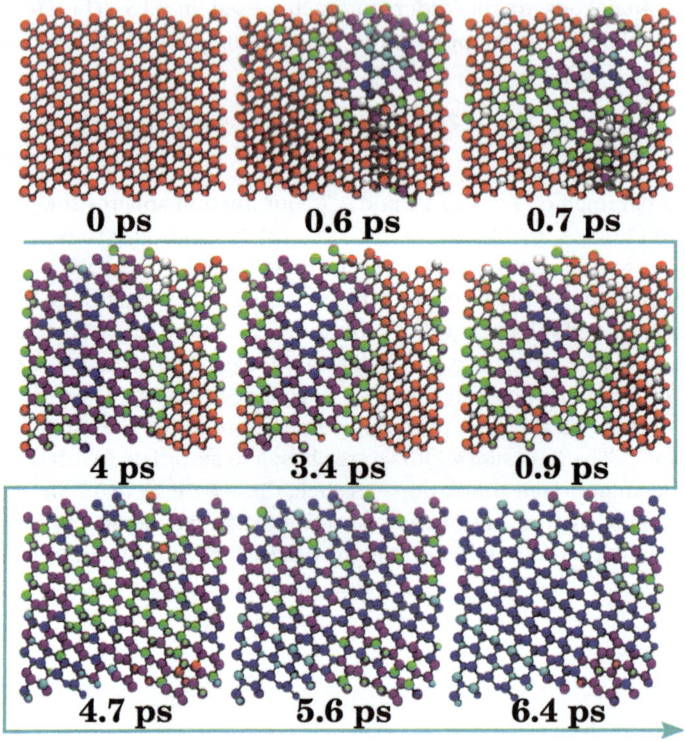

Fig. 2 Coordination-coloured snapshots from a representative trajectory, showing the sequence from rocksalt (B1) to a mixed wurtzite/zincblende (B4/B3) configuration. A nucleus forms in B1 within 500 fs, followed by shearing along $(111)_{B1}$ at 0.6–0.7 ps. The nucleus grows further over the intermediates characterised by 5- and 4-fold coordination (0.9–4.7 ps), until the system transforms into a 4-coordinate motif, which is composed primarily of hexagonal wurtzite (5.6–6.4 ps). Colour code: red: B1; green: *iH*, *iT*; dark blue: B4 (wurtzite); light blue: B3 (zincblende); purple: 4-coordinated, but not corresponding to B3 or B4.

region and B1. The transformation into mixed B4/B3 entails the complete resorption of areas of 5-fold coordination (Fig. 2, 4.7–6.4 ps). The transformation of the remaining B1 motifs (4 ps) is associated with a sharp volume expansion, which causes the box volume to fluctuate, under the appearance of local 5-fold motifs (Fig. 2, 4.0–4.7 ps).

From a configuration of the type seen in Fig. 2 at 0.6 ps, a less severe compression leads to the growth of a 5-fold region at the expense of any 4-connected one. This accompanies the formation of a mostly 5-coordinated intermediate containing residual B1, denoted B1–*iH*. This configuration persists for as long as 500 ps in subsequent molecular dynamics simulations. The presence of an intermediate, which is not vital to the transformation and corresponds to a 5-membered intermediate along the transition pathway, was noted in previous work.[12] The nature of this intermediate is the hexagonal *iH* form, whilst the tetragonal analogue *iT* only appears as a motif at interfaces during the reconstruction.

The B4/B3 → B1 mechanism slightly deviates from the reverse process (Fig. 3). In this trajectory type, the starting 4-coordinate material is first collectively distorted

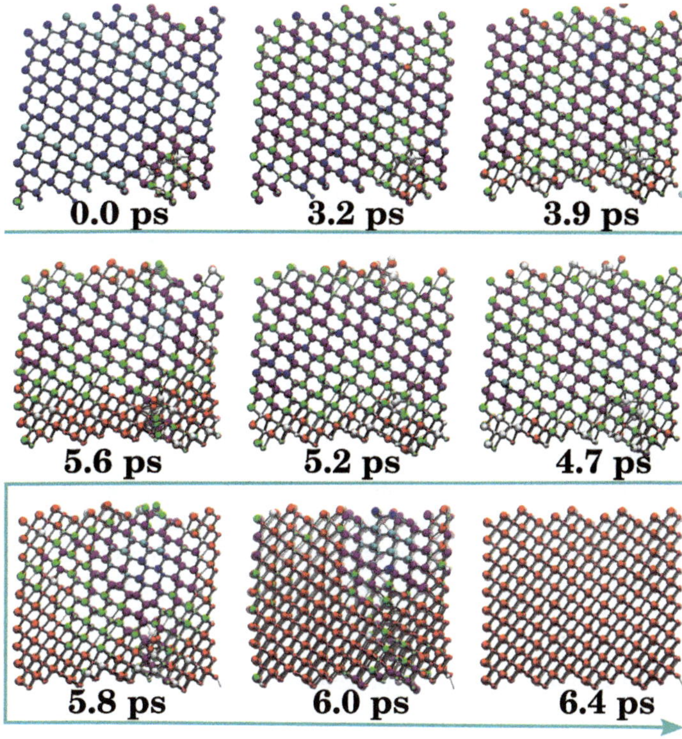

Fig. 3 Coordination-coloured snapshots of a representative B4/B3 → B1 trajectory. Mixed (B4/B3) wurtzite/zincblende (0.0 ps) is followed by a 4-coordinated distorted configuration (0.0–3.2 ps), purple atoms. Subsequent growth of 5-coordinated areas (green), mainly of type *iH*, is observed (3.2–5.2 ps), from which B1 (red) develops (5.2–6.0 ps) until the structure locks-in to 6-coordinate B1 (6.4 ps). Colours as in Fig. 2.

(Fig. 3, 0.0–3.2 ps, purple atoms). From the 5-coordinated areas of the *iH* motifs (Fig. 3, 3.9–5.2 ps), the B1 structure develops (Fig. 3, 5.2–6.4 ps). Thus, over half of the transformation time is spent slowly squeezing the system until it can transform into 5-coordinated *iH* regions, indicating the presence of a deep but fairly gently-sloping energy barrier separating the B4/B3 basin from the intermediate configurations. In all the configurations, the *iH* motifs are relatively long-lived. This is not the same as the reverse B1 → B4/B3 transition, in which the *iH* regions were less extensive and promptly eliminated at the interfaces (Fig. 2), with the majority of the simulation time being dominated by coexisting 4- and 6-coordinated motifs.

Consistent with previous calculations,[12] different intermediates are associated with different types of trajectories. The presence of B3 adds further structures to an already complex scenario. In previous work, low transition pressures (less than 10 GPa) were observed to favour the appearance of *iH*, which benefits from a less severe $[0001]_{B4}$ compression. The results presented here appear to corroborate that view.

Metashooting

In order to better understand the competitive nature of different mechanisms, the free energy (FE) was evaluated by coarse-graining the potential energy surface onto a suitable set of collective variables. Free energy surfaces have found extensive uses for the description of biological systems (in particular peptide calculations[33]), but are less frequently used in the solid state.

In metashooting, mechanistic details acquired from the transition path sampling procedure remain unspoiled. As the process is driven by TPS trajectories, metastable basins are filled in a balanced way – starting from the two limiting configurations, and finishing with the characterisation of all the relevant intermediate configurations. Repeated iterations of the metashooting algorithm (see Methods) enable the free energy landscapes corresponding only to the courses of the TPS-derived trajectories to be built up in a stepwise and equally distributed manner. As a result, regions around the transition states are considered only at the end of the analysis, avoiding any potential bias that may scramble the mechanistic details.

The algorithm should first fill up the two starting basins of attraction from the final TPS trajectories (B4/B3 and B1). Visiting these regions should become discouraged as a result of the bias potential. The progress of metashooting can be monitored by visualising the Gaussian deposition. Eventually, the metashooting algorithm should fully characterise the central (around the metastable basins) and peripheral regions of the CV space over the course of the trajectory. Convergence will be seen in a similar way to a standard metadynamics scheme – that is, the underlying free energy landscape should no longer significantly change with further iterations of the procedure. At this point, meaningful analysis of the intermediates and energy barriers associated with each process can be carried out.

Harnessing the free energy in this way should have a number of advantages. The landscape can be thought of as the surface upon which the trajectories travel, and the individual characteristics of the trajectories are determined by which regions of the surface are visited. Thus, its elucidation allows for a complete quantitative analysis of the underlying thermodynamics and kinetics of the process; measurements of the activation energy and changes in the free energy between different steps of the procedure, as well as the starting and finishing products, can trivially be taken from the plot. Moreover, the intermediates and transition states that are vitally important to the transformation will appear obvious along the reaction scheme. Such insight into trajectories using only path-based analyses of transformations is not trivial.

The free energy profile was constructed from the converged TPS trajectories. After 250 iterations of the metashooting procedure, the free energy had fully converged into a detailed landscape.

Progression of metashooting

The evolution of the free energy landscape according to the metashooting algorithm is first discussed, followed by an analysis of the final profile generated by the procedure.

The first few iterations of the procedure (Fig. 4) accumulated Gaussian functions only within the two limiting basins (B4/B3 and B1). A small 'shoulder-peak' formed in the 4-coordinate basin as a result of oscillations between different forms of the low-coordination material, including B3, B4 and other forms.

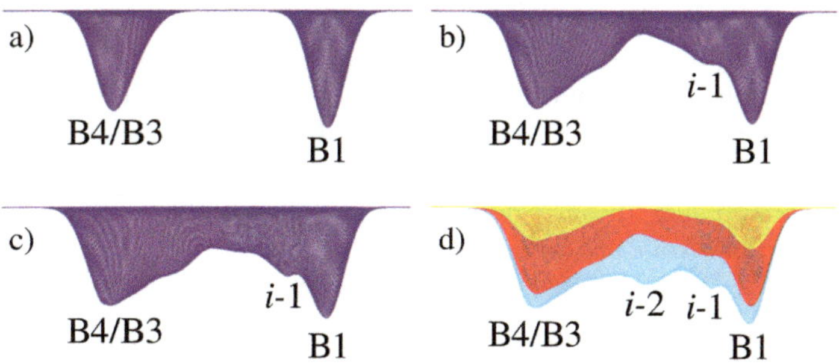

Fig. 4 Free energy landscapes at different stages of metashooting. (a) Filling of the limiting basins centred on the B4/B3 and B1 motifs (1st metashooting cycle), (b) appearance of a minimum *i*-1 after 50 cycles, (c) intermediate areas are explored after 200 cycles. (d) Stages 10 (yellow), 75 (red) and 225 (blue). In the latest cycles only the intermediate *i*-2 appears, which corresponds to B1–*iH*. The 3D free energy is plotted as a function of the 1st and 3rd coordination numbers. See main text for further details.

After 50 iterations (Fig. 4b), a notable 'shoulder-peak' adjacent to the B1 basin was observed. After 150 iterations, this was resolved into a *bona fide* minimum, *i*-1. By iteration 200, a second minimum centred halfway between the two limiting basins started to appear (Fig. 4c), which was fully resolved by iteration 225 (Fig. 4d), denoted *i*-2. By 250 iterations (Fig. 5), the plot was fully converged – the energy barriers and positions of the minima remained unchanged with any further iterations of the metashooting procedure.

Energy profile and intermediate configurations

There are at least seven points of interest on the free energy surface – the two initial basins of attraction (corresponding to B4/B3 and B1), two intermediate

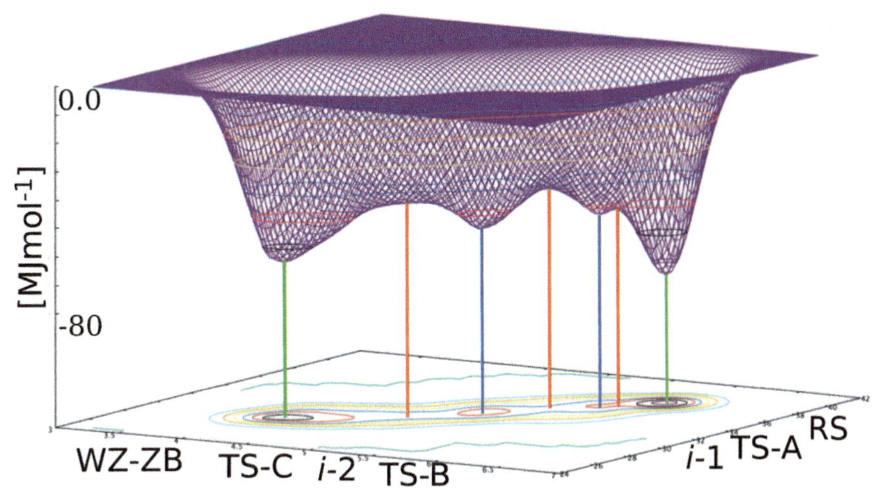

Fig. 5 Three-dimensional plot of the converged free energy surface for the transition after 250 iterations of the metashooting procedure. Two main basins, WZ-ZB (B4/B3) and RS (B1), two intermediate basins (*i*-1 and *i*-2), and three transition states (TS-A, TS-B, TS-C) are visible.

basins and three maxima, corresponding to the transition state regions. The structures and roles of these turning points in the transformation can be ascertained by obtaining the 1st and 3rd coordination numbers of the turning points on the graph and matching these up to structures along the hundreds of trajectories from both the 'plain' TPS runs and the metashooting trajectories (Table 1). It stands to reason that these regions must play important roles in the transformation mechanism.

Free energy intermediates and maxima

i-1 is the metastable basin adjacent to the B1 basin. Every trajectory congregates at *i*-1 before proceeding along a single pathway to or from the rocksalt structure. Despite the small and shallow dimensions of the basin, it clearly plays a central role in the transformation. It is separated from B1 by TS-A. The configuration associated with this maximum is an almost entirely B1 structure at the moment of deformation of the rocksalt motifs (Fig. 6). For the first few Zn–O sites this entails overcoming an activation energy, as they leave their 6-fold coordination.

In *i*-1 a portion of the pristine B1 fragments contains 4- and 5-coordinated motifs. With respect to the way the square pattern is opened, the locally 5-coordinated motifs correspond to *iH*. This small region is extremely short-lived but is present in every successful trajectory sampled, thus explaining its place on the free energy profile. Interestingly, this mechanistic pattern cannot propagate as it is. The next free energy barrier TS-B is overcome by including an additional $(111)_{B1}$ shearing mechanism, as seen in both the forward and reverse TPS trajectories. This configuration, which includes an enlarged seed region and the onset of shearing, corresponds to the free energy maximum TS-B. The TS-B configuration could easily be seen in the TPS trajectories, however the fact that this part of the transformation corresponds to two distinct events would have been completely missed without the metashooting analysis.

The second intermediate region, *i*-2, is a deep energy well placed halfway between the two limiting basins. Trajectories visiting configurations characterised by the 1st CN in the interval [5.0–5.3] and the 3rd CN within [32–34] can in principle contribute to this basin. Accordingly, three different types of trajectory can be distinguished:

(1) Successful B4/B3 → B1 trajectories, along which extended 5-coordinated regions are present (see Fig. 3, 3.9–5.6 ps).

Table 1 1st and 3rd coordination spheres (as defined by the switching function) at the centre of the regions defining the turning points

Turning point	1st CN	3rd CN
RS (B1)	5.98	37.90
TS-A	5.75	36.66
i-1	5.68	36.21
TS-B	5.43	34.86
i-2	5.15	33.13
TS-C	4.87	31.62
WZ-ZB (B4/B3)	4.10	29.22

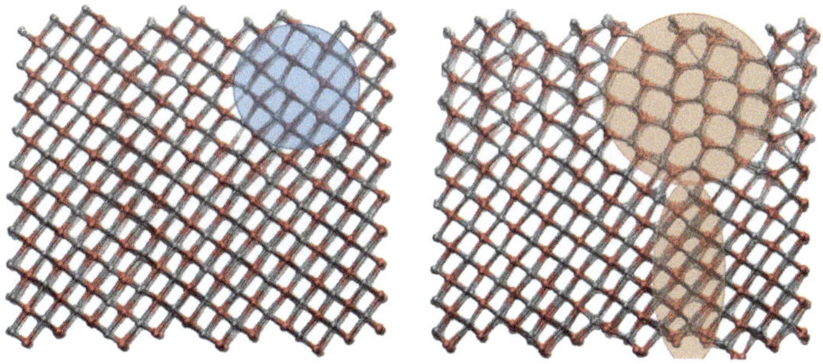

Fig. 6 (Left) Configuration corresponding to the small minimum i-1, which is just after the very first rocksalt motifs locally deform into hexagons. (Right) A representative configuration corresponding to the maximum TS-B, showing an enlarged seed accompanied by $(111)_{B1}$ layer shearing.

(2) Successful B1 → B4/B3 trajectories, in which Zn or O is 4-, 5- and 6-coordinated, giving an overall average 1st coordination sphere of five (see Fig. 2, 0.7–0.9 ps).

(3) Trajectories where the system visits and lingers in the predominantly 5-coordinate intermediate B1–*iH*. The average time spent in this basin is >200 ps.

Representative trajectories of the three types above are mapped on the free energy surface in Fig. 7. Successful trajectories tend to keep a higher 3rd coordination number while they approach the large *i*-2 basin. On the other hand, a lower 3rd CN is associated with a trajectory entering an intermediate basin. Closer to the B1 basin, the trajectories are more similar, while the B4/B3 basin can be escaped in different ways. Around TS-B, mechanistic competition (assisted by box size fluctuations[12,22]) may result in very different trajectories.

It seems likely, therefore, that the depth of the intermediate basin (and hence the energy barriers associated with it) cannot be taken as representing a proper intermediate along a particular type of trajectory. Instead, numerous possible configurations with similar average coordination number values all add weight to the free energy at these values of the collective variables, even though not every configuration is realisable in every trajectory.

Unlike TS-A and TS-B, TS-C corresponds to a very 'gentle' maximum, with a very shallow gradient either side of it. This again corroborates what was seen in the path sampling investigations – the system spends a great deal of time in a 4-coordinate state, or a mixture of 4- and 5-coordinate states, as the result of a slow, steady process, which is mapped here onto a gentle free energy gradient. Like its associated intermediate basin *i*-2, the height of TS-C is influenced by a number of configurations, all of which contain some combination of 4-coordinated ZnO, *iH* intermediate and residual B1. Crossing this maximum implies a significant coordination number (order parameter) change, and corresponds to mixed motif configurations as at 4 ps in Fig. 2 for the B1 → B4/B3 direction, or at 4.7 ps for B4/B3 → B1 (Fig. 3).

The appearance of mixed B1–*iH* agrees with previous studies on ionic compounds undergoing the same B4/B3 → B1 transition,[34–36] which may indicate the transferability of this mechanistic analysis.

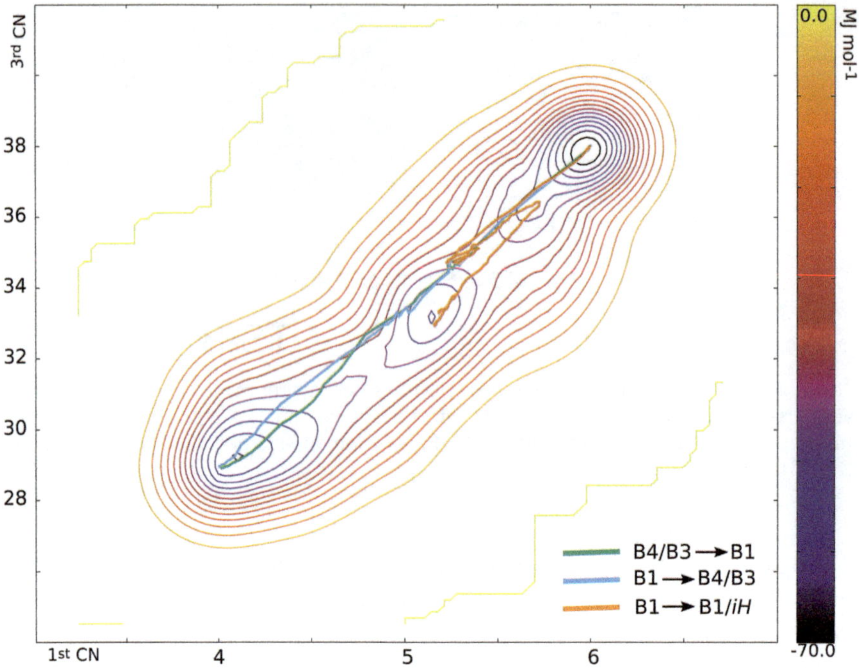

Fig. 7 Three representative TPS trajectories, mapped on the free energy surface. Successful B4/B3 → B1 trajectory (green), successful B1 → B4/B3 trajectory (light blue) and B1 → mixed B1–*iH* (orange). The latter is a long-lived intermediate configuration, which coincides with the intermediate *i*-2 basin.

Energy barriers

From the converged free energy surface, it is apparent that there are six distinguishable energy barriers associated with the pathway – three for B4/B3 → B1 (Fig. 8, E_{1-3}), and three for the reverse process (Fig. 8, E_{5-7}).

Trajectory variety has shown that the system may proceed along different pathways, therefore crossing different features of the free energy plot. This will

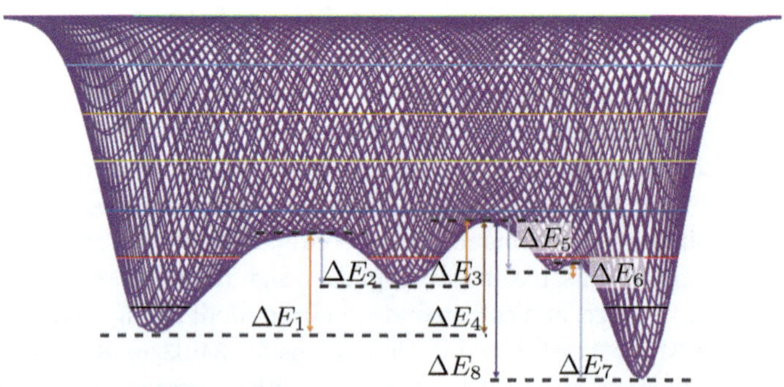

Fig. 8 Summary of energy barriers associated with the free energy profile. The activation energies are defined as the largest barriers from the starting basin. Note that these barriers are calculated as the difference from the centre of each of the turning points.

lead to different values for the total amount of energy overcome by the individual trajectories. The following quantitative analysis of the energy differences assumes measurement to and from the centre of each of the turning points, as described in Table 2, thereby indicating the energy associated with the minimum energy transition pathway that visits all maxima and minima. Every trajectory must surmount an energy barrier at least equivalent to the activation energy presented, as successful trajectories must overcome the highest energy barrier to cross to the other side, irrespective of the path taken.

It is also apparent from inspection that the energy barriers for each stage of the transformation are very different for the forward and reverse trajectories. This may go some way towards explaining why there are different pathways associated with the forward and reverse trajectories, and perhaps why there is such notable hysteresis present in such transformations.

The change in free energy is calculated as the difference between the energy values of the B1 and B4/B3 basin centres. The initial pressure was chosen based on the zero enthalpy difference between B3 and B1, which was dictated by the model used to commence the TPS cycles (see above).

The difference in free energy between the forward (B4/B3 $\rightarrow$ B1) and backward transformation is not zero at the chosen simulation pressure (p = 9.8 GPa). The inclusion of lattice dynamics[37] in first principles calculations was shown to shift the (ΔH = 0) "equilibrium" pressure to a lower value. Consistent with those findings, Fig. 8 pinpoints the role of entropy, which is positive going from B4/B3 to B1. The disordered nature of B4/B3 reinforces this effect.

Discussion and conclusions

Using transition path sampling methods, the phase transition characterised by a sequence of nucleation and growth events between mixed wurtzite/zincblende (B4/B3) and rocksalt (B1) has been described. The forward and reverse trajectories were found to be different, each with different competing intermediates and preferences for one mechanism over the other for both transformations.

A typical B1 $\rightarrow$ B4/B3 trajectory was characterised by the formation of an initial seed, followed by the gradual transformation to a mixed 4-coordinate product *via* coexisting motifs. Trajectories could also end up trapped in a long-lived intermediate 5-coordinate basin, corresponding to a mixed rocksalt-5-coordinate

Table 2 Energy barriers associated with the forward and reverse trajectories. Energy differences are measured to and from the centres of the turning points on the energy surface, corresponding to the (unobserved) minimum energy pathway trajectory

Energy difference	kJ mol^{-1}	k_BT/ZnO
ΔE_1	20 447.0	6.8
ΔE_2	10 483.8	3.5
ΔE_3	13 085.9	4.4
ΔE_5	10 376.6	3.5
ΔE_6	1357.0	0.5
ΔE_7	23 726.3	7.9
ΔE_4	35 069.9	11.6
ΔE_8	44 586.7	14.9

hexagonal structure denoted B1–*iH*. The B4/B3 → B1 transition entailed a more stepwise process, which involved a transition to a 4-coordinate mixture *via* local motifs of hexagonal *iH* before transforming to the final B1 product.

A novel method denoted 'metashooting' was developed as a combination of TPS and well-tempered metadynamics, using the 1st and 3rd average coordination spheres as the collective variables. TPS ensured rapid variations of the CVs, and balanced filling of the basins of interest. The metashooting procedure is able to decipher the underlying free energy landscape of the transformation, and structural information about the nature of appropriate maxima and minima can be extracted from the resultant plot. In addition, the energy barriers and changes in free energy can be ascertained and the successful and failed trajectories can be mapped onto the final free energy plot, in order to gain a true understanding of the energetics and structural changes inherent within each individual trajectory.

Metashooting can accommodate additional collective variables, for example higher coordination sequences, in order to further distinguish between the phases. Using the 1st and 3rd coordination spheres was convenient in this case as it was an identical measurement to the order parameter used in the path sampling calculations. In general, the CV set used on the metashooting layer can be more detailed then the order parameter utilised in the TPS steps.

It is believed that the metashooting procedure, which combines transition path sampling and metadynamics, could set a novel paradigm for future investigations into condensed matter phase transitions. Such a scheme could easily be transferred to another system, under any level of theory, to ascertain the underlying thermodynamics and kinetics of a temperature- or pressure-induced phase transformation.

Conflicts of interest

There are no conflicts to declare.

Acknowledgements

We would like to thank Davide Branduardi and Michele Parrinello for useful discussions and valuable comments. S. A. J and S. L. would like to thank ARCCA at Cardiff for the generous allocation of computational resources. They also acknowledge support from the UK Research Council for using work in the paper that was undertaken under project no. EP/M50631X/1. *Via* our membership of the UK's HPC Materials Chemistry Consortium, which is funded by EPSRC (EP/L000202), this work made use of the facilities of ARCHER, the UK's National High-Performance Computing Service, which is funded by the Office of Science and Technology through EPSRC's High End Computing Programme.

References

1 D. Chandler, *J. Chem. Phys.*, 1978, **68**, 2959–2970.
2 C. Dellago, P. G. Bolhuis and P. L. Geissler, *Lect. Notes Phys.*, 2006, **703**, 349–391.
3 A. Laio and M. Parrinello, *Proc. Natl. Acad. Sci. U. S. A.*, 2002, **99**, 12562–12566.
4 F. Glover, *Interfaces*, 1990, **20**, 74–94.

5 P. Sibani, J. C. Schön, P. Salamon and J. O. Andersson, *Europhys. Lett.*, 1993, **22**, 479–485.

6 C. Dellago, P. G. Bolhuis, F. S. Csajka and D. Chandler, *J. Chem. Phys.*, 1998, **108**, 1964–1977.

7 G. M. Torrie and J. P. Valleau, *J. Comput. Phys.*, 1977, **23**, 187–199.

8 C. Abrams and G. Bussi, *Entropy*, 2014, **16**, 163–199.

9 O. Valsson and M. Parrinello, *Phys. Rev. Lett.*, 2014, **113**, 090601.

10 O. Valsson and M. Parrinello, *J. Chem. Theory Comput.*, 2015, **11**, 1996–2002.

11 P. G. Bolhuis, C. Dellago and D. Chandler, *Faraday Discuss.*, 1998, **110**, 421–436.

12 S. E. Boulfelfel and S. Leoni, *Phys. Rev. B*, 2008, **78**, 125204.

13 S. Leoni and D. Zahn, *Z. für Kristallogr.*, 2004, **219**, 339–344.

14 D. Zahn and S. Leoni, *Phys. Rev. Lett.*, 2004, **92**, 250201.

15 S. E. Boulfelfel, D. Zahn, Y. Grin and S. Leoni, *Phys. Rev. Lett.*, 2007, **99**, 125505.

16 S. Leoni, S. E. Boulfelfel and I. A. Baburin, *Z. Anorg. Allg. Chem.*, 2011, **637**, 864–869.

17 S. E. Boulfelfel, G. Seifert, Y. Grin and S. Leoni, *Phys. Rev. B*, 2012, **85**, 014110.

18 S. E. Boulfelfel, A. R. Oganov and S. Leoni, *Sci. Rep.*, 2012, **2**, 471.

19 D. Zagorac, J. C. Schoen and M. Jansen, *J. Phys. Chem. C*, 2012, **116**, 16726–16739.

20 C. R. A. Catlow, S. A. French, A. A. Sokol, A. A. Al-Sunaidi and S. M. Woodley, *J. Comput. Chem.*, 2008, **29**, 2234–2249.

21 A. M. Saitta and F. Decremps, *Phys. Rev. B*, 2004, **70**, 035214.

22 L. Bayarjargal and B. Winkler, *Z. Kristallogr. - Cryst. Mater.*, 2014, **229**, 92–100.

23 J. Cai and N. Chen, *J. Phys. Condens. Matter.*, 2007, **19**, 266207.

24 H. Z. Liu, Y. Ding, M. Somayazulu, J. Qian, J. Shu, D. Häusermann and H. K. Mao, *Phys. Rev. B*, 2005, **71**, 212103.

25 P. G. Bolhuis, D. Chandler, C. Dellago and P. L. Geissler, *Annu. Rev. Phys. Chem.*, 2002, **53**, 291–318.

26 S. Leoni, R. Ramlau, K. Meier, M. Schmidt and U. Schwarz, *Proc. Natl. Acad. Sci. U. S. A.*, 2008, **105**, 19612–19616.

27 G. Martyna, D. Tobias and M. Klein, *J. Chem. Phys.*, 1994, **101**, 4177–4189.

28 G. Martyna, M. Tuckerman, D. Tobias and M. Klein, *Mol. Phys.*, 1996, **87**, 1117–1157.

29 D. J. Binks and R. W. Grimes, *J. Am. Ceram. Soc.*, 1993, **76**, 2370–2372.

30 G. A. Tribello, M. Bonomi, D. Branduardi, C. Camilloni and G. Bussi, *Comput. Phys. Commun.*, 2014, **185**, 604–613.

31 A. Barducci, G. Bussi and M. Parrinello, *Phys. Rev. Lett.*, 2008, **100**, 020603.

32 S. Leoni and R. Nesper, *Acta Crystallogr., Sect. A: Found. Crystallogr.*, 2000, **56**, 383–393.

33 D. Granata, F. Baftizadeh, J. Habchi, C. Galvagnion, A. De Simone, C. Camilloni, A. Laio and M. Vendruscolo, *Sci. Rep.*, 2015, **5**, 15449.

34 J. C. Schön and M. Jansen, *Comput. Mater. Sci.*, 1995, **4**, 43–58.

35 J. C. Schön, *Z. Anorg. Allg. Chem.*, 2004, **630**, 2354–2366.

36 J. C. Schön, *Process. Appl. Ceram.*, 2015, **9**, 157–168.

37 A. Seko, F. Oba, A. Kuwabara and I. Tanaka, *Phys. Rev. B*, 2005, **72**, 485.

Faraday Discussions

PAPER

First-principles stability ranking of molecular crystal polymorphs with the DFT+MBD approach†

Johannes Hoja and Alexandre Tkatchenko *

Received 19th March 2018, Accepted 4th May 2018
DOI: 10.1039/c8fd00066b

The ability to accurately calculate the relative stabilities of numerous polymorphs of a given molecular crystal is crucial for the success of any molecular crystal structure prediction (CSP) approach. We have recently presented a hierarchical CSP procedure based on van-der-Waals-inclusive density functional theory [Hoja *et al.*, 2018, arXiv:1803.07503], which yields excellent stability rankings for molecular crystals involving rigid molecules, salts, co-crystals, and highly polymorphic drug-like molecules. This approach includes many-body dispersion effects, exact exchange, as well as vibrational free energies. Here, we discuss in detail the impact of these effects on the obtained stability rankings. In addition, we assess the impact of the approximations used in our hierarchical procedure. We show that our procedure is generally robust to 1–2 kJ mol^{-1} for the systems in the latest CSP blind test but vibrational free energies for crystals involving flexible molecules would benefit from directly including many-body dispersion interactions. In addition, we also discuss the effect of temperature on the structure of molecular crystals and a simple but effective method for estimating anharmonic effects.

1 Introduction

The prediction of molecular crystal structures is an essential tool for the development of new pharmaceutical compounds and for the engineering of novel molecular materials.[1,2] In addition to drugs, molecular crystals find applications as organic semiconductors[3,4] and explosives,[5] and are used for solid-state reactions.[6] Computational molecular crystal structure predictions (CSP) face many challenges due to the fact that sampling procedures have to cover the whole conformational space of the involved molecules as well as a vast number of possible crystal-packing arrangements.[7] Furthermore, the majority of

Physics and Materials Science Research Unit, University of Luxembourg, L-1511 Luxembourg, Luxembourg. E-mail: alexandre.tkatchenko@uni.lu

† Electronic supplementary information (ESI) available: All optimized structures, phonon band structures for XXIII-N18, and phonon densities of state for several XXIII structures. See DOI: 10.1039/c8fd00066b

experimentally observed polymorphs differ in their relative stability by less than 4 kJ mol^{-1} ($\approx$1 kcal mol^{-1}) but often by even less than 1 kJ mol^{-1},[8] which amounts to only a few percent of the actual lattice energies of these crystals. Therefore, the computational methods used for the final stability ranking should ideally have an accuracy of about 1 kJ mol^{-1}. Recently, this accuracy could be reached for the lattice energy of a benzene crystal utilizing a high-level fragment coupled cluster approach.[9,10] However, this methodology cannot currently be applied to molecular crystals of pharmaceutical relevance.

The most accurate first-principles approach, which can routinely be applied to systems of pharmaceutical relevance, is density-functional theory (DFT) supplemented by a high-level method for van der Waals (vdW) dispersion interactions.[11–13] For instance, the PBE0 density functional approximation (DFA)[14] combined with the many-body dispersion (MBD) model[15,16] is capable of describing the lattice energies of the X23 benchmark set[17–19] with a mean absolute deviation (MAD) of 3.8 kJ mol^{-1}. In comparison, force-field approaches usually yield in the best case MADs that are twice as large.[20] The same is true for the semi-empirical density functional tight binding (DFTB) approach when paired with a vdW correction.[21] However, this applies to absolute values of lattice energies and not necessarily to the energy differences between polymorphs. It was recently shown for oxalic acid, glycine, and coumarin that the PBE0+MBD approach is capable of describing relative polymorph stabilities with an accuracy of about 1 kJ mol^{-1} w.r.t. experimental measurements.[17,18,22–24] The purpose of this manuscript is to present a detailed methodology that we use to achieve this accuracy and also highlight the remaining challenges towards a broadly applicable computational molecular CSP method.

The quality of molecular CSP procedures is regularly assessed by the blind tests of the Cambridge Crystallographic Data Centre[25–30] and the performance of several vdW-inclusive DFT-approaches has been subsequently discussed.[31–33] We have recently presented a hierarchical CSP approach which combined the most successful crystallographic space sampling of the latest blind test with the DFT+MBD framework.[34] The initial crystal structures are generated by a Monte Carlo parallel tempering algorithm using a tailor-made force field and dispersion-inclusive DFT within the Grace software.[30,31,35,36] Subsequently, the relative stabilities are calculated with a hierarchical procedure, which includes three effects commonly neglected in production CSP studies: (i) many-body dispersion (MBD) interactions, (ii) exact exchange (hybrid functional), (iii) vibrational free energies.[34] In this paper, we discuss in detail the impact of these three mentioned effects on the stability ranking. In addition, we also evaluate the possible impact of several approximations employed during our hierarchical approach. We focus in our discussion on system XXIII from the latest blind test, which was a former drug candidate,[37] involves a flexible molecule, and already has 5 experimentally confirmed polymorphs. Therefore, this system provides an optimal testing ground for challenging systems of pharmaceutical importance.

2 Methods

We utilized for all calculations the all-electron code FHI-aims.[38–44] We use in general two accuracy levels, which are labeled light and tight. The light level (l) corresponds to the light species default settings in FHI-aims in terms of integration grids and basis functions used to describe the Kohn–Sham orbitals. The number of used k-points (n) in each direction amounts to the smallest integer number satisfying the relation $n \times a \geq x$, with a describing the unit cell length in this direction and x being 25 Å. The tight level (t) corresponds to the tight species default settings within FHI-aims and the number of k-points is determined as for the light settings but with $x = 30$ Å. The following convergence criteria are used for all calculations: 10^{-6} eV for the total energy, 10^{-5} electrons per Å^3 for the charge density, 10^{-4} eV Å^{-1} for the forces, and 10^{-3} eV for the sum of the eigenvalues. The employed naming scheme for the structures corresponds to the labels used in ref. 34.

Full lattice relaxations and geometry optimizations were performed using the following methods: (i) PBE+TS[45,46] (pairwise Tkatchenko–Scheffler dispersion model), (ii) PBE+MBD[15,16] (many-body dispersion model at the MBD@rsSCS level), and (iii) PBE0+MBD.[14] These optimizations were performed using the light settings and the force components were converged to 0.005 eV Å^{-1}. In addition, the PBE+MBD optimizations were also carried out utilizing the tight settings. The PBE+TS structures[34] were used as the starting point for all other optimizations. The optimized structures were symmetrized by using PLATON[47] and are provided in the ESI.† The agreement of the optimized structures w.r.t. experimental structures is measured by the root-mean-square-deviation (RMSD) of 20 molecules within the crystal (RMSD$_{20}$) utilizing Mercury.[48]

The relative static lattice energies based on DFT total energies were evaluated at the PBE0+MBD level. These energies are obtained by calculating the differences between the PBE0+MBD and PBE+MBD energies at the light level and adding these differences to the PBE+MBD energies obtained at the tight level. This provides an excellent approximation of converged PBE0+MBD energies[34] and reduces the computation time by about a factor of 10 compared to tight settings. All presented stability rankings are normalized per chemical unit (per molecule for XXII, XXIII, XXVI; per dimer for XXV; per trimer for XXIV).

Vibrational free energies were calculated with PBE+TS, PBE+MBD, and PBE0+MBD within the harmonic approximation using phonopy[49] on top of structures optimized at the same level. In general, vibrational free energies were calculated using the light settings. In addition, several structures were also calculated with PBE+MBD using tight settings. The temperature used to evaluate each free energy corresponds always to the temperature at which the experimental crystal structure were measured, *i.e.*, 150 K for system XXII and 300 K for system XXIII. Displacements of 0.005 Å were used throughout for the finite displacements. Unless stated otherwise, we used supercells which extend at least 10 Å in every direction and the number of q-points for the reciprocal space sampling was calculated in the same way as the k-grid mentioned above, but using $x = 50$ Å.

Thermally expanded structures corresponding to 300 K were calculated for several structures of system XXIII within the quasi-harmonic approximation (QHA) utilizing the PBE+TS method with light settings.[34] Structures with different unit cell volumes were generated by full lattice and geometry optimizations under external hydrostatic pressures of 0.4, 0.2, −0.2, −0.4, and −0.6 GPa. Harmonic vibrational free energies at 300 K were calculated for all so-obtained structures. The unit cell volume corresponding to 300 K was determined with the Murnaghan equation of state[50] fitted to the sum of the PBE+TS (light) total energies and vibrational free energies. In addition, the room temperature structure for XXIII-A was calculated within the QHA utilizing entirely PBE+MBD with light settings for energies and vibrational free energies. Furthermore, we also calculated the vibrational free energies using PBE+MBD for the experimentally determined structures of system XXIII. For that, we fixed the lattice vectors to the experimental values and optimized only the atomic positions within the unit cells.

Finally, we estimated for several structures of system XXIII the anharmonic vibrational free energies by substituting the harmonic oscillators with Morse oscillators.[51] This substitution is made for all modes at the Γ point. To have a consistent description, cells containing 4 molecules were used throughout. For form B and structures N18, N31, and N42 $2 \times 1 \times 1$ super cells were used while the unit cell was used for all other structures. Optimized structures were displaced along normal modes in both directions. The magnitude of the displacement was chosen to correspond to energy differences of $0.5k_BT$ and k_BT within the harmonic approximation, with a temperature $T = 300$ K and k_B being the Boltzmann constant. The method used for calculating the energy of the displaced structures corresponds always to the method used for the harmonic vibrational free energies. In order to have a consistent sampling of the accessible energy window, we constrained the largest observed energy change to energies between k_BT and $1.5k_BT$. Hence, the displacement amplitudes for several low-frequency modes were reduced to sample the desired energy window. Next, a Morse potential[51,52] was fitted to the obtained data points. The Morse potential is given by

$$V(x) = D(1 - \mathrm{e}^{-a(x-x_0)})^2, \tag{1}$$

with x being the displacement amplitude and a, D, and x_0 describing the width of the potential, the well depth, and the minimum of the potential, respectively. For a vibrational mode in state v the energy can be calculated according to

$$E(v) = \hbar\omega_0\left(v + \frac{1}{2}\right) - \frac{\hbar^2{\omega_0}^2}{4D}\left(v + \frac{1}{2}\right)^2, \tag{2}$$

with

$$\omega_0 = \sqrt{\frac{2a^2D}{\mu}}, \tag{3}$$

where μ describes the reduced mass. The anharmonic vibrational free energy ($\tilde{F}_{\mathrm{vib}}$) was calculated according to

$$\tilde{F}_{\mathrm{vib},\Gamma} = -k_BT \ln Q_{\mathrm{vib}}, \tag{4}$$

with

$$Q_{\mathrm{vib}} = \prod_i \sum_v \exp\left(\frac{-E_{i,v}}{k_BT}\right), \tag{5}$$

where i describes the phonon mode number. With this approach we obtained anharmonic vibrational energies for the Γ point of the used cells. For the other q-points we relied on the harmonic approximation. The total vibrational free energies were calculated according to

$$\tilde{F}_{\mathrm{vib}} = F_{\mathrm{vib,full}} + \tilde{F}_{\mathrm{vib},\Gamma} - F_{\mathrm{vib},\Gamma}, \tag{6}$$

with $F_{vib,full}$ being the fully converged harmonic vibrational free energy and $F_{vib,\Gamma}$ being the harmonic vibrational free energy calculated only at the Γ point of the used cells.

3 Static lattice energies

First, we analyze the different stability rankings we have presented in ref. 34. Therein, PBE+TS, PBE+MBD, and PBE0+MBD energies were calculated on top of PBE+TS-optimized structures for all the systems of the latest CSP blind test.[30] The chemical diagrams of the involved molecules are visualized in Fig. 1. We will use throughout the labeling scheme of the blind test for all systems. This very diverse set of systems includes a crystal consisting of a small and rigid molecule (XXII), a highly polymorphic system composed of a flexible molecule (XXIII), a salt (XXIV), a co-crystal (XXV), and a crystal involving a large molecule (XXVI). The accurate description of all these systems using a consistent CSP method is challenging due to the varying system sizes and their different chemical natures. The best stability ranking in ref. 34 is obtained when all three of the above effects are accounted for.

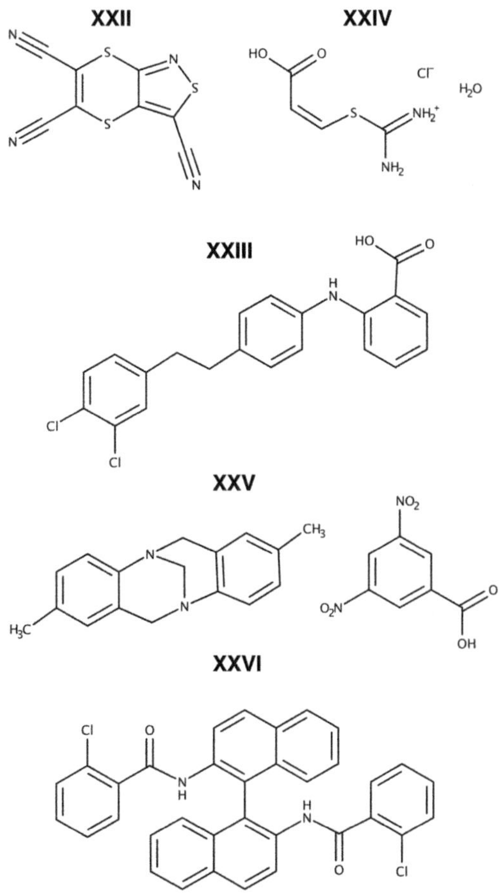

Fig. 1 Chemical diagrams for all systems of the latest CSP blind test.[30]

Let us start by discussing the importance of many-body dispersion (MBD) interactions and exact exchange for relative static lattice energies. Fig. 2 shows a correlation plot between PBE+TS and PBE+MBD energies for all systems. The most stable structure according to the PBE+MBD energies is always set to zero and all relative energies are normalized per chemical unit (per molecule for XXII, XXIII, XXVI; per dimer for XXV; per trimer for XXIV). In addition, perfect correlation is indicated with the black line and intervals of 1 kJ mol^{-1} and 1 kcal mol^{-1} around this line are marked with the dotted lines. It can be seen that for no system are all data points within the $\pm$1 kcal mol^{-1} energy window. This immediately illustrates the crucial nature of MBD interactions given the energy window for co-existing polymorphs. Furthermore, it can be seen that even re-fitting a pairwise approach would not yield a significant improvement of the correlation, given the large spread of the data points.

In order to numerically quantify the impact of MBD and exact exchange effects, we calculated and compared for every system all possible relative energies between polymorphs. To illustrate the effect of MBD interactions we discuss the deviation between the PBE+TS and PBE+MBD relative energies in terms of the mean absolute deviation (MAD) and the maximum absolute deviation (MAX). These values are shown for every system in Table 1.

It can be seen that the average impact of MBD interactions on relative stabilities varies between 1.8 and 3.2 kJ mol^{-1}. As expected, the least impact is found for the two smallest systems involving rigid molecules. The largest impact is found for the former drug candidate XXIII, which involves a quite flexible molecule, and the co-crystal XXV. Note that the maximal observed change in relative energies is for all systems larger than 7.7 kJ mol^{-1}. Therefore, MBD interactions clearly cannot be neglected for any of the studied systems.

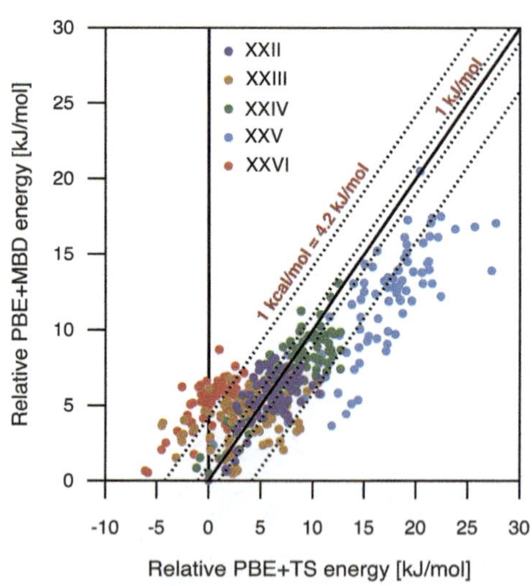

Fig. 2 Correlation between PBE+TS and PBE+MBD energies calculated on top of PBE+TS-optimized structures.[34] The most stable PBE+MBD structure is always set to zero and energies are normalized per chemical unit.

Table 1 Statistics for the impact of several interactions on all possible relative energies expressed in terms of the mean absolute deviation (MAD) in kJ mol^{-1} and the maximum absolute deviation (MAX) in kJ mol^{-1} for all studied systems;[34] PBE+MBD *vs.* PBE+TS (MBD), PBE0+MBD *vs.* PBE+MBD (PBE0), PBE0+MBD + F_{vib} *vs.* PBE0+MBD (F_{vib})

	MBD		PBE0		F_{vib}	
System	MAD	MAX	MAD	MAX	MAD	MAX
XXII	1.8	7.7	1.1	4.6	0.5	1.0
XXIII	3.2	13.4	1.1	4.2	1.9	8.4
XXIV	1.9	7.8	2.0	8.2	0.9	1.8
XXV	3.2	15.2	1.4	5.1	1.6	3.2
XXVI	2.4	8.7	0.7	2.3	1.0	2.3

Since MBD captures many-body interactions up to infinite order, we now investigate the convergence of the relative energies. The MBD energy can be decomposed into orders n (n-atom contributions) according to

$$E_n = -\frac{1}{2\pi}\int_0^\infty d\omega \sum_{n=2}^{\infty} \frac{1}{n} \mathrm{Tr}\left[(\mathbf{AT})^n\right], \tag{7}$$

where the matrix **A** includes all atomic polarizabilities and **T** describes the dipole–dipole tensor.[12,53] We illustrate this effect by using the five experimentally observed forms of system XXIII and in addition also 4 low-energy structures identified in ref. 34 (see Fig. 3). These structures are visualized in Fig. 4. The plot shows the relative PBE+TS and PBE+MBD energies, with the MBD interactions evaluated up to the shown n-atom contribution.

Already the 2-atom (pairwise) MBD energy is closer to the final PBE+MBD result than the PBE+TS result. This is due to the fact that MBD includes in addition to many-body dispersion effects also dielectric screening effects. When we now include 3-atom interactions, the relative energies change up to 3.7 kJ mol^{-1}. However, the relative energies are by far not converged yet.

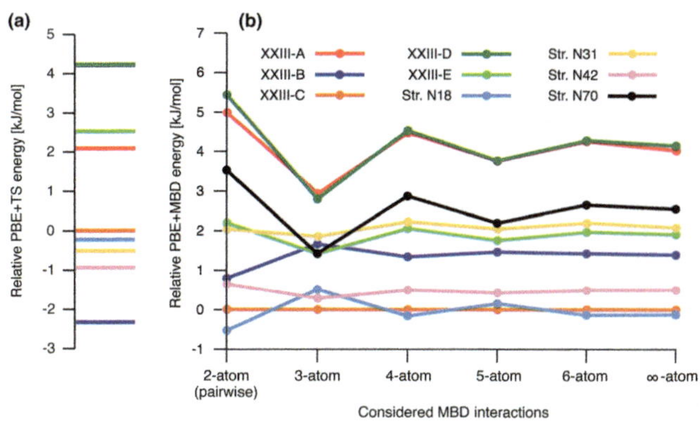

Fig. 3 Relative PBE+TS (a) and PBE+MBD (b) energies for several structures of system XXIII. The energy of form C serves as reference and is set to zero. Part (b) shows the relative PBE+MBD energies when MBD interactions are included up to the shown n-atom contribution.

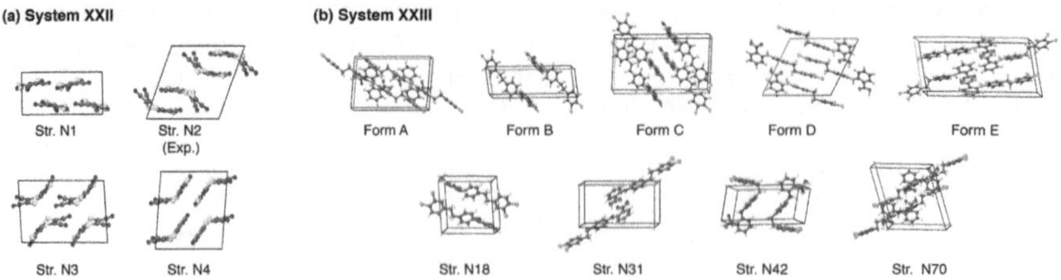

Fig. 4 Visualization of the unit cell for all explicitly discussed structures. The coordinates of all optimized structures are available in the ESI.†

Accounting, in addition, for 4-atom contributions still changes the relative energies by up to 2.4 kJ mol^{-1}. Due to the oscillatory behavior only the step from 5-atom to 6-atom contributions finally changes the relative energies by less than 1 kJ mol^{-1}.

Now let us move on to the effect of exact exchange, *i.e.*, switching from PBE to the hybrid functional PBE0 (see Table 1). This step improves for example the description of electrostatics and charge transfer. In general, we can see that this effect is less pronounced than the MBD effects with MADs varying between 0.7 and 2.0 kJ mol^{-1}. However, since the maximal changes are between 2.3 and 8.2 kJ mol^{-1} this effect is significant enough to change stability rankings. Exact exchange has as expected the largest effect on the salt, since this system has the largest electrostatic interactions. In addition, PBE0 is for example crucial for correctly identifying the experimental structure of system XXII as the most stable structure.[34,54]

4 Geometries

All stabilities in our hierarchical CSP procedure[34] have been calculated on top of PBE+TS-optimized structures. Therefore, one natural step to improve this procedure would be to include MBD interactions also in the lattice relaxations and geometry optimizations. However, this would lead on average to an increase of force calculation times of about 50% (see Section 6). Therefore, we evaluate in this section the effect of MBD and also exact exchange effects on structures and corresponding stability rankings. We have re-optimized all the structures from ref. 34 corresponding to experimentally observed polymorphs using PBE+MBD and PBE0+MBD with light settings. In addition, the structures were also optimized with PBE+MBD using tight settings to study the impact of the used basis functions and integration grids. All the optimized geometries are available in the ESI† and the errors w.r.t. the experimental measurements are shown in Table 2. Let us start by discussing the accuracy of the PBE+TS-optimized structures. It can be seen that on average we can describe cell lengths and angles within 1% of the experimental value and the mean absolute relative deviations (MARD) amount to 1.4 and 1.5%, respectively. Furthermore, all errors are less than 5%. However, we underestimate the unit cell volume on average by 2.4%. This is to be expected since molecular crystals tend to expand with temperature and our cell relaxation does not include any thermal effects. The unit cell volume can be improved by the so called quasi-harmonic approximation, which will be discussed later on. Note

Table 2 Errors of optimized structures w.r.t. experimental measurements;[30] mean relative deviation (MRD), mean absolute relative deviation (MARD) and maximum absolute deviation (MAX) in percent. The labels l and t indicate light and tight species default settings within FHI-aims, respectively. For the cell angles we only take into account values which are not exactly 90 degrees due to symmetry

	PBE+TS/l			PBE+MBD/l			PBE0+MBD/l			PBE+MBD/t		
	MRD	MARD	MAX	MRD	MARD	MAX	MRD	MARD	MAX	MRD	MARD	MAX
Lengths [%]	−0.8	1.4	4.0	−0.3	1.4	4.7	−1.3	1.6	3.8	−0.3	1.4	4.1
Angles [%]	0.8	1.5	4.3	0.8	1.6	3.2	0.8	1.7	4.3	0.7	1.4	2.7
Volume [%]	−2.4	3.1	4.7	−0.9	2.2	3.8	−3.7	3.8	5.9	−1.0	2.6	5.2

that while we are underestimating most unit cell volumes, we are overestimating the volume of systems XXII and XXIV. In addition, the quality of the molecular structure is assessed in terms of the $RMSD_{20}$ (see Method section). On average our $RMSD_{20}$ w.r.t the experimental structures amounts to 0.25 Å with a maximum of 0.48 Å (XXIII-E). The possibly disordered nature of form E[30] is most likely the reason for the observed relatively large $RMSD_{20}$. Overall, our PBE+TS optimizations yield already structures of good quality.

Now let us turn to the PBE+MBD (light) optimizations. The description of cell lengths and angles is comparable with the PBE+TS results, but the errors in the cell volume are now much smaller. PBE+MBD consistently yields larger unit cell volumes than PBE+TS. This is due to the fact that the pairwise dispersion approach tends to overestimate intermolecular interactions.[17] In terms of the $RMSD_{20}$ one can only see a minor improvement with the average and the maximum amounting to 0.24 and 0.45 Å, respectively. When we move on to PBE0+MBD, we again see a similar picture for cell lengths and angles. However, the exact exchange leads to significantly smaller unit cell volumes. Now, the unit cell volume of XXII is only slightly overestimated by 0.6%. On average, unit cell volumes are underestimated by 3.7%. Again, this underestimation originates due to the missing thermal expansion in our structures and has to be expected. In terms of the $RMSD_{20}$ the PBE0+MBD structures yield very similar results to the previously discussed methods with an average of 0.23 Å and a maximum of 0.47 Å. Finally, let us discuss the PBE+MBD optimizations with tight settings. It can be seen that the cell lengths and angles are only slightly better compared to the light settings and also the unit cell volumes are not changing significantly. Also the $RMSD_{20}$ values are again similar with an average of 0.23 Å and a maximum of 0.46 Å. Therefore, we can conclude that light settings provide already a sufficient description of the crystal structure.

Overall, all discussed methods provide quite similar structures, especially in terms of the cell lengths and angles, as well as the $RMSD_{20}$. The only major difference is the cell volume, which in general follows the following trend: PBE0+MBD < PBE+TS < PBE+MBD.

The main focus of all CSP approaches is the stability ranking or in other words the polymorphic energy landscape. Therefore, we discuss now the effect of the

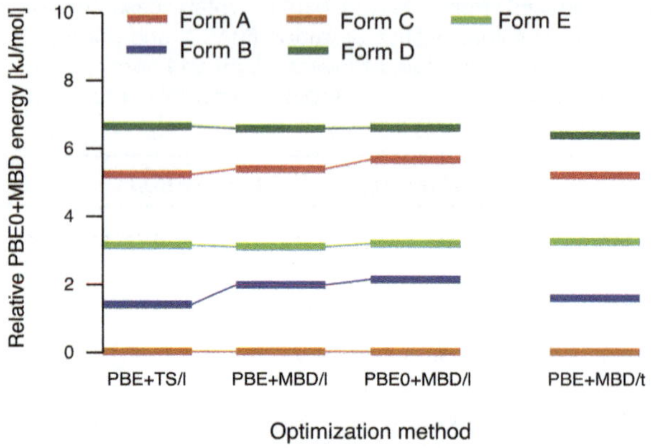

Fig. 5 Relative PBE0+MBD energies of the five experimental polymorphs of system XXIII for different optimization methods.

optimization method on the relative stabilities for the five experimentally obtained structures of system XXIII. The final static lattice energies within our CSP procedure are always calculated on the PBE0+MBD level (see Methods). Hence, Fig. 5 shows the relative PBE0+MBD energies calculated on top of the differently optimized structures. We again discuss the differences with the mean absolute deviation (MAD) and the maximum absolute deviation (MAX) evaluated based on all possible relative energies. Changing the lattice relaxations from PBE+TS (light) to PBE+MBD (light) yields a MAD of only 0.3 kJ mol^{-1} with a MAX of 0.6 kJ mol^{-1}. PBE0+MBD optimizations yield 0.4 and 0.8 kJ mol^{-1}, respectively. Utilizing tight settings for PBE+MBD optimizations instead of light settings yield a MAD of 0.3 kJ mol^{-1} with a MAX of 0.5 kJ mol^{-1}. This illustrates that the PBE0+MBD energy ranking is only mildly affected by improving the optimization method, with all changes in relative energies being less than 0.8 kJ mol^{-1}. In addition, Mortazavi *et al.* have shown that computationally very efficient density functional tight binding optimizations could be used for a pre-screening when the stability is subsequently evaluated with PBE+MBD on top of these structures.[55]

5 Vibrational free energies

In order to describe relative stabilities at finite temperatures, it is necessary to include thermal effects. The simplest way to achieve this is to add to the electronic energies (and the nuclear repulsion) the vibrational free energy. This leads then to Helmholtz free energies, which include entropic effects. The vibrational free energy $F_{\text{vib}}(T)$ can be calculated within the harmonic approximation according to

$$F_{\text{vib}}(T) = \int \mathrm{d}\omega g(\omega)\frac{\hbar\omega}{2} + \int \mathrm{d}\omega g(\omega) k_{\text{B}} T \ln\left[1 - \exp\left(-\frac{\hbar\omega}{k_{\text{B}}T}\right)\right], \tag{8}$$

with $g(\omega)$ being the phonon density of states. However, accurate calculations are not as straightforward as simple energy calculations. For accurate results it is imperative that there are no imaginary modes present at the Γ-point. Such modes indicate that this structure does not correspond to a local minimum of the potential energy surface and also drastically change the vibrational free energy.

The relative free energies are essentially determined by the low-frequency modes, since they have the largest impact on the entropic contributions. This is illustrated in the ESI.† These low-frequency modes contain intermolecular motions and therefore vdW interactions play an important role for them.

Furthermore, calculating the vibrational free energy for the Γ point of the unit cell is insufficient for accurate results, which we will illustrate using our set of 9 structures for system **XXIII**. Fig. 6 shows relative stabilities in terms of the Helmholtz free energies at 300 K, for which the vibrational free energies were calculated at the PBE+TS level. Since we are using finite displacements, we have to avoid artifacts between displacements in periodic images. It was shown that at least 9–10 Å are necessary in every direction in order to minimize artifacts.[17] Therefore, we have used for all phonon calculations in our CSP procedure[34] super cells, which extend at least 10 Å in every direction. All values in part (a) of Fig. 6 were calculated with these cells but the vibrational free energy was evaluated differently. At first, the vibrational free energy was evaluated by only using the modes at the Γ-point of the unit cell (uc), followed by the Γ point of the used supercell (sc). In all other cases the vibrational free energy was evaluated with a q-grid in reciprocal space utilizing the q-point interpolation available in phonopy.[49] The number (n) of necessary q-points in each direction was calculated according to $n \times a \geq x$, with a being the cell length in the respective direction and the used values of x are given in the plot. It can be seen that both Γ-point plots provide a completely different stability ranking compared to the converged reciprocal space sampling. This is mainly due to the fact that the three acoustic modes have zero frequency at the Γ point but contribute significantly for other q-points. Note that for forms A, D and structure N70 the supercell is identical to the unit cell, since in these unit cells all cell lengths are already larger than 10 Å. Convergence to 0.1 kJ mol^{-1} is reached at $x = 50$ Å.

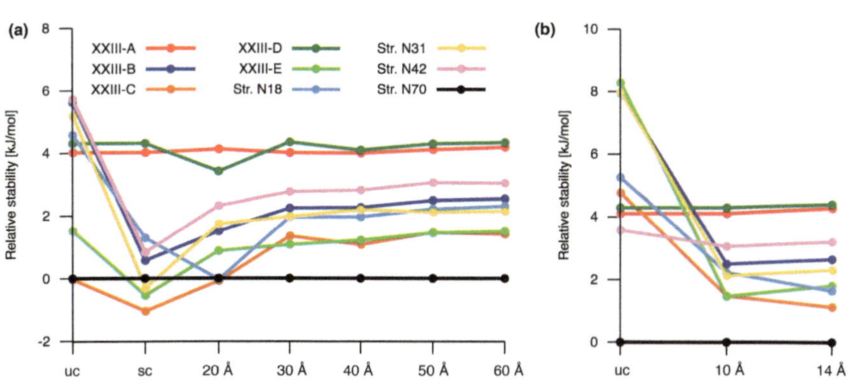

Fig. 6 Convergence of vibrational free energies. Part (a) shows the relative stabilities of 9 structures from system XXIII calculated with PBE0+MBD + F_{vib}(PBE+TS) at 300 K, where supercells with cell lengths of at least 10 Å were used for the finite difference calculation. The vibrational free energy was evaluated using only the Γ point modes of the unit cell (uc), the Γ-point modes of the used supercell (sc), or with a q-point mesh with the shown values used for x (see Methods). In part (b) a converged q-point mesh is used throughout but the finite difference calculations were performed by either using just the unit cell or supercells with minimum cell lengths of 10 and 14 Å, respectively.

Part (b) of Fig. 6 illustrates the importance of the used supercell for the finite difference calculations. Here, the vibrational free energy is always evaluated with a q-point grid corresponding to $x = 50$ Å but we used different cell sizes for the finite difference calculations. First, the calculations were performed by only using the unit cells (uc), followed by the smallest supercells which extend in every direction at least 10 Å and 14 Å. The MAD when comparing all the relative energies between the unit cell and the 10 Å supercell approach amounts to 3.3 kJ mol^{-1} with a maximum of 6.8 kJ mol^{-1}. Moving from the 10 Å to the 14 Å supercells leads to a MAD of only 0.3 kJ mol^{-1} with a maximum energy change of 0.9 kJ mol^{-1}. This illustrates how crucial these supercells are for accurate vibrational free energies. In addition, using larger supercells than 10 Å has only a minor effect on the relative stabilities and does not merit the immense increase in computation time. In general, the larger the supercell, the more points in reciprocal space are evaluated exactly, which increases the quality of the q-point interpolation used in phonopy. For most of the unit cell calculations one can see in the phonon band structure that the three acoustic modes are poorly described away from Γ. This is illustrated for structure N18 in the ESI.†

Now let us discuss the impact of the vibrational free energies on stability rankings (see Table 1). The average impact in terms of the MAD varied for the five systems from 0.5 to 1.9 kJ mol^{-1}. While this effect is not that pronounced for systems involving rigid molecules, it can be crucial for the stability ranking of systems involving flexible molecules. For our highly polymorphic system XXIII vibrational free energies can change relative stabilities by up to 8.4 kJ mol^{-1}.

All the so far discussed vibrational free energies have been calculated with PBE+TS. Therefore, we now evaluate what impact MBD interactions and exact exchange have on the vibrational free energies. Let us begin by discussing the effect of MBD based on the four most stable structures of system XXII and 9 structures of system XXIII (see Fig. 7). The total energy is always evaluated with PBE0+MBD while the vibrational free energy is calculated by using PBE+TS and PBE+MBD, respectively. For system XXII we observe only minor changes in the relative stabilities; relative energies change on average by 0.4 kJ mol^{-1} and the

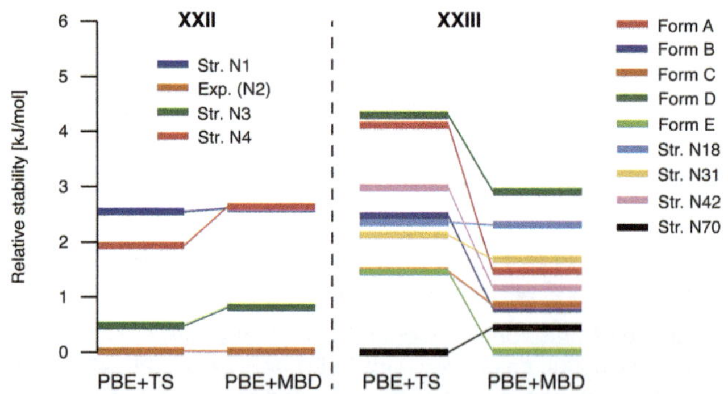

Fig. 7 Comparison of PBE+TS and PBE+MBD vibrational free energies for several structures of systems XXII and XXIII. The relative stability consists of the PBE0+MBD energy plus the vibrational free energy evaluated with the respective method. The vibrational free energies were calculated at the experimental measuring temperature of the respective crystal structures (150 K for XXII and 300 K for XXIII).

Table 3 Relative free energies in kJ mol^{-1}. The total energy was always calculated with PBE0+MBD and the structure optimization as well as the vibrational free energy were performed with the labeled method. The vibrational free energies were always calculated at the experimental measuring temperature of the respective crystal structures (150 K for XXII and 300 K for XXIII)

Method	ΔXXII[a]	ΔXXIII-1[b]	ΔXXIII-2[c]	ΔXXIII-3[d]
PBE+TS/l	2.5	0.2	4.3	4.1
PBE+MBD/l	2.6	1.5	2.4	0.9
PBE0+MBD/l	2.0	−0.8	0.3	1.1
PBE+MBD/t	2.2	1.0	2.4	1.3

[a] ΔXXII = *F*(XXII-N1) − *F*(XXII-N2). [b] ΔXXIII-1 = *F*(XXIII-D) − *F*(XXIII-A). [c] ΔXXIII-2 = *F*(XXIII-D) − *F*(XXIII-N70). [d] ΔXXIII-3 = *F*(XXIII-A) − *F*(XXIII-N70).

maximum change amounts to 0.7 kJ mol^{-1}. In contrast, the effect of MBD interaction on the vibrational free energies is much stronger for system XXIII with an average change of 1.2 kJ mol^{-1} and a maximum modification of 3.1 kJ mol^{-1}. This is consistent with ref. 56, which shows that MBD interactions change the relative free energy between two polymorphs of aspirin by about 3 kJ mol^{-1}. In general, MBD effects are more pronounced for cells with little symmetry involving especially flexible molecules.

Furthermore, let us discuss the influence of exact exchange on vibrational free energies. Therefore, two structures from system XXII, for which exact exchange significantly modified the relative stability, were selected as well as 3 structures from system XXIII, for which no supercell is necessary for the accurate evaluation of the vibrational free energy. This benchmark set provides 4 relative free energies (see Table 3). The structure optimization and the calculation of the vibrational free energy are done with the respectively mentioned methods, while the total energy is evaluated with PBE0+MBD. In addition, the PBE+MBD calculations were also performed using the tight settings in order to evaluate if a larger basis set would significantly change the result. The difference between PBE0+MBD and PBE+MBD (light) amounts to 1.3 kJ mol^{-1} in terms of the MAD and the maximal observed change is 2.3 kJ mol^{-1}. The MAD between PBE+TS and PBE0+MBD amounts to 2.1 kJ mol^{-1} with a maximum of 3.9 kJ mol^{-1}. Hence, exact exchange still modifies relative vibrational free energies, however, this effect is not large enough to justify the massive additional computational cost for production calculations. In addition, increasing the basis set yields only a minor modification of the relative stabilities with a maximum change of 0.5 kJ mol^{-1}.

6 Computational cost

The computational cost of single-point energy calculations, lattice optimizations, and vibrational free energies highly depends on the size and shape of the involved unit cells. Here we discuss the computation times for form A of system XXIII. This monoclinic unit cell contains 172 atoms, which is only slightly less than the average unit cell size studied in ref. 34. Table 4 shows the relative computation times for one single-point energy and one force calculation for all the discussed methods. The values are normalized to the PBE+TS single-point energy

Table 4 Relative computation times for one single-point energy and one force calculation of structure XXIII-A (172 atoms in the unit cell). All computation times were normalized to the PBE+TS/l energy calculation, which amounts to 2.1 CPU hours on Intel Xeon E5-2680 v4 cores (2.4 GHz)

Method	Energy	Force
PBE+TS/l	1.0	1.4
PBE+MBD/l	1.0	2.1
PBE0+MBD/l	16.9	18.2
PBE+MBD/t	9.8	13.2

calculation with light settings (l), which needs 2.1 CPU hours on Intel Xeon E5-2680 v4 cores (2.4 GHz). It can be seen that the calculation of the many-body dispersion energy (MBD) adds only a negligible cost on top of the PBE+TS calculation. However, PBE+MBD forces need about 50% more computation time than PBE+TS forces when computed at the light level. PBE0 calculations are at the light level about 17 times more expensive than PBE calculations. Furthermore, utilizing tight settings (t) increases the computation time of a PBE+MBD energy calculation by a factor of about 10 compared to light settings. When utilizing tight settings, PBE+TS and PBE+MBD force calculations are essentially equally expensive since almost all computation time is spent on the underlying DFT calculation.

The number of needed optimization steps in a geometry and lattice optimization highly depends on the starting structure, *i.e.* how similar the initial molecular conformation and the unit cell shape are to the optimized structure. Typically, 50–100 steps are necessary to converge a structure with the used settings. Therefore, the computation time amounts to about 50–100 times the time needed for one force calculation. The calculation of the harmonic vibrational free energy requires for XXIII-A 258 finite displacements and therefore also 258 force calculations. In this particular case the unit cell is sufficient for the finite differences calculation and no supercell is needed. Therefore, the calculation of the vibrational free energy of XXIII-A using PBE+TS with light settings requires about 750 CPU hours.

7 Beyond the harmonic approximation

So far we have calculated all free energies on top of the fully optimized structures. However, molecular crystals expand with increasing temperature. This structural change could lead to a modification of the relative stabilities. The simplest approach of including the effect of thermal expansion on a first-principles level is the so called quasi-harmonic approximation (QHA).[57–62] In this approach the vibrational free energy is calculated within the harmonic approximation for several unit cell volumes, and the volume at a certain temperature is determined by an equation-of-state (EOS) fit. We have calculated the cells corresponding to 300 K within the QHA for 9 structures from system XXIII.[34] The thermally expanded structures are available in the ESI.† Note that in this case we use PBE+TS for structures and vibrational free energies, as well as for the total energies. This approach yields unit cell volumes which agree on average within

1% with the experimental structures. The largest error was found for form E, where the unit cell volume is underestimated by 1.7%.

The thermal expansion for form A is illustrated in Fig. 8(a) at the PBE+TS level. The dots correspond to the harmonic free energies calculated at various volumes and at three different temperatures. The solid curves correspond to the Murnaghan equation-of-state fit, and the respective minimum is always marked with a red triangle. In addition, the light blue line indicates the volume of the fully optimized structure. Already the inclusion of zero-point energy (0 K) leads to a volume increase of 0.4%. At 150 and 300 K the thermal volumetric expansions in comparison with the optimized structure amount to 1.3 and 2.5%, respectively. The resulting volume at 300 K agrees within 0.7% with the experimentally measured volume at that temperature. The QHA is sometimes performed by isotropically scaling the cell.[33] Since we perform lattice relaxations using thermal pressures, we get an estimate for the directional thermal expansion. In this example the thermal expansion is quite different in all directions and amounts for the cell lengths a, b, and c to 0.8%, 1.4%, and 0.2%, respectively. In addition, we have also performed the QHA for form A utilizing exclusively PBE+MBD (see Fig. 8(b)). In this case the thermal expansion amounts at 300 K to 3.3%, which overestimates the experimental cell volume by 1.9%. However, when we pair the PBE+MBD phonons with PBE0+MBD total energies, the resulting volume agrees in this case within 0.2% with the experimental volume. PBE+MBD prefers in general larger volumes than PBE+TS which leads here to an overestimation of the cell volume at 300 K. In contrast, the effect of exact exchange significantly reduces the cell volume. Overall, the QHA entirely done at the PBE+TS level provides a good description of the unit cell volume at room temperature.

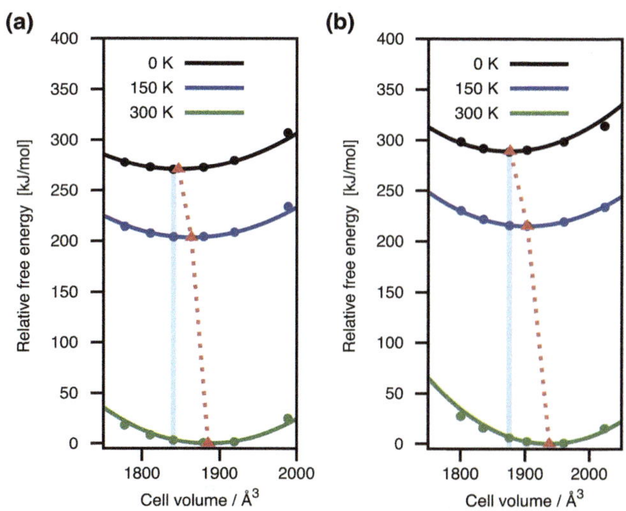

Fig. 8 Illustration of the QHA thermal expansion for form A of system XXIII as described by PBE+TS (a) and PBE+MBD (b). The dots correspond to the harmonic free energies calculated at various volumes and at three different temperatures. The solid curves correspond to the Murnaghan equation-of-state fits, and the respective minimum is always marked with a red triangle. The light blue line indicates the volume of the fully optimized structure.

Next, let us focus on how the QHA affects the relative stabilities. Therefore, the left part of Fig. 9 compares the relative stabilities of the optimized XXIII structures with the thermally expanded ones. The total energies are again calculated with PBE0+MBD and the vibrational free energies with PBE+TS. It can be seen that the QHA only mildly modifies the relative stabilities, with an average change amounting to 0.4 kJ mol^{-1} and a maximum of 1.4 kJ mol^{-1}. At this level we find all the so far experimentally observed structures within an energy window of only 2.6 kJ mol^{-1}, which in general corresponds quite well with the expected energy window of co-existing polymorphs. In comparison, Brandenburg and Grimme[33] also studied the experimentally confirmed polymorphs of system XXIII with the QHA utilizing a minimal basis set Hartree Fock method[63] (HF-3c) as well as density functional tight binding[21] (DFTB) calculations for vibrational free energies, and TPSS-D3 (ref. 64 and 65) for total energies. In their case the spread between the experimental structures is much larger, amounting to 8 kJ mol^{-1}. Experimental evidence[30] suggests that form D should be the most stable one at room temperature. However, in our ranking form E is the most stable one and form D is in fact the second-least stable one. In addition, we find one structure (N70) which is described as more stable than any experimental structure.

Harmonic PBE+MBD vibrational free energies at the optimized structures describe form D as the least stable and form E as the most stable form, with a spread of 3 kJ mol^{-1}. Therefore, we investigated if we could reproduce with PBE+MBD phonons the expected stability ranking, assuming that we have knowledge of the exact unit cell. These calculations were performed using the experimental cell vectors and the results are shown in the right part of Fig. 9. In this case all the experimental structures differ by less than 0.8 kJ mol^{-1}, which means that given our expected accuracy all the structures are in fact degenerate.

All the so far discussed vibrational free energies have been determined within the harmonic approximation. However, in reality these modes are not entirely harmonic. Therefore, we have estimated anharmonic vibrational free energies by replacing the harmonic oscillators with Morse oscillators[51,52] (see Methods). These

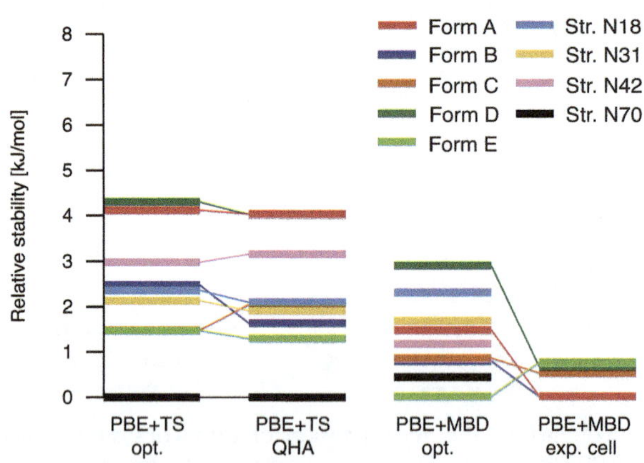

Fig. 9 Relative stabilities for 9 structures of system XXIII for optimized and thermally expanded structures. The relative stability consists of the PBE0+MBD total energy plus the vibrational free energy evaluated with the respective method. The vibrational free energy was calculated in the harmonic approximation at 300 K.

anharmonic free energies were calculated for all optimized and thermally expanded structures with PBE+TS and PBE+MBD, respectively. The resulting stability rankings are shown in Fig. 10.

It can be seen that in the case of optimized structures, form D is always described as the least stable structure and that the spread of the relative energies is larger than for harmonic vibrational free energies. Let us now focus on the PBE+TS QHA results. Here, form D is highly stabilized by the thermal expansion and all the experimental structures are getting closer in terms of energies. Now, all the experimental structures are found within an energy window of only 1.5 kJ mol^{-1}. Form C is described as the most stable experimental structure. However, we underestimate the unit cell volume of form C much more than for form D. Since a further increase in the unit cell volume would lead to higher energies, we can expect further destabilization of form C if we have even more accurate unit cells. Interestingly, form N70 is still described as the most stable structure (see the ESI of ref. 34 for a discussion). Overall, the result in the QHA with PBE+TS anharmonic free energies already agrees well with the experimental expectation in terms of the stability ranking. Given our expected accuracy, all experimentally confirmed polymorphs are essentially degenerate. Finally, let us discuss the anharmonic free energies calculated with the exact cell vectors with PBE+MBD. In this case all the experimental structures except form E are found within an energy window of only about 0.2 kJ mol^{-1}. This picture agrees very well with the available experimental observations. However, form E is described as 2.2 kJ mol^{-1} less stable than the other polymorphs. This large difference occurs probably because the optimized PBE+MBD structure has a quite large RMSD w.r.t. the experimental structure (0.45 Å), most likely due to potential disorder in the experimental form E. We note that this approach for calculating anharmonic free energies does not take into account the interactions between phonon modes. More accurate anharmonic free energies can be obtained by using for example the vibrational self-consistent field approach[66–68] or path-integral molecular dynamics.[69]

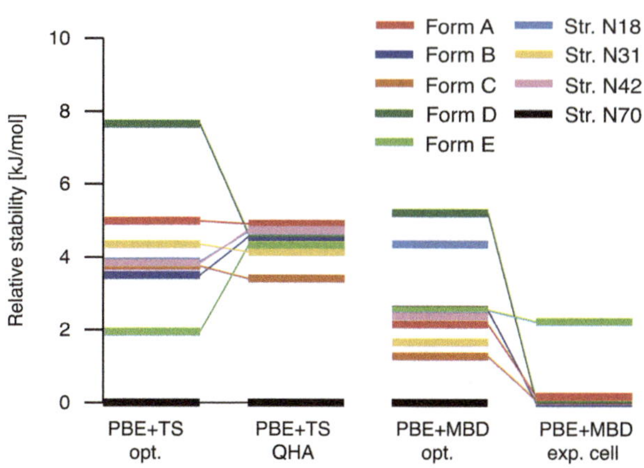

Fig. 10 Relative stabilities for 9 structures of system XXIII for optimized and thermally expanded structures. The relative stability consists of the PBE0+MBD total energy plus the vibrational free energy evaluated with the respective method. The vibrational free energy was calculated using Morse oscillators at 300 K.

8 Conclusion

We have presented an in-depth discussion of how crucial various commonly neglected effects are for the calculation of the relative stabilities of molecular crystal polymorphs using density functional theory. Many-body dispersion effects, exact exchange, and vibrational free energies all contribute significantly to the obtained relative stabilities. Therefore, none of the studied effects can safely be neglected, especially for systems involving flexible molecules. We have shown that our suggested hierarchical CSP approach presented in ref. 34, which yields excellent results for the systems of the latest blind test, is already quite robust. Increasing the basis set for structures and vibrational free energies does not significantly change the resulting stability ranking. In addition, the evaluation of vibrational free energies at the PBE+TS level instead of PBE0+MBD results in an average absolute error of about 2 kJ mol^{-1}. Many-body dispersion interactions affect the vibrational free energies on average by about 1 kJ mol^{-1} but can change relative stabilities in system XXIII by up to 3 kJ mol^{-1}. In the case of aspirin polymorphs, these interactions are crucial to obtaining the correct stability.[56] Given the recent development and efficient implementation of analytical nuclear gradients for MBD,[70,71] geometry optimizations and vibrational free energies could be performed with PBE+MBD instead of PBE+TS in the case of complicated polymorphic energy landscapes. This would increase the computation time for forces on average by about 50%.

Furthermore, we have studied the effect of thermal expansion within the quasi-harmonic approximation. The resulting thermally expanded structures obtained at the PBE+TS level provide accurate unit cell volumes. However, the relative stabilities remain very similar to the ones of the fully optimized structures. In addition, we also estimated anharmonic free energies by using Morse oscillators. These results are for system XXIII in good agreement with the available experimental information on relative stabilities. Including the QHA as well as the anharmonic free energy estimates changes the relative stabilities on average by about 1 kJ mol^{-1}. Therefore, the harmonic approximation on top of fully optimized structures will often provide sufficient results.

Our CSP procedure is applicable to molecular crystals large enough to be of pharmaceutical relevance. However, due to the computational cost it is intended for the final stability ranking of a limited structure set. Therefore, more efficient but still accurate methods are desirable. The computationally most expensive part is the calculation of vibrational free energies. Hence, a more efficient approach for vibrational free energies would significantly increase the applicability of our procedure in terms of the treatable number of structures for large crystals. This could potentially be achieved with a high-level machine learning approach[72,73] based on a database of PBE+TS or PBE+MBD vibrational free energies. In addition, density-functional tight binding (DFTB) with MBD could be utilized for efficient geometry optimizations[55] with final energy rankings evaluated using PBE0+MBD. Furthermore, our PBE0+MBD total energies could potentially also serve as reference or benchmark data for the development of accurate tailor-made force fields or machine learning approaches for lattice energies.

In summary, reliable and practical molecular crystal structure predictions are now possible for pharmaceutically relevant systems.[34] We have shown that the presented CSP procedure is generally robust and the used approximations lead on average to changes in relative stabilities of only 1–2 kJ mol^{-1}.

Conflicts of interest

There are no conflicts to declare.

Acknowledgements

J. H. and A. T. acknowledge support from the Deutsche Forschungsgemeinschaft under the program DFG-SPP 1807 and the European Research Council (ERC-CoG BeStMo). This research was carried out by using computational resources provided by the Fritz Haber Institute of the Max Planck Society (FHI-aims and Mercury) and the HPC facilities of the University of Luxembourg[74] – see http://hpc.uni.lu.

References

1 S. Datta and D. J. W. Grant, *Nat. Rev. Drug Discovery*, 2004, **3**, 42–57.
2 G. R. Desiraju, *J. Am. Chem. Soc.*, 2013, **135**, 9952–9967.
3 V. Coropceanu, J. Cornil, D. A. da Silva Filho, Y. Olivier, R. Silbey and J.-L. Brédas, *Chem. Rev.*, 2007, **107**, 926–952.
4 J. E. Anthony, *Angew. Chem., Int. Ed.*, 2008, **47**, 452–483.
5 A. G. Davies, A. D. Burnett, W. Fan, E. H. Linfield and J. E. Cunningham, *Mater. Today*, 2008, **11**, 18–26.
6 G. Kaupp, *Curr. Opin. Solid State Mater. Sci.*, 2002, **6**, 131–138.
7 S. L. Price, *Chem. Soc. Rev.*, 2014, **43**, 2098–2111.
8 A. J. Cruz-Cabeza, S. M. Reutzel-Edens and J. Bernstein, *Chem. Soc. Rev.*, 2015, **44**, 8619–8635.
9 J. Yang, W. Hu, D. Usvyat, D. Matthews, M. Schutz and G. K.-L. Chan, *Science*, 2014, **345**, 640–643.
10 G. J. O. Beran, *Angew. Chem., Int. Ed.*, 2015, **54**, 396–398.
11 G. J. O. Beran, *Chem. Rev.*, 2016, **116**, 5567–5613.
12 J. Hermann, R. A. DiStasio Jr and A. Tkatchenko, *Chem. Rev.*, 2017, **117**, 4714–4758.
13 S. Grimme, A. Hansen, J. G. Brandenburg and C. Bannwarth, *Chem. Rev.*, 2016, **116**, 5105–5154.
14 C. Adamo and V. Barone, *J. Chem. Phys.*, 1999, **110**, 6158–6170.
15 A. Tkatchenko, R. A. DiStasio Jr, R. Car and M. Scheffler, *Phys. Rev. Lett.*, 2012, **108**, 236402.
16 A. Ambrosetti, A. M. Reilly, R. A. DiStasio Jr and A. Tkatchenko, *J. Chem. Phys.*, 2014, **140**, 18A508.
17 A. M. Reilly and A. Tkatchenko, *J. Chem. Phys.*, 2013, **139**, 024705.
18 A. M. Reilly and A. Tkatchenko, *J. Phys. Chem. Lett.*, 2013, **4**, 1028–1033.
19 A. Otero-de-la Roza and E. R. Johnson, *J. Chem. Phys.*, 2012, **137**, 054103.
20 J. Nyman, O. S. Pundyke and G. M. Day, *Phys. Chem. Chem. Phys.*, 2016, **18**, 15828–15837.
21 J. G. Brandenburg and S. Grimme, *J. Phys. Chem. Lett.*, 2014, **5**, 1785–1789.

22 N. Marom, R. A. DiStasio Jr, V. Atalla, S. Levchenko, A. M. Reilly, J. R. Chelikowsky, L. Leiserowitz and A. Tkatchenko, *Angew. Chem., Int. Ed.*, 2013, **52**, 6629–6632.

23 A. M. Reilly and A. Tkatchenko, *Chem. Sci.*, 2015, **6**, 3289–3301.

24 A. G. Shtukenberg, Q. Zhu, D. J. Carter, L. Vogt, J. Hoja, E. Schneider, H. Song, B. Pokroy, I. Polishchuk, A. Tkatchenko, A. R. Oganov, A. L. Rohl, M. E. Tuckerman and B. Kahr, *Chem. Sci.*, 2017, **8**, 4926–4940.

25 J. P. M. Lommerse, W. D. S. Motherwell, H. L. Ammon, J. D. Dunitz, A. Gavezzotti, D. W. M. Hofmann, F. J. J. Leusen, W. T. M. Mooij, S. L. Price, B. Schweizer, M. U. Schmidt, B. P. van Eijck, P. Verwer and D. E. Williams, *Acta Crystallogr., Sect. B: Struct. Sci.*, 2000, **56**, 697–714.

26 W. D. S. Motherwell, H. L. Ammon, J. D. Dunitz, A. Dzyabchenko, P. Erk, A. Gavezzotti, D. W. M. Hofmann, F. J. J. Leusen, J. P. M. Lommerse, W. T. M. Mooij, S. L. Price, H. Scheraga, B. Schweizer, M. U. Schmidt, B. P. van Eijck, P. Verwer and D. E. Williams, *Acta Crystallogr., Sect. B: Struct. Sci.*, 2002, **58**, 647–661.

27 G. M. Day, W. D. S. Motherwell, H. L. Ammon, S. X. M. Boerrigter, R. G. Della Valle, E. Venuti, A. Dzyabchenko, J. D. Dunitz, B. Schweizer, B. P. van Eijck, P. Erk, J. C. Facelli, V. E. Bazterra, M. B. Ferraro, D. W. M. Hofmann, F. J. J. Leusen, C. Liang, C. C. Pantelides, P. G. Karamertzanis, S. L. Price, T. C. Lewis, H. Nowell, A. Torrisi, H. A. Scheraga, Y. A. Arnautova, M. U. Schmidt and P. Verwer, *Acta Crystallogr., Sect. B: Struct. Sci.*, 2005, **61**, 511–527.

28 G. M. Day, T. G. Cooper, A. J. Cruz-Cabeza, K. E. Hejczyk, H. L. Ammon, S. X. M. Boerrigter, J. S. Tan, R. G. Della Valle, E. Venuti, J. Jose, S. R. Gadre, G. R. Desiraju, T. S. Thakur, B. P. van Eijck, J. C. Facelli, V. E. Bazterra, M. B. Ferraro, D. W. M. Hofmann, M. A. Neumann, F. J. J. Leusen, J. Kendrick, S. L. Price, A. J. Misquitta, P. G. Karamertzanis, G. W. A. Welch, H. A. Scheraga, Y. A. Arnautova, M. U. Schmidt, J. van de Streek, A. K. Wolf and B. Schweizer, *Acta Crystallogr., Sect. B: Struct. Sci.*, 2009, **65**, 107–125.

29 D. A. Bardwell, C. S. Adjiman, Y. A. Arnautova, E. Bartashevich, S. X. M. Boerrigter, D. E. Braun, A. J. Cruz-Cabeza, G. M. Day, R. G. Della Valle, G. R. Desiraju, B. P. van Eijck, J. C. Facelli, M. B. Ferraro, D. Grillo, M. Habgood, D. W. M. Hofmann, F. Hofmann, K. V. J. Jose, P. G. Karamertzanis, A. V. Kazantsev, J. Kendrick, L. N. Kuleshova, F. J. J. Leusen, A. V. Maleev, A. J. Misquitta, S. Mohamed, R. J. Needs, M. a. Neumann, D. Nikylov, A. M. Orendt, R. Pal, C. C. Pantelides, C. J. Pickard, L. S. Price, S. L. Price, H. A. Scheraga, J. van de Streek, T. S. Thakur, S. Tiwari, E. Venuti and I. K. Zhitkov, *Acta Crystallogr., Sect. B: Struct. Sci.*, 2011, **67**, 535–551.

30 A. M. Reilly, R. I. Cooper, C. S. Adjiman, S. Bhattacharya, A. D. Boese, J. G. Brandenburg, P. J. Bygrave, R. Bylsma, J. E. Campbell, R. Car, D. H. Case, R. Chadha, J. C. Cole, K. Cosburn, H. M. Cuppen, F. Curtis, G. M. Day, R. A. DiStasio Jr, A. Dzyabchenko, B. P. van Eijck, D. M. Elking, J. A. van den Ende, J. C. Facelli, M. B. Ferraro, L. Fusti-Molnar, C.-A. Gatsiou, T. S. Gee, R. de Gelder, L. M. Ghiringhelli, H. Goto, S. Grimme, R. Guo, D. W. M. Hofmann, J. Hoja, R. K. Hylton, L. Iuzzolino, W. Jankiewicz, D. T. de Jong, J. Kendrick, N. J. J. de Klerk, H.-Y. Ko, L. N. Kuleshova, X. Li, S. Lohani, F. J. J. Leusen, A. M. Lund, J. Lv, Y. Ma,

N. Marom, A. E. Masunov, P. McCabe, D. P. McMahon, H. Meekes, M. P. Metz, A. J. Misquitta, S. Mohamed, B. Monserrat, R. J. Needs, M. A. Neumann, J. Nyman, S. Obata, H. Oberhofer, A. R. Oganov, A. M. Orendt, G. I. Pagola, C. C. Pantelides, C. J. Pickard, R. Podeszwa, L. S. Price, S. L. Price, A. Pulido, M. G. Read, K. Reuter, E. Schneider, C. Schober, G. P. Shields, P. Singh, I. J. Sugden, K. Szalewicz, C. R. Taylor, A. Tkatchenko, M. E. Tuckerman, F. Vacarro, M. Vasileiadis, A. Vazquez-Mayagoitia, L. Vogt, Y. Wang, R. E. Watson, G. A. de Wijs, J. Yang, Q. Zhu and C. R. Groom, *Acta Crystallogr., Sect. B: Struct. Sci.*, 2016, **72**, 439–459.

31 M. A. Neumann, F. J. J. Leusen and J. Kendrick, *Angew. Chem., Int. Ed.*, 2008, **47**, 2427–2430.

32 S. R. Whittleton, A. Otero-De-La-Roza and E. R. Johnson, *J. Chem. Theory Comput.*, 2017, **13**, 441–450.

33 J. G. Brandenburg and S. Grimme, *Acta Crystallogr., Sect. B: Struct. Sci., Cryst. Eng. Mater.*, 2016, **72**, 502–513.

34 J. Hoja, H.-Y. Ko, M. A. Neumann, R. Car, R. A. DiStasio Jr and A. Tkatchenko, 2018, arXiv:1803.07503.

35 M. A. Neumann and M.-A. Perrin, *J. Phys. Chem. B*, 2005, **109**, 15531–15541.

36 M. A. Neumann, *J. Phys. Chem. B*, 2008, **112**, 9810–9829.

37 L. J. Simons, B. W. Caprathe, M. Callahan, J. M. Graham, T. Kimura, Y. Lai, H. LeVine, W. Lipinski, A. T. Sakkab, Y. Tasaki, L. C. Walker, T. Yasunaga, Y. Ye, N. Zhuang and C. E. Augelli-Szafran, *Bioorg. Med. Chem. Lett.*, 2009, **19**, 654–657.

38 V. Blum, R. Gehrke, F. Hanke, P. Havu, V. Havu, X. Ren, K. Reuter and M. Scheffler, *Comput. Phys. Commun.*, 2009, **180**, 2175–2196.

39 A. Marek, V. Blum, R. Johanni, V. Havu, B. Lang, T. Auckenthaler, A. Heinecke, H.-J. Bungartz and H. Lederer, *J. Phys.: Condens. Matter*, 2014, **26**, 213201.

40 T. Auckenthaler, V. Blum, H.-J. Bungartz, T. Huckle, R. Johanni, L. Krämer, B. Lang, H. Lederer and P. Willems, *Parallel Comput.*, 2011, **37**, 783–794.

41 V. Havu, V. Blum, P. Havu and M. Scheffler, *J. Comput. Phys.*, 2009, **228**, 8367–8379.

42 X. Ren, P. Rinke, V. Blum, J. Wieferink, A. Tkatchenko, A. Sanfilippo, K. Reuter and M. Scheffler, *New J. Phys.*, 2012, **14**, 053020.

43 A. C. Ihrig, J. Wieferink, I. Y. Zhang, M. Ropo, X. Ren, P. Rinke, M. Scheffler and V. Blum, *New J. Phys.*, 2015, **17**, 093020.

44 F. Knuth, C. Carbogno, V. Atalla, V. Blum and M. Scheffler, *Comput. Phys. Commun.*, 2015, **190**, 33–50.

45 J. P. Perdew, K. Burke and M. Ernzerhof, *Phys. Rev. Lett.*, 1996, **77**, 3865–3868.

46 A. Tkatchenko and M. Scheffler, *Phys. Rev. Lett.*, 2009, **102**, 073005.

47 A. L. Spek, *Acta Crystallogr., Sect. D: Biol. Crystallogr.*, 2009, **65**, 148–155.

48 C. F. Macrae, I. J. Bruno, J. A. Chisholm, P. R. Edgington, P. McCabe, E. Pidcock, L. Rodriguez-Monge, R. Taylor, J. van de Streek and P. A. Wood, *J. Appl. Crystallogr.*, 2008, **41**, 466–470.

49 A. Togo and I. Tanaka, *Scr. Mater.*, 2015, **108**, 1–5.

50 F. D. Murnaghan, *Proc. Natl. Acad. Sci. U. S. A.*, 1944, **30**, 244–247.

51 P. M. Morse, *Phys. Rev.*, 1929, **34**, 57–64.

52 J. P. Dahl and M. Springborg, *J. Chem. Phys.*, 1988, **88**, 4535–4547.

53 A. Tkatchenko, A. Ambrosetti and R. A. DiStasio Jr, *J. Chem. Phys.*, 2013, **138**, 074106.

54 F. Curtis, X. Wang and N. Marom, *Acta Crystallogr., Sect. B: Struct. Sci., Cryst. Eng. Mater.*, 2016, **72**, 562–570.
55 M. Mortazavi, J. G. Brandenburg, R. J. Maurer and A. Tkatchenko, *J. Phys. Chem. Lett.*, 2018, **9**, 399–405.
56 A. M. Reilly and A. Tkatchenko, *Phys. Rev. Lett.*, 2014, **113**, 055701.
57 R. E. Allen and F. W. De Wette, *Phys. Rev.*, 1969, **179**, 873–886.
58 J. Hoja, A. M. Reilly and A. Tkatchenko, *Wiley Interdiscip. Rev.: Comput. Mol. Sci.*, 2017, **7**, e1294.
59 A. Erba, J. Maul and B. Civalleri, *Chem. Commun.*, 2016, **52**, 1820–1823.
60 Y. N. Heit, K. D. Nanda and G. J. O. Beran, *Chem. Sci.*, 2016, **7**, 246–255.
61 J. Nyman and G. M. Day, *Phys. Chem. Chem. Phys.*, 2016, **18**, 31132–31143.
62 G. J. O. Beran, J. D. Hartman and Y. N. Heit, *Acc. Chem. Res.*, 2016, **49**, 2501–2508.
63 R. Sure and S. Grimme, *J. Comput. Chem.*, 2013, **34**, 1672–1685.
64 J. Tao, J. P. Perdew, V. N. Staroverov and G. E. Scuseria, *Phys. Rev. Lett.*, 2003, **91**, 146401.
65 S. Grimme, J. Antony, S. Ehrlich and H. Krieg, *J. Chem. Phys.*, 2010, **132**, 154104.
66 B. Monserrat, N. D. Drummond and R. J. Needs, *Phys. Rev. B: Condens. Matter Mater. Phys.*, 2013, **87**, 144302.
67 N. D. Drummond, B. Monserrat, J. H. Lloyd-Williams, P. L. Ríos, C. J. Pickard and R. J. Needs, *Nat. Commun.*, 2015, **6**, 7794.
68 E. A. Engel, B. Monserrat and R. J. Needs, *Phys. Rev. X*, 2015, **5**, 021033.
69 M. Rossi, P. Gasparotto and M. Ceriotti, *Phys. Rev. Lett.*, 2016, **117**, 115702.
70 M. A. Blood-Forsythe, T. Markovich, R. A. DiStasio Jr, R. Car and A. Aspuru-Guzik, *Chem. Sci.*, 2016, **7**, 1712–1728.
71 T. Bučko, S. Lebègue, T. Gould and J. G. Ángyán, *J. Phys.: Condens. Matter*, 2016, **28**, 45201.
72 S. Chmiela, A. Tkatchenko, H. E. Sauceda, I. Poltavsky, K. T. Schütt and K.-R. Müller, *Sci. Adv.*, 2017, **3**, e1603015.
73 K. T. Schütt, F. Arbabzadah, S. Chmiela, K. R. Müller and A. Tkatchenko, *Nat. Commun.*, 2017, **8**, 13890.
74 S. Varrette, P. Bouvry, H. Cartiaux and F. Georgatos, *Proc. of the 2014 Intl. Conf. on High Performance Computing & Simulation (HPCS 2014)*, Bologna, Italy, 2014, pp. 959–967.

Faraday Discussions

PAPER

Crystal structure prediction of flexible pharmaceutical-like molecules: density functional tight-binding as an intermediate optimisation method and for free energy estimation†

Luca Iuzzolino,[a] Patrick McCabe,[b] Sarah L. Price[a] and Jan Gerit Brandenburg‡*[a]

Received 1st February 2018, Accepted 27th February 2018
DOI: 10.1039/c8fd00010g

Successful methodologies for theoretical crystal structure prediction (CSP) on flexible pharmaceutical-like organic molecules explore the lattice energy surface to find a set of plausible crystal structures. The initial search stages of CSP studies use relatively simple lattice energy approximations as hundreds of thousands of minima have to be considered. These generated crystal structures often have poor molecular geometries, as well as inaccurate lattice energy rankings, and performing reasonably accurate but computationally affordable optimisations of the crystal structures generated in a search would be highly desirable. Here, we seek to explore whether semi-empirical quantum-mechanical methods can perform this task. We employed the dispersion-corrected tight-binding Hamiltonian (DFTB3-D3) to relax all the inter- and intra-molecular degrees of freedom of several thousands of generated crystal structures of five pharmaceutical-like molecules, saving a large amount of computational effort compared to earlier studies. The computational cost scales better with molecular size and flexibility than other CSP methods, suggesting that it could be extended to even larger and more flexible molecules. On average, this optimisation improved the average reproduction of the eight experimental crystal structures ($RMSD_{15}$) and experimental conformers ($RMSD_1$) by 4% and 23%, respectively. The intermolecular interactions were then further optimised using distributed multipoles, derived from the molecular wave-functions, to accurately describe the electrostatic components of the intermolecular energies. In all cases, the experimental crystal structures are close to the top of the

[a]Department of Chemistry, University College London, 20 Gordon Street, London WC1H 0AJ, UK. E-mail: gerit.brandenburg@chemie.uni-goettingen.de
[b]The Cambridge Crystallographic Data Centre, 12 Union Road, Cambridge CB2 1EZ, UK

† Electronic supplementary information (ESI) available: Full details of results for each molecule. See DOI: 10.1039/c8fd00010g

‡ Current address: Institute for Physical Chemistry, University of Goettingen, Tammannstr. 6, 37077 Goettingen, Germany.

lattice energy ranking. Phonon calculations on some of the lowest energy structures were also performed with DFTB3-D3 methods to calculate the vibrational component of the Helmholtz free energy, providing further insights into the solid-state behaviour of the target molecules. We conclude that DFTB3-D3 is a cost-effective method for optimising flexible molecules, bridging the gap between the approximate methods used in CSP searches for generating crystal structures and more accurate methods required in the final energy ranking.

Introduction

In computational materials science, approaches that are based on the fundamental laws of quantum mechanics (QM) are now integral to almost any materials design initiative in academia or industry.[1–3] The prevalence of polymorphism in organic molecules[4] and its importance in determining the physical properties of organic solids[1,5–8] has led to an interest in using simulation methods to predict the range of possible crystal structures and their properties, possibly to design molecular materials with desired characteristics, like high porosity.[9] CSP studies aim to predict all the possible putative polymorphs (PPMs) of a molecule, starting from the chemical diagram only.[10] This is particularly important for pharmaceutical development when performed as a complement to solid form screening, which attempts to establish the range of solid forms that could either be developed into the pharmaceutical product or must not appear during manufacturing and storage.[8,11] Several successful CSP studies on large but relatively rigid systems, such as organic porous cages,[12,13] have been reported. For inorganic solids, *ab initio* random structure searches have been used successfully.[14–16] Pharmaceuticals are more challenging since they tend to have a considerable conformational flexibility, being designed for their biological activity, and so most molecules in drug development can crystallise in a variety of conformations.[8] For pharmaceuticals with few intramolecular degrees of freedom, several successful CSP studies have been conducted.[17–24] However, the CSP search space grows exponentially with conformational flexibility and hence more efficient methods are needed for application to modern pharmaceuticals. The challenge of dealing with molecular flexibility is compounded by small changes in some bond or torsion angles having a significantly greater effect on the overall molecule shape in large pharmaceuticals than in smaller molecules.[25] Hence, the number of conformational degrees of freedom that may change from isolated molecule (*ab initio*) values because of the crystal packing forces can be larger than expected[26] from combining the flexibility of fragment model molecules. Whilst optimising all atomic positions simultaneously with the crystal lattice positions avoids the problem of selecting the crystal packing-dependent conformational degrees of freedom, the need to balance the inter- and intramolecular forces sufficiently accurately makes this exceptionally demanding. This has been demonstrated by the challenges posed by the largest molecules in the recent Blind Tests of CSP.[1,27,28] These molecules, though large enough to generate interest in CSP methods from industry, are still small and of limited flexibility compared with pharmaceuticals in development. Hence, if CSP is to fulfil its promise for pharmaceutical development, innovations that allow CSP studies on larger molecules within a restricted time frame[11] are needed.

All CSP methods begin with a search step in which the lattice energy (E_{latt}) surface is explored to locate all the possible local minima, each corresponding to a separate possible crystal structure.[1,29] Since the surface that must be explored is extremely wide, especially for flexible molecules, a large number of candidate crystal structures are produced, at least of the order of hundreds of thousands.[22,29] Hence, cheaper but less accurate energy models must be utilised in the search. They can take the form of tailor-made force fields,[30] exp-6 atomistic intermolecular potentials combined with quantum-mechanical calculations of the conformational energy,[31,32] or transferrable empirical force fields.[33]

The lattice energy evaluations used in the crystal structure generation ("search") stage are rarely accurate enough to give a reliable energy ranking, since the lattice energy differences between observed polymorphs are usually less than 5 kJ mol^{-1}. Hence, thousands of generated crystal structures must be optimised and re-ranked to obtain a meaningful list of PPMs. Since this final step requires the use of accurate and expensive models, such as periodic dispersion-corrected density functional theory (DFT-D)[34–40] on all crystal structures (the Ψ_{crys} method), or the calculation of a high-quality wave-function for each molecular conformer (the Ψ_{mol} method),[8,41,42] the computational cost can easily become unfeasible. A possible solution is the use of an intermediate optimisation method, which can bridge the gap between the simple models used to generate the crystal structures and the more accurate ones used to generate an accurate final energy ranking. For flexible molecules, several intermediate methods are used in different CSP workflows, such as optimising some torsion angles with a transferable force-field to improve the molecular conformation geometries,[43,44] single-point energy calculations with a more accurate description of the intra- and intermolecular interactions[18,20] or partial lattice energy optimisations.[1,22]

Recently, semi-empirical QM methods have been applied to large chemical and biochemical systems;[45,46] these methods do not have the transferability problems of standard force fields but are significantly less demanding than other QM methods.[45,47,48] In this work, the interest is focused on dispersion-corrected density functional tight binding methods (DFTB3-D3), which are up to three orders of magnitude cheaper than DFT-D, and have been shown to provide reasonably accurate lattice energies for a benchmark of small molecular crystals.[49] We present the first use of semi-empirical QM methods within a full CSP workflow. We aim to verify whether DFTB3-D3 can be used as an intermediate re-ranking step in CSP. There are at least four ways in which it could provide useful insights:

(1) Producing more accurate molecular and crystalline geometries and packings than the original crystal structures generated in the search, which could be used as starting points for better lattice energy evaluations.

(2) Obtaining a more accurate energy ranking than the one obtained after the crystal structure search, in order to limit the number of structures that need to be taken to the final optimisation stage.

(3) Producing fewer minima that are artefacts of the constraints imposed by the space group and conformational variables, by optimising all crystal structures with $P1$ symmetry.

(4) Computing the phonon modes of plausible crystal structures to estimate thermodynamic free energies in the harmonic approximation.

We test the suitability of DFTB3-D3 for these purposes on structures generated in a set of searches[25] performed with the Ψ_{mol} method. E_{latt}, the energy needed to completely separate all the molecules in the static crystal structure to isolated molecules in their most stable conformation,[10] can be described as:

$$E_{latt} = U_{inter} + \Delta E_{intra}$$

ΔE_{intra} is the energy penalty for modifying the conformation of the molecule from its isolated gas-phase minimum, calculated with a good quality molecular electronic structure method, and U_{inter} is the intermolecular energy, which sums all the interactions between the molecules in the crystal.[10] The search method variant used in this work treated the most flexible torsion angles as search variables, and the other conformational degrees of freedom correspond to the isolated molecule *ab initio* values, an approach that is comparable to using multiple rigid conformations.[43] Five flexible molecules for which full CSP studies had already been performed (Fig. 1) are used to test the application of DFTB3-D3 methods: molecule XX was the first "pharmaceutical-like" molecule used in a Blind Test,[50] XXIII and XXVI are from the sixth blind test,[1] GSK269984B was used in an early pharmaceutical study,[23] while the CSP study of mebendazole was performed for an on-going collaboration.[51] These CSP methods were successful in predicting all the observed polymorphs with one molecule in the asymmetric unit cell (*i.e.* $Z' = 1$), and so provide a stringent test for the suitability of an intermediate DFTB3-D3 step to extend CSP methods to even larger, more flexible molecules.

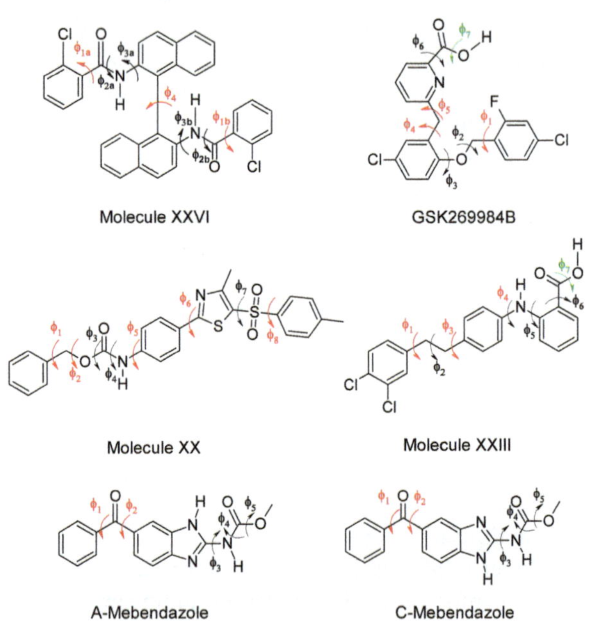

Fig. 1 Chemical diagrams of the six molecules used in this work. The torsion angles in red were treated as explicitly flexible in the crystal structure searches, those in black were constrained to a set of values defined by the CSD Conformer Generator, and those in green were constrained to a set of values after an *ab initio* gas-phase scan.

Methods

Crystal structure generation

The method for generating the crystal structures used statistical information retrieved from the Cambridge Structural Database (CSD) to guide the determination and the analysis of the conformational search space.[25] For each molecule in Fig. 1, the CSD distributions of each rotatable bond were retrieved from the CSD knowledge base conformational libraries;[52] the effect of each of these rotamers on the overall shape of each molecule was subsequently determined *via* ultra-fast shape recognition.[53] This information was used to divide the torsion angles into two groups: those that were treated as explicitly flexible in the searches, and those that were constrained to a set of values, determined by the CSD Conformer Generator.[54] Molecular electronic structure calculations at the PBE0 6-31G(d,p) level of theory using Gaussian09 [55] were used to limit the set of potential conformations to be considered and to calculate grids of ΔE_{intra} values for the explicitly flexible torsion angles. These calculations were also used to determine the fixed atomic charges used by CrystalPredictor 1.8 [32] to estimate U_{inter} in combination with the empirically fitted FIT repulsion–dispersion intermolecular potential.[56] The number of crystal structures generated varied from ~350 000 for the C tautomer of mebendazole to ~2 200 000 for molecule XXVI. The outputs from this search method and from the original CSP studies are compared in a previous publication.[25] Thus the molecular geometries prior to refinement were mostly determined by PBE0 6-31G(d,p) isolated molecule optimisations, with the torsion angles shown in red in Fig. 1 at values specifically produced in the search.

Density functional tight-binding method

As an established semi-empirical method, we used a third order density functional tight-binding Hamiltonian with self-consistent charge redistribution as implemented in dftb+.[57–59] We used the 3OB Slater–Koster files,[60,61] and an additional damping of all hydrogen-containing pair-potentials, and we corrected for missing London dispersion interactions with the atom-pairwise D3 scheme in the rational damping variant.[62,63] The Brillouin zone was then sampled with a Γ-centred mesh with a density of 0.025 Å^{-1}. The original DFTB3-D3 parametrisation focused on energetic properties only.[49] However, it has been shown that the geometries of large molecular complexes and molecular crystals are not reproduced sufficiently accurately.[64,65] A re-parametrisation of the DFTB3-MBD model has been shown to improve the geometrical properties substantially,[66] so we followed a similar strategy and optimised the D3 damping parameter to minimise the errors in centre-of-mass distances on a set of small molecular dimers (S66).[67,68] This results in a DFTB3-D3 parameter set ($S_6 = 1.0$, $S_8 = 0.0$, $a_1 = 0.841$, $a_2 = 3.834$ Bohr) with 8 kJ mol^{-1} mean absolute error on the S66 binding energies and 1.3% error in the centre-of-mass distances, compared to 3.8 kJ mol^{-1} and 3.2%, respectively with the original parametrisation. This re-parametrisation also improves the mass density errors of a set of small molecular crystals (X23)[69,70] from 11.6% to 4.2%. Additional atom pair-wise corrections to DFTB3-D3, such as geometrical hydrogen-bonding corrections[71,72] and halogen-bonding corrections,[73] were not utilised.

Structure relaxations have been conducted with a quasi-Newton optimisation in internal redundant coordinates as implemented in the CRYSTAL14 programme.[74] As DFTB3-D3 is used as an intermediate optimisation method, it is important to test the relaxation thresholds. This was tested on the POLY59 benchmark set[75] and an optimum trade-off between the accuracy in relative energy ranking and computational cost was obtained with thresholds of 0.12 a.u. and 0.003 a.u. for the root-mean-square (RMS) displacement and RMS gradient.

Phonon calculations were used as a final step to estimate the free energies for the experimental crystal structures and a few diverse low energy structures (given in ESI Table 10†). Phonon calculations using symmetric finite displacements required geometry optimisations with one order of magnitude tighter thresholds, corresponding to the default settings in CRYSTAL14. The Brillouin zone sampling for the phonon calculations was initially formed by the common practice of constructing supercells with minimum cell lengths of 10 Å.[70] As we are interested in relative polymorph energies, we additionally ensured that a similar number of atoms are covered in each supercell to enhance possible error compensations. The Helmholtz free energy, A, can be defined as:

$$A = E_{\text{latt}} + F_{\text{vib}}$$

where F_{vib} is the vibrational contribution to the free energy, which includes the thermal contribution to the free energy and the zero point energies calculated by summation.[41,65]

DFTB3-D3 intermediate optimisation of all the CrystalPredictor generated structures

All the unique generated crystal structures within 40 kJ mol^{-1} of the global minimum in CrystalPredictor energy were optimised with DFTB3-D3. In order to perform the DFTB3-D3 optimisations, the symmetry of each crystal structure had to be reduced to the $P1$ space group. 9215 crystal structures were optimised for molecule XXVI, 16 744 for GSK269984B, 28 249 for molecule XXIII, 26 650 for molecule XX, 4165 for the A-tautomer of mebendazole and 4284 for its C-tautomer. The symmetries of all the crystal structures that were successfully optimised were reintroduced with Platon,[76] and then they were clustered to remove duplicates using the Crystal Packing Similarity tool within the CSD Python API;[77] details on the clustering procedure can be found in ESI Section 1.1.†

The energy ranking calculated *via* DFTB3-D3 proved to be very different, as shown in Fig. 2 for XXVI, with the calculated E_{latt} values having a larger energy spread than after the search stage. However, the poor energies of the structures corresponding to the experimental forms showed that, contrary to the original hope, DFTB3-D3 had not produced a more accurate energy ranking. Hence, all the unique optimised crystal structures within a large window of 50 kJ mol^{-1} of the global minimum in DFTB3-D3 energy were taken forward. Clustering to remove duplicates and applying the energy cut-off each reduced the number of structures by a molecule-dependent proportion, which is broken down in ESI Table 1.† This resulted in 3346 crystal structures for molecule XXVI, 5328 for GSK269984B, 13 490 for molecule XXIII, 19 146 for molecule XX, 3078 for the A tautomer of mebendazole and 3352 for its C tautomer, requiring optimisation with a more accurate method.

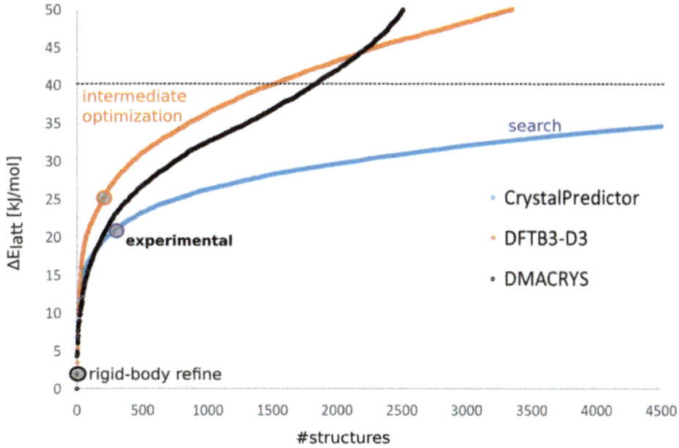

Fig. 2 The energy distributions of the computer-generated crystal structures of molecule XXVI showing how the ~9000 structures below 40 kJ mol^{-1} generated by the CrystalPredictor search (blue) are optimised by DFTB3-D3 (orange) to provide only ~3000 structures below 50 kJ mol^{-1}, and their corresponding energies calculated by DMACRYS using the $\Psi_{\text{mol}}^{\text{PBE0+FIT}}$ model (black). The coloured circles indicate the energies of the experimental structures relative to the global minimum in E_{latt} at that stage. Similar plots for the other molecules are shown in ESI Fig. 1–5.†

Re-ranking using improved molecular wave-functions

The lattice energies with DFTB3-D3 were clearly inadequate, hence a further lattice energy evaluation with a more accurate description of the wave-function was required. We assumed that the DFTB3-D3 optimisation had improved the molecular geometries, and this was exploited by reverting to the Ψ_{mol} approach used to generate the crystal structures, but using distributed multipoles instead of atomic charges, and a more accurate evaluation of ΔE_{intra}. For all the DFTB3-D3 optimised crystal structures, the molecular wave-function of each distinct molecular conformation was calculated at the PBE0 6-31G(d,p) level of theory using Gaussian09, and distributed multipoles up to hexadecapoles were derived from the charge density using GDMA 2.2.[78] Finally the intermolecular interactions were optimised with DMACRYS.[79] U_{inter} was modelled using an electrostatic component calculated from distributed multipoles combined with a repulsion–dispersion component calculated with the empirically-fitted exp-6 FIT potential. The crystal structures optimised with this specific method, denoted $\Psi_{\text{mol}}^{\text{PBE0+FIT}}$, were ranked in terms of E_{latt}, with ΔE_{intra} estimated as the difference between the energy of each crystalline conformer and the PBE0 6-31G(d,p) energy of the DFTB3-D3 optimised gas-phase minimum calculated with Gaussian09. This was used to avoid the absolute values of E_{latt} being affected by a ΔE_{intra} component calculated relative to a value optimised with a different wave-function. Although the energies of the DFTB3-D3 optimised gas-phase minima are 12–25 kJ mol^{-1} higher than their PBE0 6-31G(d,p) optimised counterparts, this systematic error does not change the energy differences between polymorphs, which are the main assessment criteria of CSP studies. For each molecule, all the crystal structures within 50 kJ mol^{-1} were clustered to remove duplicates.

Rigid-molecule harmonic phonon calculations were carried out on the same structures as for the DFTB3-D3 phonons with DMACRYS using supercells generated following the methodology of Nyman and Day.[80] This computationally cheap approach effectively assumes that the molecular modes do not affect ΔF_{vib}, in contrast to the DFTB3-D3 calculations that allow the molecular and lattice modes to couple. The rigid-molecule $\Psi_{mol}^{PBE0+FIT}$ model assumes that the molecular modes make an identical contribution to F_{vib} for all structures. This is a necessary assumption since the pure molecular modes could only be calculated at a local isolated-molecule minimum, and many conformations are far from such minima (although they would not necessarily be defined as conformational polymorphs).[81]

Assessment of the results

Crystal structures were compared with the experimental crystal structures and sets of the most significant computer-generated PPMs from the previous CSP studies identified in a previous work[25] using the Crystal Packing Similarity tool in the CSD Python API, trying to match 15-molecule clusters with 20% distance and 20° angle tolerances. In a few cases, the tolerances had to be slightly increased to find matches with some of the computer-generated structures.

The generated crystal structures that upon optimisation ended up matching the experimental crystal structures and the significant computer-generated PPMs[25] were optimised with CrystalOptimizer[82] at the PBE0 6-31G(d,p) level of theory, as a function of a set of torsion and bond angles (shown in ESI Fig. 6†) that were selected according to rules defined by Nyman and Day[26] based on chemical intuition derived from small model molecules. These calculations were used as a stringent test to verify how the geometries and energies obtained using DFTB3-D3 and $\Psi_{mol}^{PBE0+FIT}$ compare with those obtained through optimising the most flexible conformational degrees of freedom using a more accurate and computationally expensive Ψ_{mol} model.

Results

Intra- and intermolecular geometries

The quality of reproduction of each experimental $Z' = 1$ crystal structure and the molecular conformation within it, before and after optimisations with DTFB3-D3 and $\Psi_{mol}^{PBE0+FIT}$ is shown in Table 1, and contrasted with the reproductions achieved with the more computationally demanding CrystalOptimizer.

DFTB3-D3 optimisations are effective at improving all the crystalline molecular conformations from the search generated structure, with a 23% improvement in the average $RMSD_1$. The improvement in the overall crystal packing is more variable, with some becoming slightly worse and some slightly better with a mere 4% improvement in the average $RMSD_{15}$. Keeping the molecule rigid at the DFTB3-D3 geometry but using the distributed multipoles and intramolecular energy from a superior electronic structure method improves the intermolecular packing, leading to an overall improvement of the average $RMSD_{15}$ of 22% after the $\Psi_{mol}^{PBE0+FIT}$ optimisations. Both the molecular and crystalline geometries are of a slightly worse quality than those obtained after the CrystalOptimizer optimisations, which improve the search structure by an average $RMSD_1$ of 31% and an average $RMSD_{15}$ of 38%. However, DFTB3-D3 refinements could be carried out at a much lower computational cost for these types of molecules, and yet successfully reproduced the structures.

Table 1 Accuracy of the reproduction of the experimental crystal structures ($RMSD_{15}$) and experimental conformers ($RMSD_1$). See ESI Table 2 for a visual comparison. Note that the molecular conformers were treated as rigid in the final DMACRYS optimisation, and so the $RMSD_1$ was not affected

Experimental crystal structure (CSD refcode)	Reproduction of the experimental conformers ($RMSD_1$)			Reproduction of the experimental crystal lattices ($RMSD_{15}$)			
	Search/Å	DFTB3-D3/Å	CrystalOptimizer/Å	Search/Å	DFTB3-D3/Å	$\Psi_{mol}^{PBE0+FIT}$/Å	CrystalOptimizer/Å
XXVI (XAFQIH)	0.2	0.133	0.123	0.533	0.425	0.376	0.294
GSK269984B (BIFHOP)	0.093	0.071	0.094	0.219	0.329	0.212	0.21
XXIII a (XAFPAY)	0.236	0.216	0.212	0.689	0.68	0.663	0.443
XXIII b (XAFPAY01)	0.278	0.18	0.159	0.626	0.528	0.415	0.373
XXIII d (XAFPAY03)	0.31	0.278	0.237	0.511	0.544	0.583	0.53
XX (OBEQIX)	0.227	0.137	0.101	0.455	0.469	0.39	0.218
Mebendazole A (TUXPEJ)	0.201	0.176	0.143	0.465	0.408	0.402	0.321
Mebendazole C (YULGIW)	0.096	0.067	0.066	0.825	0.763	0.313	0.28
Average	**0.205**	**0.157**	**0.142**	**0.540**	**0.518**	**0.419**	**0.334**
Standard deviation	**0.073**	**0.067**	**0.055**	**0.169**	**0.135**	**0.134**	**0.103**

Lattice energy rankings

The energy rankings of the crystal structures matching the experimental ones and their stabilities relative to the most stable one at that stage are shown in Table 2, and compared to those obtained in the original studies.

DFTB3-D3 energies are inadequate for CSP and in most cases the experimental crystal structures are ranked worse than after the search. On the other hand, the final optimisation of just U_{inter} (*i.e.* keeping the DFTB3-D3-optimised molecular geometries rigid) with the $\Psi_{\text{mol}}^{\text{PBE0+FIT}}$ model leads to a large improvement in the relative energies, with the crystal structures matching their experimental counterparts having rankings similar to those in the original CSP studies, which had been performed at a much larger computational cost. In three cases (molecule XX, GSK269984B, and mebendazole form A) the experimental crystal structure is still the global minimum. Experimental form B of molecule XXIII, form C of mebendazole, and XXVI are all within a few kJ mol^{-1} of the global minimum, while forms A and D of molecule XXIII are higher in energy, although still within a sensible energy window considering the issues with predicting the stability of the polymorphs of XXIII with any method as shown in Fig. 5.[1]

A question that arises is whether the energy landscapes, such as the two examples in Fig. 3, differ significantly in the range of putative polymorphs produced. Any comparison with the original studies is not straightforward because of the developments in the algorithms, variations in settings and the potential energy surfaces being used. A selection of the lowest energy PPMs and a few conformationally diverse structures from the original CSP studies had been selected for testing the workflow for generating the crystal structures[25] and we find that the vast majority of these structures remain (see ESI Tables 3–7†) on the crystal energy landscapes (Fig. 3 and ESI Fig. 7–9†). Out of the 180 targeted PPMs for the five molecules, 21 were not found in the set of $\Psi_{\text{mol}}^{\text{PBE0+FIT}}$ optimised crystal structures. However, in eight of those cases the PPMs were already missing after the search, and for a further nine the recent search structure was a very poor match.[25] Only four PPMs that had been considered as 'certainly found' in the search[25] were lost in this study, and either these were towards the high energy end of the sample, or similar types of packing are present in the crystal energy landscape. Some PPMs were found, but only above 20 kJ mol^{-1} from the global minimum. This can reflect the huge sensitivity of E_{latt} to quite subtle changes in conformation, particularly when hydrogen bonds can be inter- or intramolecular as in GSK269984B, and a crystal structure containing a molecular conformation optimised at the DFTB3-D3 level may converge to a less favourable local minimum when the intermolecular interactions are optimised with the $\Psi_{\text{mol}}^{\text{PBE0+FIT}}$ method. The huge dependence of E_{latt} on subtle changes is further highlighted by the differences between the energies obtained in previous CSP studies and those obtained *via* re-optimisation of the key crystal structures with CrystalOptimizer, as shown in ESI Tables 3–7.† These optimisations also seem to lead to different geometries compared to the $\Psi_{\text{mol}}^{\text{PBE0+FIT}}$ optimised crystal structures, highlighting how different routes can lead to different outcomes.

Table 2 Energy rankings and energy differences with respect to the global minimum at each stage of the workflow, which can be compared with their rankings in the previous CSP studies (in italics)

Experimental crystal structure	Ranking after search	ΔE_{latt} after search/kJ mol^{-1}	Ranking after DFTB3-D3	ΔE_{latt} after DFTB3-D3/ kJ mol^{-1}	Ranking after $\Psi_{mol}^{PBE0+FIT}$	ΔE_{latt} after $\Psi_{mol}^{PBE0+FIT}$/ kJ mol^{-1}	Ranking in previous CSP study	ΔE_{latt} in previous CSP study/kJ mol^{-1}
XXVI (XAFQIH)	262	20.5	208	25.44	2	1.96	*2*	*0.49*
GSK269984B (BIFHOP)	54	6.63	1803	37.2	1	0	*1*	*0*
XXIII a (XAFPAY)	4311	20.57	3670	33.8	232	12.31	*280*	*13.52*
XXIII B (XAFPAY01)	55	7.72	805	23.49	3	0.82	*1*	*0*
XXIII D (XAFPAY03)	1318	16.28	3748	33.99	114	10.25	*85*	*9.21*
XX (OBEQIX)	10	3.86	10	2.94	1	0	*1*	*0*
Mebendazole A (TUXPEJ)[a]	2	1.37	14	4.32	1	0	*1*	*0*
Mebendazole C (YULGIW)[a]	92	9.64	101	8.96	15	5.78	*4*	*2.55*

[a] The crystal structures of the two tautomers of mebendazole are ranked together.

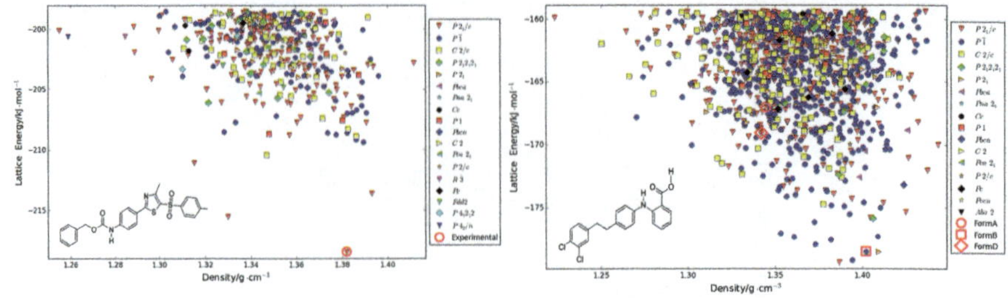

Fig. 3 Crystal energy landscapes of molecules (left) XX and (right) XXIII, with E_{latt} calculated by $\Psi_{mol}^{PBE0+FIT}$, plotted against the density of each crystal structure. The structures corresponding to the three experimental forms with $Z' = 1$ only are given, with the relative energies of the $Z' = 2$ polymorphs of XXIII being shown in Fig. 5.

Phonon calculations

The calculation of phonons by DFTB3-D3 methods proved difficult to automate for reasons that would apply equally to significantly more expensive periodic electronic structure methods. For certain specific crystal structures, it was necessary to perform optimisations to an even tighter convergence on the energies and gradients in order to remove small imaginary frequencies (<100 i cm^{-1}). The recipe of only needing supercells to have a minimum length of 10 Å proved insufficient to converge the Brillouin sampling so that the difference in F_{vib} between structures could be converged below 1 kJ mol^{-1}, mainly because of the differences in entropy. Thus the contrasting calculations in F_{vib} (Fig. 4) contain significant numerical uncertainties, which do not cancel between different structures of the same molecule. Overall, the DFTB3-D3 phonons do not cause significant re-ranking; the most notable exception is for molecule XXVI, where the experimental crystal structure becomes lower in free energy than the global minimum.

It is clear that the inclusion of the coupling of the molecular modes with the lattice modes in the DFTB3-D3 calculations leads to a wider range of free energy differences than with rigid-body phonons, implying that there is a considerable effect from the variations of the modes in different conformations and their couplings. The rigid-body phonons, which used 4 or 5 linear supercells to converge the Brillouin zone sampling, may be subject to greater errors in ignoring the coupling of the modes for this type of molecule.

Computational cost

The savings in computational cost are quite large with this methodology, when compared with the original CSP studies, as detailed in ESI Tables 11–13.† The smallest saving (30%) was for the two tautomers of mebendazole, which is the least complex molecule in terms of size and conformational flexibility, and the greatest was for XXVI (90%). Some of this reduction in computational cost would be due to the use of different computer clusters, but the saving from replacing CrystalOptimizer and sometimes intermediate refinement methods with DFTB3-D3 structure refinement and $\Psi_{mol}^{PBE0+FIT}$ seems to increase as a function of molecular size and flexibility. The rigid-molecule phonons were much cheaper to calculate than the DFTB3-D3 phonons.

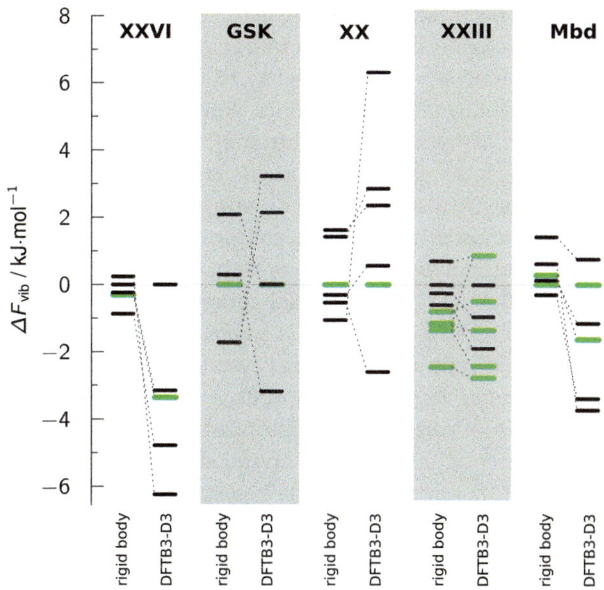

Fig. 4 The relative vibrational contributions to the Helmholtz free energy ΔF_{vib}, which can be added to E_{latt} to calculate A. For each molecule, ΔF_{vib} is calculated relative to the structure which is the E_{latt} global minimum within the $\Psi_{mol}^{PBE0+FIT}$ method, for which $\Delta F_{vib} = 0$. The experimental structures are in green. The rigid-body modes are pure lattice modes, calculated from the Ψ_{mol} method. Full details about the crystal structures and their energies at the different stages can be found in ESI Table 10.†

Discussion

This study pioneers the assessment of the DFTB3-D3 method for CSP studies of pharmaceuticals. In principle, the free energy calculated using a dispersion corrected PBE0 functional for each crystal structure (*i.e.* Ψ_{crys} using the same wavefunction as in the $\Psi_{mol}^{PBE0+FIT}$ method) should be more accurate than using the PBE functional that is commonly used for the final lattice energy evaluation in CSP studies,[1] with both being more accurate than DFTB3-D3 calculations. However, the enormous computational cost makes this impractical for use in accurate CSP studies of large flexible molecules with current computer resources. Full DFTB3-D3 optimisations are significantly better than fixed cell optimisations of selected torsion angles with a transferable force-field, a previously used intermediate step in CSP for flexible molecules,[43,44,50] where the energies were discarded. In practice, theoretical accuracy, which is very dependent on the molecule and range of crystal structures, has to be compromised by the huge number of structures that need to be considered, and the need to make such studies affordable for larger pharmaceutical studies. Thus the time saving produced by an intermediate DFTB3-D3 step is an important consideration for extending CSP to be used more widely in pharmaceutical development. DFTB3-D3 scales as $N \ln(N)$, in contrast to N^3 for periodic DFT-D methods,[58] where N is the number of atoms in the unit cell. The cost of the Ψ_{mol} approach is more variable, as by using databases of previous calculations and methods of interpolation[82] the cost of this approach decreases as more structures are optimised. However, the Ψ_{mol} approach scales badly with the number of conformational degrees of freedom that are allowed to vary in response to the packing forces of specific crystals. The other conformational variables cannot be fixed, but have to be simultaneously *ab initio* optimised as intramolecular energy barriers can be drastically lowered by small changes in other parts of the molecule.[83] Pharmaceutical molecules, where packing forces acting on large aromatic rings at different ends of the molecule strain the conformation of linking groups, and inter- and intramolecular dispersion need to be balanced, are very demanding of the quality of the modelling of the potential energy surface.

The numerical results shown in Table 2 suggest that DFTB3-D3 is of similar accuracy to PBE0 partial optimisation using CrystalOptimizer for reproducing the experimental conformers, as judged by $RMSD_1$ values. Holding the DFTB3-D3 conformations as rigid during the final optimisation with the $\Psi_{mol}^{PBE0+FIT}$ model gives a slightly worse reproduction of the crystal structures in terms of $RMSD_{15}$, but at a reduced cost. Comparisons with the experimental conformations (see ESI Table 2†) show that in most cases the DFTB3-D3 reproduction is good, probably within the amplitude of the vibrational motions of the terminal groups. The only exception is XXIII form D, where the optimisations have found a conformer close to the search generated structure, but visually different from the experimental one. The differences between crystal structures, as judged by $RMSD_{15}$, powder pattern similarity, or any other measure, raise the issue of clustering: how can one eliminate duplicates while being certain not to eliminate potential polymorphs? Many polymorphs are very similar in their packing.[84] Determining which structures correspond to distinct free energy minima at crystallisation temperatures requires realistic and expensive modelling of molecular motions within the crystals and would only ever be applied to the most promising PPMs resulting from a CSP study. The question of whether structures could remain distinct during crystallisation and growth is closely linked to the possibility of disorder within crystal structures, and indeed CSP can help rationalise disordered pharmaceuticals.[8,11] For example, molecules that can pack in hydrogen-bonded layers that can stack in different ways with similar energies could form polytypic packings or have stacking faults.[20] Closely related structures can manifest as static or dynamic disorder depending on the energy barriers.[17] Hence, decisions about the extent of the search and the clustering criteria can have a major effect on the number of structures that need to be considered and their interpretation. It appears that using an electronic structure level optimisation (DFTB3-D3) has not significantly reduced the number of minima (only by 1–4%, see ESI Table 1†) for the molecules considered in this study, hinting that few structures are artefacts of the Ψ_{mol} approach.

The disappointment is that the relative energies generated by DFTB3-D3 are too poor (see ESI Section 2.3.1†) to allow a confident, drastic reduction in the number of structures to be investigated with more accurate and expensive methods. This is probably due to the limitations of this methodology, such as the approximate Hamiltonian used in the molecular overlap regime where atoms are in van der Waals contact, and approximating the long-range electrostatic potential by atomic monopoles, although they are iterated to self-consistency to allow some polarisation of the molecules by the field of the other molecules.[59] There is also an underestimation of the Pauli repulsion that can lead to an overestimate of the density of molecular crystals;[45] the unbalanced description of Pauli repulsion in the parametrization of DFTB seems to originate from the underlying basis confinement.[60] Some observed deficiencies in hydrogen bond and halogen bond geometries and energies have led to the proposal of additional corrections,[71,73] but these do not seem to improve the results for large molecular assemblies.[46] The long range electrostatic interactions are described by atomic charges in both DFTB3-D3 and in the crystal generation stage, but from atomic multipoles in the final energy evaluation, and the quality of the electrostatic model is very critical in crystal structure prediction, particularly for polar hydrogen bonded molecules;[85] this explains why the final energy ranking appears to be much more accurate than the initial and intermediate ones. Several other differences between the various methodologies employed exist in the ways in which intra- and intermolecular interactions are treated and how they are combined to optimise the crystal structures and calculate E_{latt}. All these differences play a part in determining the quality of the geometries and of the energy ranking, and it appears that despite the clear advantages of periodic electronic methods a high-quality crystal energy landscape cannot be calculated with a simplified wave-function.

DFTB3-D3 and the Ψ_{mol} approach are less applicable in the presence of strong induction forces, suggesting they should not be used for salts or zwitterions, though there have been successful studies which make allowance for these deficiencies.[86,87] The affordable density functionals such as PBE also have known problems with these types of systems.[88–90] Furthermore, it is very challenging to have computational methods that are equally reliable for intramolecular dispersion effects[91] and intermolecular dispersion extending through the range of separations sampled in crystal structures, but this balance underlies the observed preference for flexible molecules to pack in extended crystal structures.[92] The large dispersion contributions make the energies very density dependent, as well as being sensitive to conformational distortion and the stronger more directional interactions such as hydrogen bonding.

The phonon calculations within the Ψ_{crys} approach, which are only affordable using DFTB3-D3, are conceptually different from the rigid-body Ψ_{mol} approach, in including effects for molecules having many molecular modes that are of comparable frequency to the lattice modes.[41] The use of DFTB3-D3 accounts for differences in the molecular modes within different crystalline geometries. This leads to an even larger spread in F_{vib} values than with the rigid-body model, which already shows a range of ΔF_{vib} that is comparable with polymorph energy differences,[4] and was estimated to reorder the calculated stability of about 9% of over 500 observed pairs of polymorphs.[80] This raises a question as to whether thermal effects are likely to be more significant in determining the relative thermodynamic stability of pharmaceutical polymorphs than for more rigid molecules, particularly when there is a larger range of densities amongst the PPMs. DFTB3-D3 calculates all phonon modes within the specific crystal in a reasonable timeframe, and these can be added to E_{latt} calculated with $\Psi_{\text{mol}}^{\text{PBE0+FIT}}$ or a better periodic DFT-D method.[70,93,94] The disadvantages of a poorer potential energy surface need to be balanced with the problems of converging energies with the size of the supercell to equivalent accuracy for different polymorphs demonstrated in this study.[95] Even if it were possible to calculate completely reliable harmonic modes, which the current results in Fig. 5 show has yet to be achieved, thermal expansion is likely to be important for both absolute free energies[65] and the differences in thermal expansion between polymorphs, which will affect the relative free energies.[41,96] These estimates assume that the (quasi)-harmonic model is equally adequate for all polymorphs, which is questionable for larger molecules where methyl groups may be rotating and phenyl rings may have different large amplitude motions, depending on the packing. Hence, the potential energy surface should be explored more explicitly, for example by molecular dynamics simulations,[97,98] as none of the methodologies currently available can fully describe all the possible thermal effects that can affect crystal structures and their relative free energy ranking.

Many of the challenges in CSP studies are illustrated by the contrast between the crystal energy landscapes of XX and XXIII in Fig. 3, which from the chemical diagrams look like similar extended pharmaceutical molecules. The crystal energy landscape of molecule XX represents the simplest outcome of a CSP study, with a clear prediction of one crystal structure being more stable than any of its competitors. This structural refinement has led to a larger energy gap than previously found with two different Ψ_{mol} approaches,[50] using CrystalOptimizer (FCC) and a force-field (RCM) for modifying the molecular conformation in response to the packing forces. Both had the experimentally-known crystal structure as the global minimum in E_{latt} but with a margin of less than 1 kJ mol^{-1} from the second most favourable PPM. Comparing the quality of this landscape with those of the previous studies is impossible as only three structures among the lowest ten found with the RCM method were among the top ten when optimised with the FCC method. ESI Table 5† shows that seven of the ten lowest-energy FCC crystal structures are among the ten lowest energy $\Psi_{\text{mol}}^{\text{PBE0+FIT}}$ crystal structures, while for RCM this is only true for the crystal structure that matches the experimental form (see ESI Table 8†). The sensitivity of energies to the E_{latt} model is further shown by a generally very effective Ψ_{crys} methodology developed

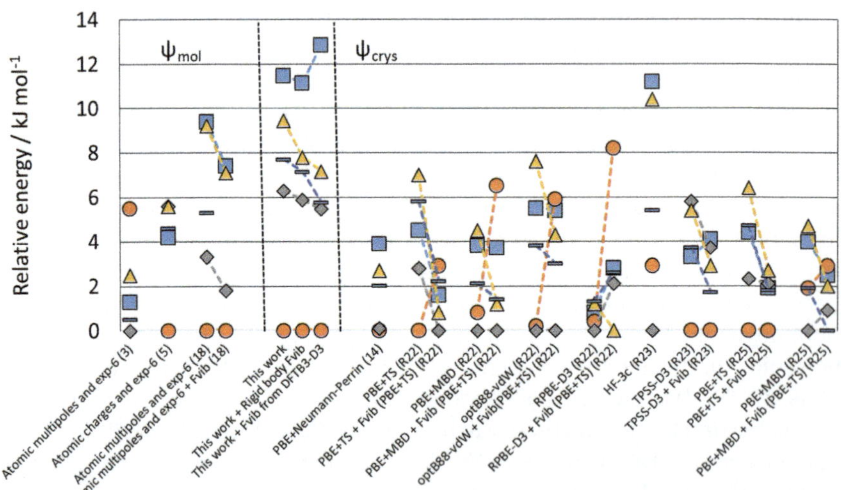

Fig. 5 Relative energies of the polymorphs of XXIII calculated in this study compared with those reported by the participants of the 6th Blind Test.[1] Note that values linked by dashed lines denote the changes from adding free energy estimates, with the group identifier in parentheses, and R denoting groups which only re-ranked the crystal structures.

by Neumann *et al.*[21,24,99–101] having the experimental crystal structure ranked 7th,[27] although that search included some $Z' = 2$ crystal structures. Some of these $Z' = 2$ structures also appear to be highly competitive when optimised with DFTB3-D3 and $\Psi_{mol}^{PBE0+FIT}$, as shown in ESI Table 9.† Nevertheless in our study the structure matching the only known experimental form is calculated to be the most stable by approximately 3 kJ mol^{-1}, and there are relatively few alternatives to be considered as possible metastable polymorphs.

In contrast, the crystal energy landscape of XXIII has successfully found the three $Z' = 1$ polymorphs, but amongst a plethora of competitive $Z' = 1$ structures. Two further $Z' = 2$ polymorphs are known from the industrial polymorph screening. Only form B is among the lowest energy crystal structures, while forms A and D are much higher in energy amongst many other structures of similar energy and density. Comparing these results with the sensitivity of the known polymorphs to different energy models, Fig. 5 suggests that the energy differences would have been reduced if the final energy optimisation had been performed using a good periodic DFT-D method, like TPSS-D3,[102] PBE-MBD[98] or B86bPBE-XDM.[94,103] However, this contrasts with experimental stabilities from slurrying data, where form A is the most stable at 257 K and form D at 293 K.[1] Overall, the sensitivity of the energies of the known forms of XXIII highlights the challenges of modelling the thermodynamics of the organic solid state. The contrast between the crystal energy landscapes of XX and XXIII emphasises the need for more CSP studies to be performed in collaboration with experimentalists in industry and academia.

Conclusions

Periodic DFTB3-D3 is suitable for optimising the crystal structures of flexible molecules, but does not provide a sufficiently accurate energy ranking. Experimental conformers and crystal structures are reproduced by DFTB3-D3 with average $RMSD_1$ and $RMSD_{15}$ values of 0.16 Å and 0.52 Å, respectively. DFTB3-D3 calculations suggest that the phonon modes may cause a more significant re-ranking of the energies of polymorphs of flexible pharmaceuticals than for smaller, more rigid molecules. The demands of performing CSP for flexible pharmaceuticals, and hence the most cost-efficient strategy, are very dependent on the molecule and the number of competitive crystal structures.

Conflicts of interest

There are no conflicts to declare.

Acknowledgements

We thank the CCDC for organising the Blind Tests of CSP and for partial financial support of L. I. along with the M3S Centre for Doctoral Training (EPSRC GRANT EP/G036675/1). We also thank Profs Pantelides and Adjiman (Imperial College) for the use of CrystalPredictor and CrystalOptimizer. J. G. B acknowledges support by the Alexander von Humboldt foundation within the Feodor Lynen program. The CSP computational software is developed under EP/K039229/1.

References

1 A. M. Reilly, R. I. Cooper, C. S. Adjiman, S. Bhattacharya, A. D. Boese, J. G. Brandenburg, P. J. Bygrave, R. Bylsma, J. E. Campbell, R. Car, D. H. Case, R. Chadha, J. C. Cole, K. Cosburn, H. M. Cuppen, F. Curtis, G. M. Day, R. A. DiStasio Jr, A. Dzyabchenko, B. P. van Eijck, D. M. Elking, J. A. van den Ende, J. C. Facelli, M. B. Ferraro, L. Fusti-Molnar, C.-A. Gatsiou, T. S. Gee, R. de Gelder, L. M. Ghiringhelli, H. Goto, S. Grimme, R. Guo, D. W. M. Hofmann, J. Hoja, R. K. Hylton, L. Iuzzolino, W. Jankiewicz, D. T. de Jong, J. Kendrick, N. J. J. de Klerk, H.-Y. Ko, L. N. Kuleshova, X. Li, S. Lohani, F. J. J. Leusen, A. M. Lund, J. Lv, Y. Ma, N. Marom, A. E. Masunov, P. McCabe, D. P. McMahon, H. Meekes, M. P. Metz, A. J. Misquitta, S. Mohamed, B. Monserrat, R. J. Needs, M. A. Neumann, J. Nyman, S. Obata, H. Oberhofer, A. R. Oganov, A. M. Orendt, G. I. Pagola, C. C. Pantelides, C. J. Pickard, R. Podeszwa, L. S. Price, S. L. Price, A. Pulido, M. G. Read, K. Reuter, E. Schneider, C. Schober, G. P. Shields, P. Singh, I. J. Sugden, K. Szalewicz, C. R. Taylor, A. Tkatchenko, M. E. Tuckerman, F. Vacarro, M. Vasileiadis, A. Vazquez-Mayagoitia, L. Vogt, Y. Wang, R. E. Watson, G. A. de Wijs, J. Z. Yang, Q. Zhu and C. R. Groom, *Acta Crystallogr., Sect. B: Struct. Sci., Cryst. Eng. Mater.*, 2016, **72**, 439–459.

2 S. Curtarolo, G. L. W. Hart, M. B. Nardelli, N. Mingo, S. Sanvito and O. Levy, *Nat. Mater.*, 2013, **12**, 191.

3 N. Marzari, *Nat. Mater.*, 2016, **15**, 381.

4 A. J. Cruz-Cabeza, S. M. Reutzel-Edens and J. Bernstein, *Chem. Soc. Rev.*, 2015, **44**, 8619–8635.

5 J. Bernstein, *Polymorphism in Molecular Crystals*, Clarendon Press, Oxford, 2002.

6 Y. Abramov, *Org. Process Res. Dev.*, 2013, **17**, 472–485.

7 A. Y. Lee, D. Erdemir and A. S. Myerson, *Annu. Rev. Chem. Biomol. Eng.*, 2011, **2**, 259–280.

8 S. L. Price and S. M. Reutzel-Edens, *Drug Discovery Today*, 2016, **21**, 912–923.

9 A. Pulido, L. Chen, T. Kaczorowski, D. Holden, M. Little, S. Chong, B. Slater, D. McMahon, B. Bonillo, C. Stackhouse, A. Stephenson, C. Kane, R. Clowes, T. Hasell, A. Cooper and G. Day, *Nature*, 2017, **543**, 657–664.

10 S. L. Price, *Chem. Soc. Rev.*, 2014, **43**, 2098–2111.

11 S. L. Price, D. E. Braun and S. M. Reutzel-Edens, *Chem. Commun.*, 2016, **52**, 7065–7077.

12 J. T. A. Jones, T. Hasell, X. F. Wu, J. Bacsa, K. E. Jelfs, M. Schmidtmann, S. Y. Chong, D. J. Adams, A. Trewin, F. Schiffman, F. Cora, B. Slater, A. Steiner, G. M. Day and A. I. Cooper, *Nature*, 2011, **474**, 367–371.

13 E. O. Pyzer-Knapp, H. P. G. Thompson, F. Schiffmann, K. E. Jelfs, S. Y. Chong, M. A. Little, A. I. Cooper and G. M. Day, *Chem. Sci.*, 2014, **5**, 2235–2245.

14 C. J. Pickard and R. J. Needs, *J. Phys.: Condens. Matter*, 2011, **23**.

15 R. T. Strong, C. J. Pickard, V. Milman, G. Thimm and B. Winkler, *Phys. Rev. B: Condens. Matter Mater. Phys.*, 2004, **70**, 045101.

16 M. Martinez-Canales, C. J. Pickard and R. J. Needs, *Phys. Rev. Lett.*, 2012, **108**, 045704.

17 D. E. Braun, J. A. McMahon, L. H. Koztecki, S. L. Price and S. M. Reutzel-Edens, *Cryst. Growth Des.*, 2014, **14**, 2056–2072.

18 R. M. Bhardwaj, L. S. Price, S. L. Price, S. M. Reutzel-Edens, G. J. Miller, I. D. H. Oswald, B. Johnston and A. J. Florence, *Cryst. Growth Des.*, 2013, **13**, 1602–1617.

19 M. Baias, J. N. Dumez, P. H. Svensson, S. Schantz, G. M. Day and L. Emsley, *J. Am. Chem. Soc.*, 2013, **135**, 17501–17507.

20 L. S. Price, J. A. McMahon, S. R. Lingireddy, S. F. Lau, B. A. Diseroad, S. L. Price and S. M. Reutzel-Edens, *J. Mol. Struct.*, 2014, **1078**, 26–42.

21 M. A. Neumann, J. V. de Streek, F. P. A. Fabbiani, P. Hidber and O. Grassmann, *Nat. Commun.*, 2015, **6**, 7793.

22 M. Vasileiadis, C. C. Pantelides and C. S. Adjiman, *Chem. Eng. Sci.*, 2015, **121**, 60–76.

23 S. Z. Ismail, C. L. Anderton, R. C. Copley, L. S. Price and S. L. Price, *Cryst. Growth Des.*, 2013, **13**, 2396–2406.

24 J. Kendrick, G. A. Stephenson, M. A. Neumann and F. J. J. Leusen, *Cryst. Growth Des.*, 2013, **13**, 581–589.

25 L. Iuzzolino, A. M. Reilly, P. McCabe and S. L. Price, *J. Chem. Theory Comput.*, 2017, **13**, 5163–5171.

26 J. Nyman and G. Day, *Phys. Chem. Chem. Phys.*, 2016, **18**, 31132–31143.

27 D. A. Bardwell, C. S. Adjiman, Y. A. Arnautova, E. Bartashevich, S. X. M. Boerrigter, D. E. Braun, A. J. Cruz-Cabeza, G. M. Day, R. G. Della Valle, G. R. Desiraju, B. P. van Eijck, J. C. Facelli, M. B. Ferraro, D. Grillo, M. Habgood, D. W. M. Hofmann, F. Hofmann, K. V. J. Jose, P. G. Karamertzanis, A. V. Kazantsev, J. Kendrick, L. N. Kuleshova, F. J. J. Leusen, A. V. Maleev, A. J. Misquitta, S. Mohamed, R. J. Needs, M. A. Neumann, D. Nikylov, A. M. Orendt, R. Pal, C. C. Pantelides, C. J. Pickard, L. S. Price, S. L. Price, H. A. Scheraga, J. van de Streek, T. S. Thakur, S. Tiwari, E. Venuti and I. K. Zhitkov, *Acta Crystallogr., Sect. B: Struct. Sci.*, 2011, **67**, 535–551.

28 G. M. Day, T. G. Cooper, A. J. Cruz-Cabeza, K. E. Hejczyk, H. L. Ammon, S. X. M. Boerrigter, J. Tan, R. G. Della Valle, E. Venuti, J. Jose, S. R. Gadre, G. R. Desiraju, T. S. Thakur, B. P. van Eijck, J. C. Facelli, V. E. Bazterra, M. B. Ferraro, D. W. M. Hofmann, M. Neumann, F. J. J. Leusen, J. Kendrick, S. L. Price, A. J. Misquitta, P. G. Karamertzanis, G. W. A. Welch, H. A. Scheraga, Y. A. Arnautova, M. U. Schmidt, J. van de Streek, A. Wolf and B. Schweizer, *Acta Crystallogr., Sect. B: Struct. Sci.*, 2009, **65**, 107–125.

29 C. C. Pantelides, C. S. Adjiman and A. V. Kazantsev, *Top. Curr. Chem.*, 2014, **345**, 25–58.

30 M. A. Neumann, *J. Phys. Chem. B*, 2008, **112**, 9810–9829.

31 M. Habgood, I. J. Sugdan, A. V. Kazantsev, C. S. Adjiman and C. Pantelides, *J. Chem. Theory Comput.*, 2015, **11**, 1957–1969.

32 P. G. Karamertzanis and C. C. Pantelides, *Mol. Phys.*, 2007, **105**, 273–291.

33 S. Kim, A. M. Orendt, M. B. Ferraro and J. C. Facelli, *J. Comput. Chem.*, 2009, **30**, 1973–1985.

34 R. G. Parr and W. Yang, *Density-Functional Theory of Atoms and Molecules*, Oxford University Press, Oxford, 1989.

35 W. Kohn, *Rev. Mod. Phys.*, 1999, **71**, 1253–1266.

36 S. Grimme, A. Hansen, J. G. Brandenburg and C. Bannwarth, *Chem. Rev.*, 2016, **116**, 5105–5154.

37 L. Kronik and A. Tkatchenko, *Acc. Chem. Res.*, 2014, **47**, 3208–3216.

38 J. Hoja, A. Reilly and A. Tkatchenko, *Wiley Interdiscip. Rev.: Comput. Mol. Sci.*, 2017, **7**, e1294.

39 G. J. O. Beran, *Chem. Rev.*, 2016, **116**, 5567–5613.

40 Y. N. Heit, K. D. Nanda and G. J. O. Beran, *Chem. Sci.*, 2016, **7**, 246–255.

41 H. K. Buchholz, R. K. Hylton, J. G. Brandenburg, A. Seidel-Morgenstern, H. Lorenz, M. Stein and S. L. Price, *Cryst. Growth Des.*, 2017, **17**, 4676–4686.

42 S. L. Price and J. G. Brandenburg, in *Non-Covalent Interactions in Quantum Chemistry and Physics*, Elsevier, 2017, pp. 333–363.

43 G. M. Day, W. D. S. Motherwell and W. Jones, *Phys. Chem. Chem. Phys.*, 2007, **9**, 1693–1704.

44 G. M. Day and T. G. Cooper, *CrystEngComm*, 2010, **12**, 2443–2453.

45 A. S. Christensen, T. Kubař, Q. Cui and M. Elstner, *Chem. Rev.*, 2016, **116**, 5301–5337.

46 J. G. Brandenburg, M. Hochheim, T. Bredow and S. Grimme, *J. Phys. Chem. Lett.*, 2014, **5**, 4275–4284.

47 A. V. Akimov and O. V. Prezhdo, *Chem. Rev.*, 2015, **115**, 5797–5890.

48 S. Grimme, C. Bannwarth and P. Shushkov, *J. Chem. Theory Comput.*, 2017, **13**, 1989–2009.

49 J. G. Brandenburg and S. Grimme, *J. Phys. Chem. Lett.*, 2014, **5**, 1785–1789.

50 A. V. Kazantsev, P. G. Karamertzanis, C. S. Adjiman, C. C. Pantelides, S. L. Price, P. T. A. Galek, G. M. Day and A. J. Cruz-Cabeza, *Int. J. Pharm.*, 2011, **418**, 168–178.

51 M. K. Corpinot, L. Iuzzolino, S. L. Price and D.-K. Bučar, unpublished work.

52 R. Taylor, J. Cole, O. Korb and P. McCabe, *J. Chem. Inf. Model.*, 2014, **54**, 2500–2514.

53 P. J. Ballester and W. G. Richards, *J. Comput. Chem.*, 2007, **28**, 1711–1723.

54 J. C. Cole, O. Korb, P. McCabe, M. G. Read and R. Taylor, *J. Chem. Inf. Model.*, 2018, **58**, 615–629.

55 M. J. Frisch, G. W. Trucks, H. B. Schlegel, G. E. Scuseria, M. A. Robb, J. R. Cheeseman, G. Scalmani, V. Barone, B. Mennucci, G. A. Petersson, H. Nakatsuji, M. Caricato, X. Li, H. P. Hratchian, A. F. Izmaylov, J. Bloino, G. Zheng, J. L. Sonnenberg, M. Hada, M. Ehara, K. Toyota, R. Fukuda, J. Hasegawa, M. Ishida, T. Nakajima, Y. Honda, O. Kitao, H. Nakai, T. Vreven, J. A. Montgomery Jr, J. E. Peralta, F. Ogliaro, M. J. Bearpark, J. Heyd, E. N. Brothers, K. N. Kudin, V. N. Staroverov, R. Kobayashi, J. Normand, K. Raghavachari, A. P. Rendell, J. C. Burant, S. S. Iyengar, J. Tomasi, M. Cossi, N. Rega, N. J. Millam, M. Klene, J. E. Knox, J. B. Cross, V. Bakken, C. Adamo, J. Jaramillo, R. Gomperts, R. E. Stratmann, O. Yazyev, A. J. Austin, R. Cammi, C. Pomelli, J. W. Ochterski, R. L. Martin, K. Morokuma, V. G. Zakrzewski, G. A. Voth, P. Salvador, J. J. Dannenberg, S. Dapprich, A. D. Daniels, Ö. Farkas, J. B. Foresman, J. V. Ortiz, J. Cioslowski and D. J. Fox, *Gaussian 09 Revision E.01*, Gaussian Inc., Wallingford, CT, 2009.

56 D. S. Coombes, S. L. Price, D. J. Willock and M. Leslie, *J. Phys. Chem.*, 1996, **100**, 7352–7360.

57 G. Seifert and J.-O. Joswig, *Wiley Interdiscip. Rev.: Comput. Mol. Sci.*, 2012, **2**, 456–465.
58 B. Aradi, B. Hourahine and T. Frauenheim, *J. Phys. Chem. A*, 2007, **111**, 5678–5684.
59 M. Elstner, *J. Phys. Chem. A*, 2007, **111**, 5614–5621.
60 M. Gaus, A. Goez and M. Elstner, *J. Chem. Theory Comput.*, 2013, **9**, 338–354.
61 M. Kubillus, T. Kubař, M. Gaus, J. Řezáč and M. Elstner, *J. Chem. Theory Comput.*, 2015, **11**, 332–342.
62 S. Grimme, J. Antony, S. Ehrlich and H. A. Krieg, A consistent and accurate ab initio parametrization of density functional dispersion correction (DFT-D) for the 94 elements H-Pu, *J. Chem. Phys.*, 2010, **132**, 154104.
63 S. Grimme, S. Ehrlich and L. Goerigk, *J. Comput. Chem.*, 2011, **32**, 1456–1465.
64 R. Sure, J. G. Brandenburg and S. Grimme, *ChemistryOpen*, 2016, **5**, 94–109.
65 J. G. Brandenburg, J. Potticary, H. A. Sparkes, S. L. Price and S. R. Hall, *J. Phys. Chem. Lett.*, 2017, **8**, 4319–4324.
66 M. Mortazavi, J. G. Brandenburg, R. J. Maurer and A. Tkatchenko, *J. Phys. Chem. Lett.*, 2018, **9**, 399–405.
67 J. Řezáč, K. E. Riley and P. Hobza, *J. Chem. Theory Comput.*, 2011, **7**, 2427–2438.
68 B. Brauer, M. K. Kesharwani, S. Kozuch and J. M. L. Martin, *Phys. Chem. Chem. Phys.*, 2016, **18**, 20905–20925.
69 A. Otero-de-la-Roza and E. R. Johnson, *J. Chem. Phys.*, 2012, **137**, 054103.
70 A. M. Reilly and A. Tkatchenko, *J. Chem. Phys.*, 2013, **139**, 024705.
71 M. Korth, M. Pitoňák, J. Řezáč and P. Hobza, *J. Chem. Theory Comput.*, 2010, **6**, 344–352.
72 V. M. Miriyala and J. Rezac, *J. Comput. Chem.*, 2017, **38**, 688–697.
73 J. Rezac and P. Hobza, *Chem. Phys. Lett.*, 2011, **506**, 286–289.
74 R. Dovesi, R. Orlando, A. Erba, C. M. Zicovich-Wilson, B. Civalleri, S. Casassa, L. Maschio, M. Ferrabone, M. De La Pierre, P. D'Arco, Y. Noël, M. Causà, M. Rérat and B. Kirtman, *Int. J. Quantum Chem.*, 2014, **114**, 1287–1317.
75 J. G. Brandenburg and S. Grimme, *Acta Crystallogr., Sect. B: Struct. Sci., Cryst. Eng. Mater.*, 2016, **72**, 502–513.
76 A. L. Spek, *Acta Crystallogr., Sect. D: Biol. Crystallogr.*, 2009, **65**, 148–155.
77 C. R. Groom, I. J. Bruno, M. P. Lightfoot and S. C. Ward, *Acta Crystallogr., Sect. B: Struct. Sci., Cryst. Eng. Mater.*, 2016, **72**, 171–179.
78 A. J. Stone, *J. Chem. Theory Comput.*, 2005, **1**, 1128–1132.
79 S. L. Price, M. Leslie, G. W. A. Welch, M. Habgood, L. S. Price, P. G. Karamertzanis and G. M. Day, *Phys. Chem. Chem. Phys.*, 2010, **12**, 8478–8490.
80 J. Nyman and G. M. Day, *CrystEngComm*, 2015, **17**, 5154–5165.
81 A. J. Cruz-Cabeza and J. Bernstein, *Chem. Rev.*, 2014, **114**, 2170–2191.
82 A. V. Kazantsev, P. G. Karamertzanis, C. S. Adjiman and C. C. Pantelides, in *Molecular System Engineering*, ed. C. S. Adjiman and A. Galindo, Wiley-VCH Verlag GmbH & Co., Weinheim, 2010, vol. 6, pp. 1–42.
83 O. G. Uzoh, P. T. A. Galek and S. L. Price, *Phys. Chem. Chem. Phys.*, 2015, **17**, 7936–7948.
84 S. J. Coles, T. L. Threlfall and G. J. Tizzard, *Cryst. Growth Des.*, 2014, **14**, 1623–1628.

85 G. M. Day, W. D. S. Motherwell and W. Jones, *Cryst. Growth Des.*, 2005, **5**, 1023–1033.
86 D. E. Braun, S. R. Lingireddy, M. D. Beidelschies, R. Guo, P. Muller, S. L. Price and S. M. Reutzel-Edens, *Cryst. Growth Des.*, 2017, **17**, 5349–5365.
87 D. E. Braun, M. Orlova and U. J. Griesser, *Cryst. Growth Des.*, 2014, **14**, 4895–4900.
88 J. G. Brandenburg, T. Maas and S. Grimme, *J. Chem. Phys.*, 2015, 142.
89 J. Klimes, D. R. Bowler and A. Michaelides, *Phys. Rev. B: Condens. Matter Mater. Phys.*, 2011, **83**, 195131.
90 L. Goerigk, H. Kruse and S. Grimme, *ChemPhysChem*, 2011, **12**, 3421–3433.
91 A. Fokin, T. Zhuk, S. Blomeyer, C. Perez, L. Chernish, A. Pashenko, J. Antony, Y. Vishnevskiy, R. Berger, S. Grimme, C. Logemann, M. Schnell, N. Mitzel and P. Schreiner, *J. Am. Chem. Soc.*, 2017, **139**, 16696–16707.
92 H. P. G. Thompson and G. M. Day, *Chem. Sci.*, 2014, **5**, 3173–3182.
93 S. Grimme, *Chem.–Eur. J.*, 2012, **18**, 9955–9964.
94 S. R. Whittleton, A. Otero-de-la-Roza and E. R. Johnson, *J. Chem. Theory Comput.*, 2017, **13**, 441–450.
95 R. Guo, K. Refson and S. L. Price, personal communication.
96 Y. Heit and G. Beran, *Acta Crystallogr., Sect. B: Struct. Sci., Cryst. Eng. Mater.*, 2016, **72**, 514–529.
97 M. Rossi, P. Gasparotto and M. Ceriotti, *Phys. Rev. Lett.*, 2016, **117**, 115702.
98 A. Shtukenberg, Q. Zhu, D. Carter, L. Vogt, J. Hoja, E. Schneider, H. Song, B. Pokroy, I. Polishchuk, A. Tkatchenko, A. Oganov, A. Rohl, M. Tuckerman and B. Kahr, *Chem. Sci.*, 2017, **8**, 4926–4940.
99 M. A. Neumann, F. J. J. Leusen and J. Kendrick, *Angew. Chem., Int. Ed.*, 2008, **47**, 2427–2430.
100 A. Asmadi, M. A. Neumann, J. Kendrick, P. Girard, M. A. Perrin and F. J. J. Leusen, *J. Phys. Chem. B*, 2009, **113**, 16303–16313.
101 J. van de Streek and M. A. Neumann, *CrystEngComm*, 2011, **13**, 7135–7142.
102 J. G. Brandenburg, J. E. Bates, J. Sun and J. P. Perdew, *Phys. Rev. B*, 2016, **94**, 115144.
103 S. R. Whittleton, A. Otero-de-la-Roza and E. R. Johnson, *J. Chem. Theory Comput.*, 2017, **13**, 5332–5342.

Faraday Discussions

PAPER

Repulsion–dispersion parameters for the modelling of organic molecular crystals containing N, O, S and Cl†

Christina A. Gatsiou, Claire S. Adjiman and Constantinos C. Pantelides*

Received 9th March 2018, Accepted 1st May 2018
DOI: 10.1039/c8fd00064f

In lattice energy models that combine *ab initio* and empirical components, it is important to ensure consistency between these components so that meaningful quantitative results are obtained. A method for deriving parameters of atom–atom repulsion dispersion potentials for crystals, tailored to different *ab initio* models, is presented. It is based on minimization of the sum of squared deviations between experimental and calculated structures and energies. The solution algorithm is designed to avoid convergence to local minima in the parameter space by combining a deterministic low-discrepancy sequence for the generation of multiple initial parameter guesses with an efficient local minimization algorithm. The proposed approach is applied to derive transferable exp-6 potential parameters suitable for use in conjunction with a distributed multipole electrostatics model derived from isolated molecule charge densities calculated at the M06/6-31G(d,p) level of theory. Data for hydrocarbons, azahydrocarbons, oxohydrocarbons, organosulphur compounds and chlorohydrocarbons are used for the estimation. A good fit is achieved for the new set of parameters with a mean absolute error in sublimation enthalpies of 4.1 kJ mol^{-1} and an average rmsd$_{15}$ of 0.31 Å. The parameters are found to perform well on a separate cross-validation set of 39 compounds.

1 Introduction

Crystal Structure Prediction (CSP) has become a valuable tool for determining and understanding the solid states of an increasingly wide range of compounds and mixtures. As highlighted in the most recent (2016) blind test,[1] CSP methods for organic molecules are now applicable to a wide range of real systems, including salts, hydrates and large flexible molecules, with at least one successful prediction being achieved for each target system in the test.

Molecular Systems Engineering Group, Centre for Process Systems Engineering, Department of Chemical Engineering, Imperial College London, London SW7 2AZ, UK. E-mail: c.pantelides@imperial.ac.uk

† Electronic supplementary information (ESI) available: Further information on calculation methods, training and cross-validation data sets, detailed description of the new potentials and results of crystal packing calculations with all parameter sets. See DOI: 10.1039/c8fd00064f

Notwithstanding this significant progress, the results of the blind test also demonstrated some outstanding challenges for CSP methods specifically relating to energy ranking. In some cases, experimentally-observed structures for a given organic compound were deemed to have too high a lattice energy to occur in nature, and were consequently discarded from further consideration. Even methods that are generally successful in identifying the experimentally-observed crystal structures often fail to predict the experimentally-observed relative stability order.[2] While this may partly be due to the neglect of entropic contributions, the problem primarily arises from the insufficient accuracy of the lattice energy model in evaluating the energy differences between predicted structures, which can be as low as 1 kJ mol^{-1}.[3]

A variety of models have been developed and used in CSP for the calculation of the lattice energy, ranging from empirical force fields to electronic structure methods. Dispersion-corrected Density Functional Theory (DFT) models[4–6] have emerged as accurate methods for calculating energies,[1] with benchmark studies suggesting an uncertainty of 3 kJ mol^{-1} for absolute lattice energies.[7,8] However, the computational cost of these models precludes their use for the minimization of large numbers (tens or hundreds of thousands) of structures, a critical step in current CSP methodologies. On the other hand generic force fields such as COMPASS,[9] CVFF[10] and others are computationally efficient but of limited accuracy.[11,12] In view of these factors, the development of force fields that bridge the gap between these extremes has been an important step for all successful CSP methods. In this context, the models used in the sixth blind test included tailor-made potentials fitted to periodic *ab initio* data,[13] potentials derived from symmetry-adapted perturbation theory based on DFT (SAPT(DFT)) calculations,[14] and hybrid models combining isolated-molecule electronic structure calculations with empirical repulsion–dispersion potentials.[15–17] Further details of the different lattice energy models that have been used in CSP through the years and their performance can be found in the blind test papers[18–22] and review papers.[11,23]

In the CSP approach developed in our group,[24] different energy models are used for the global search step in which thousands of possible structures are generated (*CrystalPredictor*[16,17,25,26]) and in the refinement step in which the ranking of structures is finalized (*CrystalOptimizer*[27,28]). Recent developments in the model for the global search[16,17] have led to significant improvements in the accuracy of the force fields for large flexible molecules (such as molecule XXVI in the sixth blind test); in general, all experimentally-relevant structures are typically identified within 20 kJ mol^{-1} of the global minimum and carried through to the refinement stage.[16,17] The latter requires a higher degree of accuracy in the lattice energy. The intramolecular energy is calculated using local approximate models derived from the isolated-molecule *ab initio* energy[27] computed at a chosen level of theory. The intermolecular energy is derived by combining electrostatic interactions represented by distributed multipoles[15,29,30] derived from the isolated molecule charge density calculated at the same level of theory, and repulsion–dispersion interactions modelled by an empirical exp-6 potential with parameters taken from the literature such as the FIT potential,[31–37] W01 (ref. 38) or other parameterizations.[39]

The values of the repulsion–dispersion potential parameters used in most current methodologies were originally estimated from experimental crystal data analysed using an atomic charge representation of electrostatics, with charges fitted to the HF/6-31G(d,p) charge density; intramolecular energy contributions were neglected.[10,18,39–41] As a result, such parameters are not necessarily consistent with more sophisticated electrostatic models such as those based on multipoles. In addition, only parameters relating to the most common atoms (C, N, O, H, Cl, F) were derived from extensive datasets. For atoms such as sulphur or bromine, parameters from a variety of literature sources are currently being used, a practice that is likely to result in inaccuracies. Furthermore, even within the set of atoms considered in the FIT and W01 parameterizations, some cross interactions (*e.g.*, O–N) were not fitted to experimental data; instead, they were approximated using geometric combining rules that are known to be of limited validity and may lead to inaccurate predictions.[42]

Overall, as has been stated elsewhere,[13] force fields should be carefully parameterized so that the different terms complement each other. In practical terms, repulsion–dispersion potentials need to be reparameterized for the specific electrostatics and intramolecular energy models with which they are intended to be used. There has been a recent effort[42] in this direction through the partial reparameterization of the FIT set using the B3LYP/6-31G** and B3LYP/6-311G** levels of theory and multipole expansions for the electrostatics. Focusing on O and N parameters, the work led to encouraging results, with experimental unit cells being reproduced to within 3% and hydrogen bond geometries and sublimation enthalpies between 7.4 and 9.0%.

In this paper, we present a systematic approach for the reparameterization of the repulsion–dispersion term using experimental structural and sublimation enthalpy data. The proposed algorithm can easily be used to derive empirical parameters consistent with different electrostatic models or levels of theory. Here, its applicability is demonstrated by deriving a set of exp-6 repulsion–dispersion potential parameters for C, H, N, O, S and Cl consistent with electrostatic interactions described by distributed multipoles derived from the M06/6-31G(d,p) level of theory. For the first time, this leads to parameters for sulphur that are consistent with the remaining parameter set and this is shown to lead to a significant improvement in performance.

2 The lattice energy model

To relate computational models of crystal structures to experimental data, it is generally assumed that the experimentally-observed crystals correspond to local minima of the Gibbs free energy. Moreover, given the relatively small contribution of entropic effects at low temperatures and the complexity associated with their calculation, the lattice energy is usually employed as an approximation of the free energy. Thus, the objective of CSP is the identification of all low-energy local minima in the lattice energy surface.

2.1 Defining the crystal structure

In frequently-used codes in CSP, such as the *CrystalOptimizer*[27,28] code on which we focus, the lattice energy of a crystal is a function of the following independent variables:

- The unit cell lengths and angles, collectively denoted by X.

• The positions of the centres of mass and the orientation of chemical entities within the unit cell, collectively denoted by β.

• The molecular conformational degrees of freedom (CDFs), *i.e.* bond lengths, angles and torsions, collectively denoted by θ.

It is also useful to introduce two dependent quantities, *i.e.* variables that are functions of the independent variables:

• $Q(\theta)$ denotes the set of parameters in the distributed multipoles[43] describing the molecule's electrostatic field for a given molecular conformation θ.

• $Y(\theta,\beta)$ denotes the Cartesian coordinates of all atoms in the unit cell, which depend on the molecular positions and orientations β and the molecular conformations θ.

2.2 Lattice energy model for CSP

The lattice energy of a crystal U_{latt} can be expressed as the sum of intramolecular and intermolecular energy contributions:

$$U_{\text{latt}}(X,\beta,\theta) = \Delta U^{\text{intra}}(\theta) + U^{\text{e}}(Y,X,Q) + U^{\text{rd}}(Y,X) \tag{1}$$

where ΔU^{intra} is the intramolecular energy contribution (*i.e.*, the electronic energy after subtraction of the gas-phase internal energies of the chemical species in the crystal weighted by their stoichiometric coefficients) and U^{e} and U^{rd} represent, respectively, the intermolecular electrostatic and repulsion–dispersion contributions.

The intramolecular contribution $\Delta U^{\text{intra}}(\theta)$ in the above expression is typically computed *via* a QM evaluation of the conformational energy at given θ.

The intermolecular electrostatic model is based on a distributed multipole representation of electrostatics, placing an expansion up to the hexadecapole moment at each atomic position (as in *DMACRYS*[15] and hence *CrystalOptimizer*[27]). The GDMA program[30] is used to derive the parameters $Q(\theta)$ from the isolated molecule wavefunction;[29] the latter is also determined by the QM calculation used for the computation of $\Delta U^{\text{intra}}(\theta)$ at given θ. The intermolecular electrostatic energy, U^{e}, is calculated as the sum of interactions between multipoles on atomic sites of different molecules. Such a representation of electrostatics has been shown to be successful in predicting highly-directional (anisotropic) lone-pair interactions, π–π stacking in aromatic rings and hydrogen bond geometries in organic crystals.[44–46]

The repulsion–dispersion component of the intermolecular energy, U^{rd}, is calculated as the sum of all repulsion–dispersion interactions within a predefined cutoff. Each interaction, U_{ij}^{rd}, between two atoms of types i and j occurring in two different molecules and separated by a distance r is described through an isotropic atom–atom Buckingham potential:

$$U_{ij}^{\text{rd}}(r) = A_{ij}\exp\left(-r/B_{ij}\right) - \frac{C_{ij}}{r^6} \tag{2}$$

where the interatomic separation r between any two atoms can be derived from X and Y. In many current CSP methodologies, the values of the adjustable parameters A_{ij}, B_{ij}, C_{ij} for atoms of the same type ($i = j$) are usually taken from the FIT set;[18,31–36] for unlike atom types ($i \neq j$), the parameters are usually computed *via* combining rules:

$$A_{ij} = \sqrt{A_{ii}A_{jj}};\;\; B_{ij} = \frac{B_{ii} + B_{jj}}{2};\;\; C_{ij} = \sqrt{C_{ii}C_{jj}}, \tag{3}$$

although more recently,[42] parameters A_{ij} and C_{ij} for some atom types have been regressed directly from data. The most common elements C, N, O, S, F, Cl are treated as transferable atom types in the FIT parameterization, but three distinct atom types are employed for hydrogen, namely hydrogen bonded to carbon, H_C; hydrogen bonded to nitrogen, H_N; and hydrogen bonded to oxygen, H_O.

Finally, in calculating U^e and U^{rd}, summations of intermolecular atom–atom interactions between the central unit cell and other cells in the crystal must be computed. This is done *via* a combination of direct and Ewald summations within a predefined cutoff which is set to 15 Å by default.[15]

2.3 Lattice energy model for parameterization

While the intramolecular energy plays an important role in CSP for flexible molecules, accounting for conformational changes leads to significant increases in the computational cost of energy evaluations. As a result, in keeping with previous work on the parameterization of force fields for CSP,[31–36] we rely on a dataset that contains structures of relatively rigid molecules with no flexible torsions. For such a dataset, intermolecular interactions within the solid will not significantly distort the conformation of the molecule, and the experimental values of the molecular conformation variables, θ^{exp}, can be expected to be close to those for the global minimum in the QM energy function for the isolated molecule, θ^{gas}. However, due to experimental error and theoretical approximations, there are typically some small differences between θ^{exp} and θ^{gas}. Thus, for the purpose of the calculations, we fix the CDFs to the values of θ^{gas} consistent with the chosen level of the theory. The intramolecular energy, ΔU^{intra}, is then equal to zero and as a result the lattice energy model takes the simplified form:

$$U_{latt}(X,\beta;\theta^{gas}) = U^e(Y,X,Q(\theta^{gas})) + U^{rd}(Y,X,p) \tag{4}$$

where p denotes the set of model parameters A_{ij}, B_{ij}, C_{ij} for all atom–atom interactions.

To assess the validity of the lattice energy model, and in particular of the parameter values p, we seek a local energy minimum close to the given experimental structure, (X^{exp},β^{exp}). The lattice energy computed for a given p, denoted by $U^*_{latt}(p)$, is given by:

$$U^*_{latt}(p) = \min_{X,\beta} U^{latt}(X,\beta;p,\theta^{gas}|X^{exp},\beta^{exp}) \tag{5}$$

where the notation "X^{exp},β^{exp}" denotes that the values X^{exp},β^{exp} are used as the starting point for the local minimization of the lattice energy; the semicolon preceding p and θ^{gas} indicates that these quantities remain fixed throughout the lattice energy minimization. The above minimization problem is solved subject to all appropriate symmetry considerations, and its solution determines the geometry of the predicted structure, denoted by (X^*,β^*).

The quantities U^*_{latt} and (X^*,β^*) form the basis for comparison between predicted and experimental structures, as discussed in the next section.

3 Parameter estimation

An extensive set of experimental crystal property data is required to determine appropriate values for the repulsion–dispersion parameters. More specifically, the properties that can be most directly related to the lattice energy model are the geometries of experimentally-observed crystal structures determined by diffraction methods, and sublimation enthalpies.

3.1 Quantifying deviations in crystal geometry

The structural features predicted by the model can be assessed by comparison to different types of experimental structure data as available in the Cambridge Structural Database (CSD).[47] As experimental X-ray diffraction (XRD) data are subject to relatively small uncertainties, they are valuable for parameter estimation. However, there are well-known difficulties in determining hydrogen positions[39] and in determining the bond lengths of polar or high-order bonds (such as C≡N, C═C, C≡C) due to the fact that interatomic nuclei distances in XRD are defined as distances between electron densities.[48,49] Thermal effects may be another source of error in structural data: typically, unit cell lengths in organic crystals may increase by around 3% over a temperature range of several hundred degrees.[50]

In order to minimize the potential effects of experimental errors, our study applied the following criteria in selecting appropriate experimental crystallographic data from the CSD:

1. Relatively rigid molecules should be chosen.
2. Solvate and hydrate crystals are avoided.
3. Structures at high pressures (above ambient pressure) and above room temperature are rejected.
4. Single-crystal diffraction data are preferable over powder diffraction data.
5. Structures resolved *via* neutron scattering are preferred when available.
6. Structures resolved with XRD with an X-ray discrepancy factor (R) exceeding 5% are normally rejected, unless their sublimation enthalpies are also available or other criteria are met.

Several distance metrics are available for quantifying the deviation between two given crystal structures.[51] In our work, the deviation between experimental and predicted structures for a crystal i is described by a metric $G_i(p)$ that consists of two contributions. The first contribution measures the relative deviations between the experimental and predicted unit cell parameter vectors, X_i^{exp} and X_i^* respectively:

$$\frac{1}{N_{X_i}}\sum_{j=1}^{N_{X_i}}\left(\frac{X_{ij}^{\text{exp}}-X_{ij}^*(p)}{X_{ij}^{\text{exp}}}\right)^2 \tag{6}$$

where N_{X_i} is the number of unit cell parameters not constrained by symmetry, and the jth such parameter in crystal i is indicated by subscript j. The second contribution measures the deviations between atomic positions within the asymmetric unit. Given the fractional coordinates $Y'_{x,i,j}$, $Y'_{y,i,j}$ and $Y'_{z,i,j}$‡ of an atom j in the asymmetric unit of a crystal i, its relative position vector with respect to a reference atom,§ denoted by the subscript "ref", is given by:

$$\delta Y'_{x,i,j}=Y'_{x,i,j}-Y'_{x,i,\text{ref}};\;\delta Y'_{y,i,j}=Y'_{y,i,j}-Y'_{y,i,\text{ref}};\;\delta Y'_{z,i,j}=Y'_{z,i,j}-Y'_{z,i,\text{ref}} \tag{7}$$

‡ Fractional coordinates can be computed directly from the Cartesian coordinates Y and the unit cell parameters X, *i.e.* $Y' = Y'(X,Y)$.

§ The first heavy atom listed in the SHELX file.

The contribution of atomic positions to the deviation measure $G_i(p)$ between experimental and predicted structures for crystal i is then given by:

$$\frac{1}{3\left(N_{\mathrm{asym}_i}-1\right)}\sum_{j=1}^{N_{\mathrm{asym}_i}-1}\left(\delta Y'^{\mathrm{exp}}_{x,i,j}-\delta Y'^{*}_{x,i,j}(p)\right)^2+\left(\delta Y'^{\mathrm{exp}}_{y,i,j}-\delta Y'^{*}_{y,i,j}(p)\right)^2 + \left(\delta Y'^{\mathrm{exp}}_{z,i,j}-\delta Y'^{*}_{z,i,j}(p)\right)^2 \tag{8}$$

where N_{asym_i} is the number of atoms in the asymmetric unit cell for crystal i.

The calculation of the relative position vectors is described in detail in the ESI.† In order to avoid introducing an inherent offset in the deviation metric $G_i(p)$, the fractional atomic coordinates in the experimental crystal are computed based on the predicted (rather than the experimental) molecular conformation. This is appropriate provided that the differences between θ^{gas} and θ^{exp} are small, a condition that needs to be verified when constructing the experimental dataset.

Overall, the function $G_i(p)$ used as a measure of similarity between experimental and predicted structures is defined as:

$$G_i(p)=\frac{1}{N_{X_i}}\sum_{j=1}^{N_{X_i}}\left(\frac{X^{\mathrm{exp}}_{ij}-X^{*}_{ij}(p)}{X^{\mathrm{exp}}_{ij}}\right)^2+\frac{1}{3\left(N_{\mathrm{asym}_i}-1\right)} \times\sum_{j=1}^{N_{\mathrm{asym}_i}-1}\left(\delta Y'^{\mathrm{exp}}_{x,i,j}-\delta Y'^{*}_{x,i,j}(p)\right)^2+\left(\delta Y'^{\mathrm{exp}}_{y,i,j}-\delta Y'^{*}_{y,i,j}(p)\right)^2 + \left(\delta Y'^{\mathrm{exp}}_{z,i,j}-\delta Y'^{*}_{z,i,j}(p)\right)^2 \tag{9}$$

It is worth noting that the above expression aims to balance the contributions of various terms *via* the application of appropriate scaling factors. Thus, relative (rather than absolute) deviations are used for the unit cell parameters. No such scaling is needed for the position vectors as they are already normalized. Moreover, since the two summations may comprise very different numbers of terms, the first one is divided by the number of unit cell parameters not constrained by symmetry, N_{X_i}, and the second one by the number of fractional coordinates describing the asymmetric unit, $3(N_{\mathrm{asymi}}-1)$.

The quantity $G_i(p)$ is directly incorporated in the objective function of the parameter estimation problem. In analysing the quality of the results of the parameter estimation, the root mean squared deviation of the 15 molecule coordination sphere, rmsd_{15},[52] is used as an additional measure of similarity between predicted and experimental crystals.

3.2 Quantifying deviations in sublimation enthalpy

The accuracy of the computed lattice energies can be assessed from sublimation enthalpies reported in thermophysical properties databases such as NIST[53] and DETHERM.[54] Uncertainties in measured sublimation enthalpies can be very large due to several factors such as polymorphism, a lack of standards for compounds with low vapour pressures, chirality, systematic errors associated with the measurement techniques and further errors that can be introduced in measurements by adjustment to a reference temperature (which is necessary when comparing different measurements).[55] Based on 451 measurements of 80 compounds, Chickos *et al.*[55] estimated the average uncertainty of sublimation enthalpies to be 4.9 kJ mol^{-1}. Consequently, our study does not make use of any sublimation enthalpy measurement with a reported error exceeding 4.9 kJ mol^{-1}.

A calculated enthalpy of sublimation at temperature T and pressure P, $\Delta H^*_{\text{sub}}(T,P)$, can be obtained from the predicted lattice energy U^*_{latt} by the following commonly-used approximation:[56–58]

$$\Delta H^*_{\text{sub}}(T,P) \approx -U^*_{\text{latt}}(T=0\ \text{K},P)) - 2RT, \tag{10}$$

in which the zero-point energy is neglected. A derivation of eqn (10) is given in the ESI.†

In previous studies by Arnautova *et al.*[59] and Gavezzotti,[60] computed lattice energies were directly compared to experimental sublimation enthalpies without any correction, an approach justified by the authors on the basis of uncertainty in the sublimation enthalpy data. We prefer to apply the correction shown in eqn (10) in order to reduce any systematic error in the comparison between predicted and experimental values. Thus, we employ the following deviation function as a measure of the difference between the experimental and predicted sublimation enthalpies for a given crystal i:

$$E_i(p) = \left(\frac{\Delta H^{\text{exp}}_{\text{sub},i} - \Delta H^*_{\text{sub},i}(p)}{\Delta H^{\text{exp}}_{\text{sub},i}}\right)^2 \tag{11}$$

3.3 Obtaining globally optimal parameter estimates

Parameter estimation aims to determine the values of parameters p for which the calculated quantities best match the experimentally-observed quantities. This can be expressed mathematically as the minimization of a combined deviation metric $R(p)$:

$$\min_p R(p) \equiv \sum_{i\in I_{\text{G}}} w_i^{\text{G}} G_i(p) + \sum_{i\in I_{\text{E}}} w_i^{\text{E}} E_i(p) \tag{12}$$

where I_{G} and I_{E} are the sets of crystal structures for which we have reliable experimental data on the geometry and sublimation enthalpy respectively. In general, I_{E} is a subset of I_{G}, reflecting the fact that reliable sublimation enthalpy measurements may not be available for all crystal structures under consideration. The constants w_i^{G} and w_i^{E} are non-negative weights that can be adjusted to reflect differences in the reliability between different experimental data and to yield a desirable trade-off between geometry and energy reproduction.

In general, we seek to establish the values of p that lie within given lower and upper bounds, p^{l} and p^{u} respectively, and lead to a globally minimal value of $R(p)$. To compute $R(p)$ at a given set of parameter values p, we need to perform the lattice energy minimization (5) for each crystal structure $i \in I_{\text{G}}$ starting from the corresponding experimental structure. Once the optimal solution is obtained, we can compute the geometry deviation $G_i(p)$ and, if $i \in I_{\text{E}}$, the sublimation enthalpy deviation $E_i(p)$. The quantities $G_i(p)$ and $E_i(p)$ can then be used to compute $R(p)$ *via* eqn (12). Overall, this is a relatively expensive computation as it involves a large number of crystal structure minimizations, but these can be carried out in parallel.

A more serious challenge for the reliable solution of problem (12) is that the objective function $R(p)$ may have multiple local minima, in which case the solution obtained *via* the application of standard local minimization algorithms may depend on the starting points (*i.e.* the initial guesses) for the parameters p. In order to increase the likelihood of identifying the globally optimal parameter vector, we adopt a step-wise approach involving the generation of many starting points in the space of parameters p, followed by a gradient-based local minimization from selected starting points:

1. Compute the reference quantity $R(p^{\mathrm{FIT}})$ where p^{FIT} are the values of the parameters in the existing FIT parameterization.

2. Generate N points p^k, $k = 1, \ldots, N$, in the parameter space by means of a Sobol' sequence[61,62] within the defined bounds $[p^{\mathrm{l}},p^{\mathrm{u}}]$. The advantages of low-discrepancy (such as Sobol') sequences over other methods (*e.g.* Monte Carlo sampling or uniform grids) for efficiently searching multivariable domains are well established.[63]

3. For each point p^k, evaluate the corresponding $R(p^k)$ as described above; discard all points for which $R(p^k)$ exceeds the reference value $R(p^{\mathrm{FIT}})$ by more than a specified factor α (*e.g.* 5) as these are unlikely to yield parameter estimates that are better than the FIT one.

4. Starting from p^k as the initial guess, perform the local minimization (eqn (12)) to obtain a (locally) optimal parameter estimate $p^{*,k}$ with a corresponding objective function value $R(p^{*,k})$. As described in the ESI,† a sequential quadratic programming (SQP) algorithm with a Broyden–Fletcher–Goldfarb–Shanno (BFGS)[64] update of the Hessian matrix is used for this purpose.

5. Choose the solution $p^{*,k}$ with the lowest corresponding value $R(p^{*,k})$ as the best parameter estimate p^*.

A flowchart of the overall parameter estimation methodology is shown in Fig. 1.

3.4 Cross-validation of optimal parameter estimates

To further assess the quality of the optimal parameter estimates p^* obtained by the procedure described in the previous section, we examine its ability to reproduce experimental geometry and sublimation enthalpy data both for the structures used for the estimation, and for a separate set of structures reserved for cross-validation purposes.

The quality of geometry reproduction is assessed based on the usual $\mathrm{rmsd}_{15,i}$ measure between an experimental and predicted crystal i calculated using the *COMPACK*[52] code. We also consider the rmsd_{15} averaged over a set of crystal structures.

For assessing the quality of sublimation enthalpy prediction for each individual crystal i, we use the Absolute Deviation (AD_i) defined as:

$$\mathrm{AD}_i = \left|\Delta H^{\mathrm{exp}}_{\mathrm{sub},i} - \Delta H^{*}_{\mathrm{sub},i}\right| \tag{13}$$

as well as the average and maximum values of AD over a set of N_{str} crystal structures:

$$\mathrm{AAD} = \frac{1}{N_{\mathrm{str}}}\sum_{i=1}^{N_{\mathrm{str}}}\left|\Delta H^{\mathrm{exp}}_{\mathrm{sub},i} - \Delta H^{*}_{\mathrm{sub},i}\right| \tag{14}$$

and

$$\mathrm{maxAD} = \max_{i=1,\ldots,N_{\mathrm{str}}}\left|\Delta H^{\mathrm{exp}}_{\mathrm{sub},i} - \Delta H^{*}_{\mathrm{sub},i}\right| \tag{15}$$

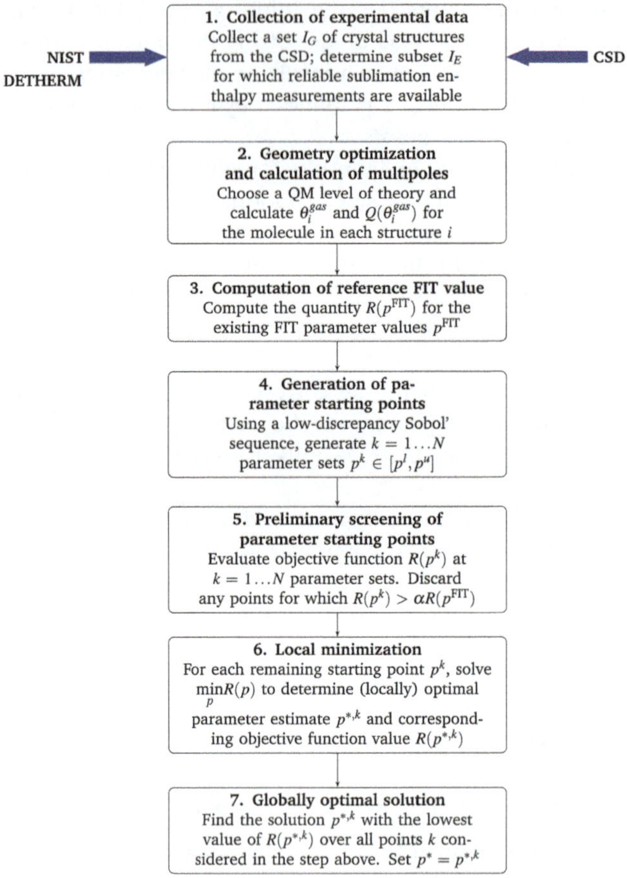

Fig. 1 Flowchart of the methodology for parameter estimation.

We generally aim to derive parameter estimates for which the average values AAD and rmsd_{15} do not exceed 5 kJ mol^{-1} and 1 Å respectively.[22]

4 Parameterization strategy

For a parameter estimation study involving N_A distinct types of atoms, there are $N_A(N_A + 1)/2$ possible interaction pairs that need to be considered, and consequently $3N_A(N_A + 1)/2$ repulsion–dispersion parameters that need to be estimated from experimental data. Overall, even for relatively small N_A, this is a difficult and computationally expensive problem to solve.

An approach commonly used for reducing the parameter space is to use combining rules for the parameters describing interactions between atoms of different types. The use of such combining rules for the dispersion coefficient does have some physical justification at the level of the Unsöld approximation to the r_{ij}^{-6} dispersion. However, other rules are known to be limited in their accuracy,[65] and their use may lead to lower quality of predictions in crystal structure calculations.[32,42,66] Therefore, in this study we prefer not to use combining rules, relying instead on the estimation of the cross-interaction parameters directly from experimental data.

On the other hand, and in common with the approach adopted by Pyzer-Knapp *et al.*,[42] we do hold the exponential B_{ij} parameters fixed at the FIT values: the shape of the potential is very sensitive to even small changes in these parameters, which renders their numerical handling as part of the parameter estimation procedure problematic.

Finally, in order to make the problem computationally manageable and ensure compatibility and transferability of the different sets of parameters, the parameter estimation is carried out in a sequential manner as shown schematically in Fig. 2. Similarly to the work of Sun,[67] optimal parameters for carbon and hydrogen (H_c) are determined first based on experimental data for hydrocarbons. These are then fixed for the subsequent steps, and interaction parameters for other atom pairs are estimated from data on appropriate compounds.

5 Illustration of methodology

The methodology proposed in this paper is illustrated for the estimation of parameters for the C···C, C···H_C and H_C···H_C interactions using experimental data for hydrocarbon crystals (*cf.* the top box in Fig. 2). As indicated in the flowchart in Fig. 1, we start by collecting relevant experimental data based on the general criteria described in Section 3.1. The experimental crystallographic data and energetic data collected for hydrocarbons are listed in Table S1 of the ESI.† This consists of 19 compounds and exhibits a diversity of hybridizations not present in the training sets for the FIT parameterization.

We then perform an isolated-molecule gas-phase QM calculation for each molecule using *Gaussian09* (ref. 68) with the M06 (ref. 69) functional and a 6-31G(d,p) basis set. The resulting wavefunctions are used to generate multipole moments *via* the distributed multipole analysis method[43] as implemented in the *GDMA 2.2* program.[43]

We now fix the three B_{ij} parameters to their FIT values, and aim to determine optimal estimates for the 6 A_{ij} and C_{ij} parameters; the latter constitute our parameter vector p. For the purposes of this estimation, we will allow p to vary within ±40% of the FIT values. As explained in Section 3.3, in order to increase

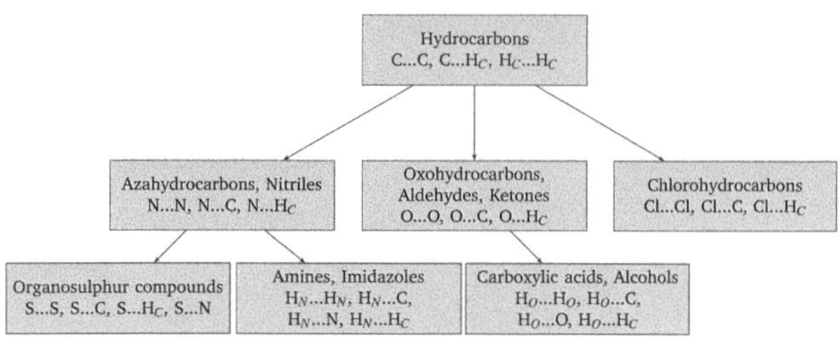

Fig. 2 Parameterization tree. Each box indicates the type of molecules for which experimental crystal structures are used for the estimation (first line) and the interaction parameters being estimated from these data (second line). The values of the parameters estimated in each box are thereafter held constant in the estimations corresponding to the boxes below, as indicated by the direction of the arrows.

the probability of obtaining globally optimal parameter estimates, we use a Sobol' sequence to generate 1000 starting points p^k, $k = 1, \ldots, 1000$.

Before proceeding with actually performing local minimizations of function $R(p)$ (*cf.* eqn (12)) starting from some of the above points, it is instructive to gain some insight as to the appropriate choices of the weighting factors w_i^G and w_i^E. For simplicity, we use the same weight values (w^G or w^E) for all 19 structures i. We consider 15 distinct combinations of these weights as shown in Table 1, and evaluate the objective function $R(p^k)$ for each such combination. The 19 000 local lattice energy minimizations required for evaluating $R(p^k)$, $k = 1, \ldots, 1000$ are performed with *DMACRYS* at 0 Pa pressure.

For each weight combination, having evaluated the objective function $R(p^k)$ at the 1000 Sobol' points p^k, we select the Sobol' point with the lowest value of the objective function. This serves as an approximation to the optimal parameter estimate p^* that might be obtained using that particular combination of weights. The solutions arising from different weight combinations are compared in Fig. 3 based on the geometry and sublimation deviation functions, $G(p^*)$ and $E(p^*)$, averaged over the 19 crystal structures under consideration.

The maximum and minimum values of $G_i(p^*)$ across all 19 crystal structures i are also depicted on the horizontal axis of Fig. 3 while the maximum and minimum $E_i(p^*)$ are shown against the vertical axis. We note that several weight ratios lead to the same solution; in fact, only 4 distinct solutions are obtained, as depicted by the solid symbols in Fig. 3. Based on these results, the weights $w^G = 1$ and $w^E = 1$ appear to offer both the best trade-off between geometry and energy errors, and the smallest variation of these errors across the different structures being considered. Accordingly, these values are chosen to be used throughout our parameter estimation methodology in this paper.

The values of $R(p^k)$ for the weights $w^G = w^E = 1$ at the 1000 points p^k are plotted in Fig. 4. The corresponding reference value $R(p^{FIT})$ for the FIT parameter values p^{FIT} is also shown. We note that the FIT parameter point, p^{FIT}, lies among the lowest points in this figure, which suggests that the FIT parameter values are

Table 1 Combinations of objective function weights considered

w^G	w^E
100	1
50	1
20	1
15	1
10	1
5	1
2	1
1	1
1	2
1	5
1	10
1	15
1	20
1	50
1	100

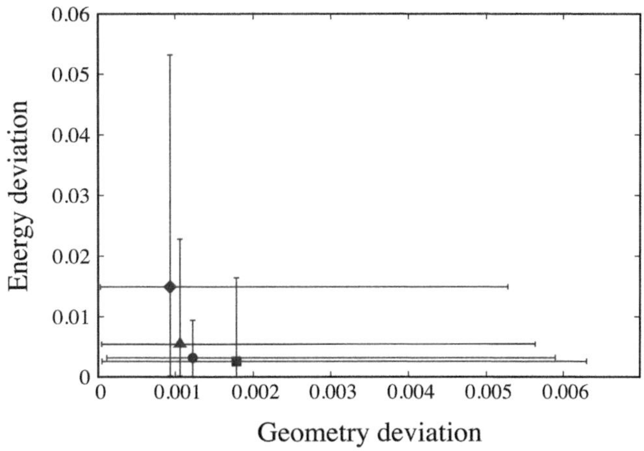

Fig. 3 Geometry and energy deviation functions, $G(p^*)$ and $E(p^*)$, at (approximately) optimal points p^* obtained using different weight ratios $w^G : w^E$. Solid points indicate $G_i(p^*)$ and $E_i(p^*)$ values averaged across all 19 structures i. There are 4 distinct such points: diamond (◆): weight ratio 100 : 1; triangle (▲): weight ratios 50 : 1, 20 : 1, 15 : 1; circle (●): weight ratios 10 : 1, 5 : 1, 2 : 1, 1 : 1; and square (■): weight ratios 1 : 2, 1 : 5, 1 : 10, 1 : 15, 1 : 20, 1 : 50, 1 : 100. The horizontal and vertical line segments through each point indicate the range of values of, respectively, $G_i(p)$ and $E_i(p)$ across all 19 structures.

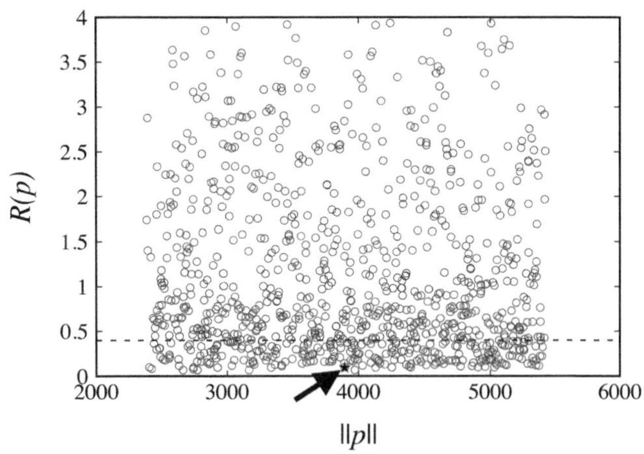

Fig. 4 Distribution of objective function values, $R(p^k)$, obtained with weights $w^G = w^E = 1$, at starting parameter points p^k, $k = 1, \ldots, 1000$ (denoted by open circles, ○). Symbol ★, highlighted with the arrow, indicates the corresponding objective function value $R(p^{FIT})$ evaluated at the FIT parameter values. The dashed line signifies the cutoff applied for the subsequent local minimization step.

already good estimates, a result that was expected given the similarity between the training set used in this work for hydrocarbons and that used by Williams *et al.*[31]

The FIT reference value $R(p^{FIT})$ allows us to prune out any Sobol' points that are unlikely to result in an improved fit of the experimental data over the already available FIT parameters (*cf.* step 5 of the algorithm in Fig. 1). Here we apply a cutoff ratio α of approximately 5, which excludes all but 250 of the original

Sobol' points from further consideration. We then solve the optimization problem expressed by eqn (12) starting from each of the remaining 250 points. The initial objective function values $R(p^k)$ and the final objective function values after local minimization, $R(p^{*,k})$, are shown in Fig. 5. It can be seen that the minimization does result in a significant reduction of the objective function $R(p)$. The results suggest a "flat" objective function with many broad and shallow minima; however, it is also possible that some of the points are not true local minima, but simply arise due to premature termination of the challenging local minimization.

The parameter values p^* resulting in the lowest objective function value are given in Table 2 where they are also compared with the original FIT parameters. A comparison of the corresponding objective function values is presented in Table 3. The objective function value for p^* is 63% lower than the corresponding value for the FIT parameters. The new parameters also achieve 23% and 72% reductions in the geometry and energy deviation functions respectively. Overall, the new parameters yield a significantly better reproduction of both the geometries and energies of the crystals in the training set even in the case of such

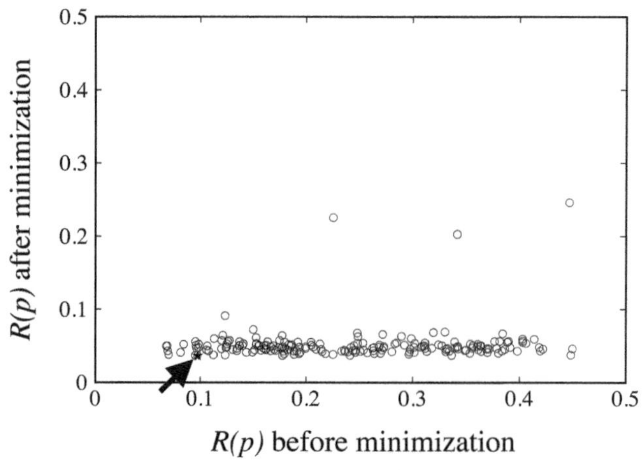

Fig. 5 Local minimization results for the 250 initial Sobol' points. Objective function value after minimization *vs.* objective function value before local minimization (denoted by open circles, ∘). The best solution is marked with a star (★).

Table 2 Exp-6 potential parameters for hydrocarbons obtained as a result of the refinement

Atom type		FIT[a]			This work		
i	j	A_{ij}/eV	B_{ij}/Å	C_{ij}/eV Å^6	A_{ij}/eV	B_{ij}/Å	C_{ij}/eV Å^6
C	C	3832.147	0.278	25.287	2299.288	0.278	18.020
C	H_C	689.537	0.272	5.979	449.3172	0.272	6.265
H_C	H_C	124.072	0.267	1.414	173.559	0.267	1.957

[a] Parameter values taken from the CPOSS website.[37]

Table 3 Comparison of objective function values between p^{FIT} and p^* for the hydrocarbon test set. The geometry and energy deviation functions averaged over the 19 structures under consideration are also compared

Parameter set	$R(p)$	$\frac{1}{N_{str}}\sum_{i=1}^{N_{str}} G_i(p)$	$\frac{1}{N_{str}}\sum_{i=1}^{N_{str}} E_i(p)$
FIT p^{FIT}	0.098	0.036	0.063
Best solution p^*	0.037	0.028	0.010

a relatively simple and well-studied parameter set. Further details of the quality of fit for each individual structure are presented in Table S4 of the ESI.†

6 Results and discussion

In this section, the stagewise procedure depicted in Fig. 2 is applied to derive optimal estimates of the repulsion–dispersion interaction parameters for a range of atoms. The parameters obtained are consistent with a distributed multipole electrostatic model derived from charge densities calculated at the M06/6-31G(d,p) level of theory.

6.1 Choice of training set and cross-validation set

The crystal structures of 106 organic molecular crystals including hydrocarbons, azahydrocarbons, oxohydrocarbons, organosulphur compounds, chlorohydrocarbons, amines, imidazoles, carboxylic acids and alcohols were collected from the CSD[47] for the purpose of the parameter estimation reported in this work. The training set was selected to contain molecules with each element in different bonding situations. For example, the training set of hydrocarbons contains alkanes and cycloalkanes with sp^3 hybridized carbons that form single bonds, aromatics with sp^2 hybridized carbons forming sigma bonds and the delocalization of electrons between the ring carbons and hydrocarbons with an alkynyl group of sp hybridized carbons that form a triple bond. Similar considerations underpinned the selection of molecules containing N, O and S, as summarized in Table 4. The molecular diagrams, CSD refcodes, numbers of molecules in the unit cell Z, space groups, R-factors and experimental temperatures are summarized in Table S1 of the ESI.†

Gas-phase molecular conformations are computed at the M06/6-31G(d,p) level of theory. The rigidity of the molecules is ascertained through $rmsd_1$ comparisons between computed and experimental conformations with *COMPACK*, giving an average $rmsd_1$ of only 0.05 Å. In a previous study by Arnautova *et al.*,[59] the C–H bond lengths of the experimental conformations were adjusted to the average experimental value from neutron diffraction data to overcome issues of unreliable hydrogen positions determined by X-ray diffraction. We found the C–H bond lengths in the gas-phase conformations to be approximately 1.087 Å, which is very close to the average experimental value of 1.089 Å derived from neutron data given in the CSD. Therefore, no modification was applied to the C–H bond lengths.

Experimentally determined enthalpies of sublimation were identified for 53 of the 106 molecules, which is significantly more than the number used in the

Table 4 Atom types and classes of compounds used for the estimation of the new potential parameters

Atom type	Bonding structures represented in dataset	Compounds classes, functional groups	Representative structures
C	Carbon bonded to four other atoms	Alkanes	$R{-}(CH_2)_nH$
	Carbon bonded to three other atoms	Alkenes	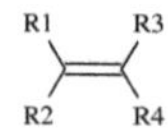
		Benzene derivatives	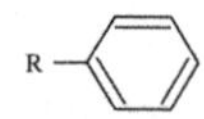
	Carbon bonded to two other atoms	Alkynes	R–C≡C–R′
H_C	Hydrogen bonded to carbon		
N	Nitrogen to three other atoms (no bonded hydrogen)	Amines	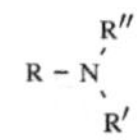
	Nitrogen to two other atoms (no bonded hydrogen)	Imines	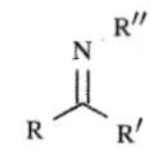
		Azo compounds	R′–N=N–R′
		Pyridine derivatives	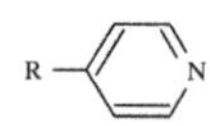
	Nitrogen with triple bond (no bonded hydrogen)	Nitriles	R–C≡N
O	Oxygen bonded to one other atom (no bonded hydrogen)	Ketones	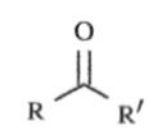
		Aldehydes	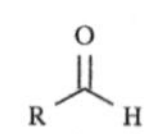
	Oxygen bonded to two other atoms (no bonded hydrogen)	Ethers	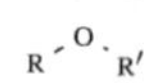
		Esters	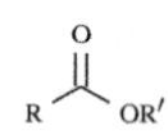
S	Sulphur bonded to one other atom (no bonded hydrogen)	Thiones	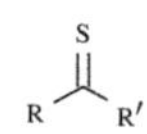
		Thial	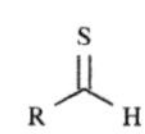
	Sulphur bonded to two other atoms (no bonded hydrogen)	Sulfides	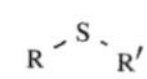
		Disulfides, polysulfides	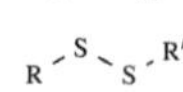
H_N	Hydrogen bonded to nitrogen	Amines	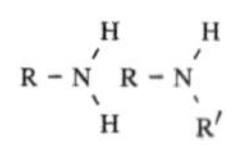
		Imines	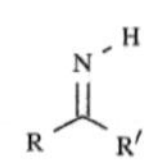
H_O	Hydrogen bonded to oxygen	Alcohols	R–OH
		Carboxylic acids	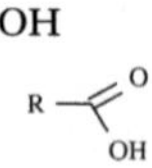
Cl	Chlorine bonded to carbon	Chloroalkanes	R–Cl
		Chlorinated aromatics	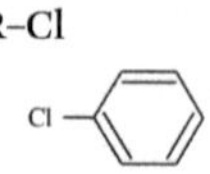

parameterization of FIT. The sublimation enthalpies used in this work are given in Table S1 of the ESI.†

6.2 Optimized parameters

The optimal parameter estimates for the 24 atom–atom interactions occurring in the datasets are listed in Table 5, where they are also compared with the corresponding FIT parameters. The optimized parameters for sulphur interactions are further compared with two additional sets of parameters, namely those by No *et al.*[70] (denoted as S-No) and by Abraha *et al.*[71] (denoted as S-Ab). These alternative parameter sets are presented in Table 6.

The optimized parameters differ significantly from the original set. However, because of the high degree of correlation between the A_{ij} and C_{ij} parameters, large changes in their individual values do not necessarily result in a large change in the potential. Fig. 6 shows the differences in equilibrium distances and well depths of the potentials. For most pairwise interactions, these differences are relatively small. A striking exception is the O···O interaction for which we observe a significant change in the potential well depth and location of the minimum (*cf.* Fig. 7a), with the new parameters indicating stronger and shorter-range interactions. On the other hand, the differences between our potentials for the interactions of oxygen with other atoms and the corresponding FIT potentials are not very pronounced. This is

Table 5 Exp-6 potential parameters obtained in the FIT parameterization and in this work

Atom type		FIT[a]			This work		
i	j	A_{ij}/eV	B_{ij}/Å	C_{ij}/eV Å^6	A_{ij}/eV	B_{ij}/Å	C_{ij}/eV Å^6
C	C	3832.147	0.278	25.287	2299.288	0.278	18.020
C	H_C	689.537	0.272	5.979	449.3172	0.272	6.265
H_C	H_C	124.072	0.267	1.414	173.559	0.267	1.957
C	N	3179.515	0.271	19.007	3355.322	0.271	24.171
N	N	2638.029	0.265	14.286	3687.092	0.265	17.577
H_C	N	572.105	0.266	4.494	516.587	0.266	2.696
C	O	3022.85	0.265	17.16	1352.853	0.265	10.164
H_C	O	543.916	0.260	4.057	304.084	0.260	2.481
O	O	2384.466	0.253	11.645	963.846	0.253	18.632
C	S	3990.989	0.290	38.957	2819.374	0.290	46.499
N	S	3311.305	0.282	29.281	3760.816	0.282	31.148
H_C	S	718.118	0.284	9.211	802.131	0.284	9.891
S	S	4156.415	0.303	60.016	3401.359	0.303	95.872
C	H_N	446.952	0.242	2.374	478.038	0.242	3.561
H_C	H_N	80.422	0.238	0.561	40.211	0.238	0.842
N	H_N	370.834	0.237	1.784	297.752	0.237	0.892
H_N	H_N	52.129	0.215	0.223	78.194	0.215	0.111
H_O	C	446.952	0.242	2.374	519.271	0.242	2.488
H_O	H_C	80.422	0.238	0.561	128.676	0.238	0.224
H_O	O	352.562	0.232	1.611	231.169	0.232	0.644
H_O	H_O	52.129	0.215	0.223	61.708	0.215	0.089
C	Cl	6060.173	0.281	45.040	2627.877	0.290	39.868
H_C	Cl	1090.436	0.276	10.650	1371.320	0.276	17.708
Cl	Cl	9583.584	0.285	80.224	7940.587	0.285	67.687

[a] Parameters from (ref. 37).

Table 6 Exp-6 potential parameters for S-Ab and S-No

Atom type		S-Ab			S-No		
i	j	A_{ij}/eV	B_{ij}/Å	C_{ij}/eV Å^6	A_{ij}/eV	B_{ij}/Å	C_{ij}/eV Å^6
C	S	2623.650	0.308	55.295	3064.776	0.308	46.911
N	S	2176.830	0.299	41.562	2542.830	0.299	35.260
H_C	S	472.092	0.301	13.074	551.467	0.301	11.092
S	S	1796.306	0.345	120.918	2451.128	0.345	87.029

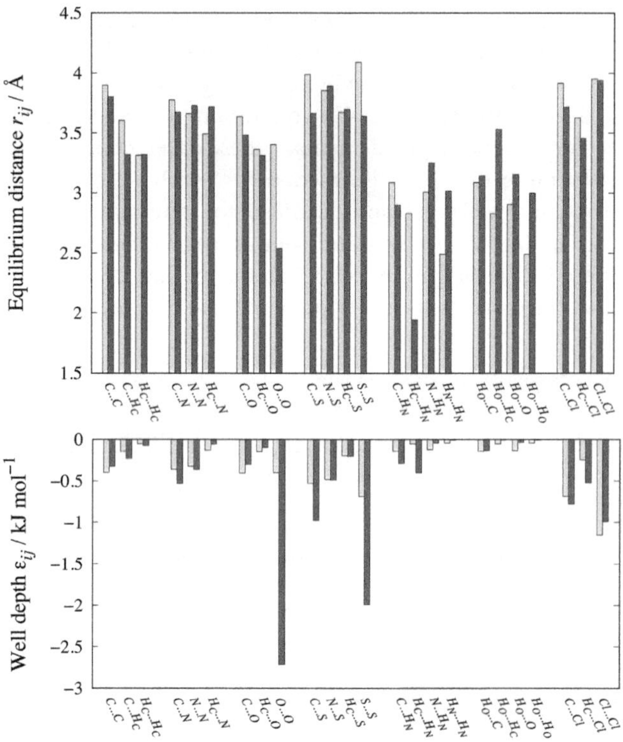

Fig. 6 Comparison of equilibrium distances and well depths for 24 interactions as calculated with FIT (light grey bars) and our new parameter set (dark grey bars). The data along with the results for the S-Ab and S-No parameter sets are given in Table S3 of the ESI.†

an indication of the inaccuracies that may be introduced by relying on combining rules for estimating cross-interaction parameters.

Shorter equilibrium distances and deeper wells are also observed for all S interactions apart from the $H_C \cdots S$ one (*cf.* Fig. 6 and 7b and c). We note that significant differences also exist among the FIT, S-Ab and S-No parameter sets as shown in Fig. S3 of the ESI.†

Both oxygen interactions and sulphur interactions were fitted to a number of sublimation enthalpy data. We note that no such data were used in the derivation of the FIT set. The resulting improvement in sublimation enthalpy prediction demonstrates the impact of using energetic data to fit the potentials.

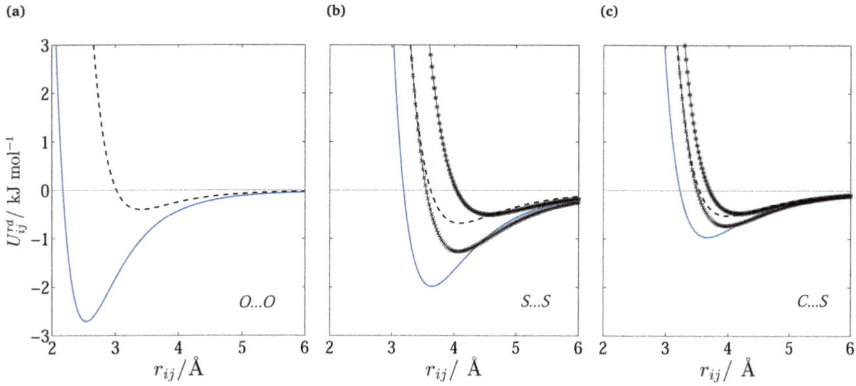

Fig. 7 Repulsion–dispersion potential curves for (a) O···O, (b) S···S and (c) C···S interactions. The optimized potential curves (continuous curves) are compared with the FIT potential curves (dashed curves) and with S-No[70] (✱) and S-Ab[71] (×).

6.3 Performance of new parameters: sublimation enthalpy

Overall, the new parameter values result in a good fit for the experimental structures in the training set. As shown in Table 7, the sublimation enthalpy AAD is

Table 7 Comparison of accuracy of computed sublimation enthalpies and geometries with different parameter values over the training set

	Absolute sublimation enthalpy error (kJ mol^{-1})							
	FIT		This work		S-Ab		S-No	
Training set	AAD	maxAD	AAD	maxAD	AAD	maxAD	AAD	maxAD
Hydrocarbons	2.66	4.83	1.47	4.08				
Azahydrocarbons	3.43	6.80	1.28	3.56				
Oxohydrocarbons	9.46	18.22	6.88	16.66				
Organosulphur compounds	20.97	38.71	5.29	10.44	7.94	16.74	25.73	46.85
Amines, imidazoles	11.00	19.40	9.17	13.67				
Carboxylic acids, alcohols	13.98	24.93	5.94	10.50				
Chlorohydrocarbons	7.30	8.90	0.15	0.46				
Overall	8.78	38.71	4.11	16.66				
	Absolute rmsd$_{15}$ error (Å)							
	FIT		This work		S-Ab		S-No	
	Average	Max	Average	Max	Average	Max	Average	Max
Hydrocarbons	0.349	0.922	0.285	0.985				
Azahydrocarbons	0.223	0.747	0.231	0.631				
Oxohydrocarbons	0.342	0.728	0.306	0.808				
Organosulphur compounds	0.321	1.047	0.326	1.047	0.298	0.996	0.430	1.195
Amines, imidazoles	0.289	0.581	0.291	0.640				
Carboxylic acids, alcohols	0.336	0.868	0.434	0.875				
Chlorohydrocarbons	0.246	0.529	0.263	0.735				
Overall	0.305	1.047	0.309	1.047				

4.1 kJ mol^{-1}, which is 53% smaller than the corresponding FIT value while the average rmsd$_{15}$ of 0.31 Å is practically identical to that achieved by FIT. In fact, for hydrocarbons and azahydrocarbons, the achieved sublimation enthalpy AADs of less than 2 kJ mol^{-1} are similar to the errors observed with some dispersion-corrected DFT methods.[7] There is also a decrease in the spread of the sublimation enthalpy errors, with a maxAD value of 16.7 kJ mol^{-1} compared to 38.7 kJ mol^{-1} with FIT.

In addition to the training set, we employ a cross-validation set consisting of 39 crystal structures for representative molecules for all the classes of compounds considered here (see Table S2 in the ESI†). For six of these structures, the sublimation enthalpy is also available. The criteria for the choice of the cross-validation set were similar to those for the training set (*cf.* Section 3.1) although, due to the limited availability of such data for rigid molecules, the requirements regarding experimental error were less stringent in some cases. The sublimation enthalpy results reported in Table 8 indicate similar improvements to those for the training set (*cf.* Table 7). However, the sublimation enthalpy AAD of 7.87 kJ mol^{-1} is larger than the corresponding value for the training set. This may be due to the relative sparsity of energetic data, which leads to reduced statistical significance for the parameters. We also note that some of the experimental sublimation enthalpy values in the cross-validation set are less reliable than the enthalpies reported for the compounds in the training set.

Fig. 8a shows parity plots between experimental and computed sublimation enthalpies for both the training and cross-validation sets. A significant improvement over the FIT set is evident across both sets. Fig. 8b focuses on the

Table 8 Comparison of accuracy of computed sublimation enthalpies and geometries for different parameter values over the cross-validation set

	Absolute sublimation enthalpy error (kJ mol^{-1})							
	FIT		This work		S-Ab		S-No	
Training set	AAD	maxAD	AAD	maxAD	AAD	maxAD	AAD	maxAD
Hydrocarbons	8.65	13.49	8.31	16.82				
Azahydrocarbons	9.51	9.51	7.49	7.49				
Amines, imidazoles	11.45	11.45	4.91	4.91				
Carboxylic acids, alcohols	23.26	23.26	9.87	9.87				
Overall	11.69	23.26	7.87	16.82				
	Absolute rmsd$_{15}$ error (Å)							
	FIT		This work		S-Ab		S-No	
	Average	Max	Average	Max	Average	Max	Average	Max
Hydrocarbons	0.201	0.286	0.169	0.285				
Azahydrocarbons	0.402	0.747	0.317	0.433				
Oxohydrocarbons	0.147	0.188	0.197	0.289				
Organosulphur compounds	0.239	0.340	0.275	0.499	0.221	0.390	0.276	0.450
Amines, imidazoles	0.327	0.574	0.275	0.377				
Carboxylic acids, alcohols	0.526	1.828	0.579	1.537				
Chlorohydrocarbons	0.246	0.316	0.235	0.311				
Overall	0.294	1.828	0.282	1.537				

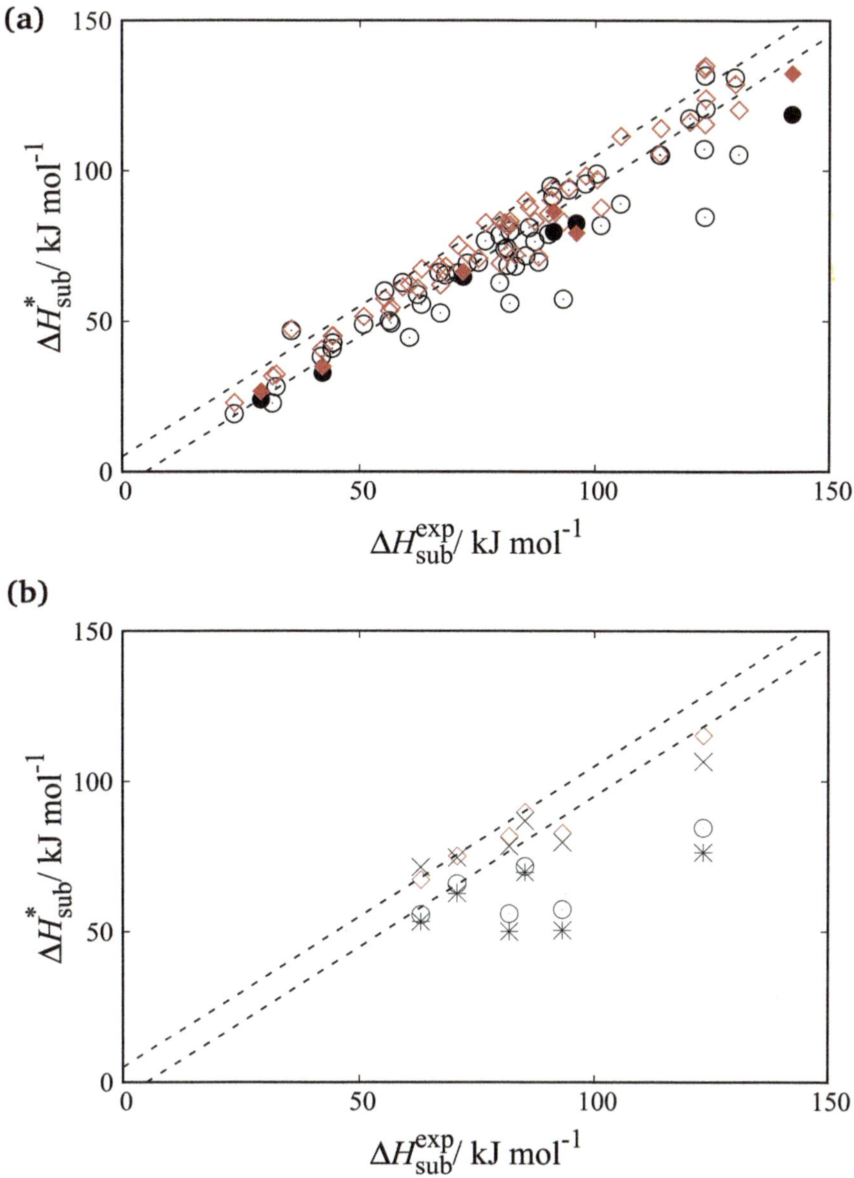

Fig. 8 Parity plots of experimental and predicted sublimation enthalpies with different parameter sets. The diagonal dashed lines (- - -) indicate experimental error bounds. (a) Training and cross-validation sets (empty and filled symbols respectively). Circles (•) and diamonds (◇) indicate values calculated with FIT parameters and the new parameters respectively. (b) Sublimation enthalpies for organosulphur compounds. FIT parameters (○), S-Ab (×), S-No (✱) and this work (◇).

organosulphur compounds. It is clear that the accuracy of sublimation enthalpy predictions varies substantially across the previously derived parameter sets (FIT, S-No, and S-Ab). The improved predictive accuracy of the new parameters is important as the study of molecular crystals containing sulphur has so far been hampered by the limited availability of reliable transferable parameters.

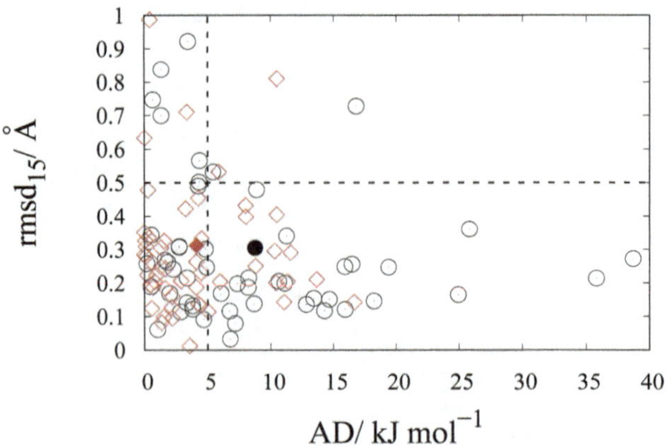

Fig. 9 Rmsd$_{15}$ relative to the corresponding absolute error in sublimation enthalpy, AD, for all the compounds in the training set calculated for FIT (○) and this work (◇). Filled symbols (● and ◆) represent the corresponding averages. The dashed lines (- - -) indicate the acceptable maximum errors in sublimation enthalpy (5 kJ mol^{-1}) and rmsd$_{15}$ (0.5 Å).

6.4 Performance of new parameters: rmsd$_{15}$

As mentioned in previous sections, the rmsd$_{15}$ between the optimized unit cells and the experimentally observed ones is also used to assess the ability of our optimal parameter estimates to accurately predict unit cell geometries. An rmsd$_{15}$ below 0.5 Å is generally considered to indicate a good degree of agreement between two structures. In our case, the average rmsd$_{15}$ over the entire training set was 0.31 Å. The rmsd$_{15}$ averages for each separate set, shown in Table 7, are below 0.5 Å. Overall, we can conclude that there is a good agreement with the experimental structures for each class although a few individual structures show rmsd$_{15}$ values nearer 1.

As indicated by the entries in the last rows of Tables 7 and 8, our new parameter set does not have a clear overall advantage over FIT in terms of rmsd$_{15}$. This is partly explained by the fact that we used a different metric for geometry in our objective function. In our methodology, improved geometry reproduction could be achieved by setting $w^{G} \gg w^{E}$), thereby placing greater importance on the geometry terms than the energy terms. However, the results presented here demonstrate that using equal weights leads to a significant increase in the accuracy of the computed sublimation enthalpies while maintaining good accuracy in geometry reproduction. In particular, as shown in Fig. 9, the new parameters lead to fewer points showing large errors in the sublimation enthalpy (>5 kJ mol^{-1}) or/and in the rmsd$_{15}$ (>0.5 Å).

7 Concluding remarks

The fundamental premise of the work reported in this paper is that the values of the repulsion–dispersion parameters estimated from experimental data depend on the model of electrostatic interactions¶ used in this estimation. Accordingly,

¶ And, in the case of flexible molecules, the model used for intramolecular energy contributions.

this electrostatic model needs to be the same as that used in the subsequent CSP calculations. The approach proposed here allows the systematic reparameterization of the repulsion–dispersion model to derive transferable parameters that are consistent with the rest of the lattice energy model. This is particularly important in the context of current CSP methodologies which are relying on increasingly sophisticated representations of electrostatic and intramolecular contributions based on *ab initio* quantum mechanical calculations.

The estimation of the repulsion–dispersion parameters in the lattice energy model has been formulated mathematically as a weighted least-squares minimization of the deviation between the computed and experimentally measured crystal geometries and sublimation enthalpies. Important aspects of our formulation are the weighting and scaling schemes used for balancing the contributions of different elements of the experimental dataset. The proposed solution algorithm aims to mitigate the large computational cost associated with this problem and to reduce the risk of the minimization being trapped in local minima.

The feasibility and effectiveness of the proposed approach were demonstrated by the estimation of a set of atom–atom interaction parameters for the isotropic Buckingham potential, consistent with distributed multipole electrostatics derived from charge densities computed *via* isolated-molecule quantum mechanical calculations at the M06/6-31G(d,p) level of theory.‖ Interaction parameters between unlike atom types were also included directly in the estimation instead of being approximated *via* combining rules. The study reported here focused on organic molecular crystals with N, O, S and Cl atoms, and made use of a training dataset containing 106 crystal geometries and the sublimation enthalpies for 53 of these structures. A separate set of 39 structures (including 6 sublimation enthalpy values) was used for cross-validation of the derived parameter estimates. Overall, compared to the commonly used FIT parameter set, the new parameter values were found to result in significantly improved sublimation enthalpy predictions while maintaining a comparable quality of geometry reproduction.

In the work reported here, the B_{ij} parameters of the Buckingham potential were kept fixed at the corresponding FIT values. This was mainly done in order to avoid the numerical problems caused by the unphysical predictions of this potential at low interatomic separation distances. Including these parameters in the estimation set could result in improved predictive accuracy. However, with the availability of systematic ways of estimating reliable and consistent parameter values for any form of repulsion–dispersion potential, it may be worth considering alternative potential forms that do not exhibit such unphysical behavior in the first place.

Data statement

All underlying data supporting this article are available in the ESI.†

Conflicts of interest

There are no conflicts to declare.

‖ Parameter values consistent with QM calculations at the HF/6-31G(d,p) and MP2/6-31G(d,p) levels of theory are reported elsewhere.[72]

Acknowledgements

The authors gratefully acknowledge financial support from the Engineering and Physical Sciences Research Council (EPSRC) grants EP/J014958/1 and EP/J003840/1 and access to computational resources and support from the High Performance Computing Cluster at Imperial College London. We are grateful to Professor S. L. Price for supplying the DMACRYS code.

Notes and references

1 A. M. Reilly, A. M. Cooper, C. S. Adjiman, S. Bhattacharya, D. Boese, J. G. Brandenburg, P. J. Bygrave, R. Bylsma, J. E. Campbell, R. Car, D. H. Case, R. Gadha, J. C. Cole, K. Cosburn, H. M. Cuppen, F. Curtis, G. M. Day, R. A. Distasio Jr, A. Dzyabchenko, B. P. van Eijck, D. M. Elking, J. A. van den Ende, J. C. Facelli, M. B. Ferraro, L. Fusti-Molnar, C. A. Gatsiou, T. S. Gee, R. de Gelder, L. M. Ghiringhelli, H. Goto, S. Grimme, R. Guo, D. W. M. Hofmann, J. Hoja, R. K. Hylton, L. Luzzolino, W. Jankiewicz, D. T. de Jong, J. Kendrick, N. J. J. de Klerk, H. Y. Ko, L. N. Kuleshova, X. Li, S. Lohani, F. J. J. Leusen, A. M. Lund, J. Lv, Y. Ma, N. Marom, A. E. Masunov, P. McCabe, D. P. McMahon, H. Meekes, M. P. Metz, A. J. Misquitta, S. Mohamed, B. Monserrat, R. J. Needs, M. A. Neumann, J. Nyman, S. Obata, H. Oberhofer, A. R. Oganov, A. M. Orendt, G. I. Pagola, C. C. Pantelides, C. J. Pickard, R. Podeszwa, L. S. Price, S. L. Price, A. Pulido, M. Read, K. Reuter, E. Schneider, C. Schober, G. P. Shields, P. Singh, I. J. Sugden, K. Szalewicz, C. R. Taylor, A. Tkatchenko, M. E. Tuckerman, F. Vacarro, M. Vasileiadis, Á. Vásquez-Mayagoittia, L. Vogt, Y. Wang, R. E. Watson, G. A. de Wijs, J. Yang, Q. Zhu and C. R. Groom, *Acta Crystallogr., Sect. B: Struct. Sci., Cryst. Eng. Mater.*, 2016, **72**, 439–459.

2 M. Vasileiadis, A. V. Kazantsev, P. G. Karamertzanis, C. S. Adjiman and C. C. Pantelides, *Acta Crystallogr., Sect. B: Struct. Sci.*, 2012, **68**, 677–685.

3 A. T. Hulme and S. L. Price, *J. Chem. Theory Comput.*, 2007, **3**, 1597–1608.

4 S. Grimme, J. Antony, S. Ehrlich and H. Krieg, *J. Chem. Phys.*, 2010, **132**, 154104.

5 M. A. Neumann and M. A. Perrin, *J. Phys. Chem. B*, 2005, **109**, 15531–15541.

6 A. Tkatchenko and M. Scheffler, *Phys. Rev. Lett.*, 2009, **102**, 073005.

7 A. Otero-De-La-Roza and E. R. Johnson, *J. Chem. Phys.*, 2012, **137**, 054103.

8 J. Moellmann and S. Grimme, *J. Phys. Chem. C*, 2014, **118**, 7615–7621.

9 H. Sun, *J. Phys. Chem. B*, 1998, **102**, 7338–7364.

10 P. Dauber-Osguthorpe, V. A. Roberts, D. J. Osguthorpe, J. Wolff, M. Genest and A. T. Hagler, *Proteins: Struct., Funct., Bioinf.*, 1988, **4**, 31–47.

11 G. M. Day, *Crystallogr. Rev.*, 2011, **17**, 3–52.

12 G. M. Day, W. D. S. Motherwell and W. Jones, *Phys. Chem. Chem. Phys.*, 2007, **9**, 1693–1704.

13 M. A. Neumann, *J. Phys. Chem. B*, 2008, **112**, 9810–9829.

14 M. Jeziorska, W. Cencek, K. Patkowski, B. Jeziorski and K. Szalewicz, *J. Chem. Phys.*, 2007, **127**, 124303.

15 S. L. Price, M. Leslie, G. W. A. Welch, M. Habgood, L. S. Price, P. G. Karamertzanis and G. M. Day, *Phys. Chem. Chem. Phys.*, 2010, **12**, 8478–8490.

16 M. Habgood, I. J. Sugden, A. V. Kazantsev, C. S. Adjiman and C. C. Pantelides, *J. Chem. Theory Comput.*, 2015, **11**, 1957–1969.

17 I. Sugden, C. S. Adjiman and C. C. Pantelides, *Acta Crystallogr., Sect. B: Struct. Sci., Cryst. Eng. Mater.*, 2016, **72**, 864–874.

18 J. P. M. Lommerse, W. D. S. Motherwell, H. L. Ammon, J. D. Dunitz, A. Gavezzotti, D. W. M. Hofmann, F. J. J. Leusen, W. T. M. Mooij, S. L. Price, B. Schweizer, M. U. Schmidt, B. P. van Eijck, P. Verwer and D. E. Williams, *Acta Crystallogr., Sect. B: Struct. Sci.*, 2000, **56**, 697–714.

19 W. D. S. Motherwell, H. L. Ammon, J. D. Dunitz, A. Dzyabchenko, P. Erk, A. Gavezzotti, D. W. M. Hofmann, F. J. J. Leusen, J. P. M. Lommerse, W. T. M. Mooij, S. L. Price, H. Scheraga, B. Schweizer, M. U. Schmidt, B. P. van Eijck, P. Verwer and D. E. Williams, *Acta Crystallogr., Sect. B: Struct. Sci.*, 2002, **58**, 647–661.

20 G. M. Day, W. D. S. Motherwell, H. L. Ammon, S. X. M. Boerrigter, R. G. Della Valle, E. Venuti, A. Dzyabchenko, J. D. Dunitz, B. Schweizer, B. P. van Eijck, P. Erk, J. C. Facelli, V. E. Bazterra, M. B. Ferraro, D. W. M. Hofmann, F. J. J. Leusen, C. Liang, C. C. Pantelides, P. G. Karamertzanis, S. L. Price, T. C. Lewis, H. Nowell, A. Torrisi, H. A. Scheraga, Y. A. Arnautova, M. U. Schmidt and P. Verwer, *Acta Crystallogr., Sect. B: Struct. Sci.*, 2005, **61**, 511–527.

21 G. M. Day, T. G. Cooper, A. J. Cruz-Cabeza, K. E. Hejczyk, H. L. Ammon, S. X. M. Boerrigter, J. S. Tan, R. G. Della Valle, E. Venuti, J. Jose, S. R. Gadre, G. R. Desiraju, T. S. Thakur, B. P. van Eijck, J. C. Facelli, V. E. Bazterra, M. B. Ferraro, D. W. M. Hofmann, M. A. Neumann, F. J. J. Leusen, J. Kendrick, S. L. Price, A. J. Misquitta, P. G. Karamertzanis, G. W. A. Welch, H. A. Scheraga, Y. A. Arnautova, M. U. Schmidt, J. van de Streek, A. K. Wolf and B. Schweizer, *Acta Crystallogr., Sect. B: Struct. Sci.*, 2009, **65**, 107–125.

22 D. A. Bardwell, C. S. Adjiman, Y. A. Arnautova, E. Bartashevich, S. X. M. Boerrigter, D. E. Braun, A. J. Cruz-Cabeza, G. M. Day, R. G. Della Valle, G. R. Desiraju, B. P. van Eijck, J. C. Facelli, M. B. Ferraro, D. Grillo, M. Habgood, D. W. M. Hofmann, F. Hofmann, K. V. J. Jose, P. G. Karamertzanis, A. V. Kazantsev, J. Kendrick, L. N. Kuleshova, F. J. J. Leusen, A. V. Maleev, A. J. Misquitta, S. Mohamed, R. J. Needs, M. A. Neumann, D. Nikylov, A. M. Orendt, R. Pal, C. C. Pantelides, C. J. Pickard, L. S. Price, S. L. Price, H. A. Scheraga, J. van de Streek, T. S. Thakur, S. Tiwari, E. Venuti and I. K. Zhitkov, *Acta Crystallogr., Sect. B: Struct. Sci.*, 2011, **67**, 535–551.

23 S. M. Woodley and R. Catlow, *Nat. Mater.*, 2008, **7**, 937–946.

24 C. C. Pantelides, C. S. Adjiman and A. V. Kazantsev, in *General Computational Algorithms for Ab Initio Crystal Structure Prediction for Organic Molecules*, ed. S. Atahan-Evrenk and A. Aspuru-Guzik, Springer International Publishing, Cham, 2014, pp. 25–58.

25 P. G. Karamertzanis and C. C. Pantelides, *J. Comput. Chem.*, 2005, **26**, 304–324.

26 P. G. Karamertzanis and C. C. Pantelides, *Mol. Phys.*, 2007, **105**, 273–291.

27 A. V. Kazantsev, P. G. Karamertzanis, C. C. Pantelides and C. S. Adjiman, in *Process Systems Engineering*, Wiley-VCH Verlag GmbH and Co. KGaA, 2014, vol. 6, pp. 1–42.

28 A. V. Kazantsev, P. G. Karamertzanis, C. S. Adjiman and C. C. Pantelides, *J. Chem. Theory Comput.*, 2011, **7**, 1998–2016.

29 A. J. Stone and M. Alderton, *Mol. Phys.*, 1985, **56**, 1047–1064.
30 A. J. Stone, *J. Chem. Theory Comput.*, 2005, **1**, 1128–1132.
31 D. E. Williams and S. R. Cox, *Acta Crystallogr., Sect. B: Struct. Sci.*, 1984, **40**, 404–417.
32 D. S. Coombes, S. L. Price, D. J. Willock and M. Leslie, *J. Phys. Chem.*, 1996, **100**, 7352–7360.
33 T. Beyer and S. L. Price, *J. Phys. Chem. B*, 2000, **104**, 2647–2655.
34 S. R. Cox, L.-Y. Hsu and D. E. Williams, *Acta Crystallogr., Sect. A: Cryst. Phys., Diffr., Theor. Gen. Crystallogr.*, 1981, **37**, 293–301.
35 D. E. Williams and D. J. Houpt, *Acta Crystallogr., Sect. B: Struct. Sci.*, 1986, **42**, 286–295.
36 L.-Y. Hsu and D. E. Williams, *Acta Crystallogr., Sect. A: Cryst. Phys., Diffr., Theor. Gen. Crystallogr.*, 1980, **36**, 277–281.
37 CPOSS Website, The FIT and WILL01 empirical "repulsion–dispersion" potentials, 2014, http://www.chem.ucl.ac.uk/cposs/dmacrys/fit_and_will01_empirical_potentials.htm, last accessed, 04 March 2018.
38 D. E. Williams, *J. Comput. Chem.*, 2001, **22**, 1154–1166.
39 Y. A. Arnautova, A. Jagielska, J. Pillardy and H. A. Scheraga, *J. Phys. Chem. B*, 2003, **107**, 7143–7154.
40 A. T. Hagler, E. Huler and S. Lifson, *J. Am. Chem. Soc.*, 1974, **96**, 5319–5327.
41 A. T. Hagler and S. Lifson, *J. Am. Chem. Soc.*, 1974, **96**, 5327–5335.
42 E. O. Pyzer-Knapp, H. P. G. Thompson and G. M. Day, *Acta Crystallogr., Sect. B: Struct. Sci., Cryst. Eng. Mater.*, 2016, **72**, 477–487.
43 A. J. Stone, *J. Chem. Theory Comput.*, 2005, **1**, 1128–1132.
44 S. Brodersen, S. Wilke, F. J. J. Leusen and G. Engel, *Phys. Chem. Chem. Phys.*, 2003, **5**, 4923–4931.
45 G. M. Day, W. D. S. Motherwell and W. Jones, *Cryst. Growth Des.*, 2005, **5**, 1023–1033.
46 W. T. M. Mooij and F. J. J. Leusen, *Phys. Chem. Chem. Phys.*, 2001, **3**, 5063–5066.
47 F. H. Allen, *Acta Crystallogr., Sect. B: Struct. Sci.*, 2002, **58**, 380–388.
48 R. G. Little, D. Pautler and P. Coppens, *Acta Crystallogr., Sect. B: Struct. Crystallogr. Cryst. Chem.*, 1971, **27**, 1493–1499.
49 P. Coppens, T. M. Sabine, G. Delaplane and J. A. Ibers, *Acta Crystallogr., Sect. B: Struct. Crystallogr. Cryst. Chem.*, 1969, **25**, 2451–2458.
50 D. E. Williams, *J. Comput. Chem.*, 2001, **22**, 1–20.
51 R. Hundt, J. C. Schön and M. Jansen, *J. Appl. Crystallogr.*, 2006, **39**, 6–16.
52 J. A. Chisholm and S. Motherwell, *J. Appl. Crystallogr.*, 2005, **38**, 228–231.
53 P. J. Linstrom and W. G. Mallard, *J. Chem. Eng. Data*, 2001, **46**, 1059–1063.
54 U. Westhaus, T. Dröge and R. Sass, *Fluid Phase Equilib.*, 1999, **158**, 429–435.
55 J. S. Chickos, *Netsu Sokutei*, 2003, **30**, 116–124.
56 L. Lorenzo Maschio, B. Civalleri, P. Ugliengo and A. Gavezzotti, *J. Phys. Chem. A*, 2011, **115**, 11179–11186.
57 A. Otero-de-la Roza and E. R. Johnson, *J. Chem. Phys.*, 2012, **137**, 054103.
58 A. M. Reilly and A. Tkatchenko, *J. Chem. Phys.*, 2013, **139**, 024705.
59 Y. A. Arnautova, J. Pillardy, C. Czaplewski and H. A. Scheraga, *J. Phys. Chem. B*, 2003, **107**, 712–723.
60 A. Gavezzotti, *Theoretical Aspects and Computer Modeling of the Molecular Solid State*, John Wiley & Sons, 1997, p. 97.

61 I. M. Sobol, *USSR Computational Mathematics and Mathematical Physics*, 1967, vol. 7, pp. 86–112.
62 P. Bratley and B. L. Fox, *ACM Trans. Math Software*, 1988, **14**, 88–100.
63 P. G. Karamertzanis, PhD thesis, Imperial College London, 2004.
64 C. G. Broyden, *IMA J. Appl. Math.*, 1970, **6**, 76–90.
65 A. J. Stone and C.-S. Tong, *J. Comput. Chem.*, 1994, **15**, 1377–1392.
66 H. H. Y. Tsui and S. L. Price, *CrystEngComm*, 1999, **1**, 24–32.
67 H. Sun, *J. Phys. Chem. B*, 1998, **102**, 7338–7364.
68 M. J. Frisch, G. W. Trucks, H. B. Schlegel, G. E. Scuseria, M. A. Robb, J. R. Cheeseman, G. Scalmani, V. Barone, B. Mennucci, G. A. Petersson, H. Nakatsuji, M. Caricato, X. Li, H. P. Hratchian, A. F. Izmaylov, J. Bloino, G. Zheng, J. L. Sonnenberg, M. Hada, M. Ehara, K. Toyota, R. Fukuda, J. Hasegawa, M. Ishida, T. Nakajima, Y. Honda, O. Kitao, H. Nakai, T. Vreven, J. A. Montgomery Jr, J. E. Peralta, F. Ogliaro, M. Bearpark, J. J. Heyd, E. Brothers, K. N. Kudin, V. N. Staroverov, R. Kobayashi, J. Normand, K. Raghavachari, A. Rendell, J. C. Burant, S. S. Iyengar, J. Tomasi, M. Cossi, N. Rega, J. M. Millam, M. Klene, J. E. Knox, J. B. Cross, V. Bakken, C. Adamo, J. Jaramillo, R. Gomperts, R. E. Stratmann, O. Yazyev, A. J. Austin, R. Cammi, C. Pomelli, J. W. Ochterski, R. L. Martin, K. Morokuma, V. G. Zakrzewski, G. A. Voth, P. Salvador, J. J. Dannenberg, S. Dapprich, A. D. Daniels, Ö. Farkas, J. B. Foresman, J. V. Ortiz, J. Cioslowski and D. J. Fox, *Gaussian 09*, Gaussian, Inc., Wallingford CT, 2009.
69 Y. Zhao and D. G. Truhlar, *Theor. Chem. Acc.*, 2008, **120**, 215–241.
70 K. T. No, O. Y. Kwon, S. Y. Kim, K. H. Cho, C. N. Yoon, Y. K. Kang, K. D. Gibson, M. S. Jhon and H. A. Scheraga, *J. Phys. Chem.*, 1995, **99**, 13019–13027.
71 A. Abraha and D. E. Williams, *Inorg. Chem.*, 1999, **38**, 4224–4228.
72 C.-A. Gatsiou, PhD thesis, Imperial College London, 2016.

Faraday Discussions

DISCUSSIONS

Crystal structure evaluation: calculating relative stabilities and other criteria: general discussion

Matthew Addicoat, Claire S. Adjiman, Mihails Arhangelskis, Gregory J. O. Beran, David Bowskill, Jan Gerit Brandenburg, Doris E. Braun, Virginia Burger, Jason Cole, Aurora J. Cruz-Cabeza, Graeme M. Day, Volker L. Deringer, Rui Guo, Alan Hare, Julian Helfferich, Johannes Hoja, Luca Iuzzolino, Samuel Jobbins, Noa Marom, David McKay, John B. O. Mitchell, Sharmarke Mohamed, Marcus Neumann, Sten Nilsson Lill, Jonas Nyman, Artem R. Oganov, Pablo Piaggi, Sarah L. Price, Susan Reutzel-Edens, Ivo Rietveld, Michael Ruggiero, Matthew R. Ryder, German Sastre, J. Christian Schön, Christopher Taylor, Alexandre Tkatchenko, Seiji Tsuzuki, Joost van den Ende, Scott M. Woodley, Grahame Woollam and Qiang Zhu

DOI: 10.1039/C8FD90031K

J. Christian Schön opened the discussion of the paper by Gregory J. O. Beran: Concerning the multi-tier approach to improving the quality of the computations, where you did a careful analysis of different types of approximations, it is still very disconcerting that there appears to be no inherently systematic approach to gain an estimate of the errors involved in the calculations – often it seems that one does not really know where the error is hiding (*e.g.* fortuitous error cancellations). This contrasts with the analytical convergence criteria and error bounds one gets when using *e.g.* perturbation theory, and is quite bothersome (especially since many people do not even try to provide error estimates in the discussion sections of their papers). Do you see any progress in this area?

Gregory J. O. Beran replied: Quantification of uncertainty is a major challenge in electronic structure theory (and many other areas of chemical simulation). Our preferred, albeit imperfect, approach is to investigate the sensitivity of the prediction to the modeling choices. For example, one can explore how the structures, thermochemical properties, relative polymorph stabilities, or other properties of interest converge with respect to the basis set and level of theory. A few earlier studies from our group exemplify this. For aspirin forms I and II, the relative energy rankings proved quite insensitive to the electronic structure method and basis set, leading to relatively high confidence in the relative energetics.[1] The close similarities in crystal packing between the two forms allow for substantial error cancellation in the relative energies. In contrast, the energy rankings for the five polymorphs of oxalyl dihydrazide are very sensitive to the electronic structure treatment, making the quantitative stability ranking less certain.[2] In that case, the basis set superposition error (BSSE) associated with the intramolecular hydrogen bonding in the alpha polymorph is harder to correct than the BSSE for the purely intermolecular hydrogen bonding found in the other polymorphs. Only in the limit of the large basis sets does one achieve a balanced description of both types of interactions.

Perhaps the best example of sensitivity analysis/uncertainty quantification from our work occurs in our recent study, which predicted the methanol phase diagram.[3] In that case, by examining how the predicted phase boundaries change upon altering the relative free energies between two forms, we showed that our calculations reproduced the experimental phase diagram to within $\sim$0.5 kJ mol^{-1}. Furthermore, the sensitivity analysis was used to investigate which features of the calculations were most important and where one should focus the computational effort in order to obtain accurate predictions. The Electronic Supplementary Information for ref. 3 goes into considerable detail on these issues.

1 S. Wen and G. J. O. Beran, *Cryst. Growth Des.*, 2012, **12**, 2169–2172.
2 S. Wen and G. J. O. Beran, *J. Chem. Theory Comput.*, 2012, **8**, 2698–2705.
3 C. Cervinka and G. J. O. Beran, *Chem. Sci.*, 2018, **9**, 4622–4629.

Aurora J. Cruz-Cabeza asked: Could you please give an estimate as to how your Tier 1–4 calculations compare in terms of computational expense?

Gregory J. O. Beran answered: Several factors play a role in the overall computational timings. Generally speaking, MP2 scales $O(N^5)$ with system size N, while large-scale DFT calculations typically scale $O(N^3)$ or better. However, things become more nuanced for the calculations performed here. The fragment-based hybrid many-body interaction (HMBI) energy approach decomposes the full crystal calculation to a series of quantum mechanical (QM) monomer and short-range dimer calculations, while longer-range dimers and many-body interactions are approximated at a lower level of theory (*e.g.* with the AMOEBA polarizable force field or in some cases here, periodic Hartree–Fock, HF). Evaluation of the QM dimers forms the computational bottleneck. At long range, the model switches to a force field or Hartree–Fock description of the dimer intermolecular interactions.

Furthermore, the HMBI approach exploits space group symmetry and requires evaluation only of the symmetrically unique dimers. Thus, the computational cost grows $O(N^5)$ with the size of the monomer, but it grows only linearly with the number of molecules in the asymmetric unit. The fragment-based calculations employ atom-centered Gaussian-orbital basis sets, and the convergence of the MP2 correlation energy with basis set size is slow—cubic, rather than the exponential convergence DFT would have in the same basis sets.

For fragment-based phonon calculations like those with HMBI, it turns out that the same QM dimer calculations are needed for either a Γ-point-only calculation or a supercell lattice dynamics calculation.[1] The extra work associated with the supercell occurs purely at the low level of theory (*e.g.* polarizable force field). In addition, the finite displacements are done on a smaller scale (in individual monomers or dimers, which often have fewer atoms than the whole unit cell), which further reduces the cost. So while MP2 frequencies are considerably more expensive than DFT ones, it might prove more efficient to use the fragment approach to capture phonon dispersion for sufficiently large cells.

For plane-wave DFT, the basis set grows with the size of the unit cell. This becomes especially problematic for large supercell phonon calculations. On the other hand, one can typically perform less k-point sampling in larger cells. DFT codes typically also exploit the symmetry of the cell to reduce the number of k-points.

Overall, one expects the cost of fragment-based MP2 calculations to grow more rapidly than plane-wave DFT as the molecules get larger, while the plane-wave DFT calculation cost grows more steeply with unit cell size. Acetic acid, imidazole and acetaminophen each have four molecules in the unit cell and one in the asymmetric unit. Carbon dioxide and ice are not discussed here, since their small molecular size and the extraordinarily high symmetry in carbon dioxide make them anomalously fast compared to larger organic molecules of interest to the community.

For acetic acid or imidazole, a single-point energy MP2/aug-cc-pVTZ + AMOEBA calculation is more expensive than the plane-wave B86bPBE-XDM (80 Ry plane-wave cut-off) one by a factor of $\sim$1.5.[2] The relative cost increases by a factor of $\sim$6–8$\times$ for a force or frequency calculation. The DFT basis set used in those calculations was probably larger than necessary, and for acetaminophen we reduced the plane-wave cut-off to 60 Ry. We also reduced the k-point mesh from $5\times5\times5$ to $1\times3\times3$. As a result, MP2 is $\sim$49$\times$, $\sim$242$\times$, and $\sim$153$\times$ more expensive than DFT for an energy, force, and frequency calculation. For comparison, if we kept the 80 Ry plane-wave cut-off, the ratios would be $\sim$3$\times$, $\sim$16$\times$, and $\sim$0.2$\times$. With the very large DFT basis set, the frequency calculation is actually cheaper with fragment MP2/aug-cc-pVTZ, though again, that DFT basis set would probably be unnecessarily large.

Based on these test calculations, we estimated overall computational costs in CPU hours and relative costs for the Tier 1–4 calculations. This includes mapping out the energy–volume curves at 10 different geometries (assuming 100 optimization cycles per geometry optimization) and computing phonon frequencies at three volumes (for the reference frequencies and two more to compute the Grüneisen parameters). The results are summarized in Table 1. From this table, one sees that the Tier 1 calculations cost roughly 1–2 orders of magnitude more than the Tier 4 ones. On the other hand, for Tiers 3 and 4, the computational cost is dominated by the DFT geometry optimizations and phonon calculations. Single-point energy refinement at the MP2/aug-cc-pVTZ level in Tier 3 increases the computational cost only marginally—no more than $\sim$20% here. In contrast, switching to Tier 2 starts to increase the computational cost much more dramatically, especially for acetaminophen.

Even the single-point refinement calculations in the paper, which extrapolate to the complete basis set limit, correct the known deficiencies in MP2 with MP2C, and replace AMOEBA with periodic HF, do not increase the cost of Tier 3 too much. For the smaller two crystals, this refinement incurs only 10–20% higher cost compared to the pure DFT Tier 4 calculations. For acetaminophen, it roughly doubles the cost compared to the Tier 4 calculation. When combined with the

Table 1 Approximate computational costs for Tiers 1–4 using either HMBI (MP2/aug-cc-pVTZ + AMOEBA) or B86bPBE-XDM. The estimates assume 10 geometry optimization/energy calculations to map out the energy volume curves, 3 phonon calculations to obtain the reference phonons and Grüneisen parameters, and 100 geometry optimization cycles per geometry optimization

	CPU hours				Relative time			
Crystal	Tier 1	Tier 2	Tier 3	Tier 4	Tier 1	Tier 2	Tier 3	Tier 4
Acetic acid[a]	37 600	30 700	5330	5270	7.1	5.8	1.0	1
Imidazole[a]	92 800	74 100	13 100	13 000	7.1	5.7	1.0	1
Acetaminophen[b]	1 290 000	745 000	8000	6600	195	113	1.2	1

[a] DFT calculations employed an 80 Ry plane-wave cut-off and a 5×5×5 *k*-point mesh. [b] DFT calculations employed a 60 Ry plane-wave cut-off and a 1×3×3 *k*-point mesh.

quality of the predictions discussed in the paper, Tier 3 provides a valuable compromise between accuracy and computational cost.

1 S. Hirata, *J. Chem. Phys.*, 2008, **129**, 204104.
2 All timings were performed on a 16-core node with dual 2.4 GHz Intel Xeon E5-2630 v3 processors, 64 GB of RAM, and several terabytes of local disk storage.

Volker L. Deringer said: You have nicely discussed the different "tiers" of methodology, and how the acetaminophen example was only amenable to the computationally more affordable tiers. How do you expect this to change over time, when more and cheaper CPU time will become available? Do you expect "Tier 1" methods to be routinely in reach within a time frame of a couple of years – and what impact will that have on the field?

Gregory J. O. Beran responded: Future advances in computer power will certainly help, but as the timings in Table 1 indicate, the computational cost of a Tier 1 calculation is roughly 200 times that of a Tier 3 or Tier 4 one, and that ratio will continue growing with increasing molecular size. For many applications, the difference between Tier 1 and 3 results may not be worth the difference in computational cost. In the longer term, we think it will likely prove more beneficial to move beyond the quasi-harmonic approximation and find better ways to handle the dynamics and anharmonicity in the crystal, perhaps using machine learning or other approaches to develop accurate potentials that can be sampled more cheaply. High-quality energies/forces from fragment methods could potentially provide inputs for developing such potentials.

David McKay communicated: The four tiers of computational methodology are set out in decreasing order of computational cost, with Tier 1 having HMBI energies, geometries and frequencies, then DFT coming in from right to left, *i.e.* frequencies, then geometries, then energies. Would the results of methods including DFT be improved (closer to the all-HMBI Tier 1) if, rather than considering the cost of the methods, you considered the competency of the methods at each stage, perhaps giving a Tier 2′ of HMBI, DFT and HMBI for energies, geometries and frequencies, respectively? Perhaps also a Tier 3′ of DFT, DFT, HMBI.

Gregory J. O. Beran communicated in reply: Given the reality of needing to reduce computational cost, as evidenced by the timing data presented in response to the previous question, we did not investigate approaches of the sort you mention. What follows is therefore speculation: in the molecular quantum chemistry literature, there are usually not substantial differences between the quality of harmonic vibrational frequencies for intramolecular modes predicted with DFT *versus* MP2, assuming sufficiently large basis sets. The empirical factors people use to scale the frequencies to improve agreement with experiment are generally similar in magnitude (typically around 0.9–1.0, depending on the specific model chemistry). It is less obvious how well that insight translates to the soft lattice phonons, which contribute significantly to thermal expansion and thermal contributions to the enthalpy and entropy. On the other hand, the soft lattice modes are largely governed by the crystal packing and cell volume. So as long as the model gets those reasonably correctly, it should perform well. Our thermal expansion data and plot of the phonon frequencies for ice demonstrate generally similar performance for either HMBI or DFT on the low-frequency modes, so one might think there could be some freedom to mix and match the different models along the lines you suggest. However, in the end, the practical matters of computational effort drive us toward Tiers 3 and 4 discussed in the paper.

Mihails Arhangelskis enquired: My question concerns modelling unit cell expansion with quasi-harmonic calculations. How was anisotropic unit cell expansion accounted for? In the case of low symmetry (monoclinic and, especially, triclinic) unit cells I would imagine that a large number of structure distortions would have to be considered in order to map the temperature expansion in each crystallographic direction. Presumably, one would have to study expansion about each principal axis separately, while also considering the variations in unit cell angles. I would be curious to know whether such calculations are practical for low symmetry cells, or whether there this can be done in a more efficient way.

Gregory J. O. Beran responded: For the quasi-harmonic calculation, we perform the expansion/compression along a one-dimensional semi-anisotropic coordinate. Specifically, the energy *versus* volume curve is generated by optimizing the structure under various positive and negative hydrostatic pressures. This allows the volume to expand or contract by deforming the unit cell in whatever manner minimizes the energy change. A closely related approach would optimize the cell subject to a volume constraint for various different volumes. However, algorithms to perform such constant volume optimizations are not currently implemented in the software packages used in our work. Either way, the volume dependence of the phonons can subsequently be determined along the same one-dimensional coordinate. The semi-anisotropic approach used in our work is an approximation relative to a fully anisotropic expansion model which would allow all independent degrees of freedom to vary independently, but it should provide a more realistic description of how the unit cell varies than a simple isotropic scaling of the lattice parameters. For detailed analysis of the impact of including different degrees of anisotropy in quasi-harmonic calculations, see the recent work by Abraham and Shirts.[1] In the examples they looked at, the difference between an isotropic and a fully anisotropic model was up to ~1 kJ mol^{-1}. The differences involving quasi-anisotropic approaches like ours are considerably smaller. Regarding low-symmetry cells: the computational bottleneck for the hybrid many-body interaction fragment model is the number of dimer calculations that need to be evaluated. For careful calculations of the sort employed here, we typically employ conservative cut-offs that evaluate dimer interactions quantum mechanically to 9–10 Å, before smoothly switching to an MM treatment for longer-range interactions. See ref. 2 for more details. Exploiting space group symmetry typically reduces the number of dimers by a factor corresponding to the ratio between the total number of molecules in the unit cell (Z) and the number of molecules in the asymmetric unit (Z'). For example, in the orthorhombic polymorph of acetic acid ($Pna2_1$ space group, $Z = 4$, $Z' = 1$) with the 9–10 Å cut-off mentioned above, a typical single-point energy calculation on crystalline acetic acid might require over 212 dimers without space group symmetry, or 53 dimers with symmetry (savings factor = 4). See ref. 3 for more details. Four-fold symmetry is common to many of the most frequently occurring space groups, and such symmetry savings factors are common. Therefore, calculations on a unit cell with $P1$ symmetry will require several-fold more computational effort than calculations on a comparable cell with four-fold symmetry.

One nice feature of the fragment approach, however, is that the computational effort for the expensive quantum mechanical calculations grows linearly with the number of molecules in the unit cell. The quantum mechanical dimer interactions occur within a finite radius (*e.g.* the aforementioned 9–10 Å cut-off), regardless of whether those molecules occupy the central unit cell or a periodic image cell. In contrast, the computational cost of the periodic density functional grows cubically with the basis set size, and increasing the unit cell dimensions increases the number of plane-waves which fit in the unit cell for a given plane-wave cut-off. This behavior could make a fragment approach particularly attractive for modeling a defect in a larger supercell, for example.

1 N. S. Abraham and M. R. Shirts, 2017, arXiv:1712.00936 [cond-mat.mtrl-sci].
2 S. Wen, K. D. Nanda, Y. Huang and G. J. O. Beran, *Phys. Chem. Chem. Phys.*, 2012, **14**, 7578–7590.
3 Y. N. Heit and G. J. O. Beran, *J. Comput. Chem.*, 2014, **35**, 2205–2214.

Marcus Neumann asked: In the calculation of the Grüneisen parameter you apply hydrostatic pressure in order to estimate the unit cell changes under thermal pressure. How well is this approximation validated? What is the expected free energy error of the approximation?

Gregory J. O. Beran answered: We have not investigated this issue in detail, though we provide some discussion regarding the isotropic *versus* the quasi-anisotropic hydrostatic approach in our answer to Mihails Arhangelskis' previous question. Abraham and Shirts recently investigated various ways of performing the quasi-harmonic approximation that include varying degrees of anisotropy.[1] They found that fully anisotropic expansion allowed the crystal to adopt structures up to 1 kJ mol^{-1} lower in energy than isotropic expansion. Free energy differences between polymorphs were even smaller, at a fraction of a kJ mol^{-1}. Our approximation, which lies intermediate between the isotropic and fully anisotropic models, seems like it should perform reasonably well.

1 N. S. Abraham and M. R. Shirts, 2017, arXiv:1712.00936 [cond-mat.mtrl-sci].

Joost van den Ende remarked: In Fig. 4 and Fig. S6 you show the dependence of the molar volume and electronic energy of CO_2 as a function of the different density functionals + dispersion correction used. Intuitively, I would think that using empirical dispersion corrections works better for larger molecules which fall “better” in the training set of the used dispersion correction. Therefore I would argue that CO_2 might not be the best test system to analyse the performance of empirical dispersion corrections. As a very experienced user of electronic structure calculations, do you think there is a size dependence of the studied molecules on the precision of empirical dispersion corrections?

Gregory J. O. Beran responded: Previous work by Gohr *et al.* suggests that the dispersion correction is quite important even for carbon dioxide.[1] For example, the lattice energy of phase I is 8.7, 22.7, and 23.5 using PBE, PBE-D2, and PBE-D3(BJ), respectively. Omitting the dispersion correction also leads to substantial overestimation of the molar volume and underestimation of the bulk modulus. Similar evidence for the reasonably good performance of various dispersion corrections for carbon dioxide, acetic acid and imidazole can be found in ref. 2.

More generally, modern dispersion corrections (Grimme’s D3, Distasio and Tkatchenko’s many-body dispersion, and Becke and Johnson’s XDM) are largely non-empirical and adapt well to a variety of systems. So I do not think that small molecules used for some of the benchmarks are a significant problem. Our paper did use the more empirical D2 for a few results, since D3 was not available in the Quantum Espresso software used, and it generally performed less well. We do not advocate general use of the D2 correction.

1 S. Gohr, S. Grimme, T. Soehnel, B. Paulus and P. Schwerdtfeger, *J. Chem. Phys.*, 2013, **139**, 174501.
2 A. Otero-de-la-Roza and E. R. Johnson, *J. Chem. Phys.*, 2012, **137**, 054103.

Johannes Hoja said: You have shown that the quasi-harmonic approximation at the Tier 3 level provides a good description of the thermal expansion and accurate sublimation enthalpies for small molecule crystals. Do you have an estimate of the level of molecular flexibility up to which one could obtain reliable sublimation enthalpies and entropies using the quasi-harmonic approximation?

Gregory J. O. Beran replied: This is a really important question and one we intend to investigate in the future. The quasi-harmonic calculations do appear to perform well for acetaminophen in the paper, which has slightly more flexibility than the completely rigid small molecules considered in earlier parts of the paper. Of course, many pharmaceuticals and other organic molecules of interest exhibit far more flexibility than any of the species considered in our paper. For sufficiently flexible species, it might make more sense to treat them with molecular dynamics (using a high-quality potential) or some other technique which can capture the anharmonicity in these species. On the other hand, our paper shows (*e.g.* Fig. 2) that zero-point vibrational energy can contribute appreciably to the molar volume. Zero-point vibrational effects are not captured in classical molecular dynamics. One might also expect that the anharmonicity of the intramolecular vibrational modes will play an increasingly important role in larger molecules. The use of a one-dimensional Morse approximation for the anharmonicity described in your paper (DOI: 10.1039/c8fd00066b) might help address that problem. Future work will hopefully provide further clarity on the practical limits of the quasi-harmonic approximation for more flexible species.

Michael Ruggiero remarked: To follow up on the previous point, in Tier 1 the unit cell was constrained to an isotropic expansion/contraction, while in Tiers 2–4 the cell was allowed to optimize anisotropically. Does this have any influence on the results?

Gregory J. O. Beran responded: This is a good point. The Tier 1 results come from an earlier publication,[1] and our computational procedures have evolved since then. In that work, we optimized the geometries on-the-fly (instead of using the 1D energy–volume curves). While that allows the structures to relax in a fully anisotropic fashion, the Grüneisen parameters in that work were computed from volumes that were scaled isotropically. For ice and carbon dioxide, the experimental expansion is quite isotropic, so this is not a significant issue. However, acetic acid and imidazole expand somewhat more anisotropically, so there is probably some error introduced by computing the Grüneisen parameters under an isotropic approximation. Specifically, one would anticipate that the energy changes associated with those volume changes are too large, and that the frequency changes for the lowest-energy phonon modes will be overestimated. Hopefully the errors introduced by this approximation are small, and indeed the Tier 1 thermal expansion curves are fairly parallel to the experimental ones.

1 Y. N. Heit and G. J. O. Beran, *Acta Cryst. B*, 2016, **72**, 514–529.

Michael Ruggiero said: In the poster presented, the results using phonon dispersion are very compelling. In the cases where phonon dispersion isn't utilized and yields good results, in your opinion do you think that this is due to a cancellation of errors, a system dependent case where phonon dispersion isn't actually needed, or some combination of both?

Gregory J. O. Beran replied: For the Tier 3 hybrid approach, which we advocate due to its relatively low computational cost and good quality of results, the thermal expansion is pretty consistently underestimated across all the examples. As Jessica McKinley showed in her poster at the meeting (for another example, see the Electronic Supplementary Information for ref. 1), including phonon dispersion helps correct this. In contrast, the Tier 1 results, which do include phonon dispersion, track the experimental thermal expansion quite well. It is true that some of the calculations (most notably the Tier 4 DFT calculations, especially those with the D2 dispersion correction) expand the crystals more significantly with increasing temperature, even without including phonon dispersion. We suspect that this is due to error cancellation. As shown in Fig. 2, 5 and S7 from the main paper (DOI: 10.1039/c8fd00048d), the expansion is suppressed somewhat once the MP2-level single-point energies are applied on top of those energies.

1 Y. N. Heit, K. D. Nanda and G. J. O. Beran, *Chem. Sci.*, 2016, 7, 246–255.

Michael Ruggiero asked: You've shown the importance of phonon dispersion for accurately determining thermodynamic properties, however such simulations are often prohibitively expensive. Is there any indication or marker for how much phonon dispersion to include to best marry computational cost and accuracy, or is it highly system dependent? To follow up, when thermodynamic values are computed accurately without phonon dispersion, what happens when dispersion is ultimately included? Are the non-dispersion simulations accurate due to a cancellation of errors, or do such methods simply not require dispersion?

Gregory J. O. Beran answered: We are pursuing additional strategies for capturing phonon dispersion effectively with reasonable computational cost and hope that it will be feasible to capture these effects in much larger crystals in the future. For example, Brandenburg *et al.* reported promising results in modeling the thermal expansion of carbamazepine with a density functional tight binding model.[1] Regarding the accuracy of the thermochemical predictions without phonon dispersion, one can obtain a reasonable estimate for the impact on the sublimation enthalpies or entropies by comparing the temperature dependent curvature of the Tier 1 *versus* Tier 3/4 results in Fig. 6 and 7 of the main paper (DOI: 10.1039/c8fd00048d). Visual inspection suggests that the sublimation enthalpy errors at higher temperatures that are due to neglecting phonon dispersion are somewhere around 1–2 kJ mol^{-1} for carbon dioxide, acetic acid and imidazole, for example. The errors are considerably smaller for ice, which expands much less due to the strong three-dimensional hydrogen bond network. In other words, the magnitude of the errors will depend on the nature of the crystal packing. It is possible, of course, that fortuitous error cancellation will sometimes lead to reasonable-looking predictions of thermochemical properties despite modeling the thermal expansion poorly.

1 J. G. Brandenburg, J. Potticary, H. A. Sparkes, S. L. Price and S. R. Hall, *J. Phys. Chem. Lett.*, 2017, **8**, 4319–4324.

Sarah L. Price said: Chick Wilson's paper[1] shows the thermal ellipsoids of a methyl group libration in form I paracetamol (acetaminophen) over the range of temperatures of the structures you have been modelling. The amplitude of the librations is huge, even as extrapolated to 0 K. Do you think that this may have affected your results for predicting the thermal expansion, as would the quasi-harmonic expansion be appropriate?

1 C. C. Wilson, Zero point motion of the librating methyl group in *p*-hydroxyacetanilide, *Chem. Phys. Lett.*, 1997, **280**, (5–6), 531–534.

Gregory J. O. Beran responded: This is an interesting question that merits future investigation. At the moment, we lack the data to answer this definitively. In other cases we have examined, neglecting phonon dispersion leads to an underestimation of the thermal expansion. It would be interesting to perform the phonon dispersion calculations and see how much of the thermal expansion the model captures. The magnitude of the residual disagreement between the predicted and experimental temperature dependence of the molar volumes could point toward the importance of effects such as the high-amplitude librations you point to.

Sarah L. Price remarked: Your work has been very important in showing that even zero-point lattice modes have a significant effect in expanding the lattice, implying that lattice energy minimisation with the perfect model would give errors for even the lowest-temperature crystal structures. The difference in methyl rotational amplitudes between polymorphs, depending on how tightly the methyl group is packed, could well affect the relative free energies. It would be unfeasible if we needed to calculate both quantum mechanical zero-point effects and had to do classical MD to get a full description of how the molecules are moving to get accurate thermodynamic and spectral properties.

Gregory J. O. Beran replied: This raises a good point. The quasi-harmonic approximation has clear limitations (*e.g.* high temperatures or highly flexible molecules). If one can obtain a good potential, switching to a dynamical treatment would be an appealing way to circumvent those limitations. However, classical molecular dynamics simulations do not capture the zero-point vibrational effects, which appear to contribute appreciably to molecular crystal volumes (*e.g.* Fig. 2 in the main article (DOI: 10.1039/c8fd00048d)). There have been advances in path integral molecular dynamics techniques in recent years,[1] so perhaps it will become feasible at some point to use those types of approaches to capture the nuclear quantum effects more effectively.

1 T. E. Markland and M. Ceriotti, *Nature Rev. Chem.*, 2018, **2**, 0109.

Seiji Tsuzuki remarked: I have comments on the DFT calculations of the intermolecular interaction energy. You showed that the performance of the DFT calculations for evaluating the sublimation energy is not very good. The calculated sublimation energy depends on the calculated depth of the intermolecular interaction energy potential. The depth of the potential depends strongly on the choice of dispersion correction method and functionals used for the DFT calculations. I believe that the choice of PBE functionals is not so good. I believe that the performance of the calculations of the sublimation energies will be improved if you choose an appropriate dispersion correction method and functionals.

Gregory J. O. Beran responded: Our paper considers multiple different density functionals and van der Waals dispersion corrections. Our most important results use the B86bPBE-XDM model, which Erin Johnson and co-workers have demonstrated works well for molecular crystals and other systems where non-covalent interactions are important.[1,2] In addition, one of the important results in the paper is that when using hybrid approaches like Tier 3 that use wavefunction-based single-point energy refinement of DFT geometries/refinement, the final results are relatively insensitive to the choice of density functional/dispersion correction (see Fig. 4 in the paper).

During the discussion I also shared two predicted values for the sublimation enthalpy of alpha methanol, which were taken from our recent paper.[3] These numbers compared the sublimation enthalpy computed with our fragment method at the CCSD(T)/CBS + periodic HF level of theory. The difference between the two values was whether we optimized the geometries and computed the phonons using MP2-level fragment calculations (*i.e.* Tier 1) or with plane-wave PBE-D3 (Tier 3). Because the single-point energies defining the energy–volume curve are at the CCSD(T) level, the results are not directly dependent on the PBE-D3 energy well depth. In that case, the Tier 1 sublimation enthalpy was 45.0 kJ mol^{-1}, which is in excellent agreement with the experimentally reported value of 45.7 ± 0.3 kJ mol^{-1}.[3] The Tier 3 value of 42.8 kJ mol^{-1} has an error that is several kJ mol^{-1} larger. This result is consistent with our other results presented in the *Faraday Discussions* paper (DOI: 10.1039/c8fd00048d), which demonstrate that while the hybrid wavefunction/DFT calculations perform well, they are not quite as good as those done entirely with higher-level wavefunction methods.

1 S. R. Whittleton, A. Otero-de-la-Roza and E. R. Johnson, *J. Chem. Theory Comput.*, 2017, **13**, 441–450.
2 S. R. Whittleton, A. Otero-de-la-Roza and E. R. Johnson, *J. Chem. Theory Comput.*, 2017, **13**, 5332–5342.
3 C. Cervinka and G. J. O. Beran, *Chem. Sci.*, 2018, **9**, 4622–4629.

Seiji Tsuzuki enquired: You used the HF method for the analysis of many-body interactions. The HF method overestimates the Coulombic interactions (electrostatic and induction interactions). The induction interaction is a major source of the many-body interactions in general. Therefore, the HF method probably overestimates the many-body interactions. On the other hand, DFT calculations can evaluate Coulombic interactions more accurately owing to the consideration of electron correlation. Why did you not use DFT calculations instead of HF calculations for the evaluation of the many-body interactions?

Gregory J. O. Beran answered: Our experience has been that when focusing on the long-range pairwise and many-body induction interactions, HF typically performs better than typical generalized gradient approximation (GGA) and can even be better than hybrid density functionals. There are examples of fragment methods that combine periodic DFT with monomer and dimer corrections computed with wavefunction methods,[1–3] However, those work best in cases where polarization effects are relatively weak. This can be understood in terms of the delocalization errors and problems in the many-body repulsion inherent in density functionals. Our earlier benchmark calculations on three-body interactions demonstrate this nicely.[4] Additional discussion of repulsion issues can be found in a paper by Gillan.[5] In the context of our fragment approaches, it is important to emphasize that the energies are dominated by the 1-body and shorter-range 2-body terms, which are computed with higher levels of theory. HF is only used for the non-pairwise additive contributions involving three or more molecules. In practice, such contributions are typically no more than 10–20%, and often they are 5–10%.

1 O. Bludsky, M. Rubes and P. Soldan, *Phys. Rev. B*, 2008, **77**, 092103.
2 S. Tsuzuki, H. Orita, K. Honda and M. Mikami, *J. Phys. Chem. B*, 2010, **114**, 6799–6805.
3 C. R. Taylor, P. J. Bygrave, J. N. Hart, N. L. Allan and F. R. Manby, *Phys. Chem. Chem. Phys.*, 2012, **14**, 7739–7743.
4 J. Rezac, Y. Huang, P. Hobza and G. J. O. Beran, *J. Chem. Theory Comput.*, 2015, **11**, 3065–3079.
5 M. J. Gillan, *J. Chem. Phys.*, 2014, **141**, 224106.

Jan Gerit Brandenburg communicated: The authors nicely analyzed how to combine different electronic structure methods to compute different components of the free energy (relaxed geometry, lattice energy, lattice dynamics). I wondering if their ranking (good to less good) is well justified in following the hierarchy of HMBI, DFT, FF. This will most likely depend on the type of many-body methodology that is used for the clusters and the type of embedding. As a first point, the many-body method employed is mostly MP2, which is certainly appropriate for CO_2 and ice I_h. However, for unsaturated π-stacked systems, MP2 is known to have deficiencies and DFT might be even more accurate. The literature data on the studied systems is summarized in Table 2. Could the authors comment on the accuracy of the employed HMBI variants for the lattice energy? The second issue is the force field describing the embedding. While I am confident that both geometries and energies from the HMBI approach will be only slightly perturbed by the embedding, this might not hold for the low-energy frequency modes. Have the authors studied this, for instance by enlarging the region that is treated with the many-body method?

1 A. Otero-de-la-Roza and E. R. Johnson, *J. Chem. Phys.*, 2012, **137**, 054103.
2 A. M. Reilly and A. Tkatchenko, *J. Chem. Phys.*, **2013**, 139, 024705.
3 A. Zen, J. G. Brandenburg, J. Klimeš, A. Tkatchenko, D. Alfè and A. Michaelides, *Proc. Natl. Acad. Sci. USA*, 2018, **115**, 1724–1729.
4 S. Wen and G. J. O. Beran, *J. Chem. Theory Comput.*, 2011, **7**, 3733–3742.
5 M. J. Gillan, D. Alfè, P. J. Bygrave, C. R. Taylor and F. R. Manby, *J. Chem. Phys.*, 2013, **139**, 114101.
6 M. Cutini, B. Civalleri, M. Corno, R. Orlando, J. G. Brandenburg, L. Maschio and P. Ugliengo, *J. Chem. Theory Comput.*, 2016, **12**, 3340–3352.

Gregory J. O. Beran communicated in reply: Lattice energies: We agree about the limitations of MP2 for describing non-covalent interactions. This issue in the context of molecular crystals is discussed in some detail in our earlier review.[1] There are multiple strategies for addressing these MP2 deficiencies, but one of our favored approaches and one we adopt in the current paper is MP2C.[2] This employs intermolecular perturbation theory to correct the treatment of the dispersion energy in MP2. Based on its excellent performance for non-covalent interactions and $O(N^5)$ scaling with system size N, David Sherrill and co-workers anointed MP2C the "bronze standard" for non-covalent interactions.[3] We have used MP2C to study the polymorphs of oxalyl dihydrazide,[4] and we have a new MP2D model[5] which recasts MP2C in a framework based on Grimme's D3 dispersion correction. MP2D extends

Table 2 Lattice energies of the studied crystals from MP2 and DFT compared to semi-experimental estimated and high-level results. Energies are given in kJ mol^{-1}

	Semi-exp.[a]	DMC[b]	MP2[c]	DFT[d]
CO_2	−28.4	−28.2	−29.1	−21.6
Ice I_h	−58.8	−59.2	−58.7	−72.8
Acetic acid	−72.8	—	−61.2	−71.8
Imidazole	−86.8	—	−88.6	−87.5

[a] Experimental sublimation enthalpies with estimations of zero-point and thermal effects from DFT computations as summarized in ref. 1–3. [b] Diffusion Monte Carlo data from ref. 3. [c] MP2 data from ref. 4–6, non converged basis set for acetic acid and imidazole. [d] B86bPBE-XDM for CO_2, acetic acid, imidazole, PBE-D3 for ice I_h.

the range of applicability of MP2C to also include intramolecular interactions and facile geometry optimizations, which could prove useful in molecular crystals.

To compare against the lattice energies you cite, we extracted our best Tier 3 lattice energies at the MP2/aug-cc-pVTZ + AMOEBA and MP2C/CBS + periodic HF levels. Specifically, we evaluated the crystal energy at the minimum of the energy *versus* volume curves that result from single-point energy refinement on the B86bPBE-XDM geometries. As can be seen in Table 3, MP2/aug-cc-pVTZ + AMOEBA performs well for carbon dioxide, reasonably for ice, and poorly for acetic acid and imidazole. Imidazole in particular demonstrates the problem of MP2 over-binding the van der Waals dispersion interactions involving π systems.

Refinement of the single-point energies by replacing MP2 with MP2C, extrapolating to the complete basis set limit, and improving the many-body treatment from the AMOEBA force field to periodic Hartree–Fock leads to considerable improvement. Three of the four MP2C/CBS + periodic HF lattice energies lie within 2 kJ mol^{-1} of the semi-experimental value. For acetic acid, the error is larger at 4 kcal mol^{-1}. More detailed analysis of basis set effects and post-MP2 correlation contributions are found in our earlier studies.[6–8] Aside from acetic acid, the MP2C lattice energies are appreciably better than the results we obtained with our own B86bPBE-XDM calculations, which are also reported in Table 3.

For additional comparison, consider the sublimation enthalpy data in our paper. For acetic acid at 50 K (the lowest temperature we have experimentally derived values for), the predicted and experimental sublimation enthalpies are 63.4 and 66.0 kJ mol^{-1}, or an error of 2.6 kJ mol^{-1}. Looking at all four crystals, the errors in the low-temperature sublimation enthalpies are in the range of ~1–3 kJ mol^{-1} with MP2C (Fig. 6). The sublimation enthalpies computed with B86bPBE-XDM in our paper exhibit errors of ~3–8 kJ mol^{-1}. Overall, the accuracy of sublimation enthalpies and lattice energies appears fairly consistent, especially given that the experimental data may well have uncertainties of a couple of kJ mol^{-1} or more.

From a pragmatic point of view, dispersion-corrected DFT has advanced tremendously over the last decade, and it performs very well for many molecular crystal problems.[9–11] In many cases its accuracy is more than sufficient to make useful predictions. However, in cases where one is not confident in the DFT results (*e.g.* when particularly small energy differences are obtained or when one observes high sensitivity of the results to the specific functional or dispersion correction), then the fragment approaches provide an alternative that can allow one to push toward even higher accuracy. This philosophy fits nicely with the Tier

Table 3 Lattice energies for the four small molecule crystals (kJ mol^{-1}). Errors *vs.* the semi-experimental values are given in parentheses

	Semi-exp.	MP2/aug-cc-pVTZ + AMOEBA	MP2C/CBS + periodic HF	B86bPBE-XDM
CO_2	28.4	27.7 (−0.7)	26.3 (−2.1)	23.9 (−4.5)
Ice	58.8	55.4 (−3.4)	57.4 (−1.4)	67.9 (9.1)
Acetic acid	72.8	64.5 (−8.3)	68.7 (−4.1)	73.5 (0.7)
Imidazole	86.8	101.2 (14.4)	88.8 (2.0)	90.8 (4.0)

3 approach: if DFT modeling of a given system proves inconclusive, the results can be refined with moderate amounts of additional computational work.

Embedding: First let us clarify that we do not embed in the traditional sense (*e.g.* there is no electrostatic embedding). Rather, the quantum mechanical monomers and dimers are computed in isolation, and those terms are used to replace the corresponding polarizable MM contributions from the super-system calculation.[6] That said, one parameter in our model is the distance interval over which one should transition smoothly from the short-range dimers treated at the high level (*e.g.* MP2 or MP2C) to long-range dimers modeled at the low level (AMOEBA or periodic HF). We typically choose a conservative interval of 9–10 Å for this transition, which ensures that the important pairwise interactions are well described by the high-level method. We have examined these cut-offs in some detail previously.[12] For the many-body contribution, which is evaluated solely at the low level of theory, replacing AMOEBA with periodic HF provides an improved description and insight into the approximation made using AMOEBA. In addition to the data shown in the *Faraday Discussions* paper (DOI: 10.1039/c8fd00048d), the positive impact resulting from this model improvement is also demonstrated in our recent work on the methanol phase diagram.[13] Switching from AMOEBA to periodic HF for many-body treatment brings the predictions of the both the sublimation enthalpy and the molar volume of alpha methanol into excellent agreement with experimental results.

1 G. J. O. Beran, *Chem. Rev.*, 2016, **116**, 5567–5613.
2 A. Hesselmann, *J. Chem. Phys.*, 2008, **128**, 144112.
3 L. A. Burns, M. S. Marshall and C. D. Sherrill, *J. Chem. Phys.*, 2014, **141**, 234111.
4 S. Wen and G. J. O. Beran, *J. Chem. Theory Comput.*, 2012, **8**, 2698–2705.
5 J. Řezáč, C. Greenwell and G. J. O. Beran, *J. Chem. Theory Comput.*, 2018, **14**, 4711–4721.
6 G. J. O. Beran and K. Nanda, *J. Phys. Chem. Lett.*, 2010, **1**, 3480–3487.
7 S. Wen and G. J. O. Beran, *J. Chem. Theory Comput.*, 2011, **7**, 3733–3742.
8 Y. Huang, Y. Shao and G. J. O. Beran, *J. Chem. Phys.*, 2013, **138**, 224112.
9 M. A. Neumann, F. J. J. Leusen and J. Kendrick, *Angew. Chem., Int. Ed.*, 2008, **47**, 2427–2430.
10 S. Grimme, A. Hansen, J. G. Brandenburg and C. Bannwarth, *Chem. Rev.*, 2016, **116**, 5105–5154.
11 J. Hoja, A. M. Reilly and A. Tkatchenko, *WIREs Comput. Mol. Sci.*, 2017, **7**, e1294.
12 S. Wen, K. Nanda, Y. Huang and G. J. O. Beran, *Phys. Chem. Chem. Phys.*, 2012, **14**, 7578–7590.
13 C. Červinka and G. J. O. Beran, *Chem. Sci.*, 2018, **9**, 4622–4629.

Ivo Rietveld opened the discussion of the paper by Grahame R. Woollam: To what level do the results of the crystal structure prediction help explain the experimental results: the observed stability behaviour of the polymorphs and the conditions under which those polymorphs can be obtained?

Grahame Woollam responded: The experimentally observed stability relationship is only reproduced in the crystal structure landscape when the energy contribution of disorder is taken into account. The disordered model of Form I is found to be lower in energy than all of the other predicted structures and there is no indication of a missing thermodynamically more stable form, which represents the outcome of the experimental polymorphism screens. Experimentally Form II is metastable to Form I, converting during competitive ripening studies. The calculated internal energy difference for Forms II and I is 0.54 kcal mol^{-1}, which compares well to the experimentally measured (by DSC) exothermic crystallisation enthalpy of 0.33 kcal mol^{-1} (following the melt of Form II). Concerning the conditions to prepare Form II, it cannot be formed by a simple recrystallisation procedure; we need to do something extreme to prepare the metastable form. The toluene/TBME method gives it; in retrospect, we can access both conformers from solution. The kinetically driven process results in a structure that accommodates both conformers in the frozen-in disordered Form II structure. From the crystal structure landscape, we see that the Form II disorder model moves up in relative energy in between the single components of ranks 1 and 58. Conversely, the Form I disorder model moves down to a lower relative energy than either single component (ranks 11 and 45) by more than $kT \ln 2$.

Graeme M. Day enquired: You have shown in this paper that, by properly modelling disorder in Form I, the prediction results improve and disordered Form I gets resolved to below the global lattice energy minimum of the landscape of ordered crystal structures. To be confident in now having the correct energy ranking, wouldn't we have to look at every structure on the landscape and assess whether it has disordered configurations? Have you looked at modelling the disorder in some of the other predicted structures to see if they gain a similar amount of stabilisation? In terms of assessing disorder as part of a workflow, could you also comment on how automated the procedure could be made: identifying structures with disorder, creating symmetry adapted ensembles and assessing the resulting free energy? Identifying structures that could be disordered seems to be the trickiest step.

Grahame Woollam replied: The contrast of the *ab initio* crystal structure landscape featuring the single component of experimentally metastable Form II as rank 1 prompted a closer look into the structures containing two molecules in the asymmetric unit. If we accept that in the case of loratadine the ordered structures containing two molecules in the asymmetric unit provided insight into the relevance and importance of configurational variation, it is even more informative that only families 1 and 2 contained $Z' > 1$ (Forms II and I, respectively). As such, we have explored all families that may exist in other ordered arrangements with lower symmetry that were not generated by the CSP procedure.

This is a pertinent point; we are actively working in this area. A stepwise approach is required; the largely automated workflow follows the order identified by Graeme M. Day. For the moment, we are working pragmatically by benchmarking using systems with experimentally determined disorder. Indeed, identifying *a priori* that a molecule could feature thermodynamically favourable disorder is a challenge. However, going back to the CSP of loratadine, the $Z' = 2$ structures were informative; Form I conformers 1 and 2 were found 18 and 19 times, respectively. Of those 18 times, conformer 1 (AAAA) coincided with a second conformer 1 within a $Z' = 2$ structure just once (*i.e.* AAAA and AAAA). Whereas eight out of the 38 $Z' = 2$ structures contain both Form I conformers 1 and 2 (AAAA and AABA), suggesting that the low-energy structures containing both conformers 1 and 2 result in a thermodynamically stable product, owing to the number of times they were generated *ab initio* (thermodynamically favourable). Additionally, all of the AAAA and AABA structures are within 0.90 kcal mol^{-1} of the lowest-energy structure.

Conversely, Form II conformers 1 and 2 appear together in a $Z' = 2$ structure just once (ABAA and BBAA) in rank 12, highlighting that the structure is stable at only 0.61 kcal mol^{-1} higher energy than rank 1; however, not as significant as the Form I AAAA and AABA configuration. As such, a thorough investigation of the $Z' > 1$ structures may provide the necessary information to get started on the symmetry adapted ensembles to assess the resulting free energies.

Jason Cole said: Graeme M. Day made the point that it would be interesting to analyse the possibility of lowering the energies of all structures in the predicted landscapes by using the same substitution of conformations. I was wondering whether this had been attempted for any individual examples, other than the experimental cases, to get an understanding of what the overall distortion to the CSP landscape might be expected as a consequence: *i.e.* how common is the motif in the modelled structures that gets energetically improved by considering the disorder?

Grahame Woollam answered: There has been no disorder modelling of the predicted structures outside of Form I (family 2, energy rank structures 11 and 45) and Form II (family 1, energy rank structures 1 and 58). Families 3–5 and the individual structures have not been assessed; the response to Graeme M. Day's question on this paper applies here. For families 1 and 2 the motifs don't change following the introduction of configurational variation; they remain as rods and mosaics coming from the single component systems and matching the experimental structures of Forms II and I, respectively.

J. Christian Schön remarked: The discussion about disorder brings to my mind the issue of time scales and self-averaging. It appears to me that frozen-in disorder is just a matter of time scales, and similarly when we discuss dynamical disorder. It is really important to take the observational time scales into account when deciding whether one should add a configurational entropy term to the free energy. Now, in macroscopic (and often even in mesoscopic) systems, one can never fully equilibrate all the disordered states, since we cannot explore enough of the configuration space of the energy landscape for such systems. However, we can hope to use the self-averaging concept: here we would divide the macroscopic system into many subsystems that are just at the size limit of being a "macroscopic" system, in the sense that we would still recognize them as miniature representatives of the bulk system. If the equilibration time scales of these sub-systems are short enough that we can assume the sub-blocks are pretty close to equilibrium on the observational time scale, then we can approximate the full free energy by a sum over the free energies of each such sub-system, where each sub-system will have a slightly different free energy (as if they were members of a Boltzmann ensemble), but their average will be a good approximation to the total free energy of the macroscopic system (that is where the term "self-averaging" comes from, of course). Here, it is important to keep in mind that when trying to identify systems with "controlled disorder", such as solid solutions (the disorder only applies to the distribution of *e.g.* two cations on the positions of one cation in the aristotype of the solid solution structure, but not to the underlying cation–anion–lattice structure!) during a crystal structure prediction using global optimization techniques, we need to analyze all the minimum structures found and place them into structure families – often there are several structure families within one chemical system (*e.g.* a rocksalt family and a wurtzite family in mixed alkali–halide compounds[1]). This can lead to several solid solution phases present in the same diagram with different lifetimes,[2] *i.e.* on short observational time scales the system can be in one solid solution phase or the other (without knowing that another feasible – perhaps more stable – solid solution phase exists in the system)! Proceeding in this fashion for molecular crystals by identifying families of local minimum structures sounds very promising.

On another point, you also discussed the asymmetric unit in your study, but I am not quite sure that this makes sense – once you insert disorder, the symmetry of the structure is lost, and thus the asymmetric unit becomes the full set of atoms/molecules in the cell. So how can you proceed the way you do once this symmetry is lost?

1 J. C. Schön, I. V. Pentin and M. Jansen, *J. Phys. Chem. B*, 2007, **111**, 3943–3952.
2 M. Jansen, I. V. Pentin and J. C. Schön, *Angew. Chem., Int. Ed.*, 2012, **51**, 132–135.

Grahame Woollam responded: In our paper we used symmetry-adapted ensemble theory to compute the configurational free energy of Form I. The free energy of this form is associated with disorder that is a thermodynamic equilibrium property. The approach is indeed only valid if the system under consideration has the time to equilibrate. Since neither crystal structure prediction nor symmetry-adapted ensemble theory provide rate constants for configurational changes, we take a pragmatic approach to decide whether the method is applicable. We first use symmetry-adapted ensemble theory to compute occupancy factors for each molecular conformation. If the calculated occupancies match the experimental occupancies to within a few percent, we take this as a confirmation that the system is indeed close to equilibrium and that symmetry-adapted ensemble theory applies. If, on the other hand, the computed and experimental occupancies strongly disagree, we conclude that the disorder is not an equilibrium property. Experimental occupancies are determined for the atoms in the asymmetric unit. In symmetry-adapted ensemble theory, individual configurations indeed break the crystal symmetry, but not so the actual ensemble. By averaging over all configurations of the ensemble and their symmetry copies, a computed asymmetric unit with computed occupancies is obtained that can be compared to the experimental results. It is important to realise that molecular crystals can get into a state close to thermodynamic equilibrium even if the rate constants for configurational changes in the bulk are so slow that the disorder appears static. When crystals grow, there is a surface layer in which molecules are significantly more mobile than in the bulk and are subjected to intermolecular interactions that resemble the crystal field of the bulk. The surface layer can hence equilibrate towards a bulk-like equilibrium state. In such cases, the standard textbook formulas for the calculation of the configurational free energy apply at the temperature of the crystal growth process.

*P*1 structures are commonly used to construct or depict a supercell for simplicity. However, the supercells of loratadine do maintain symmetry elements; this is a key aspect of the SAE model, and the disordered components interacting with one another.

Aurora J. Cruz-Cabeza queried: I would like to know whether the disorder in loratadine Form I is static or dynamic.

Grahame Woollam answered: The ratio of Form I conformers 1 and 2 changed with increasing temperature; conformer 1 increased from 0.51 at 10 K through to 0.55 at 273 K, with the change marginally outside of 3 times e.s.d. At 353 K (just 54 K below the melting point of Form I) the increase in ratio was discontinuous.

Aurora J. Cruz-Cabeza asked: If disorder in Form I is dynamic, then rotation of the group would need to be considered through molecular dynamics. This may be a limitation for your static model of disorder and poses a challenge for the computation of entropies and free energies.

Grahame Woollam replied: Dynamic disorder is temperature dependent disorder; could we consider that Form I is dynamic based on the information above? Static disorder is considered random and free to orient; while the carbamate tail has a number of configurations available, the "forward" and "back" positions are favoured in the crystal systems with alternating frequency. The disorder models don't represent the model as random and free to orient, but instead thermodynamically favourable through alternating configurations in microdomains. Substitutional disorder, which is largely seen in inorganic chemistry, is where chemically inequivalent species replace the host molecules in a substitutional manner, as the name suggests. If this is considered to be like static disorder, perhaps Form I could be considered a chemically equivalent solid solution. Definitions are taken from Disorder in crystal structures: new approaches in finding the best model, Markus Neuburger, Basel, 2012.[1] The QM static calculation may be regarded as a microstate of what an MD simulation could provide (with improved accuracy). The static calculation may represent a frame from the MD trajectory(?). However, we do not capture the information over time scales and temperature using the static approach. The experimental occupancy measured in Form I changes with temperature, although marginally. As such, the difference is subtle, which may ultimately be within the error associated with the MD force field calculations.

1 M. Neuburger, Disorder in crystal structures: new approaches in finding the best model, PhD thesis, University of Basel, 2012, DOI: 10.5451/unibas-005832349.

Aurora J. Cruz-Cabeza said: The disorder present in loratadine Form II is static. The molecules are "frozen" in the conformations available during crystallisation. I wonder if crystallisations of Form II have been carried out at different temperatures and whether or not a change in occupancies has been observed.

Grahame Woollam responded: Many crystallisations have been attempted to prepare Form II; the conditions are specific and prescribed, and as such the form is crystallised with an onset *ca.* 20 °C and isolated at 4 °C. No other preparations have yielded Form II; instead Form I is prepared from the same solvent and solute composition if isolated at higher temperatures. The best Form I crystals, for example (and those used in the vT-SCXRD experiments), were obtained from toluene/TBME isolated at room temperature. The Form II single crystal structure presented in the paper was solved at 10 K using Synchrotron radiation. A 100 K structure solution was made using equipment available in-house; however the quality of the data did not allow the same conclusions to be drawn for Form II as for Form I. This relates to the size and quality of the metastable crystals formed under severe kinetically driven conditions.

Rui Guo asked: If loratadine Form II is indeed frozen-in during the specific crystallization setup, can we assume that the global minimum found in the CSP is still out there, waiting to be found? Are there any temperature-varied experiments on Form II, similar to those for Form I in Table 3?

Grahame Woollam answered: If the question relates to the rank 1 structure, which is an ordered component of Form II, given our time frame and crystallisation conditions, it is hypothesised that Form II is an accommodating structure that is adopted due to its ability to feature the two configurations of the cycloheptane head group. The structures of family 1 are less numerous on the crystal landscape, and from a solution mediated process we would expect to convert to the more stable Form I as the crystal landscape is dominated by this family of structures; as such, many nucleation avenues lead to this dynamically favoured form. It may be postulated that the Form II structure could anneal in the solid state to a fully ordered single-component structure given sufficient time, the correct temperature and perhaps humidity to enable relaxation. However, the saddle point for conversion of the cycloheptane head group from the BBAA to the ABAA conformation in this accommodating structure is increased (as calculated from the Form II disorder model structure) to 2.5 kcal mol^{-1}, as compared to 2.0 kcal mol^{-1} in the Form I disorder model structure. Regarding vT-SCXRD, the response to the previous question applies.

Rui Guo queried: With regard to the supercells you used to calculate the energy gain from configuration disorder, how sensitive is this energy to the size and shape of the supercells?

Grahame Woollam replied: When disordered sites are close enough to interact strongly, the isolated site model is not applicable and in principle one needs to consider all combinations of disordered configurations in a cell or supercell. However, many of these combinations are symmetry equivalent, and by limiting lattice energy calculations to one representative per symmetry equivalent set and working out the correct multiplicities, the computational effort can be greatly reduced. This approach is called symmetry-adapted ensemble theory. Typically the number of molecules per cell or supercell would be the right choice to feature in the calculation of energy gain, though with Forms I and II of loratadine an alternative selection allowed for the conservation of computational time. In both cases it was clear from geometrical calculations that the disordered sites interact along one-dimensional rods, with little interaction between the rods. Therefore, the energies refer to the number of molecules per rod, with two rods per supercell. The evaluation of the energy differences was again based on the energies of the entire supercell. As such the energy calculations are not sensitive to the size and shape of the cell/supercell.

Rui Guo asked: Did you use the supercells of the same size and shape for both forms of loratadine? Isn't it reasonable to assume that a more balanced choice of supercells should be based on their specific ways of molecular packing, which as you have shown clearly for the two forms are indeed rather different?

Grahame Woollam answered: The supercells were constructed independently; both feature doubling of the shortest axis (incidentally the *b* axis in both systems, and as such 1×2×1 supercells resulted) in order to capture the interactions between neighbouring disordered molecules – translated and featuring the stacking of the tricyclic ring systems.

Jason Cole enquired: Have any efforts been made to explore incommensurate refinements or possible very large supercells with respect to the crystallography to resolve or more accurately model the underlying statistical disorder in the forms?

Grahame Woollam responded: No refinements were made to the experimental structures to resolve an incommensurate structure or very large supercell.

Jason Cole asked: Were satellite peaks observed in the structural data, as these would be indicative of a more ordered supercell?

Grahame Woollam replied: There were no interpretable additional peaks observed for any structures.

Sten Nilsson Lill enquired: In relation to the comment by Aurora J. Cruz-Cabeza, have you investigated the barriers and dynamics involved for interconversion of the conformers seen in the disordered parts of your crystal structures I and II?

Grahame Woollam answered: There have not been any MD calculations to investigate the flipping of the alternating configurations in the crystal. However, the saddle point for conversion of the cycloheptane head group from the BBAA to the ABAA conformer in the Form II disorder model structure was calculated to be 2.5 kcal mol^{-1}, as compared to 2.0 kcal mol^{-1} in the Form I disorder model structure. This encourages the idea that the alternative cycloheptane head configuration is frozen-in. In terms of varying the ethyl carbamate tail in Form I, Table 5 demonstrates the energy gain or penalty from having a favourable or unfavourable assembly, with the 1×2×1 supercell featuring the configuration of conformer 2, 1, 2, 1 in the abcd configuration, *i.e.* a 50 : 50 occupancy at 0 K (see Fig. 14).

Susan Reutzel-Edens remarked: You have indicated that Form II, the polymorph with "frozen-in" disorder, was obtained under a very specific set of conditions, so perhaps it is unsurprising that only a limited range of disorder was ever observed for this polymorph. As it is not unusual to find limited conditions in which to nucleate a specific polymorph, I am curious as to whether you have considered seeding to possibly grow Form II crystals under diverse conditions to assess the accessible range of disorder experimentally. Can you please comment?

Grahame Woollam responded: This is a valid point; perhaps the limited range of crystallisation conditions provide a limited range of disorder observed experimentally, because the structure accommodates those configurations available from the solution to the interface/bulk of the crystal. On the flip side, it could be that the accommodating structure is only formed under such conditions, whereby the available solution conformers are provided at a specified rate during desupersaturation through both the cooling crystallisation and self-induced drown-out concomitantly to form such a metastable structure, which is fleeting in any other organic solvent systems, yet long-lasting in the solid state.

Sharmarke Mohamed said: This work has shown that the crystal form landscapes generated from crystal structure prediction methods could be used to rationalise the observed crystal disorder for conformationally flexible molecules. It is promising to see that GRACE is capable of locating the two ordered components of each disordered structure of loratadine. Could the authors comment on the capabilities of GRACE in modelling cases of dynamic proton disorder (*i.e.* in ionisable molecules leading to salts or cocrystals) where you have covalent bond breaking and formation as a result of proton disorder? This is not just a question that is of general academic interest (although the problem of proton disorder clearly continues to be a challenge for crystallographers), but as was evident in the discussions, there are clear regulatory requirements in formulating active pharmaceutical ingredients in solid forms with well-defined covalent bonding. The relative energies of the neutral or ionised forms of molecular complexes can be significant and so it is interesting to know whether GRACE can suggest the correct molecular complex (salt or cocrystal) as well as the energy barrier/stability for proton migration.

Grahame Woollam replied: The prediction of proton positions and proton disorder in molecular crystals of pharmaceutical interest is still a challenge. The problem is not specific to GRACE, but rather related to the underlying level of theory in the computation of the potential energy hypersurface. According to the experience of one of the authors (Marcus Neumann), calculations with the PBE functional and the Neumann-Perrin dispersion correction have a substantial failure rate in identifying the correct proton position in cases where the proton position is ambiguous.

Doris E. Braun remarked: In Table 4 of your manuscript an enthalpy of recrystallisation of 0.33 kcal mol^{-1} is given, which is also compared to the internal energy difference obtained for Forms II and I (0.54 kcal mol^{-1}). Does the value really correspond to the enthalpy of recrystallization? How was it measured/integrated?

Grahame Woollam responded: The value relates to the exothermic transition (crystallisation) measured by DSC following the melt of Form II and prior to the subsequent melt of Form I.

Doris E. Braun commented: The enthalpy of fusion rule[1] is often applied to derive the energy differences between polymorphs, which gives an energy difference (enthalpy of transformation) of 1.74 kcal mol^{-1} for the two loratadine polymorphs.

1 A. Burger and R. Ramberger, On the polymorphism of pharmaceuticals and other molecular crystals: I. Theory of thermodynamic rules, *Microchim. Acta*, 1979, 2(3–4), 259–271.

Grahame Woollam replied: For the 10 K min^{-1} heating rate and conditions applied, this value is correct and represents the difference in enthalpy of fusion between the polymorphic pair, assuming full and complete melting, without interference of the recrystallisation exotherm.

Doris E. Braun asked: By integrating the melting peak of Form II (5.09 kJ mol^{-1}) and the recrystallisation peak of Form I (−6.83 kJ mol^{-1}, if complete recrystallisation occurred), the value should be exothermic and not endothermic.

Grahame Woollam answered: It is indeed exothermic.

Virginia Burger enquired: I have a question about how each of the partially occupied conformers is identified within a disordered polymorph. Is disorder identified within the electron density map and are the individual conformers then fitted to the map and assigned occupancies based on their electron densities? Are the actual partially occupied conformers visible within the electron density map? Or could, for example, polymorph Form I have been assigned three (or more) partially occupied conformers instead of just two?

Grahame Woollam responded: We see two distinct conformations in the electron density.

Virginia Burger asked: In the electron density for polymorph Form I, does the crystallographer see two distinct configurations (for conformers 1 and 2), or is the data just "blurry"?

Grahame Woollam answered: We see two distinct conformations in the electron density.

Michael Ruggiero opened the discussion of the paper by Samuel A Jobbins: What happens when the system requires some large-scale reorientation, perhaps occurring along a non-intuitive pathway? What limitations are there to the *metashooting* technique? Do you limit yourself by inputting the transition state directly? How prone is the algorithm to "missing" the right mechanism?

Samuel Jobbins responded: Thanks for your questions. The transition mechanism is deduced by implementing "plain" transition path sampling, which is completed prior to the *metashooting* procedure being carried out. Thus, by the time the *metashooting* takes place, the most likely subset of the transition path ensemble has been obtained (which is what is sampled using the *metashooting*). The shooting steps that take place in the *metashooting* part of the work are to generate the correct velocity distributions for new trajectories on the biased energy surface, in order to ensure control in the variation of the collective variable.

Large-scale reorientations may occur in the transition path sampling runs. When one generates an initial trajectory using the geometrical–topological procedure we have followed, it is quite easy to deliberately avoid unfavourable regions of the trajectory space. If any large-scale reorientation is to occur, the path sampling algorithm should find it – however, it may take more iterations of the algorithm before the converged subset of trajectories is found.

Inputting the transition state directly should not itself not be a limitation – in fact, we consider it a positive of the method. The transition state for both the transition path sampling and the *metashooting* must be dynamical, as it is mandatory that it links the two basins of attraction. The initial path sampling intermediate is generated from the geometric trajectory, and the initial *metashooting* intermediate may be any point along a (converged) trajectory that links the two basins. From here, shooting to the two basins and filling up from the "outside-in" allows for accurate weighting of the entire underlying free energy landscape.

Finally, the algorithm should not "miss" the ideal trajectory. Transition path sampling, by definition, should always find the most favourable subset of transition pathways in trajectory space. It is these that are biased upon in the *metashooting* technique, and the trajectories that we report in the paper. Further information can be found in papers on transition path sampling.[1–3]

1 C. Dellago, P. G. Bolhuis, F. S. Csajka and D. Chandler, *J. Chem. Phys.*, 1998, **108**, 1964–1977.
2 P. G. Bolhuis, D. Chandler, C. Dellago and P. L. Geissler, *Annu. Rev. Phys. Chem.*, 2002, **53**, 291–318.
3 C. Dellago, P. G. Bolhuis and P. L. Geissler, *Advances in Chemical Physics*, ed. I. Prigogine and S. A. Rice, John Wiley and Sons, 2002, pp. 1–78.

Michael Ruggiero asked: How do you determine which trajectories to follow? Since you start from a pre-determined intermediate, what if your initial transition state is inaccurate? Is it possible to seed the system with specific information, *e.g.* vibrational normal modes, to help "steer" the algorithm?

Samuel Jobbins replied: Thank you for your intriguing questions. The trajectories utilised in the *metashooting* component of the work are extracted from the prior transition path sampling scheme, which converged upon the most likely subset of reactive trajectories linking two basins of interest after more than two hundred iterations. For more information on the path sampling procedure and how it converges upon the most plausible subset of the transition path ensemble, I'd recommend reading the papers which illustrate the method in detail.[1–3]

In our work, we generated an initial trajectory using a geometrical–topological technique involving the transformation of periodic nodal surfaces, as used in previous work by our group.[4,5] Intermediates could then be generated by interpolation, and a structure which linked the two basins (by propagation with MD forwards and backwards in time) was selected to be the initial intermediate. One of the advantages of shooting-off from a generated intermediate in this way is that it does not matter at all whether or not the initial transition state is a plausible starting structure – the starting trajectory need only link the two basins of interest. In fact, it is highly likely that none of the intermediate states in our initial trajectory have any physical relevance at all. The TPS algorithm will gradually steer the system away from this initial unfavourable regime towards a more favourable one. Thus, the initial regime should not matter – however, generating an energetically reasonable first trajectory certainly decreases the number of TPS iterations required before initial trajectory decoherence, and minimises the computing expenditure required to move towards more relevant regions of trajectory space.

Another advantage of this geometrical–topological approach is that, if desired, certain characteristics can be hard-coded into the initial trajectory. The non-linear mapping from the two limiting configurations proceeds according to a set of particular deformation modes that are defined within the model. As a result, specific features may be deliberated encoded (or avoided!) in the initial pathway to greatly reduce the number of path sampling iterations required. Additionally, a direct appreciation of the favourability of a particular model can be ascertained by monitoring the conservation or disappearance of initial transition motifs whilst the path sampling procedure is being performed.

1 C. Dellago, P. G. Bolhuis, F. S. Csajka and D. Chandler, *J. Chem. Phys.*, 1998, **108**, 1964–1977.
2 P. G. Bolhuis, D. Chandler, C. Dellago and P. L. Geissler, *Annu. Rev. Phys. Chem.*, 2002, **53**, 291–318.
3 C. Dellago, P. G. Bolhuis and P. L. Geissler, *Advances in Chemical Physics*, ed. I. Prigogine and S. A. Rice, John Wiley and Sons, 2002, pp. 1–78.
4 S. E. Boulfelfel and S. Leoni, *Phys. Rev. B.*, 2008, **78**, 125204.
5 H. G. von Schnering and R. Nesper, *Z. Phys. B: Condens. Matter*, 1991, **83**, 407–412.

Marcus Neumann said: You mentioned that it's difficult to pick the order parameter, but is that also true if the start point and end point of a transformation are known? What if the start point and end point are predicted structures in the crystal energy landscape of a flexible organic molecule? Are there any general rules for the construction of the order parameter?

Samuel Jobbins answered: Thanks for your questions. Selecting the collective variable(s)/order parameter(s) is often the most difficult part of the job! There are a number of general rules for constructing suitable collective variable(s):

(1) The collective variable(s) must distinguish between the initial and final configurations of interest, and (if appropriate) all relevant intermediate and transition state regions, meaning no overlap between regions of importance;

(2) The collective variable(s) must be able to describe all of the events of importance in a given transformation, particularly those that are "slow", and tolerate fluctuations in thermal energy;

(3) The number of collective variables should not be too large – otherwise the calculation will be prohibitively expensive.

It is immediately apparent, however, that points (2) and (3) can be quite mutually exclusive, as characterising all the necessary reaction coordinates to describe a particular system can require quite a number of collective variables! In addition, omitting a relevant collective variable can lead to basin over-filling or failure to sample relevant regions of configuration space. (Indeed, this is why we have included the 3rd coordination sphere in the work as well as the 1st, as biasing only the 1st coordination sphere proved too coarse to distinguish between competing intermediates and produced an incorrect energy scheme.[1,2]) In my opinion, the best approach is to characterise the system using as few collective variables as possible, but by cleverly designing them such that they are truly "collective" and that each order parameter encapsulates as much information about the system and the transformation of interest as possible.

Knowing the start and end points of a simulation can obviously help to intuitively deduce a relevant set of collective variables. Sometimes it is quite evident – for example, when simulating a chemical reaction, a certain bond length might be the only variable of interest. However, in more complex systems with many degrees of freedom, ascertaining a fundamental set of collective variables can be a very non-trivial task. For example, in the case of flexible organic molecules, a reasonable start would be to address which particular bonds, angles, torsions, *etc.* play a vital role in the transformation under scrutiny. Some combination of these important reaction coordinates might then prove sufficient to correctly generate a free energy profile using metadynamics or the *metashooting* procedure.

One final point – in the scheme described in the paper, we used TPS initiated from a model first trajectory. As such, knowing the start and end points of the simulation are mandatory if one wishes to utilise transition path sampling and the *metashooting* technique in its current formulation. But, as I've said, whether or not this helps to determine the collective variables depends highly on the nature of the system and the event under scrutiny.

1 A. Laio and M. Parrinello, *Proc. Natl. Acad. Sci. USA*, 2002, **99**, 12562–12566.
2 A. Laio and F. L. Gervasio, *Rep. Prog. Phys.*, 2008, **71**, 126601.

Julian Helfferich remarked: We have so far discussed the collective variables, but I would like to address the methods. How far does your *metashooting* procedure depend on the methods you have chosen, metadynamics and the shooter method? Could you exchange them for other, potentially cheaper, biasing and transition path methods?

Samuel Jobbins responded: Thank you for your question. The *metashooting* procedure is by its very nature dependent on its two component techniques – metadynamics and transition path sampling. The TPS steps in the *metashooting* procedure control the variation in the collective variable, ensuring that it remains within the trajectory space corresponding to the reactive transition path ensemble. The metadynamics component deposits the bias within only this relevant region, allowing for the underlying energy surface corresponding to the transition to be explored efficiently and in precise detail. It is conceivable that the two core methods could be "switched out" for other biasing/transition path methods. However, we have illustrated that the union of TPS and metadynamics currently implemented can produce exquisite and accurate detail, particularly of the intermediate and transition state regions. Whether or not changing the methods for others would have an effect on the accuracy or the expense of the work, or whether unifying other techniques in this manner would be as robust or effective, remain open questions. For myself, rather than changing the methods utilised, I would advocate tuning and optimising the parameters involved in the two techniques of the procedure, in order to maximise the efficacy and speed of the *metashooting* analysis.

Pablo Piaggi said: I would like to comment on the performance of metadynamics. First of all, there is a common misconception that the metadynamics algorithm becomes slower as the simulation progresses. This is only true in the most rudimentary implementation, in which an increasing number of Gaussians have to be summed in order to calculate the bias potential. In modern implementations, a grid is used to construct the bias potential and therefore there is no need to sum all the Gaussians.[1] As a consequence, the cost of the algorithm is constant in time. Furthermore, other methods, such as adaptive biasing force, are also based on collective coordinates (CVs) and typically the calculation of the CVs is the bottleneck of these methods.

1 https://plumed.github.io/doc-v2.4/user-doc/html/_metadyn.html

Samuel Jobbins replied: I wholly concur with this contribution. Thank you for bringing it to attention!

Pablo Piaggi remarked: It has been shown[1] that in finite size systems the barriers can be different to those in an infinite system. In the case under consideration the transformation will probably proceed by nucleation and growth, and I suspect the size of the simulation box is smaller than the critical cluster. Did you assess whether the size of the system affects the results?

1 M. Salvalaglio, C. Perego, F. Giberti, M. Mazzotti and M. Parrinello, Molecular-dynamics simulations of urea nucleation from aqueous solution, *Proc. Natl. Acad. Sci. USA*, 2015, **112**, E6–E14.

Samuel Jobbins responded: Thanks for your question. We have not directly tested whether or not the size of the system would affect the results – however, we strongly believe that it would, as it is likely that the precise nature of the motifs seen in the transition path sampling runs is highly dependent on the simulation system size. Addressing this would be a useful validation of the method, and indeed would help to determine whether or not the system can be suitably scaled-down and studied by *ab initio* methods. However, based on previous work,[1,2] we believe that the system size utilised produced representative motifs and nucleation/growth events, whilst keeping the number of atoms in the system tractable for efficient MD/metadynamics simulations.

1 S. E. Boulfelfel and S. Leoni, *Phys. Rev. B.*, 2008, **78**, 125204.
2 D. Zahn, Y. Grin and S. Leoni, *Phys. Rev. B.*, 2005, 72, 064110.

Sarah L. Price enquired: It is really spectacular that you have identified intermediate basins and transition states in this solid state transformation of ZnO. Your calculations were done with the simple exp-6 interatomic potentials that Robin Grimes derived in 1993, and were later validated for ZnO in 2007 in your group, but this validation is unlikely to have included these local basins and transition state regions of the potential energy surface. Please can you comment on the extent to which your results could be an artefact of the potentials used?

Samuel Jobbins answered: Thank you for your question and for your high praise about the work! The structures represented by the extrema on the free energy surface that are described in this work correspond very closely to the previously reported structures. Notably, the five-coordinated tetragonal and hexagonal forms of ZnO have been discussed at great length in the literature. Indeed, there still remains considerable debate as to the roles of these two intermediates – for example, one *ab initio* study suggested that the tetragonal intermediate was vital to the WZ-RS transformation in ZnO,[1] however an experimental study utilising high resolution angular dispersive X-ray diffraction appeared to indicate that the hexagonal pathway was more favourable for zinc oxide transformations.[2]

Subsequent work has shown that the two intermediates are in direct competition with one another, and that a number of external factors are relevant, such as pressure, temperature and the arrangement of local coordination motifs.[3] In addition, the work of Zagorac *et al.* found a number of the metastable ZnO phases using *ab initio* techniques, some of which correlate directly with the configurations observed in this work.[4] Finally, in the group's previous work, the existence and energy rankings of intermediates found using this force field were validated using density functional theory calculations.[5]

Therefore, all of these works, both theoretical and experimental, report the same (or very similar) metastable configurations as those presented in this paper. As a result, I believe I can say with some certainty that the chances of the intermediate basins having little physical significance is very low – however, as always, further study (perhaps using a different potential model or a higher level of theory) would be a very suitable future endeavour to ensure this!

1 A. M. Saitta and F. Decremps, *Phys. Rev. B.*, 2004, **70**, 035214.
2 H. Liu, Y. Ding, M. Somayazulu, J. Qian, J. Shu, D. Häusermann and H. Mao, *Phys. Rev. B.*, 2005, **71**, 212103.
3 J. Cai and N. Chen, *J. Phys.: Condens. Matter*, 2007, **19**, 266207.
4 D. Zagorac, J. C. Schön, J. Zagorac and M. Jansen, *Phys. Rev. B.*, 2014, **89**, 075201.
5 S. E. Boulfelfel and S. Leoni, *Phys. Rev. B.*, 2008, **78**, 125204.

David McKay queried: The speaker responded to a previous question on the computational expense of metadynamics by saying that particular intermediate states are targeted to avoid spending time in unwanted minima. He also remarked that regarding accuracy, to verify these results he would like to be able to repeat the process using the DFT potential in metadynamics and transition path sampling. Rather than repeating the whole process with DFT, is there any scope for only using DFT as a correction to key structures along the potential energy surface?

Samuel Jobbins replied: Many thanks for your interesting question. It may be conceivable to try to treat the entire transition pathway (including the key structures) using density functional theory on top of the existing results. In particular, a few iterations of the *metashooting* procedure could be carried out using density functional theory, in order to see if there is significant deviation from the obtained results and how the energetics of the new trajectories compare with older ones.

However, both the transformation mechanism (as deduced by transition path sampling) and the free energy surface (produced by *metashooting*) have been ascertained by describing the atomic interactions using a classical Buckingham-style empirical pair potential.[1] One would hope that the scheme and (in particular) the intermediate regions produced by the potential would correlate well with any results that would be obtained using a higher level of theory. Previous work by our group has shown that the intermediates generated by the force field can be stabilised using density functional theory and have the correct energy rankings.[2] However, to be absolutely sure about this work, the entire process could be repeated again with a different potential model or higher level of theory.

1 D. J. Binks and R. W. Grimes, *J. Am. Ceram. Soc.*, 1993, **76**, 2370–2372.
2 S. E. Boulfelfel and S. Leoni, *Phys. Rev. B.*, 2008, **78**, 125204.

David McKay asked: Another consideration when going to more accurate methods is system size, which in this case is large for DFT. Does the study at force-field level provide any pointers as to how the system size could be reduced for DFT calculations, while retaining a reasonable model system?

Samuel Jobbins responded: It is of course very true that the mechanism deduced by transition path sampling is dependent on the size of the system. The system size of 1200 Zn–O pairs was chosen as this is small enough for computational tractability but large enough for critical nucleation and growth events to be visualised.[1,2] To scale down for a density functional theory-type study, it is conceivable that one could ascertain the size of these critical nuclei and the number of atoms required for their formation, and construct a new model system with a sufficient number of atoms to produce such features (with some additional neighbouring atomic sites, so that an interface is present when considering the periodic boundary conditions).

However, one must be very careful when scaling down the system. Making the system too small can completely suppress the formation of key intermediates, and the system may get "locked-in" to a regime of unrealistic trajectories, both in terms of the details which emerge and the time scales over which they occur. I would not be surprised if 1200 Zn–O pairs actually approached the minimum size required to generate such meaningful events, however it would be interesting to experiment with this to properly determine the minimum system size that could be studied with density functional theory.

1 S. E. Boulfelfel and S. Leoni, *Phys. Rev. B.*, 2008, **78**, 125204.
2 D. Zahn, Y. Grin and S. Leoni, *Phys. Rev. B.*, 2005, **72**, 064110.

J. Christian Schön said: We have seen similar intermediates (*e.g.* the 5-5 structure, the ionic analogue of the hexagonal BN structure) on the energy landscapes of related systems,[1–3] which were also stable minima when locally optimized on the *ab initio* energy landscape. Thus, I would believe they also exist in your system. On another issue, I would hesitate to make the system smaller – you might be in danger of locking the system into a limited number of (perhaps unrealistic) pathways between the regions you want to connect. If you recall the water-freezing simulations by Ohmine and co-workers *ca.* 15 years ago,[4] they found that even the large "crystalline" nuclei that had formed inside the liquid phase after about 100 ns vanished again (I think they needed about 200 ns before the whole system crystallized). So, a smaller system size might lead to unrealistically fast freezing rates. Do you ever see paths where seeds of the new phase grow and then vanish again?

You also mentioned that you could measure activation energies – can you also deduce activation entropies, or more generally measure the generalized barriers[5] *via* the probability flows on the landscape of the system between metastable configurations?[6,7] Perhaps *via* effective transition rates?

1 J. C. Schön and M. Jansen, *Comp. Mater. Sci.*, 1995, **4**, 43–58.
2 J. C. Schön, *Z. Anorg. Allg. Chem.*, 2004, **630**, 2354–2366.
3 D. Zagorac, J. C. Schön and M. Jansen, *J. Phys. Chem. C*, 2012, **116**, 16726–16739.
4 M. Matsumoto, S. Saito and I. Ohmine, *Nature*, 2002, **416**, 409–413.
5 J. C. Schön, M. A. C. Wevers and M. Jansen, *J. Phys. Cond. Matt.*, 2003, **15**, 5479–5486.
6 M. A. C. Wevers, J. C. Schön and M. Jansen, *J. Phys. Cond. Matt.*, 1999, **11**, 6487–6499.
7 S. Neelamraju, J. C. Schön, K. Doll and M Jansen, *Phys. Chem. Chem. Phys.*, 2012, **14**, 1223–1234.

Samuel Jobbins answered: Thanks for your correspondence and queries. I appreciate the vote of confidence with respect to the existence of the intermediates and transition states found in the work! I am very confident that they are meaningful metastable minima, as they have been seen in a great deal of previous work, including your own. I also am inclined to agree with your comment regarding system size – it would be interesting to see how small one could go whilst still observing plausible mechanistic details, however my intuition tells me that this limit would not be much smaller than the system size that we have used in this work! During the initial transition path sampling scheme, there are multiple examples of "failed" trajectories, in which a transformation begins, but then "fails" and reverts back to the existing structure. Such failed trajectories were often seen in the very early stages of transformation, particularly from rocksalt to zincblende, around the time of the formation of the cubic seed. Another frequent example of trajectory "failure" was when the transition became "stuck" in the intermediate 5-coordinate basin.

Pairing TPS and metadynamics allows us to globally map the free energy. In transition path sampling, the system still spends the most time around the main basins of attraction, while *metashooting* can weight even elusive intermediates accurately. TPS is a method that allows the generation of reactive trajectories without bias, as single trajectories follow their natural dynamics without external perturbation. Reaction rates are thus accessible by combining TPS with methods like Umbrella Sampling. A more precise outcome is provided by Transition Interface Sampling (TIS).[1] Additionally, the separation of enthalpic and entropic contributions can be addressed in metadynamics by introducing an instantaneous measure of entropy, based on pair correlation functions. All of these measures require a precise knowledge of the free energy, including possible intermediates, which is exactly what is provided by our *metashooting* methodology.

1 T. S. van Erp, D. Moroni and P. G. Bolhuis, *J. Chem. Phys.*, 2003, **118**, 7762–7774.

J. Christian Schön asked: Regarding your shooting algorithm: since there are many competing basins in your system, how can you sit "in the middle" and shoot, and still reach the two basins between which you try to establish a transformation route? Where is the "middle", exactly?

Samuel Jobbins responded: The shooter algorithm here corresponds directly to the original algorithm reported in the transition path sampling literature. More information about the transition path sampling method can be found in ref. 1–3.

When referring to the "middle", we simply mean a dynamical mid-point of the current trajectory that links the two basins of interest. In the beginning of the transition path sampling work, the mid-point is an interpolated intermediate from the geometric trajectory that successfully links the two basins of attraction. Subsequent mid-points are dynamical intermediates from the latest reactive trajectory. In the *metashooting* component, the transformation route has already been established by transition path sampling, and so shooting from the mid-point still reaches the two basins of interest along the converged subset of trajectories. By the time *metashooting* is carried out, the transition path ensemble has been sampled and the most likely subset of trajectories has been found – it is these trajectories which are biased and evaluated by the *metashooting* procedure.

1 C. Dellago, P. G. Bolhuis, F. S. Csajka and D. Chandler, *J. Chem. Phys.*, 1998, **108**, 1964–1977.
2 P. G. Bolhuis, D. Chandler, C. Dellago and P. L. Geissler, *Annu. Rev. Phys. Chem.*, 2002, **53**, 291–318.
3 C. Dellago, P. G. Bolhuis and P. L. Geissler, *Advances in Chemical Physics*, ed. I. Prigogine and S. A. Rice, John Wiley and Sons, 2002, pp. 1–78.

J. Christian Schön enquired: Does your procedure measure the probability flows directly? Also, we should not confuse the local "entropy"/"free energy" measured as a function of the order parameter during the metadynamics phase of the simulations with the "entropic" part of the generalized barrier that is related to the probability flows and thus contributes to the rate constants of the processes being analyzed. After all, a local free energy is only defined on observational time scales that are larger than the equilibration time and shorter than the escape time from the region, so you might be able to measure local densities of states – a static quantity – but it might not be permissible to compute the free energy, which is a cale dependent quantity!

Samuel Jobbins replied: I'm not sure I fully agree with this analysis. The Laio *et al.* paper on metadynamics describes in considerable detail how, at the infinite time limit, the deposited bias is approximately equal to the negative of the free energy, plus an irrelevant additive constant.[1] There are a number of ways that the error in this can be estimated, and also a number of methods which can be implemented to reduce this error. Utilising the well-tempered prescription of metadynamics also helps to overcome any potential issues here associated with over-filling (and hence over-estimating the free energy barriers). However – to reiterate – Laio *et al.* clearly show that the free energy (an equilibrium property) can be very well estimated using non-equilibrium dynamics in metadynamics. This is indeed one of the great selling points of the method, and as such both enthalpic and entropic contributions should be accounted for in the present scheme.

More generally, one can consider the situation in this way: large free energy barriers represent a major limitation for the exploration of reactive events – even more so if *ab initio* methods are used. In "plain" molecular dynamics, the escape time in the presence of energy barriers much larger than k_BT is, in principle, infinite. Metadynamics implements a coarse-grained representation of the true high-dimensional potential energy surface of interest, which will contain a large number of low-lying saddle points. In the lower dimensional coordinate space of the collective variables, the free energy surface is smooth. In this regime, the basin escape time is finite, allowing for a global mapping of the free energy, which converges within a finite time.

1 A. Laio and F. L. Gervasio, *Rep. Prog. Phys.*, 2008, **71**, 126601.

John B. O. Mitchell asked: Systems like chemical reactions, with few degrees of freedom, have tightly defined mechanisms corresponding to a specific trajectory or a narrow range of trajectories across the potential energy surface. Other energy landscapes, like those for protein folding, with very many degrees of freedom, have much looser families of trajectories connecting regions. Would you classify the phase transition you are looking at as a single mechanism, or a broad family of region-connecting trajectories? Do you think your methodology is better suited for tightly defined mechanisms or for broader families of trajectories?

Samuel Jobbins answered: Thanks for your intriguing question. The underlying energy landscape corresponding to a condensed matter phase transition can be very coarse, owing to the many degrees of freedom in the system. This is of course dependent on the number of atoms in the system, but for 1200 Zn–O pairs the complexity of the (complete) energy landscape is probably intermediate between the two examples you have cited here! In this particular transition, there are a great number of transition pathways possible, each of which has a particular probability (as determined by the energy of the transformation). The initial transition path sampling procedure samples reactive pathways in trajectory space and ultimately converges upon the most likely subset of reactive trajectories linking the two basins (prior to the *metashooting* technique actually being utilised). For more information on transition path sampling, I'd recommend taking a look at ref. 1–3.

These lowest energy reactive trajectories correspond to the transition mechanisms reported in the paper. The *metashooting* procedure is then only applied to this lowest energy subset of reactive trajectories in the coarse-grained, lower dimensional space of the collective variables, which is why the final free energy profile looks relatively smooth and well defined. I see no reason why the methodology should not work on more simple or complex cases. Systems with fewer degrees of freedom would be easier to characterise and define in terms of reaction coordinates, and would likely converge more quickly than their larger counterparts. More complicated systems may be more expensive to compute, however if an appropriate order parameter is chosen, transition path sampling should still be able to find the corresponding favourable reactive subset in trajectory space. These trajectories can then be biased using *metashooting* to generate the energy landscape corresponding to the "path of least resistance" for a given system.

1 C. Dellago, P. G. Bolhuis, F. S. Csajka and D. Chandler, *J. Chem. Phys.*, 1998, **108**, 1964–1977.
2 P. G. Bolhuis, D. Chandler, C. Dellago and P. L. Geissler, *Annu. Rev. Phys. Chem.*, 2002, **53**, 291–318.
3 C. Dellago, P. G. Bolhuis and P. L. Geissler, *Advances in Chemical Physics*, ed. I. Prigogine and S. A. Rice, John Wiley and Sons, 2002, pp. 1–78.

Marcus Neumann said: In a periodic supercell, many paths leading from a highly symmetric start point to a highly symmetric end point are related by symmetry. Does your method somehow exploit these energy relationships?

Samuel Jobbins replied: The initial trajectory is generated using a geometrical–topological approach, in which periodic nodal surfaces are wrapped around the limiting structures and transformed according to a pre-defined set of deformation modes. In this respect, symmetry can be incorporated into the initial trajectory. Such a scheme has been used in previous work by our group.[1,2] However, in its current implementation, the actual *metashooting* procedure does not take symmetry into account at all when sampling the trajectory space or applying the metadynamics scheme. However, this could certainly be worth looking into in the future, as it may well help to increase the efficiency of the technique.

1 S. E. Boulfelfel and S. Leoni, *Phys. Rev. B.*, 2008, **78**, 125204.
2 H. G. von Schnering and R. Nesper, *Z. Phys. B: Condens. Matter*, 1991, **83**, 407–412.

Scott M. Woodley communicated: What is the definition of your energy? You state in your paper that you use CP2K and mention the use of Buckingham potentials with parameters derived from Grimes *et al.*, but is this the Shell Model or rigid ions? If the latter, how would you expect your results to change if you added oxygen polarisation into your model?

Samuel Jobbins communicated in reply: In the molecular dynamics and transition path sampling runs, the total energy (as computed by CP2K) was used as the energy function. In the *metashooting*, the free energy is reconstructed as the negative of the deposited bias potential, as explained in the original paper on metadynamics. We indeed use the Buckingham potential of Binks and Grimes,[1] however in this work the oxygen polarisation term was not included. Initial calculations showed that just using the core charges and van der Waals interactions was sufficient to describe the relevant bulk phases of ZnO, with reasonable values for the geometric parameters and transition pressures. I personally suspect that the oxygen shells would actually make little difference to the transition mechanism – however, there may be some subtle effects, and there would be considerable differences if one, for example, wanted to calculate phonon dispersion curves. Perhaps repeating the procedure with the shells (or with another potential model) might be a good future endeavour to help validate the methodology.

1 D. J. Binks and R. W. Grimes, *J. Am. Ceram. Soc.*, 1993, **76**, 2370–2372.

Noa Marom communicated: You showed an example of solid state transformation from one crystal structure to another. Can *metashooting* also be used for amorphous to crystalline transformations?

Samuel Jobbins communicated in response: Thanks for your question! In principle, any system that can be described with a suitable set of collective variables/order parameters for the metadynamics and TPS stages of *metashooting* can be studied using our technique. In fact, we are currently in the process of studying a phase transition between two configurations of a metal–organic framework, which is believe to proceed *via* an amorphous intermediate. Once the most likely subset of trajectories linking the two configurations is found, we intend to apply the *metashooting* method to quantify the underlying free energy surface. Of course, as usual, the difficult part of this is determining which reaction coordinates are relevant to the transition! My advice for treating an amorphous system would be to attempt to exploit any short-range order present in the system, and to use mean coordination sequences or Steinhardt parameters to attempt to globally characterise the system.

Noa Marom communicated: Do you envision using machine learning to identify the best collective variables for a system?

Samuel Jobbins communicated in reply: What a fascinating idea! In principle, the idea of using a machine learning algorithm to substantially reduce the dimensionality of a problem to a handful of tractable collective variables is a very enticing prospect indeed (and arguably is a textbook example of the merits of machine learning). In practice, however, there are often so many possible variables to utilise in each system that it is quite hard to envisage a machine learning algorithm being able to successfully capture the often subtle chemical/physical nuances in every possible system. Often, a great deal of chemical intuition (and sometimes luck!) is required to ensure that only appropriate variables are chosen, and that all of the important variables are considered. I, however, am only just beginning to enter the machine learning world and would love to know if this could be done. Do you have any suggestions about how this could be done in practice? There are so many different systems (and hence collective variables) to study – how would one train such an algorithm to ensure that it could make useful suggestions for such a disparate selection of systems?

Marcus Neumann opened the discussion of the paper by Alexandre Tkatchenko:† Fig. 9 and Fig. 10 seem to suggest that thermal expansion affects the lattice free energy only weakly if phonons are treated in the harmonic approximation, and much more strongly when the anharmonicity of phonon modes is taken into account. Is this interpretation correct?

Johannes Hoja responded: This seems to be the case for this particular system. However, for a general statement further investigations involving a variety of systems would be necessary.

John B. O. Mitchell enquired: Thinking about the things you added to your model, I'm not surprised that you added mode-by-mode analysis of the vibrations. I am looking for an accuracy of 3 kJ mol^{-1} in my applications, and would also expect to require that. However, I am surprised by how much difference the many-body dispersion makes, compared to using just pairwise dispersion. This might be disturbing, since omission of many-body effects seems to lead to significant increases in error. Is this widely true across drug-like molecules? Are there other

† Alexandre Tkatchenko's paper was presented by Johannes Hoja, *University of Luxembourg*.

non-obvious sources of error that we need to be aware of that are currently missing from most models?

Johannes Hoja replied: For the five systems discussed in our paper, the mean absolute deviation between the PBE+TS and PBE+MBD energies varies between 1.8 and 3.2 kJ mol^{-1}. Therefore, we observe significant effects for all systems but their magnitude seems to depend on the molecular size and flexibility. Fig. 2 in our paper shows a correlation plot between PBE+TS and PBE+MBD. Given the spread of the data points, even fitting the best possible pairwise model to the MBD data would not provide sufficient correlation. Note that the differences between PBE+TS and PBE+MBD originate from many-body dispersion interactions and also from the included dielectric screening effects. Therefore, Fig. 3(b) illustrates purely the effect of many-body interactions. It can be seen that contributions of up to about 5-atom interactions are necessary to converge the relative stabilities. A discussion about many-body dispersion effects for molecular dimers, supramolecular systems and molecular crystals can be found in ref. 1 and 2. Non-obvious sources of error for static energy calculations could be revealed by benchmarks on larger and more flexible crystals using Quantum Monte Carlo (QMC).

1 A. M. Reilly and A. Tkatchenko, *Chem. Sci.*, 2015, **6**, 3289–3301.
2 A. Ambrosetti, D. Alfè, R. A. DiStasio Jr. and A. Tkatchenko, *J. Phys. Chem. Lett.*, 2014, **5**, 849–855.

Graeme M. Day asked: I have a question about the validation of crystal structure prediction methods: when evaluating methods used to rank structures, how do we know when we have the correct ranking? Target systems for benchmarking should be chosen carefully and should be experimentally well characterised.

Are the blind test molecules the best systems to revisit when assessing new methods? It does not seem to me that the blind test molecules are selected in a way that makes them good benchmarks. Apart from a few exceptions, such as the polymorphic molecule XXIII, most of the blind test molecules have not been experimentally studied very much at all. They are included in the blind tests simply because their crystal structures have not been published. So, once they are no longer blind targets, I don't think that they are valuable for validation. Should the community agree on some better sets of molecules to use for testing?

Johannes Hoja answered: I agree that the blind test systems are not ideal benchmarks (with the exception of system XXIII). It would be better to create a set of experimentally well-studied systems, for which extensive polymorph screening has been done and for which reliable experimental relative stabilities are available (maybe at different thermodynamic conditions). Ideally, this set should cover a variety of different types of molecular crystals, *i.e.* crystals with different molecular size and flexibility, salts, co-crystals, solvates, and also disordered systems.

Susan Reutzel-Edens said: As a follow-up to Graeme's general question about the use of blind test molecules for benchmarking purposes, the evaluation of systems from the literature that are well characterized experimentally is likely to produce more meaningful results than for systems chosen simply for being blind test targets – unless, of course, experimental validation is forthcoming. In this work, what experimental validation data is available to assess the various energy models that were used for the CCDC blind test molecules?

Johannes Hoja responded: The available experimental data for the systems of the latest blind test[1] consists of the measured crystal structures (5 polymorphs for system XXIII and one form for the other systems). In addition, slurring experiments indicate that form A of system XXIII is the most stable polymorph at 257 K, while form D is the most stable polymorph at 293 K. Furthermore, all 5 polymorphs of XXIII co-exist under ambient conditions. Therefore, we can assume that these 5 polymorphs must have very similar stabilities.

1 A. M. Reilly *et al.*, *Acta Cryst. B*, 2016, **72**, 439–459.

German Sastre enquired: My question concerns large molecules (*e.g.* molecule XXVI in your Fig. 1) containing a large number of dihedral angles, where large variations imply very similar energies. What is the balance between intramolecular dispersion contributions and intermolecular dispersion contributions? There are not many comments in your paper regarding the effect of geometry (in particular dihedral angles) on energy. What degree of accuracy in the dispersion is necessary in order to obtain correct dihedrals?

Johannes Hoja replied: In our methods, we do not distinguish between intra- and intermolecular dispersion interactions. We evaluated the quality of our calculated structures by comparing the cell parameters and the root-mean-square deviation to those of the experimentally measured structures. In these metrics, PBE+TS, PBE+MBD and PBE0+MBD yield structures of quite similar quality (see Table 2). All the optimized structures are available in the Supporting Information. We also explored what impact the structure optimization method has on the resulting PBE0+MBD energies of the five XXIII polymorphs (see Fig. 5). It can be seen that the optimization method used has only a small impact on the PBE0+MBD relative stabilities; the largest observed change in relative energy amounts to 0.8 kJ mol^{-1}.

The similarity of the structures is illustrated in Fig. 1 for system XXVI. Fig. 1 shows an overlay of the two molecules within the optimized unit cells obtained with PBE+TS (blue), PBE+MBD (red), and PBE0+MBD (green). To put this into perspective for dihedral angles, we compared the dihedral angles defined by atoms 1–4 in the plot. PBE+TS yields a dihedral angle of 51.3 degrees, while the PBE+MBD and PBE0+MBD dihedral angles amount to 52.2 degrees and 51.4 degrees, respectively. For this dihedral angle all three methods agree within 1 degree.

Rui Guo remarked: I noticed that in the paper, the construction of supercells are completely based on the unit cell parameters. I'd like to add that this practice largely mirrors that routinely used for metals or inorganic materials where the interacting species in the unit cell, metals or ions, have isotropic interactions. For organic molecular crystals, where intermolecular interactions or even the shapes of the molecules can be very anisotropic, a more informed approach, focusing more on molecular packing and intermolecular interactions, may lead to better

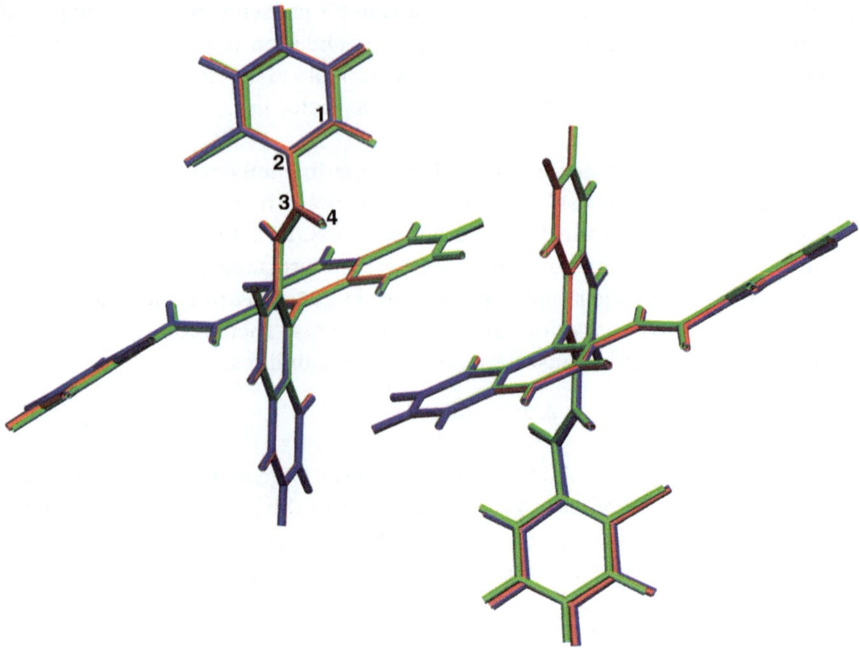

Fig. 1 System XXVI: overlay of the two molecules within the optimized unit cells obtained with PBE+TS (blue), PBE+MBD (red), and PBE0+MBD (green).

computing efficiency. As an example, succinic acid form β has a molecular packing with hydrogen-bonded chains running along the [10–1] direction, and weak CH···O contacts between the chains along the *c* axis. I recently calculated its phonon dispersion curves, and found that due to the specific molecular packing, extending the unit cell along the *c* axis for a supercell converges the integrated Helmholtz free energy much faster than extending along the *a* or *b* axes. It can be rationalized that extending the *c* axis preserves the network of weak CH–O contacts in the supercell, while extending the other two axes would have no such effect. A paper on this subject is currently being prepared with Prof. Keith Refson.

Johannes Hoja answered: Utilizing a more informed approach for the generation of supercells is a great idea for reducing the computational cost. In our paper, we employed supercells with minimal lengths of 10 Å in each direction. For a selected structure set we also evaluated the vibrational free energies using larger supercells with a minimal length of 14 Å. Increasing the supercell size changed the relative stabilities on average by 0.3 kJ mol^{-1}, and the maximum observed change amounts to 0.9 kJ mol^{-1}. For the structure showing the largest energy change upon increasing the supercell size, we plotted the phonon dispersion curves in the Supporting Information for the unit cell and the two calculated supercells. It can be seen that supercells of appropriate size are crucial for the correct description of the acoustic modes.

Luca Iuzzolino asked: With your methodology for optimising and ranking crystal structures, how many do you think you could tackle with a computational expense feasible for a practically useful CSP study?

Johannes Hoja replied: I believe that the lattice relaxation with PBE+TS and a subsequent PBE0+MBD energy calculation would be feasible for up to around 100 structures. Vibrational free energies within the harmonic approximation are in my opinion reasonable for tens of structures, while the quasi-harmonic and Morse free energies should be done for a few selected structures.

Luca Iuzzolino said: The methods you use to calculate the relative energy differences between polymorphs are very accurate but also very computationally expensive. In this study they were applied to a few crystal structures; do you think they would be widely applicable to CSP studies where a large number of putative forms need to be considered? Otherwise you will have to trust the energy ranking produced at the search stage to limit the number of crystal structures that are optimised to a feasible number, and this may lead to a loss of important forms if they are not well-ranked.

Johannes Hoja answered: Our presented approach is intended for the final stage of a crystal structure prediction procedure, assuming an already limited structure set. Therefore, one would need to trust the preceding methods not to miss important forms. However, our method could also function as reference data for the creation of tailor-made force fields or machine-learning models that can be used at the search stage.

Artem R. Oganov remarked: Last week I saw John Perdew, who was excited about his new SCAN functional. The recipe is to combine it with the rVV10 van der Waals functional when one needs to describe van der Waals systems. What is better – SCAN+rVV10 or PBE(PBE0)+MBD? Or perhaps SCAN+MBD (which I haven't seen used yet)?

Johannes Hoja responded: The performance of several density-functional approximations (including PBE, PBE0, SCAN) when paired with the MBD, VV10, and D3 van der Waals (vdW) models are discussed in ref. 1. SCAN+vdW can provide accurate interaction energies when the damping function used is optimized for the studied system. However, the effective range of the SCAN functional is much more system dependent than that of PBE or PBE0. Therefore, the PBE+vdW and PBE0+vdW methods provide a more universal description since the same range-separation parameter can be used for a variety of systems. Note that in contrast to MBD, VV10 is a pairwise method and provides therefore no description of many-body dispersion interactions.

1 J. Hermann and A. Tkatchenko, *J. Chem. Theory Comput.*, 2018, **14**, 1361–1369.

Artem R. Oganov enquired: In your work, you used PBE (or PBE0) as exchange–correlation functional, supplemented with many-body dispersion. Other people recommend SCAN exchange–correlation functional, supplemented with rVV10 functional for the description of van der Waals bonding. Which scheme is better? SCAN seems to be better than PBE; is this your impression too? What about the comparison between many-body dispersion and rVV10 functional?

Johannes Hoja replied: A comparison between PBE and SCAN with different van der Waals models (MBD, VV10, D3) can be found in ref. 1 (see the previous question).

1 J. Hermann and A. Tkatchenko, *J. Chem. Theory Comput.*, 2018, **14**, 1361–1369.

Gregory J. O. Beran said: Your results seem to show that as you refine the energies by increasing the sophistication of your computational models, the energy gaps between structures often decrease. We have seen similar behaviors in our own work. Pessimistically, this suggests that even though the field is increasingly able to predict the energetics accurately, that does not necessarily translate to an increased ability to rank/discriminate between structures. It also ties back to Sally Price's comment in the introductory lecture about how the field has transitioned away from the search for *the* unique crystal structure, whose energetic stability would ideally be well-separated from the other candidate structures/polymorphs, to a more nuanced understanding of how complicated crystal energy landscapes can be.

Johannes Hoja answered: I agree that we will often be faced with complicated crystal energy landscapes, where several structures are in a very narrow energy window. Such cases could lead to the identification of new crystal forms or highlight once again the necessity of also including kinetic effects. However, accurate first-principles calculations are a prerequisite for the understanding of such complicated energy landscapes.

Virginia Burger queried: You mentioned that someone is trying to crystallize the low energy polymorph (structure N70). Did you learn anything from your theoretical analysis that could guide the experimental search?

Johannes Hoja responded: Structure N70 has several similarities with the experimentally obtained Form A and both structures differ mainly by the stacking of the sheets. At the static lattice energy level structure N70 is about 1 kJ mol^{-1} more stable than Form A and is additionally significantly stabilized by vibrational free energies, *i.e.* temperature effects. Therefore, Form A could be dynamically favored over structure N70. We believe that, for example, extremely low melting beginning from Form A could potentially lead to the crystallization of structure N70.

Virginia Burger remarked: Regarding benchmarking: XtalPi has been testing our CSP work on axitinib, which has around five polymorph crystal structures, many more polymorphs and lots of experimental data, and we have found it to be a more useful benchmark case because of all the experimental data. Finally, you mentioned that simulations could provide more accurate free energy results. We have run such simulations for molecule XXIII from the blind test and have found that the more stable experimental structures have the lowest free energies at room temperature, which is exciting. We are putting together a paper about this now.

Johannes Hoja replied: Indeed, molecular dynamics simulations would provide a better treatment of anharmonic effects than our quasi-harmonic/Morse free energies. However, the accuracy of such an approach will also highly depend on the quality of the underlying energy evaluation method.

Jan Gerit Brandenburg communicated: During the discussion of the author's excellent contribution, the importance of many-body dispersion effects was mentioned. I would like to clarify that the statements made crucially depend on the definition of the "body", *i.e.* a dipole excitation *vs.* an effective atomic excitation.

Alexandre Tkatchenko communicated in reply: Indeed, the absolute magnitude of the many-body dispersion energy will depend on at least two critical aspects: (1) the definition of the fragment and its polarizability (*i.e.* infinitesimal volume of electron density, localized orbital, or atomic density fragment), and (2) the definition of the coupling potential between such fragments. However, the point we make is rather independent of such considerations. We have demonstrated that MBD (where atomic fragments are employed) provides a quantitative description of the relative energies of molecular crystal polymorphs when combined with non-empirical hybrid exchange–correlation functionals (such as PBE0). Independently of which atom-pairwise model is used to describe vdW dispersion interactions, the correlation between MBD and such effective pairwise models leads to a spread that is larger than 4 kJ mol^{-1}. This is at least 4 times larger than the desired accuracy of 1 kJ mol^{-1}. We have also demonstrated that the convergence of the many-atom expansion within the MBD method is qualitatively different for various polymorphs. We expect these qualitative findings to hold independently of the method employed to describe many-body dispersion interactions.

David McKay communicated: Section 6 and Table 4 nicely describe the computational costs of the methods involved in the study. PBE0-MBD is the most successful method here, but is 17 times more expensive than PBE-TS. Could this cost increase be elaborated on? I have previously shied away from hybrid-DFT functionals in the solid state, as we tend to see these as prohibitively expensive. I assume there is a cut-off distance for the Fock exchange calculation – is this a parameter that must be converged, and is the cost kept low here with a short distance?

Johannes Hoja communicated in response: Hybrid density functional approximations like PBE0 are significantly more expensive than generalized gradient approximations like PBE because of the necessity to calculate the Hartree–Fock exchange. A detailed description of the periodic Hartree–Fock exchange implementation in FHI-aims utilizing numeric atom-centered basis functions is available in ref. 1. In our calculations we did not use any special settings or cut-off distances for the evaluation of the Hartree–Fock exchange. Note that the relative timings provided in our paper are for a particular molecular crystal of average size, which has 172 atoms in the unit cell. In general, the relative cost will depend on the calculated system and also on the HPC infrastructure used.

1 S. V. Levchenko, X. Ren, J. Wieferink, R. Johanni, P. Rinke, V. Blum and M. Scheffler, *Comput. Phys. Commun.*, 2015, **192**, 60–69.

Volker L. Deringer communicated: Towards the end of your paper you've taken the method to the limit of what can be done (in terms of computing cost), and in the conclusions you suggest that machine-learning methods might help with this in the future. There seem to be several possibilities indeed: one could fit potentials to MBD-corrected input data, or "learn" only the MBD correction, or possibly even thermal properties directly. Could you elaborate on this a bit – what do you think are the most promising directions there?

Alexandre Tkatchenko communicated in reply: Indeed, currently the cost of explicit PBE0+MBD calculations plus vibrational free energies is quite high. However, valuable and reliable reference data can already be produced with this methodology for 10–100 crystalline polymorphs of a given molecule. I see two possible approaches that combine machine learning with PBE0+MBD:

(1) Directly learn the forces/Hessians for different polymorphs and reconstruct a "molecule in a crystal" force field from a finite number of polymorphs.

(2) Learn the PBE0 energy (which should be rather local for closed-shell molecules) and learn the MBD Hamiltonian parameters and compute the MBD energy explicitly.

Machine learning ZPE and thermal properties should also be rather straightforward. We have demonstrated that for gas-phase molecules this can be done with quantitative accuracy.[1] Applying such models to crystals requires some non-trivial extension, but this should be doable.

1 W. Pronobis, A. Tkatchenko and K.-R. Müller, *J. Chem. Theory Comput.*, 2018, **14**, 2991–3003.

Seiji Tsuzuki communicated: You showed that the contributions of the many-body dispersion interactions and vibrational energies to the lattice energies of crystals are not negligible. Do the contributions of many-body dispersion and vibrational energies have a significant impact on the relative stability of polymorphs, or are the contributions of these terms to the relative stability not significant due to cancellation?

Alexandre Tkatchenko communicated in response: As our paper shows, both the many-body dispersion and vibrational free energies are crucial for obtaining the correct description of free energy differences between polymorphs. The many-body energy decomposition shows that many-body effects vary significantly for different polymorphs. In the case of vibrational free energies, the low-frequency phonons depend heavily on the crystal structure and hence lead to substantially different entropy contributions for different polymorphs.

David Bowskill communicated: This approach determines the geometry of organic crystals using the PBE functional and then applies the MBD correction, followed by a proportion of exact exchange using the PBE0 functional. By applying these corrections, the static crystal structure and associated energy no longer reside at local minima. Do you have any evidence of how much this will affect the predicted crystal energies? For example, would full geometry optimisation using the PBE0 functional with MBD correction produce noticeably different results? In addition, how much does this affect the harmonic approximation, which assumes that the molecules are positioned at minima in the crystal energy landscape?

Johannes Hoja communicated in reply: In our paper we studied how the PBE0+MBD energies depend on the optimization method used (see Fig. 5). It can be seen that the structure optimization method has only a small effect on the relative PBE0+MBD stabilities, with a maximum change of 0.8 kJ mol^{-1} in the relative stabilities. The structures used for the vibrational free energy calculations in the harmonic approximation were always optimized with exactly the same method as is used for the harmonic approximation to ensure that we are in a local minimum.

Marcus Neumann opened the discussion of the paper by Jan Gerit Brandenburg: What is the root mean square deviation between the lattice energy differences calculated with DFTB3-D3 and higher level methods for the compounds that you have studied?

Jan Gerit Brandenburg answered: The root mean square (rms) deviation was calculated between the energies of the crystal structures after DFTB3-D3 and the final optimisations. The distributed multipoles and intramolecular energy were calculated with an improved charge distribution (PBE0 6-31G(d,p)). The FIT repulsion–dispersion potential was calculated relative to the global minimum structure in E_{latt} after the final optimisations (see Table 4). The rms values were calculated for all the CSP-generated crystal structures that were found to be within 50 kJ mol^{-1} after the final optimisations, as well as for a more limited set of crystal structures that matched the most important putative polymorphs found in the original CSP studies (see the Electronic Supplementary Information); this set included all the experimentally known forms with $Z' = 1$. The results in Table 4 confirm the conclusion of the paper that we cannot rely on DFTB3-D3 lattice energies.

Related to this is a comparison with periodic DFT-D methods for selected crystal structures from the 6th crystal structure prediction blind test (POLY59 set[1]). The absolute lattice energies (X23 set) mean unsigned errors of DFTB3-D3 are competitive to PBE-TS, but for relative lattice energies, periodic dispersion corrected density functional (DFT-D) profits a lot from error compensation, which does not hold for the tight-binding Hamiltonian.

Table 4 Root mean square (rms) deviations calculated between the energies of crystal structures

Molecule	~$rms_{\Delta E_{latt}}$ for all crystal structures/kJ mol^{-1}	~$rms_{\Delta E^{latt}}$ for crystal structures matching the most important PPMs/kJ mol^{-1}
XXVI	21	13
GSK269984B	34	21
XX	15	5
XXIII	21	10
Mebendazole	15	7
Overall	21	13

1 J. G. Brandenburg and S. Grimme, *Acta Cryst. B*, 2016, **72**, 502–513.

J. Christian Schön enquired: While looking like a nice compromise between *ab initio* calculations and empirical potentials, there is one big problem with DFTB: up to now, not enough pairwise tight-binding parameters are available. According to the practitioners, it takes 1/2 to 1 year to generate a new set of parameters for your problem of interest, and you need such parameters for every pair of atom types (n atoms $\rightarrow$ $n(n+1)/2$ pairs). What is the story on this at the moment? We have found some parameters for organic molecules in the literature, but it would be incredibly helpful if there were also such parameters *e.g.* for interactions between metals and organic molecules.

Jan Gerit Brandenburg replied: There are different groups working on DFTB parametrization. While we used the 3OB parameter set, a current list is available.[1] A different strategy from the group of Stefan Grimme is the development of a tight-binding Hamiltonian that relies on element parameters only and can thus be parameterized much more rapidly. Elements up to Rn are available.[2]

1 http://www.dftb.org/parameters/download
2 S. Grimme *et al.*, *J. Chem. Theory Comput.*, 2017, **13**, 1989–2009.

Matthew Addicoat asked: Following up on parameters, you may have been referring to the QUASINANO parameters, which are single-element parameters up to element 90 (Th).[1,2] Did you try using the MIO parameters at all? I've had significant experience in both COFs and ionic liquids where we used the MIO parameters with Lennard-Jones dispersion – in one paper a reviewer did ask us to repeat our calculations with 3OB – we did so and got poorer results.[3]

In your paper, you did significant work on re-parameterising the dispersion component, but didn't add any extra hydrogen bonding or halogen corrections. Three of your molecules do have fluorine, chlorine or both. Did you see any problems or poor behaviour with those molecules?

1 M. Wahiduzzaman, A. F. Oliveira, P. Philipsen, L. Zhechkov, E. van Lenthe, H. A. Witek and T. Heine, *J. Chem. Theory Comput.*, 2013, **9**, 4006–4017.
2 A. F. Oliveira, P. Philipsen and T. Heine, *J. Chem. Theory Comput.*, 2015, **11**, 5209–5218.
3 M. A. Addicoat, R. Stefanovic, G. B. Webber, R. Atkin and A. J. Page, *J. Chem. Theory Comput.*, 2014, **10**, 4633–4643.

Jan Gerit Brandenburg answered: The QUASINANO set mentioned is part of the effort to generate a DFTB parametrization for the whole periodic table. However, it is still necessary to combine them with pair-parameters and those are only available up to Ca.[1] A more extensive full parametrization is provided by the GFN-xTB variant.[2]

A systematic comparison of QUASINANO, MIO, and 3OB parameter sets for DFTB3 combined with the D3 dispersion correction has not been published. All our initial tests on established benchmark sets for molecular dimers and crystals (S66, L7, S12L, X23 test sets) have shown that the use of 3OB parameters is better suited to describe the noncovalent interactions of mostly organic molecules. Our optimized damping parameter sets with corresponding statistical evaluation have been distributed in a private communication to ADF, where the QUASINANO set is used, and will be provided by us upon request. A closely related test has recently been done by combining DFTB3(3OB) and DFTB3(MIO) with the TS and MBD dispersion correction models. In the supporting information of ref. 3, we directly compare the intermolecular interactions and distances of molecular dimers (S66 set) obtained from different parametrizations (see Fig. S7). In agreement with our previous experience, the 3OB parameters are recommended for this system class.

Hydrogen- and halogen-bonded molecules can be tougher to describe with a simple point-charge electrostatic model. Ref. 71–73 in our paper try to correct some of the systematic errors with additional pair-potentials. However, we found that these additional corrections do improve the interaction energies in the reference sets, but lead to only slight improvements in larger complexes.[4] Thus, we did not use these in our paper (DOI: 10.1039/c8fd00010g) and could not detect major problems with the target molecules including halogens. However we did not compare the energetic ranking in detail, as this was not sufficiently accurate for any of our targets, and we recommend a subsequent energy evaluation with a more accurate method.

1 A. F. Oliveira *et al.*, *J. Chem. Theory Comput.*, 2015, **11**, 5209–5218.
2 S. Grimme *et al.*, *J. Chem. Theory Comput.*, 2017, **13**, 1989–2009.
3 M. Mortazavi *et al.*, *J. Phys. Chem. Lett.*, 2018, **9**, 399–405.
4 J. G. Brandenburg *et al.*, *J. Phys. Chem. Lett.*, 2014, **5**, 4275–4284.

Christopher Taylor enquired: You mentioned that you obtained significantly improved results when performing a single-point calculation and using an electrostatic multipole and empirical force field interaction model (*i.e.* using the DMACRYS program). Can you clarify this – do you mean that you ran only a single-point calculation in DMACRYS (*i.e.* the resulting structures are the same as those from your DF-TB calculations, and not an energy minimum at the multipole-and-force-field level of theory)? Or are you referring to the single-point *ab initio* calculation required to obtain said multipoles (*i.e.* the structures are fully minimised in DMACRYS after having been generated through DF-TB)?

Jan Gerit Brandenburg replied: We did indeed follow the latter strategy, *i.e.* we used the rigid DFTB3-D3 molecular geometries to compute the distributed charge multipoles. Subsequently the DFTB3-D3 crystal geometry was relaxed with the DMACRYS code by combining the multipole electrostatic with the empirical FIT potential. This relaxation of the intermolecular degrees of freedom does not notably increase the computational cost and we thus recommend including this step. However, as a more general strategy, we have shown that even the intermolecular geometries of DFTB3-D3 (or DFTB3-MBD) can be used directly for computed density functional based single-point energies.[1]

1 M. Mortazavi *et al.*, *J. Phys. Chem. Lett.*, 2018, **9**, 399–405.

Seiji Tsuzuki said: I have two questions on the accuracy of DFT-B. The first question concerns the accuracy of the charge distributions obtained by the DFT-B method. The electrostatic interactions are important in determining the arrangement of molecules in crystals, since the electrostatic interaction is a highly directional interaction. It can be attractive or repulsive, depending on the orientation of the molecules. Therefore, accurate evaluation of the electrostatic interactions is important for crystal structure prediction. The accuracy of the calculated charge distributions can be judged by comparing the calculated dipole moment with the experimental one. *Ab initio* calculations show that a split valence basis set including polarization functions and electron correlation corrections are necessary for accurate calculation of the charge distributions. On the other hand, tight-binding calculations are semi-empirical calculations and use very small basis sets. I suspect that the DFT-B calculations cannot reproduce the experimental dipole moment well. Did you compare the dipole moments of molecules calculated by DFT-B with experimental values? I would like to know how large the errors of the calculated dipole moments are.

The second question concerns the accuracy of the calculated torsional potential by DFT-B. To reproduce the small energy difference between rotamers by *ab initio* calculations, a split valence basis set with polarization functions and electron correlation corrections are necessary. Is it possible to reproduce the experimental torsional potential by DFT-B? Have you compared the torsional potentials calculated by DFT-B with those obtained by high level *ab initio* calculations?

Jan Gerit Brandenburg responded: The dipole moments are indeed a way to test the electrostatic interactions from DFTB3-D3. While similar studies have been conducted before, we have computed the dipole moment on a set of small semi-rigid molecules with experimental references; experimental values gathered in ref. 1. While DFT methods in converged basis set expansions (augmented triple-zeta or larger) indeed produce higher quality dipole moments, the results from DFTB3-D3 are reasonably good and actually better than those from a very recent tight-binding parametrization, GFN-xTB.[2] Correlations with the reference data (Pearson correlation and mean unsigned errors) are shown in Fig. 2. However, semi-empirical methods rely strongly on error compensations, *i.e.* the true test should be the total interaction energy that has been tested for molecular crystals in ref. 3.

It is correct that torsion potentials are crucial to describe the relative stabilities of molecular conformers. Individual torsions have sometimes been compared directly to high-level references. However, this is not trivial as any benchmark of full torsion potentials has to be based on multi-reference wavefunction methods. Thus, it is easier and more instructive to directly compare the relative stability of local equilibrium structures, *i.e.* molecular conformers. We have not done this for the target molecules in our paper (DOI: 10.1039/c8fd00010g), but in ref. 4 several established conformational benchmark sets were analyzed with different tight-binding approaches (including DFTB3-D3), showing reasonably good performance.

1 A. L. Hickey and C. N. Rowley, *J. Phys. Chem. A*, 2014, **118**, 3678–3687.
2 S. Grimme *et al.*, *J. Chem. Theory Comput.*, 2017, **13**, 1989–2009.
3 J. G. Brandenburg and S. Grimme, *J. Phys. Chem. Lett.*, 2014, **5**, 1785–1789.
4 S. Grimme, C. Bannwarth and P. Shushkov, *J. Chem. Theory Comput.*, 2017, **13**, 1989–2009.

Gregory J. O. Beran asked: You noted that DFTB3-D3 provides useful predictions for the free energy contributions. Does it reproduce each individual phonon mode frequency well, or is it a case where good error cancellation occurs when summing over all the vibrational modes when computing free energies?

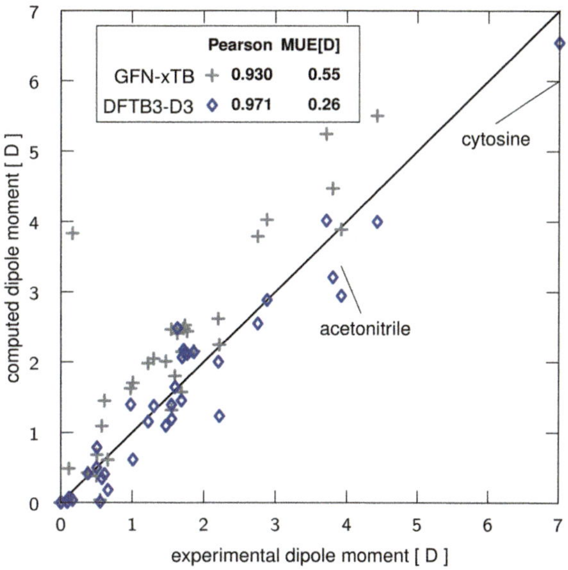

Fig. 2 Correlation with the reference data (Pearson correlation and mean unsigned errors).

Jan Gerit Brandenburg answered: In our paper (DOI: 10.1039/c8fd00010g), we computed free energy contributions based on the DFTB3-D3 model and contrasted it with rigid-body models. Some comparisons to higher-level DFT results and experimental measurements have been conducted.[1,2] However, our analysis is indeed a rather coarse-grained one, as only integrated quantities (like free energy contributions) are analyzed. In this regard, Prof. Beran is correct and we do rely on compensation of errors.

In earlier studies, the covalent modes from DFTB have been analyzed.[3] While the authors recommended a scaling factor, it is not clear how this would translate to soft non-covalent modes. Clearly, further detailed analysis for molecular crystals is needed in order to make reliable predictions. We would emphasize that methods need to be tested and validated for flexible pharmaceutical crystal structures to test the description of the coupling between the molecular and intermolecular modes, which seems likely have a significant effect on the relative free energies.

1 J. G. Brandenburg *et al.*, *J. Phys. Chem. Lett.*, 2017, **8**, 4319–4324.
2 H. K. Buchholz *et al.*, *Cryst. Growth Des.*, 2017, **17**, 4676–4686.
3 H. A. Witek and K. Morokuma, *J. Comput. Chem.*, 2004, **25**, 1858–1864.

Marcus Neumann remarked: Some years ago we introduced phonon free energies in GRACE based on our tailor-made force fields, and found out that those free energies actually had a negative impact on our energy rankings. Phonon free energies strongly depend on the vibrational frequencies of the soft modes, and small errors in the force calcuations result in large errors in the phonon free energies. The energy ranking performance of DFTB3-D3 does not seem to be better than the performance of tailor-made force fields, so I would expect the same for its use in phonon free energy calculations. What evidence do you have that phonon free energies calculated with DFTB3-D3 actually improve the energy ranking performance?

Jan Gerit Brandenburg replied: DFTB3-D3 based lattice energies are not satisfactory for a reliable energy ranking and might not outperform the tailor-made force field within GRACE. On the other hand, the equilibrium geometries seem to be of good quality, as shown by the small deviations w.r.t. experimental crystal structures. How this translates into the properties of second derivatives is not clear *a priori*, especially as the low-energy modes are most important for the entropy estimates. From the ranking of our computed energy landscapes, we have some indication of an improved ranking, *e.g.* for target XXVI the experimental structure is rank 2 based on lattice energies and moves to rank 1 upon inclusion of the DFTB3-D3 based free energy contributions. A direct comparison of free energy contributions from DFTB3-D3 with DFT based references has been recently done on a carbamazepine polymorph.[1] This has to be extended to broader tests. The contributions of paper 11895 (DOI: 10.1039/c8fd00048d) and paper 11896 (DOI: 10.1039/c8fd00066b) might be ideally suited to conduct this analysis.

1 J. G. Brandenburg *et al.*, *J. Phys. Chem. Lett.*, 2017, **8**, 4319–4324.

Johannes Hoja asked: You mention in your paper that tighter convergence criteria had to be used in order to remove imaginary frequencies in the DFTB3-D3 phonon calculations. What were the tightest convergence settings necessary in order to avoid imaginary frequencies?

Jan Gerit Brandenburg responded: The "tightened" optimization criteria correspond to the standard thresholds of CRYSTAL17, *i.e.* maximal gradient component of 0.00045 a.u., rms gradient of 0.0003, maximal displacement of 0.0018 a.u., and rms displacement of 0.0012 a.u. Additionally, the components of the dynamical matrix are computed by symmetric finite displacements (in contrast to the default of just one displacement).

Ivo Rietveld opened the discussion of the paper by Claire S. Adjiman: The potentials in the calculations of the crystal structures are adjusted using sublimation enthalpy data. How do you guarantee that the sublimation enthalpy data is accurate and not affected by impurities, and that the data is for the correct polymorph? How is the precision of the sublimation data dealt with, in particular when the vapor pressure is low? Sublimation data cannot be used for new polymorphs that are obtained purely by prediction. How is the available sublimation data generalised within the code to improve the stability prediction of unknown polymorphs?

Claire S. Adjiman answered: We have carefully selected data from the literature, paying attention to the experimental methods used, but ultimately we cannot guarantee the accuracy of all data points. The error in the data can be taken into account by adjusting the weight given to each data point (*e.g.* in the maximum likelihood approach, each data point is weighted based on the corresponding variance). We investigated the impact of such weights on the parameters for the hydrocarbons (*c.f.* Fig. 3 in the paper) and found that using equal weights for the structural data and for the sublimation enthalpies led to parameters which gave a good structure reproduction relative to what is achieved with a weight of 0 for sublimation enthalpies, and a good quality of reproduction of the sublimation enthalpies. We subsequently used equal weights. We do not use any information on predicted or unknown polymorphs for parameter estimation. In our data set, we have only one molecule with two polymorphs, tetracyanoethylene, and we ensured that the parameters yielded the correct energy ranking of these two polymorphs. In the future, we would like to include information on relative energies in the objective function.

Jonas Nyman asked: Have you benchmarked your new force field against the currently very popular X23 benchmark? Are some of the structures in the X23 set in your training and/or validation sets? Also, have the parametrizations been automated in such a way that you could fit other functional forms, such as a Lennard-Jones potential, or a damped Buckingham with additional higher-order C_8/C_{10} dispersion terms?

Claire S. Adjiman replied: We have not benchmarked our force field against the X23 data set. The following X23 structures are included in our training set: 1,4-cyclohexanedione (CYHEXO), acetic acid (ACETAC07), adamantane (ADAMAN08), anthracene (ANTCEN16), benzene (BENZEN06), imidazole (IMAZOL06), naphthalene (NAPHTA31), pyrazine (PYRAZI01), succinic acid (SUCANH14), triazine (TRIZIN01), and trioxane (TROXAN). The parameter estimation code currently does not allow the direct use of alternative potentials. However, it can be modified relatively easily to work for the Lennard-Jones potential or a higher-order Buckingham potential.

Sharmarke Mohamed remarked: The work presented in your paper is a welcome contribution for practitioners of crystal structure prediction, who for a long time have used model intermolecular potentials that have been fitted to crystal structures that are often not representative of the systems under study. In work conducted within my group, I have found that the FIT force field actually performs well in modelling the intermolecular interactions in two-component molecular salt systems. However this force field is not appropriate for three-component molecular systems (*e.g.* cocrystal hydrates or ionic cocrystals). The *ab initio* model used to fit the potential parameters is an important factor in this and so it is very encouraging that the authors have developed and tested a method for deriving potential parameters that can be tailored to different *ab initio* models. I am curious, however, whether the authors feel that the lack of complex molecular systems (*e.g.* hydrates, cocrystals, salts) in their dataset of 106 crystal structures will present challenges in the use of these potential parameters for studies of multicomponent systems. Have the authors tested the transferability of these new parameter sets to multicomponent systems that display competing intermolecular interactions? Could the authors also comment on the typical cost in time (*i.e.* how many weeks?) that would need to be spent on using the method proposed in this paper to derive system-specific parameters (*i.e.* potential parameters for salt hydrates) using a specific *ab initio* model?

Claire S. Adjiman responded: Thank you for the positive comments. We have not yet tested the proposed parameters for such multicomponent systems. In our view, the dataset needs to be expanded to include hydrates and salts, as well as flexible molecules, to increase the transferability. We are currently working in this direction. The main cost in deriving parameters is to gather reliable data on crystal structures with sufficient diversity. The computations are highly parallelisable and their cost depends principally on the number of data points in the training set. Using one core, the regression of the parameters corresponding to one box in Fig. 2 of the paper requires approximately 150 days of wall time for a single core. This is equivalent to about 20 h per local minimisation, and all minimisations are independent of each other.

Sten Nilsson Lill asked: Your choice of functional (M06) is very sensitive to the numerical grid accuracy. Did you apply a tight grid in your calculations to ensure you had a stable final solution?

Claire S. Adjiman answered: Yes, we used a very tight grid for our calculations.

Seiji Tsuzuki enquired: I have a question about the exchange–repulsion potential in your force field. You showed that the performance of the calculation of sublimation energy is improved by the improvement of the force field, but the calculated sublimation energies still have large errors in some cases. It is well known that the exchange–repulsion potentials of hydrogen and halogen atoms have strong anisotropy. The importance of the anisotropy of hydrogen atoms in reproducing the crystal packing of hydrocarbon molecules by force field calculations was suggested by Williams in the 1970s. I believe that it is necessary to consider the anisotropy of exchange–repulsion interactions to improve the performance of the calculation of sublimation energies. Why don't you use anisotropic exchange–repulsion potentials for the calculation of the sublimation energies?

Claire S. Adjiman responded: I agree that it is often important to include anisotropy in the calculations. We have designed our parameter estimation code so that the parameters can easily be re-estimated as one makes changes to the lattice energy model, either in the *ab initio* components of the model (intramolecular energy and electrostatics) or in the form of the repulsion–dispersion potential. In this paper, we focused on re-parameterising the most common potential, the isotropic Buckingham, for one specific level of theory for the DFT calculations (M06/6-31G(d,p)). In the future, we intend to re-use the methodology and code to develop anisotropic potentials.

Luca Iuzzolino said: You have shown that the new repulsion–dispersion parameters improve the absolute energies of the crystal structures that were considered in this study. Is this also true for the relative energy differences between polymorphic forms of the same molecules?

Claire S. Adjiman replied: The relative stability of TCNE polymorphs in the training set is correctly reproduced. We have not yet tested other polymorphic systems.

Sarah L. Price remarked: It is about time that something was done to improve the FIT potential. In 1996, David Coombes fitted a polar hydrogen potential to allow modelling of some hydrogen-bonded structures when used with in combination with Williams' old parameters from papers from the 1980s. Williams had fitted his parameters to few crystal structures and very few sublimation energies. The fluorine parameters have given some problems, probably because of the sublimation energy used in derivation, but generally it is surprising how many hundreds of CSP studies we have been able to do with the FIT potential, given how it was derived. I look forward to trying your new sets of parameters, with distributed multipoles for a good quality of charge density. However, I notice that you haven't got many cross-terms for two heteroatoms, such as nitrogen–oxygen, which we would need for amide hydrogen bonding. What are the difficulties? Would you advise trying combination rules for the missing parameters?

Claire S. Adjiman answered: Thank you for the positive comments. In the first instance, I would indeed advise using standard combining rules. The main limiting step is the creation of the training and validation sets, curating the data to ensure high reliability and adequate coverage of different types of structures. We are currently working to expand the set of parameters published here to include fluorine as well as heteroatom cross-interactions.

Alan Hare asked: Following up on the question of intramolecular smoothing: is the discontinuity in the potential true or apparent? While sometimes there can be an abrupt separation between two proximate, distinct functions, in others – like melting "squares of chocolate", as the poster reminded me! – a single, continuous function may really straddle the divide, or the "break between squares"… which can itself sometimes be transformed into a simpler function. On reflection, am I right in thinking that in the intramolecular case – with its multiplicity of rotations – the discontinuity is truly an abrupt one? If so, I imagine that smoothing across it may involve the use of a hyperbolic tangent (or possibly sigmoidal) function close to the "cut-over", so to speak. Or do you have another way of handling it altogether?

Claire S. Adjiman responded: This relates to the poster presented by Isaac Sugden. The functions are discontinuous because that there are two different values for the intramolecular energy at the boundary between two local approximate models (LAM points). This is because in the early versions of this approach each LAM has a well-defined region of use and we switch abruptly between LAMs as we cross boundaries. I have sketched this roughly to help visualise what happens (see Fig. 3). In an intuitive sense, how abrupt the discontinuity is depends on how different the two function values are. Regardless, the discontinuities are always expected to be present. These discontinuities are not physical – they are artefacts of the local models used. In this work, we corrected this by replacing the use of a single LAM by an average of all the LAMs. The average is weighted so that the LAM nearest to the current point of interest carries the most weight. We compute the weight of each LAM so that it decays exponentially as the distance between the point of interest and the LAM point increases.

Jan Gerit Brandenburg communicated: The authors made a nice effort in parameterizing the FIT potential for use in organic molecular crystal structure prediction containing the elements H, C, N, O and Cl. While hydrogen has parameters that depend on its binding motif, other elements have a single set of parameters. Considering the London dispersion interaction, this might be a crude approximation, especially for carbon, which has quite a large variety in its dynamic polarizability. Fig. 4 shows computed (TD-DFT) dispersion coefficients for carbon in different environments, which are then used as references in the D3 dispersion correction scheme.[1] As the range of C_6^{CC} is from 20 to about 50 a.u., this might be relevant for the FIT type potential. The D3 coefficients are computationally easily accessible and yield high-quality inter-molecular dispersion coefficients (see Fig. 2 in ref. 2). Are the authors considering using environment dependent parameters for elements other than H?

1 S. Grimme, J. Antony, S. Ehrlich and H. Krieg, *J. Chem. Phys.*, 2010, **132**, 154104.
2 E. Caldeweyher and J. G. Brandenburg, *J. Phys.: Condens. Matter*, 2018, **30**, 213001.
3 S. Grimme, A. Hansen, J. G. Brandenburg and C. Bannwarth, *Chem. Rev.*, 2016, **116**, 5105–5154.

Claire S. Adjiman communicated in reply: It is indeed an approximation not to take into account the environment of other elements. We strive for a balance between the accuracy of the model and the statistical significance of the parameter values. As we increase the number of atom types and hence the number of parameters, we are likely to achieve greater accuracy, at the cost of decreased statistical significance as data scarcity is an issue. The set of atom types chosen has so far provided a good balance between accuracy and significance, with enough data available in most cases for training and validation of the models. We note that the parameters are effective values rather than a reflection of intrinsic

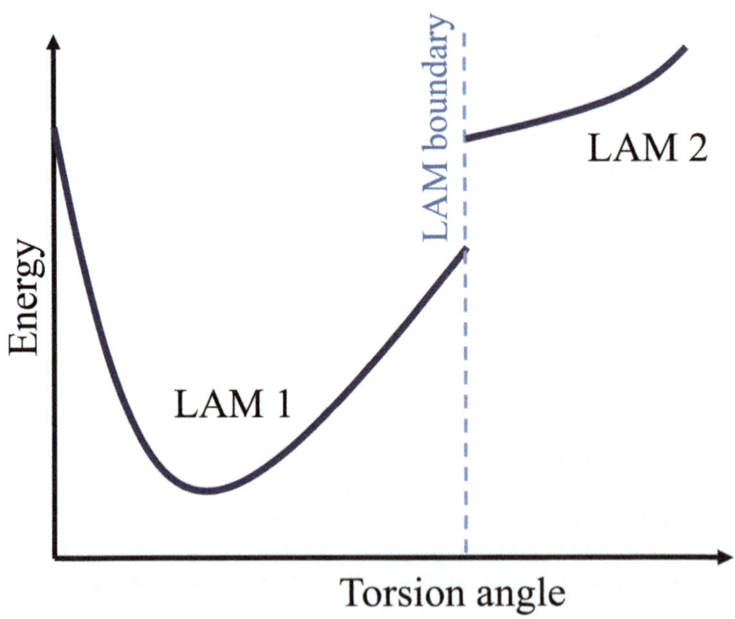

Fig. 3 Abrupt switching between LAMs at LAM boundary.

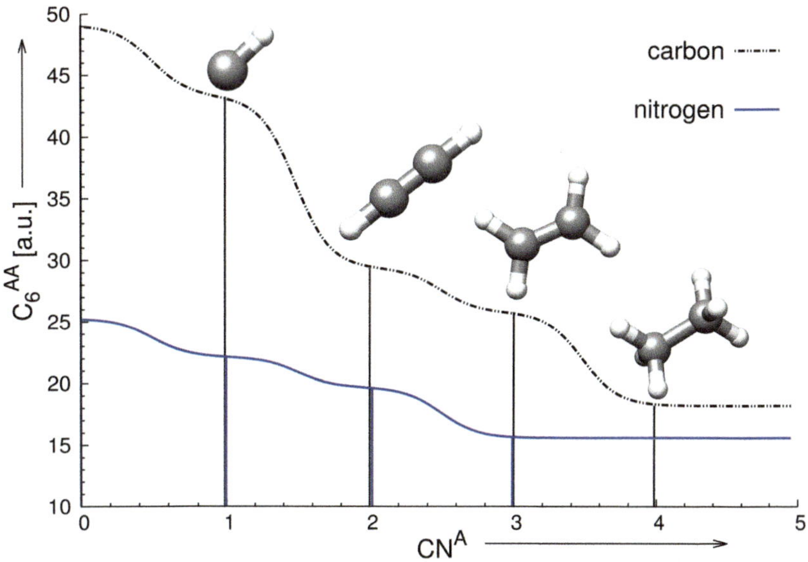

Fig. 4 CN dependent atom pairwise dispersion coefficient for homoatomic pairs C_6^{AA} as a function of the coordination number CN_A. The reference points $C_{6,ref}^{AA}$ are drawn as vertical sticks in the respective colors (black for carbon, blue for nitrogen). Figure reproduced with permission from ref. 3.

repulsive–dispersive interactions: as such, we include multiple atom environments in the data set for training and obtain parameters that provide an average interaction. In addition, as discussed in the paper, the repulsion–dispersion term in our model of lattice energy includes all the aspects that are not taken into account in the model, such as entropic contributions or polarizability. This is very different in concept to the D3 coefficients, for which the objective is to obtain an intrinsic value of the dispersive interactions only. Thus, the proposed repulsion–dispersion potential contains an empirical correction of the potential. One could in principle modify our model to use D3 coefficients and introduce a different empirical correction.

Marcus Neumann opened a general discussion of the papers by Alexandre Tkatchenko, Jan Gerit Brandenburg and Claire S. Adjiman: Calculating the relative stabilities of crystal polymorphs is at the very heart of organic crystal structure prediction. Yet very little reliable information, especially for larger molecules, is available in the scientific literature. Experimental error bars are missing in most cases. If we really want to make progress in the energy ranking accuracy, a reliable experimental energy benchmark must be established. Ideally, 20–100 polymorphic enthalpy and free energy differences should be measured, each by different methods and different researchers.

Having talked to several experimentalists in academia about the issue, it seems that getting funding for such an endeavour is very difficult because accurate thermodynamic measurements by themselves are not considered innovative. As a community, we should try to convince a national or transnational funding agency to support the accurate measurement of polymorphic energy differences.

Claire S. Adjiman responded: I concur that it is important to generate such a benchmark data set. This could be done through collaborative projects where one student works across experimental and computational groups, measuring some data and assessing/improving the state-of-the-art in modelling. This would provide a coherent scientific problem and expose the students to different tools/learnings, and by working in tandem such projects may be more compelling for both researchers and funders. Of course one can imagine doing this across a larger consortium of collaborators, where we might explore a broader range of properties. ITN or COST would be good ways forward. There is also the example of the UNIFAC consortium, which has been obtaining data and developing models for fluid phase equilibria for many years, first in academia and now as an independent company.[1]

1 http://unifac.ddbst.de/unifac_.html.

Ivo Rietveld responded: As an experimentalist, I would like to add a few points. First of all, it is hard to give a PhD student such a subject, because he or she should at least have to deal with a scientific problem that will lead to a proper scientific thesis (and answer…?). Just data as such will not be enough, but in the end we also look for precise thermodynamic rankings, so it is a matter of matching the right substances with the right scientific questions. That will remain a sort of case-by-case issue.

Second of all, precision is difficult, also experimentally. The experimentalists may claim small error bars, and certainly with calorimetry and X-ray diffraction we are good at determining differences, as long as we do it on the same equipment(!), but again everything depends on the system, and whether the chemical compound can be measured with the precision that is required for computer calculations. Therefore, from an experimental point of view, the database is probably easier to define not in terms of substances, but in terms of transition values of which we are very confident. For ROY, for example (with which I am in fact not very familiar), there may be phases that are extremely well documented, whereas others are not because they are an experimental nightmare.

I think therefore that this question may be resolved by some sort of COST (or other!) network, where the uptake into the database of experimental systems that we are studying for other reasons anyway can be discussed. A decision to take up a substance or transition in the database within this COST network will initiate additional measurements by other experimentalists and calculations by computationists to come to consolidated transition values between very specific polymorphs. This latter part will likely need extra funding, but we may be able to leverage funding within such a COST network. This COST network should also function as quality control of the experimental values and computational values.

For the precise measurements, we may be able to apply for grants that emphasise the industrial development of scientific findings, particularly if all the large European pharmaceutical companies can be involved in the grant application. Some validation work may be done by a number of post-docs and papers may be published in the same vein as the polymorph prediction blind tests. Why not discuss the experimental and computational caveats and how we interpret, say, the experimental and computational results of 10 polymorphic transitions towards a rigorous database? More importantly: who is in?

Alexandre Tkatchenko responded: I strongly concur with Marcus and others as well. Perhaps the effort of encouraging our experimental colleagues would have a higher chance of success if all the participants (theorists and experimentalists from academia and industry) show strong interest and push each other. For a start, it would be good if industry decided to disclose existing thermodynamic data for interesting polymorphic systems. I am sure this data is lying around in different companies. High-level quantum-chemical calculations are also very relevant and should be done. One current problem with non-canonical CCSD(T) and QMC is that accuracy is very hard to control to the required level of 1 kJ mol^{-1}, especially for larger molecules. So far, the accuracy of these methods in periodic calculations has been assessed by comparing to experiments, and never for energy differences between polymorphs. In this case, glycine, oxalic acid and coumarin (mentioned by Qiang Zhu) would be good systems to which to start applying QMC/CCSD(T). Regarding challenging systems, axitinib and ROY are good ones going forward. However, in the case of ROY, my understanding is that it is hard to ensure that experiments measure pristine and non-defect structures.

J. Christian Schön responded: I strongly concur with Marcus Neumann. I have tried for many years to convince my experimental solid state chemistry colleagues to perform systematic analyses of their syntheses of well-defined systems, producing quantitative outcomes as function of synthesis parameters, in order to provide data to sharpen our theoretical tools for modeling chemical syntheses. But in the end, nobody was willing to give a student this endeavor – I guess it is not "sexy" enough (no new compounds or crystal structures being produced, *etc.*) and it might be too difficult to publish such a study. (I have heard similar complaints from people in the phase diagram community – they have a hard time getting funding agencies interested unless the chemical system has important applications.) I can see that a similar problem would appear in the accurate measurement of free energy differences, and I would strongly support such a measurement campaign.

Qiang Zhu responded: I agree. I think the first step is to create a good polymorph data set that can present the real challenges in energy ranking. Currently, the X23 set has been studied a lot,[1,2] but it seems to me that many methods produce nice results at about the same level. Do they really reflect the challenge in energy ranking? I have two systems which strongly oppose this conclusion – coumarin[3] and ROY (DOI: 10.1039/c8fd00039e). These might be very small molecules by Marcus Neumann's standards, but most methods still fail to reproduce the ranking nicely. As a community, we need to come up with a plan to create a list of challenges and then work it out. This requires a synergy between experiment and theory.

1 A. Otero-de-la-Roza and E. R. Johnson, *J. Chem. Phys.*, 2012, **137**, 054103.
2 A. M. Reilly and A. Tkatchenko, *J. Phys. Chem. Lett.*, 2013, **4**, 1028–1033.
3 A. G. Shtukenberg, Q. Zhu, D. J. Carter, L. Vogt, J. Hoja, E. Schneider, H. Song, B. Pokroy, I. Polishchuk, A. Tkatchenko, A. R. Oganov, A. L. Rohl, M. E. Tuckerman and B. Kahr, *Chem. Sci.*, 2017, **8**, 4926–4940.

Jan Gerit Brandenburg remarked: The electronic structure community has made some substantial developments in the past years, with CCSD(T) being applicable to large and sometimes periodic systems (mostly relying on embedding or localization schemes). We have contributed to this endeavour by extending the applicability of Quantum Monte Carlo to organic crystals.[1] This approach can potentially be used to compute the static lattice energies of large molecular crystals and their polymorphs (~200 atoms per unit cell) in a reasonable time frame.

This could ideally be complemented by experimental relative stabilities to independently test the capabilities of CSP methods to predict relative lattice energies (*vs.* high-level theory) and relative free energy differences (*vs.* experiment). This is closely connected to Marcus Neumann's previous question and similarly requires funding to afford the substantial CPU time.

As an add-on concerning the CPU time: the computation of a lattice energy for benzene and anthracene crystals with sub-chemical accuracy (regarding the thresholds we can control) requires 6000 and 74 000 CPU hours. The time will clearly depend on the system size (roughly cubic scaling with the number of atoms), the target accuracy (*e.g.* reducing it from 0.5 kJ mol^{-1} to 0.1 kJ mol^{-1} increases the CPU time by an order of magnitude), and the employed time step (doubling it halves the CPU time), which varies a lot with the target molecule. Once a decision on target crystal structures has been made, more quantitative estimates are possible.

1 A. Zen, J. G. Brandenburg, J. Klimeš, A. Tkatchenko, D. Alfè and A. Michaelides, *Proc. Natl. Acad. Sci. USA*, 2018, **115**, 1724–1729.

Matthew R. Ryder replied: I agree that this approach could yield valuable insights into the capabilities of different CSP methods to predict lattice energies.

J. Christian Schön addressed Johannes Hoja and Jan Gerit Brandenburg: It looks like we are getting good results on the energy evaluation front for molecules and molecular crystals. But what about the quality/computational expense of your methods for applications to molecules on surfaces? Can you handle this as "just another application" of your methods, or would one still need years of new developments/adaptations? What about the size of the (flexible) molecules – do you envision computing proteins on this level of accuracy?

Johannes Hoja answered: Our methods can be and have already been used to study the adsorption of molecules on surfaces (chemisorption, physisorption, and intermediate situations). Based on the Tkatchenko–Scheffler dispersion model, the vdW^{surf} method was developed to also incorporate screening effects from the substrate electrons.[1] Examples of applications of the DFT+vdW^{surf} and DFT+MBD methods can be found in ref. 2–5. These calculations provide accurate adsorption energies and geometries, and can readily be carried out on state-of-the-art computing facilities. Several DFT+MBD single-point energies could be obtained for isolated small proteins.

1 V. G. Ruiz, W. Liu, E. Zojer, M. Scheffler and A. Tkatchenko, *Phys. Rev. Lett.*, 2012, **108**, 146103.
2 V. G. Ruiz, W. Liu and A. Tkatchenko, *Phys. Rev. B*, 2016, **93**, 035118.

3 R. J. Maurer, V. G. Ruiz, J. Camarillo-Cisneros, W. Liu, N. Ferri, K. Reuter and A. Tkatchenko, *Progr. Surf. Sci.*, 2016, **91**, 72–100.
4 Y. Jiang, S. Yang, S. Li and W. Liu, *Sci. Rep.*, 2016, **6**, 39529.
5 S. Yang, Y. Jiang, S. Li and W. Liu, *Carbon*, 2017, **111**, 513–518.

Jan Gerit Brandenburg answered: Molecules adsorbed on surfaces are indeed technically just another application. The computational expense grows (typically cubic scaling) with the size of the system and one has to converge the size of the unit cell to remove artificial interactions of the molecule with itself, *i.e.* to converge to the low-coverage adsorption, as well as the thickness of the surface layer. However, one has to carefully control the accuracy of the vdW corrected density functional as it is less established compared to organic crystals. This holds in particular for conducting surfaces, as the vdW corrections rely on local response models that might not be justified for conductors. Even the water adsorption on a wide gap hBN surface can be a challenge.[1] As the corresponding measurements are not trivial and reference quality calculations are rare, we need more experience for a better quality assessment.

VdW-inclusive DFT can be applied to small protein crystals like crambin within standard codes.[2] GPU accelerated program codes have been used for studying protein ligand binding affinities,[3] and linear scaling codes can be used to study the ion channel gramicidin A.[4]

1 Y. S. Al-Hamdani, M. Rossi, D. Alfè, T. Tsatsoulis, B. Ramberger, J. G. Brandenburg, A. Zen, G. Kresse, A. Grüneis, A. Tkatchenko and A. Michaelides, *J. Chem. Phys.*, 2017, **147**, 044710.
2 M. Delle Piane *et al.*, *Chem. Sci.*, 2016, **7**, 1496–1507; CRYSTAL17 code.
3 S. Ehrlich *et al.*, *ChemPhysChem*, 2017, **18**, 898–905; TerraChem code.
4 M. Todorović *et al.*, *J. R. Soc. Interface*, 2013, **10**, 20130547; CONQUEST code.

J. Christian Schön addressed Johannes Hoja and Jan Gerit Brandenburg: What is a “large” molecule in the context of your energy calculations?

Johannes Hoja replied: The calculation time for molecular crystals does not directly depend on the size of the involved molecules, but rather on the number of atoms present in the unit cell of the studied molecular crystal. The largest unit cell in our study of the latest blind test systems consists of about 1000 atoms.

Jan Gerit Brandenburg replied: From my experience this mainly depends on the type of application. The examples given in the answer to the previous question range from 1000 to 10 000 atoms per unit cell and would be considered large.

J. Christian Schön addressed Johannes Hoja and Jan Gerit Brandenburg: Where would you place the current limits of your calculations, *i.e.* what kind of computational facilities (and what kind of time scales/levels of global optimization) are needed for which size of system?

Johannes Hoja responded: I believe that for production calculations the upper limit of PBE+TS or PBE+MBD lattice relaxations and subsequent PBE0+MBD energy calculations would be around 100 structures, and that vibrational free energies in the harmonic approximation could be calculated for tens of structures. The computation time actually required highly depends on the sizes of the sampled unit cells, and for the phonon calculations, also on the unit cell shape and symmetry. In general, a cluster with more than 500 cores would be advisable for these kinds of calculations. For example, a PBE+TS lattice optimization with subsequent PBE0+MBD energy evaluation and PBE+TS vibrational free energies (with settings as described in the paper) for Form A of system XXIII (172 atoms per unit cell) would require about 1100 CPU hours.

Jan Gerit Brandenburg responded: Different codes scale differently and not all can fully use HPC facilities; for instance CRYSTAL17 has been shown to scale well up to 32 000 cores.[1] A DFT based re-ranking of crystal energy landscapes (or lists of putative polymorphs) has been conducted for the 6th CSP blind test;[2] the CPU time for local optimizations of 5000 structures are presented in Table S9 using 71 000 to 3 200 000 CPU hours. 100 000 CPU hours would amount to approximately 16 d elapsed time on a 256-core machine.

1 A. Erba *et al.*, *J. Chem. Theory Comput.*, 2017, **13**, 5019–5027.
2 A. M. Reilly *et al.*, *Acta Cryst. B*, 2016, **72**, 439–459.

J. Christian Schön enquired: Experimental colleagues tell me that there is a net transfer of charge between the molecule and the metal surface. Does this complicate the calculations for you, or can you predict this charge transfer quantitatively?

Johannes Hoja answered: Van der Waals-inclusive DFT calculations capture charge transfer effects in principle, but I am not aware of any benchmark calculations.

Jan Gerit Brandenburg answered: Technically this is not an issue, as the electron density is relaxed self-consistently. One can partition the self-consistent density to molecules, which would correspond to the charge transfer when compared to neutral molecules. However, some of the most frequently used DFT approaches of the generalized gradient type have a systematic error of too strong electron density delocalization, *i.e.* for a quantitative comparison to experiments one probably has to use a hybrid functional that reduces this error.

Marcus Neumann commented: Regarding a previous question about the CPU time requirements for PBE(0)+MBD+Fvib, I would like to mention that we recently implemented the approach in GRACE and are in the possession of some early CPU time benchmark results. The actual QM calculations were carried out with FHIaims through a driver implemented in GRACE. For blind test compound XXVI, we found that it is possible to process about 4 crystal structures on 384 modern Intel XEON cores in 1 day.

J. Christian Schön addressed Johannes Hoja and Jan Gerit Brandenburg: What kind of codes support your methods – only FHI-AIMS?

Johannes Hoja replied: To my knowledge the Tkatchenko–Scheffler (TS) and the many-body dispersion (MBD) methods are implemented in FHI-aims, Quantum Espresso, VASP, CASTEP, ADF, Q-Chem, and DFTB+. Furthermore, a general many-body dispersion library is currently in development.[1]

1 https://github.com/azag0/libmbd

Jan Gerit Brandenburg replied: Both D3 as well as MBD dispersion corrections are implemented to be used in conjunction with DFTB3 or DFT in freely available codes.[1,2] Additionally, several codes support the corrections; a non-complete list: VASP (D3 and MBD), FHI-aim (D3 and MBD), ADF (D3 and MBD), DFTB+ (D3 and MBD), CRYSTAL17 (D3) CP2K (D3), Orca (D3), Turbomole (D3), Psi4 (D3), Gaussian (D3), QuantumEspresso (older D2 and TS), CASTEP (older D2 and TS), and Siesta (older D2).

1 http://www.thch.uni-bonn.de/tc/dftd3
2 http://www.fhi-berlin.mpg.de/~tkatchen/MBD

Conflicts of interest

Julian Helfferich and Marcus Neumann work for a company that develops software for organic crystal structure prediction. There are no other conflicts to declare.

Faraday Discussions

PAPER

Computational modelling of solvent effects in a prolific solvatomorphic porous organic cage†

David P. McMahon,[a] Andrew Stephenson,[b] Samantha Y. Chong, [b] Marc A. Little, [b] James T. A. Jones,‡[b] Andrew I. Cooper *[b] and Graeme M. Day *[a]

Received 16th February 2018, Accepted 22nd March 2018
DOI: 10.1039/c8fd00031j

Crystal structure prediction methods can enable the *in silico* design of functional molecular crystals, but solvent effects can have a major influence on relative lattice energies, sometimes thwarting predictions. This is particularly true for porous solids, where solvent included in the pores can have an important energetic contribution. We present a Monte Carlo solvent insertion procedure for predicting the solvent filling of porous structures from crystal structure prediction landscapes, tested using a highly solvatomorphic porous organic cage molecule, **CC1**. Using this method, we can understand why the predicted global energy minimum structure for **CC1** is never observed from solvent crystallisation. We also explain the formation of three different solvatomorphs of **CC1** from three structurally-similar chlorinated solvents. Calculated solvent stabilisation energies are found to correlate with experimental results from thermogravimetric analysis, suggesting a future computational framework for *a priori* materials design that factors in solvation effects.

Introduction

A fundamental goal for computational chemistry is the *in silico* design of functional materials. Computational methods can reduce our reliance on trial and error experiments, saving laboratory time. Because materials properties are defined by structure, *in silico* methods for screening hypothetical functional molecules must involve a structural hypothesis; this can be done using analogy,

[a]*Computational Systems Chemistry, School of Chemistry, University of Southampton, SO17 1BJ, UK. E-mail: g.m.day@soton.ac.uk*

[b]*Department of Chemistry and Materials Innovation Factory, University of Liverpool, Crown St., Liverpool L69 7ZD, UK. E-mail: aicooper@liverpool.ac.uk*

† Electronic supplementary information (ESI) available. CCDC 1575910–1575914. For ESI and crystallographic data in CIF or other electronic format see DOI: 10.1039/c8fd00031j

‡ Present Address: Defence Science and Technology Laboratory, Dstl, Porton Down, Salisbury, Wiltshire UK.

empirical rules, or *ab initio* solid-state structure prediction. For molecular crystals, empirical rules are unreliable because the crystal packing is determined by the summation of many weak interactions, rather than a single, dominant interaction. A small change in molecular structure often alters the crystal structure completely. This means that intuitive design strategies for molecular crystals will frequently fail or, at least, fail to capture the true complexity of the potential crystallisation landscape for a given molecule. Computational methods developed for crystal structure prediction (CSP) give us the potential to predict structure from the molecular building blocks alone.[1,2] So far, CSP methods have largely focused on pharmaceutical molecules,[3–5] but they have also been applied in areas of materials chemistry, such as organic semiconductors[6–9] and porous organic molecular crystals.[10–13]

An important factor that is usually overlooked in CSP is the effect of solvent on the relative crystal structure stabilities and, hence, the structure that is predicted to form. Alternative crystal structures are often predicted to be close in lattice energy and the interaction of solvent molecules during crystallisation can affect the stabilisation of one potential packing arrangement with respect to another. Frequently, the influence of solvent is limited to interactions with the surface of a crystal. The total free energy of a crystal has contributions from the bulk and surface energies, such that structures with the lowest surface energies can be more stable than those with the lowest bulk energy. The balance between bulk and surface energies depends on crystallite size and shape, with surface terms increasing for small crystallite sizes.[14] While the bulk energy of a crystal is unaffected by the solvent environment, surface energies can be strongly influenced. Thus, different solvents can stabilise different structures and solvent stabilisation of crystal surfaces can have an important influence at the critical nucleation stage, leading to the growth of metastable phases. These effects on polymorph stabilities have been modelled for known polymorphs of inorganic[15,16] and organic[17] materials using explicit[15,16] or continuum[17] models of solvent on the dominant crystal faces.

The influence of solvent becomes especially important for porous crystals, which tend to crystallise with solvent in the pore channels. Solvated polymorphs, or "solvatomorphs", are of intrinsic interest because solvatomorphs can have different physical properties.[18–21] For porous crystals, different porous polymorphs can be isolated by desolvating solvatomorphs with different crystal structures.[10,22,23] This can be used to an advantage: for example, we showed previously that 1,4-dioxane acts as a directing solvent to induce the formation of a specific highly-porous crystal packing of organic cage molecules.[23] The solvent molecules in a porous molecular crystal can also be an integral part of the crystal structure, and hence the removal of solvent from solvatomorphs with retention of the crystal packing of the host structure is not always possible.[24,25]

It has been shown that the inclusion frameworks of organic molecules often correspond to local minima on the lattice energy surface, even in the absence of the guest solvent molecules.[26] They can therefore be located using CSP methods, usually at energies significantly above the global lattice energy minimum. These structures can be identified because they contain voids of the right size to accommodate molecular guests. It is not usually clear, *a priori*, which solvent will stabilise which particular crystal packing.[27] As such, predicting the influence of solvent on relative crystal lattice energies is of strong value because it could allow crystallisation conditions to be explored *in silico* to target specific polymorphs. For example, we recently used solvent stabilisation calculations to explain why a hydrogen bonded molecule, triptycene-tris(benzimidazolone), crystallises as a series of porous polymorphs and not as a dense, non-porous phase, which is the predicted global lattice energy minimum structure.[10]

Here, we investigate the effect of solvent on the crystallisation of an ethylenediamine-derived [4 + 6] imine cage, **CC1** (Fig. 1a).[22,28] **CC1** is a good candidate for studying solvent effects on polymorphism: unlike many imine cages, where bulky vertices control the crystal packing, **CC1** is nearly spherical and it has many structures of similar energy on its predicted crystal energy landscape.[12] The **CC1**-α′ polymorph is formed by desolvation of an ethyl acetate (EtOAc) solvate, **CC1**·2.5(EtOAc), while **CC1**-β′ is formed by desolvation of a dichloromethane (DCM) solvate, **CC1**·2.5(DCM).[22] The desolvated crystal structures are related directly to their respective solvatomorphs and the two different polymorphs have different properties: **CC1**-β′ is porous to nitrogen while **CC1**-α′ is not.[22] The phase behaviour of **CC1** is not limited to these polymorphs:

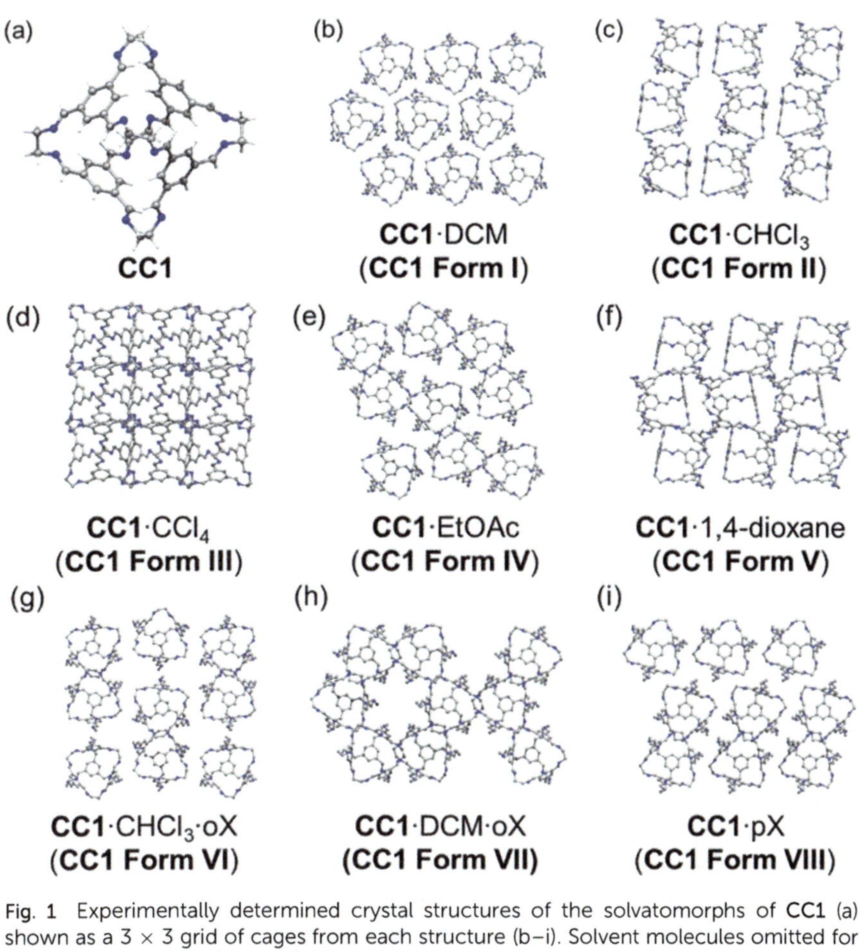

Fig. 1 Experimentally determined crystal structures of the solvatomorphs of **CC1** (a) shown as a 3 × 3 grid of cages from each structure (b–i). Solvent molecules omitted for clarity. oX = *ortho*-xylene and pX = *para*-xylene.

for example, recrystallisation from DCM/mesitylene affords **CC1**·4(mesitylene) ($P\bar{1}$),[29] and recrystallisation from DCM/1,4-dioxane affords 2(**CC1**)·7(1,4-dioxane) ($P2_1/c$).[23] Hence, the solid form landscape of **CC1** is particularly rich and non-intuitive from an experimental perspective.

We present here the preparation and characterisation of five new solvatomorphs of **CC1** (Fig. 1), which exist as either single solvates (one cocrystallised solvent) or as bisolvates (two different cocrystallised solvents). By using a Monte Carlo solvent filling procedure in conjunction with lattice energy minimisation, we show that all the solvatomorphs are solvent stabilised compared to the hypothetical global minimum on the CSP landscape. In addition, we show for three different chlorinated solvents that the experimentally obtained crystal solvatomorph is correctly predicted by solvent stabilisation calculations; that is, the observed solvatomorph is predicted to be more stabilised by the solvent used for its crystallisation than any of the alternative **CC1** structures that were tested. This is further corroborated by thermogravimetric analysis (TGA) of the experimental solvates, which shows a correlation between the experimental stability of the solvatomorph and the computed solvent stabilisation energy.

Methods

Crystallisation and characterisation

Crystals of **CC1** were obtained using either slow evaporation, vapour diffusion, or layering methods, details of which are provided in the ESI.† **CC1** is soluble in chlorinated solvents such as DCM and $CHCl_3$ but poorly soluble in non-polar solvents, such as xylenes, and also in some polar solvents, such as EtOAc. Hence, combinations of solvents were used in some cases for the crystallisation of **CC1** where limited solubility did not allow a single solvent to be used.

Single crystal X-ray data sets were measured on a Rigaku MicroMax-007 HF rotating anode diffractometer (Mo-Kα radiation, $\lambda = 0.71073$ Å, Kappa 4-circle goniometer, Saturn 724+ detector); at beamline I19, Diamond Light Source, UK using silicon double crystal monochromated synchrotron radiation ($\lambda = 0.6889$ Å, Saturn 724+ detector);[30] or at beamline 11.3.1, Advanced Light Source, Berkeley, USA, using silicon monochromated synchrotron radiation ($\lambda = 0.7749$ Å, APEX-II detector). Solvated single crystals were immersed in a protective oil, mounted on a MiTeGen loop, and flash cooled under a dry N_2 gas flow. Absorption corrections, using the multi-scan method, were performed with the program SADABS.[31] Structures were solved with SHELXD,[32] or by direct methods using SHELXS,[33] and refined by full-matrix least squares on $|F|^2$ by SHELXL.[34] Supplementary CIFs, that include structure factors, are available free of charge from the Cambridge Crystallographic Data Centre (CCDC). 2(**CC1**)·7.75(DCM) #1575913, 2(**CC1**)·11.32($CHCl_3$) #1575914, 2(**CC1**)·10(CCl_4) #1575912, CC1·2(*o*X)·$CHCl_3$ #1575911, and CC1·*o*X·DCM #1575910. For full refinement details, see the ESI (Tables S1 and S2, Fig. S1–S7†).

Powder X-ray diffraction data was collected in transmission mode using the Mythen-II position sensitive detector and capillary spinner on the I11 beamline at the Diamond Light Source ($\lambda = 0.827157$ Å). The sample was contained in a 0.5 mm diameter borosilicate glass capillary. The structure of CC1·1.47(*p*X) was solved using the simulated annealing routine implemented in Topas-Academic.[35] For PXRD refinement details, see the ESI (Fig. S8 and S9†).

TGA was carried out using a TA Q5000IR analyser with an automated vertical overhead thermobalance. Samples were heated at a rate of 10 °C min^{-1}. T_{onset} was calculated using the in-built feature in the software 'TA Universal Analysis'. TGA curves and calculated T_{onset} values are in the ESI.†

Crystal structure prediction

As a reference crystal energy landscape for **CC1** (Fig. 3), we use the structures from Case *et al.*,[36] which was generated by quasi-random sampling with one molecule in the asymmetric unit ($Z' = 1$) in 16 common space groups ($P1$, $P2_1$, $C2$, $P2_12_12_1$, $P2_12_12$, $C222_1$, $P4_12_12$, $R3$, $P\bar{1}$, Cc, $P2_1/c$, $C2/c$, $Pna2_1$, $Pbcn$, $Pbca$, and $Pnma$). This search used 5000 accepted trial crystal structures in each space group, in which molecular clashes were relieved using the SAT-expand method described in ref. 36. This previously published set of predicted crystal structures for **CC1** was supplemented by additional crystal structure searching where we identified that the sampling in our previous set was not complete. Specifically, we added structures in $C2/c$ with $Z' = 1$ and sampled those space groups with two independent molecules ($Z' = 2$) where this symmetry is observed in one of the known **CC1** structures. These additional crystal structure searches were performed using the Global Lattice Energy Explorer (GLEE) software,[36] which performs a quasi-random sampling of lattice dimensions, molecular positions and orientations. The quasi-random sampling used to generate the CSP landscape has the advantage over other global optimisation strategies that the sampling of higher energy regions of the landscape is given equal importance to sampling the low energy region near the global minimum. We find (as discussed below) that the **CC1** frameworks of observed solvates can be high in energy and so could be easily missed by more aggressive global optimisation algorithms.

Lattice energy minimisation calculations used the DMACRYS[37] software in which molecular geometries are held rigid at the gas phase B3LYP/6-31G** geometry. Intermolecular interactions are described using an empirically parameterised atom–atom repulsion–dispersion model (W99 (ref. 38)) and a distributed multipole electrostatic model, using a distributed multipole analysis[39] (DMA) of the B3LYP/6-31G** electron density with multipoles up to hexadecapole on each atom. Ewald summation was used to calculate the charge–charge, charge–dipole, and dipole–dipole interactions; all other intermolecular interactions were subject to a 30 Å cutoff. Full details are provided in the ESI.†

Solvent insertion calculations

A Monte Carlo (MC) approach was used to insert solvent molecules into **CC1** host frameworks (Fig. 2). MC calculations were performed using Towhee-7.10 (ref. 40) on the CSP structures corresponding to all of the observed forms of **CC1**, as well as the global lattice energy minimum structure determined from CSP. 10 000 cycle MC simulations were run in the NVT ensemble using a combination of translational, rotational and configuration-bias regrowth and reinsertion moves (the distribution of moves is given in the ESI†). For bisolvates, we also included two-molecule centre-of-mass switch moves. Simulations were run for a range of solvent loading ratios (**CC1** : solvent = 1 : N, N = 1,2,3…) until no further solvent could be added without disrupting the **CC1** packing.

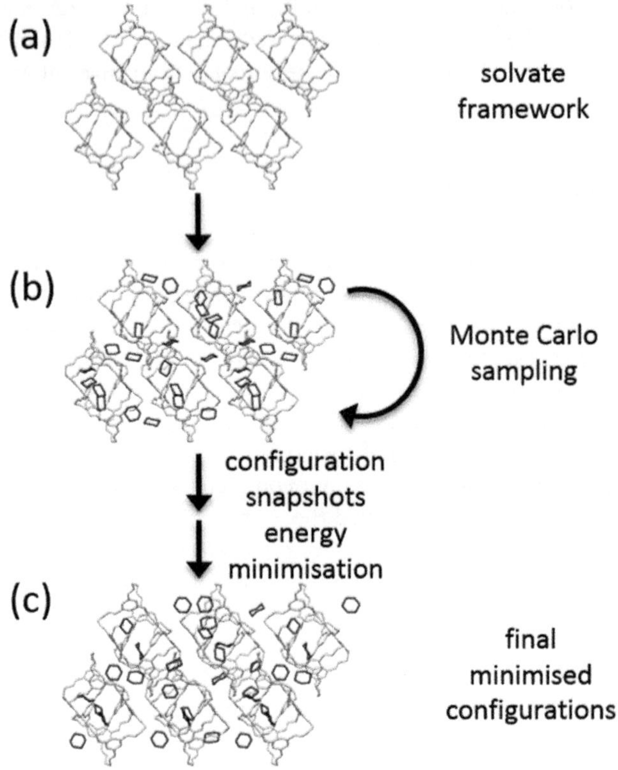

Fig. 2 Scheme of the Monte Carlo solvent loading calculations. Starting from a **CC1** host generated by crystal structure prediction (a), Monte Carlo steps are used to sample configurations of solvent molecules (b), and snapshots from the Monte Carlo sampling are subjected to lattice energy minimisation (c).

We attempted insertion of each solvent ($CHCl_3$, DCM, CCl_4, EtOAc, 1,4-dioxane, *p*-xylene) and solvent mixtures ($CHCl_3$: *o*-xylene and DCM : *o*-xylene) into the **CC1** framework that is observed to form with that solvent. For mixed solvent systems (bisolvates), the experimental solvent ratios were used during the solvent insertion calculations. It is a future challenge to extend our current methods to assess mixed solvent systems without experimental information on the solvent ratio in the solvated crystal structure. We also used the Monte Carlo approach to load five of the single solvents ($CHCl_3$, DCM, CCl_4, EtOAc, 1,4-dioxane) into all of the solvent frameworks and the CSP global minimum crystal structure.

An artificially high temperature of $T = 5000$ K was used to ensure good sampling of solvent configurations within the crystal structures. Frames sampled at 10 cycle intervals were lattice energy minimised with the same energy model used in the CSP calculations (W99 + DMA). No space group symmetry was enforced during solvent sampling or lattice energy minimisation. Solvated structures that led to large distortions of the **CC1** framework were discarded. This was tested by comparison of the **CC1** arrangement within the solvated structure to the original CSP structure using the COMPACK algorithm.[41]

The lattice energies of the solvated crystal structures were then corrected for the cost of removing N molecules of solvent from the pure phase:

$$E_{\text{latt}} = E_{\text{latt}}^{\text{CC1+solvent}} - N\left(E^{\text{solvent}} - \frac{3}{2}RT\right) + E_{\text{conf}} \tag{1}$$

where E^{solvent} was calculated from NVT MC simulations on pure solvent at 300 K, in which the solvent densities were fixed at the experimental density.

The $\frac{3}{2}NRT$ correction accounts for the change in the kinetic energy contribution of the solvent molecules to the internal energy. We assume solvent molecule rotational motion to be free in the bulk solvent phase, and to become librational vibrations in the solvate crystals and use equipartition estimates of the energy. E_{conf} is an energy correction between the conformations of **CC1**, which is added to any structure with a higher energy conformation.

Results and discussion

Crystallisation

The various experimental solvatomorphs of **CC1** are shown in Fig. 1. **CC1** crystallises from $CHCl_3$ as 2(**CC1**)·11.32($CHCl_3$) in the triclinic space group, $P\bar{1}$. In this structure, ordered $CHCl_3$ molecules were found in the cage cavities (Fig. S1†). In addition, a mixture of ordered and disordered $CHCl_3$ molecules was found in extrinsic 1-D channels running along the *c*-axis. By contrast, **CC1** crystallises from a CCl_4/$CHCl_3$ mixture as 2(**CC1**)·10(CCl_4) in the cubic space group, *F*23. In this structure, ordered CCl_4 molecules were found in the cage cavities, with the Cl atoms pointing toward the four tetrahedrally-arranged cage windows (Fig. S2 and S3†). 2(**CC1**)·10(CCl_4) was formed by layering CCl_4 onto a solution of **CC1** dissolved in $CHCl_3$, yet the refined crystal structure indicates that only CCl_4 crystallised with **CC1**, and not $CHCl_3$, which was also confirmed by ^{1}H NMR (Fig. S4†). In the published crystal structure of **CC1**·2.5(DCM) (*R*3), the dichloromethane (DCM) molecules were poorly resolved.[22] By repeating this crystallisation, we were able to isolate better quality crystals and the new structure was refined as 2(**CC1**)·7.75(DCM). There was still some disorder evident in 2(**CC1**)·7.75(DCM), but it was possible to accurately determine solvent positions and occupancies (Fig. S5†). Interestingly, the crystal structures obtained from these three chlorinated solvents ($CHCl_3$, CCl_4, and DCM) are all quite different (Fig. 1b–d), despite the relatively weak interactions between the solvents and the cage molecules.

In **CC1**·2.5(EtOAc) (*C*2/*c*), one EtOAc molecule was located in the **CC1** cavity with additional disordered EtOAc molecules being found throughout the narrow 1-D channel along the *b*-axis.[28] In 2(**CC1**)·7(1,4-dioxane) ($P2_1/c$), 1,4-dioxane sits in the window of two adjacent cage molecules and directs **CC1** to pack window-to-window. Extrinsic pockets also exist between **CC1** molecules filled with two 1,4-dioxane molecules.[23,28]

By using anti-solvent vapour diffusion crystallisations with *ortho*-xylene (*o*X) and *para*-xylene (*p*X), we isolated three additional solvatomorphs with different **CC1** crystal structures: **CC1**·DCM·*o*X ($P2_12_12_1$); **CC1**·$CHCl_3$·2(*o*X) ($P2_1/n$); and **CC1**·1.47(*p*X) ($P2_1$) (Fig. 1). **CC1**·$CHCl_3$·2(*o*X) was obtained from a mixture of chloroform and *o*X, crystallising in the form of thin needles (Fig. S6†). In this structure, $CHCl_3$ occupies the cage centres and *o*X occupies the interstitial spaces between the cage molecules to form extended, solvent-filled 1-D channels along the crystallographic *a*-axis. When DCM and *o*X were used, **CC1**·DCM·*o*X was obtained (Fig. S7†). As with the $CHCl_3$ analogue, the chlorinated solvent is in the cage centre. In this case, *o*X is in 1-D channels along the crystallographic *b*-axis and the resultant packing motif of **CC1** is very different. In **CC1**·DCM·*o*X, the channels are formed by arrangement of 6 **CC1** molecules such that the *o*X sits in a relatively large hexagonal-shaped channel (Fig. 1h). By contrast, in **CC1**·$CHCl_3$·*o*X, the *o*X molecules sit in smaller channels surrounded by just 4 cages (Fig. 1g). This leads to a different **CC1** : solvent ratio; in both cases, there is one chlorinated solvent molecule per cage, but in the $CHCl_3$·*o*X solvate, there are two *o*X molecules per cage, while in the DCM·*o*X solvate there is only one. We were unable to obtain suitable single crystals for the *p*X solvate of **CC1**, **CC1**·1.47(*p*X), but its structure was solved by powder X-ray diffraction (PXRD) in the monoclinic space group, $P2_1$ (Fig. S8 and S9†). There are no solvent-filled channels between cages in this solvatomorph (Fig. 1i); instead, the *p*X solvent sits inside the cage cavity and also in the interstitial voids, pointing into the cage window. The intrinsic site is partially occupied, with an occupancy of less than 50% in contrast to the extrinsic window site, which shows full occupancy. The total number of *p*X guests per cage is 1.471(2).

Crystal structure prediction

For many of the chiral organic cages that we have studied previously, we predicted a large energy gap between the global lattice energy minimum and the bulk of the predicted structures;[12,13,42] for example, this energy gap is 26 kJ mol^{-1} for **CC3-R**. In comparison, for this structurally related cage, **CC1**, we predicted a lattice energy surface with no large energy gaps between structures (Fig. 3). Less dense structures on the **CC1** landscape exhibit, in general, higher relative lattice energies, reflecting the energetic cost of void space in the crystal structure. The shapes and sizes of the extrinsic voids in each of the predicted structures are different, but it is non-trivial to determine what effect different crystallisation solvents will have on the energies of these structures. Therefore, it is not possible to simply determine which crystal packing will be formed from a given crystallisation solvent system using the CSP results for **CC1** alone.

The two unsolvated experimental polymorphs, **CC1**-α' and **CC1**-β', are both present in the predicted set of structures, 6.3 and 6.2 kJ mol^{-1} above the global minimum in the energy landscape, respectively, which is within the usual energetic range of polymorphism[43] in organic molecular crystals. For comparison, we calculated lattice energies of the various experimental solvates (Fig. 1) after artificial removal of the lattice solvent and subsequent lattice energy minimisation. All of these 'artificial desolvates' are located on the CSP landscape. The resulting structures are plotted on the CSP crystal energy landscape (Fig. 3, unfilled coloured circles). These results confirm that all 7 of the experimental solvate packings for **CC1** can be located using CSP methods without including solvent molecules in the calculation. In the case of the high-energy **Form II** structure, the **CC1** framework was only found after additional sampling in the observed space group ($C2/c$, $Z' = 1$); this is a particularly high energy structure (53.8 kJ mol^{-1} above the global minimum) and there are many local minima at such high energies, making the location of all local minima very challenging.

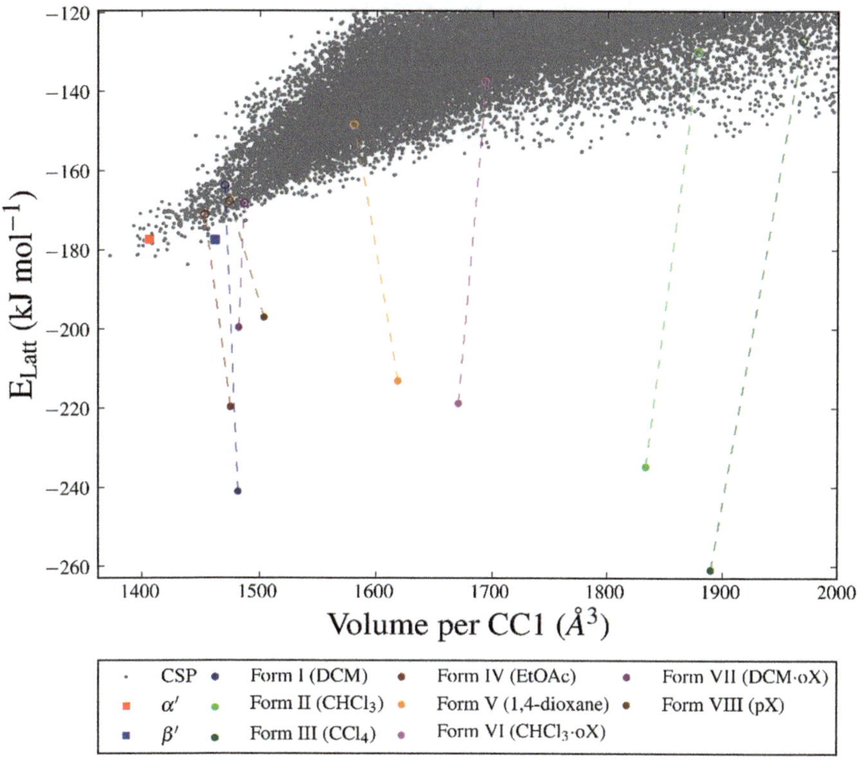

Fig. 3 CSP energy landscape for **CC1** (grey points) also showing the calculated lattice energies of the desolvated structures (**Forms I–VIII**, unfilled circles) and after solvent stabilisation (filled circles). Each solvent-filled structure is represented by the lowest energy configuration from the Monte Carlo sampling. The experimentally determined α′ and β′ polymorphs are shown as squares.

The significant challenge, then, is to identify which of the many predicted structures will be stabilised by which solvent or solvent mixture. The first step, tackled here, is to understand the relevant energetics: that is, the energy range on the CSP landscape where solvates are found and the energetic stabilisation that arises from the inclusion of solvent molecules. The calculated lattice energies for the artificially desolvated **CC1** structures range from 12.6 to 56.7 kJ mol^{-1} above the global lattice energy minimum – that is, the lowest energy solvent-free crystal packing predicted for **CC1** (Fig. 3). Some of these solvate frameworks are therefore much higher in energy than typical organic polymorphs; hence, these packing arrangements of **CC1** are only observed because of the energetic influence of solvent templating.

Solvent stabilisation for observed solvatomorphs

We first investigated the solvent stabilisation energies for the experimentally-observed solvatomorphs by inserting the relevant solvent into the respective solvent-free structure on the CSP energy landscape. To do this, a Monte Carlo procedure was used to insert solvent into the CSP predicted matches to each known **CC1** solvate structure at a range of solvent : **CC1** loadings.

The sampling performed in the Monte Carlo approach is important in evaluating the solvent stabilisation energies because of the strong dependence of the energy on the positions and orientations of the solvent molecules in the **CC1** framework. Fig. 4 shows the range of energies for the sampled configurations of DCM in **Form I** at all sampled solvent loadings (all sampled configurations of each solvent system are also shown in Fig. S11†). At $N = 4$, where the **CC1** framework is fully solvated, the different solvent configurations produced by the Monte Carlo sampling cover a range of approximately 40 kJ mol^{-1} and only a few of these are close in energy to the lowest energy configuration.

To evaluate the energetic stabilisation, we assume ordered solvent in the voids of the **CC1** structures and, so, we take the most stable calculated solvent arrangement in each of the host **CC1** frameworks **Forms I–VII**. We find that all the structures are stabilised significantly by solvent inclusion in their voids; this stabilisation is calculated to be 29.4–133.8 kJ mol^{-1} for the various solvates. Importantly, these large stabilisation energies lower the lattice energy of the known solvates below the hypothetical CSP global minimum energy structure on the crystal energy landscape (Fig. 3). This demonstrates that the solvated **CC1** structures are more stable than the unsolvated CSP global minimum and rationalises why the observed solvated structures of **CC1** are formed from organic solvents in preference to the lowest energy possible packing of pure **CC1**. The extent of the calculated solvent stabilisation with respect to the CSP global minimum ranges from 13.3 (**Form VIII**, *p*X) to 77.1 (**Form III**, CCl_4) kJ mol^{-1}.

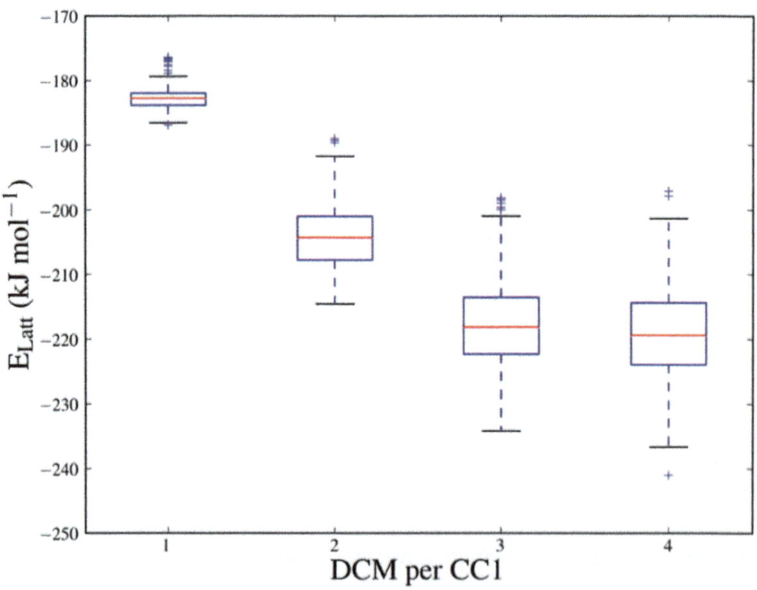

Fig. 4 Box plot of the energies of solvated structures of DCM in **CC1 Form I** at a range of solvent loadings (**CC1** : DCM = 1 : *N*, *N* = 1–4). Red lines indicate the median energy across the set of configurations for each value of *N*. The upper and lower limits of the box indicate the 1st and 3rd quartiles. The whiskers show the range of the calculated energies within 1.5× the interquartile range of the box limits and outliers are denoted by a cross.

Table 1 Observed ($N_{observed}$) and calculated ($N_{predicted}$) solvent loadings (**CC1** : solvent, 1 : N) of the known solvatomorphs of **CC1**

	Solvent	$N_{observed}$	$N_{predicted}$
Form I	DCM	3.88	4
Form II	$CHCl_3$	5.66	5
Form III	CCl_4	5	5
Form IV	EtOAc	2.5	2
Form V	1,4-Dioxane	3.5	3
Form VI	$CHCl_3 \cdot o$X	1 : 1	1 : 1
Form VII	DCM·oX	1 : 2	1 : 2
Form VIII	pX	1.47	1

The data in Fig. 4 also show how the **CC1** solvate system lowers its energy quickly as the first solvent molecules are added, but then starts to level off, approaching a minimum in stabilisation energy as more solvent is added (for DCM in **Form I**, around $N = 3$ and 4). Beyond $N = 4$, any further addition of DCM completely disrupts the **CC1 Form I** framework structure. For each solvatomorph, we predict the solvent loading as the composition with the lowest energy before the **CC1** arrangement is disrupted.

For the most stabilised solvent structures, the Monte Carlo predicted solvent loading agrees remarkably well with the experimental composition (Table 1). For all cage-solvent pairings, the predicted ratio of **CC1** to solvent is within 1 of the experimental value, which represents the closest possible agreement, given the use of integer sampling ratios used in the computational work. This suggests that it is possible to predict, *a priori*, the most stable solvation level for a given solvent within a predicted structure on a CSP landscape.

Preferential solvent stabilisation between CC1 frameworks

While the results summarised in Fig. 3 demonstrate that all the solvates tested are indeed predicted to be more stable than the desolvated CSP global minimum structure, they do not answer a key question: is the observed experimental solvatomorph the most stable structure for that specific solvent?

To assess the predictive potential of our methods, we considered the five solvents DCM, $CHCl_3$, CCl_4, EtOAc and 1,4-dioxane, using the Monte Carlo procedure to insert each solvent into each of the artificially desolvated **CC1 Forms I–V**, as well as the CSP global minimum structure (CSP min). The solvated lattice energies were then calculated for the most stable loading of each solvent into each solvate framework (Table 2).

First, these results clearly rationalise why the CSP global lattice energy minimum structure is not observed from crystallisation in the presence of any of these five solvents. The global minimum structure from the CSP landscape can accommodate each of the solvents studied here, and it is energetically stabilised by most solvents (final row, Table 2), apart from oX and pX, where the lattice energy is increased relative to the neat, solvent-free **CC1** structure (Table S8†). However, for all the solvents tested, there is always at least one competing solvate, including the experimentally observed solvate in each case, that is more stable (Tables 2, S8 and S9†); as such, the global minimum predicted structure is never predicted to be favoured.

Table 2 Comparison of the E_{Latt} (kJ mol^{-1}) for a selection of different solvated **CC1** crystal packing arrangement (**Forms I–V**) combinations

	DCM	$CHCl_3$	CCl_4	EtOAc	1,4-Dioxane
Form I	−241.1[a]	−204.1	−203.8	−220.1	−178.4
Form II	−226.4	−234.8[a]	N/A	−181.9	N/A[b]
Form III	−186.4	−227.6	−260.9[a]	−199.5	−198.9
Form IV	−225.7	−227.9	−237.5	−219.7[a]	−232.8
Form V	N/A[b]	N/A[b]	N/A[b]	N/A[b]	−213.2[a]
CSP min	−217.9	−214.2	−222.1	−206.8	−205.4

[a] Green highlighting indicates that the experimentally determined solvent is the most stabilising; yellow indicates that another solvent is more stabilising. [b] N/A denotes structures into which solvent could not be inserted without significantly distorting the **CC1** packing (see ESI).

For the three chlorinated solvents, DCM, $CHCl_3$, and CCl_4, we found excellent agreement between prediction and experiment; for each solvent, the observed solvate structure is the most stable of the **CC1** structures after computational solvation (Table 2, green highlighted values).

For EtOAc, we find that **Form I** and **Form IV** (the observed EtOAc solvatomorph) are effectively equi-energetic, with **Form I** being slightly more stable by 0.4 kJ mol^{-1}. EtOAc is the largest and most flexible of the solvents tested in this cross comparison and it is possible that the rigid-molecule approach used in our calculations performs less well in terms of distinguishing the observed solvate structure than for essentially rigid solvent molecules (*i.e.*, DCM, $CHCl_3$, and CCl_4). Nonetheless, from these calculations, one might predict that EtOAc would lead to either **Form I** or **Form IV**, the latter of which is observed by experiment.

In the case of the 1,4-dioxane solvate, we found that none of the four alternative solvents could solvate the **CC1** framework of the observed structure (**Form V**); for each of DCM, $CHCl_3$, CCl_4 and EtOAc, solvent insertion leads to a rearrangement of the **Form V CC1** framework. The observed packing, in which 1,4-dioxane lies flat between the cages, cannot be replicated by the other solvents. The fact that no other solvent can form this solvated crystal structure agrees well with experiment – in a previous study,[23] more than 40 other solvents were trialled in the laboratory, but none was found to direct the **CC1** window-to-window packing observed in the 1,4-dioxane solvate, **Form V**. Comparing the inclusion of 1,4-dioxane within the set of **CC1** frameworks (column 5; Table 2), we find that 1,4-dioxane forms a more stable solvate in the **Form IV CC1** arrangement than the observed **Form V**. This is due to a failure of the Monte Carlo sampling to locate the most stable configuration of 1,4-dioxane within **Form V**; by calculating the lattice energy using the positions of 1,4-dioxane from the experimentally determined crystal structure, we obtain an energy for the **Form V** 1,4-dioxane solvate of −241.9 kJ mol^{-1}. This configuration is 28.7 kJ mol^{-1} more stable than the most stable configuration determined by our solvent insertion methodology and it renders **Form V** the lowest energy of the tested pairings between 1,4-dioxane solvent and the **CC1** host frameworks. This 1,4-dioxane result shows that the solvent filling approach could be successful, but illustrates that further developments in solvent configuration sampling methods are needed to make these methods more robust in the future.

Solvent insertion into predicted crystal structures shows good results in terms of qualitatively identifying those crystal structures in a CSP landscape that can be solvent stabilised and assessing the preferential solvent stabilisation between possible host frameworks. One aspect that we are as yet unable to address with the current method is the possibility of structural transformations that can occur upon desolvation of a solvated crystal structure. For example, desolvation of **CC1**·2.5(EtOAc) to afford **CC1**-α' results in a 10% contraction of the unit cell volume and a transformation of the space group from $C2/c$ to $P2_1/c$.[22] Artificial, *in silico* desolvation and lattice energy minimisation of the experimental **CC1**·2.5(EtOAc) solvate leads to a polymorph which is 6.3 kJ mol^{-1} higher on the CSP landscape than the experimentally determined structure. Likewise, the DCM solvate, 2(**CC1**)·7.75(DCM), transforms during desolvation which results in a change from 2 to 1 molecule in the asymmetric unit ($Z' = 2 \rightarrow Z' = 1$), which is related to a conformational change in one of the **CC1** molecules.[22]

Comparison with thermogravimetric analysis

To further validate the results of our computational methods, we compared the magnitudes of the calculated solvent stabilisation energies with an experimentally determined measure of the solvate stability. Specifically, we assessed the thermal stability of each single solvate by measuring the difference ($T_{onset} - T_{bp}$) by thermogravimetric analysis (TGA), where T_{onset} and T_{bp} are the temperatures

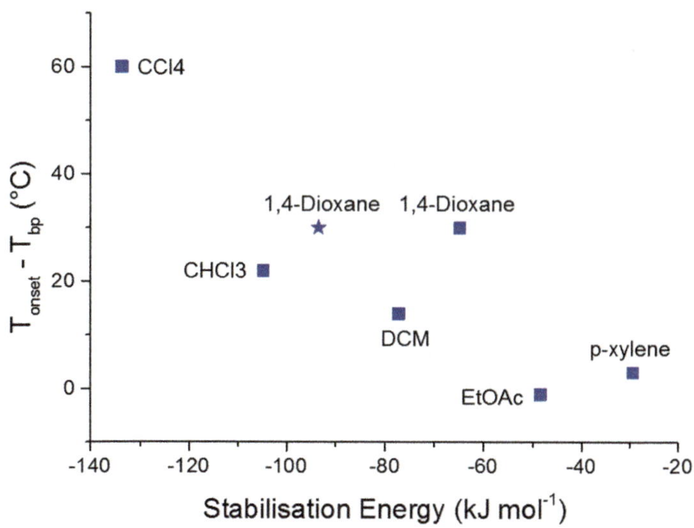

Fig. 5 Correlation between ($T_{onset} - T_{bp}$) and the calculated solvent stabilisation energy for the observed **CC1** solvates. The blue squares show the lowest energies calculated from the Monte Carlo simulations. The calculated stabilisation energy from the lattice energy minimised experimental structure of the 1,4-dioxane solvate is also shown as a blue star.

that mark the onset of guest release and the boiling point of the guest solvent, respectively (Table S10 and Fig. S15†). This quantity, $(T_{onset} - T_{bp})$, has been used previously to probe thermal stability of host-guest systems, including clathrates.[44,45]

A broad correlation is observed between $(T_{onset} - T_{bp})$ and the calculated solvent stabilisation energies (Fig. 5). TGA shows that 2(**CC1**)·10(CCl_4) is by far the most thermally stable of the **CC1** solvates ($T_{onset} - T_{bp} = 60$ °C) and this solvate also has the largest calculated stabilisation energy (−133.8 kJ mol^{-1}). This encouraging comparison extends to all three chlorinated solvents and EtOAc, where the stability orders based on TGA and the computational results are the same. The 1,4-dioxane solvate also fits this correlation, particularly when we use the energy calculated using the experimentally-determined solvent position within the **Form V CC1** framework (blue star in Fig. 5), rather than the Monte Carlo minimum energy solvent configuration.

Conclusions

In summary, we have shown how a Monte Carlo solvent insertion procedure can be used to rationalise the stabilities of the different solvated crystal structures of a highly solvatomorphic porous organic cage, **CC1**. All the observed solvated structures, with eight different crystal packing motifs, are stabilised significantly with respect to their corresponding desolvated **CC1** framework by up to 133.8 kJ mol^{-1}. As a result, the observed solvates are calculated to be more energetically stable than the lowest energy **CC1** crystal structure calculated using crystal structure prediction methods. Thus, we rationalise why these relatively high energy **CC1** crystal packing arrangements are observed, rather than the predicted global minimum, when this molecule is crystallised from solvent. This presents a challenge for CSP by demonstrating that experimentally relevant crystal structures can occupy high energy regions on the landscape, so that crystal structure search algorithms should have the capability to provide complete structure sets, even high above the global lattice energy minimum.

This computational approach is also shown to have promise in predicting the preferential stabilisation that different solvents provide to different porous crystal structures: for example, the calculations explain why three structurally-similar chlorinated solvents force **CC1** to adopt three different solvated crystal packing motifs. In each case, the solvated structure is most stable in the correct experimental packing motif and that packing motif is stabilised the most by the solvent from which it is crystallised.

These calculations could be used in a predictive sense by solvating the crystal structures produced by CSP with a range of solvents to identify a solvent that might be used to preferentially stabilise a targeted porous packing arrangement. Such an approach would benefit from a more efficient global optimiser to find the lowest energy solvent configuration within each framework, especially given the failure in this work to locate the most favourable 1,4-dioxane configuration within the **Form V** structure. The assumption of ordered solvent within porous crystal structures should also be relaxed in future developments of the method; where multiple, energetically similar solvent configurations are possible, the solvent is likely to be disordered and the related entropic stabilisation should be considered. If the target is to desolvate the resulting structure to create a porous material, the methods should also be developed to assess the possible changes in the crystal packing motif that may occur upon desolvation, as seen in the conversion of **CC1**-α to **CC1**-α'.[22]

We hope that this study will motivate further development in these methods, which could have an important impact on the *in silico* prediction of solvated structures and, hence, the *a priori* design of new porous molecular crystals.

Conflicts of interest

There are no conflicts to declare.

Acknowledgements

We acknowledge funding from the Engineering and Physical Sciences Research Council (EPSRC) (grants EP/K018132/1, EP/N004884/1 and EP/K018396/1) and the European Research Council under the European Union's Seventh Framework Programme (FP/2007-2013)/ERC through grant agreement number 321156 (ERC-AG-PE5-ROBOT). We acknowledge use of the IRIDIS High Performance Computing Facility and associated support services at the University of Southampton, in the completion of this work. We thank Diamond Light Source for access to beamlines I19 (MT15777) and I11 (EE12336). We thank the Advanced Light Source, supported by the Director, Office of Science, Office of Basic Energy Sciences, of the US Department of Energy under contract number DE-AC02-05CH11231, and S. J. Teat for his assistance. We also thank Dr Marc Schmidtmann for assistance with the X-ray crystallography.

Notes and references

1 G. M. Day, *Crystallogr. Rev.*, 2011, **17**, 3–52.
2 S. L. Price, *Chem. Soc. Rev.*, 2014, **43**, 2098–2111.
3 A. V. Kazantsev, P. G. Karamertzanis, C. S. Adjiman, C. C. Pantelides, S. L. Price, P. T. A. Galek, G. M. Day and A. J. Cruz-Cabeza, *Int. J. Pharm.*, 2011, **418**, 168–178.
4 M. A. Neumann, J. van de Streek, F. P. A. Fabbiani, P. Hidber and O. Grassmann, *Nat. Commun.*, 2015, **6**, 7793.
5 D. E. Braun, J. A. McMahon, L. H. Koztecki, S. L. Price and S. M. Reutzel-Edens, *Cryst. Growth Des.*, 2014, **14**, 2056–2072.
6 R. G. Della Valle, E. Venuti, A. Brillante and A. Girlando, *J. Chem. Phys.*, 2003, **118**, 807–815.
7 J. E. Campbell, J. Yang and G. M. Day, *J. Mater. Chem. C*, 2017, **5**, 7574–7584.
8 Y. Yang, B. Rice, X. Shi, J. R. Brandt, R. Correa da Costa, G. J. Hedley, D.-M. Smilgies, J. M. Frost, I. D. W. Samuel, A. Otero-de-la-Roza, E. R. Johnson, K. E. Jelfs, J. Nelson, A. J. Campbell and M. J. Fuchter, *ACS Nano*, 2017, **11**, 8329–8338.
9 F. Musil, S. De, J. Yang, J. E. Campbell, G. M. Day and M. Ceriotti, *Chem. Sci.*, 2018, **9**, 1289–1300.
10 A. Pulido, L. Chen, T. Kaczorowski, D. Holden, M. A. Little, S. Y. Chong, B. J. Slater, D. P. McMahon, B. Bonillo, C. J. Stackhouse, A. Stephenson,

C. M. Kane, R. Clowes, T. Hasell, A. I. Cooper and G. M. Day, *Nature*, 2017, **543**, 657–664.

11 J. D. Evans, D. M. Huang, M. Haranczyk, A. W. Thornton, C. J. Sumby and C. J. Doonan, *CrystEngComm*, 2016, **18**, 4133–4141.

12 E. O. Pyzer-Knapp, H. P. G. Thompson, F. Schiffmann, K. E. Jelfs, S. Y. Chong, M. A. Little, A. I. Cooper and G. M. Day, *Chem. Sci.*, 2014, **5**, 2235–2245.

13 J. T. A. Jones, T. Hasell, X. Wu, J. Bacsa, K. E. Jelfs, M. Schmidtmann, S. Y. Chong, D. J. Adams, A. Trewin, F. Schiffman, F. Cora, B. Slater, A. Steiner, G. M. Day and A. I. Cooper, *Nature*, 2011, **474**, 367–371.

14 A. Navrotsky, L. Mazeina and J. Majzlan, *Science*, 2008, **319**, 1635–1638.

15 A. S. Barnard and L. A. Curtiss, *Nano Lett.*, 2005, **5**, 1261–1266.

16 W. Sun, S. Jayaraman, W. Chen, K. A. Persson and G. Ceder, *Proc. Natl. Acad. Sci. U. S. A.*, 2015, **112**, 3199–3204.

17 A. M. Belenguer, G. I. Lampronti, A. J. Cruz-Cabeza, C. A. Hunter and J. K. M. Sanders, *Chem. Sci.*, 2016, **7**, 6617–6627.

18 B. Stöger, P. Kautny, D. Lumpi, E. Zobetz and J. Fröhlich, *Acta Crystallogr., Sect. B: Struct. Sci.*, 2012, **68**, 667–676.

19 C. B. Aakeröy, N. R. Champness, C. Janiak, R. Champness and C. Janiak, *CrystEngComm*, 2010, **12**, 22–43.

20 J. Karpinska, A. Erxleben and P. McArdle, *Cryst. Growth Des.*, 2011, **11**, 2829–2838.

21 L. R. Nassimbeni, *Acc. Chem. Res.*, 2003, **36**, 631–637.

22 J. T. A. Jones, D. Holden, T. Mitra, T. Hasell, D. J. Adams, K. E. Jelfs, A. Trewin, D. J. Willock, G. M. Day, J. Bacsa, A. Steiner and A. I. Cooper, *Angew. Chem., Int. Ed.*, 2011, **50**, 749–753.

23 T. Hasell, J. L. Culshaw, S. Y. Chong, M. Schmidtmann, M. A. Little, K. E. Jelfs, E. O. Pyzer-Knapp, H. Shepherd, D. J. Adams, G. M. Day and A. I. Cooper, *J. Am. Chem. Soc.*, 2014, **136**, 1438–1448.

24 M. Mastalerz, M. W. Schneider, I. M. Oppel and O. Presly, *Angew. Chem., Int. Ed.*, 2011, **50**, 1046–1051.

25 K. E. Jelfs, X. Wu, M. Schmidtmann, J. T. A. Jones, J. E. Warren, D. J. Adams and A. I. Cooper, *Angew. Chem.*, 2011, **123**, 10841–10844.

26 A. J. Cruz-Cabeza, G. M. Day and W. Jones, *Chem. - Eur. J.*, 2009, **15**, 13033–13040.

27 A. J. Florence, A. Johnston, S. L. Price, H. Nowell, A. R. Kennedy and N. Shankland, *J. Pharm. Sci.*, 2006, **95**, 1918–1930.

28 T. Tozawa, J. T. A. Jones, S. I. Swamy, S. Jiang, D. J. Adams, S. Shakespeare, R. Clowes, D. Bradshaw, T. Hasell, S. Y. Chong, C. Tang, S. Thompson, J. Parker, A. Trewin, J. Bacsa, A. M. Z. Slawin, A. Steiner and A. I. Cooper, *Nat. Mater.*, 2009, **8**, 973–978.

29 T. Hasell, X. Wu, J. T. A. Jones, J. Bacsa, A. Steiner, T. Mitra, A. Trewin, D. J. Adams and A. I. Cooper, *Nat. Chem.*, 2010, **2**, 750–755.

30 H. Nowell, S. A. Barnett, K. E. Christensen, S. J. Teat and D. R. Allan, *J. Synchrotron Radiat.*, 2012, **19**, 435–441.

31 G. M. Sheldrick, *SADABS Programs Scaling Absorpt. Correct. Area Detect. Data*, University of Göttingen, Göttingen, Germany, 1997.

32 G. M. Sheldrick, *Acta Crystallogr., Sect. D: Biol. Crystallogr.*, 2010, **66**, 479–485.

33 G. M. Sheldrick, *Acta Crystallogr., Sect. A: Found. Crystallogr.*, 2008, **64**, 112–122.

34 G. M. Sheldrick, *Acta Crystallogr., Sect. C: Struct. Chem.*, 2015, **71**, 3–8.
35 A. Cohelo, *Coelho Software (v. 5)*, TOPAS-Academic, Brisbane, Australia, 2012.
36 D. H. Case, J. E. Campbell, P. J. Bygrave and G. M. Day, *J. Chem. Theory Comput.*, 2016, **12**, 910–924.
37 S. L. Price, M. Leslie, G. W. A. Welch, M. Habgood, L. S. Price, P. G. Karamertzanis and G. M. Day, *Phys. Chem. Chem. Phys.*, 2010, **12**, 8478–8490.
38 D. E. Williams, *J. Comput. Chem.*, 2001, **22**, 1154–1166.
39 A. J. Stone and M. Alderton, *Mol. Phys.*, 1985, **56**, 1047–1064.
40 M. G. Martin, *Mol. Simul.*, 2013, **39**, 1212–1222.
41 J. A. Chisholm and S. Motherwell, *J. Appl. Crystallogr.*, 2005, **38**, 228–231.
42 A. G. Slater, P. S. Reiss, A. Pulido, M. A. Little, D. L. Holden, L. Chen, S. Y. Chong, B. M. Alston, R. Clowes, M. Haranczyk, M. E. Briggs, T. Hasell, G. M. Day and A. I. Cooper, *ACS Cent. Sci.*, 2017, **3**, 734–742.
43 J. Nyman and G. M. Day, *CrystEngComm*, 2015, **17**, 5154–5165.
44 J. L. Atwood, L. J. Barbour and A. Jerga, *Science*, 2002, **296**, 2367–2369.
45 C. M. Kane, A. Banisafar, T. P. Dougherty, L. J. Barbour and K. T. Holman, *J. Am. Chem. Soc.*, 2016, **138**, 4377–4392.

Faraday Discussions

PAPER

Towards the systematic crystallisation of molecular ionic cocrystals: insights from computed crystal form landscapes†

Sharmarke Mohamed, *[a] Ahmad A. Alwan, [a] Tomislav Friščić, [b] Andrew J. Morris [c] and Mihails Arhangelskis [b]

Received 18th February 2018, Accepted 20th March 2018
DOI: 10.1039/c8fd00036k

The underlying molecular and crystal properties affecting the crystallisation of ionic cocrystals (ICCs) with the general formula A^-B^+N (A^- = anion, B^+ = cation and N = neutral acid molecule; 1 : 1 : 1 stoichiometry) are reported for a limited set of known crystal structures determined following the cocrystallisation of either 4-aminopyridine (which forms salts) or 4-dimethylaminopyridine (which forms salts and ICCs) with the same set of monoprotic acids with a single hydroxy or halogen substitution at the *ortho* or *para* position. Periodic density functional theory calculations (PBE + D2) on the energetic driving force for ICC crystallisation for a set of known crystal structures with well characterised acid, salt and ICC structures show that all but 1 of the 7 experimental ICC structures surveyed were more stable than the sum of their component salt and acid structures with 4 displaying relative stabilities (ΔE_{ICC}) ranging from 2.47–8.02 kJ mol^{-1}. The majority of molecular ICCs that are more stable with respect to their component salt and acid structures display the formation of discrete intermolecular $O{-}H_{acid}\cdots O_{anion}$ hydrogen bonds with the $D_1^1(2)$ graph set between the carboxylic acid OH donor and the carboxylate oxygen acceptor of the anion. Computed crystal form landscapes for model 1 : 1 salts derived from acid–base pairs (involving 4-dimethylaminopyridine) known to form molecular ICCs show that on average the most stable predicted polymorphs of the 1 : 1 salts have efficient packing of the ions with packing coefficients in the range 65–80% and this is comparable to the packing coefficients of the most stable predicted polymorphs of 1 : 1 salts (involving 4-aminopyridine) that have no ICCs reported. This suggests that the cocrystallisation of equimolar amounts of the 1 : 1 salt and the acid to form a 1 : 1 : 1 molecular ICC is a complicated phenomenon that cannot be explained on the basis of inefficiencies in the crystal packing of the salt ions.

[a]*Department of Chemistry, Khalifa University of Science and Technology, PO Box 127788, Abu Dhabi, United Arab Emirates. E-mail: sharmarke.mohamed@ku.ac.ae*

[b]*Department of Chemistry, McGill University, 801 Sherbrooke St. W., H3A 0B8 Montreal, Canada*

[c]*School of Metallurgy and Materials, University of Birmingham, Edgbaston, Birmingham B15 2TT, UK*

† Electronic supplementary information (ESI) available. See DOI: 10.1039/c8fd00036k

1. Introduction

The chemical composition of crystal forms is an important intrinsic property affecting the physicochemical properties[1–3] of solid materials. For active pharmaceutical ingredients (APIs), the identity and composition of the species in the lattice is also an important regulatory[4] consideration for the classification of the solid form as a polymorph, solvate, salt or cocrystal. Each of these types of solid form offers the opportunity to enhance one or more physicochemical properties of the API.[5–7] In the domain of multicomponent crystal forms, the past decade has seen significant effort by crystal engineers to develop predictive theoretical[8] and empirical rules[9] for the synthesis of cocrystal solid forms. Cocrystals have received significant attention because of their proven utility[10–12] as vehicles for enhancing the performance characteristics and solid-state properties of a range of ionisable and non-ionisable molecules. The hierarchy of hydrogen bond synthons[13–15] observed in cocrystal solid forms suggests that theoretical models capable of predicting and accurately ranking the stabilities of different cocrystal polymorphs as a function of coformer identity and relative composition, could prove useful in experimental efforts in cocrystal selection. There is already increasing evidence[8,16,17] that cocrystal formation may be predicted on thermodynamic grounds given a sufficiently accurate theoretical model for calculating the relative energy of the cocrystal with respect to its component molecules.

Computational methods of crystal structure prediction (CSP) have been shown to be a useful aid[18–21] in experimental efforts targeted at the discovery of new polymorphs[22–24] of organic compounds. The crystal form landscape (CFL)[18,25] calculated from a typical CSP study is usually depicted as a scatter plot of lattice energy *versus* density and shows all the unique lattice energy minima corresponding to hypothetical polymorphs of the system at a nominal 0 K temperature. This CFL in principle contains enough structural information about the packing preferences and stabilities of crystal polymorphs, that an accurate algorithm for translating this CFL into the experimental conditions necessary for targeting each polymorph would provide a critical leap forward in the industrial solid-form screening strategy. The continued exponential increase in computing power is already making it possible to apply machine learning techniques to a variety of problems in science.[26–28] Although the calculation of crystal properties from calculated CFLs is still non-routine and state-of-the-art,[29] recent work combining CSP with the estimation of physical properties has led to the calculation of energy–structure–function maps that have facilitated the discovery of previously unknown highly porous molecular crystals.[30,31] The periodic blind tests[32–37] in organic crystal structure prediction organised by the Cambridge Crystallographic Data Centre (CCDC) have shown a steady improvement in the capabilities of existing CSP algorithms in predicting the crystal structures of flexible organic molecules as well as those crystal structures that could be classified as multi-component crystal forms. Dispersion-corrected density-functional theory (DFT-D)[38–40] methods have been shown[41] to be reliable in producing final relative lattice energies for predicted structures that are often in accord with the experimental stability ordering for polymorphs[42] or show that the predicted global lattice energy minimum structure corresponds to the observed experimental structure.[43] In the domain of multicomponent crystal form synthesis, CSP methods have been shown to be a useful complementary tool in experimental efforts for the discovery of cocrystals,[44] hydrates[45] and salt hydrates.[25,46]

The term ionic cocrystal (ICC)[47–52] has recently entered the crystal engineering lexicon to describe a multicomponent solid form where the crystal lattice consists of a salt and an additional chemical species which may be a neutral molecule or another salt complex. The existence of ICC solid forms has been known for a long time however, with some reports[52] suggesting that it was Kobell who in 1843 first reported[53] an ICC consisting of sodium chloride and glucose. Despite the long history of ICC solid forms, the systematic crystallisation of ICCs using tested crystal engineering principles or theoretical methods is an underexplored area of solid state chemistry research. Here, we define a molecular ICC as any solid form with the general formula[52] A^-B^+N (where A^- = anion, B^+ = cation and N = neutral molecule or salt) where the ions of the salt are derived from molecular species. In the case where N is a neutral molecule, the systematic manipulation of the chemical identity of this molecule under a range of crystallisation conditions opens up the possibility of synthesising several types of ICCs with a range of physicochemical properties,[54] thereby further widening the utility of multicomponent solid forms within industrial solid-state screening laboratories. The charged intermolecular interactions between the ions are expected to be much stronger than the interactions between the neutral molecule and the salt complex. This raises interesting questions about the role of this additional neutral molecule: what is the structural role of the additional neutral molecule within the lattice? Does the inclusion of an additional neutral molecule within the lattice imply that ICCs with lattice stoichiometries such as 1 : 1 : 1 are thermodynamically more stable than the sum of the lattice energies of the 1 : 1 salt and the neutral acid molecule? Are ICCs more likely to be observed for acid–base combinations where the putative packing modes available to the charged ions of the hypothetical salt have low packing coefficients? All of these are interesting yet unanswered questions that will be probed in this contribution. When compared to cocrystals, salt solid forms are generally known to be more prone to crystallising in lattices with molecular stoichiometries that deviate from those expected (*i.e.* 1 : 1 or 2 : 1) on the basis of the number of ionisable functional groups on the acid and base, thereby leading to salt solvates.[55] The literature is replete[51,52,56,57] with examples of ICCs comprised of coordination complexes of neutral and ionised forms of organic compounds bound to a metal centre. However, there remain unanswered questions about whether organic molecular ICCs are largely a serendipitous[58–61] result of the crystallisation experiment whose observation is difficult to predict *a priori* or whether there are underlying trends in the molecular properties of the constituent molecules/ions of the ICC, which make them amenable to purposeful design using computational methods of CSP.[62]

Here, we compare the potential of two relatively simple organic bases (**1-2**) to form salts and molecular ICCs using the results of previously reported cocrystallisation experiments involving the combination of **1** or **2** with three aromatic monoprotic acids (**3–5**) with a single substitution at the *ortho* or *para* position (Table 1). These molecules (**1–5**) were carefully chosen to provide a contrast in the behaviour of 4-aminopyridine (**1**, 4-AP) which forms salts with the acids **3–5** and 4-dimethylaminopyridine (**2**, 4-DMAP) which forms both salt (1 : 1) and molecular ICC structures (1 : 1 : 1). Table 1 summarises the numbering system for the 1 : 1 salts (denoted **I–VI**) that are expected following the cocrystallisation of **1** or **2** with

Table 1 Grid depicting the numbering system (**I–VI**) for the 1 : 1 salts resulting from the combination of **1** or **2** with the acids **3–5**. The asymmetric units of the known experimental solid forms following cocrystallisation of the bases **1** or **2** with the acid coformers **3–5** under varying conditions are shown. Cambridge Structural Database (CSD)[63] reference codes for the known crystal structures are shown in bold capital letters. For all acid–base combinations, the calculated[70] aqueous ΔpK_a was in the range 4.2–6.2, illustrating high probability of ionisation to form the salt.[71–73]

	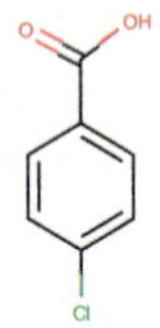 3 4-Chlorobenzoic Acid	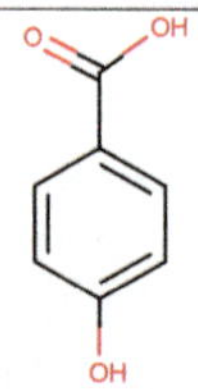 4 4-Hydroxybenzoic Acid	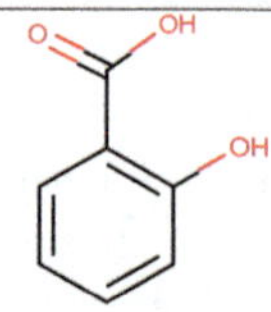 5 2-Hydroxybenzoic acid
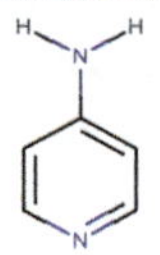 1 4-Aminopyridine	**I**=4-Aminopyridinium 4-chlorobenzoate (1:1) **NO SALT/ICC KNOWN**	**II**=4-Aminopyridinium 4-hydroxybenzoate (1:1) **Known Solid Forms:** 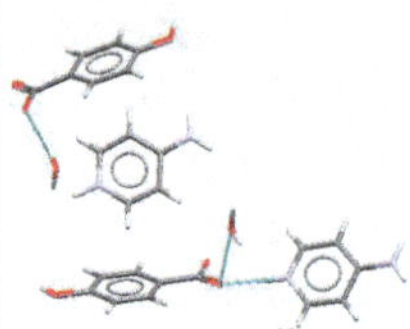 **MOYQOH**[66] 4-Aminopyridinium 4-hydroxybenzoate monohydrate	**III**=4-Aminopyridinium 2-hydroxybenzoate (1:1) **Known Solid Forms:** 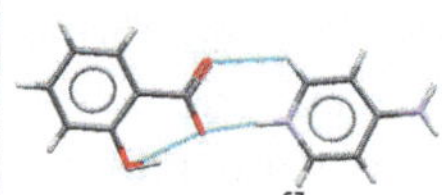 **DUSYOG**[67] 4-Aminopyridinium 2-hydroxybenzoate
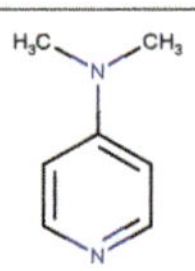 2 4-Dimethylaminopyridine	**IV**=4-Dimethylaminopyridinium 4-chlorobenzoate (1:1) **Known Solid Forms:** 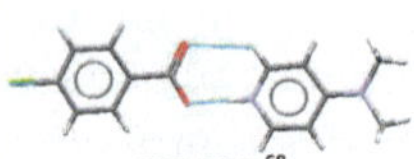 **CUKNED**[68] 4-Dimethylaminopyridinium 4-chlorobenzoate 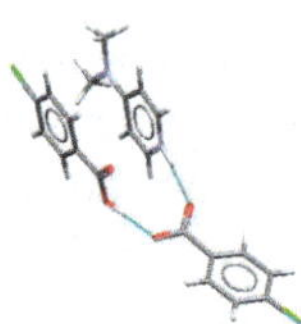 **CUKNON**[68] 4-Dimethylaminopyridinium 4-chlorobenzoate 4-chlorobenzoic acid	**V**=4-Dimethylaminopyridinium 4-hydroxybenzoate (1:1) **Known Solid Forms:** 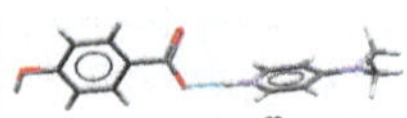 **SOLGUX**[69] 4-Dimethylaminopyridinium 4-hydroxybenzoate 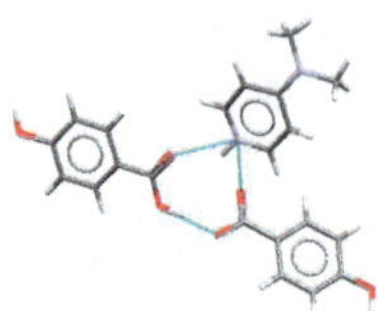 **CUKNUT**[68] 4-Dimethylaminopyridinium 4-hydroxybenzoate 4-hydroxybenzoic acid	**VI**=4-Dimethylaminopyridinium 2-hydroxybenzoate (1:1) **Known Solid Forms:** 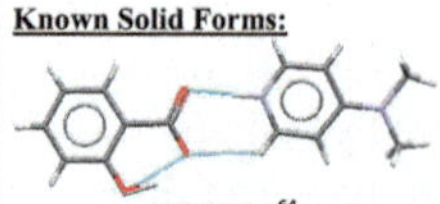 **KUJDEA**[64] 4-Dimethylaminopyridinium 2-hydroxybenzoate 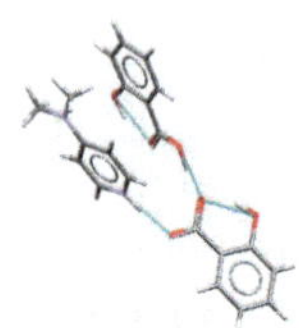 **KUJDIE**[64] 4-Dimethylaminopyridinium 2-hydroxybenzoate 2-hydroxybenzoic acid

the acids **3–5** as well as the previously reported multicomponent solid forms reported in the Cambridge Structural Database (CSD)[63] for each combination of acid and base. The objectives of this work are to probe the underlying molecular and crystal properties that give rise to ICC crystallisation. Given that ICC solid forms could be synthesised *via* the cocrystallisation of stoichiometric amounts of a salt with an acid,[54,64,65] we can consider ICC formation to be a type of "cocrystallisation" experiment between a salt and an acid coformer. Computed CFLs of model 1 : 1 salts are therefore of interest in understanding whether ICC crystallisation could be inferred on the basis of the packing modes available to the salt ions.

Calculated DFT-D relative lattice energies of a series of experimentally characterised 1 : 1 : 1 ICCs and their corresponding 1 : 1 salt and acid crystal structures were used to quantify the thermodynamic driving force for ICC crystallisation. The calculated packing efficiencies of the ions in the predicted polymorphs of the 1 : 1 salts were used to quantify the degree to which ICC crystallisation is driven by packing frustration of the salt ions.

2. Methodology

2.1. Equilibrium solubility measurements

The saturation temperatures (T_s) corresponding to saturated solutions of **1–5** were determined in a methanol solvent assuming a total solvent volume of 1 ml. Turbidity measurements as implemented in the Crystal16 parallel crystallisation unit (Technobis Crystallization Systems) were used to determine the solution saturation temperatures. Four heating and cooling cycles were set up for each compound **1–5** at a range of solution concentrations starting with a saturated solution of the compound at room temperature. Solution concentrations differed by approximately 5–10 mg ml^{-1} for each system. Each heating and cooling cycle had a minimum temperature of −15 °C and a maximum temperature of 50 °C. A constant heating/cooling rate of 1 °C min^{-1} was used for all experiments. Nitrogen purge gas was supplied to the reactors at a constant pressure of 0.4 bar. The contents of each vial were stirred at a rate of 500 rpm using a single magnetic flea stirrer bar using the bottom-stirring option of the Crystal16 unit. The resulting solubility data were analysed according to the measured clear and cloud points. The Van't Hoff equation was used to estimate the mole fractions (x_m) of each solute at a reference saturation temperature (T_s) of 298 K.

2.2. Crystal structure prediction

The crystal form landscapes (CFLs) for the 1 : 1 salt (**I–VI**) complexes (Table 1), were calculated using the Polymorph Predictor module of BIOVIA Materials Studio 8.0,[74] which uses a Monte Carlo simulated annealing method[75,76] to generate trial crystal structures. For the salt complexes, we assumed a 1 : 1 ratio of the ions in the asymmetric unit. Hypothetical crystal structures were generated in the following space groups: $P\bar{1}$, $P2_1$, $P2_1/c$, $P2_12_12_1$, $Pbca$, and $C2/c$. Motion groups were defined for each molecule/ion in the asymmetric unit and these motion groups were kept rigid throughout the cell refinement stage. For all systems except **II** and **V**, the assumed input geometry for the ions in the asymmetric unit corresponded to the calculated global minimum conformation obtained from gas phase geometry optimisation at the MP2/6-31G(d,p) level of theory using GAUSSIAN09.[77] For **II** and **V**, several input conformations of the 4-hydroxybenzoate anion (see Section 1.1.1 in the ESI†) were used in separate rigid body searches with the corresponding 4-aminopyridinium (**II**) and 4-dimethylaminopyridinium (**V**) cation geometries fixed to their gas phase global minimum conformations. For all CSP searches, torsion angles were not optimised during the cell refinement stage. For each of the above space groups used to generate trial crystal structures, the search space for the Monte Carlo simulation was limited to 7000 steps. The minimum temperature that the simulation could reach was 300 K and the maximum temperature was 1 × 10^5 K. Initial estimates of the lattice energies of predicted crystal structures were made within the Polymorph Predictor module of BIOVIA Materials Studio 8.0 using the Dreiding[78] force field with atomic charges derived from fitting to the molecular electrostatic potential of the *ab initio* wavefunction of the optimised conformation for the molecule/ion. The ChelpG[79] scheme as implemented in GAUSSIAN09 was used to fit the atomic charges to the molecular electrostatic potential. Clustering of the predicted crystal structures was performed within BIOVIA Materials Studio 8.0 in order to remove duplicates of the predicted lattice energy minima using a tolerance of 0.13, a cut-off of 7 Å for the distance-distribution analysis using 140 bins and a maximum cluster size of 500.

2.3. Final energy rankings of predicted structures using empirical force fields

For all systems except **II** and **V**, the 1000 most stable structures produced following clustering were passed to DMACRYS[80] for lattice energy minimisation using a distributed multipole model[81,82] for the electrostatic contribution towards the lattice energy. For systems **II** and **V**, since several rigid body searches were performed as a result of different input conformations for the 4-hydroxybenzoate anion, only the most stable 200 structures from each rigid body search were passed to DMACRYS. Multipoles were calculated up to rank 4 (hexadecapole) for all atoms by performing a distributed multipole analysis of the *ab initio* charge density using GDMA2.2.[83] The dispersion–repulsion contributions towards the lattice energy were estimated using a Buckingham exp-6 function using the FIT potential parameters for C, H_C (hydrogen attached to carbon), N, O and Cl from the work of Williams[84–86] as well as the parameters for H_N[87] (hydrogen attached to nitrogen) and H_O[88] (hydrogen attached to oxygen) that were subsequently determined by fitting to crystal structures containing the N–H···O=C interactions and carboxylic acid structures respectively. For each search, a final clustering step was performed to remove all duplicate structures with lattice energies within 0.2 kJ mol^{-1}, cell volumes within 1.0 Å^3 and a powder pattern similarity index[89] of at least 0.97. The Crystal Packing Similarity module[90] of Mercury 3.9 (ref. 89) was used to match predicted lattice energy minima with the observed salt structure retrieved from the CSD. This was done by estimating the root-mean squared deviation for matching at least 15 molecules ($RMSD_{15}$) in the coordination spheres of the experimental and predicted structures using a tolerance of 20% for the distances and 20° for the angles. In order to facilitate the crystal packing similarities, the lattice energies of the experimental structures were minimised using the same conformation and energy model used in the CSP. The Kitaigorodsky Packing Index (KPI)[91] values for the predicted polymorphs on the CFLs of **I–VI** were calculated using PLATON[92] and the distribution of the KPI values for the predicted polymorphs within 20 kJ mol^{-1} of the global minimum structure were used to rationalise the packing efficiencies of the ions. Computationally demanding DFT structural optimisations (atomic positions and unit cell parameters) were performed for a limited set of systems using the most stable structures (up to rank 20) following DMACRYS lattice energy minimisation. This was done in order to investigate the extent of re-ranking in the relative stability of the predicted polymorphs as a function of the DFT method used. Further details about the methodology for these calculations and a discussion of the results can be found in Section 1.4 of the ESI.†

2.4. Solid-state periodic DFT-D calculations for estimating the thermodynamic driving force for ICC crystallisation

Experimental crystal structures corresponding to molecular ICCs with a 1 : 1 : 1 stoichiometry and their corresponding 1 : 1 salt and acid crystal structures were retrieved from the CSD and optimized using CASTEP 8.0 (ref. 38) as implemented in BIOVIA Materials Studio 8.0 using the Perdew–Burke–Ernzerhof (PBE) generalised gradient approximation (GGA) exchange–correlation functional[93] and ultrasoft pseudopotentials.[94] The D2 (G06) semi-empirical dispersion correction of Grimme[95] was used in all cases. Brillouin zone integrations were performed on a symmetrised Monkhorst–Pack *k*-point grid[96] generated using the highest quality setting on Materials Studio 8.0 corresponding to a *k*-point separation of 0.07 $Å^{-1}$. The plane-wave basis set cut-off was set at 340 eV and the self-consistent field (SCF) convergence on the total energy was set at 1×10^{-6} eV per atom. The BFGS[97] algorithm was used for the geometry optimisation of all atomic positions and unit cell parameters, within the constraints of the space group symmetry. The structural optimisation was considered complete when the following convergence criteria were satisfied: maximum energy change of 1×10^{-5} eV per atom, maximum force of 3×10^{-2} eV $Å^{-1}$, maximum stress of 5×10^{-2} GPa and maximum displacement of 1×10^{-3} Å. In some cases, the BFGS algorithm did not converge after reaching 100 iterations and geometry optimisation was performed stepwise by first optimising only the atomic positions (fixed unit cell) followed by a full geometry optimisation on all atomic positions and the unit cell. Such a stepwise approach has been shown[16,98] to aid convergence in the optimisation of organic crystal structures when using DFT methods. Where a system displayed crystal polymorphism, the most stable reported structure (as determined using thermal methods) was used to carry out the DFT-D lattice energy calculations.

3. Results and discussion

3.1. Molecular and crystal properties of systems known to form ICCs: insights from the CSD

A search of the Cambridge Structural Database (CSD)[63] for all three-component ($Z'' = 3$)[99] crystal structures consisting of the general formula A^-B^+N (Fig. 1) where the cation is a pyridinium moiety, the anion is an aromatic carboxylate and N is any neutral molecule has revealed that there are a total of 185 crystal structures in the database that match the search query. Approximately 52% of these structures were salt hydrates where the additional chemical species N is a molecule of water. Molecular ICCs where the additional molecule N is the same acid used to form the anion (A^-B^+A) account for 27%, whilst ICCs where the additional molecule N is the base (A^-B^+B) account for only 2.7%. The remaining 18% of structures are ternary ICCs where N is a distinct molecular species to the acid or base used to form the salt.

With the exception of salt hydrates, given that molecular ICCs with 1 : 1 : 1 stoichiometry are predominantly those where the additional molecule N is the same acid that forms the anion (A^-B^+A), we conducted a careful analysis of the molecular properties of the acid–base pairs used to crystallise the ICCs with the view of gaining a better understanding of the underlying molecular properties that favour ICC crystallisation. All the acids that form ICCs have molecular

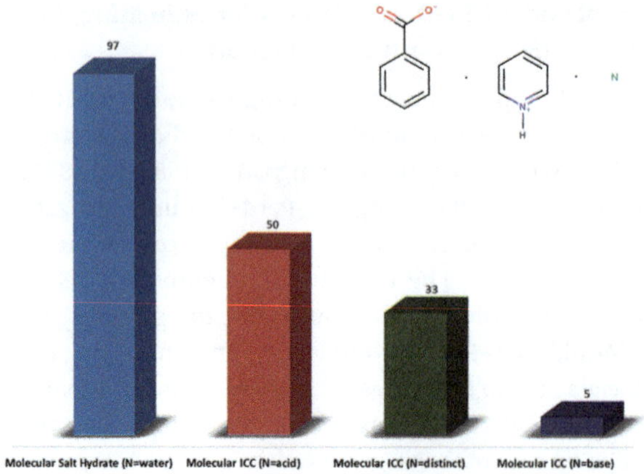

Fig. 1 Bar graph depicting the frequencies of different types of organic molecular ICCs retrieved from the CSD. All categories of structures depicted above consist of a 1 : 1 : 1 stoichiometry of anion : cation : neutral molecule (N) in the lattice, where N is a molecule of water, acid, base or a distinct molecular species to the acid and base. The inset on the top right-hand corner shows the search query used in the CSD search. The Cambridge Structural Database Version 5.37 (November 2015 + 3 updates up to and including the May 2016 release) was searched using Conquest V1.18 to retrieve the crystal structures of all pyridinium carboxylate salts containing 3 residues. The following filters were used in the search: 3D coordinates determined, not disordered, no errors, not polymeric, no powder structures and only organics.

structures with one or more hydrogen bond acceptor groups and in 41% of the acids surveyed, there were more acceptor groups than donor groups. By contrast, all the basic molecules used to form ICCs have more acceptors than donors in line with expectation. The polar surface area (PSA) of the acid molecules, which is a good measure of the potential of a molecule to engage in intermolecular hydrogen bonding, is on average 79 Å^2 across the sample space compared to just 19 Å^2 across all the basic molecules known to form ICCs (Fig. 2). This suggests that there is a significant mismatch in the available molecular surface area that could be used to form intermolecular hydrogen bond heterosynthons in acid–base pairs that are prone to crystallising as ICCs.

The hydrogen bond synthons observed in molecular ICCs of the type A^-B^+A (A^- = anion, B^+ = cation and A = acid; 1 : 1 : 1) retrieved from the CSD display two common types of interactions, which are denoted as Type 1 or Type 2 (Fig. 3). In both cases, the acid engages in discrete $O–H_{acid}\cdots O_{anion}$ [graph set: $D_1^1(2)$] hydrogen bonding interactions with the oxygen of the anion. The key difference between the motifs is that the Type 1 motif occurs in ICCs where the OH donor of the acid and the N^+–H donor of the cation are both engaged in hydrogen bonding interactions with the same oxygen of the carboxylate group, whilst in Type 2 interactions, these donor groups interact with different hydrogen bond acceptors on the anion. The Type 2 motif is found in 39% of all ICC structures surveyed whilst the Type 1 motif occurs in just 18% of structures. Approximately 63% of all ICCs that display a Type 1 motif were found to have cation–anion heterosynthons with the $R_2^2(7)$ graph set, whereas structures with Type 2 motifs all displayed discrete $D_1^1(2)$ heterosynthons between the cation and anion.

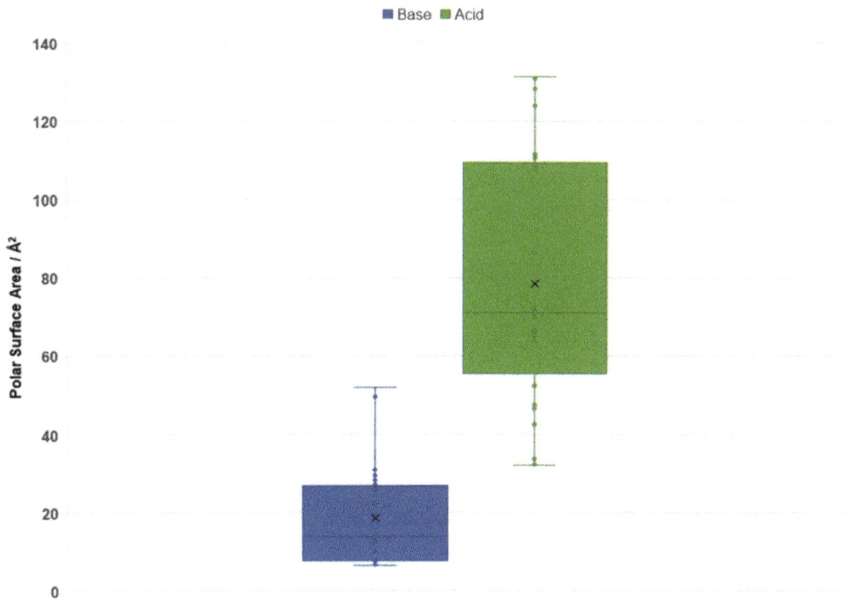

Fig. 2 Box-plots depicting the distribution of the calculated polar surface areas (PSA) of the acid (green) and base (blue) molecules that form ICCs according to the data retrieved from the CSD. The PSA for each acid or base was estimated using Spartan'16 (ref. 100) *via* the summation of the area (in $Å^2$) occupied by nitrogen, oxygen and hydrogens attached to either nitrogen or oxygen. The Corey–Pauling–Koltun (CPK)[101] model was used to represent the atoms. The mean value for each distribution is shown as a cross within the box-plot for each category. See Table S2 in the ESI† for more details.

3.2. Estimating the thermodynamic driving force for ICC crystallisation

Dispersion-corrected density functional theory methods (DFT-D) were used to estimate the relative lattice energies of molecular ICCs with 1 : 1 : 1 stoichiometry (A^-B^+A) with respect to the lattice energies of the corresponding 1 : 1 salt (A^-B^+) and acid (A) structures known for each system. These relative lattice energies

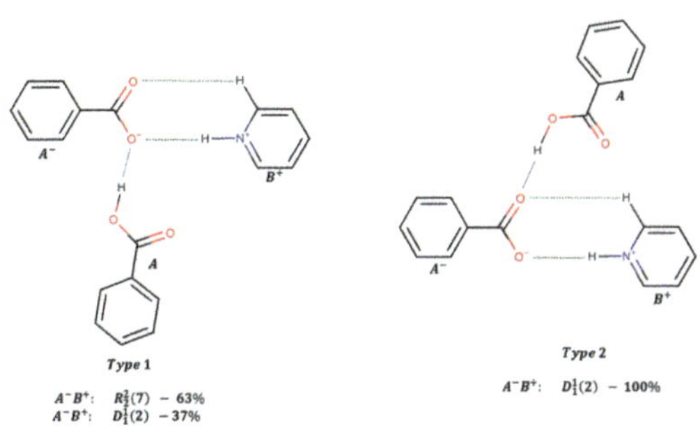

Fig. 3 Illustration of the common Type 1 and Type 2 hydrogen-bond heterosynthons found in the crystal structures of molecular ICCs with the formula A^-B^+A (A^- = anion, B^+ = cation and A = acid) derived from the cocrystallisation of substituted carboxylic acid and pyridine molecules. The hydrogen bond heterosynthons are depicted using graph-set notation.[102]

provide an indication of the thermodynamic driving force for ICC crystallisation given that the cocrystallisation of equimolar amounts of a salt and acid is known[54,64,65] to lead to the formation of molecular ICCs. Only 7 (of the total 50) well-characterised molecular ICC crystal structures retrieved from the CSD were found to have known 1 : 1 salt and acid crystal structures and hence only these systems were used to carry out the relative stability estimates. In each case, the thermodynamic driving force for ICC crystallisation (ΔE_{ICC}) was estimated according to the equation

$$\Delta E_{\mathrm{ICC}}(\mathrm{A}^- : \mathrm{B}^+ : \mathrm{A}) = E_{\mathrm{ICC}}(\mathrm{A}^- : \mathrm{B}^+ : \mathrm{A}) - [E_{\mathrm{SALT}}(\mathrm{A}^- : \mathrm{B}^+) + E_{\mathrm{acid}}(\mathrm{A})]$$

where $E_{\mathrm{ICC}}(\mathrm{A}^- : \mathrm{B}^+ : \mathrm{A})$, $E_{\mathrm{SALT}}(\mathrm{A}^- : \mathrm{B}^+)$ and $E_{\mathrm{ACID}}(\mathrm{A})$ are the total lattice energies of the 1 : 1 : 1 ICC, 1 : 1 salt and acid crystal structures respectively expressed per formula unit.

The relative energy estimates (Fig. 4) revealed that for 6 out of the 7 systems surveyed, the 1 : 1 : 1 ICC crystal structure was more stable than the sum of the energies of its corresponding 1 : 1 salt and acid crystal structures. In the case of **HAXFER**,[103] a molecular ICC formed from 3,5-dimethylpyridine and 3,5-dinitrobenzoic acid, the ICC is approximately 8 kJ mol^{-1} more stable than the sum of the lattice energies of the 1 : 1 salt and acid structures. The ICCs **CUKNUT**[68] and **NIMDIW**,[104] which were both crystallised using mono-substituted hydroxybenzoic acids also have clear thermodynamic stabilities of approximately 4 kJ mol^{-1} with respect to their component salt and acid structures. In all three cases (**HAXFER**, **CUKNUT** and **NIMDIW**), careful inspection of the hydrogen bond synthons in the molecular ICCs when compared to those observed in the salt structures shows that the additional acid molecule in the ICC is engaged in

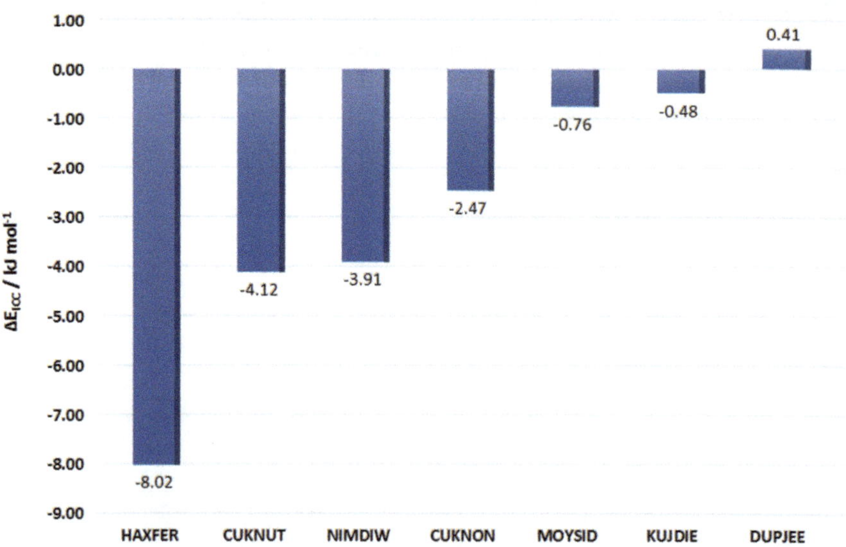

Fig. 4 Relative stabilities (ΔE_{ICC}) of the 1 : 1 : 1 experimental ICCs (labelled according to CSD refcodes) with respect to the stabilities of the known experimental 1 : 1 salt and acid crystal structures. All relative stabilities were calculated using PBE + D2 lattice energies of the experimental ICC, salt and acid crystal structures obtained following full structural optimisation with CASTEP.

discrete hydrogen bonding interactions with the carboxylate of the anion. In all cases, there is a significant perturbation to the crystal packing of the salt ions upon inclusion of the acid in the lattice leading to the ICC structure (Fig. 5). For example, in **HAXFER** which displays the highest relative stability with respect to its constituent salt and acid structures, the $R_2^2(7)$ dimer motif between the ions in the salt (**PUHRIT**[105]) is lost in the ICC, potentially due to the significant change in the relative packing of the ions as a result of acid inclusion into the lattice. Instead, discrete $D_1^1(2)$ acid-anion hydrogen bonds are formed. By comparison, for the **DUPJEE**[106] structure which has a lattice energy that is comparable in value to the sum of the lattice energies of the salt and acid crystal structures, the formation of the ICC does not lead to a change in the hydrogen bonded synthons involving the ions. This is evidenced by the persistence of the $R_2^2(7)$ dimer motif between the salt ions, which is a constant feature (Fig. 5) in both the salt (**BOVTEM**[107]) and ICC (**DUPJEE**) crystal packings. For three of the observed molecular ICCs (**MOYSID**,[66] **KUJDIE**[64] and **DUPJEE**[106]), the calculated energy differences are too small (<1 kJ mol^{-1}) to be indicative of a clear energetic driving force for the formation of a molecular ICC with 1 : 1 : 1 stoichiometry. The crystallisation of **MOYSID**[66] may be an interesting case study in the kinetics of ICC crystallisation as the authors report[66] the appearance of single crystals of **MOYSID** (3-aminopyridinium 3,5-dihydroxybenzoate 3,5-dihydroxybenzoic acid; 1 : 1 : 1) from a methanol solution of the acid and base following solvent evaporation at low temperatures inside a domestic refrigerator. If a solution with the same composition is allowed to undergo slow solvent evaporation under ambient conditions, single crystals of the 1 : 1 molecular salt 3-aminopyridinium 3,5-dihydroxybenzoate (**MUCMUT**[66]) are observed instead. Both the ICC and salt are reported to crystallise from the same methanol solution prepared *via* the cocrystallisation of equimolar amounts of 3-aminopyridine and

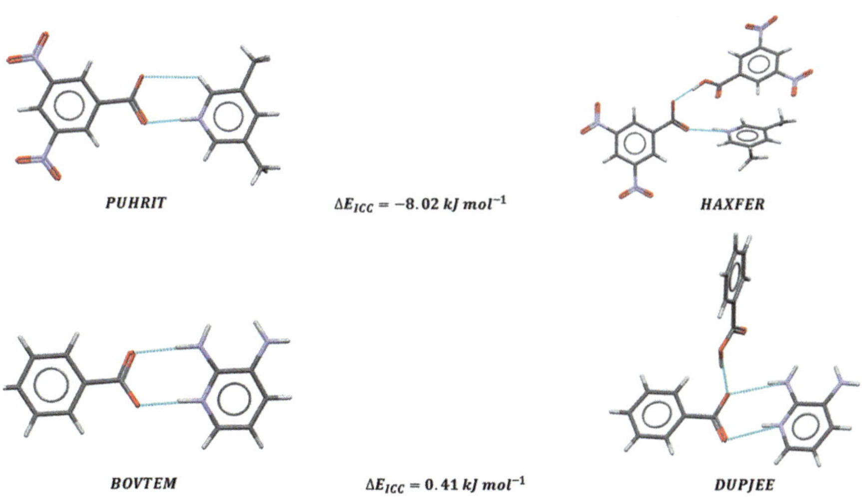

Fig. 5 Comparison of the differences induced in the hydrogen-bonded heterosynthons of the ions upon switching from the 1 : 1 salts **PUHRIT** and **BOVTEM** to three-component ICCs **HAXFER** and **DUPJEE**, respectively. The quoted ΔE_{ICC} values correspond to the calculated DFT-D relative lattice energies of these ICCs with respect to the known salt and acid crystal structures for each system as reported in Fig. 4.

3,5-dihydroxybenzoic acid. The small relative energy ($\Delta E_{ICC} = -0.76$ kJ mol^{-1}) of **MOYSID** with respect to the known salt and acid solid forms, may therefore be indicative of a kinetic ICC crystallisation product. The lattice energies of a wider set of experimental ICCs and their corresponding salt and acid structures as well as the experimental conditions used to target the observed solid forms must be surveyed before drawing any general conclusions about the thermodynamic driving force for ICC crystallisation. But the evidence from this limited set of structures is that DFT-D relative energies can be diagnostic enough to guide the selection of acid–base pairs with the potential to form stable molecular ICCs when coupled with CSP or X-ray diffraction techniques for determining the crystal structures.

3.3. Solubility differences of acid–base pairs known to form molecular ICCs

Differences in the solubilities in acid–base pairs are known to affect the stoichiometry of cocrystal solid forms.[108] The measured solubility differences of acid–base pairs are therefore critical parameters to consider when conducting multicomponent crystallisation screens. The Crystal16 platform was used to measure the saturation temperatures (T_s) as a function of the solution concentrations for **1–5** (Table 1) in methanol solvent. Methanol was chosen as the solvent of choice given that all the ICCs (**CUKNON**, **CUKNUT** and **KUJDIE**) of interest in this study (see Table 1) were crystallised from this solvent. The intention behind the solubility screen was to obtain a quantitative initial assessment of the likely solution concentrations of the acid–base pairs in methanol solvent at a given saturation temperature. This was deemed as a necessary first step in conducting more detailed studies on the effects of component solution concentrations on the stoichiometry of ICCs. To the best of our knowledge, the use of ternary phase diagrams for exploring the range of solution concentrations for targeting ionic cocrystals and the correlation of such concentrations with the stoichiometry of the resulting solid form are not typically reported in the literature. For a narrow temperature range, the Van't Hoff equation can be used to describe the variation in the mole fraction (x_m) of a given solute (**1–5**) as a function of the experimental saturation temperature, (T_s), according to the equation

$$\ln x_m = -\frac{\Delta H_{fus}}{R}\left(\frac{1}{T_s} - \frac{1}{T_0}\right)$$

where ΔH_{fus} is the enthalpy of fusion, R is the gas constant and T_0 is the melting temperature. For a series of experimental x_m values corresponding to different solution concentrations of **1–5**, the saturation temperatures of these solutions were measured using turbidity sensors. Linear fitting of the experimental solubilities and saturation temperatures from the Crystal16 platform was used to calculate the mole fraction (x_m) corresponding to a saturated solution of **1–5** at a reference saturation temperature of $T_s = 298$ K. Fig. 6 shows the x_m values at $T_s = 298$ K for **1–5** in methanol solvent. Comparison of the relative solubilities of the acid–base pairs in methanol shows that 4-chlorobenzoic acid (**3**) had the lowest solubility whilst 4-hydroxybenzoic acid (**4**) and 2-hydroxybenzoic acid (**5**) have comparable solubilities in methanol. The solubility of 4-dimethylaminopyridine (**2**) was at least a factor of 3 larger than the solubilities of **4** and **5**. By contrast 4-aminopyridine (**1**) had a marginally higher solubility than **4** and **5**. For

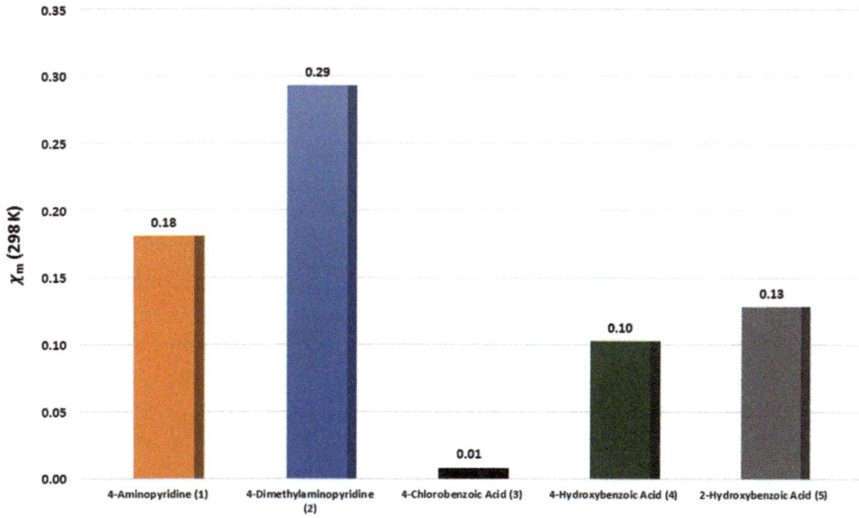

Fig. 6 Equilibrium solubilities expressed as mole fractions, x_m, for the molecules **1–5** in methanol solvent. The mole fractions, x_m, were derived for all molecules **1–5** in methanol solvent at a reference saturation temperature of 298 K. All mole fractions, x_m, shown above were derived from solubility data in methanol solvent using the Van't Hoff equation at a defined saturation temperature of 298 K.

the limited set of systems studied, acid–base pairs known to form ICCs (combinations involving **2** and the acids **3–5**) therefore have greater differences in relative solubilities at a given saturation temperature when compared to acid–base pairs that only form 1 : 1 salts or hydrates of 1 : 1 salts (combinations involving **1** and the acids **4** and **5**) from solution crystallisation experiments. The extent to which a comprehensive experimental screen for ICCs of **1** with the acid coformers **3–5** was performed and how exhaustive[109] such screens were in being confident that ICC crystallisation is not favourable, remain clear caveats in this analysis. However, given that chemists often rely on specific molar stoichiometric ratios of acid–base pairs when performing cocrystallisation experiments rather than relative concentrations of the acid or base at a given saturation temperature, the significant differences in the solubilities of **2** and each of the acids **3–5** show that it is important to consider the solution concentrations of the acid–base pair as this will determine the shape of the ternary phase diagram[110] and hence the composition of the resulting multicomponent solid form under a given set of experimental conditions. Several molecular ICCs[106,111–113] with 1 : 1 : 1 stoichiometry of the type A^-B^+A retrieved from the CSD were synthesised from solution crystallisation experiments involving 1 : 1 molar ratios of the acid and base. It is therefore conceivable that a 1 : 1 molar ratio of acid : base translates to a slightly supersaturated solution of the acid compared to the result assumed by the molar ratios, especially in acid–base pairs (like those involving **2** and the acids **3–5**) where the differences in the solubilities are significant. This is worthy of further investigation by looking at a wider set of relative solubilities for acid–base pairs known to form molecular ICCs.

3.4. Model crystal form landscapes of 1 : 1 salts: packing efficiencies of ions and inferences about ICC crystallisation

Model crystal form landscapes (CFLs) of the 1 : 1 salts **I–VI** (Table 1) were calculated in an attempt to see if the packing efficiencies of the salt ions could provide a strong indication as to the potential of the salt system to form a molecular ICC. The hypothesis being tested here is that molecular salts with inefficient packing of the ions are likely to form 1 : 1 : 1 ICCs (A^-B^+A) *via* the inclusion of a neutral acid molecule (A) into the salt lattice following the cocrystallisation of equimolar amounts of A and the 1 : 1 salt (A^-B^+), provided there is a thermodynamic driving force (Fig. 4) for the crystallisation of the ICC. This is worthy of investigation since the CSD survey (Fig. 1) revealed a number of ICC crystal structures that were described as "acid solvates"[106,112–114] according to their entry in the CSD or in one case, the title of the structure report itself.[106] The implication here is that the neutral acid molecule occupies the empty space in the ICC lattice created by the inefficient packing of the salt ions, in the same way that solvent molecules such as water may be said to occupy the voids in the crystal lattice of salt hydrates.[46,115] The 4-dimethylaminopyridinium 2-hydroxybenzoate 2-hydroxybenzoic acid ICC (**KUJDIE**) with 1 : 1 : 1 stoichiometry can be prepared from the solution crystallisation of equimolar amounts of 2-hydroxybenzoic acid (**5**) and 4-dimethylaminopyridinium 2-hydroxybenzoate (1 : 1) salt (**VI, KUJDEA**) in methanol solvent.[64] By contrast, the cocrystallisation of equimolar amounts of **5** with 4-aminopyridine (**1**) leads to a 1 : 1 salt of 4-aminopyridinium 2-hydroxybenzoate (**III, DUSYOG**)[61] with no ICC of the salt and **5** reported in the CSD (Table 1). Of the 1 : 1 salt systems **I–III** (Table 1) that could potentially be formed *via* the cocrystallisation of equimolar amounts of **1** and each of the acids **3–5**, only the combination **1** and **5** has a known anhydrate 1 : 1 salt structure (**III, DUSYOG**). The combination of **1** and **3** does not have a known salt or ICC structure in the CSD and cocrystallisation of **1** and **4** leads to a salt monohydrate (**MOYQOH**). Fig. 7 contrasts the CFLs of the salt systems **III** (Fig. 7a) and **VI** (Fig. 7b), which are plotted for each predicted lattice energy minimum in terms of the Kitaigorodsky Packing Index (KPI, %) as a function of the relative lattice energy using the predicted global minimum structure on each CFL as the reference point.

The CFL of **III** shows that the experimental structure is the third ranked structure in lattice energy (Fig. 7a) using the empirical FIT potential coupled with a distributed multipole electrostatic model. The predicted rank 3 structure corresponding to the experimental salt structure of **III** (**DUSYOG**) displays a KPI of 69.3%. By contrast the CFL of **VI** shows that the experimental salt structure (**KUJDEA**) is the global minimum in lattice energy. The CFL of **VI** displays fewer energetically competitive packings of the salt ions, with the rank 2 structure approximately 4 kJ mol^{-1} higher in energy when compared to the predicted global minimum structure. Structural optimisation (atomic positions and unit cell parameters) of the most stable predicted polymorphs (up to rank 20) for **III** and **VI** using the PBE functional[93] combined with a plane-wave implementation of the Grimme D2 (ref. 95) dispersion correction (PBE + D2), has shown (see Section 1.4 of the ESI† for more details) that there is significant re-ranking in the stabilities of the predicted polymorphs when the results of the empirical force field calculations are compared with those obtained from the PBE + D2 method. For **III**, the experimental rank 3 structure from the empirical force field calculations is the

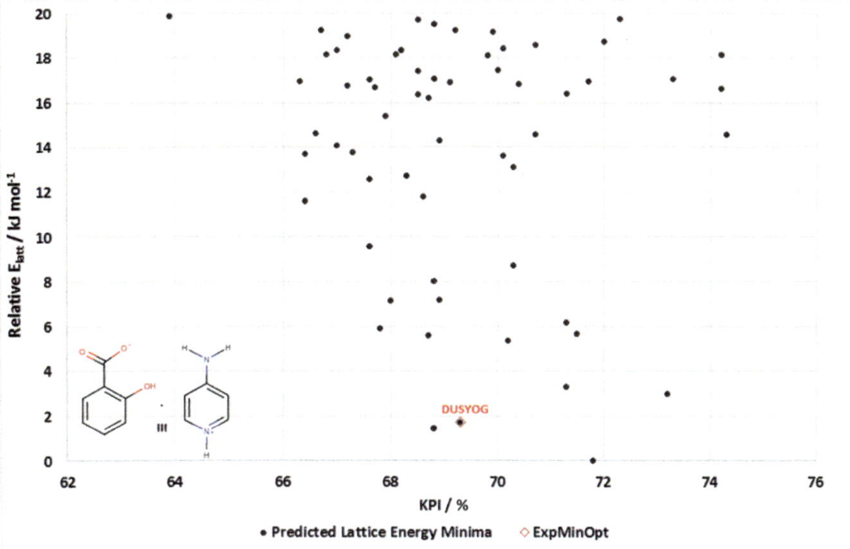

(a)

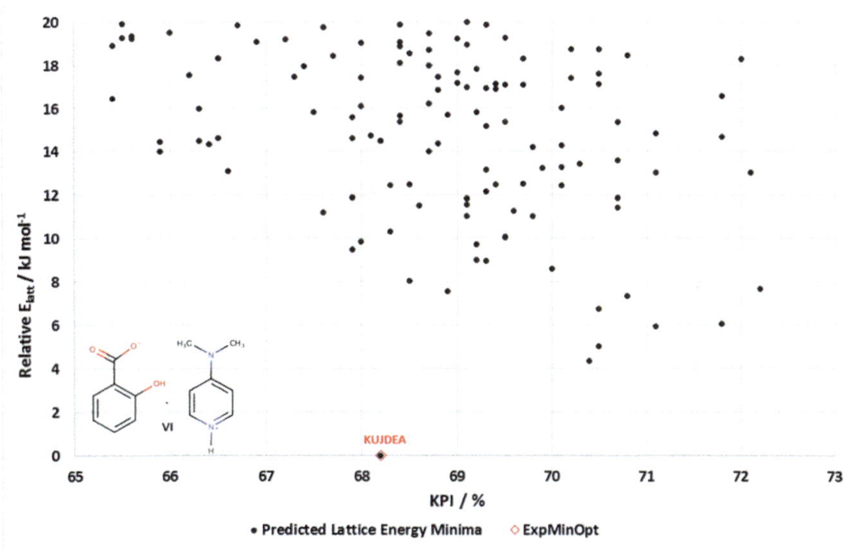

(b)

Fig. 7 Comparison of the calculated crystal form landscapes (CFLs) of the 1 : 1 salts **III** (a) and **VI** (b). The CFLs are plotted as relative lattice energies with respect to the calculated global minimum structure as a function of the calculated Kitaigorodsky Packing Index (KPI, %) for each lattice energy minimum. On each CFL, the predicted polymorphs are shown in dark-grey circles whilst the polymorph corresponding to the experimental structure (ExpMinOpt) is indicated in an open red diamond shape. See Tables S3–S8 in the ESI† for a detailed listing of the most stable predicted lattice energy minima for all the 1 : 1 salt complexes **I–VI**.

global minimum in lattice energy using the PBE + D2 method, with the second most stable structure (originally the global minimum using the empirical force field) approximately 1.89 kJ mol^{-1} higher in energy. Optimisations performed with the PBE functional coupled with the TS dispersion correction (PBE + TS)[116] as well as optimisations performed using the uncorrected LDA functional, reveal that in both cases the experimental structure of **III** is ranked as the second most stable structure. For **VI**, the re-ranking observed is much more pronounced and sensitive to the DFT method used. PBE + D2 and PBE + TS optimisations reveal that the experimental structure of **VI** changes from rank 1 using the empirical force field to rank 4 with relative lattice energies of 3.61 kJ mol^{-1} and 2.26 kJ mol^{-1} respectively. The results of the DFT optimisations show that the relative energies of predicted polymorphs for molecular organic salts are sensitive to the type of DFT method used. This contrasts with recently reported results on the relative energies of metal pentazolate[117] framework structures, where the rankings of the structures were consistent across a range of dispersion-correction schemes used.

The KPI of the predicted global minimum structure of **VI** is 68.2% and is comparable in value to that for the predicted rank 3 structure on the CFL of **III** which has a KPI of 69.3%. The comparable KPIs for the predicted lattice energy minima of **III** and **VI** that correspond to the experimental structures **DUSYOG** and **KUJDEA** respectively, suggest that the observation of an ICC in **VI** is a complicated phenomenon that has more to do with the finely balanced competing intermolecular interactions in both systems rather than arguments based on packing efficiency alone. This is supported by a closer inspection (Fig. 8) of the

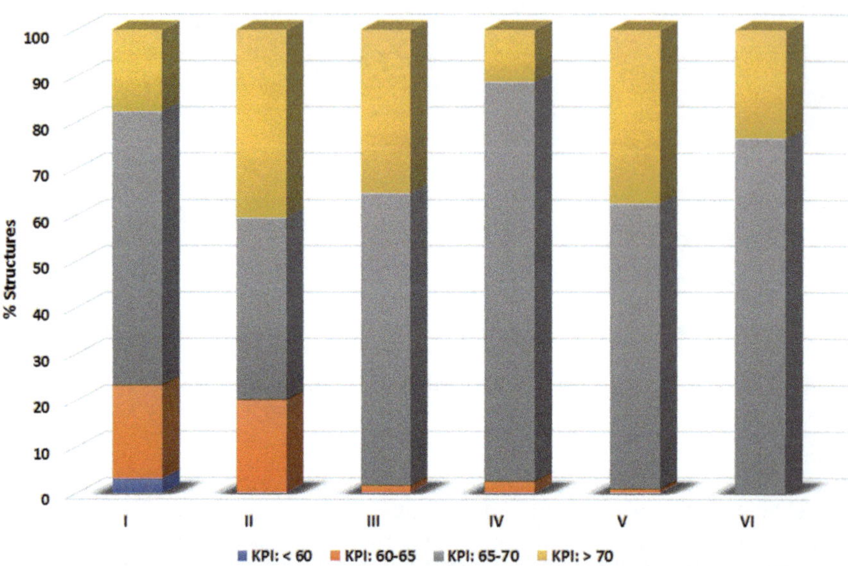

Fig. 8 Distribution of the calculated Kitaigorodsky Packing Index (KPI) values for the 1 : 1 molecular salts **I–VI** (Table 1) for the predicted lattice energy minima within 20 kJ mol^{-1} of the predicted global minimum structure for each system. The *y*-axis shows the percentage of structures within a particular KPI range (see legend). For all 1 : 1 salt systems, the majority of the predicted lattice energy minima within 20 kJ mol^{-1} have packing coefficients with KPI = 65–70 or KPI > 70.

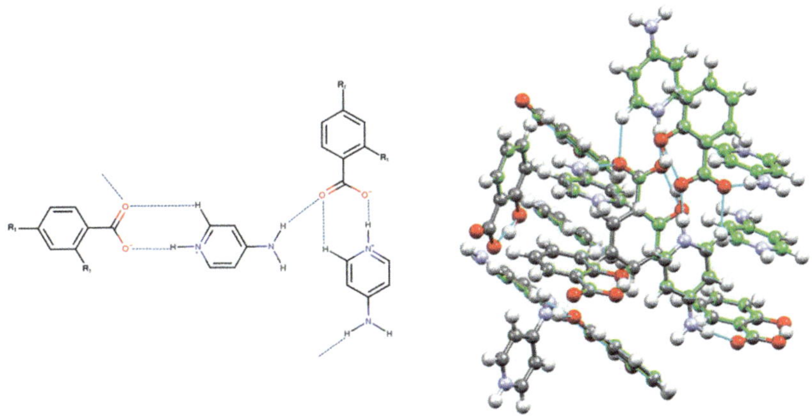

Fig. 9 Hydrogen bond motif found in the experimental (left) structure of III (DUSYOG) and an overlay (right) of the predicted rank 3 structure from the CFL of III on the experimental structure (DUSYOG) of III whose lattice energy was minimised using the same energy model used in the CSP (denoted ExpMinOpt). The calculated ExpMinOpt is coloured by element and the predicted rank 3 structure is coloured green. All 15/15 molecules in the coordination spheres of the two structures matched with a powder X-ray diffraction similarity index of 1.

distribution of the calculated KPIs of all the predicted polymorphs for salts **I–III** (which do not form ICCs) and salts **IV–VI** (which do form ICCs). Within a relative lattice energy range of 20 kJ mol^{-1} with respect to the global minimum structure, the majority of the predicted polymorphs of all the salts have KPIs that are in the range 65–80%, which is within the range we would expect for stable molecular crystals.[118] Moreover, the most stable polymorphs on each CFL have KPIs (see ESI: Tables S3–S8†) in the range 65–70%, which suggests no packing frustration of the salt ions. However, approximately 20% of the predicted salt polymorphs for **I** and **II** within a relative lattice energy range of 20 kJ mol^{-1} with respect to the global minimum display low KPIs (60–65%) when compared to the KPIs of the lattice energy minima corresponding to the experimental salts of **III** and **VI**. This may be significant given that anhydrate salt structures for **I** and **II** have yet to be reported.

Careful contrast of the hydrogen bond motifs adopted by the predicted low energy structures in the CFLs of **III** and **VI** show that all predicted polymorphs of **III** within 5 kJ mol^{-1} of the global minimum structure display N–H···O hydrogen bonds between the 4-amino group of the cation and the oxygen of the carboxylate/hydroxyl group on the anion (Fig. 9). The most stable predicted polymorphs of **I** and **II** also display similar N–H···O hydrogen bonds between the ions as observed in the experimental structure of **III** (Fig. 9). These N–H···O hydrogen bonds in the experimental structure of **III** are analogous to the Type 2 $O–H_{acid}···O_{anion}$ hydrogen bonds (Fig. 3) found in some of the ICCs reported in Table 1.

4. Conclusions

The cocrystallisation of an acid and base can lead to a variety of multicomponent solid forms such as a salt, cocrystal or salt hydrate. A number of qualitative descriptors have been proposed to account for the factors that affect the synthesis of cocrystal solid forms. Some of the proposed molecular descriptors/properties affecting the crystallisation of cocrystals include the complementarity of hydrogen bond synthons, the relative acidity (ΔpK_a) and the solubility regime of the acid–base pair on the ternary phase diagram. Whilst there are many reported examples of molecular ionic cocrystals (ICCs) in the literature, to date there has been no systematic investigation as to the factors that control the crystallisation of molecular ICCs, let alone the proposal of empirical or theoretical rules for predicting the crystallisation of such solid forms. The empirical evidence from the crystal structures of a small set of 50 molecular ICCs reported in the CSD as well as a limited grid of salt and molecular ICC solid forms resulting from the cocrystallisation of pyridines with a range of substituted monoprotic acids, was used to probe the underlying molecular properties of acid–base pairs that may be important drivers for ICC crystallisation. Acid–base pairs known to form ICCs have statistically significant differences in the polar surface areas available for intermolecular hydrogen bonding. Solubility differences in acid–base pairs known to form ICCs were also more pronounced than those known to form only salts. Solid-state DFT calculations show that comparisons of the relative lattice energies of 1 : 1 : 1 ICCs with the lattice energies of the known 1 : 1 salt and acid crystal structures reveal a thermodynamic preference for ICC crystallisation ranging from 2.47–8.02 kJ mol^{-1} for 4 of out of the 7 experimental ICC systems surveyed. However, the energy differences were less than 1 kJ mol^{-1} for the remaining 3 systems. This illustrates the challenges of relying on just the relative lattice energies when attempting to draw definitive conclusions about the crystallisation of multicomponent solid forms, which display many more finely balanced intermolecular interactions than single component crystal forms. Finally, the hypothesis that ICCs are favoured when the 1 : 1 salt ions do not have favourable modes of packing with themselves was disproven by the results of computed crystal form landscapes, which show that the predicted polymorphs of 1 : 1 salts known to form ICCs are just as likely to have a range of poor, reasonable and good packing coefficients when compared to anhydrous 1 : 1 salts that do not have reported ICCs. The description of ICCs as "acid solvates" is discouraged due to the observation of discrete intermolecular $O–H_{acid}\cdots O_{anion}$ hydrogen bonds between the carboxylic acid OH donor and the carboxylate oxygen acceptor of the anion in a significant number of the ICC crystal structures surveyed, suggesting that the additional acid molecule is integral to the ICC structure rather than just occupying empty space within the crystal. Future work will address some of the above unanswered questions from this study. These include the role of solution concentrations of acid–base pairs in affecting the stoichiometry of molecular ICCs. We will also conduct a wider study looking at the functional groups responsible for directing the finely balanced set of halogen and hydrogen bonds that are evident in the crystal structures of molecular ICCs. This will pave the way for the discovery of useful supramolecular design motifs that may be used by crystal engineers to facilitate the systematic crystallisation of molecular ICCs with useful solid-state properties.

Conflicts of interest

The authors confirm that there are no conflicts of interest to declare.

Acknowledgements

We are grateful for the support of the research computing department at Khalifa University and we acknowledge the use of the high-performance computing clusters at the Masdar Institute and at the Nuclear Engineering Department of Khalifa University. This research was enabled in part by support provided by Calcul Québec (http://www.calculquebec.ca) and Compute Canada (www.computecanada.ca). We acknowledge the support of the National Science and Engineering Council (NSERC) Discovery Grant (RGPIN-2017-06467) and E. W. R. Steacie Memorial Fellowship (SMFSU 507347-17). Khalifa University is acknowledged for awarding SM a 3-year competitive internal research award (CIRA-2018-068) to support his work in supramolecular and materials chemistry.

Notes and references

1 R. K. Khankari and D. J. W. Grant, *Thermochim. Acta*, 1995, **248**, 61–79.
2 L. B. Ta, H. M. Hobgood, A. Rohatgi and R. N. Thomas, *J. Appl. Phys.*, 1982, **53**, 5771–5775.
3 J. F. Remenar, S. L. Morissette, M. L. Peterson, B. Moulton, J. M. MacPhee, H. R. Guzmán and Ö. Almarsson, *J. Am. Chem. Soc.*, 2003, **125**, 8456–8457.
4 A. V. Trask, *Mol. Pharmaceutics*, 2007, **4**, 301–309.
5 W. Curatolo, *Pharm. Sci. Technol. Today*, 1998, **1**, 387–393.
6 L.-F. Huang and W.-Q. Tong, *Adv. Drug Delivery Rev.*, 2004, **56**, 321–334.
7 R. J. Bastin, M. J. Bowker and B. J. Slater, *Org. Process Res. Dev.*, 2000, **4**, 427–435.
8 N. Issa, P. G. Karamertzanis, G. W. A. Welch and S. L. Price, *Cryst. Growth Des.*, 2009, **9**, 442–453.
9 C. C. Seaton, K. Chadwick, G. Sadiq, K. Guo and R. J. Davey, *Cryst. Growth Des.*, 2009, **10**, 726–733.
10 S. Hiendrawan, B. Veriansyah, E. Widjojokusumo, S. N. Soewandhi, S. Wikarsa and R. R. Tjandrawinata, *Int. J. Pharm.*, 2016, **497**, 106–113.
11 J. Li, L. Wang, X. Fu, Z. Deng, Y. Q. Ye, Q. Ren and H. Zhang, *Eur. J. Pharm. Sci.*, 2016, **85**, 47–52.
12 N. Shan, M. L. Perry, D. R. Weyna and M. J. Zaworotko, *Expert Opin. Drug Metab. Toxicol.*, 2014, **10**, 1255–1271.
13 G. R. Desiraju and T. Steiner, *The weak hydrogen bond: in structural chemistry and biology*, Oxford University Press, Oxford, 2001.
14 C. B. Aakeroy and D. J. Salmon, *CrystEngComm*, 2005, **7**, 439–448.
15 T. R. Shattock, K. K. Arora, P. Vishweshwar and M. J. Zaworotko, *Cryst. Growth Des.*, 2008, **8**, 4533–4545.
16 C. R. Taylor and G. M. Day, *Cryst. Growth Des.*, 2018, **18**, 892–904.
17 H. S. Chan, J. Kendrick, M. A. Neumann and F. J. Leusen, *CrystEngComm*, 2013, **15**, 3799–3807.
18 S. L. Price, D. E. Braun and S. M. Reutzel-Edens, *Chem. Commun.*, 2016, **52**, 7065–7077.
19 S. L. Price and S. M. Reutzel-Edens, *Drug Discovery Today*, 2016, **21**, 912–923.
20 M. A. Neumann, J. van de Streek, F. P. A. Fabbiani, P. Hidber and O. Grassmann, *Nat. Commun.*, 2015, **6**, 7793.

21 S. Mohamed, *Integrating Computed Crystal Energy Landscapes in Crystal Form Discovery and Characterisation in Understanding Intermolecular Interactions in the Solid State: Approaches and Techniques*, ed. D. Chopra, Monographs in Supramolecular Chemistry, Royal Society of Chemistry, Cambridge, 2018, pp. 1–31, ISBN: 978-1-78801-079-5.

22 A. T. Hulme, S. L. Price and D. A. Tocher, *J. Am. Chem. Soc.*, 2005, **127**, 1116–1117.

23 J.-B. Arlin, L. S. Price, S. L. Price and A. J. Florence, *Chem. Commun.*, 2011, **47**, 7074–7076.

24 A. G. Shtukenberg, Q. Zhu, D. J. Carter, L. Vogt, J. Hoja, E. Schneider, H. Song, B. Pokroy, I. Polishchuk, A. Tkatchenko, A. R. Oganov, A. L. Rohl, M. E. Tuckerman and B. Kahr, *Chem. Sci.*, 2017, **8**, 4926–4940.

25 S. Mohamed, *Acta Crystallogr., Sect. E: Crystallogr. Commun.*, 2016, **72**, 1348–1352.

26 Y. Liu, T. Zhao, W. Ju and S. Shi, *J. Materiomics*, 2017, **3**, 159–177.

27 F. Musil, S. De, J. Yang, J. E. Campbell, G. M. Day and M. Ceriotti, *Chem. Sci.*, 2018, **9**, 1289–1300.

28 R. Burbidge, M. Trotter, B. Buxton and S. Holden, *Comput. Chem.*, 2001, **26**, 5–14.

29 E. L. Willighagen, R. Wehrens, W. Melssen, R. de Gelder and L. M. C. Buydens, *Cryst. Growth Des.*, 2007, **7**, 1738–1745.

30 A. Pulido, L. Chen, T. Kaczorowski, D. Holden, M. A. Little, S. Y. Chong, B. J. Slater, D. P. McMahon, B. Bonillo, C. J. Stackhouse, A. Stephenson, C. M. Kane, R. Clowes, T. Hasell, A. I. Cooper and G. M. Day, *Nature*, 2017, **543**, 657.

31 G. M. Day and A. I. Cooper, *Adv. Mater.*, 2017, 1704944.

32 A. M. Reilly, R. I. Cooper, C. S. Adjiman, S. Bhattacharya, A. D. Boese, J. G. Brandenburg, P. J. Bygrave, R. Bylsma, J. E. Campbell, R. Car, D. H. Case, R. Chadha, J. C. Cole, K. Cosburn, H. M. Cuppen, F. Curtis, G. M. Day, R. A. DiStasio Jr, A. Dzyabchenko, B. P. van Eijck, D. M. Elking, J. A. van den Ende, J. C. Facelli, M. B. Ferraro, L. Fusti-Molnar, C.-A. Gatsiou, T. S. Gee, R. de Gelder, L. M. Ghiringhelli, H. Goto, S. Grimme, R. Guo, D. W. M. Hofmann, J. Hoja, R. K. Hylton, L. Iuzzolino, W. Jankiewicz, D. T. de Jong, J. Kendrick, N. J. J. de Klerk, H.-Y. Ko, L. N. Kuleshova, X. Li, S. Lohani, F. J. J. Leusen, A. M. Lund, J. Lv, Y. Ma, N. Marom, A. E. Masunov, P. McCabe, D. P. McMahon, H. Meekes, M. P. Metz, A. J. Misquitta, S. Mohamed, B. Monserrat, R. J. Needs, M. A. Neumann, J. Nyman, S. Obata, H. Oberhofer, A. R. Oganov, A. M. Orendt, G. I. Pagola, C. C. Pantelides, C. J. Pickard, R. Podeszwa, L. S. Price, S. L. Price, A. Pulido, M. G. Read, K. Reuter, E. Schneider, C. Schober, G. P. Shields, P. Singh, I. J. Sugden, K. Szalewicz, C. R. Taylor, A. Tkatchenko, M. E. Tuckerman, F. Vacarro, M. Vasileiadis, A. Vazquez-Mayagoitia, L. Vogt, Y. Wang, R. E. Watson, G. A. de Wijs, J. Yang, Q. Zhu and C. R. Groom, *Acta Crystallogr., Sect. B: Struct. Sci., Cryst. Eng. Mater.*, 2016, **72**, 439–459.

33 D. A. Bardwell, C. S. Adjiman, Y. A. Arnautova, E. Bartashevich, S. X. M. Boerrigter, D. E. Braun, A. J. Cruz-Cabeza, G. M. Day, R. G. Della Valle, G. R. Desiraju, B. P. van Eijck, J. C. Facelli, M. B. Ferraro, D. Grillo, M. Habgood, D. W. M. Hofmann, F. Hofmann, K. V. J. Jose,

P. G. Karamertzanis, A. V. Kazantsev, J. Kendrick, L. N. Kuleshova, F. J. J. Leusen, A. V. Maleev, A. J. Misquitta, S. Mohamed, R. J. Needs, M. A. Neumann, D. Nikylov, A. M. Orendt, R. Pal, C. C. Pantelides, C. J. Pickard, L. S. Price, S. L. Price, H. A. Scheraga, J. van de Streek, T. S. Thakur, S. Tiwari, E. Venuti and I. K. Zhitkov, *Acta Crystallogr., Sect. B: Struct. Sci.*, 2011, **67**, 535–551.

34 G. M. Day, T. G. Cooper, A. J. Cruz-Cabeza, K. E. Hejczyk, H. L. Ammon, S. X. M. Boerrigter, J. S. Tan, R. G. Della Valle, E. Venuti, J. Jose, S. R. Gadre, G. R. Desiraju, T. S. Thakur, B. P. van Eijck, J. C. Facelli, V. E. Bazterra, M. B. Ferraro, D. W. M. Hofmann, M. A. Neumann, F. J. J. Leusen, J. Kendrick, S. L. Price, A. J. Misquitta, P. G. Karamertzanis, G. W. A. Welch, H. A. Scheraga, Y. A. Arnautova, M. U. Schmidt, J. van de Streek, A. K. Wolf and B. Schweizer, *Acta Crystallogr., Sect. B: Struct. Sci.*, 2009, **65**, 107–125.

35 G. M. Day, W. D. S. Motherwell, H. L. Ammon, S. X. M. Boerrigter, R. G. Della Valle, E. Venuti, A. Dzyabchenko, J. D. Dunitz, B. Schweizer, B. P. van Eijck, P. Erk, J. C. Facelli, V. E. Bazterra, M. B. Ferraro, D. W. M. Hofmann, F. J. J. Leusen, C. Liang, C. C. Pantelides, P. G. Karamertzanis, S. L. Price, T. C. Lewis, H. Nowell, A. Torrisi, H. A. Scheraga, Y. A. Arnautova, M. U. Schmidt and P. Verwer, *Acta Crystallogr., Sect. B: Struct. Sci.*, 2005, **61**, 511–527.

36 W. D. S. Motherwell, H. L. Ammon, J. D. Dunitz, A. Dzyabchenko, P. Erk, A. Gavezzotti, D. W. M. Hofmann, F. J. J. Leusen, J. P. M. Lommerse, W. T. M. Mooij, S. L. Price, H. Scheraga, B. Schweizer, M. U. Schmidt, B. P. van Eijck, P. Verwer and D. E. Williams, *Acta Crystallogr., Sect. B: Struct. Sci.*, 2002, **58**, 647–661.

37 J. P. M. Lommerse, W. D. S. Motherwell, H. L. Ammon, J. D. Dunitz, A. Gavezzotti, D. W. M. Hofmann, F. J. J. Leusen, W. T. M. Mooij, S. L. Price, B. Schweizer, M. U. Schmidt, B. P. van Eijck, P. Verwer and D. E. Williams, *Acta Crystallogr., Sect. B: Struct. Sci.*, 2000, **56**, 697–714.

38 S. J. Clark, M. D. Segall, C. J. Pickard, P. J. Hasnip, M. J. Probert, K. Refson and M. C. Payne, *Z. Kristall.*, 2005, **220**, 567–570.

39 M. A. Neumann and M.-A. Perrin, *J. Phys. Chem. B*, 2005, **109**, 15531–15541.

40 M. C. Payne, M. P. Teter, D. C. Allan, T. A. Arias and J. D. Joannopoulos, *Rev. Mod. Phys.*, 1992, **64**, 1045–1097.

41 M. A. Neumann, F. J. Leusen and J. Kendrick, *Angew. Chem., Int. Ed.*, 2008, **47**, 2427–2430.

42 H. C. S. Chan, J. Kendrick and F. J. J. Leusen, *Angew. Chem., Int. Ed.*, 2011, **50**, 2979–2981.

43 A. Asmadi, M. A. Neumann, J. Kendrick, P. Girard, M.-A. Perrin and F. J. J. Leusen, *J. Phys. Chem. B*, 2009, **113**, 16303–16313.

44 M. H. D. B. Alsirawan, V. R. Vangala, J. Kendrick, F. J. J. Leusen and A. Paradkar, *Cryst. Growth Des.*, 2016, **16**, 3072–3075.

45 D. E. Braun, P. G. Karamertzanis and S. L. Price, *Chem. Commun.*, 2011, **47**, 5443–5445.

46 S. Mohamed, D. P. Karothu and P. Naumov, *Acta Crystallogr., Sect. B: Struct. Sci., Cryst. Eng. Mater.*, 2016, **72**, 551–561.

47 D. Braga, F. Grepioni, L. Maini, S. Prosperi, R. Gobetto and M. R. Chierotti, *Chem. Commun.*, 2010, **46**, 7715–7717.

48 D. Braga, F. Grepioni, G. I. Lampronti, L. Maini and A. Turrina, *Cryst. Growth Des.*, 2011, **11**, 5621–5627.
49 L. Casali, L. Mazzei, O. Shemchuk, K. Honer, F. Grepioni, S. Ciurli, D. Braga and J. Baltrusaitis, *Chemical Communications*, 2018, **54**, 7637–7640.
50 A. J. Smith, S.-H. Kim, N. K. Duggirala, J. Jin, L. Wojtas, J. Ehrhart, B. Giunta, J. Tan, M. J. Zaworotko and R. D. Shytle, *Mol. Pharm.*, 2013, **10**, 4728–4738.
51 A. R. Buist and A. R. Kennedy, *Cryst. Growth Des.*, 2014, **14**, 6508–6513.
52 N. K. Duggirala, A. J. Smith, Ł. Wojtas, R. D. Shytle and M. J. Zaworotko, *Cryst. Growth Des.*, 2014, **14**, 6135–6142.
53 F. v. Kobell, *J. Prakt. Chem.*, 1843, **28**, 489–491.
54 S. L. Childs, L. J. Chyall, J. T. Dunlap, V. N. Smolenskaya, B. C. Stahly and G. P. Stahly, *J. Am. Chem. Soc.*, 2004, **126**, 13335–13342.
55 C. B. Aakeröy, M. E. Fasulo and J. Desper, *Mol. Pharm.*, 2007, **4**, 317–322.
56 D. P. Karothu, I. Jahovic, G. Jovanovski, B. Kaitner and P. Naumov, *CrystEngComm*, 2017, **19**, 4338–4344.
57 K. Honer, E. Kalfaoglu, C. Pico, J. McCann and J. Baltrusaitis, *ACS Sustainable Chem. Eng.*, 2017, **5**, 8546–8550.
58 M. Rafilovich and J. Bernstein, *J. Am. Chem. Soc.*, 2006, **128**, 12185–12191.
59 K. Shankland, C. K. Leech, S. Mohamed, S. A. Barnett and D. A. Tocher, *Acta Crystallogr., Sect. E: Struct. Rep. Online*, 2007, **63**, o3574.
60 D.-K. Bučar, R. W. Lancaster and J. Bernstein, *Angew. Chem., Int. Ed.*, 2015, **54**, 6972–6993.
61 S. Mohamed, S. A. Barnett and D. A. Tocher, *Acta Crystallogr., Sect. E: Struct. Rep. Online*, 2007, **63**, o3575.
62 S. Mohamed and L. Li, *CrystEngComm*, 2018, DOI: 10.1039/c8ce00758f.
63 C. R. Groom, I. J. Bruno, M. P. Lightfoot and S. C. Ward, *Acta Crystallogr., Sect. B: Struct. Sci., Cryst. Eng. Mater.*, 2016, **72**, 171–179.
64 B. Lou, S. R. Perumalla and C. C. Sun, *Cryst. Growth Des.*, 2015, **15**, 24–28.
65 A. Sikorski and D. Trzybiński, *J. Mol. Struct.*, 2013, **1049**, 90–98.
66 B. Sarma, N. K. Nath, B. R. Bhogala and A. Nangia, *Cryst. Growth Des.*, 2009, **9**, 1546–1557.
67 H.-K. Fun, M. Hemamalini and V. Rajakannan, *Acta Crystallogr., Sect. E: Struct. Rep. Online*, 2010, **66**, o2010–o2011.
68 B. Lou, S. R. Perumalla and C. C. Sun, *J. Mol. Struct.*, 2015, **1099**, 516–522.
69 A. Mukherjee and G. R. Desiraju, *Cryst. Growth Des.*, 2014, **14**, 1375–1385.
70 Calculator Plugins were used for structure property prediction and calculation, MarvinSketch, Version 16.3.7.0, ChemAxon, http://www.chemaxon.com, 2016.
71 A. J. Cruz-Cabeza, *CrystEngComm*, 2012, **14**, 6362–6365.
72 S. Mohamed, D. A. Tocher, M. Vickers, P. G. Karamertzanis and S. L. Price, *Cryst. Growth Des.*, 2009, **9**, 2881–2889.
73 S. L. Childs, G. P. Stahly and A. Park, *Mol. Pharm.*, 2007, **4**, 323–338.
74 Dassault Systèmes BIOVIA, *BIOVIA Materials Studio, Version 8.0.100.21*, Dassault Systèmes, San Diego, 2014.
75 H. R. Karfunkel and R. J. Gdanitz, *J. Comput. Chem.*, 1992, **13**, 1171–1183.
76 R. L. Akkermans, N. A. Spenley and S. H. Robertson, *Mol. Simul.*, 2013, **39**, 1153–1164.
77 M. J. Frisch, G. W. Trucks, H. B. Schlegel, G. E. Scuseria, M. A. Robb, J. R. Cheeseman, G. Scalmani, V. Barone, B. Mennucci, G. A. Petersson,

H. Nakatsuji, M. Caricato, X. Li, H. P. Hratchian, A. F. Izmaylov, J. Bloino, G. Zheng, J. L. Sonnenberg, M. Hada, M. Ehara, K. Toyota, R. Fukuda, J. Hasegawa, M. Ishida, T. Nakajima, Y. Honda, O. Kitao, H. Nakai, T. Vreven, J. A. Montgomery Jr, J. E. Peralta, F. Ogliaro, M. J. Bearpark, J. Heyd, E. N. Brothers, K. N. Kudin, V. N. Staroverov, R. Kobayashi, J. Normand, K. Raghavachari, A. P. Rendell, J. C. Burant, S. S. Iyengar, J. Tomasi, M. Cossi, N. Rega, N. J. Millam, M. Klene, J. E. Knox, J. B. Cross, V. Bakken, C. Adamo, J. Jaramillo, R. Gomperts, R. E. Stratmann, O. Yazyev, A. J. Austin, R. Cammi, C. Pomelli, J. W. Ochterski, R. L. Martin, K. Morokuma, V. G. Zakrzewski, G. A. Voth, P. Salvador, J. J. Dannenberg, S. Dapprich, A. D. Daniels, Ö. Farkas, J. B. Foresman, J. V. Ortiz, J. Cioslowski and D. J. Fox, *Gaussian 09 Revision D.01*, Wallingford, CT, USA, 2009.

78 S. L. Mayo, B. D. Olafson and W. A. Goddard, *J. Phys. Chem.*, 1990, **94**, 8897–8909.

79 C. M. Breneman and K. B. Wiberg, *J. Comput. Chem.*, 1990, **11**, 361–373.

80 S. L. Price, M. Leslie, G. W. A. Welch, M. Habgood, L. S. Price, P. G. Karamertzanis and G. M. Day, *Phys. Chem. Chem. Phys.*, 2010, **12**, 8478–8490.

81 A. J. Stone, *Chem. Phys. Lett.*, 1981, **83**, 233–239.

82 A. J. Stone and M. Alderton, *Mol. Phys.*, 1985, **56**, 1047–1064.

83 A. J. Stone, *J. Chem. Theory Comput.*, 2005, **1**, 1128–1132.

84 L.-Y. Hsu and D. E. Williams, *Acta Crystallogr., Sect. A: Cryst. Phys., Diffr., Theor. Gen. Crystallogr.*, 1980, **36**, 277–281.

85 S. R. Cox, L.-Y. Hsu and D. E. Williams, *Acta Crystallogr., Sect. A: Cryst. Phys., Diffr., Theor. Gen. Crystallogr.*, 1981, **37**, 293–301.

86 D. E. Williams and S. R. Cox, *Acta Crystallogr., Sect. B: Struct. Sci.*, 1984, **40**, 404–417.

87 D. S. Coombes, S. L. Price, D. J. Willock and M. Leslie, *J. Phys. Chem.*, 1996, **100**, 7352–7360.

88 T. Beyer and S. L. Price, *J. Phys. Chem. B*, 2000, **104**, 2647–2655.

89 C. F. Macrae, I. J. Bruno, J. A. Chisholm, P. R. Edgington, P. McCabe, E. Pidcock, L. Rodriguez-Monge, R. Taylor, J. v. Streek and P. A. Wood, *J. Appl. Crystallogr.*, 2008, **41**, 466–470.

90 J. A. Chisholm and S. Motherwell, *J. Appl. Crystallogr.*, 2005, **38**, 228–231.

91 A. Kitaigorodsky, *Molecular crystals and molecules*, Elsevier, 2012.

92 A. Spek, *Acta Crystallogr., Sect. D: Biol. Crystallogr.*, 2009, **65**, 148–155.

93 J. P. Perdew, K. Burke and M. Ernzerhof, *Phys. Rev. Lett.*, 1996, **77**, 3865–3868.

94 D. Vanderbilt, *Phys. Rev. B*, 1990, **41**, 7892–7895.

95 S. Grimme, *J. Comput. Chem.*, 2006, **27**, 1787–1799.

96 H. J. Monkhorst and J. D. Pack, *Phys. Rev. B*, 1976, **13**, 5188–5192.

97 B. G. Pfrommer, M. Côté, S. G. Louie and M. L. Cohen, *J. Comput. Phys.*, 1997, **131**, 233–240.

98 J. van de Streek and M. A. Neumann, *Acta Crystallogr., Sect. B: Struct. Sci.*, 2010, **66**, 544–558.

99 B. P. van Eijck and J. Kroon, *Acta Crystallogr., Sect. B: Struct. Sci.*, 2000, **56**, 535–542.

100 Wavefunction Inc., *Spartan'16, Version 2.0.7*, 18401 Von Karman Ave., Suite 370, Irvine CA 92612, 2017.

101 W. L. Koltun, *Biopolymers*, 1965, **3**, 665–679.
102 M. C. Etter, J. C. MacDonald and J. Bernstein, *Acta Crystallogr., Sect. B: Struct. Sci.*, 1990, **46**, 256–262.
103 L. B. Jerzykiewicz, T. Lis, Z. Malarski and E. Grech, *J. Crystallogr. Spectrosc. Res.*, 1993, **23**, 805–812.
104 J. R. Jebamony and P. Thomas Muthiah, *Acta Crystallogr., Sect. C: Cryst. Struct. Commun.*, 1998, **54**, 539–540.
105 L. B. Jerzykiewicz, Z. Malarski, L. Sobczyk, T. Lis and E. Grech, *J. Mol. Struct.*, 1998, **440**, 175–185.
106 M. Hemamalini and H.-K. Fun, *Acta Crystallogr., Sect. E: Struct. Rep. Online*, 2010, **66**, o479–o480.
107 K. Balasubramani and H.-K. Fun, *Acta Crystallogr., Sect. E: Struct. Rep. Online*, 2009, **65**, o1519.
108 A. Jayasankar, L. S. Reddy, S. J. Bethune and N. Rodríguez-Hornedo, *Cryst. Growth Des.*, 2009, **9**, 889–897.
109 D. E. Braun, P. G. Karamertzanis, J.-B. Arlin, A. J. Florence, V. Kahlenberg, D. A. Tocher, U. J. Griesser and S. L. Price, *Cryst. Growth Des.*, 2011, **11**, 210–220.
110 R. A. Chiarella, R. J. Davey and M. L. Peterson, *Cryst. Growth Des.*, 2007, **7**, 1223–1226.
111 G. Smith, U. D. Wermuth, P. C. Healy and J. M. White, *Aust. J. Chem.*, 2003, **56**, 707–713.
112 T. Wu, X.-P. Zhang and X.-Z. You, *RSC Adv.*, 2013, **3**, 26047–26051.
113 H. D. Arman and E. R. T. Tiekink, *Z. Kristallogr. – Cryst. Mater.*, 2013, **228**(6), 289–294.
114 M. Hemamalini and H.-K. Fun, *Acta Crystallogr., Sect. E: Struct. Rep. Online*, 2010, **66**, o2535.
115 D. E. Braun, T. Gelbrich, V. Kahlenberg and U. J. Griesser, *Mol. Pharm.*, 2014, **11**, 3145–3163.
116 A. Tkatchenko and M. Scheffler, *Phys. Rev. Lett.*, 2009, **102**, 073005.
117 M. Arhangelskis, A. D. Katsenis, A. J. Morris and T. Friščić, *Chem. Sci.*, 2018, **9**, 3367–3375.
118 C. P. Brock and J. D. Dunitz, *Chem. Mater.*, 1994, **6**, 1118–1127.

Faraday Discussions

PAPER

Predicting the structures and associated phase transition mechanisms in disordered crystals *via* a combination of experimental and theoretical methods

Michael T. Ruggiero, †* Johanna Kölbel, Qi Li and J. Axel Zeitler

Received 19th February 2018, Accepted 14th March 2018
DOI: 10.1039/c8fd00042e

Disordered materials make up a large proportion of condensed phase systems, but the difficulties in describing their structures and molecular dynamics limit their potential applications. Disordered crystalline systems, also known as plastic crystals, offer a unique perspective into these factors because the systems retain a degree of crystallinity, reducing the degrees of freedom that must be explored when interpreting the results. However, while disordered crystals do diffract X-rays, it is difficult to fully resolve meaningful crystalline structures, with the best scenario resulting in lattice parameters. In this study, we use a combination of experimental terahertz time-domain spectroscopy, and theoretical solid-state *ab initio* density functional theory and molecular dynamics simulations to fully elucidate the structures and associated dynamics of organic molecular solids. The results highlight that this combination provides a complete description of the energetic and mechanistic pathways involved in the formation of disordered crystals, and highlights the importance of low-frequency dynamics in their properties. Finally, with structures fully determined and validated by the experimental results, recent progress into anharmonic calculations, namely the quasi-harmonic approximation method, enables full temperature and pressure-dependent properties to be understood within the framework of the potential energy hyper-surface structure.

1 Introduction

Disordered (amorphous) solids are an important class of materials that have found use in a wide range of fields, from the semiconductor[1] to pharmaceutical industries.[2] The utility of amorphous materials lies in their metastable nature,[3,4]

Department of Chemical Engineering and Biotechnology, University of Cambridge, Philippa Fawcett Drive, Cambridge, CB3 0AS, UK. E-mail: Michael.Ruggiero@uvm.edu; mtr34@cam.ac.uk

† Present address: Department of Chemistry, University of Vermont, 82 University Place, Burlington, VT 05403, USA.

which ultimately leads to increased molecular dynamics compared to their crystalline counterparts.[5] However, the actual structures and associated motions of the molecules within disordered solids remain a topic of debate, as traditional techniques, such as X-ray crystallography, provide little information that can be directly interpreted without significant approximations.[6]

Recently, the importance of low-frequency ($\leq$330 cm^{-1} or 10 THz) collective vibrations in bulk properties and phase transformation phenomena in molecular solids has been suggested,[7] however the types of motions occurring in disordered materials have not yet been completely understood. Experimentally, low-frequency (terahertz) vibrational spectroscopy is a powerful tool for investigating solids, as the terahertz radiation is able to excite large-amplitude collective motions of entire molecules within the solid.[8] This leads to both the internal conformational and intermolecular potential energy surfaces being probed simultaneously, providing a valuable measure of the bulk structure and dynamics. Terahertz time-domain spectroscopy (THz-TDS) is the far-IR extension of the more traditional FTIR technique, but unlike FTIR, which probes internal vibrations, there are no standard functional group specific terahertz vibrations, meaning each solid has a unique spectral fingerprint at terahertz frequencies.[9,10] Therefore, the interpretation of THz-TDS spectra requires significant computational modelling, and (at a minimum) the structures of the associated solids are necessary for initialising the simulations.[11,12]

Disordered (or plastic) crystals are a class of materials that exhibit pseudo-order, in that the molecules occupy crystalline sites but lack true translational symmetry due to uncoupled motions within their crystallographic sites.[13,14] This leads to a scenario where the material diffracts X-rays (like a traditional crystal), but rather than yielding meaningful atomic-level structural details the data only provides a broad description of occupied sites within the solid, *i.e.* an average "sphere" of electron density at each lattice site.[15] However, the psuedo-order limits the possible degrees of freedom that the molecules within the solid have, enabling more detailed investigation into the bulk structure and dynamics to be performed.[16,17]

Because disordered crystals are ultimately disordered, diffraction methods often rely on other techniques to fully elucidate the static-ordered structures.[18] While this has been successfully achieved in a number of cases, the static picture of a disordered crystal is not truly physical, since the molecules are free to move in their crystallographic sites uncoupled from their neighbours.[19] In order to obtain a more physical description, molecular dynamics (MD) simulations can be used as a means of predicting the nature of the motions occurring in the solid,[20,21] as well as predicting the vibrational spectra through determination of either the molecular dipole moments (for infrared (IR) spectroscopies), or the molecular polarisabilities (for Raman spectroscopies).[22,23] However, traditional MD simulations are performed using user-defined force constants, which leads to biased vibrational results. A solution to this is *ab initio* molecular dynamics (AIMD), in which the forces are recalculated using density functional theory (DFT) at each individual time-step, leading to a more accurate description of the vibrational dynamics occurring within the system.[24–27]

AIMD simulations also enable structural transformations to be investigated.[28–30] As many plastic crystals often undergo a transformation to an ordered crystal at low temperatures, it is possible to analyse the ordered form experimentally with diffraction methods to obtain an atomic-level structure. By using

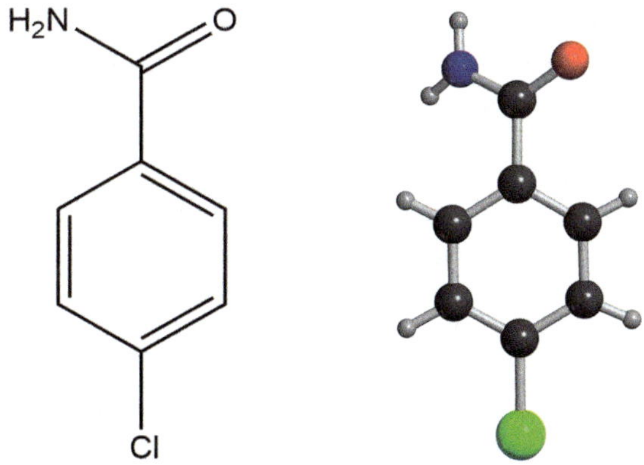

Fig. 1 Skeletal and three-dimensional models of a single PCB molecule.

this structure as a starting point for the AIMD simulations, prediction of the transformation pathway as well as the disordered structure can be achieved.

In this study, a combination of experimental THz-TDS and theoretical static solid-state-DFT (s-DFT) and AIMD simulations are used to predict the structures, dynamics, and phase-transition pathways in the order–disorder transformation of parachlorobenzamide (PCB, Fig. 1).[31] The static structures of both the ordered and plastic phases of the solid have been previously reported experimentally using synchrotron-based X-ray diffraction,[32,33] which enables the proposed methodology to be validated by experimental results. Overall, this combination of techniques allows the static structures, as well as the associated dynamical aspects, to be fully understood, providing insight into the energetic and mechanistic forces that drive the order–disorder phase transition in plastic crystals.

2 Methods

2.1 Experimental methods

Parachlorobenzamide was purchased from Sigma-Aldrich (98%, Poole, UK) and was used as-received. Samples were prepared for THz-TDS measurements by mixing each sample with polyethylene to a 10% w/w concentration and subsequently grinding with a mortar and pestle to homogenise the sample and reduce particle size. A hydraulic press was then used to press the mixture into a 13 mm diameter freestanding pellet with a thickness of 2 mm. A corresponding pellet containing pure polyethylene was also made to act as a standard blank for absorption measurements. All experimental THz-TDS spectra were obtained using a commercial TeraPulse 4000 spectrometer (TeraView Ltd, Cambridge, UK). Variable-temperature measurements were performed using a liquid nitrogen cryostat (Janis, Massachusetts, USA) equipped with an externally controlled heating element (Lakeshore 330, Ohio, USA), permitting measurements from 80–400 K. The time-domain waveforms were symmetrically zero-padded and Fourier transformed to yield terahertz power spectra. The absorption spectra presented are a result of division of the sample spectra by that of a blank spectrum.

2.2 Computational methods

2.2.1 AIMD simulations. All AIMD simulations were performed using the freely-available CP2K software package,[34,35] which incorporates periodic boundary conditions. A $1 \times 3 \times 2$ supercell (36 molecules, 576 atoms) was generated from the experimental structure of α-PCB in order to minimise interactions with neighbouring unit cells and fully capture the physical phase transformation process. Simulations were performed within the isothermal–isobaric (NPT) ensemble, with the temperature controlled using a Nosé–Hoover chain thermostat,[36,37] and pressure maintained using the extended Nosé–Hoover barostat of Martyna *et al.*[38,39] All atoms were modelled using the double-ζ DZVP-GTH basis set[40] and the Goedecker–Teter–Hutter (GTH) pseudopotentials.[41] The Perdew–Burke–Ernzerhof (PBE) density functional[42] and Grimme D3-dispersion correction[43,44] were used for all AIMD simulations. A triclinic cell was chosen so that all lattice components were free to relax throughout the simulations. A 1.0 fs time-step was used for all simulations, and each trajectory was run for a total of 20 000 fs (20 ps). The convergence criteria were set to $\Delta E \leq 10^{-9}$ hartree for all simulations.

IR spectra were generated by performing NVT simulations starting from the equilibrated NPT supercells. A time-step of 0.5 fs was used for all vibrational analyses, for a total trajectory length of 30 ps. Both solids underwent an initial equilibration period prior to the production simulation, which was set to 3.0 ps and 5.0 ps for the α- and γ-forms, respectively. The molecular dipoles were determined *via* calculation of the Wannier centres, which was performed every 2.5 fs. The resulting trajectories were analysed using the TRAVIS post-processing software package,[45] and IR spectra were calculated by performing the Fourier transform of the dipole moment autocorrelation function.[24,27]

2.2.2 S-DFT simulations. All s-DFT simulations were performed using the fully-periodic CRYSTAL17 software package.[46] In order to allow comparison to the AIMD simulations, the computational parameters were kept as similar as possible between the two methods. The atomic orbitals were represented using the atom-centred def2-SVP basis set,[47] and the D3-corrected PBE density functional was used for all calculations. The experimental α-PCB structure[32,33] and the AIMD-generated unit cell of the γ-form (*vide infra*) were used to initialise the respective simulations. In both cases the simulations were performed in the absence of any space group symmetry in order to mimic the AIMD methodology and remove any spurious constraints. Both structures were fully optimised to a convergence of $\Delta E \leq 10^{-8}$ hartree, and vibrational analyses were performed using numerical differentiation within the harmonic approximation, with an energy convergence of $\Delta E \leq 10^{-11}$ hartree.[48,49] Infrared intensities were determined using the Berry phase approach.[50] Quasi-harmonic approximation (QHA) simulations were performed by simulating the optimised structures and vibrational properties at four volumes corresponding to a −2%, 0, +2%, and +4% contraction/expansion relative to the equilibrium structure.[51–54]

3 Results and discussion

3.1 Terahertz time-domain spectroscopy

In order to determine the vibrational response of PCB over a wide temperature range, THz-TDS spectra were acquired from 100–350 K and the results are shown

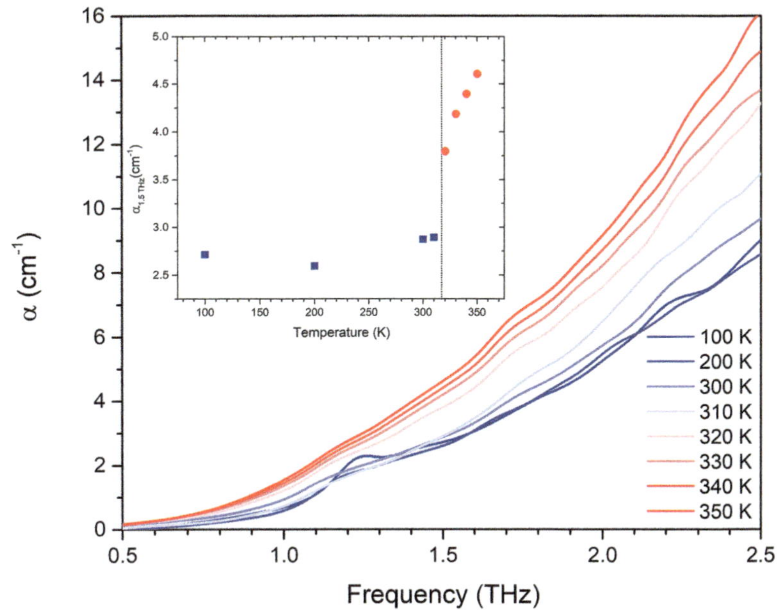

Fig. 2 Experimental THz-TDS spectra of PCB at various temperatures ranging from 100 K (blue) to 350 K (red). There is a marked transition between 310 K and 320 K, corresponding to the order–disorder transition, with all spectra corresponding to the α-form coloured blue, while all spectra corresponding to the γ-form are coloured red. The inset shows the absorption at 1.5 THz as a function of temperature, with the experimental transition temperature of 317 K noted by a dotted vertical line.

in Fig. 2. At low temperatures, there is at least one distinct feature at 1.23 THz, with a number of potential features present at higher frequencies, which is characteristic of a well-ordered crystalline solid. As the temperature is increased, the main absorption feature significantly broadens and shifts to lower frequencies, likely due to thermal expansion leading to a softening of the vibrational potential energy and an increased number of thermally-activated vibrational relaxation pathways, the latter of which leads to spectral broadening. It is important to note that while many materials experience these phenomena upon heating, the magnitude of the broadening observed here is rather large in comparison to most molecular crystals.[27,55] The origins of the spectral broadening might be related to the ordered solid deviating from an ideal crystalline geometry, as minor displacements from equilibrium positions would lead to slightly different vibrational potentials for each unit molecule. This would manifest in the experimental spectra as broadened features, as observed in the continuous phase transition of crystalline *N*-methyl-4-carboxypyridinium chloride.[7] Additionally, a decrease in the intensity of the spectra can also be attributed to the increased vibrational motion with increasing temperature, as the IR intensity is proportional to the change in dipole moment with respect to the atomic displacement, $I \propto \frac{\partial \mu}{\partial Q}$.[27] Overall, it is likely that the experimental spectral broadening is due to a combination of all of these phenomena, which results in the large effect observed.

Upon increasing the temperature above 310 K, there is a clear change in the vibrational response of the material, with the spectra exhibiting a noticeable increase in intensity across the entire spectral region that continues to increase with rising temperature, a common trait of disordered solids and in contrast to crystalline solids, which typically undergo a decrease in absorption as temperature is increased. While there is apparently little difference in the shape of the individual spectra (*i.e.* both spectra appear to be featureless with an increasing baseline), there is an obvious difference in the temperature-dependence of the spectra. This is highlighted in the Fig. 2 inset, which shows the clear change in the absorption coefficient at 1.5 THz that is observed between 310 K and 320 K. This discontinuity in the spectral trends corresponds to the order–disorder phase transition in PCB, which is in good agreement with the published transition temperature of 317 K.[32,33,56] However, unlike fully-disordered materials, which do not exhibit any resolvable low-frequency vibrational features, the spectra of the disordered γ-PCB solid seemingly exhibit at least one absorption at ~1.72 THz. It is important to note that experimentally the discontinuity in the THz-TDS measurements is reproducible upon heating and cooling, indicating that the kinetics of the transformation are faster than the experimental method can probe (minutes, due to the equilibration time required for temperature changes).

3.2 Structural elucidation and phase transformation pathway

In order to determine the structure, and associated dynamics, of the disordered γ-PCB solid, AIMD simulations were performed. Initially, a $1 \times 3 \times 2$ supercell of the α-PCB solid was generated and used as the starting geometry for the AIMD simulations. Fully periodic NPT simulations were performed using CP2K, with the pressure set to 1 bar and the temperature varied from 100–400 K, with a time-step of 1.0 fs. In the high temperature simulations (above 300 K), the system quickly undergoes a large-scale transformation, with all of the PCB molecules experiencing large amplitude hindered-rotational type motions about their long axis (see Fig. 3). It is important to note that the AIMD simulations underestimate the transition temperature by ~20 K, however, given the limitations of the exact implementation of thermal energy in the model, this represents a relatively minor error (20 K ≈ 0.04 kcal mol^{-1}).

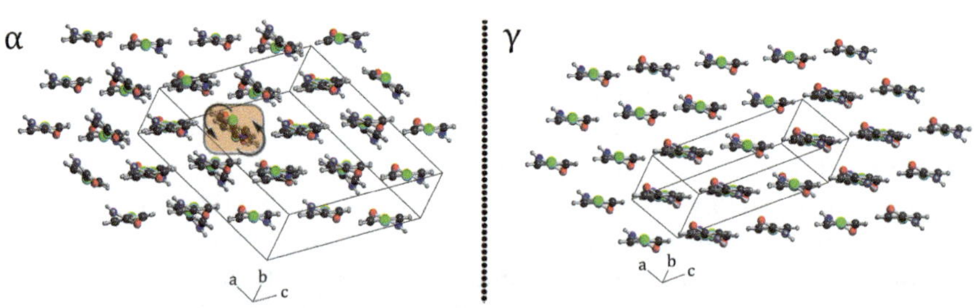

Fig. 3 Bulk static structures of the two forms of PCB, the low-temperature ordered α-form (left) and the high-temperature plastic γ-crystal (right). The main structural difference between the two crystals is highlighted in orange, along with arrows denoting the hindered-rotational motion that drives the transformation between the two forms and corresponds to the vibrations discussed later in the text.

The phase transformation can be clearly observed through analysis of the angle between the planes formed by the two distinct molecular environments in the α-form. In low-temperature AIMD simulations (below 300 K) the molecules do not undergo any large-scale rearrangement, rather they simply oscillate within their crystallographic sites about their equilibrium positions, typical of an ordered molecular crystal (Fig. 4). However, at higher temperatures, there still exist oscillations, but with a much larger magnitude compared to the low-temperature case, as well as an overall trend towards a vanishing angle between the planes of the molecules (*i.e.* all molecules trending towards forming planar sheets). Additionally, the increased molecular motion in the high-temperature simulation results in an initial expansion of the simulation cell, which is a direct example of vibrational dynamics leading to thermal expansion. After ~10 ps, the high-temperature simulation relaxes to a new bulk structure that generally exhibits an overall planar arrangement of all of the molecules within the system, however the individual molecules continue to undergo rather large-amplitude hindered-rotational type motions that are seemingly uncoupled from the bulk. Moreover, in the high-temperature simulation, the molecules are generally not as well-ordered as in the low-temperature simulation, with many deviations from the static crystal structure readily apparent. However, there do exist a number of local ordering events that make extracting a molecular basis possible, and this process was used to generate a crystallographic unit cell for the γ-form of PCB.

Six molecules (two stacked rows of three molecules each, with alternating dipole moments) were extracted from the equilibrated AIMD simulation and a unit cell was fit around them. The system was loaded into the fully-periodic CRYSTAL17 software package and optimised within the constraints of the input lattice vectors, in order to minimise any possibility of relaxation back into the α-form. Upon successful optimisation, the system was then allowed to fully relax with no constraints (lattice vectors and atomic positions). The resulting structure (Fig. 3) was found to be triclinic with $P\bar{1}$ space group symmetry, lattice parameters

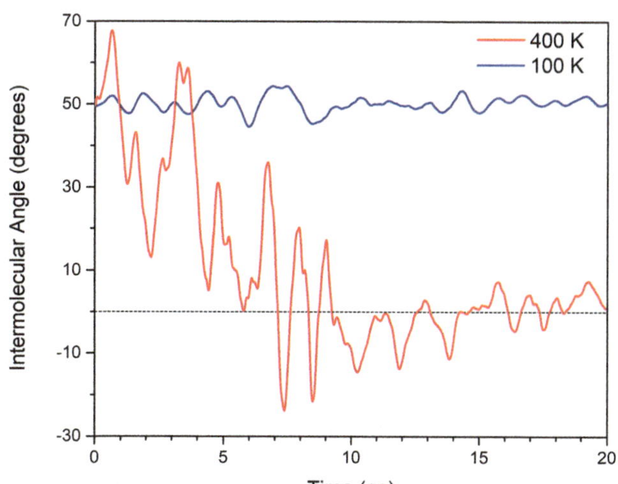

Fig. 4 AIMD simulated angles between adjacent molecular planes at 100 K (blue) and 300 K (red). A horizontal dashed line is drawn through the origin as a guide.

of $a = 4.934$ Å, $b = 5.383$ Å, $c = 13.794$ Å, $\alpha = 93.931°$, $\beta = 116.421°$, and $\gamma = 93.498°$, and a $Z = 2$ ($Z' = 1$). This structure is in good general agreement with the previously suggested static-structure of γ-PCB (bond, angle, and dihedral angle RMSDs of 0.035 Å, 0.656°, and 2.552°, respectively), although with slightly contracted lattice vectors due to the s-DFT simulations effectively simulating a 0 K crystal.

In both simulations, the individual molecules undergo predominately rocking motions with respect to their neighbours, which is evident in the periodic fluctuations present in Fig. 3. Even in the plastic crystal, in which the molecules generally move freely within their sites, there is a clear correlated rocking-type motion present. Interestingly, these oscillations have periods of about 1–2 ps, which is on the order of the period of terahertz vibrations (1 THz = 10^{12} Hz = 1 ps). With the structures of the two forms known, a more rigorous analysis of the vibrational dynamics occurring within the solids can be performed.

3.3 Vibrational dynamics

The vibrational dynamics of both PCB solids were predicted using the AIMD simulations, and the results are shown in Fig. 5. In both cases, the AIMD simulations generally reproduce the experimental spectra, yielding a number of features that correspond to those experimentally observed. However, the AIMD spectra provide little specific atomic-level detail surrounding the nature of each individual vibrational mode, leaving further analyses necessary. In this regard, s-DFT simulations are able to determine the precise atomic-motions that produce each vibrational transition, and this method was employed once the static structure of the γ-form was determined. The static crystal structures of both forms of PCB were fully optimised with no constraints using CRYSTAL17, and

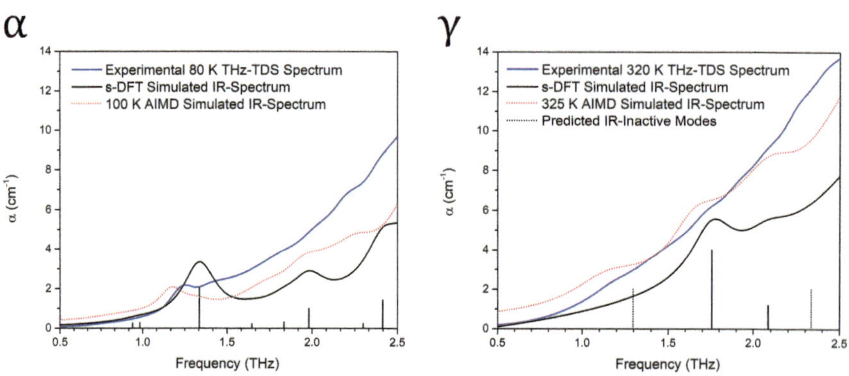

Fig. 5 Experimental THz-TDS spectra (blue) and s-DFT simulated IR spectra (black) for both PCB solids. The α-PCB spectrum (80 K) is predicted to have a number of IR-active modes which primarily correspond to hindered-rotational motions of the individual molecules within the solid. The γ-form contains fewer IR-active modes (due to the smaller static unit cell compared to the α-form), however there are additional IR-inactive modes (dotted lines) predicted that might be activated in the experimental disordered structure. In both panels, the simulated AIMD spectra are also shown as dotted red lines, and are in good general agreement with the experimental spectra. Note that the experimental spectra have been offset by +1.5 cm^{-1} for clarity.

subsequently the phonon properties were calculated using a finite difference method within the harmonic approximation. The results of the vibrational simulations are shown in Fig. 5 for both crystals.

In the α-form, there are a number of low-frequency vibrations predicted that result in good agreement with the experimental spectrum. Overall, the predicted spectrum is slightly over-estimated in terms of frequency, which is likely due to the simulations being performed at 0 K while the experimental spectrum was recorded at 80 K, as in some cases it has been shown that spectra continue to shift and sharpen upon cooling to temperatures as low at 4 K.[57] Indeed, the s-DFT simulations are also blue-shifted compared to the 100 K AIMD simulations, further indicating that the spectrum is strongly influenced by temperature. Additionally, AIMD vibrational analysis performed at higher temperatures confirms that the spectra become much more diffuse and red-shifted as the transition temperature is approached (Fig. 6). The temperature-dependent AIMD-simulated vibrational spectra generally reproduce the experimental results, predicting a red-shift of most features, as well as an increase in absorption intensity between ~1.25 THz and 1.85 THz, which originates from the combination of spectral shifting and broadening.

The vibrational mode-types are mainly hindered rotations of the individual molecules with respect to one another, along both the long molecular axis (*i.e.* the Cl-ring bond, see Fig. 3) as well as librations orthogonal to the plane of the molecule, with the exception of the weak predicted feature at 1.645 THz which is a translational-type motion. The well-resolved experimental transition at 1.23 THz, and predicted at 1.337 THz, involves a coupled hindered rotation of all of the molecules within the solid along their long-axes (see Fig. 3), which at its maximum displacement yields a totally planar arrangement of all of the molecules within the solid. Thus, this mode corresponds well to the phase-transformation pathway predicted from the AIMD simulations. Interestingly, this mode is strongly influenced by temperature both experimentally (see Fig. 2) and theoretically (Fig. 6), and this assignment helps to explain the significant

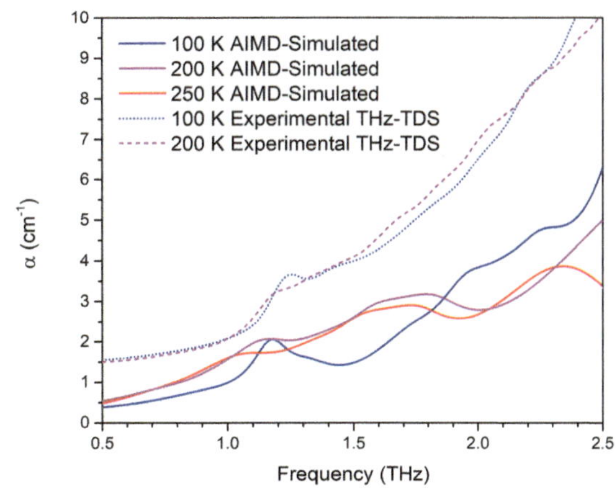

Fig. 6 AIMD-simulated IR spectra of α-PCB at 100, 200, and 250 K (blue, purple, and red, respectively). The spectra exhibit a significant red-shift and spectral broadening as the temperature is increased.

broadening and shifting that occurs with increasing temperature, as the vibrational potential likely becomes much softer and more anharmonic (leading to more relaxation pathways) with increasing thermal expansion (and thermal energy).

In the γ-form, there are fewer IR-active modes predicted due to fewer molecules within the crystallographic unit cell. The resolvable features at 1.72 THz can be assigned to an intense mode predicted at 1.755 THz. This is a surprising result given most disordered materials do not show discernible low-frequency absorption features, and it is even more surprising that the static-model is able to reproduce the disordered experimental results. Investigation of the vibrational mode type highlights that the vibrational motion is an asymmetric rocking of two molecules about their long axes with respect to one another. This directly corresponds to the coupled oscillations present in the AIMD simulations between pairs of neighbouring molecules (see Fig. 4), and also corresponds well to the oscillation period, indicating that this particular feature is indeed the local hindered-rotational motion suggested. Furthermore, the observation of this mode experimentally confirms that this vibration is not a long-range collective vibration, which would lead to a very broad experimental absorption, but rather is predominantly a local effect, being primarily dependent on a single molecular neighbour.

However, the s-DFT predicted spectrum fails to reproduce the correct IR density of states in the terahertz region as shown by the experimental and AIMD results. This is not surprising, as the static model is not a true representation of the disordered solid. However, there are a number of IR-inactive modes predicted which could become activated upon the removal of crystalline symmetry, and interestingly these transitions are predicted at similar frequencies to those predicted by the AIMD simulations. These modes, shown in Fig. 5 as dotted sticks, also correspond to hindered-rotational motions, with the lowest-frequency mode (1.296 THz) corresponding well with the mode type of the 1.337 THz vibration in the α-form. Indeed, the occurrence of a transition in both the experimental and AIMD-predicted spectra at ~1.17 THz confirms that in the absence of true crystalline symmetry this vibration can become activated.

In order to further predict the role that thermal expansion plays in the structures and vibrational dynamics of the PCB crystals, QHA simulations were performed. The QHA method performs volume-constrained geometry optimisations (all lattice vectors and atomic positions were allowed to relax while maintaining a constant volume) and frequency analyses at four distinct volumes corresponding to the equilibrium (no volume constraint) structure, and −2, +2, and +4% contracted/expanded volumes. Based on these results, the temperature dependence of the vibrational features can be estimated within the context of thermal expansion effects. The results indicate that all of the low-frequency vibrations in both crystals are dependent on the volume to a large extent, with most modes experiencing shifts on the order of 10 cm^{-1} between the two extreme volumes. This is also reflected in the temperature-dependent AIMD simulations (see Fig. 6), which also predict significant red-shifting upon increasing the temperature. Remarkably, the vibrational modes that involve motions corresponding to the phase-transformation mode (*i.e.* hindered rotations about the molecular plane) involve even larger shifts, for example the mode at 1.337 THz in the α-form is predicted to have a 26 cm^{-1} red-shift. These data highlight the importance of unit cell dimensions in the low-frequency vibrational potentials in molecular solids.

3.4 Thermodynamic analysis

The accurate modelling of the vibrational spectrum by the s-DFT simulations lends confidence in the theoretical methodology employed, and enables the prediction of thermodynamic parameters to be performed. It is important to note that while the s-DFT predicted IR (and Raman) activities might not be completely accurate for certain modes in the disordered-crystal, thermodynamic functions depend more simply on the correct identification of the vibrational states, regardless of their symmetry. The harmonic canonical vibrational partition function was determined, and related quantities such as the entropy and free energy were calculated. All the energies discussed have been corrected for basis set superposition error using the counterpoise correction method.[58] When strictly considering the electronic energy, which has no intrinsic temperature or pressure dependence, the α-crystal is the more stable system by 0.56 kJ mol^{-1}. However, when the vibrational results are used to determine the temperature-dependent entropic and free energy parameters, the results suggest that the γ-solid becomes the more thermodynamically stable crystal above 301 K, close to the experimental transition temperature of 317 K (Fig. 7).

In an attempt to increase the accuracy of the calculated thermodynamic constants, the QHA vibrational results were used to determine the temperature-dependent free energies. Because these simulations explicitly take thermal expansion into account, they are able to provide a more physical description of the thermodynamic forces involved in these materials. The simulated QHA free

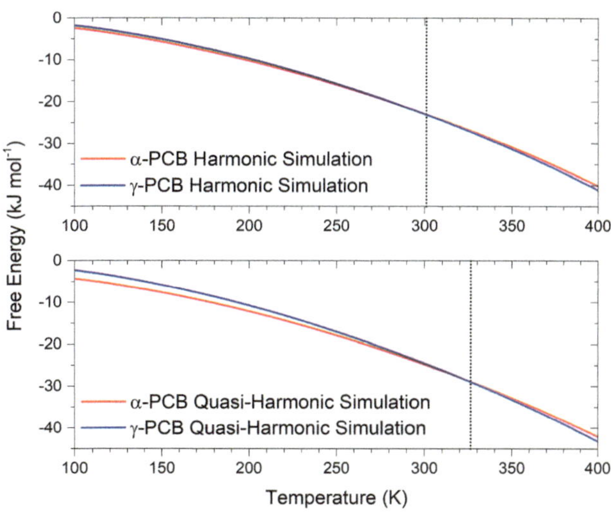

Fig. 7 Harmonic (top) and quasi-harmonic (bottom) free energy curves for the α- and γ-PCB crystals (red and blue, respectively). The vertical dotted line is drawn at the intersection of the two curves, occurring at 301 K and 326 K for the harmonic and quasi-harmonic simulations, respectively. In both charts the data have been offset relative to the value of the α-form free energy at 50 K.

energy curves, also shown in Fig. 7, not only show a larger deviation in energy between the two solids, but also provide a more accurate description of the phase-transition temperature, predicted at 326 K, only 9 K higher than the experimentally observed temperature. The greater accuracy achieved when explicitly considering the thermal expansion properties of the solids, while not surprising, highlights the importance of taking into account these forces when attempting to describe the thermodynamics of weakly bound molecular crystals.

3.5 Interpretation and further developments of proposed method

The combination of experimental and theoretical results enables a great deal of information surrounding the phase transition behaviour in molecular solids to be understood. Thus, such a methodology is generally applicable to a large variety of materials, particularly those that undergo diffusionless transformations. The sensitivity of THz-TDS measurements to the bulk packing arrangement, and any minor deviations from ideal crystalline ordering, leads to it being a powerful probe of the processes and dynamics occurring within solid materials. While X-ray diffraction is often used for interpreting this type of behaviour, it is in reality a complement to the THz-TDS experiment, as the former provides information related to the average structure, while the latter probes molecular dynamics. The addition of computational results provides a means of interpreting the experimental spectra, indicating what types of motions are occurring within the solids. Additionally, the accurate reproduction of the experimental THz-TDS spectra indicates that the simulations are correctly reproducing the forces, including the weak intermolecular interactions, providing a high level of confidence in them. However, not all systems are well suited for study using this methodology, particularly in predicting large-scale structural reorganisations, which would require much longer AIMD simulations in order to capture the effect. Nonetheless, the proposed methodology is well suited for order–disorder transitions, particularly in molecular solids where such transformations occur near ambient conditions.

4 Conclusions

Overall, the proposed combination of experimental terahertz time-domain spectroscopy and theoretical *ab initio* molecular dynamics and static-solid-state-density functional theory methods provides a complete means to elucidate the structures, phase transition mechanisms, and bulk properties (*e.g.* thermodynamics and mechanochemical effects such as thermal expansion, and other related quantities) of molecular solids. The results shed light on the importance of low-frequency vibrational modes in many aspects of the properties of disordered crystals, and the atomic-level details provided by the s-DFT simulations enable experimental observations regarding temperature-dependent spectral trends to be fully rationalised. And while traditional harmonic approximation vibrational analyses were shown to be a suitable means of determining the thermodynamic properties of molecular crystals, the results can be significantly improved by taking into account thermal expansion effects *via* the use of quasi-harmonic approximation simulations, which provide a more physical (and accurate) description of both the temperature-dependent vibrational and thermodynamical forces present within such materials.

Conflicts of interest

There are no conflicts to declare.

Acknowledgements

The authors thank the UK Engineering and Physical Sciences Research Council (EPSRC) for funding (EP/N022769/1). Simulations were possible *via* our membership of the UK's HEC Materials Chemistry Consortium, which is funded by the EPSRC (EP/L000202), and this work used the ARCHER UK National Supercomputing Service (http://www.archer.ac.uk). JAZ and JK thank the EPSRC Cambridge Centre for Doctoral Training in Sensor Technologies and Applications (EP/L015889/1). JK thanks the German Academic Scholarship Foundation for funding.

Notes and references

1 V. Coropceanu, J. Cornil, D. A. da Silva Filho, Y. Olivier, R. Silbey and J.-L. Brédas, *Chem. Rev.*, 2007, **107**, 926–952.
2 S. B. Murdande, M. J. Pikal, R. M. Shanker and R. H. Bogner, *J. Pharm. Sci.*, 2010, **99**, 1254–1264.
3 A. Cavagna, *Phys. Rep.*, 2009, **476**, 51–124.
4 M. Oguni and N. Okamotot, *Solid State Commun.*, 1996, **99**, 53–56.
5 M. T. Ruggiero, M. Krynski, E. O. Kissi, J. Sibik, D. Markl, N. Y. Tan, D. Arslanov, W. van der Zande, B. Redlich, T. M. Korter, H. Grohganz, K. Löbmann, T. Rades, S. R. Elliott and J. A. Zeitler, *Phys. Chem. Chem. Phys.*, 2017, **19**, 30039–30047.
6 E. Shalaev and A. K. Soper, *J. Phys. Chem. B*, 2016, **120**, 7289–7296.
7 M. T. Ruggiero, J. Axel Zeitler and T. M. Korter, *Phys. Chem. Chem. Phys.*, 2017, **19**, 28502–28506.
8 E. P. J. Parrott and J. A. Zeitler, *Appl. Spectrosc.*, 2015, **69**, 1–25.
9 J. A. Zeitler, P. F. Taday, D. A. Newnham, M. Pepper, K. C. Gordon and T. Rades, *J. Pharm. Pharmacol.*, 2007, **59**, 209–223.
10 G. J. O. Beran, *Chem. Rev.*, 2016, **116**, 5567–5613.
11 M. D. King, T. N. Blanton, S. T. Misture and T. M. Korter, *Cryst. Growth Des.*, 2011, **11**, 5733–5740.
12 M. T. Ruggiero, J. Sibik, R. Orlando, J. A. Zeitler and T. M. Korter, *Angew. Chem., Int. Ed.*, 2016, **55**, 6877–6881.
13 J. Timmermans, *J. Phys. Chem. Solids*, 1961, **18**, 1–8.
14 J. M. Pringle, P. C. Howlett, D. R. MacFarlane and M. Forsyth, *J. Mater. Chem.*, 2010, **20**, 2056–2062.
15 M. Descamps, N. T. Correia, P. Derollez, F. Danede and F. Capet, *J. Phys. Chem. B*, 2005, **109**, 16092–16098.
16 F. Affouard, J. F. Willart and M. Descamps, *J. Non-Cryst. Solids*, 2002, **9**, 307–310.
17 A. Hédoux, Y. Guinet, P. Derollez, J.-F. Willart, F. Capet and M. Descamps, *J. Phys.: Condens. Matter*, 2003, **15**, 8647–8661.
18 L. Jin, K. M. Nairn, C. M. Forsyth, A. J. Seeber, D. R. MacFarlane, P. C. Howlett, M. Forsyth and J. M. Pringle, *J. Am. Chem. Soc.*, 2012, **134**, 9688–9697.

19 R. Brand, P. Lunkenheimer and A. Loidl, *J. Chem. Phys.*, 2002, **116**, 10386–10401.
20 F. Affouard and M. Descamps, *Phys. Rev. Lett.*, 2001, **87**, 035501.
21 F. Affouard, E. Cochin, R. Decressain and M. Descamps, *Europhys. Lett.*, 2001, **53**, 611–617.
22 O. B. M. H. Duparc and M. Meyer, *J. Chem. Phys.*, 1998, **93**, 1313–1324.
23 J. Adebahr, F. C. Grozema, S. W. deLeeuw, D. R. MacFarlane and M. Forsyth, *Solid State Ionics*, 2006, **177**, 2845–2850.
24 M. Thomas, M. Brehm, R. Fligg, P. Vöhringer and B. Kirchner, *Phys. Chem. Chem. Phys.*, 2013, **15**, 6608–6622.
25 M. Yamaguchi and A. Ohira, *Comput. Theor. Chem.*, 2016, **1089**, 54–58.
26 M. Thomas, M. Brehm and B. Kirchner, *Phys. Chem. Chem. Phys.*, 2015, **17**, 3207–3213.
27 M. T. Ruggiero and J. A. Zeitler, *J. Phys. Chem. B*, 2016, **120**, 11733–11739.
28 G. Kresse and J. Hafner, *Phys. Rev. B*, 1994, **49**, 14251–14269.
29 A. S. Larsen, M. T. Ruggiero, K. E. Johansson, J. A. Zeitler and J. Rantanen, *Cryst. Growth Des.*, 2017, **17**, 5017–5022.
30 S. Tsuneyuki, H. Aoki, M. Tsukada and Y. Matsui, *Phys. Rev. Lett.*, 1990, **64**, 776–779.
31 A. Kubo, R. Ikeda and D. Nakamura, *Z. Naturforsch., A: Phys. Sci.*, 1986, **41**, 270–274.
32 A. Schönleber, P. Pattison and G. Chapuis, *Z. Kristallogr. – Cryst. Mater.*, 2003, **218**, 47.
33 Y. Takaki, K. Nakata, T. Taniguchi and K. Sakurai, *Acta Crystallogr., Sect. B: Struct. Crystallogr. Cryst. Chem.*, 1978, **34**, 2579–2586.
34 J. Hutter, M. Iannuzzi, F. Schiffmann and J. VandeVondele, *Wiley Interdiscip. Rev.: Comput. Mol. Sci.*, 2013, **4**, 15–25.
35 J. VandeVondele, M. Krack, F. Mohamed, M. Parrinello, T. Chassaing and J. Hutter, *Comput. Phys. Commun.*, 2005, **167**, 103–128.
36 S. Nosé, *J. Chem. Phys.*, 1998, **81**, 511–519.
37 S. Nosé, *Mol. Phys.*, 2002, **100**, 191–198.
38 G. J. Martyna, D. J. Tobias and M. L. Klein, *J. Chem. Phys.*, 1994, **101**, 4177–4189.
39 G. J. Martyna, M. E. Tuckerman, D. J. Tobias and M. L. Klein, *Mol. Phys.*, 1996, **87**, 1117–1157.
40 J. VandeVondele and J. Hutter, *J. Chem. Phys.*, 2007, **127**, 114105.
41 S. Goedecker and M. Teter, *Phys. Rev. B*, 1996, **54**, 1703–1710.
42 J. Perdew, K. Burke and M. Ernzerhof, *Phys. Rev. Lett.*, 1996, **77**, 3865–3868.
43 S. Grimme, J. Antony, S. Ehrlich and H. Krieg, *J. Chem. Phys.*, 2010, **132**, 154104.
44 S. Grimme, S. Ehrlich and L. Goerigk, *J. Comput. Chem.*, 2011, **32**, 1456–1465.
45 M. Brehm and B. Kirchner, *J. Chem. Inf. Model.*, 2011, **51**, 2007–2023.
46 R. Dovesi, R. Orlando, A. Erba, C. M. Zicovich-Wilson, B. Civalleri, S. Casassa, L. Maschio, M. Ferrabone, M. De La Pierre, P. D'Arco, Y. Noël, M. Causà, M. Rérat and B. Kirtman, *Int. J. Quantum Chem.*, 2014, **114**, 1287.
47 F. Weigend and R. Ahlrichs, *Phys. Chem. Chem. Phys.*, 2005, **7**, 3297–3305.
48 F. Pascale, C. M. Zicovich-Wilson, F. L. Gejo, B. Civalleri, R. Orlando and R. Dovesi, *J. Comput. Chem.*, 2004, **25**, 888–897.

49 C. M. Zicovich-Wilson, F. Pascale, C. Roetti, V. R. Saunders, R. Orlando and R. Dovesi, *J. Comput. Chem.*, 2004, **25**, 1873–1881.

50 Y. Noël, C. Zicovich-Wilson, C. M. Zicovich-Wilson, R. Dovesi, B. Civalleri, P. D'Arco and R. Dovesi, *Phys. Rev. B*, 2001, **65**, 014111–014119.

51 A. Erba, J. Maul and B. Civalleri, *Chem. Commun.*, 2016, **52**, 1820–1823.

52 A. Erba, M. Shahrokhi, R. Moradian and R. Dovesi, *J. Chem. Phys.*, 2015, **142**, 044114.

53 M. T. Ruggiero, J. A. Zeitler and A. Erba, *Chem. Commun.*, 2017, **53**, 3781–3784.

54 J. G. Brandenburg, J. Potticary, H. A. Sparkes, S. L. Price and S. R. Hall, *J. Phys. Chem. Lett.*, 2017, **8**, 4319–4324.

55 M. T. Ruggiero, T. Bardon, M. Strlič, P. F. Taday and T. M. Korter, *Phys. Chem. Chem. Phys.*, 2015, **17**, 9326–9334.

56 P. G. Karamertzanis and C. C. Pantelides, *Mol. Phys.*, 2007, **105**, 273–291.

57 Y. C. Shen, P. C. Upadhya, E. H. Linfield and A. G. Davies, *Appl. Phys. Lett.*, 2003, **82**, 2350–2352.

58 M. Gutowski and G. Chalasiński, *J. Chem. Phys.*, 1998, **98**, 5540–5554.

Faraday Discussions

PAPER

How many ritonavir cases are there still out there?

Marcus A. Neumann* and Jacco van de Streek

Received 26th March 2018, Accepted 30th April 2018
DOI: 10.1039/c8fd00069g

Based on a thorough and critical analysis of the commercial crystal structure prediction studies of 41 pharmaceutical compounds, we conclude that for between 15 and 45% of all small-molecule drugs currently on the market the most stable experimentally observed polymorph is not the thermodynamically most stable crystal structure and that the appearance of the latter is kinetically hindered.

Introduction

In the summer of 1998, the sudden appearance of a previously unknown and significantly more stable polymorph of ritonavir halted the production of Norvir capsules: because the new crystal form was significantly more stable, the compound had turned nearly insoluble and therefore largely inactive when taken orally.[1] Moreover, being thermodynamically more stable, the presence of the new form now made it impossible to obtain the old form, in what turned out to be one of the most infamous cases of a "disappearing polymorph".[2] The estimated cost of 250 million USD in lost revenue made the ritonavir incident a high-profile case, but it certainly was not an isolated case. Bučar *et al.*[3] describe more than ten additional cases, including rotigotine and ranitidine hydrochloride. These disasters highlight how a thermodynamically unstable crystal structure can be kinetically stable for years. This begs the question of whether any of the drugs currently sold or under development have as-yet undiscovered more stable polymorphs. Ritonavir, rotigotine and ranitidine hydrochloride had been under development for years without a trace of the more stable polymorphs, demonstrating that it is impossible for any approach relying solely on experiments to exclude possible kinetic factors.

Crystal structure prediction (CSP) attempts to give an overview of all relevant polymorphs based solely on their thermodynamic stability, *i.e.* irrespective of possible kinetic considerations. It therefore comes as no surprise that the pharmaceutical industry is adopting a combination of experimental and

Avant-garde Materials Simulation, Alte Str. 2, 79249 Merzhausen, Germany. E-mail: marcus.neumann@avmatsim.eu

computational polymorph screens to prevent disappearing polymorph accidents in the future.[4]

This, then, suggests an obvious approach to our question "How many ritonavir cases are there still out there?": use CSP to process a large number of small molecules with at least one known crystal structure and count in how many cases, according to the CSP results, the thermodynamically most stable crystal form has been missed.

It is not that straightforward, however. First, the answer to our question requires statistics over a significant number of compounds, not merely a case study. Second, the compounds must be relevant for the problem at hand: pharmaceutical compounds tend to be large and flexible, whereas most published CSP studies report results for small compounds of limited flexibility. Third, a thorough experimental polymorph screen must have been conducted for each of the compounds to ensure that missed polymorphs are due to consistently hindered kinetics and not merely a one-off. Fourth, the error bars on the lattice energies of the CSP studies must be known (or, as a minimum, a reasonable estimate must be available) to distinguish between those cases where the prediction of a thermodynamically more stable form is due to a missed experimental form and those cases where the prediction of a thermodynamically more stable form is merely caused by an incorrect assessment of the relative stabilities of the predicted crystal structures by the energy potential used. Fifth, the structure generation step must be complete enough to render statistics regarding the possibility of additional polymorphs meaningful.

To assess the feasibility of current CSP methodologies to answer the question in the title, it pays to take a closer look at the results of the periodically published crystal structure prediction blind tests. CSP blind tests have been organised in 1999, 2001, 2004, 2007, 2010 and 2015. The results can be found in the most recent publication[5] and references therein. For each blind test, crystallographers are invited to submit unpublished experimental crystal structures, the molecular 2D diagrams for which are then sent out to the participants together with a deadline for the submission of their predictions. The target crystal structures are guaranteed to be free from disorder and certain limits on the chemical elements, space-group symmetry and number of molecules in the asymmetric unit are also specified. After the submission deadline, the experimental crystal structures are made public and the computational results are published in a joint publication. In earlier blind tests, the compounds were small, rigid, neutral molecules, but the compounds became increasingly more complex as time progressed (and methods improved). For the purpose of our discussion, we will focus on two aspects: the results for all small, rigid, neutral molecules and the results of the most recent (2015) blind test.[5]

In gathering our statistics for large, flexible molecules, we are hampered by three error sources that may be hard to disentangle: the error bars on the lattice energies, the completeness of the computational structure generation step and the completeness of the experimental polymorph screen (which we presume to be hindered by kinetics). When investigating the relationship between ease of crystallisation and molecular descriptors, it was found that the smaller and the more rigid a molecule, the easier it crystallises.[6] For small, rigid, neutral molecules we can therefore expect both the computational and the experimental polymorph screens to be essentially complete, at least significantly more complete than those for large, flexible molecules. So as a first approximation we can make the assumption that the number of missed most stable polymorphs for small, rigid, neutral molecules (both computationally and experimentally) is equal to zero, and consequently any discrepancies between the experimental and the predicted results can be attributed solely to shortcomings in the energy potential. The results for all small, rigid, neutral CSP blind test molecules should therefore tell us something about the error bar on our energy potential. Focussing exclusively on the results obtained with our own method, the results are reported in four papers[7–10] and in total provide data for 17 small molecules. All 17 experimental target structures were successfully located in the structure generation steps. Thirteen times an experimental structure is identified as the global minimum, three times an experimental structure is ranked second and once an experimental structure is fourth in the list of predicted structures. The energy differences between the predicted polymorphs are very small, of the order of 0.1 kcal mol^{-1}, suggesting error bars for our energies of the order of 0.1 kcal mol^{-1} for small, rigid, neutral molecules. However, although these results provide a lower limit for the estimated standard deviation of our energy function, small, rigid, neutral molecules are not what we are interested in. In order to answer our question in the title, we must establish how current methods perform for compounds that reflect the complexities of pharmaceutically active molecules.

The most recent CSP blind test featured four target compounds of a complexity approaching that found in pharmaceutical compounds (Fig. 1). For targets **XXIV**, **XXV** and **XXVI** (the molecules have been numbered consecutively since the first blind test in 1999), one polymorph is known, each with one molecule in the asymmetric unit. For target **XXIII**, five polymorphs are known, of which two contain two molecules in the asymmetric unit. Twenty-five research groups participated in the blind test, providing a good overview of the state-of-the-art in molecular crystal structure prediction. First, let's take a look at the results of the other 24 participants, keeping in mind the requirements that will lead us to our goal of using the state-of-the-art in molecular crystal structure prediction to answer the question in the title. Of these 24 participants, four did not have access

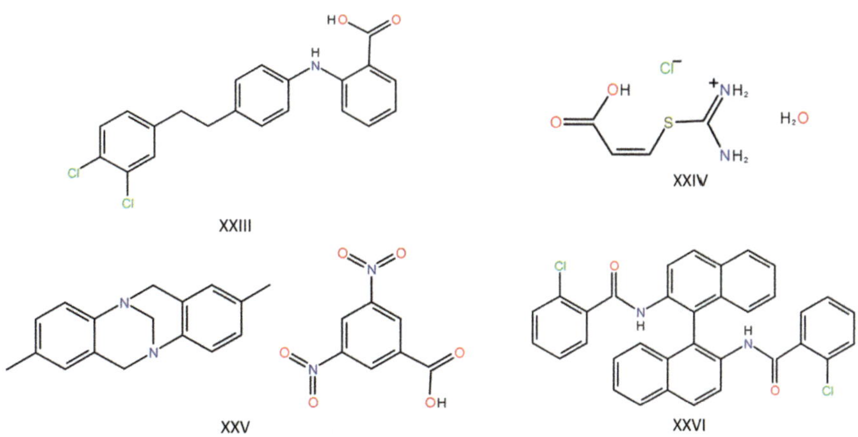

Fig. 1 The four large and flexible chemical compounds that were provided as targets in the last crystal structure prediction blind test.

to a structure generation tool of their own and relied on the list of generated structures of one of the other participants. Of the remaining 20 participating research groups (still excluding our own contribution), only seven attempted all four compounds, and only two attempted the difficult $Z' = 2$ polymorphs. In other words, of the 24 research groups involved in the development of methods for crystal structure prediction, only two submitted candidate structures for target molecule **XXIII** with two molecules in the asymmetric unit. Adding together all polymorphs, eight distinct experimental structures should be found for these four compounds. The best results were obtained by the research group of Prof. Price, who were the only group to successfully predict four of the eight crystal structures, including one of the two $Z' = 2$ polymorphs. The latest CSP blind test therefore paints a very sobering picture: large-scale prediction of the crystal structures of large, flexible molecules with one and two molecules in the asymmetric unit is hard and is currently beyond the capability of most CSP methods, either due to limitations in the available hardware or due to limitations in the available software. In addition, the blind test compounds, with the exception of molecule **XXIII**, have not been experimentally screened for polymorphism.

The results with our in-house developed software for crystal structure prediction and our in-house hardware were more promising: seven of the eight experimentally observed crystal structures were generated in our searches, missing out only one of the more challenging $Z' = 2$ targets. Furthermore, the compounds we are sent to investigate have always been thoroughly experimentally screened. These results should stand us in good stead to use our crystal structure prediction technology to address the question in the title: "How many ritonavir cases are there still out there?".

Methods

Software

The CSP studies were carried out with the software package GRACE.[11] Conceptually, a CSP study consists of two steps: the crystal structure generation step, which attempts to generate all relevant polymorphs, and the stability ranking step, which attempts to predict the relative thermodynamic stability of each generated polymorph. The way GRACE was designed is more easily understood if the ranking is described first.

Energy potential

In GRACE, the thermodynamic stability of a crystal structure is approximated as its enthalpy, $U + PV$. In other words, the temperature is set to 0. At ambient pressure, PV is negligible and set to 0. U is calculated using density functional theory (DFT) with the software package VASP.[12–14] Zero-point vibrations are not included. Even though our lattice energies are calculated in the classical 0 K approximation, for instances in which the predictions suggest the possibility of disorder, the configurational entropy at 300 K is included. The exchange–correlation functional used is the PBE functional[15] with a plane-wave basis set with an energy cut-off of 520 eV and an approximate k-point spacing of 0.07 Å^{-1}. The core electrons are described by projector-augmented wave (PAW) pseudo-potentials. Dispersion correction is indispensable for molecular crystals. VASP is only used for pure DFT calculations, to which GRACE adds a dispersion correction; the combination of DFT with a dispersion correction is referred to as DFT-D. Two choices for the dispersion correction are currently implemented in GRACE: Neumann & Perrin[16] and Grimme 2010.[17] The results in the crystal structure prediction blind tests and the results reported in the current paper were obtained with the Neumann & Perrin dispersion correction. Further details can be found elsewhere.[16]

The last CSP blind test revealed that one group had succeeded in the development of a more accurate method for the energy ranking of molecular crystals.[18] The new method, here called PBE(0) + MBD + F_{vib}, combines energy calculations with the PBE functional with a method for the calculation of Many-Body Dispersion (MBD), a PBE0 correction to include exact exchange and a calculation of the vibrational free energy in the harmonic approximation. By comparison with PBE(0) + MBD + F_{vib} we will assess the energy error of our own energy potential.

Structure generation and tailor-made force fields

DFT-D is too slow to be used in structure generation, which can involve generating millions and millions of trial crystal structures. An off-the-shelf force field can be used, but that approach would have two disadvantages. First, the large estimated standard deviation between DFT-D and the force field means that very many of the generated structures must be optimised to ensure that the global minimum (in the DFT-D potential) has been optimised. Second, the force field might be so poor for the compound under consideration that none of the local minima in the force field optimise to the polymorph of interest during the ranking stage. In GRACE, this is solved by parameterising a tailor-made force field (TMFF) for the compound under consideration, *i.e.* a force field that is not transferable but in which all force-field parameters have been optimised for that one particular compound. The reference data against which the force-field parameters are fitted are generated with the DFT-D potential. In other words, the tailor-made force field is parameterised to mimic the DFT-D potential. A detailed description of the reference data and the fitting process are published elsewhere.[19] Larger molecules are divided up into smaller fragments that are parameterised individually and from which the tailor-made force field of the original compound is assembled afterwards, ensuring that tailor-made force fields can be prepared for molecules of virtually any size.

The crystal structure generation algorithm is described in some detail elsewhere.[20] In brief, a parallel-tempering algorithm is used for a given number of space groups with a fixed number of molecules in the asymmetric unit. Crystal structures are generated in an energy window equal to four times the estimated standard deviation of the tailor-made force field with respect to the DFT-D energies. During the search, the completeness is estimated by counting how often each unique structure has been generated, and the search is continued until it is estimated that 99% of the structures in the search space have been generated.

Search space

The space groups and the Z' values that are selected depend on the chirality of the molecule, the size of the molecule, the complexity of the energy landscape and the available budget. $Z' = 1$ was included for all cases reported here, and searches with $Z' = 2$ were also included for most cases. For $Z' = 1$, if the compound is enantiomerically pure, 21 space groups are included in the search, otherwise the number of space groups is 38. For $Z' = 2$, a reduced number of space groups is searched and the structure generation step is split into two parts. In a first round, space groups $P1$, $P2_1$ and, if relevant, $P\bar{1}$ are searched. These three space groups each have at most two symmetry operators, and generally convergence is reached within the time (=budget) available even for the more problematic energy landscapes. In a second round, $Z' = 2$ structures are generated in six (for enantiomerically pure compounds) or eight space groups, namely $C2$, $P2_12_12_1$, $P4_1$, $I4_1$, $R3$ and $P4_3$ or $C2$, Pc, Cc, $P2_1/c$, $C2/c$, $P2_12_12_1$, $Pca2_1$, and $Pna2_1$ respectively. This second $Z' = 2$ round is regularly aborted for economic reasons while still incomplete, although any generated structures are included in the final results.

Hardware

Each CSP study was carried out on a high-performance computing cluster consisting of 32 INTEL Xeon E5-2650 v4 2.2 GHz 12 core CPUs (= 384 cores in total) with 2 GB memory per core connected by an InfiniBand network or equivalent hardware.

Compounds

The 41 compounds used in this paper were 41 compounds for which commercial, confidential CSP studies were commissioned. Most of the compounds were under active development at the time of the CSP study. Although no specific compounds can be shown for confidentiality reasons, the average complexity is illustrated in Fig. 2. On average, each compound had 30 non-hydrogen atoms, 23 hydrogen atoms, three trivial flexible torsions such as $-CH_3$, six non-trivial flexible torsion angles and one flexible ring. Among the compounds were five hydrates, one salt and two zwitterions. Twenty-six of the 41 compounds were enantiomerically pure. Fig. 3 and 4 show histograms for the distributions of the number of non-hydrogen atoms and the number of degrees of freedom across the 41 compounds. The statistical analysis presented in this paper only holds for the size and flexibility range of the test set.

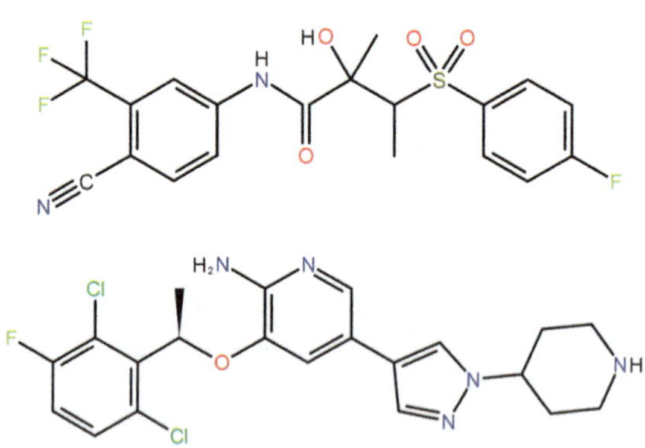

Fig. 2 Two molecules (bicalutamide and crizotinib) illustrating the average size and flexibility of the 41 compounds used in this study.

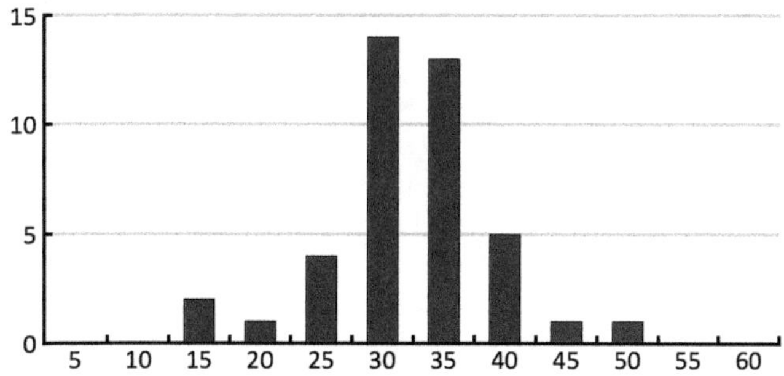

Fig. 3 Histogram of the number of non-hydrogen atoms in the 41 compounds of our test set.

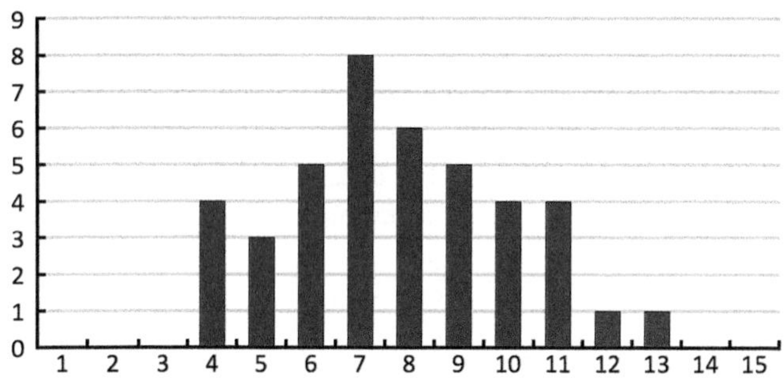

Fig. 4 Histogram of the number of non-trivial flexible torsion angles in the 41 compounds of our test set. Each flexible ring is counted as two flexible torsion angles.

Comparison to experimental data

The CSP studies were blind studies and it is only after the predictions had finished that the experimental data were made available. The experimental data may consist of single-crystal structures and X-ray powder diffraction patterns. The single-crystal structures were energy minimised and compared to the list of predicted structures to identify which predicted structures correspond to experimentally observed forms. If the experimental structure was not already present in the list, for example because the polymorph has unusual space-group symmetry, it was added to the list of predicted structures. For two of the 41 studies the most stable experimental structure crystallised with a space-group symmetry that had not been included in the structure generation step. The experimental X-ray powder diffraction patterns were compared to the simulated powder diffraction patterns of all the predicted structures to establish if there were any matches. If a match was found and confirmed by Rietveld refinement, the predicted structure was labelled as being an experimentally observed form. The results of the crystal structure prediction study were then summarised in energy landscapes (Fig. 5). For the purpose of this paper, the main result of each crystal structure prediction study is the energy gap between the predicted crystal structure with the lowest lattice energy and the experimental crystal structure with the lowest lattice energy after lattice energy minimization; for the energy landscape in Fig. 5, for example,

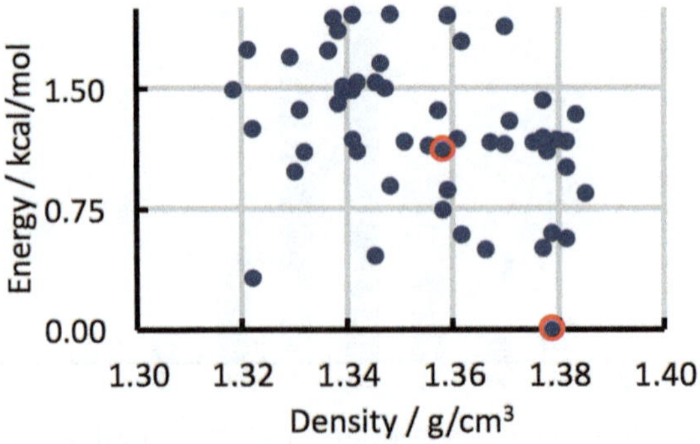

Fig. 5 Example energy landscape. Each dot represents a predicted crystal structure. Two predicted structures that have been matched with experimental structures are indicated by red circles.

this energy gap has a value of 0.0 kcal mol^{-1}. The energy gaps of the 41 studies can be summarised in a histogram (Fig. 8 in the Results and discussion). It is this histogram that is the key to the number of kinetically hindered polymorphs.

The distribution of the number of experimentally observed forms over the test set is shown in Fig. 6.

Statistical model for the number of kinetically hindered polymorphs

If in the energy landscape (Fig. 5) the global minimum in the predictions does not match an experimentally observed structure, then this can have one of two possible causes. Either the global minimum is kinetically hindered in experiment (the cases we are interested in), or the approximations in our energy potential are at fault and an experimental structure should have been the global minimum. In other words, if our energy potential were perfect, the number of kinetically hindered polymorphs would simply be the number of compounds for which the global minimum in the predictions does not match an experimentally observed structure. In practice, however, the number of predicted global minima that are experimentally unobserved is a function of both the fraction of kinetically

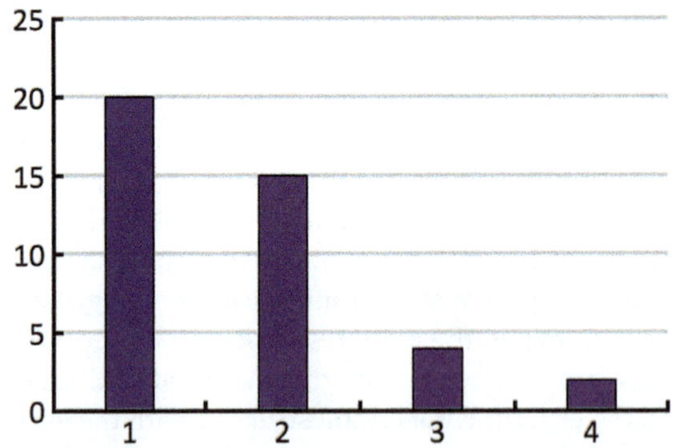

Fig. 6 Histogram of the number of confirmed experimental polymorphs per compound.

hindered structures, h, and the error in the lattice energies, σ. In order to determine h, it is therefore necessary to construct a statistical model that relates h and σ (and an additional parameter b) to the results of the 41 crystal structure prediction studies. Ideally the model would allow for fitting all its parameters to the empirical histogram shown in Fig. 8, but initial tests showed that the parameters are strongly correlated and that the small number of only 41 cases results in high statistical noise that makes a direct fitting procedure unreliable. Instead, an approach is described in the next section in which σ is determined by comparison with a more accurate energy calculation method and b is tightly bracketed by additional considerations. The agreement between the statistical model and the 41 cases is then assessed for various values of h and σ.

In order to construct a computational equivalent of the empirical histogram shown in Fig. 8 for a given set of parameters, we use a Monte Carlo procedure implemented in a computer program. The procedure generates 41 random energy landscapes (Fig. 5; the densities do not play a role and are ignored), which are combined into a histogram measuring the energy differences between the most stable predicted structure and the most stable experimental structure, analogous to Fig. 8 (see below in Results and discussion). For each set of parameter values, the random generation of 41 landscapes is repeated 100 times, generating 100 histograms. The 100 histograms are then averaged and the standard deviation for each is calculated. For each set of parameters, the average histogram is compared to the actual histogram resulting from our 41 crystal structure prediction studies (Fig. 8). To avoid confusion, we call the histogram in Fig. 8 the empirical histogram, as opposed to a single, randomly calculated histogram or the average calculated histogram for a set of parameters (we realise that the adjective "empirical" is somewhat ironic considering that the construction of the empirical histogram involved about six years of calculations on a high-performance computing cluster).

We now discuss the parameters of the model. The energy error, σ, is meant to be the root-mean-square deviation between the energies of lattice energy minima calculated at the DFT-D level and the true lattice free energies in the real world. The hindered fraction, h, is the fraction of lattice energy minima for which the appearance is strongly kinetically hindered. By definition, h is also the fraction of the thermodynamically most stable crystal structures that do not readily crystallise, *i.e.* the main quantity that we are trying to bracket in this paper. Finally, a parameter b is required that is the factor in the exponent of an exponential function describing the rapid increase of the effective number of lattice energy minima with increasing lattice energy.

The notion of the effective number of lattice energy minima, N_{eff}, requires some further discussion. Crystal energy landscapes typically contain many structures that are too similar to be observed independently. We refer to such groups of similar structures as "families" and the effective number of lattice energy minima counts the number of families rather than the number of individual family members. Currently the decision whether two predicted crystal structures belong to the same family is taken based on manual examination of the structures. The differences between family members can be subtle, such as a different orientation of a terminal ethyl group, a small unit cell change or a small molecular displacement. Or there may be structures with two molecules per asymmetric unit that almost match a structure with only one molecule in the asymmetric unit. In such cases thermal motion will either average over the different configurations or result in rapid conversion to the most stable one. But family members may also only share a rod or sheet motif held together by strong interactions. In such cases, each family member can act as a nucleation substrate for all others and the crystal growth process will typically favour the thermodynamically most stable family member. Therefore, at a given pressure and temperature, only one family member will typically be observed experimentally. The stability ranking of family members may change as a function of pressure and temperature and more than one family member may be experimentally observable. The estimation of the average number of members per family will be discussed in the Results section. Our statistical analysis is based on the number of families rather than on the number of individual lattice energy minima for three reasons:

(1) The energy error (DFT-D *versus* reality) is typically smaller for members belonging to the same family than for members belonging to different families. In each family, structures are similar and error compensation is strong. The energy error σ can be expected to be dominated by the inter-family error.

(2) Many family members are too similar to be observed independently.

(3) Even if family members can be observed independently because they have different p–T stability domains, they must be expected to be either all kinetically hindered or not kinetically hindered, because of their structural similarity and the possibility of cross-nucleation.

The expectation value of the effective number of lattice energy minima per energy interval is assumed to be $A\cdot\exp(b\cdot E)\cdot dE$, were E is the energy per chemical unit. Without loss of generality, the prefactor A is adjusted to yield an expectation value of 10 in the energy interval from 0 to 1 kcal mol^{-1} (note that 0.0 kcal mol^{-1} does not correspond to the energy of the most stable structure). The energy range from −5.0 to 5.0 kcal mol^{-1} is subdivided into 100 000 intervals. To each interval, a number of structures is randomly attributed following a Poisson distribution that yields the correct mean value according to the exponential function mentioned above. Structures are randomly marked to have kinetically hindered dynamics according to the hindered fraction h. The number of experimentally observed polymorphs, N_p, is randomly chosen according to the experimental distribution for the 41 test cases shown in Fig. 6. The first N_p non-hindered low energy structures in the 100 000 intervals are marked as experimental. Finally, the energy of each structure is randomly shifted following a normal distribution with a standard deviation that corresponds to the energy error σ.

To measure the deviation between the empirical histogram and an average histogram calculated from 100 randomly generated histograms, we use a Figure Of Merit (FOM) that is the sum over the absolute differences between the number of structures in each bin:

$$\sum_{i=1}^{N}\left|H_{1,i}-H_{2,i}\right| \quad (1)$$

where H_1 and H_2 are the two histograms to be compared and N is the number of bins in each histogram. For a given parameter set, the figure of merit can also be used to assess the deviation of each randomly calculated histogram from the average calculated histogram. As an example, Fig. 7 shows the fluctuation of the

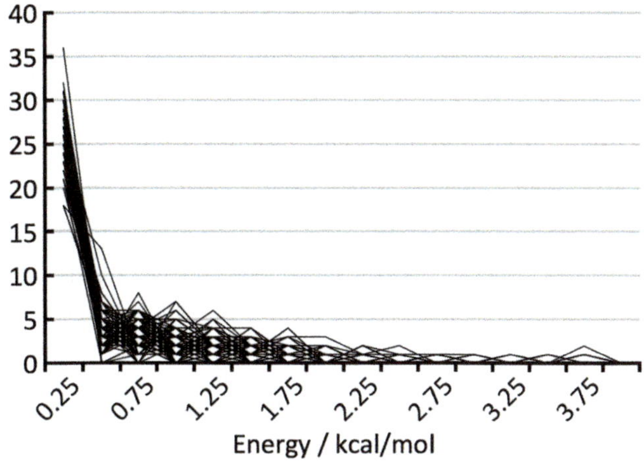

Fig. 7 Fluctuation of the 100 randomly calculated histograms for $\sigma = 0.5$ kcal mol^{-1}, $b = 1.609$ mol kcal^{-1} and $h = 0.3$ (*cf.* Fig. 9b).

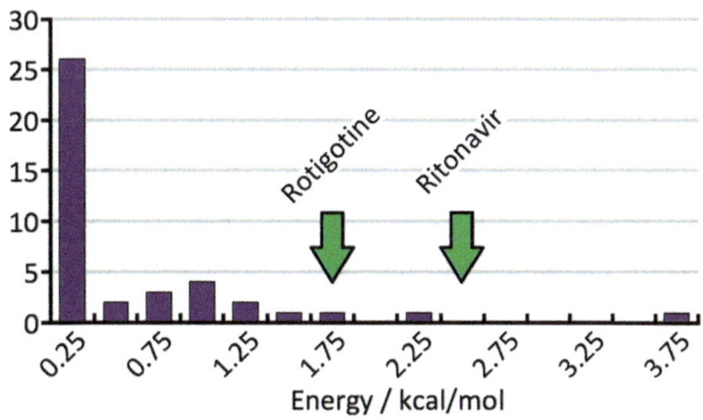

Fig. 8 Histogram of the energy gaps between the experimental structure with the lowest predicted energy and the global minimum. The arrows indicate the energy differences for the experimental polymorphs of rotigotine and ritonavir.

randomly calculated histograms around the average calculated histogram for $\sigma = 0.5$ kcal mol^{-1}, $b = 1.609$ mol kcal^{-1} and $h = 0.3$. For each parameter set an average figure of merit, $\overline{\text{FOM}}$, can be determined for the randomly calculated histograms. The fluctuation of the randomly generated histograms corresponds to the changes that would be observed if the empirical histogram was reassessed for many independent sets of 41 test cases. Any set of three parameters for which the FOM value of the empirical histogram is lower than or equal to $\overline{\text{FOM}}$ must be considered in agreement with the empirical histogram. For the parameter set of Fig. 7 we found $\overline{\text{FOM}} = 10.1$.

Results and discussion

Error bars on the energies

For the blind test compounds **XXIII, XXIV** and **XXVI** we calculated the root mean square deviation, σ_{rel}, between our own DFT-D lattice energies and the corresponding PBE(0) + MBD + F_{vib} energies. The error of the PBE(0) + MBD + F_{vib} method itself, σ_{ref}, is of the order of 1 kJ mol^{-1}, or 0.25 kcal mol^{-1}.[18] To estimate the total error of our DFT-D method we combine both errors according to:

$$\sigma_{tot} = \sqrt{\left(\sigma_{rel}^2 + \sigma_{ref}^2\right)} \tag{2}$$

The results are presented in Table 1. Compound **XXV** was not included in the comparison because PBE(0) + MBD + F_{vib} energies were available for only 4 structures, some of which are co-crystals and others are salts. Our DFT-D method is known to suffer from an increased energy error when it comes to the comparison of co-crystal and salt structures, but this is irrelevant for the present analysis because none of the 41 test cases presents this kind of ambiguity.

The average total error for our method over compounds **XXIII**, **XXIV** and **XXVI** is 0.5 kcal mol^{-1}. With 26 and 40 non-hydrogen atoms, respectively, compounds **XXIII** and **XXVI** match the size range of the 41 test compounds. The flexibility of both compounds is slightly below average. With only 11 non-hydrogen atoms, compound **XXIV** is much smaller than any compound in the test set, but being the hydrate of a chloride salt, lattice energy calculations for compound **XXIV** are much more challenging than for most other structures in the test set. Overall, it can be assumed that the average energy error obtained for compounds **XXIII**, **XXIV** and **XXVI** is fairly representative for our test set. It is important to note that the error σ_{tot} in Table 1 compares the energies of minima on DFT-D lattice energy surfaces to the true lattice free energies. Our average total error of 0.5 kcal mol^{-1} hence includes not only the inherent inaccuracy of DFT-D, but also the neglect of zero-point vibrations and entropic effects.

Energy gap histogram

Fig. 8 shows the distribution of the energy gaps between the experimental structure with the lowest predicted energy and the global minimum of the predictions. In cases where the best predicted structure matches an experimental structure, the energy gap is zero. Twenty-three times an experimentally observed structure corresponded to the global minimum of the predicted structures. The width of the bins is 0.25 kcal mol^{-1}, and the first bin not only contains the twenty-three rank-1 predictions but also three additional cases where an experimental structure falls just above the predicted global minimum. The large number of rank-1 predictions demonstrates satisfactory performance of our DFT-D method over the compounds in the test set.

Table 1 RMSD between DFT-D and PBE(0) + MBD + F_{vib} (σ_{rel}) and total error (σ_{tot}) incorporating estimated error of PBE(0) + MBD + F_{vib} calculated over N lattice energy differences

Compound	N	σ_{rel}	σ_{tot}
XXIII	46	0.44	0.51
XXIV	8	0.57	0.62
XXVI	8	0.26	0.36

For an intuitive assessment of the number of missing thermodynamically stable forms, one may argue that, considering the low number of test cases, a more stable form has certainly been missed when the energy gap exceeds two standard deviations of the energy error. According to Fig. 8, this is the case for 6 out of 41 cases. For energy gaps of between one and two standard deviations, one may roughly assign half of the cases to the energy error, and half of the cases to a missing form problem. In the case of Fig. 8 that would add another 3.5 missing forms, resulting in a total of 9.5 out of 41 (23%). We will see in the next section that this value is in good agreement with a more careful statistical analysis.

In order to support our statement that energy gaps above two standard deviations of the energy error are likely to indicate missing, more stable forms, the energy differences between the first and the late discovered polymorphs of ritonavir and rotigotine are indicated in Fig. 8 by green arrows (no crystal structure prediction studies have been carried out). If an energy landscape had been calculated for these two compounds after the discovery of the first appearing polymorph but before the discovery of the late appearing form, the energy gap would have been as indicated in Fig. 8, or greater.

Statistical analysis

To understand the shape of the histogram in Fig. 8 and to extract quantitative information, we have used the statistical model described in the Methods section. Because of the small size of the test set, many fairly different parameter sets are in agreement with the empirical histogram. It is not possible to determine the three parameters of the statistical model independently by comparison of the FOM values obtained for the various parameter sets.

As an example, Fig. 9 shows the average calculated histograms for the two parameter sets presented in Table 2. Both parameter sets are in equally good agreement with the empirical histogram. However, in one case the energy error is 0.75 kcal mol^{-1} and the hindered fraction is 0.0, whereas in the other case the energy error is 0.5 kcal mol^{-1} and the hindered fraction is 0.3. We note that because disappearing polymorph cases such as rotigotine, ranitidine and ritonavir are known to exist, our *a priori* knowledge tells us that it is not possible for h to be 0.0. We have shown both cases as a warning that the estimation of the number of missing thermodynamically most stable forms derived in this paper strongly depends on the estimated energy accuracy.

Fig. 7 shows the randomly calculated histograms for parameter set b. Some of these histograms actually have structures in the bin at 3.75 kcal mol^{-1}, and what looks like an outlier in Fig. 8 and 9 is actually in agreement with the statistical model.

When the energy error σ and the exponential factor b go up, the hindered fraction has to go down in order to maintain the same level of agreement with the empirical histogram. Above we have already derived an estimate for the energy error. In order to estimate the hindered fraction h, we now also need to bracket the exponential factor b. The exponential factor b determines the effective number of lattice minima that are found on average in a 1 kcal mol^{-1} window above the most stable structure (see Table 3). We compare these values to the full number of lattice energy minima found for the 41 test compounds. Fig. 10 shows the distribution of the full number of lattice energy minima in a 1 kcal mol^{-1}

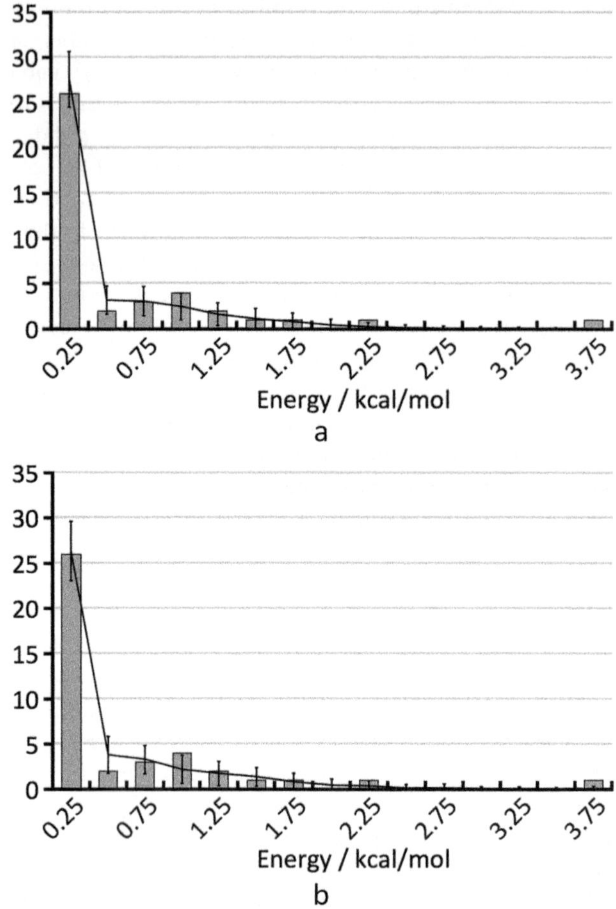

Fig. 9 The two histograms corresponding to the parameter sets from Table 2.

Table 2 Some parameter sets that provide a good match with the experimental histogram

Set	σ [kcal mol^{-1}]	b [mol kcal^{-1}]	h	FOM
a	0.75	1.609	0.0	5.5
b	0.5	1.609	0.3	6.2

window. The average value is 16.7 and goes down to about 14 if the outlier is not considered. The average full number of lattice energy minima, objectively determined from the computed crystal energy landscapes of the 41 test compounds, needs to be converted to an estimate of the effective number of lattice energy minima. This conversion is inevitably subjective at present, because there is no generally accepted measure of how similar structures need to be in order to belong to the same family. According to our experience, which is probably shared by other practitioners of CSP, families have on average about two members in the bottom 1 kcal mol^{-1} window, but certainly not less than 1.5 or more than 3. Using these last two values, we expect the effective number of lattice energy minima to

Table 3 Values for the exponential factor, b, and the corresponding average effective number of lattice energy minima, N_{eff}, in the bottom 1 kcal mol^{-1} window

b [mol kcal^{-1}]	N_{eff}
0.693	2.3
0.125	3.5
1.61	5.0
2.01	7.5
2.30	10
2.70	15
3.00	20

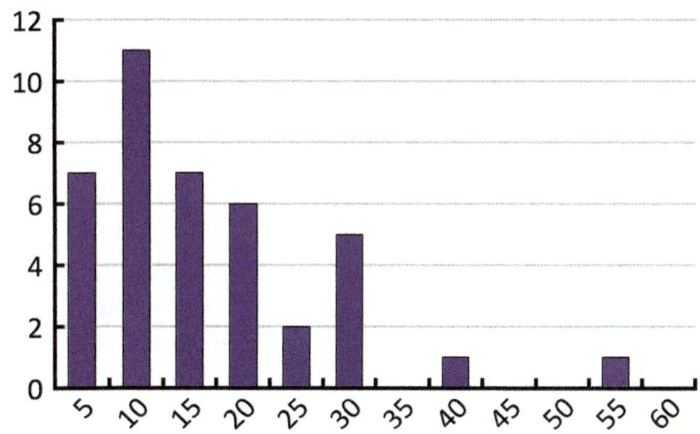

Fig. 10 Distribution of the full number of lattice energy minima in a 1 kcal mol^{-1} window above the most stable predicted structure for the 41 test compounds. One outlier with 113 crystal structures is not shown.

Table 4 Figure of merit as a function of the exponential factor b and the hindered fraction h. Boxes with a grey background correspond to cases where the figure of merit is lower than $\overline{\text{FOM}}$. The energy error σ is 0.5 kcal mol^{-1} in all cases

		b [mol kcal^{-1}]				
		1.25	1.61	2.01	2.30	2.70
h	0.0	17.5	15.1	13.7	10.0	9.1
	0.1	14.3	12.0	9.5	9.4	7.2
	0.2	10.1	8.81	7.6	6.0	8.6
	0.3	5.8	5.44	7.5	7.9	10.5
	0.4	7.2	7.47	8.3	11.9	12.9
	0.5	10.7	12.31	13.0	14.7	16.4
	0.6	16.2	16.6	17.9	19.7	21.3

be in the range of 5 to 10, which corresponds to exponential factors in the range from 1.61 to 2.30.

Table 4 presents figures of merit for $\sigma = 0.5$ kcal mol^{-1} and various values of the exponential factor b and the hindered fraction h. Parameter sets for which the figure of merit is lower than $\overline{\text{FOM}}$ are marked by a grey background in the table.

Considering only the range from 1.61 to 2.30 for the exponential factor, we conclude that the hindered fraction h must be in the range from 0.15–0.45.

Target window for rational crystallisation experiment design

One approach to reduce the risk of a late appearing thermodynamically more stable form is to use the knowledge of the crystal energy landscape as input for rational crystallisation experiment design. In such an exercise, it is often not possible to explicitly consider more than the first few structures at the bottom of the crystal energy landscape. Our experience shows that it is typically possible to suggest crystallisation experiments through chemical modification[21] for most structures in a 0.5 kcal mol^{-1} window, and the question is how likely it is to find a missing, more stable form in that region of the landscape. For parameter set b in Table 2, the probability of finding a missing most stable form in the bottom 0.5 kcal mol^{-1} window is 83%. If the energy accuracy improves to 0.25 kcal mol^{-1} as can be expected for the PBE(0) + MBD + F_{vib} method, this probability goes up to 97%.

Number of molecules in the asymmetric unit

As mentioned in the Introduction, crystal structures with two molecules in the asymmetric unit ($Z' = 2$) remain a challenge for current polymorph generation algorithms. For the 41 compounds in this work, in nine cases the polymorph with the lowest lattice energy crystallised with two molecules in the asymmetric unit, and in one case it crystallised with $Z' = 4$. For the remaining 31 compounds, the polymorph with the lowest lattice energy could be described with a single molecule in the asymmetric unit (the large, asymmetrical molecules that are typical in pharmaceutical development tend not to have internal symmetry and therefore rarely occupy special positions in the solid state, *i.e.* cases where $Z' < 1$ are rare for pharmaceutical compounds). In summary, 25% of typical pharmaceutical compounds require computational polymorph searches with more than one molecule in the asymmetric unit. Statistical studies against the Cambridge Structural Database[22] show that for crystal structures of small molecules in general, only about 10% require $Z' > 1$.[23] A possible explanation is that for enantiomerically pure compounds the presence of two molecules in the asymmetric allows for pseudo-symmetry in the crystal structure *in lieu* of true inversion centres or glide planes, which are known to facilitate close packing of small molecules.

CPU time consumption

Creating the tailor-made force field took between one and five weeks per compound. For each compound, a minimum of one full crystal structure prediction with one molecule in the asymmetric unit was completed, so the timings for $Z' = 1$ can be compared between compounds. For $Z' = 1$, the CPU resources varied between less than a day and more than two months. In total, including the predictions with two molecules in the asymmetric unit, approximately six years of continuous calculations on one 384-core cluster would be required to generate the results presented in this paper.

Future perspectives

Improving the statistics

The conclusions drawn in this work lack precision because of the large statistical errors associated with the small size of the test set. As more and more CSP studies for pharmaceutical molecules are carried out, the statistics will gradually improve. However, many CSP studies and comparisons with experimental data are carried out by scientists in pharmaceutical companies, and it would be desirable to centralise metadata about such studies, including the ranking of the experimental forms, the energy gap between the most stable predicted and experimental form and the number of local lattice energy minima within a 1 kcal mol^{-1} window above the most stable predicted form.

In addition, it is desirable to reassess the energy landscapes of the 41 test compounds at the PBE(0) + MBD + F_{vib} level of theory, and to benchmark both DFT-D and PBE(0) + MBD + F_{vib} against experimental free energy differences between crystal polymorphs for compounds that are similar to the test set in terms of size and flexibility.

Using the information to solve the problem

Knowing whether for a compound under development a more stable polymorph, à la ritonavir, exists, is a crucial first step in the right direction, but it is not a solution. However, the crystal structure prediction studies provide us with two key ingredients for the preparation of thermodynamically stable but kinetically hindered crystal structures: the crystal structure and the lattice energy of the target polymorph. Energy differences and structural differences between the currently observed forms and the kinetically hindered form can give vital clues to predict the experimental conditions under which the new form is favoured over the old forms.

Conclusions

Based on 41 CSP crystal structure prediction studies on pharmaceutical compounds for which an extensive experimental polymorph screen had been conducted, we conclude that the thermodynamically most stable form has not yet been observed experimentally in 15 to 45% of the cases. Considering the fact that late appearing forms and disappearing polymorph events have been observed in the past, it is certainly not surprising that the fraction of missing polymorphs is not zero. However, the estimated fraction of compounds that potentially present the danger of a late appearing, more stable form is probably much larger than most crystallisation scientists would expect. In this context, it is important to note that experimental observation will always substantially underestimate the fraction of missing polymorphs. If a polymorph has not been obtained experimentally during the development phase of a pharmaceutical compound despite intense experimental screening efforts, its nucleation dynamics are surely very unfavourable, and it may take years, centuries or even aeons until the missing polymorph crystallises spontaneously for the first time. Many drugs on the market may simply not have been used and crystallised for long enough to allow for a correct empirical understanding of the scope of the phenomenon.

Conflicts of interest

Both authors work for a company that develops software for crystal structure prediction. The findings of this work are likely to have a positive impact on software and contract research sales.

Notes and references

1 J. Bauer, S. Spanton, R. Henry, J. Quick, W. Dziki, W. Porter and J. Morris, *Pharm. Res.*, 2001, **18**, 859.
2 J. D. Dunitz and J. Bernstein, *Acc. Chem. Res.*, 1995, **28**, 193.
3 D.-K. Bučar, R. W. Lancaster and J. Bernstein, *Angew. Chem., Int. Ed.*, 2015, **54**, 6972.
4 Y. A. Abramov, *Org. Process Res. Dev.*, 2013, **17**, 472.
5 A. M. Reilly, *et al.*, *Acta Crystallogr., Sect. B: Struct. Sci., Cryst. Eng. Mater.*, 2016, **72**, 439.
6 J. G. P. Wicker and R. I. Cooper, *CrystEngComm*, 2015, **17**, 1927.
7 A. Asmadi, M. A. Neumann, J. Kendrick, P. Girard, M.-A. Perrin and F. J. J. Leusen, *J. Phys. Chem. B*, 2009, **113**, 16303.
8 M. A. Neumann, F. J. J. Leusen and J. Kendrick, *Angew. Chem., Int. Ed.*, 2008, **47**, 2427.
9 H. C. S. Chan, J. Kendrick and F. J. J. Leusen, *Phys. Chem. Chem. Phys.*, 2011, **13**, 20361.
10 J. Kendrick, F. J. J. Leusen, M. A. Neumann and J. van de Streek, *Chem.–Eur. J.*, 2011, **17**, 10736.
11 M. A. Neumann, *et al.*, GRACE has been developed by Avant-garde Materials Simulation since 2002, 2002.
12 G. Kresse and J. Furthmüller, *Phys. Rev. B: Condens. Matter Mater. Phys.*, 1996, **54**, 11169.
13 G. Kresse and J. Hafner, *Phys. Rev. B: Condens. Matter Mater. Phys.*, 1993, **47**, 558.
14 G. Kresse and D. Joubert, *Phys. Rev. B: Condens. Matter Mater. Phys.*, 1999, **59**, 1758.
15 J. P. Perdew, K. Burke and M. Ernzerhof, *Phys. Rev. Lett.*, 1997, **78**, 1396.
16 M. A. Neumann and M.-A. Perrin, *J. Phys. Chem. B*, 2005, **109**, 15531.
17 S. Grimme, J. Antony, S. Ehrlich and H. Krieg, *J. Chem. Phys.*, 2010, **132**, 154104.
18 A. M. Reilly and A. Tkatchenko, *Chem. Sci.*, 2015, **6**, 3289.
19 M. A. Neumann, *J. Phys. Chem. B*, 2008, **112**, 9810.
20 J. van de Streek and M. A. Neumann, *CrystEngComm*, 2011, **13**, 7135.
21 J.-B. Arlin, L. S. Price, S. L. Price and A. J. Florence, *Chem. Commun.*, 2011, **47**, 7074.
22 C. R. Groom, I. J. Bruno, M. P. Lightfoot and S. C. Ward, *Acta Crystallogr., Sect. B: Struct. Sci., Cryst. Eng. Mater.*, 2016, **72**, 171.
23 A. D. Bond, *CrystEngComm*, 2008, **10**, 411.

Faraday Discussions

PAPER

Crystal structure prediction is changing from basic science to applied technology

Jonas Nyman *[ab] and Susan M. Reutzel-Edens [b]

Received 16th February 2018, Accepted 23rd March 2018
DOI: 10.1039/c8fd00033f

Over the past three decades, the development of methods for Crystal Structure Prediction (CSP) has primarily been curiosity-driven. Because of the obvious potential for economic gain from CSP, commercial interests can be assumed to eventually take over as the main driving force of development. We argue that this transition is happening right now, not only for commercial CSP providers, but also for consumers within industry. In the context of industry-wide efforts, we describe the exploration in CSP research and algorithm development by one large pharmaceutical company, Eli Lilly and Company, and the impact that this has had on experimental solid form screening and selection. We expect that, once CSP is sufficiently reliable and automated, it will become a standard tool for analytical chemistry, on par with X-ray diffraction, calorimetry and spectroscopy.

1 From academia to industry

1.1 In this article

The purpose of this contribution is primarily to reflect on the great progress that has been achieved in our ability to predict crystal structures of organic molecules. After reminding the reader of the earliest works into crystal structure prediction (CSP), we discuss recent developments and applications in a style rarely possible in a traditional scientific paper, but hopefully suitable as the basis for a Faraday Discussion.

In this article, we make the claim that CSP is currently transitioning from being a subject of predominantly academic research to seeing widespread adoption by the pharmaceutical industry. We will point to several pieces of evidence in support of this claim.

In addition, we describe a new software program, CLilly, developed at Lilly Research Laboratories, that implements a completely automated workflow manager for CSP calculations based on a server-client approach. Finally, with

[a]*School of Pharmacy, University of Wisconsin – Madison, Madison, Wisconsin, USA. E-mail: jnyman2@wisc.edu*

[b]*Small Molecule Design & Development, Eli Lilly and Company, Indianapolis, Indiana, USA. E-mail: reutzel-edens_susan_m@lilly.com*

a look into the future, we discuss some of the consequences that arise from having access to a large and highly automated CSP capability.

1.2 Genesis

In his famous 1900 list of unsolved fundamental problems in mathematics, David Hilbert included as number 18 of 23 the problem of calculating the densest possible packing of spheres and other geometric shapes.[1] Carl Friedrich Gauss had solved the problem in 1831 for equally sized spheres in a lattice,[2] but solutions for more general cases still elude us.

Aleksander Isaakovich Kitaigorodsky attempted to predict molecular crystal structures in the 1960s by non-computational methods.[3,4] He relied on Boris Petrovich Orelkin's observation that molecules tend to pack "*in such a way that the 'projections' formed by the atoms of one molecule fill the 'hollows' between the atoms of adjacent molecules, the molecules dovetailing in a pattern of close packing*". Using such empirical principles, he predicted crystal structures by tangibly packing space-filling models of molecules as densely as possible. Kitaigorodsky also performed force field lattice energy calculations and foresaw the usefulness of computational crystal structure prediction.[4,5]

Donald Williams was another pioneer in the use of computer calculations on crystal structures in the 1960s. He realized that geometry optimisation by lattice-energy minimisation can yield crystal structures that can be used as the starting point for the indexing and refinement of experimental X-ray diffraction results.[6,7] For this purpose, he parametrized a set of force fields for organic crystals.[8–12] These force fields have since matured into the FIT and Williams force fields that have proven successful in modelling organic crystal structures.[13–15] In an almost prophetic paper, Williams discussed the prospects of computational crystal structure prediction already in 1972.[16]

Arieh Warshel and Schneior Lifson discussed,[17] in 1970, several of the particular difficulties that make the computational modelling of molecular crystal structures so challenging, including the anisotropy of intermolecular interactions,[18] the unsatisfactory performance of traditional force fields, vibrational contributions to the crystal energy,[19] and thermal expansion;[15,20] problems which are still of great importance and interest, and the cause of much frustration to the CSP community.

Hypothetical crystal structures were occasionally considered before the advent of CSP. Arnold Hagler and Leslie Leiserowitz performed lattice energy calculations on 16 crystal structures of adipamide in 1978.[21] The structures were carefully constructed to contain plausible hydrogen bonding motifs. These calculations were performed primarily for the purpose of validating force field parameters, but the authors did mention the prospect of CSP proper.

In 1983, Alexander Dzyabchenko performed a computational study of hypothetical crystal structures of benzene. That study, originally published in Russian,[22] is probably the first example of crystal structure prediction as we now use the term. He searched for lattice energy minima among randomly generated benzene crystals in four space groups. A force field by Williams was used to calculate the lattice energies. Remarkably, the global energy minimum corresponded to the known stable crystal structure and the high-pressure form of benzene was also successfully reproduced.

Despite this success, by 1988 it was still generally considered *impossible* to predict crystal structures. John Maddox, the editor of *Nature* at that time, wrote "*One of the continuing scandals in the physical sciences is that it remains in general impossible to predict the structure of even the simplest crystalline solids from a knowledge of their chemical composition.*" He continued "*one would have thought that, by now, it should be possible to equip a sufficiently large computer by a sufficiently large program, type in the formula of the chemical and obtain, as output, the coordinates of the atoms in a unit cell*".[23] In hindsight, Maddox arguably underestimated the difficulties of CSP.

A few years later, in 1991, Angelo Gavezzotti and Giuseppe Filippini performed predictions of polythienyl crystals using their own force field and Williams' program PCK83.[24,25] Through lattice dynamics they also verified that the produced structures were true energy minima and obtained vibrational spectra and entropies.

Monte Carlo simulated annealing was used early for structure predictions, with important contributions by Robert Gdanitz, Heinrich Karfunkel and Frank Leusen. Gdanitz and Karfunkel were at the time working for Ciba-Geigy, a Swiss pharma company, which later merged into Novartis. In 1992 crystal structures of hexamethylbenzene and ethylene were successfully reproduced by this method.[26] The outlook for CSP was promising, as highlighted in one review of the progress that had been made up to 1993.[27]

Building on the early success with the Monte Carlo simulated annealing method, the first true (blind) predictions of previously unknown crystal structures were demonstrated within a year.[28] James Holden and Herman Ammon described general algorithms for CSP of organic molecules of the most common elements, together with the MOLPAK and WMIN programs for trial structure generation and optimization.[29]

This was followed by a general discussion on the difficulties of practical CSP by Gavezzotti in 1994,[30] and an important study together with Giuseppe Filippini on polymorph pairwise differences in which thermal contributions were also considered.[31] At this time Gavezzotti also organised a NATO Advanced Workshop on Supramolecular Chemistry in Sestri Levante, Italy, probably the first concerted effort to convene the CSP community.

Other early works include a prediction of monosaccharides by Bouke van Eijck and of piracetam by Alexander Dzyabchenko, both in 1995.[32,33] Around that time, the ICE9 program by Anna Chaka, Rebecca Zaniewski and Gilles Klopman was able to predict structures of small rigid molecules containing only C, H and O by an alternative generation method focusing on symmetries that generate close-packed structures.[34,35]

With this, we have almost comprehensively reviewed the field of organic CSP up to 1995. By today's standards, crystal structure prediction circa 1996 was still in its infancy,[36] or in the words of Frank Leusen at that time: "still controversial". Despite the limitations of the methods, Leusen optimistically concluded that "near-future developments in computer speed and force field technology will enable the polymorph prediction of any molecular crystal".[37]

It was near this time that Molecular Simulations Ltd. introduced Polymorph Predictor, the first commercial product offering in CSP. Polymorph Predictor was implemented in the CERIUS2 molecular modeling environment and used the Monte Carlo simulated annealing approach to search for low energy structures, clustering to remove duplicate structures and finally lattice energy minimization to afford low energy structures that could represent potential polymorphs. The pharmaceutical industry was well aware of the problems of polymorphism at that time and a tool that required only a chemical diagram as input to predict how a molecule might crystallize would be a true breakthrough for the developers of drug products.[38] However, as discussed below, Polymorph Predictor would never see the uptake by the pharmaceutical industry that Molecular Simulations Ltd. had perhaps hoped for.

Academic groups, and also Marcus Neumann (who on leaving Molecular Simulations, with financial backing from a pharmaceutical company, would later develop and commercialize his own GRACE code for CSP), continued in their relentless pursuit of ever more sophisticated and automatic software to reliably and efficiently predict crystal structures of increasingly complex molecules. The progress made by the CSP community over the nearly two decades that followed has been considerable, as convincingly demonstrated in a series of blind tests organised by the CCDC, the first of which was initialised at a workshop in 1999, while the most recent sixth test was published in 2016.[39,40] With each blind test, the bar was raised as the CSP methods were tested against increasingly complex systems, including larger, more flexible molecules, hydrates, salts and co-crystals. The last few blind tests,[40–42] and other highly successful studies on *e.g.* ROY,[43] AZD8329,[44] and coumarin,[45] have shown that crystal structures of organic molecules can be predicted, although there are still a limited number of researchers who have produced results as successful as these (Fig. 1).

Some of the latest developments have included better benchmarks for calculations of the energy and structure of molecular crystals, such as X23 and POLY59.[46,47] These provide more objective ways to compare the accuracy of different modeling methods. We have also seen a generalization of the concept of a crystal (lattice) energy landscape,[48] see Fig. 2(a), to energy–structure–function maps.[49] These are landscapes that include (at least) one calculated property of the predicted structures.[50,51] In the future, these maps may rank, not static structures, but thermodynamic ensembles by their Gibbs free energy. This is schematically illustrated in Fig. 2(b).

Important questions still remain and are actively being pursued. How can we ensure that the search is complete? It is common for a CSP search not to find known structures; see Fig. 2(a), red circle. How can the reliability of predictions be improved? Currently used scoring functions have standard errors of 5–10 kJ mol^{-1}, several times larger than typical energy differences between polymorphs,[31,52] so the predicted stability ranking, even at 0 K, is quite uncertain.

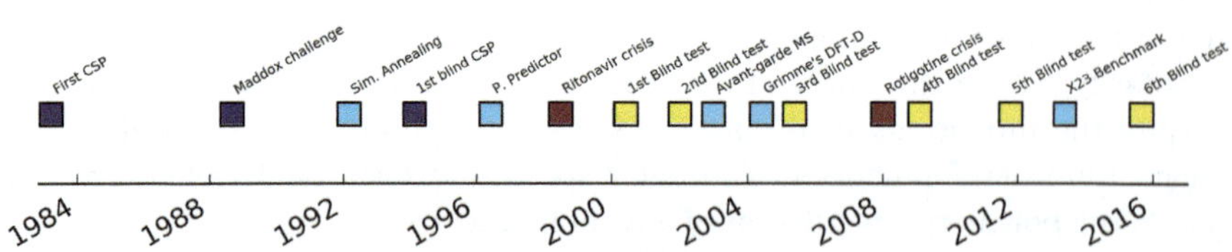

Fig. 1 Timeline of major milestones in the development of CSP methods for organic molecular crystals and key events, which have fueled the interest in and uptake of this emerging technology across the pharmaceutical industry. See the text for references to these events.

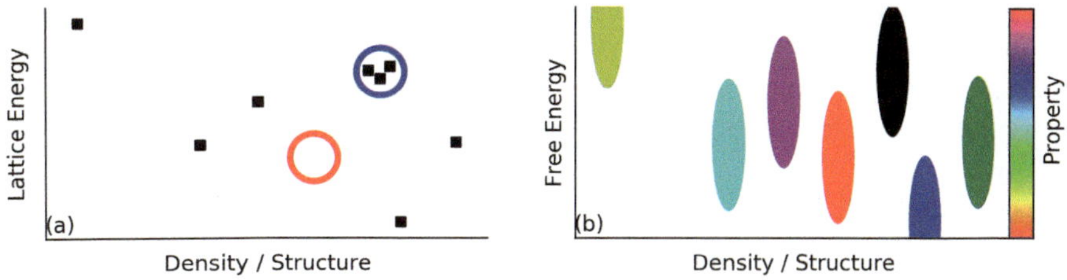

Fig. 2 (a) Crystal lattice energy landscape, with static 0 K crystal structures ranked by energy and separated horizontally by density or some other structure-dependent generalized coordinate. The red circle indicates a missed structure; the blue circle shows three lattice energy minima that suggest a disordered structure or that are likely to correspond to the same free energy minimum. Note the lack of error bars. (b) Energy–structure–function map, where thermodynamic ensembles have replaced the static crystal structures. Errors in calculated free energy and ensemble fluctuations are shown as ellipses. Colors indicate the level of some additional calculated property.

Why does CSP predict many more crystal structures than there are experimentally observed (or possibly, accessible) polymorphs?[53] Using better algorithms to calculate the free energy of the predicted forms has been suggested, but the most common method for free energy calculations, lattice dynamics, does not significantly reduce the number of predicted structures. A commonly occurring observation is that the potential energy surface is 'rugged', *i.e.*, it has several minima corresponding to *almost* identical structures; see Fig. 2(a), blue circle. Often, such similar structures are 'clustered', discarding all but one structure from each group. In reality, these structures probably correspond to the same finite-temperature *ensemble* or Gibbs energy minimum, or indicate the presence of disorder, static or dynamic. It might therefore be preferable to reduce the number of predicted structures to ensembles, either using *e.g.*, molecular dynamics simulations or combining such similar structures to a single disordered one and including the contribution from the configurational entropy to the total free energy.

Kinetic barriers to crystal nucleation and growth, along with templating effects, must also be considered for the role they play in favoring certain crystallization pathways. Here it is expected that the energy barriers and pathways for solid–solid phase transitions will be deciding factors for which higher energy structures, once formed, are likely to be kinetically trapped in the solid state. The latter would require better understanding of a solid state reaction coordinate – the lowest energy path connecting one crystal structure to another. Such modeling has only rarely been attempted.[54] New ideas are needed.

1.3 Ad industriam

1.3.1 Prospects. Drug crystal forms are of high interest to the pharmaceutical industry for the exquisite control they confer over physicochemical properties affecting drug product quality, safety and performance, as well as for their intellectual property value. As candidates progress into drug development, the selection of a suitable solid form is, not surprisingly, one of the first steps taken to transform a molecule into a medicine. Scientists and engineers responsible for solid form selection in the pharmaceutical industry would generally agree, as famously stated by Walter McCrone, that "*every compound has different polymorphic forms and that, in general, the number of forms known is proportional to the time and money that has been spent in research on that compound*".[55] Less clear is to what extent a molecule should be screened in order to find the important crystal forms.

Industry, with heightened awareness of the prevalence of polymorphism,[56] clear regulatory expectations that the solid form properties of drugs are well understood and controlled, and an unwavering desire to minimize the risk of a sudden late appearing crystal form that could threaten the market supply of a drug product (*cf.* ritonavir,[57] rotigotine[58,59]), has invested heavily in largely empirical approaches, such as high throughput screening automation, to find crystal forms. The idea was that if sufficiently diverse conditions could be surveyed in massively parallel crystallization experiments at small scale, important crystal forms would surely fall out of screens in the most rapid and material sparing of ways. There has proved, however, to be no 'sure-fire' approach to finding crystal forms. The dependence on manual screening methods has not gone away and with no obvious endpoint, solid form screens almost always extend well after the last form is found. Especially for those molecules which crystallize readily in a few forms, valuable resources are wasted simply to gain confidence that the important solid forms have not been missed.

Understandably, CSP, in reliably predicting different crystal packing arrangements and ranking them in order of their relative energies, could have tremendous implications for drug development. A crystal energy landscape would be invaluable for assessing the completeness of an experimental search, at least for the stable form, allowing solid form screens to be right-sized. Not only would it be possible to deduce the propensity for polymorphism, but solid-state properties relevant to drug product performance, such as stability, solubility, morphology and hygroscopicity, could also be modeled from CSP-derived structures before materials are even synthesized. This would allow the physical property variation across different polymorphic forms to be predicted and experimental work to be directed to only those crystal forms with desirable properties.

1.3.2 Progress. Despite the enormous upside of CSP, not to mention Molecular Simulations' strategy in the mid-1990's to market Polymorph Predictor to the pharmaceutical industry specifically, industry was slow to embrace this emerging technology. There are possibly different reasons for the slow uptake. At the time of its first release, Polymorph Predictor, a single module in a suite of CERIUS2 modeling tools, was expensive and very much unproven. Scientists at Lilly, for one, were unable to build a business case to invest in it. For others, perhaps Polymorph Predictor was tested and did not perform as advertised for their molecules. Today we know that common force fields are inadequate for CSP purposes, and if the blind test benchmarks were any indication, CSP methodologies were probably not ready to tackle large, flexible 'real-world' molecules at that time. In fact, it was not until the 5th blind test in 2010 that a molecule (target XX) approaching the size and flexibility of modern day pharmaceuticals was included and even then, only 2 of 10 groups successfully predicted the known crystal structure.[42]

Polymorph Predictor may have heightened the awareness of emerging CSP technology across the industry, but it was the CCDC-sponsored blind tests showing the progress the developers of the codes were making that we believe truly caught its attention. Today, most large pharmaceutical companies have some level of investment in CSP, however, the paths taken differ from company to company, depending at least in part on the experience, expertise and role of the sponsor(s) within each organization. Some companies continue to use Polymorph Predictor in combination with other tools for in-house CSP studies. Some have acquired the GRACE software for in-house CSP, while others have accessed CSP in collaboration with academia or through contract research (with Avant-Garde Materials Simulation and more recently, XtalPi).

Lilly was neither the first pharmaceutical company nor the last to test CSP on their molecules. Its entry into the field was also not championed by a computational chemist. Instead, a group of experimentalists, having spent decades searching for drug crystal forms, sometimes struggling to find even one, and interested in reducing the experimental footprint of solid form design, partnered with academia (Sally Price and her team at University College London). The purpose of the collaboration was to learn more about CSP, what it could offer for Lilly 'LY' molecules when used in tandem with experimental solid form screening, and how it might someday be strategically incorporated into the solid form design process at Lilly. The candidates selected for CSP were toward the smaller side of those in development and had been comprehensively screened for forms. Whenever possible, the results of these initial investigations were published for the benefit of all.

1.3.3 Perspectives. We believe the Lilly experience with a handful of chemical systems is limited compared to that of the earliest adopters of CSP in industry. This hypothesis, of course, cannot be confirmed, since the majority of CSP calculations on proprietary compounds are not likely to ever be publicly disclosed. Nonetheless, we have learned a great deal working with LY molecules about what the state-of-the-art in CSP is, the gaps, the opportunities, as well as how to interpret crystal energy landscapes in terms of diverse crystallization behaviors. For some of the compound-specific learnings, the reader is directed to our published work and others'.[60] Here, we provide an industry perspective on CSP, and highlight what we believe are important considerations and areas of need for CSP modeling to be successfully incorporated into the workflows of modern drug development.

Let us first consider the output of a CSP. Current methods are focused on generating a near complete set of low-energy crystal structures, each one a distinct energy minimum. In the Lilly experience, where CSP candidates had been comprehensively screened for forms and the crystallization behaviors were seemingly understood, the appearance of so many unobserved, yet plausible crystal packings on the crystal energy landscapes, some of which were calculated to be lower in energy than the known forms, was initially overwhelming. Yes, these structures were 0 K lattice energy minima and might not survive at relevant temperatures. But is it reasonable to expect that *most* of the structures could not persist at *e.g.*, room temperature? Perhaps not. Why then were these structures not found as polymorphs in the screens? What experiments had not been done that should have? Is it possible that some structures may, for whatever reason, never be crystallized? If so, which ones and why? CSP had raised many more questions than certainly we had answers for.

The vast number of static lattice energy minima in any given crystal energy landscape, see Fig. 2(a), has made it obvious that there is much work to be done before CSP can be fully exploited for solid form selection in industry. Except when the known structure is, based on calculated lattice energies, that much more stable than any other structure, questions will remain about whether the form with the lowest *free energy* at relevant processing and storage temperatures has been found. Even then, only part of an actual solid form landscape is likely to have been explicitly included in the search to manage the computational cost of the CSP. It is not uncommon to find important polymorphs with higher energy 'strained' conformations, non-integer or higher integer Z' packings, or infrequent space groups. Only with advances in the algorithms and access to ever more powerful computer clusters will these acknowledged sources of *under*prediction be dealt with. Improvements in the methods used to cluster similar structures and to calculate reliable free energies will also be needed to move beyond the classic lattice energy landscape to a more helpful free energy landscape. Currently used methods are just not accurate enough to reproduce free energy curves of organic crystal structures,[15,61] except perhaps in the simplest of cases.[62,63]

While important structures could be missed based on how the search space is defined, it is generally understood that CSP, in producing many energetically-feasible crystal structures as lattice energy minima, overpredicts polymorphism.[53] Reducing the landscape to free energy minima will help to direct experimental efforts to structures that are thermodynamically feasible. However, as discussed above, the kinetic barriers to crystal nucleation, growth, and solid-state phase transitions, are not sufficiently understood to identify which structures should be kinetically accessible and under what experimental conditions to obtain them.

Until high fidelity free energy landscapes can be calculated and crystallization kinetics are sufficiently well understood to be able to predict which thermodynamically feasible structures are kinetically accessible, scientists and engineers responsible for solid form selection will be challenged to identify ways to target predicted low energy structures in crystallization or to convince themselves that certain structures are not likely to crystallize readily or at all. Among various strategies taken to target crystal forms, templating has been successfully demonstrated for carbamazepine and cyheptamide, with predicted structures realized by growing them on the surfaces of structurally-similar substrates.[64,65] A new, high density dalcetrapib polymorph was crystallized for the first time at high pressure, an experiment that was similarly inspired by a CSP prediction.[66] Success stories like these are unfortunately more the exception than the rule, however, as other structurally-related substrates failed to produce new carbamazepine polymorphs and high pressure crystallization could conceivably have produced any one of several energetically-competitive high density polymorphs from the CSP.

2 Highly automated CSP calculations

2.1 The need for speed

On the typical timescale of months that it takes to complete a solid form screen experimentally, CSP can be executed concurrently, with the results of the *in silico* and *in vitro* approaches compared at or near the end to assess the completeness of the experimental search. If completed before the experimental screen, the CSP can be used to justify terminating the search once important low energy structures have been experimentally realized. However, for CSP to be used to actually *design* a solid form, the algorithms must be fast and reliable enough to intercept the decision on whether to progress a salt, a co-crystal or the parent molecule itself.

Ideally, CSP would be so fast and reliable that multiple compounds within a platform could be screened *in silico* all at once, allowing drug candidates to be selected based on their solid form properties, in addition to their biological properties. Synthetic routes could similarly be designed around key intermediates having process-relevant physical properties best suited for isolation and purification purposes. One study along these lines was recently reported by Graeme Day and Josh Campbell,[50] who performed CSP on a set of nitrogen-substituted pentacene (azapentacene) molecules, then charge transfer calculations on the resulting crystal structures. This allowed predictions of which of the azapentacenes was the most likely on crystallization to yield organic semiconductors with good properties.

We believe such extensive CSP screenings will become feasible, if not commonplace, once sufficiently automatic and efficient programs exist to deliver CSP results within days to weeks with a modest incremental cost of additional calculations. The interest in performing CSP calculations on multiple molecules simultaneously to examine how covalent chemical modifications of a candidate molecule might improve crystallizability or solid form properties is shared by many, including AstraZeneca, who have developed and are already using fast and automatic CSP calculations.[67] Sten Nilson-Lill, Anders Broo and others at AstraZeneca use GRACE for trial structure generation, but otherwise use methods developed in-house, including OPLS2005 or their own transferable force field in place of the molecule-specific force fields that are normally generated by GRACE.[68] This 'fit-for-purpose' approach, though less reliable, saves time by omitting the force field parametrisation and may allow CSP to be conducted in a useful timeframe for molecule optimization.[69]

2.2 CLilly

Having observed firsthand that CSP works, and fairly well at that, for large, flexible pharmaceutical molecules and also shown how useful the calculations can be when combined with experimental solid form screening,[70] Lilly has since brought CSP in-house. Our interest at present is to enable *solid form* selection through high fidelity CSP in as efficient a manner as possible. Today, we use GRACE, including its Force Field Factory module to generate tailor-made force fields and CSP Factory. We have also implemented a new large-scale workflow manager CLilly for CSP calculations at Lilly Research Laboratories that we hope will facilitate fast, yet comprehensive, CSP calculations.

As a CSP platform, CLilly benefits from the combined experience of the groups led by Sally Price, Graeme Day, Constantinos Pantelides and Claire Adjiman,[71–74] and implements 'best practices' developed in these groups over many years. As a workflow manager, CLilly consists of a server, see Fig. 3, a client program, a program for generating molecular conformers and one for generating trial crystal structures. These programs are written in Python 2.7 and rely only on freely available libraries. Together with various (non-free) subprograms and auxiliary Python code, CLilly automatically and efficiently performs all the steps needed in order to predict crystal structures of organic molecules.

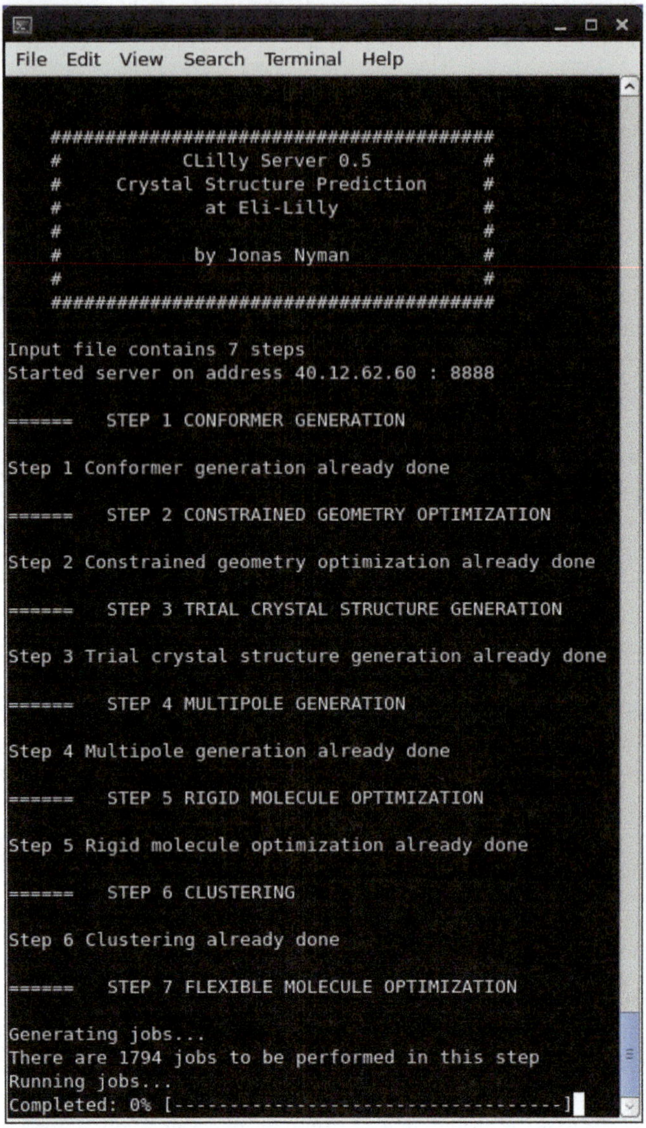

Fig. 3 The CLilly server running a CSP. The program has just been restarted and has therefore skipped the first 6 steps which have been completed previously. The server is currently waiting for clients to connect in order to perform CrystalOptimizer calculations on 1794 crystal structures. The figure has been manipulated to hide the name of the compound.

Considerable effort has been invested in making CLilly able to use the latest version of several subprograms. At the time of writing this means Gaussian 16,[75] GDMA 2.2.11 (ref. 76), DMACRYS 2.2.1.1,[73] OpenBabel 2.4.1,[77] RDKit 2016-03-3,[78] CrystalOptimizer 2.4.8.[79] and VASP 5.4.4. Some of these programs offer substantial improvements over previously used versions.

The CLilly server program, given a minimum of input from the user, proceeds through a number of consecutive steps. In each step, it reads the output from previous steps and generates a set of 'jobs' to be executed by the client program.

The client program connects to the server *via* standard TCP/IP sockets and downloads a job from the server. Communication of such jobs is performed by means of text messages and file transfers. The client then performs the calculations as instructed, normally by generating input files for, and calling, an external program. Once finished, the client collects the results and sends them back to the server, which saves them to disk and keeps track of which calculations have been completed.

CLilly has been developed to be used on Lilly's in-house Beowulf cluster, consisting of 238 compute nodes, of which 78 have Infiniband and Nvidia Tesla K80 GPUs. About 2000 CPU-cores are nearly continuously available for CSP purposes, running both GRACE and CLilly simultaneously.

The most important feature of the CLilly workflow manager is that it scales extremely well with the size of the CSP problem. Thanks to the server-client architecture, CLilly can efficiently use hundreds of compute nodes running the client program simultaneously on thousands of CPUs. This facilitates extensive calculations on large molecules or multicomponent systems. Tests conducted so far show that sampling 6.3 million trial structures is quite feasible, about an order of magnitude more than what has been used previously with similar methods.[40,80]

CLilly can currently perform any of the following tasks:

- Conformer generation with OpenBabel and RDKit.
- Molecular geometry optimization with Gaussian.
- Trial crystal structure generation with GLEE.
- Generation of distributed multipoles with GDMA.
- Crystal optimization with DMACRYS.
- Crystal optimization with CrystalOptimizer.
- Crystal optimization with VASP.
- Clustering of structures with COMPACK.[81]
- Rigid body lattice dynamics with DMACRYS and the auxiliary programs AutoLD and AutoFree.[82]

Each step is computationally highly efficient. An absolute minimum number of calculations is performed and no calculation is repeated. Furthermore, the entire workflow is restartable at any point, as shown in Fig. 3.

2.2.1 Conformer generation. Molecular flexibility is currently dealt with by using a number of discrete conformers. For each chemical species, a set of conformers is generated. The CONFAB algorithm,[77] as well as the distance geometry methods DG, KDG, ETDG and ETKDG, provided through the RDKit code,[83] are currently implemented in CLilly.

Although more sophisticated algorithms have been developed for the purpose of flexible molecule CSP,[84,85] we believe that sampling a finite number of conformers will normally be satisfactory,[69] since the difference in energy between the conformations *in crystallum* and *in vacuum* is small in most cases. Introducing a bias in favour of extended conformers, as suggested by Graeme Day and Hugh Thompson, is in progress.[86]

The generated conformers are optimized to a suitable DFT-D level of theory. The program Gaussian 16 (ref. 75) is currently used for this. This provides more accurate intramolecular energies, geometries and charge densities suitable for distributed multipole analysis.[87] It should be noted that CLilly never relies on molecular energies calculated with a force field, as these are deemed too unreliable.

2.2.2 Trial structure generation. Trial structures are quasi-randomly generated with a modified version of the GLEE code, developed by Peter Bygrave, David Case, Josh Campbell and Graeme Day.[80] Crystal structures are generated from rigid conformers, using algorithms similar to those described by Panagiotis Karamertzanis and Constantinos Pantelides.[88] The separating axis theorem is used to relieve clashes between molecules. This generation algorithm has been shown to be efficient for CSP purposes.[80,88]

2.2.3 Anisotropic intermolecular interactions. Traditional atom–atom force fields do not properly account for the anisotropy of intermolecular interactions. DMACRYS is one of very few programs that enable the use of anisotropic force fields. Electrostatic interactions are calculated from distributed multipoles, obtained from each conformer's electron density with Anthony Stone's program GDMA 2.2.11.[76,87] At this stage, CLilly also generates an intramolecular coordinate system for each molecule, which in addition to being used for the multipoles, also enables modeling of the anisotropic shape of halogen atoms.[89,90]

2.2.4 Crystal optimization with DMACRYS. The trial structures are next geometry-optimized in a rigid-molecule approach using DMACRYS. DMACRYS allows the use of the accurate FIT and Williams force fields, and distributed multipoles for calculating intermolecular interactions.[13,14,73,91,92] After this optimization, duplicate and nearly identical structures are typically removed. Currently the COMPACK program[81] is used for this purpose, but we are looking for alternative methods (see the discussion about clustering above).

2.2.5 Crystal optimization with CrystalOptimizer. Following rigid molecule optimization, CLilly uses the CrystalOptimizer program, developed by Andrei Kazantsev and Constantinos Pantelides,[93,94] to allow the molecules in the crystal structures to relax in response to lattice packing forces.[74,79] A suitable set of intramolecular degrees of freedom is chosen automatically using the AutoDOF program.[15] Extensive use is made of local approximate models (LAMs) that allow the program to interpolate between molecular conformations whose energy and electrostatic potential have already been calculated with DFT-D.[93] CLilly can synchronize these LAMs between compute nodes, so that a minimum number of relatively expensive DFT calculations is needed. This allows the optimization of thousands of crystal structures of relatively large molecules to near DFT-D accuracy at a greatly reduced computational cost.

2.2.6 Crystal optimization with periodic DFT-D. Currently, CLilly ranks the predicted crystal structures by their 0 K electronic energy, calculated with the PBE functional augmented with the D3(BJ) or MBD dispersion corrections,[95–98] as implemented in VASP 5.4.4. These have been demonstrated to be reliable scoring functions for molecular crystals.[40,99,100] There is however evidence that even this level of theory is not accurate enough to quantitatively predict the relative stability and properties of crystal structures.[100] We are preparing to test the TPSS or SCAN *meta*-GGA functionals, which have shown great promise in recent benchmark studies.[100–102]

CLilly will normally instruct VASP to use the Nvidia Tesla K80 GPUs that are available on the Lilly cluster. The GPUs not only greatly speed up periodic DFT-D calculations,[103] but also facilitate running each DFT-D calculation on a single compute node.

We are also implementing a quasi-harmonic lattice dynamical calculation method, so as to include the effects of thermal expansion in our predictions.[14,20] We are also investigating how to efficiently model configurational disorder, so that we in the future may use a more realistic estimate of the Gibbs free energy as the final scoring function.

2.2.7 **An example.** As a proof of concept that demonstrates the features of CLilly, we present here a limited CSP calculation that was performed on the compound 4-acetoamido-3-(1-acetyl-2-(2,6-dichlorobenzylidene)hydrazine)-1,2,4-triazole. This compound is probably better known to the reader as the (alphabetically) first entry in the CSD: AABHTZ. It is a typical flexible small organic molecule, chosen to be representative of the kind of molecules that CLilly has been designed for.

The ETKDG algorithm produced 81 conformers. The 24 conformers with B3LYP-D3/6-31G* energies within 50 kJ mol^{-1} of the lowest-energy conformer were used in the CSP. The same DFT-D level of theory was used throughout the CSP.

Trial structures with $Z' = 1$ were generated in the space groups $P1$, $P\bar{1}$ and $P2_1/c$. About 5000 trial structures were generated for each conformer in each space group, producing at least 630 000 trial structures. The known crystal structure belongs to $P\bar{1}$. For a comprehensive screening a much wider selection of space groups would be used.

The trial structures were first optimized with DMACRYS, applying a pressure of 100 MPa to help the optimizations converge to densely packed structures. An in-house lattice reduction algorithm was applied to structures with unusually acute or obtuse lattice angles to yield more reduced unit cells. After clustering with COMPACK, 736 structures within a 50 kJ mol^{-1} window were optimized with CrystalOptimizer. Intramolecular LAM databases were generated and synchronized between the clients and server until each client had more than 200 database entries.

For this CSP we did not perform periodic DFT-D calculations, as the purpose here is only to present features of CLilly. The crystal structures obtained from CrystalOptimizer are generally already accurate enough to unambiguously identify polymorphs and for solving powder X-ray patterns. They are also excellent starting points for lattice dynamics and molecular dynamics, or further optimization with periodic DFT-D and NMR shielding calculations *etc.*

This mini-CSP took 10 days in total. Conformer generation and optimization, trial structure generation and electrostatic multipole calculations were all completed in the first day. Rigid-molecule optimizations took 3 days while flexible-molecule optimizations required 6 days.

This CSP unfortunately proved unsuccessful; we did not reproduce the experimentally known structure. The most likely reason for this is that the molecular conformer corresponding to the conformation in the crystal structure is 53 kJ mol^{-1} higher in energy than the lowest-energy gas-phase conformer. This conformer hence fell outside our 50 kJ mol^{-1} selection criterion. This unusually large energy difference is due to the molecule's ability to form an intramolecular hydrogen bond in the vacuum, which is not present in the crystal structure. We are working on developing methods for sampling molecular conformations that are more suitable for CSP purposes.

2.3 Where do we go from here?

GRACE, CrystalPredictor, CLilly, AstraZeneca's Solid State Prediction Tool SSPT,[67] and Graeme Day's CSPy are all highly automated workflow managers for CSP. We are interested in what happens once CSP becomes automated, that is, once these software packages or others like them are deployed. When CSP becomes affordable, possibly no longer requiring a specialist to run the programs, how will that change how we work, how we think about solid forms and solid form selection?

We envision performing CSP calculations on molecules, much in the same way as we would characterize materials by standard analytical methods, such as DSC and PXRD. In this regard, the computer becomes a laboratory instrument, one of many in our arsenal, and CSP is the experiment.

With a capability like CSP in hand, new questions arise. How can we modify a molecule in order to improve the crystal landscape? Can we perform CSP, not just on lead candidates, but on a range of similar molecules in parallel and extract information from the predicted crystal landscapes to influence candidate selection? If we perform CSP on the precursors, can we find a better synthetic route? The prediction of solvents and possible cocrystals is only a matter of (huge) computing resources, and we need to prioritize. Which CSP calculations are useful and economical to run?

Finally, we would be remiss not to consider the implications of CSP on the future value of drug crystal forms as intellectual property. Crystal structures are, for all intents and purposes, predictable. However, *there is a huge difference between predicting a hypothetical crystal structure and experimentally realizing a polymorph*; CSP does not give the recipe to produce a structure that is calculated to be thermodynamically plausible and mechanically stable. Lacking efficient and accurate methods to predict nucleation, growth, and barriers to solid-state phase transitions, we are far from being able to predict which structures could be observed as experimental polymorphs, let alone properties (*e.g.*, morphology,[104] tablettability,[105] powder flow) relevant to drug product processing or performance, or the methods to produce them.

3 Conclusions

CSP is currently transitioning from exclusively an area of basic research into applied technology and this is already having an effect on the pharmaceutical industry. The value of being able to predict how a molecule will crystallize to the developers of medicines is undeniable. While the legal community continues to grapple with what CSP could mean to the future value of crystal form IP, we have pointed out that predicting crystal structures falls far short of experimentally realizing polymorphs. We believe that it is only a matter of time before regulatory bodies, too, will be confronted with the reality that CSP is here and it is here to stay.

Despite the great progress in developing CSP algorithms, there is still a need to improve their overall reliability and accuracy. As the latest blind test revealed, there is no correlation between computational cost and success rate, which implies that currently used algorithms are far from optimal. Perhaps the largest unresolved problem is the large discrepancy between the number of predicted structures and experimentally realizable forms. This is due in part to the neglect of thermal motion, as some of the predicted static 0 K structures are likely to merge into the same thermodynamic ensemble at room temperature. There are also several poorly understood kinetic factors – nucleation, growth, and transitions across energy barriers, to name a few – that we are as yet unable to model adequately. Until these issues can be resolved, the problem of *polymorph prediction* from molecular structure to material in hand will remain far from solved.

Conflicts of interest

There are no conflicts of interest to declare.

References

1 D. Hilbert, *Bull. Am. Math. Soc.*, 1902, **8**, 437–479.
2 C. F. Gauss, *Untersuchungen über die Eigenschaften der positiven Ternaeren quadratischen Formen von Ludwig August Seber*, Walter de Gruyter GmbH, 1840.
3 A. I. Kitaigorodskii, *Acta Crystallogr.*, 1965, **18**, 585–590.
4 A. I. Kitaigorodsky, *Molecular Crystals and Molecules*, Academic Press, Inc, 1973.
5 A. I. Kitaigorodsky and K. V. Mirskaya, *Mater. Res. Bull.*, 1972, **7**, 1271–1280.
6 D. E. Williams, *Science*, 1965, **147**, 605–606.
7 D. E. Williams, *Acta Crystallogr., Sect. A: Found. Crystallogr.*, 1969, **25**, 464–470.
8 D. E. Williams, *J. Chem. Phys.*, 1966, **45**, 3770–3778.
9 D. E. Williams, *Acta Crystallogr., Sect. A: Found. Crystallogr.*, 1972, **28**, 84–88.
10 D. E. Williams and T. L. Starr, *Comput. Chem.*, 1977, **1**, 173–177.
11 D. E. Williams and S. R. Cox, *Acta Crystallogr., Sect. B: Struct. Sci.*, 1984, **40**, 404–417.
12 D. E. Williams, *J. Comput. Chem.*, 2001, **22**, 1154–1166.
13 D. S. Coombes, S. L. Price, D. J. Willock and M. Leslie, *J. Phys. Chem.*, 1996, **100**, 7352–7360.
14 J. Nyman, O. Sheehan Pundyke and G. M. Day, *Phys. Chem. Chem. Phys.*, 2016, **18**, 15828–15837.
15 J. Nyman and G. M. Day, *Phys. Chem. Chem. Phys.*, 2016, **18**, 31132–31143.
16 D. E. Williams, *Acta Crystallogr., Sect. A: Found. Crystallogr.*, 1972, **28**, 629–635.
17 A. Warshel and S. Lifson, *J. Chem. Phys.*, 1970, **53**, 582–594.
18 A. J. Stone and S. L. Price, *J. Phys. Chem.*, 1988, **92**, 3325–3335.
19 B. P. van Eijck, *J. Comput. Chem.*, 2001, **22**, 816–826.
20 Y. N. Heit and G. J. O. Beran, *Acta Crystallogr., Sect. B: Struct. Sci., Cryst. Eng. Mater.*, 2016, **72**, 514–529.
21 A. Hagler and L. Leiserowitz, *J. Am. Chem. Soc.*, 1978, **100**, 5879–5887.
22 A. V. Dzyabchenko, *J. Struct. Chem.*, 1984, **25**, 416–420.
23 J. Maddox, *Nature*, 1988, **335**, 201.
24 A. Gavezzotti and G. Filippini, *Synth. Met.*, 1991, **40**, 257–266.
25 A. Gavezzotti, in *Electron crystallography of organic molecules*, Springer, 1991, pp. 77–83.
26 R. J. Gdanitz, *Chem. Phys. Lett.*, 1992, **190**, 391–396.
27 C. R. A. Catlow, J. M. Thomas, C. M. Freeman, P. A. Wright and R. G. Bell, *Proc. R. Soc. London*, 1993, **442**, 85–96.

28 H. R. Karfunkel, F. J. J. Leusen and R. J. Gdanitz, *J. Comput.-Aided Mater. Des.*, 1994, **1**, 177–185.

29 J. R. Holden, Z. Du and H. L. Ammon, *J. Comput. Chem.*, 1993, **14**, 422–437.

30 A. Gavezzotti, *Acc. Chem. Res.*, 1994, **27**, 309–314.

31 A. Gavezzotti and G. Filippini, *J. Am. Chem. Soc.*, 1995, **117**, 12299–12305.

32 B. P. van Eijck, W. T. M. Mooij and J. Kroon, *Acta Crystallogr., Sect. B: Struct. Sci.*, 1995, **51**, 99–103.

33 A. V. Dzyabchenko and V. Agafonov, *Proceedings of the Twenty-Eighth Annual Hawaii International Conference on System Sciences*, 1995, vol. 5, pp. 237–245.

34 A. M. Chaka, Ph.D. thesis, Case Western Reserve University, 1993.

35 R. C. Zaniewski, Ph.D. thesis, Case Western Reserve University, 1995.

36 D. E. Williams, *Acta Crystallogr., Sect. A: Found. Crystallogr.*, 1996, **52**, 326–328.

37 F. J. Leusen, *J. Cryst. Growth*, 1996, **166**, 900–903.

38 S. Byrn, R. Pfeiffer, M. Ganey, C. Hoiberg and G. Poochikian, *Pharm. Res.*, 1995, **12**, 945–954.

39 J. P. M. Lommerse, W. D. S. Motherwell, H. L. Ammon, J. D. Dunitz, A. Gavezzotti, D. W. M. Hofmann, F. J. J. Leusen, W. T. M. Mooij, S. L. Price, B. Schweizer, M. U. Schmidt, B. P. van Eijck, *et al.*, *Acta Crystallogr., Sect. B: Struct. Sci.*, 2000, **56**, 697–714.

40 A. M. Reilly, R. I. Cooper, *et al.*, *Acta Crystallogr., Sect. B: Struct. Sci., Cryst. Eng. Mater.*, 2016, **72**, 439–459.

41 G. M. Day, T. G. Cooper, A. J. Cruz-Cabeza, K. E. Hejczyk, H. L. Ammon, S. X. M. Boerrigter, J. S. Tan, R. G. Della Valle, *et al.*, *Acta Crystallogr., Sect. B: Struct. Sci.*, 2009, **65**, 107–125.

42 D. A. Bardwell, C. S. Adjiman, Y. A. Arnautova and E. t. Bartashevich, *Acta Crystallogr., Sect. B: Struct. Sci.*, 2011, **67**, 535–551.

43 M. Vasileiadis, A. V. Kazantsev, P. G. Karamertzanis, C. S. Adjiman and C. C. Pantelides, *Acta Crystallogr., Sect. B: Struct. Sci.*, 2012, **68**, 677–685.

44 M. Baias, J.-N. Dumez, P. H. Svensson, S. Schantz, G. M. Day and L. Emsley, *J. Am. Chem. Soc.*, 2013, **135**, 17501–17507.

45 A. G. Shtukenberg, Q. Zhu, D. J. Carter, L. Vogt, J. Hoja, E. Schneider, H. Song, B. Pokroy, I. Polishchuk, A. Tkatchenko, A. R. Oganov, A. L. Rohl, M. E. Tuckerman and B. Kahr, *Chem. Sci.*, 2017, **8**, 4926–4940.

46 A. M. Reilly and A. Tkatchenko, *J. Chem. Phys.*, 2013, **139**, 024705.

47 J. G. Brandenburg and S. Grimme, *Acta Crystallogr., Sect. B: Struct. Sci., Cryst. Eng. Mater.*, 2016, **72**, 502–513.

48 S. L. Price, *Acc. Chem. Res.*, 2009, **42**, 117–126.

49 G. M. Day and A. I. Cooper, *Adv. Mater.*, 2017, 1704944.

50 J. E. Campbell, J. Yang and G. M. Day, *J. Mater. Chem. C*, 2017, **5**, 7574–7584.

51 A. Pulido, L. Chen, T. Kaczorowski, D. Holden, M. A. Little, S. Y. Chong, B. J. Slater, D. P. McMahon, B. Bonillo, C. J. Stackhouse, *et al.*, *Nature*, 2017, **543**, 657.

52 J. Nyman and G. M. Day, *CrystEngComm*, 2015, **17**, 5154–5165.

53 S. L. Price, *Acta Crystallogr., Sect. B: Struct. Sci., Cryst. Eng. Mater.*, 2013, **69**, 313–328.

54 J. A. van den Ende, B. Ensing and H. M. Cuppen, *CrystEngComm*, 2016, **18**, 4420–4430.

55 J. Haleblian and W. McCrone, *J. Pharm. Sci.*, 1969, **58**, 911–929.

56 A. J. Cruz-Cabeza, S. M. Reutzel-Edens and J. Bernstein, *Chem. Soc. Rev.*, 2015, **44**, 8619–8635.
57 S. R. Chemburkar, J. Bauer, K. Deming, H. Spiwek, K. Patel, J. Morris, R. Henry, S. Spanton, *et al.*, *Org. Process Res. Dev.*, 2000, **4**, 413–417.
58 K. R. Chaudhuri, *Expert Opin. Drug Delivery*, 2008, **5**, 1169–1171.
59 I. B. Rietveld and R. Céolin, *J. Pharm. Sci.*, 2015, **104**, 4117–4122.
60 S. L. Price, D. E. Braun and S. M. Reutzel-Edens, *Chem. Commun.*, 2016, **52**, 7065–7077.
61 A. Grunenberg, J.-O. Henck and H. W. Siesler, *Int. J. Pharm.*, 1996, **129**, 147–158.
62 G. J. O. Beran, *Angew. Chem., Int. Ed.*, 2015, **54**, 396–398.
63 Y. N. Heit, K. D. Nanda and G. J. O. Beran, *Chem. Sci.*, 2016, **7**, 246–255.
64 J.-B. Arlin, L. S. Price, S. L. Price and A. J. Florence, *Chem. Commun.*, 2011, **47**, 7074–7076.
65 V. K. Srirambhatla, R. Guo, S. L. Price and A. J. Florence, *Chem. Commun.*, 2016, **52**, 7384–7386.
66 M. A. Neumann, J. van de Streek, F. P. A. Fabbiani, P. Hidber and O. Grassmann, *Nat. Commun.*, 2015, **6**, 7793.
67 S. O. Nilson-Lill, S. Schantz, V. Broo and A. Broo, Current computation approaches at AstraZeneca for solid state and property predictions, in *Computational Pharmaceutical Solid State Chemistry*, ed. Y. A. Abramov, Wiley, 2017.
68 M. A. Neumann, *J. Phys. Chem. B*, 2008, **112**, 9810–9829.
69 A. Broo and S. O. Nilsson Lill, *Acta Crystallogr., Sect. B: Struct. Sci., Cryst. Eng. Mater.*, 2016, **72**, 460–476.
70 S. L. Price and S. M. Reutzel-Edens, *Drug Discovery Today*, 2016, **21**, 912–923.
71 C. C. Pantelides, C. S. Adjiman and A. V. Kazantsev, in *Prediction and Calculation of Crystal Structures*, Springer, 2014, pp. 25–58.
72 G. M. Day, *Crystallogr. Rev.*, 2011, **17**, 3–52.
73 S. L. Price, M. Leslie, G. W. A. Welch, M. Habgood, L. S. Price, P. G. Karamertzanis and G. M. Day, *Phys. Chem. Chem. Phys.*, 2010, **12**, 8478–8490.
74 A. V. Kazantsev, P. G. Karamertzanis, C. C. Pantelides and C. S. Adjiman, in *20th European Symposium on Computer Aided Process Engineering*, ed. S. Pierucci and G. B. Ferraris, Elsevier, 2010, vol. 28, pp. 817–822.
75 M. J. Frisch, G. W. Trucks, H. B. Schlegel, G. E. Scuseria, M. A. Robb, J. R. Cheeseman, G. Scalmani, V. Barone, G. A. Petersson, H. Nakatsuji, X. Li, M. Caricato, A. V. Marenich, J. Bloino, B. G. Janesko, R. Gomperts, B. Mennucci, H. P. Hratchian, J. V. Ortiz, A. F. Izmaylov, J. L. Sonnenberg, D. Williams-Young, F. Ding, F. Lipparini, F. Egidi, J. Goings, B. Peng, A. Petrone, T. Henderson, D. Ranasinghe, V. G. Zakrzewski, J. Gao, N. Rega, G. Zheng, W. Liang, M. Hada, M. Ehara, K. Toyota, R. Fukuda, J. Hasegawa, M. Ishida, T. Nakajima, Y. Honda, O. Kitao, H. Nakai, T. Vreven, K. Throssell, J. A. Montgomery Jr, J. E. Peralta, F. Ogliaro, M. J. Bearpark, J. J. Heyd, E. N. Brothers, K. N. Kudin, V. N. Staroverov, T. A. Keith, R. Kobayashi, J. Normand, K. Raghavachari, A. P. Rendell, J. C. Burant, S. S. Iyengar, J. Tomasi, M. Cossi, J. M. Millam, M. Klene, C. Adamo, R. Cammi, J. W. Ochterski, R. L. Martin, K. Morokuma, O. Farkas,

J. B. Foresman and D. J. Fox, *Gaussian 16 Revision A.03*, Gaussian Inc., Wallingford CT, 2016.
76 A. J. Stone, *Distributed Multipole Analysis of Gaussian wavefunctions, GMDA version 2.2.02*.
77 N. M. O'Boyle, M. Banck, A. James, C. Morley, T. van der Meersch and G. R. Hutchison, *J. Cheminf.*, 2011, **3**, 1–14.
78 G. Landrum, RDKit: Open-source cheminformatics, http://www.rdkit.org.
79 A. V. Kazantsev, P. G. Karamertzanis, C. S. Adjiman and C. C. Pantelides, *J. Chem. Theory Comput.*, 2011, **7**, 1998–2016.
80 D. H. Case, J. E. Campbell, P. J. Bygrave and G. M. Day, *J. Chem. Theory Comput.*, 2016, **12**, 910–924.
81 J. A. Chisholm and S. Motherwell, *J. Appl. Crystallogr.*, 2005, **38**, 228–231.
82 G. M. Day, S. L. Price and M. Leslie, *J. Phys. Chem. B*, 2003, **107**, 10919–10933.
83 S. Riniker and G. A. Landrum, *J. Chem. Inf. Model.*, 2015, **55**, 2562–2574.
84 M. Habgood, I. J. Sugden, A. V. Kazantsev, C. S. Adjiman and C. C. Pantelides, *J. Chem. Theory Comput.*, 2015, **11**, 1957–1969.
85 I. Sugden, C. S. Adjiman and C. C. Pantelides, *Acta Crystallogr., Sect. B: Struct. Sci., Cryst. Eng. Mater.*, 2016, **72**, 864–874.
86 H. P. G. Thompson and G. M. Day, *Chem. Sci.*, 2014, **5**, 3173–3182.
87 A. J. Stone and M. Alderton, *Mol. Phys.*, 2002, **100**, 221–233.
88 P. G. Karamertzanis and C. C. Pantelides, *J. Comput. Chem.*, 2005, **26**, 304–324.
89 S. C. Nyburg and C. H. Faerman, *Acta Crystallogr., Sect. B: Struct. Sci.*, 1985, **41**, 274–279.
90 G. M. Day and S. L. Price, *J. Am. Chem. Soc.*, 2003, **125**, 16434–16443.
91 D. E. Williams, *J. Comput. Chem.*, 2001, **22**, 1–20.
92 E. O. Pyzer-Knapp, H. P. G. Thompson and G. M. Day, *Acta Crystallogr., Sect. B: Struct. Sci., Cryst. Eng. Mater.*, 2016, **72**, 477–487.
93 A. Kazantsev, Ph.D. thesis, Imperial College London, 2011.
94 A. Kazantsev, P. Karamertzanis, C. Pantelides and C. Adjiman, Process Systems Engineering, *Molecular Systems Engineering*, 2010, vol. 6, pp. 1–38.
95 J. P. Perdew, K. Burke and M. Ernzerhof, *Phys. Rev. Lett.*, 1996, **77**, 3865.
96 S. Grimme, J. Antony, S. Ehrlich and H. Krieg, *J. Chem. Phys.*, 2010, **132**, 154104.
97 S. Grimme, S. Ehrlich and L. Goerigk, *J. Comput. Chem.*, 2011, **32**, 1456–1465.
98 N. Marom, R. A. DiStasio, V. Atalla, S. Levchenko, A. M. Reilly, J. R. Chelikowsky, L. Leiserowitz and A. Tkatchenko, *Angew. Chem., Int. Ed.*, 2013, **52**, 6629–6632.
99 J. Hoja, A. M. Reilly and A. Tkatchenko, *Wiley Interdiscip. Rev.: Comput. Mol. Sci.*, 2017, **7**, e1294.
100 J. Moellmann and S. Grimme, *J. Phys. Chem. C*, 2014, **118**, 7615–7621.
101 J. Sun, A. Ruzsinszky and J. P. Perdew, *Phys. Rev. Lett.*, 2015, **115**, 036402.
102 J. Brandenburg, J. Bates, J. Sun and J. Perdew, *Phys. Rev. B*, 2016, **94**, 115144.
103 M. Hutchinson and M. Widom, *Comput. Phys. Commun.*, 2012, **183**, 1422–1426.
104 Y. Sun, C. J. Tilbury, S. M. Reutzel-Edens, R. M. Bhardwaj, J. Li and M. F. Doherty, *Cryst. Growth Des.*, 2018, 905–911.
105 E. Joiris, P. Di Martino, C. Berneron, A.-M. Guyot-Hermann and J.-C. Guyot, *Pharm. Res.*, 1998, **15**, 1122–1130.

Faraday Discussions

PAPER

ROY revisited, again: the eighth solved structure†

Melissa Tan,[a] Alexander G. Shtukenberg, [a] Shengcai Zhu, [b] Wenqian Xu, [c] Eric Dooryhee,[d] Shane M. Nichols,[ae] Michael D. Ward, [a] Bart Kahr*[af] and Qiang Zhu *[b]

Received 18th February 2018, Accepted 2nd May 2018
DOI: 10.1039/c8fd00039e

X-ray powder diffraction and crystal structure prediction (CSP) algorithms were used in synergy to establish the crystal structure of the eighth polymorph of 5-methyl-2-[(2-nitrophenyl)amino]-3-thiophenecarbonitrile (ROY), form R05. R05 crystallizes in the monoclinic space group $P2_1$ with lattice parameters a = 11.479(4) Å, b = 11.030(1) Å, c = 10.840(6) Å, β = 118.23(1)°. This is both the first acentric ROY polymorph, and the first with $Z' > 1$. The torsion angles defined by the S–C–N–C atom sequence of each molecule in the asymmetric unit (R05-1 and R05-2) are 44.9° and −34.0°. These values are between those previously determined for the red and orange forms of ROY. The crystal packing and intermolecular interactions in R05 are explained herein through Hirshfeld surface analysis and an updated energy stability ranking is determined using computational methods. Although the application of CSP was critical to the structure solution of R05, energy stability rankings determined using a series of DFT van der Waals (vdW)-inclusive models substantially differ from experiment, indicating that ROY polymorphism continues to be a challenge for CSP.

1 Introduction

There are ten polymorphs of 5-methyl-2-[(2-nitrophenyl)amino]-3-thiophenecarbonitrile, known simply as 'ROY' for its red, orange, and yellow crystals (Fig. 1). The abundance of colors exhibited by ROY, the peacock of

[a]Department of Chemistry, Molecular Design Institute, New York University, 100 Washington Square East, Room 1001, New York City, NY, 10003, USA. E-mail: bart.kahr@nyu.edu
[b]Department of Physics and Astronomy, High Pressure Science and Engineering Center, University of Nevada Las Vegas, Nevada, 89154, USA. E-mail: qiang.zhu@unlv.edu
[c]X-ray Science Division, Advanced Photon Source, Argonne National Laboratory, Argonne, Illinois, 60439, USA
[d]Photon Sciences Division, National Synchrotron Light Source II, Brookhaven National Laboratory, Upton, NY, 11973, USA
[e]Department of Chemistry and Chemical Biology, Harvard University, 12 Oxford St, Cambridge, MA, 02138, USA
[f]Department of Advanced Science and Engineering, Waseda University, Tokyo, 162-0056, Japan

† Electronic supplementary information (ESI) available. See DOI: 10.1039/c8fd00039e

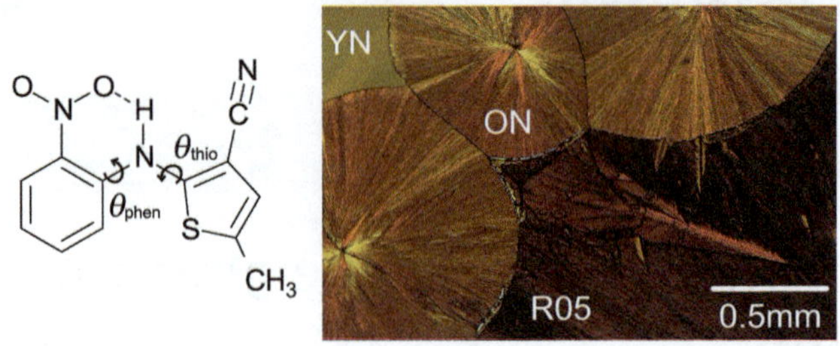

Fig. 1 ROY molecule (left). The torsion angles are indicated by θ_{thio} and θ_{phen}. Optical micrograph of spherulites of three ROY phases (YN, ON, R05) grown from the melt viewed with a polarized light microscope (right).

molecular crystals, primarily arises from the conformational flexibility exhibited in the torsion angle, θ_{thio} (Fig. 1). Of the ten known polymorphs, the crystal structures of seven have been solved,[1,2] leaving three (Y04, RPL, and R05) unknown. Here, we report the crystal structure of one of these three, R05, an abbreviation for "red 2005", the most recently discovered ROY polymorph.[3]

For those who like to keep score, ROY once held the record as the organic compound with the most associated polymorphic structures: seven in total (solvates naturally excluded).[4] In 2013, flufenamic acid[5] surpassed ROY with eight crystal structures solved. More recently still, aripiprazole,[6] the anti-depressant sold as Abilify, joined the club of eight. By determining the structure of R05, ROY – a perennial favorite of fans of polymorphs, and a wellspring for theorists – again returns to the lead, albeit a three-way tie.

R05 is distinguished from other ROY polymorphs for crystallizing in the enantiomorphous space group $P2_1$, and for having $Z' = 2$. These two independent molecules have different conformations, and to the extent that the photophysical properties of ROY can be approximated to the first order from molecular conformation, we observed from visible absorption spectra that R05 is more optically dense than any other polymorph of ROY that can be grown from the melt.

2 Crystal structure determination

2.1 Crystallization

Thin films of ROY (<5 μm) were prepared by melting 3–5 mg of powder ($C_{12}H_9N_3O_2S$, TCI, >97%, mp = 97.4–114.8 °C) between a glass slide and coverslip on a hot plate. Spontaneous crystallization of the melt at room temperature yielded seven of the ten known polymorphs: Y, YN, YT04, Y04, ON, R, and R05. Except for YT04 and Y04, each of these forms appears distinct and can easily be identified with a polarizing light microscope. R05 grows as a smooth spherulite composed of fine, red-orange needles (Fig. 1) that transform to forms Y or R by movement of the growth front. If the metastable form R05 is in contact with other polymorphs of ROY, conversion can be complete in several hours. If, however, R05 is isolated, the conversion process occurs at a slower rate unfolding over the course of several weeks.

2.2 X-ray powder diffraction

The presence of R05 was confirmed by Raman spectroscopy ($\omega_{CN} = 2217\ cm^{-1}$, $\omega_{NH} = 3276\ cm^{-1}$) and powder X-ray diffraction. Measurements were collected at New York University of an as-grown spherulite using a Bruker AXS D8 Discover GADDS microdiffractometer (CuKα radiation, $\lambda = 1.54178$ Å) equipped with a VANTEC two-dimensional detector in reflection mode. This experiment corroborated previously published data indicating characteristic Bragg peaks at $2\theta = 12.44, 12.47, 17.66, 18.38, 18.70, 23.13, 24.76, 26.44, 29.32, 31.22$, and $33.19°$.[3]

In order to achieve the higher angular resolution necessary for indexing and structure solution, additional experiments were conducted at synchrotron sources. Initially, the data were collected at the 28-ID-2 (XPD) beamline at the National Synchrotron Light Source (NSLS-II), Brookhaven National Laboratory (BNL). Thin films of ROY were prepared from the melt at NYU and screened for R05. Immediately prior to data collection, crystallites of R05 were extracted with a scalpel, yielding <1 mg of sample. The exterior wall of a 0.5 mm Kapton capillary was coated with a thin layer of Dow high vacuum grease (a silicone lubricant) and the sample was transferred onto the capillary by rolling the capillary through the powder. The measurement was performed at 250 K and $\lambda = 0.18342$ Å with the sample to detector distance set at 1200 mm. The measured 2D pattern displays almost no preferential orientation of crystallites (Fig. 2). Unfortunately, because of the very short operational wavelength employed at BNL, the angular resolution was insufficient to index the cell.

A subsequent experiment with the same sample was performed one month later on the 17-BM beamline at the Advanced Photon Source (APS), Argonne

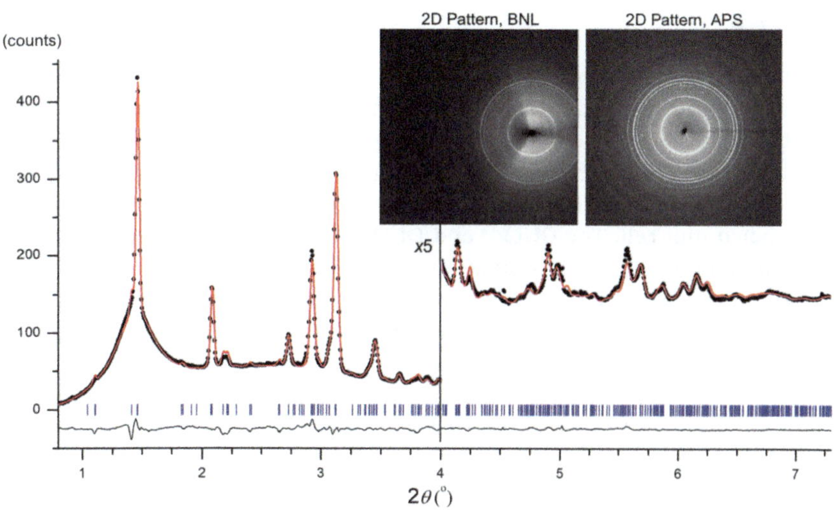

Fig. 2 Final Rietveld refinement of BNL synchrotron data ($\lambda = 0.18342$ Å, 250 K) with 2D diffraction patterns collected at BNL and APS shown in the inset. Observed intensities are depicted as black circles while calculated intensities are illustrated as a red line. The difference curve is shown beneath in black. Blue ticks indicate positions of diffraction maxima.

National Laboratory. In the interim, the capillary was stored upright in a sealed container at 4 °C. The sample was inspected at APS to determine whether conversion had begun. Although the major phase remained as form R05, a minor population of yellow crystallites had begun to grow, indicating the presence of at least one other polymorph. Two 30 second exposures were collected at 295 K (λ = 0.45336 Å, sample to detector distance = 800 mm). Both datasets were subsequently combined to create a single output file. The 2D pattern of these measurements displays evidence consistent with contamination from other phases in the form of dots and short arcs at 2θ positions that do not correspond to R05.[3] In addition, the APS pattern exhibits some preferential orientation of R05 crystallites, biasing the intensities of certain reflections (Fig. 2).

Although the instrument configurations and detectors at APS and BNL are similar, datasets collected from the same sample differ. These differences can be attributed to two effects: polymorph conversion and incident radiation. Since R05 is difficult to grow, the sample prepared at BNL was remeasured at APS. In the time between data collection at each site, the metastable form R05 had begun to transform, introducing errors in the intensity of the measured reflections of the APS data. In contrast, the dataset collected at BNL is free from contamination by other phases. Nevertheless, because the operational wavelength employed at APS was longer (λ = 0.45336 Å), the pattern affords greater angular resolution. In order to come to the structure solution of R05, it was necessary to analyze both datasets.

The diffraction pattern collected at APS was indexed using the software program McMaille v3.04.[7] The 1D pattern was compared against cleaner datasets collected at NYU and BNL to ensure that only reflections related to R05 were included in the indexing file. By this process, the unit cell for R05 was determined as monoclinic with a = 11.479(4) Å, b = 11.030(1) Å, c = 10.840(6) Å, β = 118.23(1)°, V = 1209.3(44) Å^3.

2.3 Computational crystal structure prediction

To obtain trial structures for crystal structure solution, we performed systematic crystal structure prediction (CSP) based on evolutionary algorithms, as implemented in the USPEX code.[8–11] Only the molecular geometry is used as input. The number of molecules per asymmetric unit (Z') and space group symmetries are user specified. Alternatively, the search can be constrained with an experimentally determined unit cell. The GULP[12] and DFTB+[13] codes were used to perform the structure relaxations within USPEX. In our evolutionary search, the first generation of structures was created randomly in the given space groups. All structures were relaxed at ambient pressure and 0 K with enthalpy used as a measure of fitness. The energetically worst structures (40%) were discarded and new generations were evolved by heredity and mutation operators as described elsewhere.[8] The best structure from each evolved generation was retained and runs were terminated after 50 generations.

Structural relaxations were performed in two steps. First, all structures were optimized by the GULP code[12] with the standard Dreiding force field[14] and partial charges assigned from the QEq method.[15] This was followed by further relaxation at the level of Density Functional Tight Binding with the DFTB+ code[13] and 3ob-3-1 parameter set.[16] In the final energy ranking stage, we chose the 100 lowest energy structures from the prediction process. These were re-optimized with the optB88 functional[17] as implemented in the VASP code[18] using the projector-augmented wave (PAW) method.[19] A plane wave kinetic energy cut-off of 1000 eV was used. For all geometry relaxation calculations, the Brillouin zone was sampled by uniform Γ-centered meshes with a reciprocal space resolution at least $2\pi \times 0.06$ Å, and convergence criteria of 1×10^{-5} eV per atom for total energies and 5×10^{-3} eV Å^{-1} for forces.

A crystal structure search with $Z' = 1$ at ambient pressure was initially conducted for the 30 most common space groups. Although this successfully returned all known crystal structures of ROY deposited in the CSD (QAXMEH family), there were no reasonable solutions for R05. In our second attempt, the monoclinic cell obtained from experiment ($a = 11.479$ Å, $b = 11.030$ Å, $c = 10.841$ Å, $\beta = 118.23°$) was fixed and the query was constrained to search for structures with $Z' = 1$ in $P2_1/c$, $C2$, and Cc, and $Z' = 2$ in the $P2_1$ and Pc space groups. This identified the lowest energy structure as $P2_1$ with $Z = 4$, $Z' = 2$ and a simulated PXRD pattern in good agreement with the experimental data.

2.4 Structure refinement

The synchrotron powder diffraction data collected at BNL was refined by the Rietveld method with TOPAS 4.2 (ref. 20) ($C_{12}H_9N_3O_2S$, MW = 259.29 g mol^{-1}, ρ = 1.434 g cm^{-3}, λ = 0.18342 Å, T = 250 K). The geometry of each molecule was defined by a rigid body (RB). Rotation and translation parameters were simultaneously refined with the thiophene torsion angle, θ_{thio}, of each independent molecule in the asymmetric unit. This resulted in R_{wp} = 3.12% with $\theta_{thio,1}$ = 50.68°, and $\theta_{thio,2}$ = −31.94°.

The thermal displacement parameter (B_{eq}) was subsequently refined and bond length restraints were imposed for intermolecular interactions where the measured distance between contacts was less than the expected sum of vdW radii. Following this, all parameters were refined, including both the thiophene and phenyl ring torsion angles, resulting in an improvement to the overall molecular geometry. A refinement with all restraints removed (R_p = 2.17%, R_{wp} = 2.98%, R_{int} = 1.77%, GOF = 0.207) did not significantly alter the positions of non-hydrogen atoms; however, some bond lengths became too short. For this reason, bond length restraints were reintroduced and the structure was refined to a final R_p = 3.55%, R_{wp} = 3.00%, R_{int} = 1.78%, GOF = 0.208 with $\theta_{thio,1}$ = 44.9°, $\theta_{thio,2}$ = −34.0°, $\theta_{phen,1}$ = 11.5° and $\theta_{phen,2}$ = −19.2°. This structure was deposited at the CCDC (ref code 1822444).† A comparison of the final refinement with the trial structure supplied by CSP shows close agreement. Periodic models of both structures expanded to 20-molecule clusters in COMPACK[21] resulted in a calculated highest root mean-square deviation of 0.189 Å.

3 Results and discussion

3.1 Structure analysis

All previously solved structures of ROY possess inversion symmetry and therefore contain heterochiral pairs of molecules in the unit cell. R05 is distinct from this group because it has two molecules in the asymmetric unit and lacks a center of symmetry. In the final Rietveld refinement, $\theta_{thio,1}$ = 44.9° and $\theta_{thio,2}$ = −34.9°; the molecules are twisted with opposite sense.

Color variations across different polymorphs of ROY are attributed to the conformational flexibility expressed by the torsion angle, θ_{thio} (Fig. 3).[2,22] In red form R, the aromatic rings are nearly co-planar ($\theta_{thio} = \pm 22°$, where the $\pm$ sign is due to inversion symmetry), leading to a greater degree of π conjugation, and a subsequent red shift in the visible absorption spectra.[22] In contrast, θ_{thio} is close to 90° in yellow forms Y, YT04, and YN ($\theta_{thio} = \pm(104–113)°$). In forms OP, ON, and ORP, the torsion angle falls between these extremes ranging from $\pm(39–53)°$. Thus, as is outwardly suggested by its deep red-orange color, there is a kinship between R05 and the orange colored ROY polymorphs (Table 1).

Like the previous seven structures solved, each molecule in R05 contains an intramolecular hydrogen bond between the amino and nitro groups. All known polymorphs of ROY, including R05, are characterized by a range of weak intermolecular interactions (Fig. 4). Among this group, intermolecular hydrogen bonds only exist in forms Y and Y04 between the cyano and amino group in neighboring molecules (C≡N···N–H).

Fingerprint plots derived from Hirshfeld surfaces were used to analyze packing in R05.[23–25] The Hirshfeld surface of a molecule in a crystal is constructed by partitioning space into regions where the electron distribution of a sum of spherical atoms for the molecule dominates the corresponding sum over the crystal. For each point on the surface, two distance properties are defined: d_e as the distance from a point to the nearest nucleus external to the surface, and d_i as the distance from a point to the nearest nucleus internal to the surface. 2D fingerprint plots are then derived from the Hirshfeld surface by plotting the fraction of points on the surface as a function of the pair (d_i, d_e). This concept has provided a convenient way to understand the intermolecular interactions and packing modes in molecular crystals.[25]

A detailed Hirshfeld surface analysis of R, O, OP, ON, Y, YN, and YT04 was made in 2007.[25] Here, we focus on the independent molecules in the asymmetric unit in R05 (R05-1 and R05-2), comparing each to forms R and Y. As shown in Fig. 5, the fingerprints of the two molecules in R05 are remarkably similar. Both contain a bright spot centered at $d_i = 1.8$ Å, $d_e = 1.8$ Å, which is characteristic of

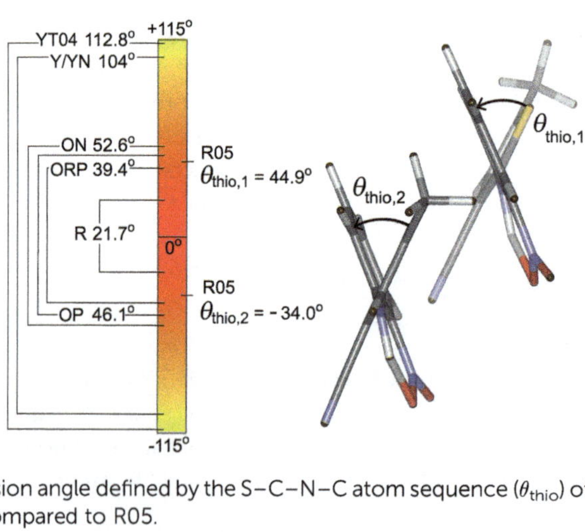

Fig. 3 The torsion angle defined by the S–C–N–C atom sequence (θ_{thio}) of all known ROY polymorphs compared to R05.

Table 1 Values of torsion angles in each ROY polymorph

Polymorph	CSD entry	θ_{thio} (°)	θ_{phen} (°)
R	QAXMEH02	±21.7	±12.4
R05	This work	44.9/−34.0	11.8/−19.2
ORP	QAXMEH05	±39.4	±6.3
OP	QAXMEH03	±46.1	±12.4
ON	QAXMEH	±52.6	±6.4
YN	QAXMEH04	±104.1	±6.0
Y	QAXMEH01	±104.7	±4.4
YT04	QAXMEH12	±112.8	±10.4

π···π stacking. In addition, the structures of R05-1 and R05-2 each feature "antennae" with internal (d_i) and external (d_e) distances of (1.3, 1.0) Å and (1.4, 1.0) Å that are associated with CH···O and CH···O intermolecular distances that are shorter than the vdW distances. The antenna at (1.2, 1.2) Å corresponds to CH···H interactions. The similarities in both molecules suggests an almost identical local environment. This allows for the independent molecules to be treated as a pseudo-centrosymmetric pair. In contrast, R and Y exhibit quite different characteristics. In both R and Y, the center spot at (1.8, 1.8) Å is weaker and more diffuse, suggesting less uniform π···π stacking as compared to R05. In addition, the "antennae" of CH···N are notably stronger in Y, which agrees with the presence of C≡N···N–H intermolecular H bonding.

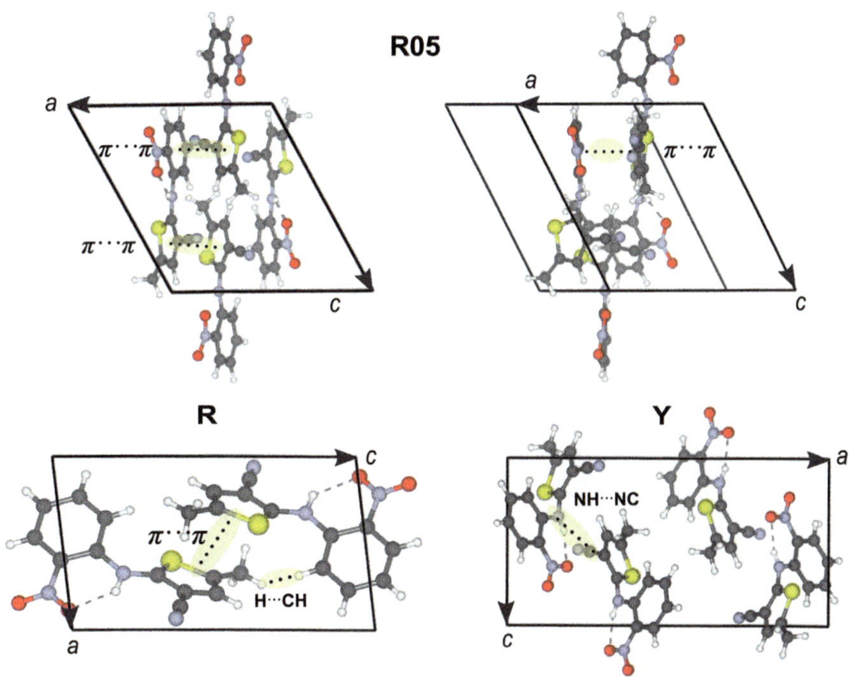

Fig. 4 Crystal structures for the selected ROY polymorphs (R, Y and R05) with some representative inter(intra)-molecular interactions highlighted.

The contributions of close intermolecular contacts shown in Fig. 6 provide another way to visualize packing differences in ROY forms R, Y, and R05. All the structures are dominated by C···H, N···H, O···H, and H···H interactions. There are large changes in C···H and O···H across the series, while the changes in N···H and H···H are smaller. R05-1 and R05-2 again display very similar distributions, with only a slight difference in C···C, which can be assigned to π···π stacking. In contrast, Y has the smallest C···C contribution and largest extent of H···H. It does exhibit π···π stacking but the oblique angle of the molecules relative to each other in the crystal structure reduces the C···C interaction and results in greater C···C contributions than in the case of form R. Interestingly, the N···H contributions from Y are also smaller than in other polymorphs, although the "antennae" of CH···N, associated with the C≡N···N–H intermolecular H bond, have the longest spikes in Y (Fig. 5). This suggests that the total number of CH···N bonds is reduced but that they are stronger. On the other hand, R is quite different from both R05-1 and R05-2. Despite having a similar θ_{thio}, R has smaller C···C contributions but larger C···H contributions. The C···C in R is from π···π between the benzene rings in chiral molecule pairs, while C···C in R05 is found between benzene–thiophene and thiophene–thiophene rings.

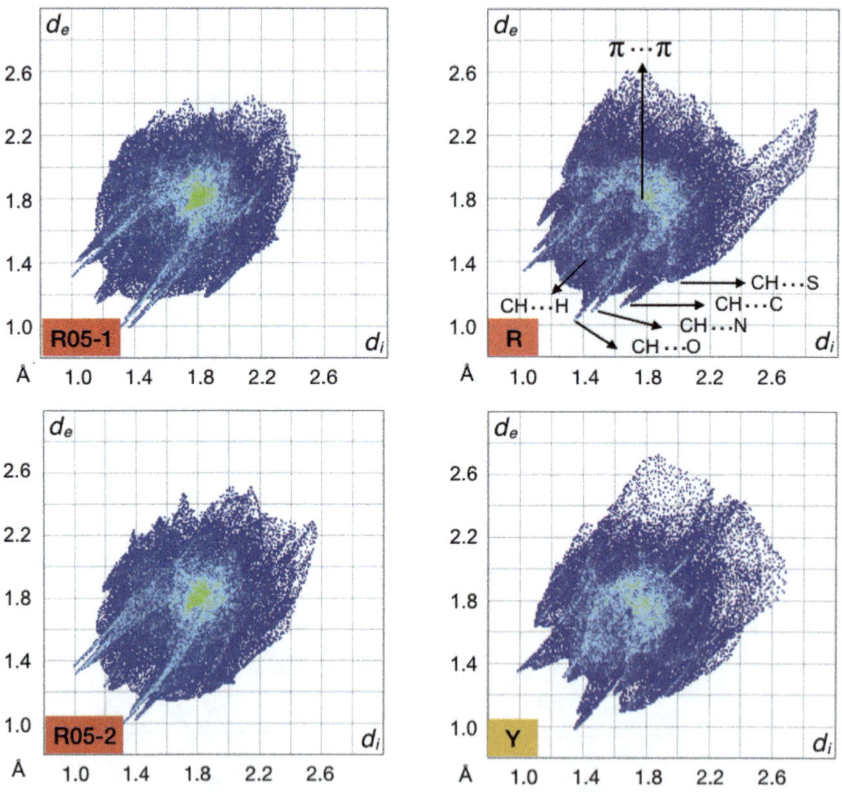

Fig. 5 2D Hirshfeld fingerprint plots for the selected ROY polymorphs (R, Y, R05-1 and R05-2). The representative intermolecular interactions are also highlighted in the plots.

3.2 Energy ranking

The rich polymorphism of ROY presents a challenge and an opportunity for computational chemists studying intermolecular interactions in organic crystals. The relative free energies of Y, YN YT04, R, OP, ON, and ORP were previously obtained by differential scanning calorimetry from melting and eutectic melting data.[2–4] The results indicate that the free energies of the seven ROY polymorphs characterized are separated by 1.5 kJ mol^{-1} at 40 °C while the enthalpies span a range of 4 kJ mol^{-1}.[22] The stability of the polymorphs at 0 K was extrapolated from the enthalpy data as follows: Y (lowest) < YT04 < R < OP < ON < YN < ORP (highest).[22] There have been several subsequent studies that have sought to determine the energetic differences between ROY polymorphs through computational techniques, each employing varying levels of theory.[26–28] So far, computational predictions have differed from experimental results.

In 2005, Dunitz and Gavezzotti[26] partitioned the intermolecular energies into different contributions (coulombic, polarization, dispersion and repulsions) between molecular pairs based on the molecular electron density distribution. The resulting ranking Y < YT04 < R < ON < OP < YN < ORP has nearly the same sequence as reported by Yu;[2] however, the energy difference (22 kJ mol^{-1}) is significantly larger than experimentally measured. Vasileiadis and co-workers[28] employed a similar semi-empirical approach based on distributed multipole analysis and found YN < Y < R < YT04 < OP < ORP < ON, with a range of 10.2 kJ mol^{-1} between YN and ON.

The first DFT study on ROY polymorphs[27] sorted the stability of the forms as ORP < ON < Y < R < OP < YN < YT04 at 0 K. Here, the energy separation between YT04 and ORP was determined to be 20 kJ mol^{-1}. This wide range in values indicated that the lattice energy alone is insufficient to determine energy rankings. Subsequent calculations were performed that incorporated the effect of the crystallization force, accounting for local changes to the electronic structure as

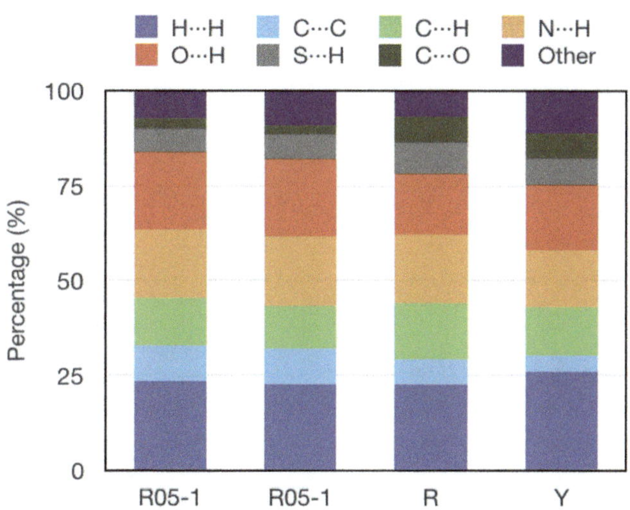

Fig. 6 Percent contributions of various intermolecular interactions to Hirshfeld area in R05-1, R05-2, R, and Y.

a result of packing.[27] In this updated model, the ranking sequence of the ROY polymorphs was determined as Y < R < ON < ORP < YN < YT04 < OP with an energy difference of 10.1 kJ mol^{-1}.

New developments in computational methodologies have improved energy rankings, describing vdW interactions in the framework of DFT.[17] Accuracy within 5 kJ mol^{-1} has been achieved with vdW-inclusive DFT.[29] Even greater accuracy (within 1 kJ mol^{-1}) can be obtained by using computationally demanding wave-function based electronic structure methods.[30] Moreover, in the most recent blind test organized by the Cambridge Crystallographic Data Center (CCDC), substantial improvements were made in the use of hierarchical approaches for ranking structures and in the application of density-functional approximations.[31] These advances, together with the new structure solution of R05, encouraged us to perform an updated benchmark of the energy rankings for ROY polymorphs at the DFT level.[27]

We selected 40 low-energy structures after merging results from all the CSP runs, and studied the lattice energy *versus* density relationship at the level of optPBE using the VASP code (see Fig. 7a). Clearly, all of the observed polymorphs cover a range of $\approx$6 kJ mol^{-1}, which is significantly smaller than the previous studies based on semi-empirical approaches.[26,28] Nevertheless, there do exist a number of hypothetical structures with close stabilities relative to the experimental structures. It is unclear whether they are real structures which have not been found by experiment, or they were misranked by the optPBE functional. Moreover, the stability ranking predicted by our calculations strongly deviates from the experimental data.[3] Form Y, the polymorph previously determined as the most stable at ambient conditions,[3] had the highest enthalpy at 0 K among all the ROY forms in our calculations. We were not surprised. In a recent study on the polymorphs of coumarin, we discovered that different vdW-inclusive models yielded inconsistent rankings.[32] Thus, we performed another benchmark test on ROY with other vdW-inclusive models, namely PBE-D3 (ref. 33) in VASP, and PBE + MBD and PBE0 + MBD[34] in FHI-aims.[35] Although the ranking varies according to the choice of computational model, all the calculations exhibit the same general trend (Fig. 7b). That is, red colored forms are the most stable, followed by orange. Yellow forms are the least stable in terms of enthalpy. This

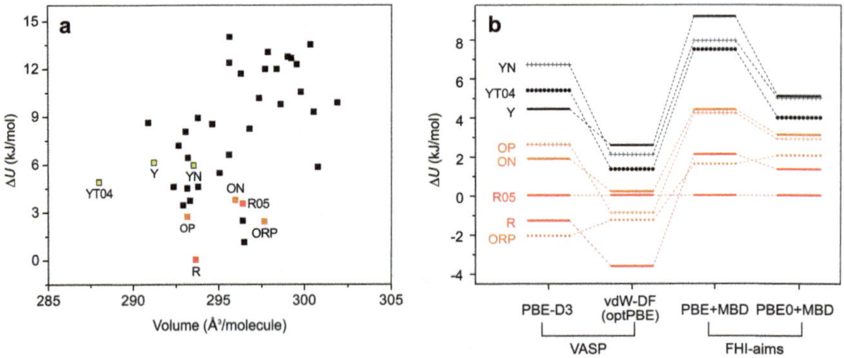

Fig. 7 (a) Lattice energy *versus* molecular volume at the optPBE level. (b) Energy rankings for the eight known polymorphs of ROY with different vdW-inclusive models.

overall ranking is contrary to the experimental data and to previous computational studies which have either ranked Y or YT04 as the most stable form. From a structural perspective, Y and YT04 differ from other forms due to the presence of weak intermolecular hydrogen bonds (C≡N···N–H). This subtlety does not seem to be recognized in our DFT models. Disagreement between our calculations and previous work in DFT on ROY[27] may also stem from the choice of functional basis set and dispersion correction models. Despite this, the energy range determined in the current study is significantly smaller as compared to that found previously,[27] suggesting an overall improvement of DFT-vdW inclusive methodology. Furthermore, we emphasize that our DFT calculations were performed at 0 K and ignore entropic contributions, while the previous stability ranking relied upon the free energy difference derived from the experimental eutectic melting data.[3] Thus, the deviation between experiment and theory might be due to a temperature effect. Calculated entropic contributions are typically small but nevertheless sufficient to switch the energy ranking for different polymorphs.[36] Indeed, our recent work on coumarin indicates that both harmonic and anharmonic vibrations might make non-negligible contributions to the total free energy. An investigation on possible origins for the discrepancy between theory and experiment remains the subject of future work.

3.3 Twisting

It is hard to make R05 from the melt. No doubt this is the reason why it was the last polymorph of ROY to have been discovered. During the process of melting and recrystallizing samples, we observed features of YN that had not been previously described in the literature. At high magnification, this form displays the tell-tale sign of twisting – optical banding. Helicoidal twisting of lamellae in molecular crystals is actually an extraordinarily common phenomenon for crystals grown under a high driving force.[37] It has been said that 25% of simple molecular crystals are twisted.[38] This is a credible estimate. Thus, since seven of the ten ROY polymorphs can be grown from the melt, it would be improbable for ROY to be twist free.

The optical properties of YN were measured with a Mueller matrix microscope that employs dual continuously rotating retarders, installed before and after the sample plane, to modulate the polarization state of the light. The operational theory of our system is similar to earlier instrument designs.[39–41] In short, the continuous rotation of each retarder creates a smoothly varying time-dependent intensity signal at the detector that can be analyzed to recover the 16 elements of the 4 × 4 Mueller matrix, **M**. Here, **M** is a polarization transfer matrix that describes how a sample transforms a four-element input polarization state vector ($\mathbf{S}_{\text{in}}$) to its output ($\mathbf{S}_{\text{out}}$). If **M** is nondepolarizing, the 16 elements of the Mueller matrix simplify to seven.[42] These seven parameters form a 4 × 4 matrix **L**. **M** is related to **L** by the matrix exponential,[39] $\mathbf{M} = \exp \mathbf{L}$ where

$$\mathbf{L} = \begin{bmatrix} -A & -\text{LD} & -\text{LD}' & \text{CD} \\ -\text{LD} & -A & \text{CB} & \text{LB}' \\ -\text{LD}' & -\text{CB} & -A & -\text{LB} \\ \text{CD} & -\text{LB}' & \text{LB} & -A \end{bmatrix}$$

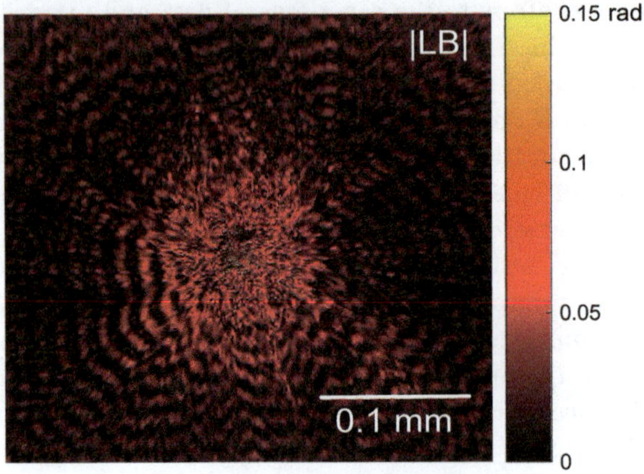

Fig. 8 The oscillation in the magnitude of the linear birefringence (|LB|) of a banded spherulite of YN measured at 620 nm is indicative of twisting.

The parameters LD, LB, CD, and CB refer to linear dichroism, linear birefringence, circular dichroism, and circular birefringence, respectively. Strictly speaking, the matrix **L** only holds for normal incidence measurements of homogenous media with plane parallel surfaces.[43] Helicoidally twisted spherulites do not fit this definition. Nevertheless, the parameters in **L** provide a convenient basis to understand the optical properties of the aggregate.[43]

Measurements were performed at normal incidence over a range of wavelengths. A precession correction algorithm was applied to the images in order to remove beam steering artifacts caused by imperfect optics.[44] Minimal depolarization allowed for **M** to be reduced to **L**. Fig. 8 plots the magnitude of linear birefringence (|LB|) of a so-called banded spherulite of YN measured at 620 nm (20×). The concentric light and dark bands indicate an in-phase helicoidal precession of radially aligned fibers. The spacing between the dark bands is equal to half the pitch.

Thus, R05 is chiral because of its crystal structure whereas YN is chiral because of its morphology. Although the coexistence of new polymorphs from the melt with twisted morphologies is commonplace, a chiral space group is not a prerequisite for a twisted form; YN crystallizes in the centric space group $P\bar{1}$ but nevertheless exhibits twisting. In contrast, the crystal structure of R05 is chiral but spherulites grown from the melt under the same conditions as YN do not twist.

4 Conclusions

Pairing high-resolution X-ray powder diffraction with crystal structure prediction techniques is an effective method for determining the structures of metastable polymorphs.[10,32,45] Indeed, the DFT optimization of candidate structures during the process of crystal structure solution allows for chemically sensible structure models to be determined even when the quality of the diffraction data is modest. In this manuscript, we employed such a multipronged approach to yield the structure of R05, the most recently discovered polymorph of ROY. R05 is acentric with two independent molecules that have opposite twists. Nevertheless, according to Hirshfeld analysis, the independent molecules have comparable local environments. Although CSP was applied with success to the problem of crystal structure solution, our DFT based vdW-inclusive calculations still do not converge to the experimentally determined energy rankings, indicating that ROY polymorphism remains a challenge for computational chemistry.

Conflicts of interest

The authors declare no conflict of interest.

Acknowledgements

This work was supported partially by the MRSEC Program of the National Science Foundation under Award Number DMR-1420073. B. K. is grateful for support from the U.S. National Science Foundation (DMR-1608374) and the National Institutes of Health (5R21GM107774-02). Q. Z. is grateful for support from the National Nuclear Security Administration under the Stewardship Science Academic Alliances program through DOE Cooperative Agreement DE-NA0001982. Calculations were performed at computing resources from XSEDE and the Center for Functional Nanomaterials (under contract no. DE-AC02-98CH10086). This research used the synchrotron X-ray resources of the Advanced Photon Source and the National Synchrotron Light Source II, U.S. Department of Energy (DOE) Office of Science User Facilities operated for the DOE Office of Science by Argonne National Laboratory and Brookhaven National Laboratory, respectively, under Contract No. DE-AC02-06CH11357 and DE-SC0012704. Q. Z. thanks Alexandre Tkatchenko and Johannes Hoja for help with MBD calculation in the FHI-aims code.

References

1 G. Stephenson, T. Borchardt, S. Byrn, J. Boyer, S. Snorek, C. Bunnell and L. Yu, *J. Pharm. Sci.*, 1995, **84**, 1385–1386.
2 L. Yu, G. A. Stephenson, C. A. Mitchell, C. A. Bunnell, S. V. Snorek, J. J. Bowyer, T. B. Borchardt, J. G. Stowell and S. R. Byrn, *J. Am. Chem. Soc.*, 2000, **122**, 585–591.
3 S. Chen, H. Xi and L. Yu, *J. Am. Chem. Soc.*, 2005, **127**, 17439–17444.
4 S. Chen, I. Guzei and L. Yu, *J. Am. Chem. Soc.*, 2005, **127**, 9881–9885.
5 V. Lopez-Mejias, J. W. Kampf and A. J. Matzger, *J. Am. Chem. Soc.*, 2012, **134**, 9872–9875.
6 S. P. Delaney, D. Pan, S. X. Yin, T. M. Smith and T. M. Korter, *Cryst. Growth Des.*, 2013, **13**, 2943–2952.
7 A. Le Bail, *Powder Diffr.*, 2004, **19**, 249–254.
8 Q. Zhu, A. R. Oganov, C. W. Glass and H. T. Stokes, *Acta Crystallogr., Sect. B: Struct. Sci.*, 2012, **68**, 215–226.
9 A. O. Lyakhov, A. R. Oganov, H. T. Stokes and Q. Zhu, *Comput. Phys. Commun.*, 2013, **184**, 1172–1182.
10 Q. Zhu, A. G. Shtukenberg, D. J. Carter, T.-Q. Yu, J. Yang, M. Chen, P. Raiteri, A. R. Oganov, B. Pokroy, I. Polishchuk, P. J. Bygrave, G. M. Day, A. L. Rohl, M. E. Tuckerman and B. Kahr, *J. Am. Chem. Soc.*, 2016, **138**, 4881–4889.

11 J. Yang, C. T. Hu, X. Zhu, Q. Zhu, M. D. Ward and B. Kahr, *Angew. Chem., Int. Ed.*, 2017, **56**, 10165–10169.
12 J. D. Gale and A. L. Rohl, *Mol. Simul.*, 2003, **29**, 291–341.
13 B. Aradi, B. Hourahine and T. Frauenheim, *J. Phys. Chem. A*, 2007, **111**, 5678–5684.
14 S. L. Mayo, B. D. Olafson and W. A. Goddard, *J. Phys. Chem.*, 1990, **94**, 8897–8909.
15 A. K. Rappe and W. A. Goddard III, *J. Phys. Chem.*, 1991, **95**, 3358–3363.
16 M. Gaus, A. Goez and M. Elstner, *J. Chem. Theory Comput.*, 2012, **9**, 338–354.
17 J. Klimeš and A. Michaelides, *J. Chem. Phys.*, 2012, **137**, 120901.
18 G. Kresse and J. Furthmüller, *Phys. Rev. B: Condens. Matter Mater. Phys.*, 1996, **54**, 11169–11186.
19 P. E. Blöchl, *Phys. Rev. B: Condens. Matter Mater. Phys.*, 1994, **50**, 17953.
20 Bruker AXS, *TOPAS v4: General profile and structure analysis software for powder diffraction data. User's Manual*, Bruker AXS, Karlsruhe, Germany, 2008.
21 J. A. Chisholm and S. Motherwell, *J. Appl. Crystallogr.*, 2005, **38**, 228–231.
22 L. Yu, *Acc. Chem. Res.*, 2010, **43**, 1257–1266.
23 F. L. Hirshfeld, *Theor. Chem. Acc.*, 1977, **44**, 129–138.
24 M. A. Spackman and J. J. McKinnon, *CrystEngComm*, 2002, **4**, 378–392.
25 J. J. McKinnon, F. P. Fabbiani and M. A. Spackman, *Cryst. Growth Des.*, 2007, **7**, 755–769.
26 J. Dunitz and A. Gavezzotti, *Cryst. Growth Des.*, 2005, **5**, 2180–2189.
27 T. Li, P. W. Ayers, S. Liu, M. J. Swadley and C. Aubrey-Medendorp, *Chem.–Eur. J.*, 2009, **15**, 361–371.
28 M. Vasileiadis, A. V. Kazantsev, P. G. Karamertzanis, C. S. Adjiman and C. C. Pantelides, *Acta Crystallogr., Sect. B: Struct. Sci.*, 2012, **68**, 677–685.
29 G. J. Beran, *Chem. Rev.*, 2016, **116**, 5567–5613.
30 J. Yang, W. Hu, D. Usvyat, D. Matthews, M. Schütz and G. K.-L. Chan, *Science*, 2014, **345**, 640–643.
31 A. M. Reilly, R. I. Cooper, C. S. Adjiman, S. Bhattacharya, A. D. Boese, J. G. Brandenburg, P. J. Bygrave, R. Bylsma, J. E. Campbell, R. Car, *et al.*, *Acta Crystallogr., Sect. B: Struct. Sci., Cryst. Eng. Mater.*, 2016, **72**, 439–459.
32 A. G. Shtukenberg, Q. Zhu, D. J. Carter, L. Vogt, J. Hoja, E. Schneider, H. Song, B. Pokroy, I. Polishchuk, A. Tkatchenko, A. R. Oganov, A. L. Rohl, M. E. Tuckerman and B. Kahr, *Chem. Sci.*, 2017, **8**, 4926–4940.
33 S. Grimme, J. Antony, S. Ehrlich and H. Krieg, *J. Chem. Phys.*, 2010, **132**, 154104.
34 A. Tkatchenko, R. A. DiStasio Jr, R. Car and M. Scheffler, *Phys. Rev. Lett.*, 2012, **108**, 236402.
35 V. Blum, R. Gehrke, F. Hanke, P. Havu, V. Havu, X. Ren, K. Reuter and M. Scheffler, *Comput. Phys. Commun.*, 2009, **180**, 2175–2196.
36 J. Nyman and G. M. Day, *CrystEngComm*, 2015, **17**, 5154–5165.
37 A. Shtukenberg, Y. Punin, A. Gujral and B. Kahr, *Angew. Chem., Int. Ed.*, 2014, **53**, 672–699.
38 F. Bernauer, *"Gedrillte" Kristalle*, Gebrüder Borntraeger, 1929.
39 R. M. A. Azzam, *Opt. Lett.*, 1978, **2**, 148–150.
40 C. Chen, I. An, G. Ferreira, N. Podraza, J. Zapien and R. Collins, *Thin Solid Films*, 2004, **455–456**, 14–23.

41 O. Arteaga, M. Baldrís, J. Antó, A. Canillas, E. Pascual and E. Bertran, *Appl. Opt.*, 2014, **53**, 2236–2245.

42 S. Cloude, *Polarisation: Applications in Remote Sensing*, Oxford Univ. Press, 2010.

43 X. Cui, S. M. Nichols, O. Arteaga, J. Freudenthal, F. Paula, A. G. Shtukenberg and B. Kahr, *J. Am. Chem. Soc.*, 2016, **138**, 12211–12218.

44 S. M. Nichols, PhD thesis, New York University, 2018.

45 A. G. Shtukenberg, C. T. Hu, Q. Zhu, M. U. Schmidt, W. Xu, M. Tan and B. Kahr, *Cryst. Growth Des.*, 2017, **17**, 3562–3566.

Faraday Discussions

DISCUSSIONS

Applications of crystal structure prediction – organic molecular structures: general discussion

Claire S. Adjiman, Jan Gerit Brandenburg, Doris E. Braun, Jason Cole, Christopher Collins, Andrew I. Cooper, Aurora J. Cruz-Cabeza, Graeme M. Day, Marta Dudek, Alan Hare, Luca Iuzzolino, David McKay, John B. O. Mitchell, Sharmarke Mohamed, Sridhar Neelamraju, Marcus Neumann, Sten Nilsson Lill, Jonas Nyman, Artem R. Oganov, Sarah L. Price, Angeles Pulido, Susan Reutzel-Edens, Ivo Rietveld, Michael T. Ruggiero, J. Christian Schön, Seiji Tsuzuki, Joost van den Ende, Grahame Woollam and Qiang Zhu

DOI: 10.1039/c8fd90032a

Marcus Neumann opened discussion of the paper by Andrew I. Cooper: In Fig. 3 of your paper you compare the lattice energies of structures of different chemical compositions. Each filled circle corresponds to a different solvate and the various solvents all have their own, temperature dependent, chemical potentials. What is the meaning of comparing the lattice energies of solvated and desolvated structures on a common energy scale?

Andrew I. Cooper replied: It shows that all of the solvates are stabilised with respect to the global minimum predicted solvent-free structure when solvent is inserted. There is some correlation between this stabilisation and the measured boiling point suppression of the solvent in the various solvates (Fig. 5 in the paper), though the onset temperature for solvent loss as measured by thermogravimetric analysis could also have a kinetic component, even at slow heating rates.

J. Christian Schön said: The problem mentioned in the previous question seems to be that in Fig. 3 you show a volume–energy correlation plot, but you are making no distinction between structures that contain different kinds of solvents inside. Of course, this means that comparisons are rather problematic – no matter which structure you have, you will reduce the energy by filling it with solvent (which adds new bonds to the energy term per atom!), and you can see which structure wins the most by filling, but since you do not have an experiment where all structures are offered all solvents at the same time, it does not make sense to compare the outcomes. (Or are you in reality doing such an experiment, perhaps by periodically sublimating and resublimating different solvents?) On the other hand, Table 2 in the paper gives the energies separately for each solvent, and one can deduce which structure will win for a given solvent. My question is: How did you pick the cage structures which you filled in your study? Or did you fill all the possible modifications you had found in your search?

Andrew I. Cooper responded: We focused on inserting all of the solvents tested into the solvent-free frameworks on the landscape that best matched the observed solvates – this gives the matrix of energies in Table 2. To do it for all structures (*i.e.*, a true *de novo* prediction, in lieu of experimental data) is in principle possible, but the computational expense is prohibitive.

I take your point about comparing different solvents on one plot – the point of this figure was less to suggest that a particular solvent "wins" (as you say, they were never in competition), but more to show that in all cases, the solvated frameworks are predicted to be more stable than the dense, non-solvated global minimum on the landscape. It rationalises the formation of the observed solvates. It would be interesting to try competitive, mixed solvent crystallisations (competitive studies) but of course, this might get complicated because we may form a variety of mixed solvates (*e.g.*, 2 $CHCl_3$ + 1 CH_2Cl_2, *etc.*).

J. Christian Schön asked: It seems to me to be rather difficult to put the solvents into the holes in the appropriate fashion – did you have a systematic procedure that allowed you to explore the full landscape for the solvent inside the structure?

Graeme M. Day responded: The bottleneck of the solvent stabilisation calculations is the need to find the lowest energy solvent configurations within the voids in the predicted crystal structures; this is where most of the computational effort was spent.

We did use a systematic method, which involved a high temperature Monte Carlo sampling with a simple, inexpensive force field model, followed by reoptimisation of structures with the more accurate force field that we used in the crystal structure prediction study. Details of the types of moves used in the Monte Carlo are given in the supporting information of the paper.

The difficulty in sampling the solvent configurations within the pores varied quite a bit between structures. For some of these structures, we had to go up to solvent loadings above 5 solvent molecules per cage molecule, with up to 8 cage molecules in the unit cell and no use of symmetry: this means sampling the positions of 40+ independent solvent molecules, which is computationally challenging.

In some cases, it was not crucial to find the lowest energy solvent configuration because the solvent molecules are unlikely to be fixed in one configuration and the final energies of many configurations are nearly iso-energetic. However, in other cases the stabilisation is much greater with one particular solvent configuration, so we always needed to do sufficient sampling to be reasonably confident that we weren't missing a particularly low energy solvent configuration.

Andrew I. Cooper added: This is a challenge – the method seems to work well in most cases, but the Monte Carlo search did fail for some solvents (*e.g.*, dioxane), as discussed in the paper, and additional searching was necessary to find the observed experimental structure.

J. Christian Schön asked: What is the structure of the solvent inside the cavities – amorphous or crystalline?

Andrew I. Cooper responded: This varies – some structures have quite well ordered solvent, in others it is highly disordered, yet others have disordered solvent with some solvent molecules ordered at specific sites in the structure. In general, we tend to see highly ordered solvent more commonly when there are specific driving forces (*e.g.*, H-bonding) to promote this.

J. Christian Schön added: As a follow-up on the previous question: if the arrangement of the solvents inside the cavities is essentially amorphous, a possible configurational entropy contribution to the free energy might be important.

Andrew I. Cooper answered: Yes, this might be important.

J. Christian Schön asked: You added the solvent afterwards to the optimized molecular crystal structures. But might it be possible that if the solvent had been available to the system throughout the whole global optimization procedure, you might have found even better candidates (with structures that would never be stable without solvent inside)?

Andrew I. Cooper replied: Yes, but this could become extremely expensive computationally since we would greatly increase the number of independent molecules in the calculation, particularly at high solvent loadings. I don't think this is practical at the current time.

Marcus Neumann said: How about the transferability of your approach to more flexible molecules? CC1 is a fairly rigid molecule and the host matrix is maintained when the solvent molecules are removed. What typically happens when solvent molecules are removed from the solvate structures of flexible molecules? Do the solvent cavities survive solvent removal or do they collapse?

Andrew I. Cooper answered: Some more flexible cage molecules do indeed collapse when the solvent is removed. An example of this can be found in ref. 1. For smaller, rigid [4+6] cages, though, the method is transferable, see for example ref. 2.

1 K. E. Jelfs, X. Wu, M. Schmidtmann, J. T. A. Jones, J. E. Warren, D. J. Adams and A. I. Cooper, Large self-assembled chiral organic cages: synthesis, structure and shape persistence, *Angew. Chem., Int. Ed.*, 2011, **50**, 10653–10656.

2 J. T. A. Jones, T. Hasell, X. Wu, J. Bacsa, K. E. Jelfs, M. Schmidtmann, S. Y. Chong, D. J. Adams, A. Trewin, F. Schiffmann, F. Cora, B. Slater, A. Steiner, G. M. Day and A. I. Cooper, Modular and predictable assembly of porous organic molecular crystals, *Nature*, 2011, **474**, 367–371.

Graeme M. Day added: We have tried a similar approach of looking for solvates of pharmaceutical molecules with moderate flexibility. The work is still in progress but one finding is that, during lattice energy minimisation, flexible molecules are able to more easily fill the voids left by removing solvent from a crystal structure. More rigid molecules have less ability for rearrangement. Therefore, the approach may work less well for more flexible molecules because the voids that could be occupied by solvent that we are looking for in structures coming from crystal structure prediction are more likely to be partially collapsed.

J. Christian Schön asked about configurational disorder related to the solvent that we have placed in the voids. We did not quantify the resulting entropy in the calculations presented here. However, we did observe that structures with large voids lead to many solvent configurations with similar energies, so probably more liquid-like, while some solvate structures lead to one solvent configuration that is energetically favoured over all others, corresponding to solvent that will be locked in one orientation in the solvate. There is some information in the ESI related to the paper.

J. Christian Schön responded: I would concur with Graeme on the configurational order issue. I think that we need to make sure that:

(a) the configurational entropy associated with a "liquid-like" (dynamical disorder inside the pore system on the observational time scale, possibly in conjunction with breathing modes *etc.* of the framework) or static disorder (self-averaging of many frozen-in disordered configurations) is taken into account when comparing free energies of the polymorphs, and

(b) we need to perform the global optimizations including various amounts of solvent as a complementary search to the standard polymorph search without solvent.

Doris E. Braun asked: You use CSP to generate porous organic cages (structures with holes). After identifying possible new frameworks, how do you choose the solvents for crystallising the structures?

Andrew I. Cooper answered: At present, we often use robotics to screen a range of solvents, in many cases correlating phases with structures predicted using CSP methods. At the moment, the *a priori* selection of solvent using these Monte Carlo insertion methods is too expensive to use on a routine basis, and here it was applied *post hoc*.

David McKay asked: There is a marked increase in the boiling point of a solvent once it is part of the discussed solvates with molecular cages. Is there also a difference observed in the boiling point of the solution made up of the molecular cage and the solvent?

Andrew I. Cooper answered: That is an excellent suggestion – we have not tried it, but we should. It may also be relevant to our research in the area of "porous liquids".[1]

1 N. Giri, M. G. Del Pópolo, G. Melaugh, R. L. Greenaway, K. Rätzke, T. Koschine, L. Pison, M. F. Costa Gomes, A. I. Cooper and S. L. James, Liquids with permanent porosity, *Nature*, 2015, **527**, 216–220.

Ivo Rietveld addressed Andrew I. Cooper and Graeme M. Day: The paper deals with crystal structures high in energy, which have large cavities that basically beg to be filled up with solvent to lower their energy.

How close are we to predicting solvate formation where the cavities would only accommodate a single solvent molecule such as 1 : 1 solvate : API systems in a given solvent?

Graeme M. Day answered: There are a few examples of crystal structure prediction studies being used to predict solvates such as 1 : 1 API : solvate systems. The usual approach to this problem is different from what we did in this paper and simply performs the full crystal structure prediction study with the molecule and solvent from the start. A couple of examples from our group are ref. 1 and 2.

This is probably the best approach where we are only interested in one solvent and we can take a good guess at the likely stoichiometry between the main molecule and the solvent. A problem that we have with solvates is that it is not possible to be confident of the amount of solvent that will be incorporated into the crystal structure. For crystal structure prediction, this means that we need to consider different stoichiometries in the prediction, then compare the energetic stability of the different compositions. This gets very computationally demanding, but has been done. See ref. 3. We took a different approach in our paper here because we are interested in the porous packings of the organic molecule and want to be able to consider what packings are accessible, irrespective of the solvent used in crystallisation. So we perform crystal structure prediction without solvent included, followed by assessing how the different structures can be stabilised with a range of possible solvents.

1 A. J. Cruz-Cabeza, G. M. Day and W. Jones, Structure Prediction, Disorder and Dynamics in a DMSO Solvate of Carbamazepine, *Phys. Chem. Chem. Phys.*, 2011, **13**, 12808–12816.
2 A. J. Cruz Cabeza, G. M. Day, W. D. S. Motherwell and W. Jones, Prediction and Observation of Isostructurality Induced by Solvent Incorporation in Multicomponent Crystals, *J. Am. Chem. Soc.*, 2006, **128**, 14466–14467.
3 A. J. Cruz-Cabeza, S. Karki, L. Fabian, T. Friscic, G. M. Day and W. Jones, Predicting Stoichiometry and Structure of Solvates, *Chem. Commun.*, 2010, **46**, 2224–2226.

Andrew I. Cooper added: This can be done already in gas phase calculations for individual cage molecules – for example, by docking individual gas molecules into cage cavities we predicted, before experiment, that a cage, CC3, would be a very favourable host for a single xenon or radon atom. Subsequent experiments showed that this translated into experiment – for example, CC3 was shown to be very effective at separating xenon from krypton.[1]

1 L. Chen, P. S. Reiss, S. Y. Chong, D. Holden, K. E. Jelfs, T. Hasell, M. A. Little, A. Kewley, M. E. Briggs, A. Stephenson, K. M. Thomas, J. A. Armstrong, J. Bell, J. Busto, R. Noel, J. Liu, D. M. Strachan, P. K. Thallapally and A. I. Cooper, Separation of rare gases and chiral molecules by selective binding in porous organic cages, *Nat. Mater.*, 2014, **13**, 954–960.

Sten Nilsson Lill asked: For selecting the solvent you want to introduce in your cavity, could you use a docking approach? It would be similar to designing as is done for an enzyme pocket so docking might be helpful.

Andrew I. Cooper replied: This is a good idea, and we've already used this in gas phase calculations to probe cage guest selectivity (*e.g.* ref. 1).

1 L. Chen, P. S. Reiss, S. Y. Chong, D. Holden, K. E. Jelfs, T. Hasell, M. A. Little, A. Kewley, M. E. Briggs, A. Stephenson, K. M. Thomas, J. A. Armstrong, J. Bell, J. Busto, R. Noel, J. Liu, D. M. Strachan, P. K. Thallapally and A. I. Cooper, *Nat. Mater.*, 2014, **13**, 954–960.

Aurora J. Cruz-Cabeza added: The previous delegate asked about the use of docking for the prediction of solvent inclusion in porous molecular crystals. I would like to point out the paper by Oliver Korb and Peter Wood entitled "Prediction of framework–guest systems using molecular docking".[1]

1 O. Korb and P. A. Wood, *Chem. Commun.*, 2010, **46**, 3318–3320.

Andrew I. Cooper responded: Thank you – I was unaware of this paper.

J. Christian Schön asked: In your introduction, you mention that the surface energy can reduce the total free energy and you can get different structures. Now, the Wulff construction of the shape of crystals *via* surface energy minimization works beautifully because no matter what shape we adopt, it is still the same crystal structure inside, and thus the bulk contribution to the energy remains unchanged. But here, you are comparing different modifications, and at macroscopic sizes, the surface energy cannot be decisive in deciding which modification you get, *i.e.* which modification is the stable bulk modification. Only for very small crystals can the surface energy contribute enough to the total energy of the system that such an effect could appear. How big or small are the crystals we are talking about? If they are considerably below the micrometre range, it could possibly make a difference. And, of course, the surface energy can affect the kind of crystal nucleus that is preferred during the nucleation phase, and once it has formed then it will be able to grow and suppress all the other crystal modifications (until a killer fluctuation creates an energetically lower nucleus and the metastable polymorph is transformed to the thermodynamically lower modification).

Andrew I. Cooper responded: In general, these crystals are micron sized. Surface effects might be important just after nucleation, but I would expect that these would have a small contribution to the total energy once the crystals are larger, especially for the highly porous crystals (measured surface areas of >3000 m^2/g), where the surface contribution is probably really insignificant. However – this might indeed be one route to form kinetic polymorphs, and it could be part of the story as to why we do not always form the lowest energy predicted polymorphs. Surface effects of this type are not captured in the CSP methods that we use here at present.

Marcus Neumann asked: As far as I understand, you generated the energy landscape of CC1, crystallized CC1 from many different solvents and then used solvent insertion to rationalize why it is the experimentally observed structures that crystallized rather than other predicted structures. Did you also try to do it the other way round: choose one or more predicted structures and try to find the solvent or solvent mixture that would make it crystallize?

Andrew I. Cooper answered: No – at the moment, the computational expense of this is prohibitive, but it could be a useful approach in the future, if these methods can be made much cheaper. Ideally, we would be able to generate multiple such "solvent corrected" landscapes, allowing us to choose the solvent *in silico* before experiment. At the moment, this is not competitive with solvent screening in the lab, which is faster, especially if we use robotics.

J. Christian Schön asked: When you discuss checks of stability, are you thinking about experiment or theory?

Marcus Neumann answered: The check is computational. All local minima found are slightly perturbed in several ways. If a minimum turns out to be a shallow side minimum of a much deeper neighbouring minimum, it is not likely to be visited very often and is excluded from the statistical completeness analysis.

Susan Reutzel-Edens asked: The porous cage structures of interest in this work were related to solvates formed by comparatively weak hydrogen bonding solvents, at least weak relative to many solvents commonly used in the pharmaceutical industry. Have you examined more strongly hydrogen bonding solvents for their ability to produce organic cage structures? Does the CC1 molecule not form solvates with such solvents or are the solvates formed not cages? Is there any way, short of crystallizing CC1 from a range of solvents in the laboratory, to know which solvents might work?

Andrew I. Cooper responded: This choice was mainly dictated by considerations of solubility – this cage dissolves best in the solvents tested here. It is not soluble in many highly polar solvents. For more hydrogen-bonding solvents, we've found for other systems (*e.g.*, hydrogen bonded frameworks) a tendency for the solvent to play a less innocent, directing role in crystallisation, and it is likely that the method presented here would break down in such cases.

Doris E. Braun said: Drug molecules often form non-stoichiometric (a variable range of solvent/water molecules) or channel solvates/hydrates. Could you use the approach for this class of solid forms as well?

Andrew I. Cooper answered: Potentially, yes – we do predict the solvate stoichiometry quite well, at least for this porous cage, CC1 (see Table 1 in the paper).

Doris E. Braun opened a discussion of the paper by Sharmarke Mohamed : You used PBE-D calculations to derive the relative stabilities of ionic cocrystals with respect to the experimental salt and acid structures. Did you start from experimental structures only? Did you do calculations on the proton positions within the salt/ionic cocrystal structures to see whether the proton moves or not?

Sharmarke Mohamed answered: The relative energy estimates (using dispersion-corrected periodic density functional theory methods) of the ionic cocrystal structures with respect to the salt and acid crystal structures were all obtained starting with the known experimental structures as reported in the Cambridge Structural Database (CSD). We did not do any additional calculations that probe the accuracy of the assigned crystallographic proton positions. We expect however that since all the ionic cocrystals whose relative energies were calculated have reported salt structures in the CSD, the potential energy barrier for proton transfer is expected to be sufficiently large so as to favor the crystallisation of solids comprising molecular ions. Indeed, despite the fact that we did not do any additional optimisations that probe the accuracy of the assigned proton positions, we would expect incorrect proton assignments to lead to optimized structures that are closer in agreement to a cocrystal of the formula 2 : 1 than an ionic cocrystal of the formula 1 : 1 : 1. We did not observe any soft barriers for proton transfer or any optimized structures resembling a 2 : 1 cocrystal.

John B. O. Mitchell said: I was interested in the solubility measurements for various molecules in your paper. Can you tell me more about them, and what are the error bars on these values?

Sharmarke Mohamed replied: The solubility measurements were obtained using turbidity probes as implemented in the Crystal16 platform. Four heating and cooling cycles were performed for each system and this allowed us to obtain consistent readings, thereby reducing the likelihood of errors in the estimates of the clear and cloud points. No error bars were generated from the measured solubility data. However, the heating/cooling rate of 1 degree/min ensured that for each heating/cooling cycle, there was ample equilibration of the contents of the vial so as to ensure that the transition points (clear and cloud points) and the solubility of the sample were accurately determined.

J. Christian Schön enquired: Some more technical questions; they may also apply to the calculations in the presentation by Prof. Cooper. Concerning energy calculations: are you using an *ab initio* code to get the shape of the molecules? And afterwards you use the molecules as rigid units, maybe with a multipole expansion of their charge distribution? I would estimate that the computational power is not large enough to do the search at the *ab initio* level. Are the shapes of the molecules allowed to change (and thus also their internal energies) during the global search? Concerning the space groups in the molecular crystals: from the databases, we know that there are only a few space groups that are found in about 95% of all molecular crystals. Does this also apply in the case of co-crystals, and is this the reason behind your choice of permitted space groups during the search? Now, would it make sense to treat the salt species elements as spherical units – this way you might get higher symmetries when generating structures, which can afterwards be relaxed into the final (presumably low-symmetry) structures? I am thinking here of the so-called "cluster-compounds" that were very popular a few years ago, where people often described their structures are "approximately cubic" (when visualizing the clusters as large spheres), *etc.* Quite generally, would it make sense to expand the allowed set of space groups you are sampling? Do we understand why this is the case?

Sharmarke Mohamed responded: In our paper, we performed crystal structure prediction (CSP) on a set of 1 : 1 molecular salt systems of interest (some that form ionic cocrystals and some that don't). We did indeed use an electronic structure package to calculate the global minimum in the molecular conformation for the ions prior to initiating the search for putative crystal packings. This was done in the gas phase with no corrections for solvent (*i.e.* by using a dielectric medium). The calculated gas phase conformational energy minimum was then used as the input for the search for putative packings of the salt ions. The conformations of the ions were kept rigid throughout the cell refinement and geometry optimization steps. This is a valid approximation in the case of the systems studied because the chosen molecules/ions are fairly rigid. Since the purpose of our CSP work was to discover the range of packing efficiencies for the low energy polymorphs of each salt and given that we already knew the experimental structure of the salt (where applicable) including the space group, we were able to restrict our search for unit cells to those that satisfy the most common space groups in the Cambridge Structural Database. If this was a purely first principles CSP work aimed at discovering previously unknown polymorphs for the salt or ionic cocrystals studied, it would have been necessary to widen the range of symmetries considered for the unit cells.

Regarding the treatment of the salt species as "spherical units", this approximation is valid for counterions that are atomic in nature (*e.g.* Cl^-) and our work did not encompass such systems. Since we are dealing with molecular ions with uneven charge distributions across the molecular surface, I don't think this is a good description for the systems studied in our work.

Sarah L. Price commented: It does seem hard to understand why a cation will pack with both the anion and neutral version of the same acid in the same crystal. What are the pK_a differences involved and the likely speciation in the solution? Are these ionic cocrystals just oddities or might there be practical uses for looking for ionic cocrystals of pharmaceutical molecules? The differences between cocrystals and salts can be only in the proton position, but this has been a scientific minefield in the pharmaceutical industry. Might an ionic cocrystal be a practical way forward for getting better solid properties?

Sharmarke Mohamed answered: Indeed, the occurrence of ionic cocrystals does raise interesting questions about the role of serendipity in crystallisation experiments as well as the degree to which these solids can be purposefully designed in order to produce materials with improved solid-state properties (*e.g.* for pharmaceutically active ingredients, APIs). Careful inspection of the literature shows that there is already increasing evidence (see ref. 1 and 2) that ionic cocrystals comprising inorganic salts and an organic molecule (*e.g.* an API) can be synthesized using a variety of methods and that such solid forms do indeed have utility in improving one or more physicochemical properties of the parent organic molecule. During my time in the pharmaceutical industry, I was fortunate enough to be involved in shaping the institutional response of my company to the draft guidance of the US Food and Drug Administration (FDA) on what constitutes a "cocrystal". I recall vividly the non-trivial implications for pharmaceutical companies that were inherent in the FDA's original draft guidance that classified cocrystals as drug product intermediates. The FDA's language and classification of cocrystals has evolved since that time and according to the most recent guidance issued by the FDA in February 2018, cocrystals are classified as "a special case of solvates and hydrates" and have the same regulatory classification as a new polymorph of the API. Definitions are therefore important and do have practical consequences for the industry. The evidence from the literature suggests that ionic cocrystals do have utility in improving the solid-state properties of pharmaceuticals.

1 L. Casali, L. Mazzei, O. Shemchuk, K. Honer, F. Grepioni, S. Ciurli, D. Braga and J. Baltrusaitis, *Chem. Commun.*, 2018, **54**, 7637–7640.
2 N. K. Duggirala, A. J. Smith, Ł. Wojtas, R. D. Shytle and M. J. Zaworotko, *Cryst. Growth Des.*, 2014, **14**, 6135–6142.

Susan Reutzel-Edens added: In response to the question raised by Sally Price on whether the formation of a salt or a co-crystal is important to industry or merely an academic curiosity, at least for the pharmaceutical industry, the determination of the ionization state is more important now than ever before. This is thanks in part to the recently revised/finalized US Food and Drug Administration co-crystal guidance (now harmonized with the European Medicines Agency), which has solidified different regulatory pathways to be taken for salts and co-crystals.

With respect to the role that an additional neutral guest molecule plays in stabilizing ionic co-crystals, the subject of Sharmarke Mohamed's Faraday contribution, the CSP community might be interested to know of a series of ICCs inspired by nature, the salt-*co*-crystals formed between cytosines/cytidines and strong acids (*e.g.* sulfonic acids). In these structures, the cytosine (or cytidine) molecule is protonated in such a way as to create a complement to the neutral molecule. Three point hydrogen bonding between the cytosine and cytosine-H^+ drives incorporation of the neutral guest; the anion is needed for charge balance. These systems may be challenging for computational chemistry, but they could be potentially very interesting substrates.

Sharmarke Mohamed replied: I agree with the comments by Susan Reutzel-Edens on the importance of the ionization state. The exact definition of the solid-form has important practical implications for the industry. However, it is important to note that regulatory guidelines always lag behind (often by several years) the fierce and often contradictory definitions put forward by practitioners (*e.g.* academics) when defining solid forms in the crystal engineering literature. You only have to look at the ongoing challenges and persistent disagreements surrounding the definition of cocrystals. For ternary systems such as ionic cocrystals, it is therefore important that the crystal engineering community adopt a unified set of definitions on what constitutes such solid forms. For example our work has shown that a simple Cambridge Structural Database survey of three-component solids with the general formula A^-B^+N and with stoichiometry 1 : 1 : 1 produces a number of crystal structures. But not all of these should be classified as ionic cocrystals since the third component N is often a solvent of crystallisation (*e.g.* water) and therefore is better described as a salt solvate. The cytosines/cytidines systems referenced by Susan Reutzel-Edens do indeed look to be interesting model systems that could help us learn more about the driving force for the crystallisation of ionic cocrystals. I would also suggest that experimentalists look at our paper and try to understand why there are so many known molecular ionic cocrystals of the general formula A^-B^+N (1 : 1 : 1) that have no corresponding 1 : 1 salts in the database. Why do molecular ionic cocrystals of stoichiometry 1 : 1 : 1 crystallize from experiments employing stoichiometric ratios of acid and base that are in fact targeting 1 : 1 salts or 1 : 1 cocrystals? Why do some molecular ionic cocrystals crystallise with solvent whilst others do not? All of these are unresolved challenges that need further investigation before we can claim to have mastered the systematic crystallisation of ionic cocrystals with desirable properties.

Doris E. Braun asked: Did you check the CSD for 2 : 1 cocrystals of related molecules of the chosen systems, *i.e.* two weak acids and one weak base?

Sharmarke Mohamed responded: We did not search the CSD for 2 : 1 cocrystals of related molecules of the chosen system.

Doris E. Braun continued: Do you think it may be possible for a system to form a cocrystal as one solid form and an ionic cocrystal as another solid-state form (hydrate or neat forms)?

Sharmarke Mohamed answered: I have not seen such cases from my review of the literature but would be interested to hear from others if such systems do in fact exist. In our paper, we deliberately defined what we mean by a "molecular ionic cocrystal" as any solid comprising molecular species (note that there is a vast literature surrounding metal coordination complexes that comprise a metal ion with ligands consisting of a combination of organic molecules and inorganic salts and these are also ionic cocrystals) that fit the general formula A^-B^+N (where N may be a molecular species or a salt complex). Based on our literature review, it is far more likely for the same molecules A and B to exist in different crystals as a salt (A^-B^+) and an ionic cocrystal (A^-B^+N). Note that in both crystals (salt and ionic cocrystal), proton transfer to form the charged ions is a constant feature, in line with chemical intuition about the relative acidity of the molecules. What your question implies however is that such proton transfer can effectively be tuned and either switched off (cocrystal) or switched on (ionic cocrystal) as a function of the experimental conditions and the presence/absence of a third chemical species in the crystal (N). This is certainly not impossible and indeed we know that proton transfer can be influenced by a range of experimental conditions (*e.g.* temperature, solvent). We also know that locating protons from X-ray diffraction methods can be tricky. So it is entirely possible that through a combination of the crystallographic misassignment of proton positions within the crystal or serendipitous experiments, that such phenomena have been observed and are documented. I would imagine however that even if such phenomena have been observed, the reproducible and targeted crystallisation of both a cocrystal and an ionic cocrystal starting from the same molecules A and B is perhaps far more difficult to achieve under ambient conditions. I therefore remain doubtful about the widespread occurrence of such a phenomenon.

Grahame Woollam asked: Andrew Maloney of CCDC analysed a subset of the CSD to investigate the contacts formed between charged species in salts; he found that the ions interacted around 65% of the time (14239 from 21956). Can we learn something from the 35% of instances in "normal salts" where a hydrogen bond or close contact is formed through an uncharged moiety in preference to the cation/anion pair, for the purpose of ionic cocrystal design?

The authors focussed on pyridine bases and carboxylic acids consisting of a 1 : 1 : 1 stoichiometry of anion : cation : neutral molecule. Molecular ICCs where the additional neutral molecule is the same acid used to form the anion accounted for 27% of the species studied. Additional insights could come from phosphoric acid cocrystals of phosphate salts such as VEGKIB, HISTPA10, HAKDED, TODNUV, WAVHOQ, and the hydrated ECEDEW, where phosphoric acid, having a number of strong H-bond donors and acceptors is able to feature both as the anionic and neutral species in the asymmetric unit. Yet to be published work has shown the benefit of developing such multicomponent systems over 1 : 1 salts. Chen *et al.* published on the Merck compound phosphoric acid cocrystal of a phosphate salt.[1]

1 A. M. Chen, M. E. Ellison, A. Peresypkin, R. M. Wenslow, N. Variankaval, C. G. Savarin, T. K. Natishan, *et al.*, *Chem. Commun.*, 2007, 419–421.

Sharmarke Mohamed responded: In the set of molecular ionic cocrystals retrieved from the Cambridge Structural Database that featured in our paper, we have also noticed a number of structures where the ion pairs are not engaged in directional hydrogen bonding interactions but instead the neutral molecule is engaged in hydrogen bonding with one of the ions. This is indeed a curious observation since chemical intuition would suggest that the charge-charge intermolecular interactions will always be stronger than the alternative interactions between a charged species and a neutral molecule. This is indeed a line of inquiry that deserves further investigation and one that we could potentially learn from. The ionic cocrystals derived from phosphoric acid may be promising model systems.

Doris E. Braun asked: Are you planning to do CSP on ionic co-crystals in the near future?

Sharmarke Mohamed responded: One of the challenges we have found in our lab with the modelling of ionic cocrystals is not just the order of magnitude increase in the computational time with the increasing size in the contents of the asymmetric unit, but also the real challenges in accurately computing the forces between the molecules and ions in the asymmetric unit. Traditional empirically-derived force fields (*e.g.* FIT) do surprisingly well in computing the relative energies of binary molecular salt systems and we have found that the known experimental salt structure is at or close to the global minimum in lattice energy (as you might expect). However, early indications are that such force fields may be unsuitable for the modelling of the relative lattice energies of ternary (and higher order) systems that comprise both charged and neutral molecular species. We therefore expect that the crystal structure prediction of ionic cocrystals will require improvements in our current capabilities for modelling the relative energies of ternary systems that comprise many finely balanced competing intermolecular interactions between charged species as well as between charged and neutral species.

J. Christian Schön opened a discussion of the paper by Michael T. Ruggiero: Concerning Fig. 6 in the paper: I am surprised at the big gap between the calculated and experimental values of alpha at a frequency of 0.5 THz (factor 3–4) – where does this offset come from?

Michael T. Ruggiero responded: Many apologies for the confusion, these data were offset by a constant to aid in visualization. We have updated our manuscript in proof to address this.

Doris E. Braun asked: You have shown for parachlorobenzamide that it is possible to model the order/disorder phase transformation. Could this method be applied to the computationally generated structures in crystal energy landscapes with the aim to reducing the number of potential forms?

Michael T. Ruggiero replied: Absolutely. If a crystal energy landscape is known, then understanding the dynamics of the system provides a great deal of information that can be utilized for crystal structure prediction. Particularly when there are multiple polymorphs that might be similar in energy, the dynamics enable the pathways responsible for undergoing transformations to be understood. So, as a hypothetical example we can imagine that there is a system that is known to undergo a solid-state phase transformation. If we have the crystal energy landscape (or a large number of structures), we can put a constraint on the search to select pairs of structures that are connected *via* a single low-energy vibrational mode. This type of search parameter would lower the number of potential matches significantly, likely down to only a handful of pairs, which then can be combined with other information to likely select the correct structures.

Doris E. Braun asked: Often we see a number of similar structures in the computed crystal energy landscapes and do not expect all of those structures to be observable as a long-lived solid form. With your calculations is it possible to identify the most probable structure of a family of related structures?

Michael T. Ruggiero replied: The kinetic *versus* thermodynamic problem is a lasting problem that is difficult to interpret, but using the methods described in this work there is certainly some progress that can be made. I think one thing that can be done using *ab initio* molecular dynamics alone is to place all probable structures into an AIMD simulation and allow the system to relax, while observing certain parameters. As described in this work and a follow up,[1] particular observables, such as intermolecular bond angles or simulation cell vectors, can be readily used to observe structural parameters that correspond to a transformation. The kinetics of those changing parameters can give some indication of the lifetime of that particular form. Additionally, one can look at the vibrational mode corresponding to the phase transitions, and look at the frequency, which is related to the shape of that potential energy curve. If two structures have very similar energies but one structure exhibits a higher-frequency phase transition mode, this would correspond to a stiffer potential, likely meaning a more stable structure.

1 M. T. Ruggiero, W. Zhang, A. D. Bond, D. M. Mittleman and J. A. Zeitler, *Phys. Rev. Lett.*, 2018, 196002.

John B. O. Mitchell asked: What is the relationship between the lower frequency normal modes and the phonon dispersion diagrams we saw in an earlier paper[1] this morning? Are you essentially slicing the diagram at the gamma point?

J. Hoja and A. Tkatchenko, *Faraday Discuss.*, 2018, DOI: 10.1039/c8fd00066b.

Michael T. Ruggiero replied: Exactly – all infrared absorption spectroscopy is performed solely at the gamma point, so in this work only phonons at gamma were calculated and used.

John B. O. Mitchell continued: What makes your peaks broad as opposed to the finer peaks seen in theoretical diagrams or in conventional IR spectra?

Michael T. Ruggiero responded: In static DFT simulations, IR intensities are determined on a mode-by-mode basis by relating the change in dipole moment with the vibrational displacement. This yields an integrated intensity for that mode, which should be imagined as a delta function. In practice, there are a few factors that determine vibrational line shapes, often the most significant being the lifetimes of the vibrational modes. For the low-frequency modes, where the vibrations are driven by long-range interactions, the uniformity of the intermolecular interactions comes into play. For example, the onset of disorder will broaden vibrational features, as each unit cell begins to feel a slightly different external potential, which will lead to slightly different vibrational frequencies, ultimately summing together to yield a broadened feature. In addition to this, low-frequency modes are strongly prone to anharmonicity, which will cause broadening, particularly with increasing temperature.

Sridhar Neelamraju asked: Are you aware of any other theoretical methods that can be used to capture transition paths for organic crystals, analogous to the transition path sampling often used for protein transitions?

Michael T. Ruggiero responded: Molecular dynamics simulations, in general, are appropriate for capturing reaction pathways. The metadynamics approach discussed previously seems like a good method, although personally we have not attempted such techniques (yet). Another technique that is available in a few solid-state density functional theory codes is the nudged elastic band method,[1,2] but we have not used this technique yet to-date. In the gas phase, the internal reaction coordinate (IRC) method has been used successfully, and there is no reason this method could not be applied to organic crystals.

1 A. Rimola, C. M. Zicovich-Wilson, R. Dovesi and P. Ugliengo, *J. Chem. Theory Comput.*, 2010, **6**, 1341–1350.
2 G. Henkelman and H. Jónsson, *J. Chem. Phys.*, 2000, **113**, 9901–9904.

Graeme M. Day asked: You mentioned the idea of looking at the lowest vibrational modes and using these to calculate phase transition pathways. A limitation here of using terahertz spectroscopy is that you observe only gamma-point modes, while the lowest frequency modes are often not at the gamma point.

Michael T. Ruggiero replied: This is an excellent point. It is true that experimentally (with terahertz spectroscopy) we are limited to modes at gamma only, and additionally we are limited to IR-active modes. However, using terahertz spectroscopy requires computational assignment –implying that all modes are available for analysis, regardless of the IR/Raman activities. The benefit of performing dual experimental and computational work in the low-frequency regime is that in a sense, the experiment can act as a tool to validate the theory. This is especially nice for the terahertz region, because the simulation of the low-frequency vibrations requires very tight numerical accuracy, implying that the accurate modeling of terahertz modes represents an overall accurate simulation for all vibrational modes, regardless of whether they are easily observed experimentally.

Graeme M. Day asked: Andrew Cooper mentioned a model of working where experimental powder diffraction patterns are matched against predicted diffraction patterns from predicted crystal structures. Agreement between experiment and one of the predicted structures can lead to the assignment of a proposed structure during solid form screening. A similar approach has been taken using electron diffraction[1] and also demonstrated using solid state NMR (*e.g.* ref. 2). How well would this approach work using terahertz spectroscopy as the experimental observable? Are the calculations good enough to be reliably matched with measurements?

1 M. D. Eddleston, K. E. Hejczyk, E. G. Bithell, G. M. Day and W. Jones, Polymorph Identification and Crystal Structure Determination by a Combined Crystal Structure Prediction and Transmission Electron Microscopy Approach, *Chem.–Eur. J.*, 2013, **19**, 7874–7882.

2 M. Baias, C. M. Widdifield, J.-N. Dumez, H. P. G. Thompson, T. G. Cooper, E. Salager, S. Bassil, R. S. Stein, A. Lesage, G. M. Day and L. Emsley, Powder Crystallography of Pharmaceutical Materials by Combined Crystal Structure Prediction and Solid-State ^{1}H-NMR Spectroscopy, *Phys. Chem. Chem. Phys.*, 2013, **15**, 8069–8080.

Michael T. Ruggiero replied: This is an excellent point, as terahertz spectroscopy can be used as a sensitive probe of the crystal structure. This is because the low-frequency modes are heavily dependent on the bulk potential energy, such that minor changes (*e.g.* due to different polymorphs or disorder) are readily probed by terahertz spectroscopy. This has been shown previously, for example ref. 1 and 2, where in each of these cases the predicted crystal structure was fully validated using a combination of PXRD and terahertz spectroscopy.

1 M. D. King, T. N. Blanton, S. T. Misture, T. M. Korter, *Cryst. Growth Des.*, 2011, **11**, 5733–5740.

2 M. T. Ruggiero, J. Sibik, R. Orlando, J. A. Zeitler, T. M.Korter, *Angew. Chemie. Int. Ed.*, 2016, **55**, 6877–6881.

Graeme M. Day continued: As a follow-up to my previous question, will the terahertz spectra differ enough between all of the predicted crystal structures to be able to identify one structure that corresponds to the experimental terahertz spectra? This is more challenging than the situation of comparing the spectra of two polymorphs because there are many more possible structures that result from crystal structure prediction, so more likelihood of similarities between spectra. This is an issue in the application of NMR crystallography (mentioned in my previous question), which leads to the need for the assignment of peaks in the spectrum to particular atoms in the molecule. This assignment provides sufficient information to pick out the correct structure (see ref. 1). I don't know if there is an equivalent approach in terahertz spectroscopy.

1 M. Baias, C. M. Widdifield, J.-N. Dumez, H. P. G. Thompson, T. G. Cooper, E. Salager, S. Bassil, R. S. Stein, A. Lesage, G. M. Day and L. Emsley, *Phys. Chem. Chem. Phys.*, 2013, **15**, 8069–8080.

Michael T. Ruggiero answered: Without question, different structures will yield different terahertz spectra. In almost all cases, even very subtle changes in the structure will result in significant contrast between the two measurements, producing a clear indication of exactly what form is present in the sample. This is because the low-frequency modes strongly depend on the bulk potential, and any structural change, even subtle ones, will result in a large net change in the weak forces that contribute to the terahertz modes. That isn't to say that this is always the case, but in situations where the spectra are very similar this is often due to coincidence rather than anything else (*e.g.* two polymorphs might not have any IR-active low-frequency modes, or might happen to have a single feature at very similar frequencies). A few examples of this can be seen in the literature, for example the terahertz spectra of aripiprazole are sufficiently different that a mixture of polymorphs can be understood using the experimental spectra.[1] In the case of very similar structures, the most drastic example I've seen has been presented by A. J. Zaczek and T. M. Korter on polymorphs of tribromobenzene, but unfortunately that result has not been published yet. Less drastic differences can be seen in the polymorphic transformation involving a martensitic-like transformation, where the structures are related by a shifting of layers, which produces a spectrum with additional features when the transformation is complete.[2] So to summarize, I do think that terahertz spectroscopy has the specificity to differentiate between similar structures, even when there might not be sufficient differences to be captured by more conventional techniques like PXRD.

1 S. P. Delaney, D. Pan, S. X. Yin, T. M. Smith and T. M. Korter, *Cryst. Growth Des.*, 2013, **13**, 2943–2952.
2 M. T. Ruggiero, J. A. Zeitler and T. M. Korter, *Phys. Chem. Chem. Phys.*, 2017, **19**, 28502.

Jonas Nyman added: Following up on Graeme M. Day's question about comparing predicted and experimental THz spectra: you said that THz spectra are characteristic for each polymorph, so in principle it is possible to solve a structure from its spectrum if you have predicted structures to compare to. Do you know if this has been done? Have you or someone else solved a crystal structure in this way?

Michael T. Ruggiero responded: Absolutely. Myself and others have used this technique a few times previously, see for example ref. 1 and 2. This is a powerful method because the low-frequency spectra are very strongly influenced by the bulk structure, so without the correct knowledge of that there is no way the correct terahertz spectrum could be simulated.

1 M. T. Ruggiero, J. Sibik, R. Orlando, J. A. Zeitler, T. M. Korter, *Angew. Chemie. Int. Ed.*, 2016, **55**, 6877–6881.

2 S. P. Delaney, D. Pan, S. X. Yin, T. M. Smith and T. M. Korter, *Cryst. Growth Des.*, 2013, **13**, 2943–2952.

J. Christian Schön remarked: If I understand your procedure correctly, you are using *ab initio* MD simulations to extract frequencies and modes. This reminds me of the studies by Oligschleger who used empirical potentials in big amorphous systems.[1,2] When starting with the system in a typical (eigen)mode, one could observe the lifetime of this mode, which is very useful information if one wants to compute *e.g.* heat conductivity from linear-response models *etc.* Looking at Fig. 4 of your paper, it seems you should also be able to see the evolution of such modes? The equilibration time seems to be on the order of tens of picoseconds –does this seem right to you? Have you thought of watching the decay of modes and trying to compute the heat transport in your system?

1 C. Oligschleger, *Phys. Rev. B*, 1999, **60**, 3182.
2 C. Oligschleger and J. C. Schön, *Phys. Rev. B*, 1999, **59**, 4125.

Michael T. Ruggiero responded: This is an excellent point. There are two distinct methods of predicting the low-frequency spectra from molecular dynamics simulations. One involves taking the Fourier transform of the dipole autocorrelation function (used for this work, described in ref. 1 and 2), and requires well-equilibrated trajectories, on the order of 20–40 picoseconds, but does not directly provide discrete normal modes. The other is what you mentioned, and involves introducing energy either by adding in kinetic energy along an eigenmode, or by introducing an oscillating packet of energy.[3] In the latter case, it is indeed somewhat straightforward to observe the relaxation of vibrational modes, and this yields information along the lines of what you refer to. Additionally, this enables the determination of anharmonic coupling parameters, as shown in ref. 4 for the terahertz region.

1 M. Thomas, M. Brehm, R. Fligg, P. Vöhringer and B. Kirchner, Phys. *Chem. Chem. Phys*, 2013, **15**, 6608–6622.
2 M. T. Ruggiero and J. A. Zeitler, *J. Phys. Chem. B.*, 2016, **120**, 11733–11739.
3 G. Katz, Z. Zybin, W. A. Goddard III, Y. Zein and R. Kosloff, *J. Phys. Chem. Lett.*, 2014, **5**, 772–776.
4 A. Pereverzev, T. D. Sewell and D. L. Thompson, *J. Chem. Phys.*, 2014, **140**, 104508.

Jonas Nyman said: On a weekly basis, I perform comparisons between experimental powder X-ray diffraction patterns and patterns simulated from CSP structures. It is common that there are several predicted candidates (possibly disorder states) that all seem to match reasonably well. It would be very interesting if THz spectroscopy and calculated phonons could be used as a second opinion that would be orthogonal/independent to PXRD to help resolve disorder *etc.*

Michael T. Ruggiero replied: Absolutely – there is no question that simulated phonons can be used in addition to PXRD for structure determination. As mentioned in my previous responses, THz-TDS is extremely sensitive to the bulk packing, therefore subtle structural changes can have a significant impact on the observed spectra. So in this case, the only limitation comes from the computational cost of performing vibrational simulations.

J. Christian Schön commented: In Fig. 2 of your paper, you claim to see strong peaks, but at first sight, I would have assumed these bumps to be just fluctuations on top of a quadratic increase in alpha with frequency. Is this broad shape of the peaks to be expected?

Michael T. Ruggiero replied: I think I should clarify that the only definitive "strong" peak observed is in the lowest temperature spectrum, at approximately 1.25 THz. In the highest temperature spectra, there is potentially a minor feature at 1.75 THz, for example. This "bump" is consistently present and upon background subtraction becomes much more significantly pronounced. It is not believed to be an artifact, since instrumental artifacts manifest as periodic oscillations and not as isolated bumps, which was confirmed by performing experiments on samples with varying concentrations. The broadness of the peaks is to be expected, as disorder has been shown to increase the linewidths of terahertz modes (see ref. 1 for a more in-depth discussion of this).

1 M. T. Ruggiero, W. Zhang, A. D. Bond, D. M. Mittleman and J. A. Zeitler, *Phys. Rev. Lett.*, 2018, 196002.

J. Christian Schön commented: Could you construct a Debye model for the phonons from your data?

Michael T. Ruggiero answered: This is a question that often arises in the terahertz community, and it can be done. However, proper fitting often requires the use of double- (and sometimes "triple") Debye models, which often does not result in much significant insight outside of obtaining a good fit to the data due to the use of fit parameters. Therefore this is not a recommended approach.

J. Christian Schön continued: Can you see the effects of the phase transition at 320 K in calorimetry measurements?

Michael T. Ruggiero replied: For this study we have not performed any calorimetry experiments ourselves, but given that this is a first-order phase transition we would fully expect to observe a clear measure of the phase transition using thermodynamic measurements. A previous study using X-ray diffraction has also observed the phase transition at the same temperature that we have.[1]

1 Y. Takari, K. Nakata, T. Taniguchi and K. Sakurai, *Acta. Cryst.*, 1978, **B34**, 2579–2586.

Ivo Rietveld commented: What kind of information can be obtained by terahertz time-domain spectroscopy in the case of plastic crystals? Have you applied the method described in the paper leading to the full temperature and pressure properties to plastic crystals?

Michael T. Ruggiero replied: This method represents a very powerful way of investigating plastic crystals. We have recently published a follow up study to this, see ref. 1, where we have done exactly what you have brought up – a temperature and pressure-dependent study of the order-disorder transition in crystalline camphor. In that work, we were able to further investigate the vibrational mode that contributes to the phase transition, observe its behaviour with temperature and pressure, and ultimately relate that mode to the kinetics of the phase transformation, for example. Additionally, the accurate simulation of the low-frequency spectra in these materials implies an accurate modelling of the thermodynamics, which we have also performed and found very good agreement with the experimental data.

1 M. T. Ruggiero, W. Zhang, A. D. Bond, D. M. Mittleman and J. A. Zeitler, *Phys. Rev. Lett.*, 2018, 196002.

Ivo Rietveld added: I would like to remark that from a thermodynamic point of view co-crystals and solvates are one and the same. The only and, say, purely practical distinction is caused by the fact that solvates contain molecules that in their pure state form a liquid under ambient conditions, whereas co-crystals contain only molecules that in their pure state are all solid under ambient conditions. I wonder therefore whether it is important to specify whether a particular molecular compound is a solvate, unless one wants to stress how such molecular compounds have been obtained.

In the case of co-crystals (and solvates), the co-formers (and/or solvents) have a clear position within the crystal structure and form an inherent part of the crystal structure. It gets more difficult for channel solvates, where mostly solvent molecules may not possess a well-defined position within the channels available to them. One could say that in such a case the crystal should be classified as a solid solution, but the boundary between co-crystal (*i.e.* solvate) and solid solution in such cases may in particular be difficult to define experimentally. And I have no idea how this may play out from a structure prediction point of view, whether it is easy to generate structures with randomly distributed molecules within a crystal structure.

Sharmarke Mohamed responded: The challenge we have as a community regarding the definition of solid forms is that we don't all refer to the same characteristics of the solid when defining it. So in the case of cocrystal *vs.* solvate, some of us focus on the underlying physical states of the molecules that make up the crystal whilst others prefer to look at other features such as the ionization state of the molecules as the basis of the classification. I suspect that we will continue to have disagreements in this area. As to whether we should continue to classify binary solids consisting of molecules with no charges into various categories (*e.g.* cocrystal, stoichiometric *vs.* non-stoichiometric hydrate, solvate), clearly some descriptor is needed in order to emphasize a particular feature of the solid. So for example, a stoichiometric hydrate implies a specific ratio of molecule to water in the crystal and also implies a specific influence of the solvent on the bulk crystal upon dehydration. The challenge with classifying cocrystals and solvates under the same category of multicomponent solid form is that such nuances would not be easily conveyed.

Andrew I. Cooper added: I do not really disagree, except that typically, a cocrystal is formed on purpose, by addition of a second component, whereas a solvate is (often) accidental. I see no real problem with having two terms, even though there is no thermodynamic distinction.

Joost van den Ende opened the discussion of the paper by Marcus Neumann: Thank you very much for writing an interesting paper about a collective statistical analysis of confidential individual CSP studies finished with a beautiful description of the conflicts of interest. As, in principle, every company doing crystal structure prediction combined with an experimental polymorph screening could do a similar analysis, I would like to request of you the following. Would it be possible to write your sentence "we use a Figure of Merit (FOM) that is the sum over the absolute differences between the number of structures in each bin" in a mathematical formula?

Marcus Neumann responded: We have modified the manuscript accordingly in proof.

Graeme M. Day asked: This is a very nice approach, but even with this large set of crystal structure prediction results you have found that there is still quite a lot of uncertainty in the final parameters of the model, including the hindered fraction h. What size of data set do you think that you really need to pin the parameters down?

Marcus Neumann answered: Independently fitting all three parameters of the model probably requires a reduction of the error bars of each bin by a factor of 2. Hence, statistics over 160 cases are required to obtain a more reliable estimate of the hindered fraction.

As the cost for computational CSP studies comes down, the number of studies performed every year goes up. I would expect the limit of 160 cases to be exceeded in 1–2 years.

Graeme M. Day said: Extending the data set of crystal structure prediction studies included in this analysis relies on the methodology being kept exactly the same in future studies. Is the crystal structure prediction methodology likely to remain unchanged or are details like the final DFT energy model that you use likely to change?

Marcus Neumann answered: It seems to me that PBE(0)+MBD+Fvib as pioneered by Alexander Tkatchenko and Johannes Hoja is a very promising method for the final energy ranking in CSP and our CSP methodology is indeed likely to change in the near future.

If CPU time availability permits, we will at least partially recalculate the energy landscapes that have entered our statistical analysis.

Alternatively, we will have to wait for a new data set to build up as more CSP contract research is performed.

John B. O. Mitchell asked: In the toy model that you produce for the landscape of polymorph energy rankings, you have a parameter for kinetically hindered structures that do not become experimental polymorphs. When the polymorph hasn't been seen, not because it's strongly kinetically hindered but merely because of insufficient "time and energy spent on research on that compound", as McCrone would have put it, should that be considered as a separate parameter in this kind of model?

Marcus Neumann replied: Most compounds sent to us for CSP are late stage compounds and have already been extensively screened experimentally. So the missing forms identified by our statistical model really do not readily crystallize.

But it is also true that the step-function-like separation into structures that do and do not crystallize is very crude. At present, the small size of the sample (41 cases) does not allow for the elaboration of a more sophisticated model.

Sharmarke Mohamed asked: The results of the paper by Neumann and van de Streek are interesting and continue to emphasize the limits of manual crystallisation screens as well as the value of integrating crystal structure prediction (CSP) methods as part of the industrial solid form screening process. The authors conclude based on the study of 41 CSP studies on pharmaceutical compounds that have gone through extensive experimental polymorph screens, that "the thermodynamically most stable form has not yet been observed experimentally in 15 to 45% of the cases". This range of 15–45% seems very wide and I was wondering if the authors could clarify its meaning and significance in the context of how prevalent this phenomenon is within the industry?

Previous CSP studies with the GRACE program have classified predicted structures based on "families" rather than space groups or other structural descriptors. If we assume the upper limit of 45% for the incidence of the above mentioned phenomenon, what proportion of these systems that display this phenomenon (where the thermodynamically most stable form has not yet been observed experimentally) have global minima that belong to a completely different "family" class to the known experimental structure? This question is of practical importance since it may provide clues as to how much effort must be expended in crystallising the thermodynamically most stable (but yet uncharacterised) form. In cases where the predicted global minimum structure is of a completely different family class to the known experimental structure, it could be argued that the experimental conditions needed to crystallise this stable but yet unknown phase will be substantially different to those used to crystallise the known form.

Marcus Neumann replied: Having examined many of the crystal energy landscapes of the 41 test compounds in detail, I can indeed confirm that a missing thermodynamically more stable form indeed typically belongs to a different family of structures compared the experimentally known forms. And yes, crystallizing the missing form is likely to require very different crystallization conditions.

Sarah L. Price said: What was the reaction of your customers in the cases where a more stable form was predicted by CSP than was already known? Your results show that there were several cases where there was a clearly more stable form that had not been found.

Marcus Neumann answered: Some customers take our predictions very seriously and substantially increase their experimental screening efforts, if a substantially more stable form is predicted to be missing.

Sarah L. Price added: The question as to whether the most thermodynamically stable computer generated structure must be experimentally realisable was one that exercised us on finding that CSP generated significantly more stable structures for both enantiopure and racemic chloromandelic acid than could be found.[1] We tried to rationalise this with a huge range of types of calculation and experimental work, but did observe that it was hard to imagine the molecules coming together to form many of the computer generated low energy structures. For example, the most stable structure would have had to avoid making or break many hydrogen bonds to get into the densest structure, and the molecules would have needed to form enantiopure hydrogen bonded sheets that then interdigitated to give a racemic crystal for others. It does sometimes seem that the statistical chance of molecules coming together to nucleate some structures in competition with less stable packings that are easier to access may be very low. Do you think that there are some structures that are calculated to be the most stable that could be dismissed as being impossible to crystallise?

1 R. K. Hylton, G. J. Tizzard, T. L. Threlfall, A. L. Ellis, S. J. Coles, C. C. Seaton, E. Schulze, H. Lorenz, A. Seidel-Morgenstern, M. Stein and S. L. Price, Are the Crystal Structures of Enantiopure and Racemic Mandelic Acids Determined by Kinetics or Thermodynamics?, *J. Am. Chem. Soc.*, 2015, **137**, 11095–11104.

Marcus Neumann responded: We have made similar observations.

When a predicted form is found to be significantly more stable than the known experimental forms, it is often intuitively clear from the crystal structures why the missing form does not readily crystallize.

The more difficult bit is to quantify the crystallizability, and possibly to conclude that a form is very unlikely to occur under typical experimental conditions within a certain time frame.

It is also important to note that while there are certainly more predicted than observed crystal packings, the two numbers approach with increasing energy ranking accuracy. In that respect, early crystal energy landscapes are misleading.

For the low-energy missing forms, it is often possible to imagine specific crystallization experiments based on the predicted crystal structure. It is too early to say how many of those attempts at rational crystallization experiment design are typically successful.

In the range of molecular complexity that we have looked at, I expect most missing, thermodynamically stable forms to be dynamically attainable under very specific conditions.

J. Christian Schön said: Very interesting analysis of the number of possibly not-yet discovered low-energy metastable kinetically stable modifications. We have performed a similar analysis at a much lower level of statistical confidence with *ab initio* energy global optimizations on the calcium carbide system,[1] and found it quite useful in estimating how far we had come in our search. Concerning your stability estimate: you should be able to take some of those systems you have been studying and try to find the energy barriers separating or surrounding "good-looking" local minima in the lower energy region of your energy landscape – this should give you an estimate of the parameter you need in your model. The exponential growth you assume in the number of minima has been seen in many systems where exhaustive search methods were applied. The exponential growth appears for the first thousands of minima, but then usually begins to level off and becomes more like a Gaussian distribution.[2,3]

1 A. Kulkarni, K. Doll, J. C. Schön and M. Jansen, *J. Phys. Chem. B*, 2010, **114**, 15573–15581.
2 J. C. Schön and P. Sibani, *Europhys. Lett.*, 2000, **49**, 196–202.
3 B. Doliwa and A. Heuer, *Phys. Rev. E*, 2003, **67**, 031506.

Marcus Neumann answered: Our statistical analysis is based on families, *i.e.* groups of rather different crystal structures, and the energy error that enters the model is the inter-family energy error. From the visual inspection of many landscapes it is clear that the energy barriers for inter-family conversion are substantial and definitely orders of magnitude higher than the small energy differences between crystal polymorphs.

I agree that the distribution of families should be a Gauss distribution if regarded over a large energy scale. However, because of the number of degrees of freedom involved, organic CSP typically focuses on the low energy fringe of the Gauss distribution.

J. Christian Schön added: Continuing the previous question/comment: Examples for such systems with an approximate exponential growth in the number of local minima in chemically relevant systems are *e.g.* lattice polymers[1] and lattice glasses.[2]

1 J. C. Schön, *J. Phys. Chem. A*, 2002, **106**, 10886–10892.
2 J. C. Schön and P. Sibani, *J. Phys. A: Math. Gen.*, 1998, **31**, 8165–8178.

Sten Nilsson Lill asked: You conclude that for about 15–45% of current drugs on the market the most stable polymorph has not yet been found. What were your expectations when you initiated this study?

Marcus Neumann answered: Since cases like rotigotine and ritonavir have occurred in the past and there are several hundred important drugs, we certainly expected the fraction of missing themodynamically stable forms to be larger than 1%.

Artem R. Oganov remarked: You make a very interesting statement: "for simple molecules, the global minimum is always the one discovered by experiment. For complex molecules (of similar complexity to most drugs) with more degrees of freedom, in 15–45% the global minimum is not yet experimentally seen". I wonder: can we state that the disappearing polymorphs phenomenon happens when there is a large number of possible conformations and the molecule crystallizes in a "wrong" conformation?

Marcus Neumann replied: Molecules adopting an unfavourable conformation in the most stable crystal structure is definitely one of the mechanisms for the disappearing polymorph phenomenon, as exemplified by the ritonavir case.

Artem R. Oganov continued: For simple molecules you expect the ground-state polymorph to be found in ~100% cases; for more complex pharmaceutical molecules you show it to be found in 55–85% cases. Imagine you increase the number of conformational degrees of freedom even further and consider proteins. Would you expect that in these cases we almost never observe the stable state and in ~100% cases deal with metastable ones? Yes, proteins fold very reliably – but perhaps into an easy-to-fold metastable state?

Marcus Neumann responded: Yes, I indeed tend to think that very large molecules such as proteins will hardly ever adopt there most stable crystal structure. For such molecules even more than for small molecules, we will have to learn which lattice energy minima are kinetically accessible.

Proteins have been designed by nature to adopt a reproducible structure in solution or cell membranes, and it remains an open question if that structure corresponds to a stable or a metastable state. The whole idea of protein crystallography is that by crystallizing a molecule from solution we can learn something about the structure of molecules in solution. Hence, proteins do not unfold and refold to adopt their most stable crystal structure.

Angeles Pulido noted: About the idea of risk: even if we put ourselves in the worst case scenario, 45% misses something. The academic brain says "we are in trouble". A pharma company thinks "do I need to worry?". Pharma companies need a good reason to stop research. How will the properties change? Of that 45%, how much do I need to worry that those properties are going to change? How much would the properties change across the families?

Marcus Neumann responded: The key property for pharmaceutical compounds is solubility. When the energy decreases by 0.5 kcal mol^{-1}, the solubility decreases by roughly a factor of 2.3. For an energy difference of 1.0 kcal mol^{-1}, the factor is 5.4. For most of the missing forms, the predicted energy difference is larger than 0.5 kcal mol^{-1}, and a decrease in solubility by a factor of more than 2.3 will cause major problems for most pharmaceutical formulations.

Grahame Woollam enquired: The CSP studies presented are primarily coming from contract research, as opposed to collaborations. Did you request that the pharmaceutical companies inform you whether they found the more stable form, outside of the three months from receiving the request to publishing the report? The conclusion that the thermodynamically most stable form has not yet been observed experimentally in 15 to 45% of the cases would have been likely to differ had the question been asked some years later, following extended experimental screening efforts.

It would be useful to ask the customer if the predicted more stable form had been crystallised in the following 1–10 years to enable a meaningful statistic to be established.

Marcus Neumann answered: I fully agree.

Unfortunately, the time tends to be quite long before we get experimental feedback on our crystal structure prediction studies. In the case of dalcetrapib, four years passed before the compound was declassified and before we found out that the high pressure experiment proposed by us had actually been successful.[1]

1 M. A. Neumann, J. van de Streek, F. P. A. Fabbiani, P. Hidber and O. Grassmann, *Nat. Commun.*, 2015, **6**, 7793.

Grahame Woollam added: In the case of dalcetrapib two of the experimental crystal forms are disordered and match several similar computer-generated structures.

Of the 41 crystal structure predictions that you performed, in how many of them do you think that disorder could play a role in stabilising/destabilising the crystal landscape?

Marcus Neumann answered: I lack the resources at the moment to check in detail but I would expect disorder to stabilize an important experimental form in 10–20% of the cases.

Marta Dudek asked: Could you give a more detailed description of how do you decide on the families of the predicted crystal structures? Which factors are accounted for? Can you think of a way by which this process could be automated?

Marcus Neumann replied: Typically, we assume that structures belong to the same family if:

The packing is essentially the same but somehow distorted.

The overall packing of the molecules is the same but there are some conformational differences with respect to side groups and no reason to assume that conversion is kinetically hindered.

The structures share thick 2D sheets.

Luca Iuzzolino communicated: In your study you state that small, rigid molecules always crystallise in the global minimum in potential energy, as they can crystallise more easily than their larger and more flexible counterparts. How can this statement be reconciled with the survey conducted by Cruz-Cabeza *et al.* in "Facts and fictions about polymorphism",[1] which showed that there is no correlation between the number of polymorphs molecules form and their size and/or level of flexibility?

A. J. Cruz-Cabeza, S. M. Reutzel-Edens and J. Bernstein, *Chem. Soc. Rev.* 2015, **44**, 8619–8635.

Marcus Neumann communicated in reply: It seems to me that the following two questions are independent enough for their answers not to require reconciliation:

(1) How likely is the crystallization of the most stable form as a function of molecular size and flexibility?

(2) How many polymorphs are likely to observed as a function of molecule size and flexibility?

Yet, I would like to add an interesting observation on the nature of the crystal energy landscapes of flexible organic molecules that may shed some extra light on the question. Predicted landscapes often exhibit a large diversity of packing motifs. Compared to this diversity, different polymorphs obtained for one and the same compound are often surprisingly similar. Entire regions of the crystal energy landscape are hard to access experimentally. Experimentally observed polymorphs often belong to only one particular region.

Jan Gerit Brandenburg began a discussion of the paper by Jonas Nyman: You showed how you want to get a workflow and apply methodology. I was worried about the overlay of your GRACE search *vs.* the CrystalPredictor search in your presentation (Fig. 1). This might be worrisome for the search algorithms. Why does this happen? Was this just one special case and are "only" high-energy structures missing?

Jonas Nyman answered: I think we should be worried about the search completeness problem. For ROY, the poor overlap between the Vasileiadis study and our CSP with GRACE is probably partly due to the suspiciously poor performance of the PBE functional, which GRACE relies heavily on. However, in 2005 Bouke van Eijck checked the search completeness for the submissions to the third blind test, and those results were very worrying indeed. Our finding just confirms van Eijck's results.

Graeme M. Day remarked: You talked about comparing the results of crystal structure prediction from different methodologies, and that Bouke van Eijck did such a study after the third blind test of crystal structure prediction.[1] His approach was to re-optimise everyone's sets of predicted structures using a consistent energy model. This means making a decision about which is the

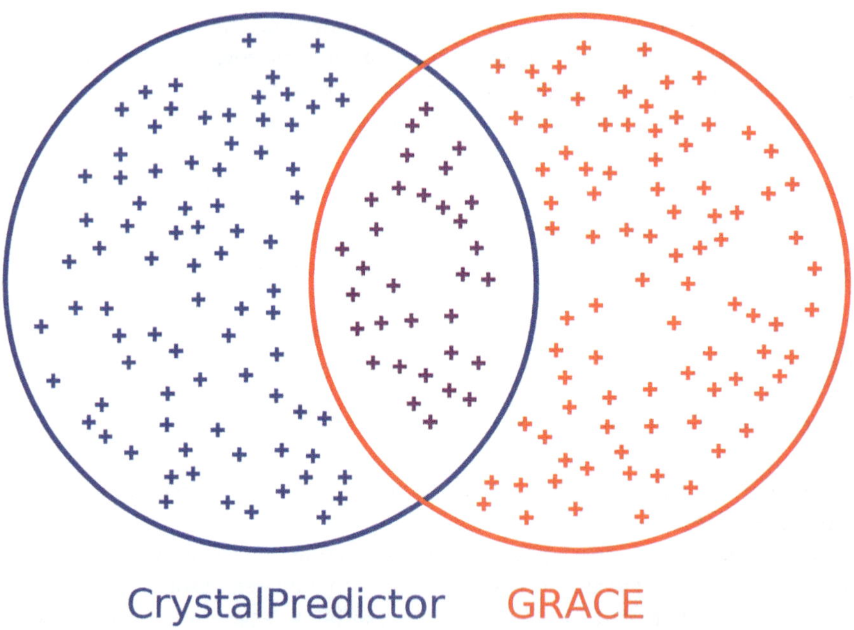

Fig. 1 Search completeness of CSPs of ROY using GRACE and CrystalPredictor+CrystalOptimizer. Taking the 100 lowest-energy structures from each landscape and comparing them to all structures from the other method shows that the two landscapes to a large extent contain structures found by only one of the methods.

appropriate energy model to use. Is this what you did, and do you think that this is a necessary step when comparing the results from different methods?

1 B. P. van Eijck, Comparing hypothetical structures generated in the third Cambridge blind test of crystal structure prediction, *Acta Cryst.*, 2005, **B61**, 528–535.

Jonas Nyman answered: Bouke van Eijck ran into great difficulties when trying to decide which structures should be considered to be the same and which should be seen as different (the clustering problem). Because of that, he had to re-optimize all the structures to a common potential energy surface. I chose not to do so for three reasons:

(1) Given the much better energy models used in the two studies I compared, I thought/hoped that the clustering problem would be less severe.
(2) The computational cost of re-optimizing all the structures with a modern energy method would make such comparisons challenging, which is unfortunate, since I would want more people to do such comparisons.
(3) Re-optimizing the structures would answer the question of whether different methods reproduce the same potential energy minima or not.

That is a relevant question for people developing CSP methods, but that is not the question I wanted to answer here. I actually wanted to see if different methods reproduce the same crystal structures or not. Bouke van Eijck's article is a sobering read, and at least in my opinion, not cited enough. He found that the submissions to the third blind test to a very large extent were in disagreement. My results only confirm his findings.

Jason Cole asked: Given your findings, do you feel there is a need for a further blind test of organic crystal structure prediction methods. If so, what form do you believe this test should take?

Jonas Nyman responded: The blind tests are the only way we can objectively judge the performance of CSP methods. So yes, I think we should have another one.

Jason Cole asked the delegates: If there were a new blind test, possibly hosted by CCDC, in your opinion:

What form do you feel would be most beneficial to the community?

How do you feel such a test should be funded?

Over what timescale would you wish to see the test occur?

Graeme M. Day responded: These are good questions because the blind tests of crystal structure prediction for organic molecules have changed each time that they were run and it is a good time to assess the questions that we want to answer from the blind tests, and the form these tests should take.

The big question that most people reading the blind test papers want to answer is:

(i) "How well do the various methods work at predicting crystal structures?"

This general question considers the whole process, which differs greatly between research groups and software packages that are developed for crystal structure prediction. Failure to predict the target crystal structure can be due to one of a number of failures, the most common of which are an incomplete search or an energy model (or other ranking criterion) that does not rank the structures correctly. Therefore, from the point of view of method development, we need to break things down, which we have tried to do in the analysis of blind test results, into

(ii) "How well do the different methods do at locating the relevant crystal structures?"

and

(iii) "How successfully do the methods rank the experimentally observed crystal structures within the lists of predictions?"

There have been suggestions to change the way that the blind tests are run, to specifically address questions (ii) and (iii). For example, by enforcing that every participant uses the same energy model, so that differences between predictions are purely down to the search method. I agree that this is a valuable test, but I'm not sure that this needs to be done as a "blind" test. I think that questions (ii) and (iii), when addressed on their own, can be tested as benchmarking validations of search methods and energy models without needing to be tests under "blind" conditions. For example, the community could develop some benchmark sets: for testing search methods, this could be a comprehensive set of generated crystal structures. How to ensure that this set is comprehensive is difficult and it might be that the set needs to evolve, allowing researchers to add new low energy structures as they are identified. The energy range of these structures is another question; the relevant range for polymorphism is quite well understood, but there are cases where higher energy structures are relevant such as desolvated solvate structures. The structures should all be geometry optimised with the same energy model, which should be of good quality, but should not be so computationally expensive that it prevents researchers from testing new methods in a reasonable amount of time. It must also be available in open source software, so that there are no large software costs involved in reproducing the benchmark set. Such sets could be put together for a range of systems with different characteristics: *e.g.* rigid molecules, flexible molecules, multi-component crystals. There is an advantage to not doing this under blind test conditions, in that we would have greater freedom to choose the best systems; blind tests end up being quite restrictive because they have to be systems where the data has not been published and is not available to the predictors. This restriction would unnecessarily rule out many useful benchmark molecules.

Benchmarks of relative energies could be well-studied polymorphs, where relative energies have been measured with sufficient accuracy. There has been a separate discussion of this topic and the need to fund such measurements for a range of systems. Again, these do not need to be blind tests because the results of such benchmarking should be reproducible: code for performing the calculations should be available to people who want to be able to reproduce the results.

The need for blind tests is really to convince the community of potential users of crystal structure prediction that the methods work, and to point them to the methods that work most reliably. Therefore, if the blind tests continue to be held, then I think that these should continue to test the "whole process" together, which means keeping a similar format to the previous tests: allow people to use whatever combination of search methods and energy models that they want. What we need to consider is what data participants need to submit when providing their predictions. This has changed since the earliest blind tests and participants are now asked to submit longer lists of predicted structures than the 3 predictions that were allowed in the early blind tests. This is so that different aspects of the predictions can be analysed. I think that this should continue, should probably involve the submission of several hundred structures per target, and needs to be made a strict criterion for submission, *i.e.* submissions that do not provide sufficient data should be excluded.

The question of funding involves a few things. As well as costs involved with the organisation (time spent gathering target structures and running the test), there are already significant hidden costs to the blind tests. The researchers spending time on the blind tests are taking their time away from other projects, which could be research council-funded projects or projects within a company. The time that researchers spend on blind tests can be significant, so participating research groups make this decision to take resources away from other research projects to take part. This includes person time and computing time. Different groups will have different capacities to do this, and will prioritise the blind tests differently, which can affect their results (*e.g.* busy groups or groups with less computing access might need to cut corners). We need the timescale and scope of the blind tests to be such that research groups can accommodate participation, while keeping the targets relevant to demonstrating advances that are being made. The move to more targets in the blind tests does not help this, so should probably be scaled back to 3 or 4 target systems. Longer timescales are helpful: if we continue with a similar format and number of targets to the most recent blind tests, then I suggest that a minimum of 12 months is needed between the release of targets and the submission deadline. An alternative is that the blind tests move away from having big, multi-target tests every few years and, instead, have smaller, more frequent tests. Perhaps each year have a single target, rotating between categories (rigid molecule, flexible molecule, multi-component crystal, *etc.*) each year. This might be easier for research groups to accommodate, although it might require a more sustained organisational effort. It would also change the way that the blind tests are reported (a large paper every few years). This is just an idea that might want to be considered. If we stick with the format of blind tests with several target systems, then a frequency of every 5 years is probably sufficient and a good balance between the need to update the community and the disruption to other research that the blind tests involve.

Funding of the organisation of blind tests is a different issue. The CCDC has been generous in hosting the blind tests until now. Having organised two of them myself, I know the amount of work involved. The community appreciates the efforts of the CCDC in their hosting of these exercises. If the CCDC no longer feels that it should bear the associated costs of acting as hosts and organisers, then it might be appropriate for the community to approach someone like the IUCr, or a funding body, to see if they are able to help.

Marcus Neumann replied: This is an interesting question and a single format for the blind test can probably not suit all purposes.

Is the blind test organized to improve CSP methodology? Or to help practitioners of CSP understand what to expect and how to best use CSP results?

In the first case, the energy ranking problem and the crystal structure generation problem should be dealt with separately. For the energy ranking part, polymorphic enthalpy and/or free energy differences should be measured but not publicly disclosed prior to the energy ranking blind test. For the structure generation part, it may be desirable that all participants use the same energy calculation method, such that the generated energy landscapes are truly comparable. However, it is important to realize that such an approach is not as objective as it may seem at first sight. Real-world CSP requires different energy methods to be used for the structure generation and the final energy ranking, with some sort of statistically-controlled connection between the two steps. This aspect of CSP would not be tested by a generation exercise in which all participants use the same energy method. Also, the most appropriate global optimization strategy depends on the properties of the energy method, such as reproducibility and the continuity or discontinuity of energies and their derivatives. Behaviour in chemically unreasonable regions also plays an important role. So the success rate of structure generation methods in such a blind test may be strongly biased by the choice of the energy calculation method.

If the blind test is meant to help practitioners of CSP, more emphasis should be put on experimental aspects. It would probably make sense to organize the blind test in several stages:

Stage I: Compounds are selected and extensive polymorph screens are carried out experimentally. Polymorphic enthalpy and free energy differences are carefully determined. Some compounds may drop out in stage I if deemed too complicated.

Stage II: The actual crystal structure prediction blind test. At the end of Stage II, the results of the predictions are compared to the experimentally available information. A workshop is held as usual, but more emphasis is put on how to crystallize predicted but not yet observed crystal forms.

Stage III: Experiments are conducted to make new forms. Another workshop is held to discuss which methods have been successful.

The experimental and organizational aspects would probably require public funding.

Organic CSP is developing fast. Having a blind test every 3–4 years would certainly be appropriate.

Sarah L. Price answered: The Blind Test has fulfilled two very valuable functions:

(1) Allowing the testing of new methods as different groups enter the field of CSP. There should continue to be new methods, which should take us beyond CSP_0, and which are likely to include the application of the CSD to CSP to exploit informatics.

(2) Allowing industry to assess the claims made, particularly by commercial companies, as to the usefulness of CSP as a complement to solid form screening and pharmaceutical development.

CSP is far from being a solved problem, as the difficulty of predicting the crystal structure is very dependent on the molecule and structure involved, and so the format should continue to stimulate improvements in the methods, rather than declare a winner. The division into search extent and ranking is a false division, which is very useful for individual groups in the post-analysis, but not one where different approaches can be usefully compared. I would propose just two or three categories, the essential two being:

A small, simple (approximately rigid) molecule, similar to XXII, to test (1), and a (model) pharmaceutical, which has been polymorph screened, with the aim of predicting the known polymorphs (*cf.* XXIII), ideally with relative stability data, to test (2).

These two categories proved the most widely discussed and useful in the last blind test. If there was to be a third, then a multi-component system (hydrate/solvate/cocrystal) would be best. However, the amount of time and resources involved in entering is such that reducing the number of targets would be helpful to many groups.

Could the next blind test be funded through CCDC's pharmaceutical crystal forms consortium? I'd hope that one company would be able to provide the outcome of the polymorphic screening data for target (2) on a dead compound that could be published. The consortium in general might provide the funding for the organisation and meeting. What does the consortium (which I think is dominated by innovators rather than screening companies) want from the blind test?

The timescale needs to be a long one, so it can be scheduled with other demands. For academic groups, having a deadline mid-summer and a meeting in early September, with the target released almost a year before the deadline, works well. The tests have had a major impact on developing CSP and it is essential that they continue. As I tried to argue in my introductory lecture, we are only at the beginning stages of having a theory and codes for genuine CSP that does not over-predict polymorphism.

Jonas Nyman responded: I think the past blind tests have served us well and I hope there will be more. I think it could make a lot of sense to split the test into two parts: the search completeness test, and the ranking test. To estimate search completeness, all groups would generate trial structures with their preferred method and then optimize them to a potential energy surface common to all the groups. The structures from all the groups are then pooled, and the search completeness for each group could be expressed as the percentage of the pool that they found.

The ranking problem test could be performed by allowing each group to use their preferred method to predict the relative thermodynamic stability between pairs of polymorphs. This would necessitate that the target system has at least two well-characterized polymorphs, but I think this would be better than the current method that assumes that the observed single form is the thermodynamically stable one. I think one year for the entire test, containing perhaps 3 or 4 target systems, is reasonable.

I do not have an opinion on who should pay for it.

Jonas Nyman continued: Is the CCDC considering organizing a new blind test? If so, would there be any changes to the format? The last blind test, with five targets, meant that trying to address all the targets was a very large commitment. Participating in a blind test can put most other research in an academic group on hold for at least half a year or so.

Jason Cole answered: The CCDC's view on new blind tests is quite simple. We would consider holding a new test if there was a strong drive and need in the CSP community for such a test, and the CSP community felt that the CCDC was the best organisation to hold such a test.

We would need to understand and define the scope of such a test with the community and also understand and define how such a test would be funded. Any new test would first be subject to input from the CSP community to design a test process that would be more beneficial to participants than the time cost of entering.

Aurora J. Cruz-Cabeza continued the discussion of the paper by Jonas Nyman: After generating a landscape computationally, you may predict that the known form is the most stable or that a thermodynamically stable form is yet to be crystallised. In the context of pharmaceuticals research and development, I wonder what the practical impact of such predictions is. For example, would such predictions result in a shortening of polymorph screenings or an extension of them? How is the technology applied in practical terms? How is this driving further experimental research?

Marcus Neumann replied: It would be dangerous to reduce the amount of experimental screening based on a crystal structure prediction study, simply because computational methods never cover the entire search space (good coverage with one molecule per asymmetric unit, often reasonable coverage with two molecules per asymmetric unit, and poor coverage above two molecules per asymmetric unit).

On the contrary, we have seen companies significantly enhance their experimental screening efforts in cases where a significantly more stable, and hence significantly less soluble crystal form was predicted to be missing.

Susan Reutzel-Edens replied: In my experience, whether crystal structure prediction can be used in pharmaceutical R&D to either shorten or extend an ongoing polymorph screen will depend on

(1) the extent to which the molecule has already been screened for solid forms and

(2) our confidence in whether or not the known (stable) form is likely to be the thermodynamically most stable of all the calculated structures at finite temperatures.

It is important to recognize that the goal of solid form screening in the pharmaceutical industry is not only to find the thermodynamically most stable (neat) form, but also important metastable forms, some of which could under the right set of conditions become the most stable form, as well as crystalline hydrates and solvates. Even the best crystal energy landscapes are limited by the scope of the search (Z' value, space group symmetry, inclusion of solvent, *etc.*) and therefore provide only a partial picture of the possible solid form diversity that is of interest to the pharmaceutical industry. Thus, only with a good faith effort in having experimentally screened for solid forms can a crystal energy landscape give at least some confidence in having found the stable form. In these cases, CSP may help to effectively "right-size" the solid form screening effort. By contrast, in those instances where a more stable structure is predicted at relevant temperatures, the challenge is on to develop strategies to target those structures in crystallization. With structure prediction falling far short of polymorph realization, it is apparent that more experimental and computational research is needed in the areas of crystal nucleation and growth.

Jonas Nyman replied: Since it is not yet possible to predict the thermodynamic stability of pharmaceutical polymorphs with quantitative accuracy, we have to take the CSP results with a (large) grain of salt. I doubt we would trust CSP results to such an extent that we would stop an experimental screening, although this scenario is highly desirable as it could save us a lot of money, time and effort. The opposite however can happen, if CSP suggests that there are additional more stable forms to be found, as is normally the case, that could motivate additional screening work, or extending the crystallization conditions with non-standard methods such as high-pressure crystallizations.

Susan Reutzel-Edens asked: From a pharmaceutical development perspective, I have one further comment with regard to whether CSP can/should be used to start/stop solid form screening. There is obviously much emphasis in CSP algorithm development on finding the most stable crystal structure and computing it as the most stable polymorph. However, industry is also very good at producing metastable forms. Here, CSP can help in identifying those forms and providing high fidelity structures suitable for modeling property variation across different lots of drug materials. In the future, solid form screens may be "right-sized" based on CSP to generate all pharmaceutically-relevant polymorphs, stable and metastable.

Luca Iuzzolino commented: You have shown in your Venn diagram (Fig. 1) that CrystalPredictor and GRACE generate very different crystal structures for ROY, as very few can be found by both methods. Where can the structures matching the known experimental polymorphs of ROY be found? Are they found by both methods? And what is the significance of this?

Jonas Nyman answered: This is interesting! Of the 7 known $Z'=1$ forms that these methods in principle could have found, CrystalPredictor+CrystalOptimizer found all of them, and GRACE missed one. That is, the experimental forms (with one exception), are in the union between the two methods. Without having done any statistical analysis, this seems to be much better than if the experimental forms were spread out randomly in the same way as the hypothetical structures. This indicates that the experimental forms are somehow easier to find than the other lattice energy minima. Presumably, this means that experimental polymorphs correspond to unusually deep or large minima, which of course is encouraging from a CSP point of view.

Luca Iuzzolino remarked: In your CSP workflow you use CrystalOptimizer as an intermediate optimisation step. You explicitly optimise only the degrees of freedom that are selected by the programme AUTODOF. In my work I realised that the absolute energies that are obtained with CrystalOptimizer are very dependent on the number of degrees of freedom are explicitly optimised. I have tested this for a set of twenty low energy crystal structures generated in the Blind Test CSP study on molecule XXVI (this includes a match to the experimental form, structure 1600). I plotted the energies obtained with various methods for selecting the independent degrees of freedom relative to the ones obtained for optimising the same crystal structures with CrystalOptimizer when all torsion and bond angles are treated as explicit variables in the optimisations (Fig. 2). The energies obtained when selecting the degrees of freedom with AUTODOF are on average approximately 5.3 kJ mol^{-1} higher than those obtained when optimising all torsion and bond angles. These are only slightly better than the energies obtained when optimising only the intuitively flexible torsion angles, which are on average approximately 6.9 kJ mol^{-1} higher). The energies can be improved by adding to the AUTODOF set a few extra torsion and bond angles, lowering the discrepancy to 3 kJ mol^{-1}. However, this is not too concerning for your workflow as the differences in the relative energies between the crystal structures are smaller than those in the absolute energies and you are only using CrystalOptimizer as an intermediate step in your workflow.

Jonas Nyman answered: You are right. Of course, optimizing more degrees of freedom will always give you better results, but often at a greatly increased computational cost. The purpose of using AutoDOF within the CLilly workflow was to select an economical set of the most important torsions and angles, to be used in an intermediate optimization step, and I think it will work reasonably well for that purpose. In the end, CLilly will re-optimize the most promising structures with full relaxation with periodic DFT-D. It is very difficult indeed to write down a small number of selection rules that picks the best degrees of freedom for any given molecule.

Claire S. Adjiman remarked: First, some comments on the comparison of ROY CSP studies presented in the presentation by Jonas Nyman and in particular the

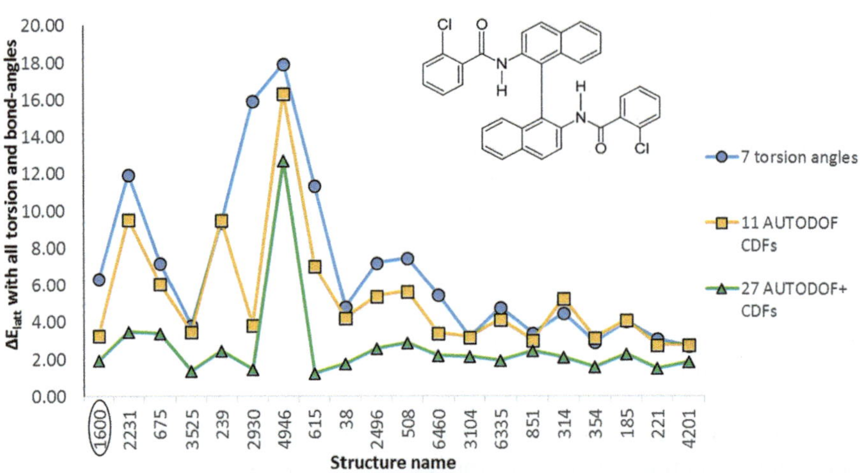

Fig. 2 Comparison between the lattice energies obtained after full optimisations with CrystalOptimizer of twenty low energy crystal structures generated in the Blind Test CSP study on molecule XXVI, in which various sets of independent conformational degrees of freedom where explicitly optimised. For each crystal structure, the energy is shown as the difference with that obtained when explicitly optimising all the torsion and bond angles for the same structure. The crystal structure that matches the experimentally known of molecule XXVI form is circled.

Venn diagram (Fig. 1) showing the overlap in the structures generated by CrystalPredictor/CrystalOptimizer and GRACE. This diagram was not shown in the paper, but highlights that there are a few common structures and many different structures in the lowest 100 structures generated with each method.

(a) I want to clarify that the results shown for CrystalPredictor/CrystalOptimizer are those obtained after the CrystalOptimizer stage as published in ref. 1.

(b) Such discrepancies have previously been observed, *e.g.*, in molecule XX of the blind test, where CrystalOptimizer and GRACE had clear differences in the ranking of the same structures re-evaluated with the lattice energy models used in the respective codes. No matter how extensive the search is, this inherently means that there will be differences in the landscapes and that we need to continue to improve the accuracy of the energy models, as discussed in detail throughout the meeting. For ROY specifically, I think it would be important to include more than the bottom 100 structures in making a comparison because the landscape is especially dense. In fact, structure ON was found at rank 129 in ref. 1, with a lattice energy 10.2 kJ mol^{-1} higher than the global minimum.

Finally, a question on CLilly. Could you expand on the drivers for developing CLilly and specifically what gaps it addresses that are not currently covered by existing codes?

1 M. Vasileiadis, A. V. Kazantsev, P. G. Karamertzanis, C. S. Adjiman and C. C. Pantelides, *Acta. Cryst.*, 2012, **B68**, 677–685.

Jonas Nyman replied: Thank you for those comments. CLilly is my attempt to automate very large calculations. When the number of structures that are considered in a CSP becomes very large, it becomes necessary to have a high degree of automation, or it becomes tedious to manually submit and monitor jobs on a cluster. CLilly is intended to provide that automation. It sets up jobs and automatically submits them to a supercomputer, and collects the results. CLilly is a very general server-client implementation so virtually any kind of calculation can be integrated. In the future, I hope to implement a better method for generating conformations, as well as methods for intermediate optimizations of crystals of flexible molecules, and better final ranking of predicted structures. I am not aware of any other program that provides the same functionality, although Graeme M. Day's group (me included) have worked on developing the CSPy codes, that to some extent implement similar functionalities. I am indebted to my former colleagues in that group for developing some of the ideas that made it into CLilly.

Joost van den Ende said: Thanks for writing a beautiful paper in which, amongst others, the history of CSP is very interestingly described. Reading the sentence "lacking efficient and accurate methods to predict nucleation, growth and barriers to solid-state phase transitions, we are far from being able to predict which structures could be observed as experimental polymorphs" made me think back about "Nucleation – A Transition State to the Directed Assembly of Materials: Faraday Discussion" which has been held a little over three years ago in Leeds. I am wondering if there will be a point in time where the fields of CSP and nucleation will merge.

Jonas Nyman answered: I think nucleation and growth are fundamental problems in chemistry that have largely resisted our attempts at modelling, and for organic molecular crystals, we are facing a huge challenge in understanding these mechanisms. Thinking of the generally very small energy differences between polymorphs, and considering the fact that the formation of cocrystals seems to be largely thermodynamically controlled (see ref. 1), we can conclude that kinetic mechanisms (probably) have a relatively small effect, though it is difficult to quantify. To properly predict polymorphs, and not just hypothetical structures, it seems to be necessary to include factors beyond thermodynamics in order to reduce the large number of predicted structures to experimentally observable polymorphs.

1 C. Taylor and G. M. Day, *Cryst. Growth Des.*, 2018, **18**, 892–904.

Qiang Zhu asked: I have a comment on the plot with the comparison of two CSP methods shown in Jonas's presentation (Fig. 1). Since two types of force fields were used in different methods, it is hard to compare the reliability of the structure search methods. CrystalOptimizer attempts to sample many structures while GRACE used Monte Carlo style methods which is more focused on the low energy structures. The latter is more sensitive to the choice of FF. If the FF cannot represent the potential energy surface well, it will fail badly. It is not fair to compare two different styles of search method in my opinion.

Jonas Nyman answered: You are wrong. What algorithms or energy models different programs use internally are irrelevant for this comparison. The figure shows to which degree different methods reproduce each other's crystal structures. Only low-energy structures were used in the comparison, so all CSP methods should be expected to reproduce these structures. It is a fair comparison.

However, there is a related question of whether different energy models have overlapping basins of attraction. That is, are crystal structures from one method reproduced when re-optimized with another force field? I think this is what you are referring to, when you object to my comparison. Bouke van Eijck investigated that issue after the third blind test, and came to a similar conclusion: CSP methods are poorly reproducible.

Qiang Zhu remarked: If you use a random structure prediction method, the results will be less sensitive to the quality of the force field as it assumes a uniform sampling. Applying a biased method like a genetic algorithm and parallel tempering, the system will be sensitive to the choice of force field (randomness). Such bias will encourage more extensive sampling in the low energy regions. If the energy landscape cannot be well represented by the energy model, the results are made to strongly deviate from the random approaches.

Artem R. Oganov said: I have a comment on the graph of CrystalPredictor *vs.* GRACE for local minima (Fig. 1): in any stochastic method, even if you do 2 runs with the same method, you will hopefully get the same solution (*i.e.* the global minimum), but very different local minima. The overlap will be quite small if you do 2 runs with the same method. You would need to run an infinitely long simulation (not a good idea in my opinion). A good CSP method gives you experimental structures (typically with very low energies) in a small amount of time. In that case overlap with high energy structures will be minimal – you want to skip local minima and get the global minimum as fast as possible.

Jonas Nyman replied: I agree that it is futile to try to find every single local minimum in the entire search space. But in the figure I showed, I had taken only the 100 lowest-energy structures from one method and looked for a corresponding structure anywhere among the structures of the second method, and then *vice versa*. The 100 lowest structures are so low in energy (a few kJ mol^{-1} from the global minimum), that we would expect any decent CSP method should find the majority of the local minima within that range, *i.e.* the search should be essentially complete in this region of the search space.

Artem R. Oganov added: Using stochastic methods, each time you run a simulation with the same input, you will sample a different set of local minima. So, it's no wonder that runs based on different methods (one of which is stochastic) give you different sets of local minima.

Jonas Nyman responded: We use either pseudo- or quasi-random numbers, so the calculations are repeatable, they are not truly stochastic. A very important requirement on any scientific method is that it is reproducible. Our models and calculations should mimic physical reality, and if/once they do, any CSP should produce the same set of predictions (in the lowest-energy part of the landscape), the predictions that correspond to minima on the true physical energy-surface. That is, for sufficiently accurate CSP methods, we would expect that each method/code yields (essentially) the same predictions. Clearly, we are not there yet.

Claire S. Adjiman added: It has been mentioned that all crystal structure prediction searches include a random element in the search. I want to point out that some methods do not include any element of randomness. This is the case for CrystalPredictor/CrystalOptimizer. In CrystalPredictor, the global search is carried out by using a Sobol′ sequence,[1–3] in which a repeatable sequence of starting points is generated. This does not depend on any seeds or any other specification. Carrying out the search twice will result in exactly the same landscape. It is possible to keep extending the search and doing so always results in an increased likelihood of finding all low energy minima. We are currently working on implementing a convergence criterion, based on the number of new structures generated for each of 50 000 minimisations/starting points.

1 I. M Sobol′, *USSR Comput. Math. Math. Phys.*, 1967, **7**, 86–112.
2 P. G. Karamertzanis and C. C. Pantelides, *J. Comput. Chem.*, 2004, **26**, 304–324.
3 I. Sugden, C. S. Adjiman and C. C. Pantelides, *Acta. Cryst.*, 2016, **B72**, 864–874.

Jonas Nyman added: I'd like to clarify that in the Venn diagram I showed (Fig. 1), I took the 100 lowest-energy structures from the first method (that have Z' and the space group common to both methods) and compared them to all structures from the second method, and then *vice versa*. That is, I only compared the lowest-energy part of the landscapes, up to a few kJ mol^{-1}. We would expect any CSP method to find essentially all the structures in that range, so it is surprising that these otherwise well-performing methods disagree to such an extent.

Artem R. Oganov responded: Actually, I am not surprised. In stochastic methods (and many, though not all, CSP methods are stochastic), every time you rerun your calculation, it will find a different set of local minima - but hopefully the same global minimum. Also, a good CSP method should sample as few local minima as possible, before finding the global minimum. So, asking that all local minima be found is very similar to asking for a very poor global optimization algorithm.

Artem R. Oganov continued: Fig. 1, showing that two runs of limited size found two different sets of local minima gives a very simple message to me – that there are many more local minima than any of your runs can sample. If you do another run, you will most likely find many additional minima.

Marcus Neumann added: In the presentation of the paper "Crystal structure prediction is changing from basic science to applied technology", a plot was shown (Fig. 1) which suggests that the crystal energy landscapes created with CrystalPredictor and GRACE are very different, each method missing a substantial number of structures.

Crystal energy landscapes calculated with different methods often seem very dissimilar, but this does not necessarily hint at a failure of the crystal structure generation for several reasons:

(1) If two approaches do not use the same energy calculation method or if one method does not use stringent lattice energy minimization convergence, structures in the two landscapes belonging to the same local attractive basin may not be considered identical by the clustering algorithm.

(2) For performance reasons, many generation algorithms are biased towards the generation of low energy structures, and landscapes can only be expected to be similar within an energy window of a few kcal mol^{-1} above the most stable predicted structure.

(3) The potential energy landscapes of flexible organic molecules are full of side minima with small attractive basins. When perturbed, *i.e.* driven over a small energy barrier, structures corresponding to such side minima typically convert to significantly more stable structures. This would also happen in the real world, triggered by thermal motions. Side minima with small attractive basins may be generated once or never in a crystal structure generation, but this is of no practical importance since the side minima do not correspond to observable structures anyhow.

Claire S. Adjiman responded: First, I want to clarify that the structures presented were those generated by CrystalOptimizer (CrystalPredictor was applied first, then the most promising structures were refined/re-ranked by CrystalOptimizer).

(1) I agree entirely.

(2) With different energy functions, I think we are likely to see many differences even near the global minimum. Having said this, in the case of ROY, and with the approach we used for the 2012 paper from which the structures were obtained, the highest-ranked structure to match an experimental structure (structure ON) was found at rank 129, and 10.2 kJ mol^{-1} above the global minimum. In the presentation by Jonas Nyman, only 100 structures were compared – I believe more should be compared.

(3) We certainly need to develop methods to eliminate such minima from the results of CSP studies.

J. Christian Schön remarked: On the issue of GRACE and CrystalPredictor giving different results and the general question of the "repeatability" of optimizations, some comments. It is quite clear that stochastic global optimization techniques will produce different results in each run, unless you drive them very close to exhaustive scanning of the landscape. Based on our experience, we feel that it pays to use two (or more) global exploration methods for crystal structure prediction *via* energy landscape searches, which are "orthogonal" to each other, in the sense that they approach the search taking different features of complex landscapes into account. Thus, we often perform *e.g.* first a simulated annealing or genetic algorithm search, and follow this up by a threshold search, since the former reacts primarily to the energetic barriers on the landscape while the latter focuses more on the local density of states and the entropic barriers at a sequence of energy lids. (As a bonus, the threshold algorithm yields information about the generalized barriers in the system, and thus the stability of the modifications.) Thus, we could find new deep-lying local minima, which did not show up during the standard simulated annealing runs although they were even expected from crystal chemical considerations: *e.g.*, the anatase type on the landscape of MgF_2,[1] which we had expected to appear in analogy to the TiO2 landscape; so far it has not been synthesized in the MgF_2 system, but we could show by MD simulations that it would grow as a kinetically stable modification during a deposition process of a Mg/F gas phase, once a seed with the anatase structure had been provided.[2] Quite generally, we feel that is would be useful for people who do not need to "defend" their method, *i.e.* prove that the algorithm they have developed is the "best", to employ a combination of different global exploration methods to solve the crystal structure prediction problem for a given system of interest, in order to make sure that all relevant candidates are found.

1 M. A. C. Wevers, J. C. Schön and M. Jansen, *J. Phys.: Condens. Matter*, 1999, **11**, 6487–6499.
2 S. Neelamraju, J. C. Schön and M. Jansen, *Inorg. Chem.*, 2015, **54**, 782–791.

Marcus Neumann responded: A good global optimization algorithm should provide some sort of completeness estimate and converge up to a user-defined degree of completeness. Subsequent and statistically independent runs should provide results that agree within the user-defined degree of completeness.

Just running several structure generation algorithms will not help if neither the overall method nor the individual algorithms implement convergence control.

However, pursuing several different structure generation strategies can turn out valuable in terms of CPU time consumption in cases where there is no clear guidance on which method performs best. But in such a case the overall method still should implement some sort of convergence control.

Sharmarke Mohamed asked: It is very encouraging to see the range of stakeholders from the pharmaceutical industry participating in this Faraday Discussion who clearly feel that crystal structure prediction (CSP) methods are a valuable asset and investment for their company. Based on my experience working in the industry, companies rightly assess the value of any technology on the basis of what value it adds to their operations. Given the evolving state-of-the-art in CSP methods, can you comment on the degree to which practitioners (*e.g.* academics) and commercial companies providing CSP methods are actively working to pass on knowledge and make their algorithms/methods available to crystallisation scientists working in the industry, rather than operating on a contractual project-by-project basis? Whilst you mentioned in your remarks that it is difficult to answer this question for the entire industry, perhaps you can give a response from the vantage point of your current efforts at Eli Lilly. The answer to this question is important if we are to see a critical mass of users of CSP methods within industry in the same way that we see an increasing interest in the hiring of computational data scientists, particle engineers and crystal engineers working on problems related to solid form screening and characterization.

Jonas Nyman replied: It is very difficult to know what commercial stakeholders are doing internally. We already see that the majority of all CSP calculations are never published. We also know, or strongly suspect, that there are pharmaceutical companies that do CSP, and have chosen not to make this publicly known. In my talk and our article, we mentioned only companies that have themselves disclosed their interest in CSP. I think it is fair to say that some are more secretive than others. At Lilly, we are applying CSP to portfolio compounds, but those results are unlikely to be published, or are published long after the work was done. We are also actively developing methods and software, often in collaborations with academic groups. For such work, we use molecules like ROY, and this we do publish as much as possible.

Ivo Rietveld opened a discussion of the paper by Qiang Zhu: The structure of the eighth polymorph of ROY was solved by generating powder diffraction patterns from calculated crystal structures.

To what extent can this method be used as a standard to solve crystal structures from powder diffraction experiments?

In your case, the unit cell was indexed from the experimental pattern. How feasible is it at present to match powder patterns from scratch without knowing the unit cell parameters?

If not, how soon do you think that this method would be feasible in the future?

Qiang Zhu replied: In our past experience, we have examined a series of joint works between theory and experiment.[1] In most of the cases where we have $Z' > 1$, CSP could solve the structures nicely without cell parameter information. However, we do have two cases where cell parameters are needed, ROY and coumarin form III. The reason lies in that our CSP method (similar to many other method(s) requires two steps:

(1) a fast screen of thousands of structures based on a classical force field;

(2) DFT energy ranking for the 100–200 lowest energy structures.

When we did our study on coumarin, form III indeed showed up in step 1, however, it was skipped in step 2 due to unfavorable energy ranking.

When we did our study on ROY, we did not really find the target structure of R05 at step 1. This is because the classical force field failed to describe the energy minimum well. ROY is also a very challenging case at the QM level. So, my short answer to your last question is the following:

This method is ready for most of the cases even when $Z' > 1$. But it relies on the accuracy of the force field. If the cell parameters are known, the dependence on the force field is less crucial and it should work in general.

1 http://www.physics.unlv.edu/~qzhu/topic2.html.

Jason Cole remarked: Your collaborators managed to index the powder pattern that led to a unit cell for your CSP search. There are existing simulation methods that can solve structures relatively easily once you have indexed the pattern, particularly when presented with synchrotron data. Did you, or your, collaborators attempt these methods before employing CSP? If so, which methods did they try and why did these methods fail?

Qiang Zhu replied: They tried, but I don't know the exact name of packages which they used. As far as I know, such methods can generate some other structures which can equally fit the PXRD well but they have relatively high energies. My interpretation is that those simulation methods are based on the degree of match in intensity, but did nothing on energy evaluation. There may exist multiple structures which may fit the experimental PXRD relatively well.

Jason Cole added: In our experience, solving a powder pattern from synchrotron data is straightforward once you have a reliable unit cell, as the search space is very constrained (see for example ref. 1 which outlines many highly complex structures that have been solved in this way, some from lab data) so I was surprised to see that this structure had been challenging.

1 E. A. Kabova, J. C. Cole, O. Korb, A. C. Williams and K. Shankland *J. Appl. Cryst.* 2017, **50**, 1421–1427.

Qiang Zhu answered: Our experimental colleagues attempted to solve the structure purely from the powder diffraction pattern by using some packages (which may include DASH). But they failed. We recently published a series of works based on the synergy between experiment and theory, for coumarin,[1], resorcinol,[2] aspirin,[3] and glycine.[4]

For the cases of coumarin and resorcinol (rigid molecules), packages like DASH may work. But we used CSP before we got the synchrotron data. For the case of ROY, aspirin and glycine, we did CSP based on the cell parameters derived from synchrotron data. Straight use of a program like DASH may generate some high energy structures which may fit the observed powder pattern equally well. Lastly, for the special case of ROY, there actually exist many rounds of conversations between experiment and theory. Our experimental colleagues sent us not only one set of parameters. In fact, they sent us four sets of cell parameters, and we had to try multiple times in order to arrive at the final solution.

1 A. G. Shtukenberg, Q. Zhu, D. J. Carter, L. Vogt, J. Hoja, E. Schneider, H. Song, B. Pokroy, I. Polishchuk, A. Tkatchenko and A. R. Oganov, *Chem. Sci.*, 2017, **8**, 4926–4940.
2 Q. Zhu, A. G. Shtukenberg, D. J. Carter, T. Q. Yu, J. Yang, M. Chen, P. Raiteri, A. R. Oganov, B. Pokroy, I. Polishchuk and P. J. Bygrave, *J. Am. Chem. Soc.*, 2016, **138**, 4881–4889.
3 A. G. Shtukenberg, C. T. Hu, Q. Zhu, M. U. Schmidt, W. Xu, M. Tan and B. Kahr, *Cryst. Growth Des.*, 2017, **17**, 3562–3566.
4 W. Xu, Q. Zhu and C. Hu, *Angew. Chem.*, 2017, **129**, 2062–2066.

J. Christian Schön said: About twenty years ago, several research groups in Bonn,[1] Birmingham,[2] and Australia[3] developed codes using such Pareto optimizations, *i.e.* the cost function was a linear combination of the potential energy/enthalpy (for $p \neq 0$) and the difference between the computed and measured powder diffractograms. Such a code is commercially available at Crystal Impact.

1 H. Putz, J. C. Schön and M. Jansen, *J. Appl. Crystallogr.*, 1999, **32**, 864–870.
2 O. J. Lanning, S. Habershon, K. D. M. Harris, R. L. Johnston, B. M. Kariuki and E. Tedesco, G. W. Turner, *Chem. Phys. Lett.*, 2000, **317**, 296–303.
3 A. A. Coelho, *J. Appl. Crystallogr.*, 2000, **33**, 899–908.

Qiang Zhu responded: I agree. Pareto optimization provides a nice solution to solve such kinds of problem. However, the main challenge for our problem is perhaps to derive the reliable PXRD data from experiment. My experimental collaborator needs to manually remove the reflections contributed by other substances. Using the raw data directly may lead to very wrong results.

J. Christian Schön asked: You mentioned that you are using information about peak indexing as part of your crystal structure prediction – in what form and at what stage of the search procedure do you enter this information?

Qiang Zhu responded: We used first principle CSP without any priors at the beginning, but we did not find any structure which matched the experimental PXRD of R05. Then the experimentalist tried very hard to collect PXRD data using a synchrotron and managed to obtain a set of cell parameters. For each cell parameter, we performed the CSP within the given fixed cell. One of the CSP runs finally yielded the right form for R05.

In short, we just used the cell parameters as the geometry constraint in the CSP. We did not optimize the PXRD match in the course of the CSP.

J. Christian Schön commented: In response to the Maddox challenge, people started restricted structure searches where they included information about cell parameters and the number of formula units per unit cell from the outset as constraints on the optimization.[1–4] How does your method compare with their approaches?

1 J. Pannetier, J. Bassas-Alsina, J. Rodriguez-Carvajal and V. Caignaert, *Nature*, 1990, **346**, 343–345.
2 C. M. Freeman, J. M. Newsam, S. M. Levine and C. R. A. Catlow, *J. Mater. Chem.*, 1993, **3**, 531–535.
3 P. A. Wright, S. Natarajan, J. M. Thomas, R. G. Bell, P. L. Gai-Boyes, R. H. Jones and J. Chen, *Angew. Chem., Int. Ed.*, 1992, **31**, 1472–1475.
4 D. K. Belashenko, *Inorg. Mater.*, 1994, **30**, 966–976.

Qiang Zhu responded: I did not go through all of these papers, but I think the method used here is conceptually similar to that used by Freeman *et al.* in 1993. The only difference is that we are using a more sophisticated way to relax the geometry and perform the energy evaluation. And the systems are more challenging here.

Sharmarke Mohamed asked: The search method used to identify polymorph R05 of ROY used the monoclinic cell obtained from experiment and fixed this using appropriate constraints. Crystal structure prediction (CSP) methods generate many candidate structures with similar unit cells but small variations in the molecular orientations or atomic positions. Have the authors considered the use of additional experimental data for reference (*e.g.* spectroscopy) so that they can compare these experimental data with similar data obtained from computational methods? For example, we know that terahertz spectra are sensitive to small variations in molecular packing and may have been a good complementary experimental method to use in determining the structure of this new polymorph of ROY. How confident are you that the combined CSP and powder X-ray diffraction method used (in the absence of other corroborating experimental data) will work for determining the crystal structures of other molecular systems of similar complexity?

Qiang Zhu replied: We didn't consider any additional data such as Raman or THz spectroscopy. It would be nice to examine this polymorph with THz spectroscopy, though I am not sure how feasible it is as a theorist. Note that we could not obtain a pure R05 phase here.

We are very confident that the structure has been solved based on the following arguments:

(1) A decent match between experimental and simulated PXRD data

(2) A lowest energy structure from our CSP search within the experimental cell parameters, and that this structure is significantly better than any other structure found in such a search (>2 kJ mol^{-1} difference)

(3) The calculated lattice energy is very close to the ground state

Christopher Collins communicated: Can the authors comment on why they obtain very low GOF values in their Rietveld refinements (ideally this shouldn't really be below 1)?

Qiang Zhu communicated in reply: The BNL data has relatively low intensity, leading to the low GOF values calculated in our refinement. In contrast, the calculated GOF for the final Rietveld refinement of data collected at APS is 2.103.

Christopher Collins communicated further: How does the Rietveld GOF compare with a GOF value obtained from a LeBail/Pawley fit?

Qiang Zhu communicated in response: The final GOF determined from the Rietveld refinement for the data collected at BNL is 0.207. The initial GOF from the Pawley fit is 0.387.

Christopher Collins communicated: Can you show the simulated PXRD pattern from the structure solution obtained from USPEX against the experimental data?

Qiang Zhu provided Fig. 3 in reply.

Christopher Collins communicated: Can the authors indicate how many parameters are used in each of the Rietveld refinements?

Qiang Zhu communicated in reply: The total number of parameters defined in the final Rietveld refinement for the data collected at BNL is 48. Of these, 11 correspond to the background profile. All the other parameters pertain to the structure. In contrast, there are 99 total parameters defined in the final Rietveld refinement for the data collected at APS. Here, 58/99 parameters are related to the background profile. The remaining parameters are related to the structure. The difference in the number of parameters defined in the BNL and APS refinements corresponds to additional peak phases present in the APS data corresponding to sample transformation.

Christopher Collins communicated further: Can the authors plot an overlay of the CSP generated structure and the experimentally refined structure?

Qiang Zhu communicated in reply: I got these results from my experimental collaborators (M. Tan, A. Shtukenberg, and B. Kahr). The result of a Crystal Structure Packing Similarity calculation in Mercury for 20 molecules is shown below in Fig. 4 (20 of 20 molecules found; RMSD = 0.195). Molecules from the CSP structure are depicted in red, while molecules from the final Rietveld refinement of the data collected at BNL are displayed in blue.

Alan Hare opened a general discussion of the papers by Marcus Neumann, Jonas Nyman, and Qiang Zhu: Years ago, it was often stated that much of the

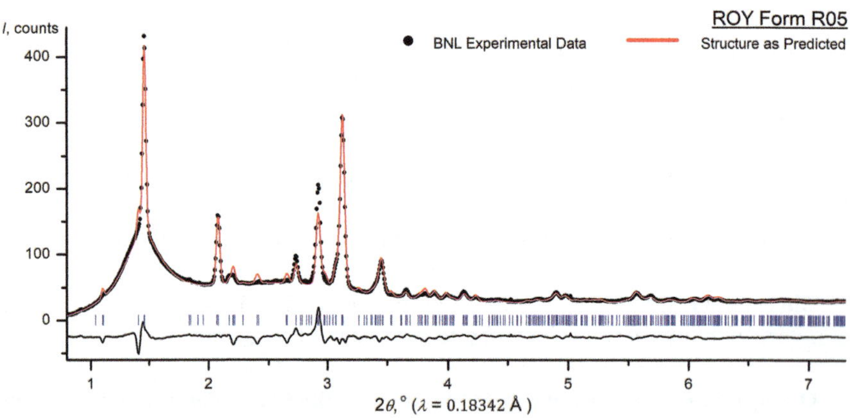

Fig. 3 PXRD comparison between the data collected at the 28-ID-2 beamline at the National Synchrotron Light Source at Brookhaven National Laboratory (referred as BNL data, in black) and the simulated data from the predicted structure (referred to as CSP, in red).

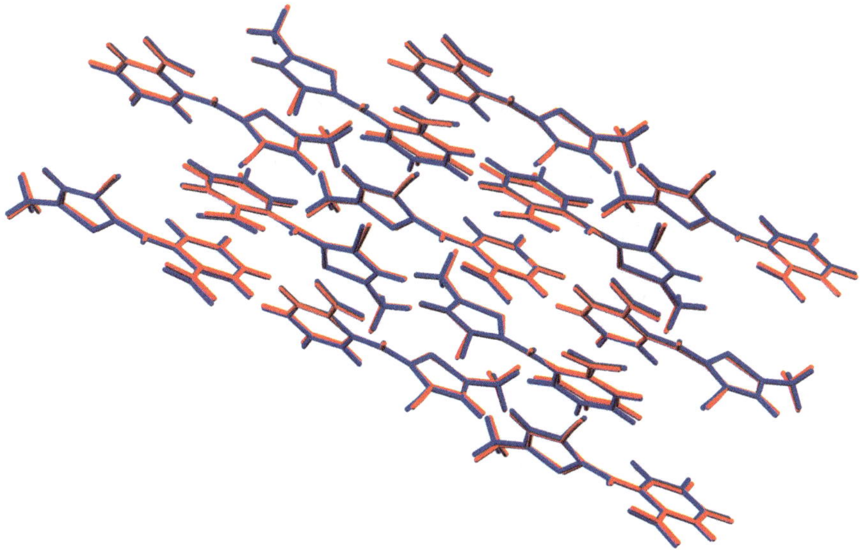

Fig. 4 The overlay of the CSP generated structure and the experimentally refined structure for ROY R05.

pharma R&D cost was due to trying to avoid toxicity. Can CSP say anything about toxicity?

(After the session, a delegate suggested that uric acid crystallisation might provide a worthwhile study. Even a short delay in treating gout with allopurinol may exacerbate the drug user's symptoms.)

Qiang Zhu replied: I don't think CSP can directly measure chemical properties like toxicity. However, it may have some relation with other properties such the solubility and eventually thermodynamical energy ranking. Some empirical relations might be valid here. That is, a metastable form may be more toxic, compared to the thermodynamical ground state. By the way, ref. 1 is a DDT paper which you may be interested in.

1 J. Yang, C. T. Hu, X. Zhu, Q. Zhu, M. D. Ward and B. Kahr, *Angew. Chem.*, 2017, **129**, 10299–10303.

Jonas Nyman responded: It's unlikely that CSP would shed much light on toxicity. I can think of only two scenarios:

(1) The toxicity is caused by a polymorph having a different solubility, causing a dramatic change in bioavailability. This is unlikely, since a new polymorph would be expected to be less soluble, but not impossible.
(2) There are examples of herbicides and insecticides (*i.e.* DDT), where different polymorphs have different efficacy/toxicity.

Sarah L. Price asked: Can toxicity kill the progress of the polymorph screening in the drug development process? Is this a common problem with industry being willing to commission or being able to evaluate the value of CSP studies?

Susan Reutzel-Edens replied: The simplest answer is yes. If compound development is terminated for toxicity reasons, then polymorph screening, along with other development activities, will be stopped. Toxicity is a surefire way to terminate a compound. In my experience, a solid form is not.

If the question here is whether different polymorphs could have different toxicities warranting a CSP search for safe polymorphs, the bioavailability of a drug, which impacts safety and efficacy, generally requires dissolution and therefore could in principle be impacted by solubility differences between polymorphs. A very interesting example of polymorphs having different toxicities in the crystalline state was recently published by Bart Kahr, Mike Ward and their teams at NYU on the contact insecticide, DDT. See ref. 1.

1 Y. Yang, C. T. Hu, X. Zhu, Q. Zhu, M. D. Ward and B. Kahr, DDT Polymorphism and the Lethality of Crystal Forms, *Angew. Chem., Int. Ed.*, 2017, **129**, 10299–10303.

Ivo Rietveld communicated: How advanced is the field in predicting temperature dependent stabilities of polymorphs?

Graeme M. Day communicated in reply: To my knowledge, there has not been a lot of work published yet on predicting the temperature dependent stabilities of polymorphs. We performed a large scale study recently on nearly 500 pairs of polymorphs to investigate the influence of temperature on their relative stabilities and properties.[1] In this work, we looked at a few polymorph pairs in detail to see if we could correctly predict the temperature dependence of the stabilities using force field-based quasi-harmonic lattice dynamics. Overall, our conclusion was that we can calculate the relative entropies quite well, but often get the overall free energy relationship wrong because of errors in the lattice energy differences (which are often small). This showed us that we can probably get improved predictions by combining the force field lattice dynamics with an improved static energy model (either from a better force field or by using quantum mechanical calculations). Another relevant study was from Cervinka and Beran[2], who used a fragment-based method to obtain the phase diagram of crystalline methanol with quite good accuracy, showing that the prediction of the temperature- and pressure-dependence of polymorph stabilities is possible, at least for relatively small molecules.

1 J. Nyman and G. M. Day, *Phys. Chem. Chem. Phys.*, 2016, **18**, 31132–31143.
2 C. Červinkaa and G. J. O. Beran, *Chem. Sci.*, 2018, **9**, 4622–4629.

Ivo Rietveld communicated further: I would like to ask whether it is possible to predict or see the amorphization of crystal structures of organic molecules under pressure and whether a precise transition pressure can be linked to it.

J. Christian Schön communicated in reply: I would think that it is a combination of exposure time and pressure that leads to the amorphization of a crystal. Similar to the glass transition temperature for which a temperature range is given, depending on the time over which the pressure is applied different levels of pressure will be needed to achieve (full) amorphization.

Seiji Tsuzuki communicated: I have a question on the force fields used for crystal structure prediction. Although several force fields have been applied for crystal structure prediction, the performance of the ranking of the stability of polymorphs depends strongly on the force field. I would like to know your opinion on which force field is the best for crystal structure prediction at present.

Jonas Nyman communicated in reply: I think it is fairly well established that general purpose force fields like COMPASS, AMBER and the like do not perform sufficiently well to be suitable for CSP. The accuracy is greatly enhanced by using multipole electrostatics, and this is normally used with the FIT and Williams type force fields, which are commonly used for CSP. However, the last blind test clearly showed that periodic DFT is now the most common method for the final, most accurate, calculation of lattice energies.

Conflicts of interest

Marcus Neumann works for a company that develops software for organic crystal structure prediction, and there are no other conflicts to declare.

Faraday Discussions

PAPER

Reducing possible combinations of Wyckoff positions for zeolite structure prediction†

Jiaze Wang,‡[a] Yuzhi Xuan,‡[a] Yue Ming[a] and Yi Li *[ab]

Received 19th February 2018, Accepted 29th March 2018
DOI: 10.1039/c8fd00040a

Zeolites are an important class of tetrahedrally coordinated inorganic materials that have been widely used in many chemical industries as catalysts, adsorbents, and ion-exchangers. To date, over 200 types of zeolite framework have been discovered. Predicting not-yet-discovered zeolite frameworks is of great importance not only for zeolite structure determination but also for the identification of promising synthetic candidates with desirable functions. However, owing to the complexity and diversity of zeolite framework topologies, zeolite structure prediction has been a challenging task for several decades. Many efforts have been devoted towards this end, among which the computer-aided assembly of zeolite framework constituent atoms (T atoms) in predefined Wyckoff positions (WPs) is a promising approach because of its high efficiency in configuration space searching. However, this approach suffers from high computational overheads caused by the large number of possible WP combinations. On the basis of the analysis of known zeolite structures, we find that the site symmetries of many WPs are incompatible with the tetrahedral coordination of T atoms. Moreover, to avoid the formation of chemically unfeasible distorted tetrahedral coordination, T atoms cannot be too "crowded" in some WPs. We define, for the first time, the T atom distribution (TAD) densities for special site symmetries as the numbers of T atoms per special point, per unit length of rotation axes, or per unit area of mirror planes, respectively. By restricting the number of T atoms in every WP so as not to exceed the highest allowed TAD density, WP combinations for zeolite structure prediction can be reduced by 1–4 orders of magnitude. Taking advantage of this discovery, the efficiency of zeolite structure prediction based on the enumeration of WP combinations can be significantly improved.

[a]*State Key Laboratory of Inorganic Synthesis and Preparative Chemistry, Jilin University, Changchun 130012, China. E-mail: yili@jlu.edu.cn*

[b]*International Center of Future Science, Jilin University, Changchun 130012, China*

† Electronic supplementary information (ESI) available. See DOI: 10.1039/c8fd00040a

‡ These authors contribute equally to this work.

I. Introduction

Zeolites are an important class of inorganic open-framework materials that are widely used in many chemical industries.[1–3] These materials feature ordered distributed micropores in molecular dimensions that are suitable for the adsorption, diffusion, and reaction of extra-framework guest species. All zeolite frameworks are built exclusively from primary TO_4 tetrahedra (T denotes tetrahedrally coordinated Si, Al, or P, *etc.*). Each TO_4 tetrahedron is connected with four neighbouring TO_4 tetrahedra by sharing their vertex O atoms, forming a three-dimensional four-connected zeolite framework. Although all zeolite frameworks are constructed from primary TO_4 tetrahedra, the different ways in which they can be connected lead to a variety of zeolite structures.[4] To date, over 200 distinct zeolite framework types have been discovered in nature or by chemical synthesis.[5]

The theoretical prediction of not-yet-discovered zeolite structures has attracted much attention since the 1970s,[6] because this could not only provide initial atomic models for zeolite structure determination but also identify a large number of synthetic candidates with superior properties for the function-led discovery of new zeolites. In particular, with the development of computer technologies, many computational methods have been developed in the past three decades to predict unknown four-connected zeolite frameworks.[4] For instance, Deem *et al.* developed a Monte Carlo algorithm to predict four-connected zeolite frameworks on the basis of a set of simple cost functions;[7–9] Treacy *et al.* proposed a symmetry constrained intersite bonding search method to predict possible four-connected zeolite frameworks from enumerated graphs;[10,11] Delgado-Friedrichs *et al.* developed a systematic enumeration approach based on mathematical tiling theory to predict four-connected crystalline networks;[12] Mellot-Draznieks *et al.* developed the *de novo* simulated annealing approach to build zeolite frameworks using secondary building units;[13,14] Li *et al.* proposed the constrained assembly of T atoms around predefined "forbidden zones" to facilitate the formation of microporous zeolite frameworks;[15–17] Woodley *et al.* utilised evolutionary algorithms to generate four-connected zeolite frameworks with "excluded regions";[18,19] and Li *et al.* developed a ternary computer coding system to enumerate zeolite frameworks built from the stacking of 6-rings.[20] Coupled with various types of structure evaluation approaches,[21–24] these computational methods have generated millions of hypothetical zeolite framework models,[25–30] among which the most promising candidates for function-led zeolite synthesis or for zeolite structure determination can be identified.

Despite all the success, *ab initio* zeolite structure prediction remains challenging, because the framework structures are too complex and the connections among primary TO_4 tetrahedra are almost arbitrary. In particular, besides the general positions, zeolite framework T atoms tend to occupy special Wyckoff positions (WPs) in the unit cells (Fig. 1). To predict the structure of an unknown zeolite framework in a given space group, one has to figure out the WP for each framework T atom. A conventional method to do this is to determine the WP for each T atom according to its randomly generated coordinates and adjust its coordinates, as well as its WP, on the fly during atom assembly.[7–9] However, this method inevitably causes fluctuation in the number of T atoms and, even worse,

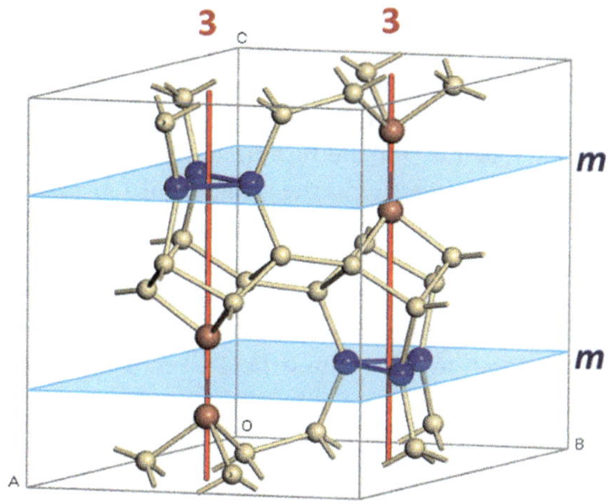

Fig. 1 Zeolite MEI with T atoms on 3-fold axes, mirror planes, and general positions. T atoms are shown in yellow, blue, and red; oxygen atoms are omitted for clarity.

a biased and incomplete search in the configuration space. An alternative method is to enumerate all the possible WP combinations for a given number of T atoms. Then, using framework-building algorithms that assemble T atoms in fixed WPs, such as FraGen,[31] all the enumerated WP combinations can be explored equally, leading to a more comprehensive search in the configuration space. Unfortunately, the number of possible WP combinations, which grows exponentially with the number of WPs and T atoms, can be prohibitively high for the subsequent framework-building computations. Therefore, reducing the possible combinations of WPs is a prerequisite for such "enumerative" zeolite structure prediction.

Here, we report our recent discovery in the reduction of WP combinations for zeolite structure prediction. We have developed a computer program to investigate the T atom distribution among various WPs in 221 known zeolite structures. We find that the site symmetries of many WPs, such as inversion centres (Fig. 2a), are incompatible with the tetrahedral coordination of zeolite T atoms, so these

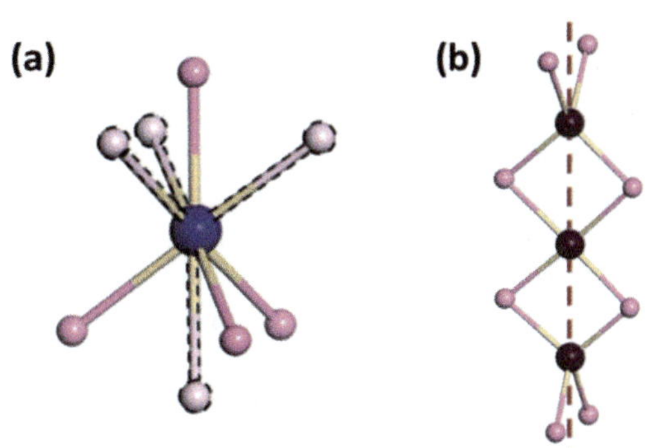

Fig. 2 (a) Inversion centre incompatible with the tetrahedral coordination of T atoms. (b) Distorted tetrahedral coordination caused by crowded T atoms on an axis. T atoms are shown in blue and brown; O atoms are shown in pink.

incompatible WPs should be avoided during the building of zeolite frameworks. Moreover, although T atoms may occupy some WPs, they should not be too "crowded" in these WPs. For instance, three close T atoms on an axis may directly bond with one another, causing the middle TO_4 tetrahedron to become highly distorted (Fig. 2b). We define, for the first time, the T atom distribution (TAD) densities for special site symmetries as the numbers of T atoms per special point, per unit length of rotation axes, or per unit area of mirror planes, respectively. By restricting the number of T atoms in every WP not to exceed the highest allowed TAD density, the possible WP combinations can be reduced by 1–4 orders of magnitude to a manageable level. With the implementation of this method, the efficiency of WP enumeration for zeolite structure prediction can be significantly improved.

II. Methods

The structure models for 221 known zeolites were obtained from the online structure database of the International Zeolite Association.[5] The 84 292 hypothetical zeolite structures were obtained from the database of Hypothetical Zeolite Frameworks.[30] Before TAD density calculations, all the structure models were geometrically optimised with the SLC potentials[32] using the program GULP.[33]

We developed a Python program to calculate the distribution of T atoms among different site symmetries in zeolite structures (Fig. 3). Starting from the CIF file of a zeolite structure, we retrieved the space group information, lattice parameters, and atomic coordinates for the corresponding zeolite structure. The locations/lengths/areas of all the special points, rotation axes, and mirror planes within the unit cell were calculated according to the lattice parameters and space group information. Then, we calculated the WP for each T atom, as well as its corresponding site symmetry, according to the coordinates of the T atom. If a T atom could be assigned to more than one WP, we chose the one with the lowest multiplicity. When the WPs for all the T atoms in the corresponding zeolite structure were obtained, we calculated the TAD density for each type of site symmetry. Here, we defined the TAD density as the number of T atoms per special point, per unit length of rotation axis, or per unit area of mirror plane, respectively. Note that the TAD density at a special point was dimensionless; it could be 1 if it was occupied by a T atom, or 0 if it was not occupied.

Framework building was performed using the program FraGen.[31] A simple cost function was defined according to the deviation of a hypothetical framework from an ideal four-coordinated zeolite structure. For each WP combination, 1000 cycles were run and 10 000 Monte Carlo moves were attempted for each cycle.

Knowing the highest TAD density for each type of site symmetry, we went on to enumerate all the combinations of WPs for zeolite structure prediction in a given unit cell. Owing to their microporous natures, the framework densities of most zeolite structures were in the range of 10–20 T per nm^3. Therefore, we were able to estimate the lower and upper limits for the number of T atoms in a given unit cell, namely N_{Tmin} and N_{Tmax}. For a space group with P types of WPs, the total number of T atoms in the unit cell, N_T, would be $\sum_{x=1}^{P} n_x m_x$, where n_x is the number of crystallographically distinct T atoms belonging to the xth WP, and m_x is the multiplicity of the xth WP. Knowing the highest TAD density for each type of site symmetry, we were able to deduce the upper limit for each n_x, *i.e.* n_{xmax}. We

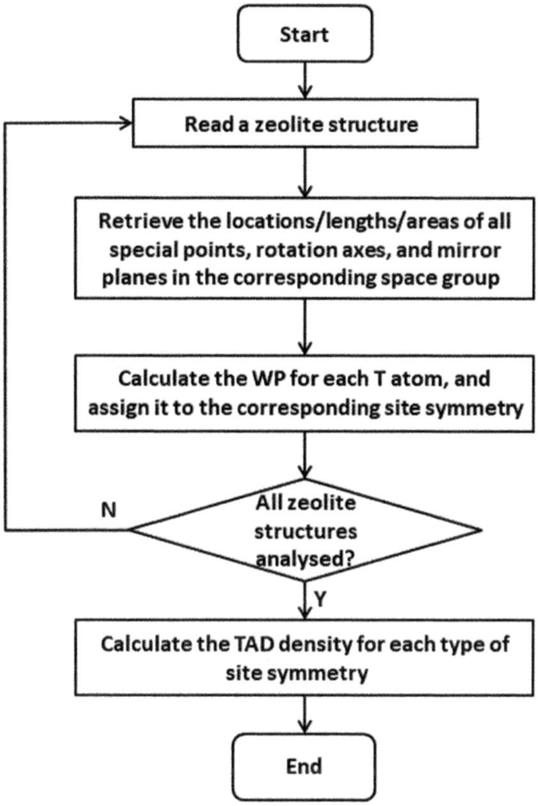

Fig. 3 Procedure to calculate the TAD densities among different site symmetries.

obtained all the possible WP combinations by enumerating all the n_x no more than $n_{x\max}$ that fulfilled $N_{\mathrm{Tmin}} \le N_{\mathrm{T}} \le N_{\mathrm{Tmax}}$. Note that $n_{x\max}$ for any special point should be 1 if it is occupiable by a tetrahedrally coordinated T atom, or 0 if it is not occupiable.

III. Results and discussion

First, we calculated the TAD densities for all the types of site symmetry in the 221 known zeolite structures. The results are listed in Table 1 and ESI Table S1†. Most of the known zeolite structures exhibit high symmetries, and nearly 70% of the known zeolites possess T atoms with special site symmetries, indicating their importance for the building of zeolite frameworks. Among all the types of site symmetry in 230 space groups, 222, $\bar{4}$, $\bar{4}2m$, $\bar{4}3m$, 2, $2mm$, 3, $3m$, m, and the general site are compatible with tetrahedral coordination, so they are all allowed to be occupied by T atoms. All these allowed site symmetries are observed in known zeolites. Besides the general positions, m, 2, and $2mm$ are the most frequently adopted site symmetries, occurring in 110, 40, and 30 known zeolite structures, respectively (Table 1). Other site symmetries, including $\bar{1}$, $\bar{3}$, $\bar{6}$, $2/m$, $4/m$, $6/m$, 32, 422, 432, 622, $\bar{3}m$, $\bar{6}2m$, mmm, $m\bar{3}m$, $4/mmm$, $6/mmm$, 4, 6, $4mm$, and $6mm$, are

Table 1 Occurrence and TAD density of each type of allowed site symmetry in 221 known zeolite structures

Site symmetry		Occurrence	Highest TAD density	Zeolite with the highest TAD density
Point	222	5	1	ASV *etc.*
	$\bar{4}$	2	1	UOZ *etc.*
	$\bar{4}2m$	8	1	SOD *etc.*
	$\bar{4}3m$	2	1	AST *etc.*
Rotation axis	2	40	2.021 T per nm	SAS
	$2mm$	30	1.603 T per nm	BCT
	3	5	2.325 T per nm	AFY
	$3m$	9	2.911 T per nm	SBN
Mirror plane	m	110	6.879 T per nm^2	IFY

incompatible with the tetrahedral coordination of T atoms, so they are all prohibited for the building of zeolite frameworks.

By calculating the number of T atoms at all the allowed points, rotation axes, and mirror planes, we obtained the TAD density for each type of site symmetry in the known zeolites. The results are shown in Table 1. As expected, to avoid the formation of distorted TO_4 tetrahedra, the TAD densities cannot be high in known zeolite structures. All the allowed special points, including 222, $\bar{4}$, $\bar{4}2m$, and $\bar{4}3m$, are found to be occupied at least twice in known zeolites. Since a point can only be occupied by an individual T atom, the highest allowed TAD densities for these point symmetries are all 1, according to our definition. The highest TAD densities for the rotation axes 2, $2mm$, 3, and $3m$, and the mirror plane m are 2.021 T per nm (in zeolite SAS), 1.603 T per nm (in zeolite BCT), 2.325 T per nm (in zeolite AFY), 2.911 T per nm (in zeolite SBN), and 6.879 T per nm^2 (in zeolite IFY), respectively. The highest TAD densities for the 3-fold axes are higher than those for the 2-fold axes. This is because the 3-fold symmetry forces two TO_4 tetrahedra to form a linear dimer aligning along the 3-fold axis, where the two T atoms and the bridging O atom form a straight line along the axis, with the rest of the vertex O atoms pointing out of the axis (Fig. 4a). Note that a linear trimer of TO_4 tetrahedra

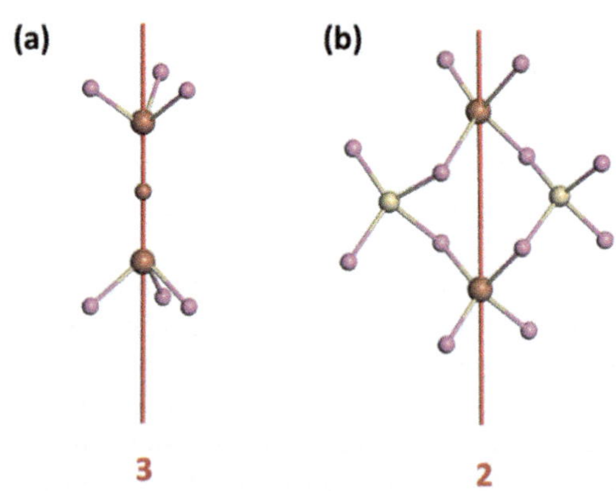

Fig. 4 (a) Linear TO_4 dimer aligned along a 3-fold axis. (b) Linking of two TO_4 tetrahedra on a 2-fold axis. T atoms are shown in red and yellow; O atoms are shown in pink.

is prohibited because it will cause distortion at the middle TO_4 tetrahedron (Fig. 2b). Therefore, a linear dimer is the closest manner for the packing of adjacent TO_4 tetrahedra, which leads to a relatively high TAD density. In comparison, 2-fold symmetry is incompatible with the alignment of a linear dimer. The closest manner to link two TO_4 tetrahedra on a 2-fold axis is *via* an additional pair of TO_4 tetrahedra to form a four-TO_4 ring (Fig. 4b). Therefore, the distributions of T atoms on 2-fold axes are generally more disperse than those on 3-fold axes.

Besides the 221 known zeolite structures, we further performed the TAD density calculations on 84 292 hypothetical zeolite structures (ESI Table S2†).[30] These hypothetical structures were predicted by Li *et al. via* a genomic approach,[20] and their chemical feasibility was confirmed by various structure evaluation methods.[21] The highest TAD densities for the 84 292 hypothetical zeolite structures do not exceed the TAD densities for the known zeolites, validating the feasibility of the highest TAD densities for the building of zeolite frameworks.

Knowing the highest TAD density for each type of site symmetry, we can enumerate all the WP combinations for zeolite structure prediction in a given unit cell. We chose sample space groups from the tetragonal, trigonal, hexagonal, and cubic crystal systems. For each of these crystal systems, we chose three space groups, including the first and the last ones according to the International Tables for Crystallography, and the most frequently observed one in known zeolite structures. As a result, we obtained 12 space groups from the four crystal systems, including $P4$ (no. 75), $I4/mmm$ (no. 139), $I4_1/acd$ (no. 142), $P3$ (no. 143), $R\bar{3}m$ (no. 166), $R\bar{3}c$ (no. 167), $P6$ (no. 168), $P6_3/mcm$ (no. 193), $P6_3/mmc$ (no. 194), $P23$ (no. 195), $Im\bar{3}m$ (no. 229), and $Ia\bar{3}d$ (no. 230). The numbers of WPs and the corresponding site symmetries in the 12 selected space groups are listed in Table 2. These 12 space groups contain many types of WP, among which special points and rotation axes occur more frequently than mirror planes. Upon removal of the WPs incompatible with the tetrahedral coordination of T atoms, such as inversion

Table 2 Numbers of WPs and the corresponding site symmetries in the selected space groups

Space group		No. of WPs[a]	No. of points[a]	No. of rotation axes[a]	No. of mirror planes[a]
Tetragonal	$P4$	4(2)	0(0)	3(1)	0(0)
	$I4/mmm$	15(10)	5(1)	6(5)	3(3)
	$I4_1/acd$	7(6)	3(2)	3(3)	0(0)
Trigonal	$P3$	4(4)	0(0)	3(3)	0(0)
	$R\bar{3}m$	9(5)	4(0)	3(3)	1(1)
	$R\bar{3}c$	6(3)	3(0)	2(2)	0(0)
Hexagonal	$P6$	4(3)	0(0)	3(2)	0(0)
	$P6_3/mcm$	12(7)	5(0)	4(4)	2(2)
	$P6_3/mmc$	12(7)	5(0)	4(4)	2(2)
Cubic	$P23$	10(8)	4(2)	5(5)	0(0)
	$Im\bar{3}m$	12(8)	4(1)	5(4)	2(2)
	$Ia\bar{3}d$	8(6)	4(2)	3(3)	0(0)

[a] Numbers in brackets represent the numbers of WPs allowed for the tetrahedral coordination of T atoms.

centres, 4-fold axes, and 6-fold axes, *etc.*, rotation axes become the major site symmetries in the 12 selected space groups (see the numbers in brackets in Table 2). Although the numbers of WPs are significantly decreased by the removal of prohibited ones, the numbers of possible WP combinations may remain prohibitively high, especially for unit cells with large numbers of crystallographically distinct T atoms.

To calculate the numbers of WP combinations, we constructed a series of unit cells in these 12 space groups. Referring to the unit cells of known zeolite structures, the lattice lengths were set to range from 1.0 nm to 2.2 nm with increments of 0.3 nm, which corresponds to the T···T distance in a linear TO_4 dimer. As a result, we obtained five unit cells for every cubic space group, and 25 unit cells for each of the remaining space groups. Meanwhile, we set the framework density in each unit cell between 8 and 18 T per nm^3, which should cover the densities of most of the known zeolite structures.

The enumerated numbers of WP combinations in the unit cells of 1.0 × 1.0 × 1.0 nm^3 and 2.2 × 2.2 × 2.2 nm^3 for each space group are listed in Table 3, and

Table 3 Numbers of WP combinations in the smallest and largest unit cells for 12 space groups

	Lattice lengths/nm		Numbers of WP combinations			
Space group	*a*, *b*	*c*	All WPs	Prohibited WPs removed	Prohibited WPs removed; rotation axes restricted	Prohibited WPs removed; rotation axes and mirror planes restricted
*P*4	1.0	1.0	1222	24	9	9
	2.2	2.2	7 272 269	1868	132	132
*I*4/*mmm*	1.0	1.0	241	22	5	4
	2.2	2.2	190 691 682	2 330 250	24 852	19 028
$I4_1/acd$	1.0	1.0	7	6	5	5
	2.2	2.2	4760	2715	312	312
*P*3	1.0	1.0	1491	1491	81	81
	2.2	2.2	10 718 396	10 718 396	6624	6624
$R\bar{3}m$	1.0	1.0	16	1	0	0
	2.2	2.2	33 341	2627	518	518
$R\bar{3}c$	1.0	1.0	2	1	0	0
	2.2	2.2	910	138	25	25
*P*6	1.0	1.0	210	34	13	13
	2.2	2.2	954 338	20 867	400	400
$P6_3/mcm$	1.0	1.0	99	11	3	3
	2.2	2.2	8 114 722	345 520	11 022	10 406
$P6_3/mmc$	1.0	1.0	193	16	5	4
	2.2	2.2	16 211 758	646 138	26 545	20 060
*P*23	1.0	1.0	415	133	121	121
	2.2	2.2	27 937 296	7 169 724	68 937	68 937
$Im\bar{3}m$	1.0	1.0	13	2	2	2
	2.2	2.2	35 550	1363	603	603
$Ia\bar{3}d$	1.0	1.0	2	0	0	0
	2.2	2.2	356	106	26	26

the results for other unit cells can be found in ESI Table S3.† For each unit cell of the selected space group, we calculate four types of WP combination. The first type is calculated using all the WPs in the selected space group, including the prohibited ones. The only restriction is that all the special points can appear in a WP combination no more than once, because they can only be occupied by no more than one T atom. As expected, the numbers of WP combinations grow rapidly with cell volumes, which are directly related to the numbers of T atoms. For instance, the number of WP combinations for space group *I*4/*mmm* increases from 241 in the unit cell of $1.0 \times 1.0 \times 1.0$ nm^3 to 190 691 682 in the unit cell of $2.2 \times 2.2 \times 2.2$ nm^3. Meanwhile, the numbers of WP combinations differ significantly among different space groups. Among the 12 selected space groups, $Ia\bar{3}d$ has the fewest WP combinations in the unit cell of $2.2 \times 2.2 \times 2.2$ nm^3, where 8 WPs lead to 356 types of combination. In the unit cell of the same size, space group *I*4/*mmm* has the most WP combinations, where 15 WPs lead to 190 691 682 types of combination (Tables 2 and 3). Note that these WP combinations are only related to space group symmetries and numbers of T atoms.

When all the prohibited WPs, such as inversion centres, 4-fold axes, and 6-fold axes, *etc.*, are removed from WP enumeration, we obtain the second type of WP combination that better reflects the true complexity for zeolite framework building. Except for space group *P*3, where there is no prohibited WP at all, the numbers of WP combinations in the rest of the 11 space groups rapidly decrease (Table 3). For instance, for space group *I*4/*mmm*, the WP combinations in the unit cell of $2.2 \times 2.2 \times 2.2$ nm^3 are reduced by over 80 times, from 190 691 682 to 2 330 250. The largest decrease in the number of WP combinations by the removal of prohibited WPs occurs in space group *P*4, where the WP combinations are reduced by nearly 4000 times, from 7 272 269 to 1868. Despite such considerable decreases in computational complexity, one should note that the numbers of WP combinations for half of the 12 space groups are still prohibitively large, and must be further decreased for zeolite-framework-building computations.

Knowing the highest allowed TAD density for each type of site symmetry derived from the known zeolites, we can further reduce the WP combinations by restricting the numbers of T atoms on rotation axes and mirror planes (the third and fourth types of WP combination; Table 3). For instance, by restricting the numbers of T atoms on all the 2-fold axes, including 2 and 2*mm*, the WP combinations in a unit cell of $2.2 \times 2.2 \times 2.2$ nm^3 in space group *I*4/*mmm* can be further reduced by over 90 times, from 2 330 250 to 24 852. Further restrictions on the numbers of T atoms on mirror planes leads to a final combination number of 19 028. Another example is space group *P*3, where there is no prohibited WP or mirror plane. Only by restricting the numbers of T atoms on the three 3-fold axes in space group *P*3 can the WP combinations in a unit cell of $2.2 \times 2.2 \times 2.2$ nm^{3}be reduced by over 1600 times, from 10 718 396 to 6624, indicating the importance of the restrictions on the numbers of T atoms on rotation axes. Combining the removal of prohibited WPs and the restrictions on the numbers of T atoms on rotation axes and mirror planes, most of the 12 selected space groups achieve dramatic decreases in the numbers of WP combinations for zeolite structure prediction. For instance, the WP combinations in the unit cell of $2.2 \times 2.2 \times 2.2$ nm^3 in space group *I*4/*mmm* are reduced by over 10^4 times. Such a dramatic decrease in computational complexity will eventually enable the zeolite-framework-building computations that are prohibited for the original numbers of WP combinations.

Note that the decrease in WP combinations differs among different crystal systems. In general, the decrease in the tetragonal system is larger than in the trigonal and hexagonal systems, and the decrease in the cubic system is the smallest. Among the 12 space groups we study in this paper, the largest decrease occurs in space group *P*4, which achieves a reduction in WP combinations of over 55 000 times in the unit cell of 2.2 × 2.2 × 2.2 nm^3; the smallest decrease occurs in space group $Ia\bar{3}d$, where the WP combinations are reduced by only 13 times in the unit cell of the same size (Table 3). Following this trend, we can expect larger decreases in WP combinations in orthorhombic and monoclinic crystal systems. To be specific, factors affecting the decrease in WP combinations include the numbers of T atoms and WPs, the multiplicity of each WP, and the numbers of prohibited and restricted WPs, *etc.* Although we cannot determine the exact correlation between the decrease in WP combinations and these influential factors, we believe that the number of restricted rotation axes plays a key role in WP combination reduction. As we show in Fig. 2b, to avoid chemically unfeasible coordination, T atoms on rotation axes have to be highly dispersed. In comparison, T atoms on mirror planes can be a little more crowded without forming unfeasible distorted tetrahedra. Therefore, our restriction on the numbers of T atoms on rotation axes is stricter than that on mirror planes. As shown in Table 3 and ESI Table S3,† restricting the numbers of T atoms on rotation axes causes a much more rapid decrease in WP combinations than restrictions on mirror planes. This also explains the fact that the first among the three space groups chosen for each crystal system always achieves the largest decrease in WP combinations, because most of its WPs correspond to rotation axes. Note that in each space group the largest decrease in WP combinations does not always occur in the largest unit cell (ESI Table S3†). For instance, in space group *P*3, the largest decrease in WP combinations occurs in the unit cell of 2.2 × 2.2 × 1.6 nm^3. As a matter of fact, among the unit cells with the same *a* and *b* in space group *P*3, the unit cells with $c = 1.6$ nm always achieve the largest decreases in WP combinations. In space group *P*3, besides the general positions, all of the remaining three WPs correspond to 3-fold axes along the *c* direction. According to the largest allowed TAD density on the 3-fold axis (2.325 T per nm), 1.6 nm of *c* allows no more than 3 T atoms on each 3-fold axis in space group *P*3, whereas 1.9 and 2.2 nm of *c* allow 4 and 5 T atoms, respectively. Meanwhile, 1.6 nm of *c* ensures that the unit cell is large enough for the assembly of a relatively large number of T atoms. Balancing the restrictions on 3-fold axes and the cell volume requirements for multiple T atoms, the unit cells with $c = 1.6$ nm in space group *P*3 eventually achieve the largest decreases in WP combinations.

With WP combinations significantly reduced, the efficiency of zeolite structure prediction can be improved dramatically. Here, we take space group $Im\bar{3}m$ as an example to illustrate the efficiency of this approach for zeolite framework building. Notice that five of the 221 known zeolite structures with lattice lengths near the range of 1.0–2.2 nm belong to this space group. We can validate the feasibility of our approach by checking whether or not we can find all of the five known zeolite structures during framework building. As shown in ESI Table S3,† space group $Im\bar{3}m$ has 40 654 WP combinations in total for five unit cells, which can be reduced to 804 using our WP combination reduction approach. Feeding the 804 WP combinations to FraGen, *i.e.* a computer program developed by us for structure generation with fixed WPs,[31] we built 490 tetrahedrally coordinated zeolite frameworks with reasonable bonding geometries (see Methods section). The whole framework-building procedure took only 248.9 CPU hours on a desktop PC with an Intel Core i7-4790 processor. Notice that all of the five known zeolite structures belonging to this space group can be found among the 490 frameworks we have built. Further geometry optimization using the program GULP[33] finally yields 320 hypothetical zeolite structures (ESI Table S4†). This proves that efficient zeolite structure prediction can be achieved by avoiding the WP combinations that cannot yield feasible zeolite structures from the very beginning.

IV. Conclusions

The computer-aided assembly of T atoms in predefined WPs is a promising method for zeolite structure prediction because of its high efficiency in configuration space searching. However, this method suffers from the large number of WP combinations in a given unit cell, prohibiting its practical applications. In this paper, we report our new approach to reduce the computational complexity for zeolite structure prediction by decreasing the number of WP combinations. *Via* TAD density calculations on known zeolite structures, we obtain the highest allowed numbers of T atoms for all types of site symmetry. Using this information to restrict the numbers of T atoms in specific WPs, the WP combinations can be dramatically reduced to a manageable level. The outcome of this study should serve as the input for zeolite-framework-building programs, where T atoms are fixed in predefined WPs. Besides zeolites, our approach to reduce the WP combinations is valid for all three-dimensional extended inorganic structures constructed from well-defined polyhedral building units.

Conflicts of interest

There are no conflicts to declare.

Acknowledgements

This work was supported by the National Natural Science Foundation of China (No. 21622102 and 21621001), the National Key Research and Development Program of China (No. 2016YFB0701100), the National 111 Project (B17020), and Program for JLUSTIRT.

References

1 M. E. Davis, *Nature*, 2002, **417**, 813–821.
2 B. M. Weckhuysen and J. Yu, *Chem. Soc. Rev.*, 2015, **44**, 7022–7024.
3 Y. Li, L. Li and J. Yu, *Chem*, 2017, **3**, 928–949.
4 Y. Li and J. Yu, *Chem. Rev.*, 2014, **114**, 7268–7316.
5 C. Baerlocher and L. B. McCusker, *Database of Zeolite Structures*, http://www.iza-structure.org/databases/.
6 J. V. Smith, *Chem. Rev.*, 1988, **88**, 149–182.
7 M. W. Deem and J. M. Newsam, *Nature*, 1989, **342**, 260–262.
8 M. W. Deem and J. M. Newsam, *J. Am. Chem. Soc.*, 1992, **114**, 7189–7198.
9 M. Falcioni and M. W. Deem, *J. Chem. Phys.*, 1999, **110**, 1754–1766.

10 M. M. J. Treacy, K. H. Randall, S. Rao, J. A. Perry and D. J. Chadi, *Z. Kristallogr.*, 1997, **212**, 768–791.
11 M. M. J. Treacy, I. Rivin, E. Balkovsky, K. H. Randall and M. D. Foster, *Microporous Mesoporous Mater.*, 2004, **74**, 121–132.
12 O. Delgado Friedrichs, A. W. M. Dress, D. H. Huson, J. Klinowski and A. L. Mackayk, *Nature*, 1999, **400**, 644–647.
13 C. Mellot Draznieks, J. M. Newsam, A. M. Gorman, C. M. Freeman and G. Férey, *Angew. Chem., Int. Ed.*, 2000, **39**, 2270–2275.
14 C. Mellot-Draznieks, S. Girard and G. Férey, *J. Am. Chem. Soc.*, 2002, **124**, 15326–15335.
15 Y. Li, J. Yu, D. Liu, W. Yan, R. Xu and Y. Xu, *Chem. Mater.*, 2003, **15**, 2780–2785.
16 Y. Li, J. Yu, Z. Wang, J. Zhang, M. Guo and R. Xu, *Chem. Mater.*, 2005, **17**, 4399–4405.
17 Y. Li, J. Yu, R. Xu, C. Baerlocher and L. B. McCusker, *Angew. Chem., Int. Ed.*, 2008, **47**, 4401–4405.
18 S. M. Woodley, P. D. Battle, J. D. Gale and C. R. A. Catlow, *Phys. Chem. Chem. Phys.*, 2004, **6**, 1815–1822.
19 S. M. Woodley, *Phys. Chem. Chem. Phys.*, 2004, **6**, 1823–1829.
20 Y. Li, X. Li, J. Liu, F. Duan and J. Yu, *Nat. Commun.*, 2015, **6**, 8328.
21 Y. Li, J. Yu and R. Xu, *Angew. Chem., Int. Ed.*, 2013, **52**, 1673–1677.
22 X. Liu, S. Valero, E. Argente, V. Botti and G. Sastre, *Z. Kristallogr.*, 2015, **230**, 291–299.
23 J. Lu, L. Li, H. Cao, Y. Li and J. Yu, *Phys. Chem. Chem. Phys.*, 2017, **19**, 1276–1280.
24 J.-R. Lu, C. Shi, Y. Li and J.-H. Yu, *Chin. Chem. Lett.*, 2017, **28**, 1365–1368.
25 D. J. Earl and M. W. Deem, *Ind. Eng. Chem. Res.*, 2006, **45**, 5449–5454.
26 M. W. Deem, R. Pophale, P. A. Cheeseman and D. J. Earl, *J. Phys. Chem. C*, 2009, **113**, 21353–21360.
27 R. Pophale, P. A. Cheeseman and M. W. Deem, *Phys. Chem. Chem. Phys.*, 2011, **13**, 12407–12412.
28 M. W. Deem, *Deem Database*, http://www.hypotheticalzeolites.net/DATABASE/DEEM/index.php.
29 M. D. Foster and M. M. J. Treacy, *Atlas of Prospective Zeolite Structures*, http://www.hypotheticalzeolites.net/.
30 Y. Li and J. Yu, *Hypothetical Zeolite Frameworks*, http://mezeopor.jlu.edu.cn/hypo/.
31 Y. Li, J. Yu and R. Xu, *J. Appl. Crystallogr.*, 2012, **45**, 855–861.
32 M. J. Sanders, M. Leslie and C. R. A. Catlow, *J. Chem. Soc., Chem. Commun.*, 1984, 1271–1273.
33 J. D. Gale and A. L. Rohl, *Mol. Simul.*, 2003, **29**, 291–341.

Faraday Discussions

ROYAL SOCIETY OF CHEMISTRY

PAPER

Materials discovery by chemical analogy: role of oxidation states in structure prediction†

Daniel W. Davies,[a] Keith T. Butler,[a] Olexandr Isayev[b] and Aron Walsh*[cd]

Received 16th February 2018, Accepted 1st March 2018
DOI: 10.1039/c8fd00032h

The likelihood of an element to adopt a specific oxidation state in a solid, given a certain set of neighbours, might often be obvious to a trained chemist. However, encoding this information for use in high-throughput searches presents a significant challenge. We carry out a statistical analysis of the occurrence of oxidation states in 16 735 ordered, inorganic compounds and show that a large number of cations are only likely to exhibit certain oxidation states in combination with particular anions. We use this data to build a model that ascribes probabilities to the formation of hypothetical compounds, given the proposed oxidation states of their constituent species. The model is then used as part of a high-throughput materials design process, which significantly narrows down the vast compositional search space for new ternary metal halide compounds. Finally, we employ a machine learning analysis of existing compounds to suggest likely structures for a small subset of the candidate compositions. We predict two new compounds, $MnZnBr_4$ and $YSnF_7$, that are thermodynamically stable according to density functional theory, as well as four compounds, $MnCdBr_4$, $MnRu_2Br_8$, $ScZnF_5$ and $ZnCoBr_4$, which lie within the window of metastability.

I. Introduction

The idea of ascribing an oxidation state to a metal can be traced back almost 200 years.[1] As the phrase suggests, it was used to describe the amount of oxygen bound to an element that was known to form multiple oxides. Since then, oxidation states have helped in the formation of many fundamental chemistry

[a]Centre for Sustainable Chemical Technologies, Department of Chemistry, University of Bath, Claverton Down, Bath BA2 7AY, UK

[b]Laboratory of Molecular Modeling, Division of Chemical Biological and Medicinal Chemistry, UNC Eshelman School of Pharmacy, University of North Carolina, Chapel Hill, North Carolina 27599, USA

[c]Department of Materials Science and Engineering, Yonsei University, Seoul 03722, Korea

[d]Department of Materials, Imperial College London, Exhibition Road, London SW7 2AZ, UK. E-mail: a.walsh@imperial.ac.uk

† Electronic supplementary information (ESI) available. See DOI: 10.1039/c8fd00032h

concepts. For example, a plot of the periodicity of accessible oxidation states (Fig. 1) by Irving Langmuir was one of the key pieces of evidence that led to the adoption of the octet rule around 100 years ago.[2] The English term itself, "oxidation state" (or equally "oxidation number"), first came into common use in the realm of electrochemistry in the 1930s,[3] and in the 1940s it gained widespread use[4] to replace the less-than-perfect system of appending -ous and -ic to the lower and higher oxidation states of metals, respectively. Ferrous became Fe(ɪɪ), ferric became Fe(ɪɪɪ), and transition metals with more than two oxidation states could now be unambiguously described. The term has remained an indispensable heuristic tool in almost all sub-disciplines of the physical sciences. It is integral to the way in which chemists think about the interactions of elements within molecules and solids.

Linus Pauling first postulated that oxidation states could be determined by approximating bonds as 100% ionic according to the electronegativities of the elements involved.[5] This simple approach did not initially gain acceptance as the use of Pauling's electronegativity scale[6] resulted in some unusual assignments. Nevertheless, his approach is reflected in the modern definition given by IUPAC: "An atom's charge after ionic approximation of its heteronuclear bonds".[7] In practice, knowledge of the chemical formula is sufficient to assign formal oxidation states in many inorganic compounds; however, there are cases where ambiguities exist (*e.g.* mixed-valence compounds, electrides, polyanions and polycations). As highlighted in a recent essay by Karen, the "atom's charge", its "heteronuclear bonds" and the "ionic approximation" are all terms that need clarification, and there are choices to be made about how each is defined.[8]

The subtlety of assigning oxidation states is still the subject of many lively discussions in both pedagogical and research contexts.[9–13] For practical purposes, we emphasise the insight of Jansen and Wedig, who point out that "It is a purely formal concept; nowhere within the definition is it claimed that a particular oxidation state can be associated with a real charge. Nevertheless, the term is certainly useful, since a specific oxidation state can be correlated to real properties".[14] It is this correlation to real properties that is useful in a materials design context. Oxidation states have had a role to play in materials design for many decades. In the 1950s and 1960s, Goodman and Pamplin were able to systematically and exhaustively design superlattices of multicomponent compounds by substitution of the cations in simple binary semiconductors, while ensuring that the octet rule remained satisfied.[15,16] This cation substitution (mutation) concept continues to inspire modern computational work on semiconductor design.[17,18]

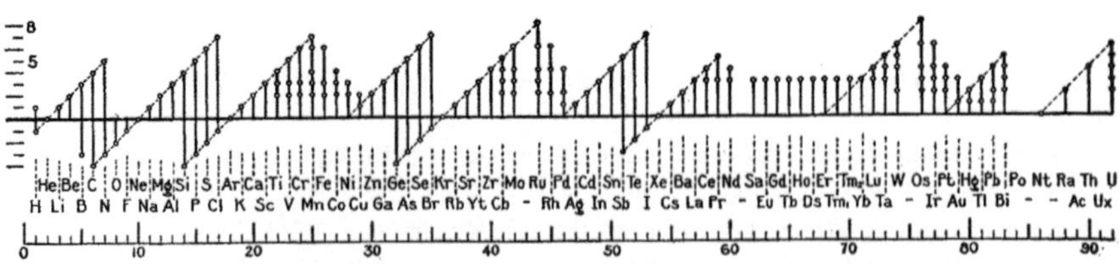

Fig. 1 Plot of accessible oxidation states reproduced from a 100-year-old paper by Irving Langmuir[2] on the octet rule.

Knowledge of accessible oxidation states for each element is advantageous because we can generate many stoichiometric combinations while ensuring that there is overall charge neutrality. For example, the formal oxidation states q of any ternary combination $A_xB_yC_z$ must sum to zero:

$$xq_A + yq_B + zq_C = 0. \tag{1}$$

We have previously demonstrated that many chemically plausible formulas can be generated in this way.[19] For example, if the stoichiometry values (x, y and z in the above equation) are limited to integers ≤ 8, the search space for ternary combinations exceeds 1×10^8, and for quaternary combinations it is over 2×10^{11}. The resulting formulas can be fed into a high-throughput screening workflow that uses machine learning structure prediction models to screen for new functional materials.[20]

In this study, we first carry out a statistical analysis of the occurrence of oxidation states in 16 735 stoichiometric, inorganic compounds in order to highlight trends and show that many elements only exhibit certain oxidation states in the presence of particular elements. We then go on to construct a screening model based on this data and apply it to the search space for ternary transition metal halides. The model we propose can be used as a general chemical filter when dealing with large composition search spaces, in order to remove those combinations of elements that are unlikely to form stable compounds. For example, we find that many transition metals are only likely to adopt their highest accessible oxidation states in the presence of sufficiently electronegative anions.

II. Results

A. Data curation

We focus on the variation of oxidation states of metals in the presence of common anions. The anions we include are the first four group VI and VII elements in their most common oxidation states, *i.e.* O^{2-}, S^{2-}, Se^{2-}, Te^{2-}, F^-, Cl^-, Br^- and I^-. These provide a reasonable range of electronegativities (Table 1) and as such we do not include group V anions. This also allows us to avoid the metalloids As and Sb. The compounds included in the dataset originate from the Inorganic Crystal Structure Database (ICSD) and were downloaded from the Materials Project (MP)[21] using their API.[22] Full details of how the dataset was refined can be found

Table 1 Anion electronegativities (χ) and number of compounds in which each anion is the most electronegative element

Anion	χ	Occurrence
F^-	3.98	1759
O^{2-}	3.44	10 546
Cl^-	3.16	924
Br^-	2.96	444
I^-	2.66	499
S^{2-}	2.58	1489
Se^{2-}	2.55	759
Te^{2-}	2.10	320

in the Methods section. In broad terms, all the compounds meet the following criteria (total number of compounds remaining in the dataset shown in brackets):

(1) Feature in both the ICSD and MP databases (34 913)

(2) Calculated to be less than 100 meV per atom above the thermodynamic convex hull by the MP (30 781)

(3) Oxidation states of all elements can be determined automatically using a bond valence analysis algorithm‡ (24 376)

(4) Contain at least one anion (as defined above) and at least one metal (16 735)

Fig. 2 shows the resulting metals that are included after this refinement has been applied. In total, 16 735 different compounds are included.

B. Occurrence of oxidation states

In the first instance, we examine the occurrence of metal oxidation states as a function of the most electronegative anion present in each compound (see Table 1). In each case, we normalise by the total number of compounds containing a given species (metal in a given oxidation state), *i.e.* we look at how the total number of instances of each species is distributed across the compounds. This is given by the ratio $\frac{N_{SX}}{N_S}$, where N_{SX} is the number of compounds containing the species S where the most electronegative anion is X, and N_S is the total number of compounds containing the species S. These values are shown graphically for all species in the ESI.†

Transition metals have the largest number of accessible oxidation states. Fig. 3 shows the distribution of some first-row d-block species. Each of these exhibits the same general trend: the likelihood of finding a metal in a higher oxidation state increases when a more electronegative anion is present in the compound

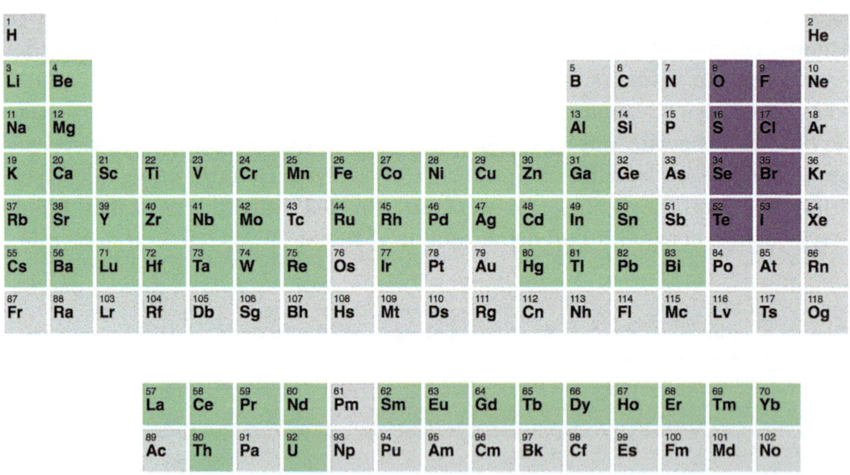

Fig. 2 Periodic table illustrating the metals included in our statistical analysis (green) and the anions considered (purple).

‡ This rules out those elements that were not included in the original study which proposed the algorithm,[23] as well as intermetallic compounds.

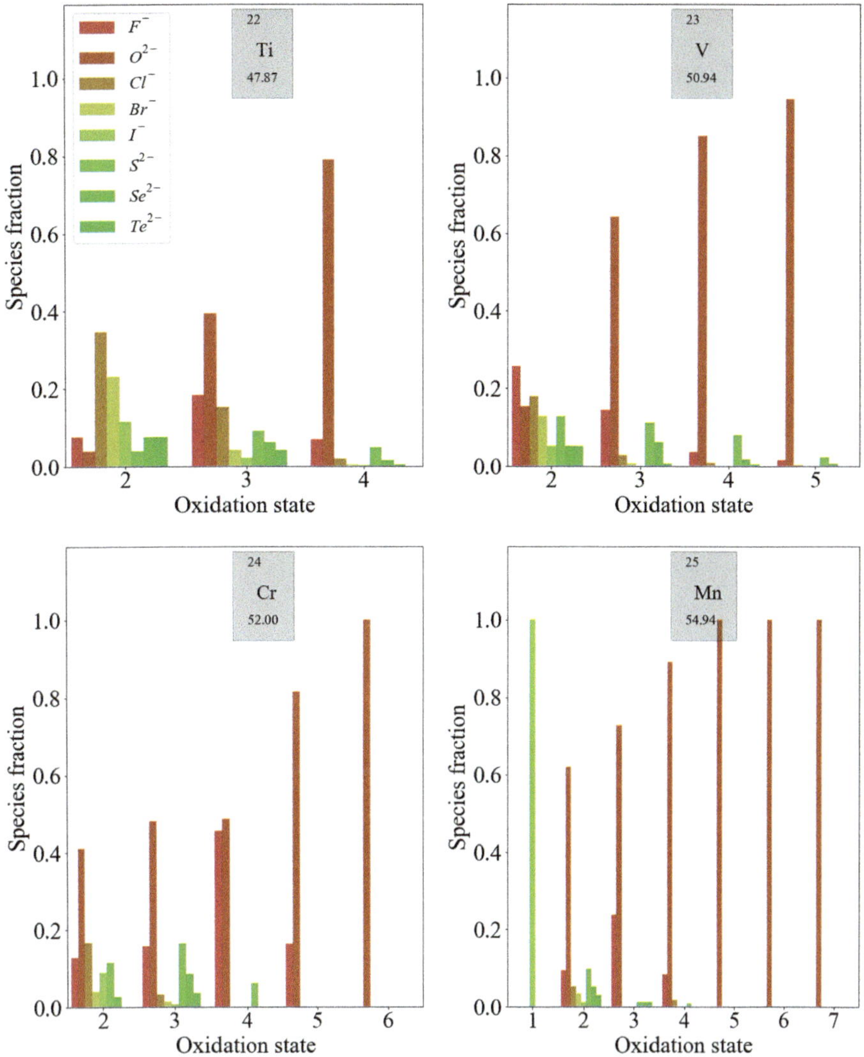

Fig. 3 Distribution of oxidation states in known inorganic crystals containing some first-row transition metal species. The color scale represents the electronegativity of the most electronegative anion present in the compound from dark red (F, most electronegative) to dark green (Te, least electronegative).

(increasing relative heights of red bars in Fig. 3). Meanwhile, metals are more likely to exhibit low oxidation states when the most electronegative anion present is of low electronegativity. More specific trends can also be extracted. For example, the higher oxidation states of Mn (Mn^{5+}–Mn^{7+}) are exclusively exhibited in oxides. This is also the case for Cr^{6+}, while Cr^{5+} is limited to oxides and fluorides.

For higher oxidation states, the likelihood of finding the metal with an anion of moderate electronegativity, such as Cl^-, Br^- or I^-, often goes to zero before the likelihood of finding it with an anion of low electronegativity, such as S^{2-}, Se^{2-}, and Te^{2-}. This is a trend that may not necessarily be expected, for example, going from V^{2+} to V^{4+}. It is also important to mention at this stage that oxides nearly always dominate each distribution, as 10 546 of the 16 735 compounds contain oxygen. This point is addressed later when using the data predictively, in order to minimise bias.

Fig. 4 shows a similar trend in the distribution of some second-row d-block species. Again, compounds containing less electronegative anions, in the absence of any more electronegative anions, are more likely to contain metal species with lower oxidation states. For the highest oxidation states of Ru (Ru^{5+} and Ru^{6+}), the presence of F^- or O^{2-} is required.

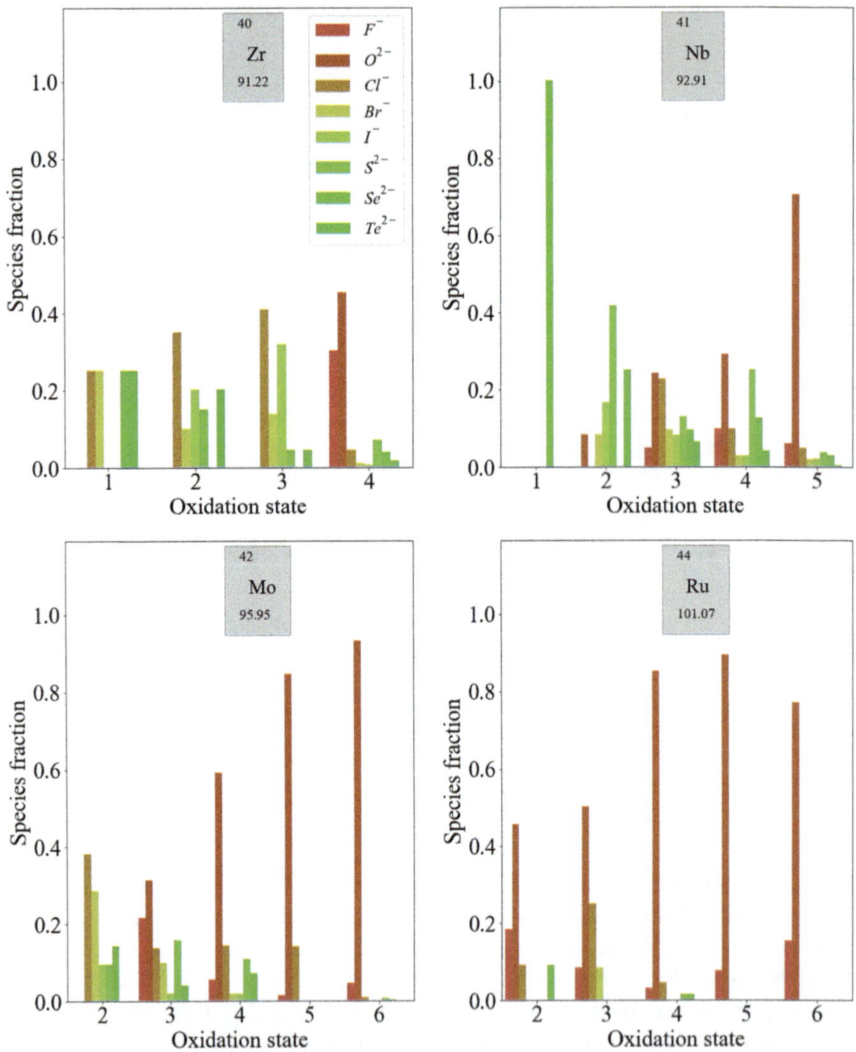

Fig. 4 Distribution of oxidation states in known inorganic crystals containing some second-row transition metal species. The color scale represents the electronegativity of the most electronegative anion present in the compound from dark red (F, most electronegative) to dark green (Te, least electronegative).

The distribution of oxidation states is more even across anions of moderate and low electronegativity for second-row transition metals compared to those in the first row. This is consistent with the established principles of chemical hardness,§ as applied to inorganic compounds by Pearson.[25] The order of chemical hardness for the halides is $F^- > Cl^- > Br^- > I^-$ and, in general, halide anions are harder than chalcogenide anions, which is consistent with the order of electronegativity in Table 1. For cations, hardness increases with increasing charge. The species in the second row have consistently lower chemical hardness than the corresponding species in the first row of the periodic table with the same oxidation state, so it should be expected that they form more compounds with softer halides.

For p-block metals, the trends are less pronounced. As shown in Fig. 5, compounds containing F^- and O^{2-} are likely to exhibit higher oxidation states of the metals. Moving from low to high oxidation states, there is less of a reduction in the fraction of compounds containing anions of lower electronegativity compared with the transition metal compounds discussed so far. The reduction in the fraction of compounds containing anions of moderate electronegativity is more pronounced for these metals. The general observation from this data, that the oxidation states of these metals are more weakly correlated to the electronegativities of the counter-ions than in the case of transition metals, is expected based on the fact that transition metals have multiple readily accessible oxidation states by virtue of their partially occupied d-bands. This is not the case for p-block metals, for which adding or removing electrons results in more significant energy differences.

The third-row transition metals and lanthanide series display similar trends to the first- and second-row transition metal series (see ESI†). For completeness, we note that the alkali and alkali-earth metals only exhibit +1 and +2 oxidation states, respectively, for the vast majority of compounds. Similarly, Sc, Y, Zn and Cd are usually not strictly classified as transition metals as there is a strong energetic preference for them to adopt the oxidation states that lead to empty (Sc^{3+}, Y^{3+}) or filled (Zn^{2+}, Cd^{2+}) valence d-orbitals, rather than these being partially filled, as the definition dictates. We also note that the later d-block metals (Ni, Cu, Pd and Ag) do not exhibit trends as clear as those for the rest of the d-block. This is due to similar effects as above, whereby particular closed (or pseudo-closed) shell configurations are favourable, for example, the d^8 electronic configuration of Ni^{2+} and Pd^{2+}.

The abundance or scarcity of particular species–anion pairings in this dataset may not always reflect what is chemically accessible. Even assuming that the dataset is sufficiently diverse, heightened interest in particular compounds or compound classes can result in their over-representation, which is a general problem in data mining. Nevertheless, we have shown that analysis of the dataset both recovers established chemical concepts and provides new insight. We now go on to develop a simple model that can be universally applied based on the dataset as a whole.

§ Chemical hardness is estimated by $\frac{I-A}{2}$ where I is the ionisation potential and A is the electron affinity. This represents half the energy gap between the highest occupied orbital and lowest unoccupied orbital. Absolute electronegativity,[24] as distinct from Pauling's definition,[6] is defined as $-\frac{I+A}{2}$ and represents the midpoint between the two orbitals.

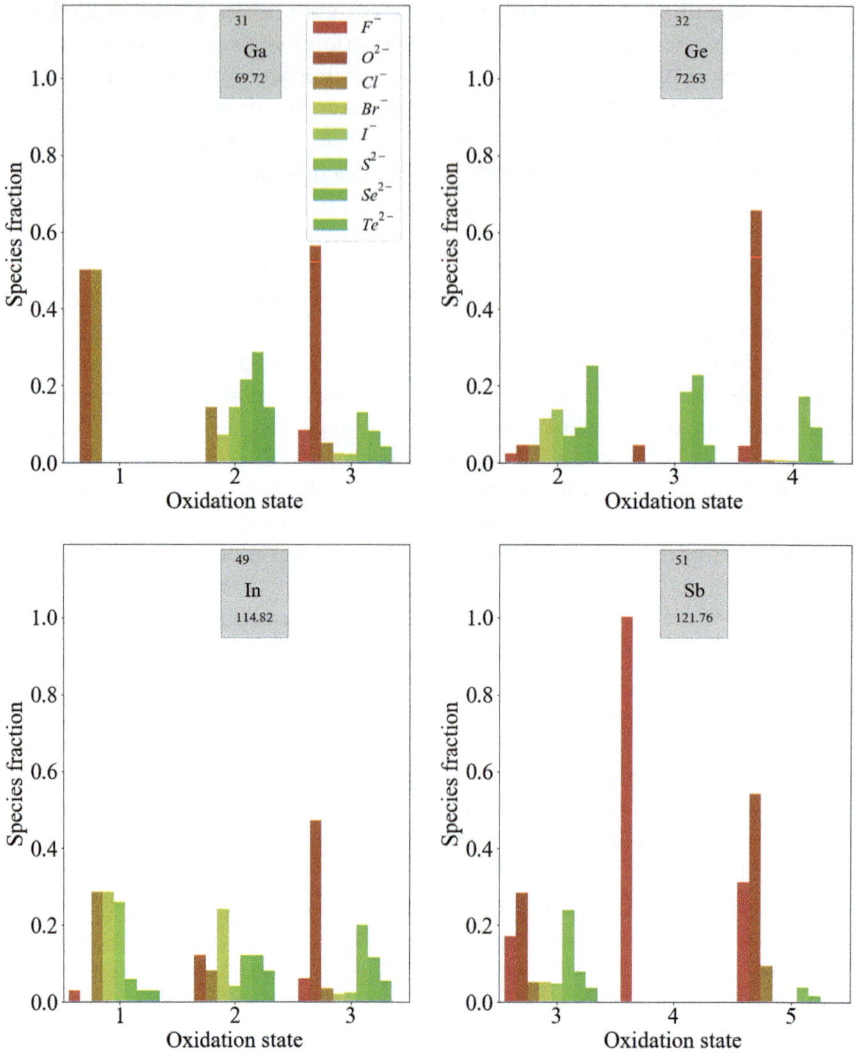

Fig. 5 Distribution of oxidation states in known inorganic crystals containing some p-block species. The color scale represents the electronegativity of the most electronegative anion present in the compound from dark red (F, most electronegative) to dark green (Te, least electronegative).

C. Probabilistic model of species combinations

There are more compounds where the most electronegative anion present is O than where it is any other anion, as shown in Table 1. To use the information from our analysis, we must ensure that the occurrence of each anion does not bias the results. To this end, we define the probability that a species is present with a given anion as:

$$P_{\mathrm{SA}} = \frac{N_{\mathrm{SX}}}{N_{\mathrm{MX}}} \quad (2)$$

where N_{MX} is the total number of compounds containing the metal element M, where X is the most electronegative anion.

We use this formula to construct a lookup table of 1320 species–anion pair probabilities (P_{SA}). The table contains 411 probabilities that equal 0, and 195 probabilities that equal 1. The former represent all the pairings that do not occur within the dataset and the latter represent pairings where for a given anion, the metal only exhibits one particular oxidation state. The P_{SA} values are also presented graphically in the ESI.† We note that this still does not mitigate against limitations that are intrinsic to the dataset. For example, there are over 100 distinct CdI_2 crystal structures in the dataset (owing to the large number of distinct polymorphs), giving rise to an anomalously high probability for the Cd^{2+}–I^- pairing.

An overall compound probability can be calculated as the product of the individual P_{SA} values. For example, for a ternary metal halide $A_aB_bX_x$, the compound probability is calculated as:

$$P_{A_aB_bX_x} = P_{AX}P_{BX} = \frac{N_{AX}}{N_{M_AX}} \times \frac{N_{BX}}{N_{M_BX}} \tag{3}$$

where M_A and M_B are the metal elements corresponding to species A and B. Stoichiometries are not factored in to the probability calculation, such that $P_{A_aB_bX_x} = P_{ABX}$. This ensures that compounds featuring elements that all have only one oxidation state are assigned a probability of 1.0. For example, Ca^{2+} and Al^{3+} are the only species in the database of Ca and Al, hence $P_{CaAl_2O_4} = 1.0$. The number of compounds in the dataset that have compound probabilities above a given threshold, t, is shown in Fig. 6. The number decreases steadily and linearly, before dropping off more rapidly as the threshold becomes more strict.

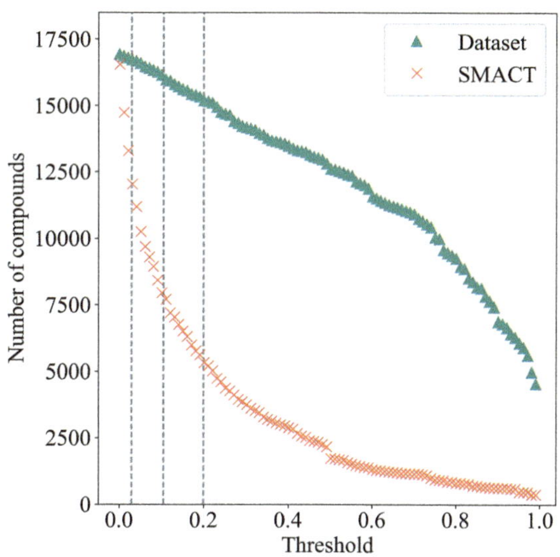

Fig. 6 Total number of allowed compounds from the entire dataset (green triangles) and of allowed compositions for ternary metal halides only generated by SMACT (red crosses) as a function of the compound probability threshold, t. Dotted vertical lines represent cut-offs that return 99%, 95% and 90% of the original dataset.

D. High-throughput compound design

We now use these compound probabilities to inform a high-throughput design workflow. Specifically, we explore the compositional landscape for ternary transition metal halide compounds. An overview of the workflow is shown in Fig. 7. The SMACT code[19] is used to generate 54 484 $A_aB_bX_x$ compositions. Of these, only 4276 are in known chemical systems (A–B–X) within the MP database. The compositions are assigned probabilities as per eqn (3), and only 18 164 are non-zero, which represents an immediate three-fold reduction in the search space.

The number of compositions produced by SMACT that pass through this probability filter as the threshold, t, is increased from zero is also shown in Fig. 6. Many compositions have low probabilities, hence, contrary to the scenario for the compound dataset, the total number drops off rapidly as the threshold increases. This separation between the two curves would, in principle, allow for a threshold to be chosen that eliminates many suggested structures but is still inclusive of the majority of the structures in the dataset. For example, choosing a threshold that

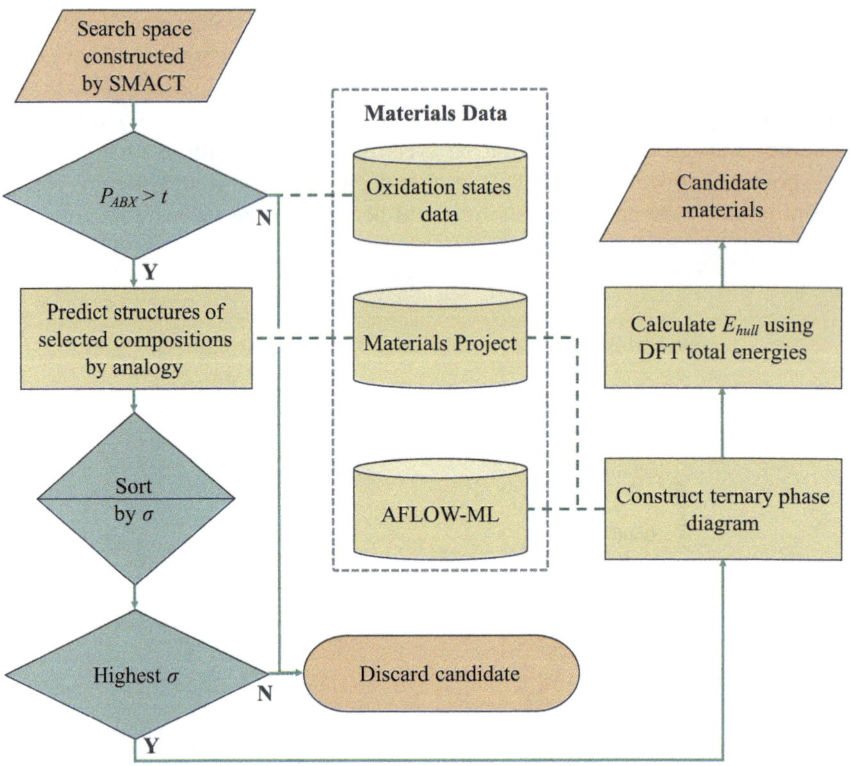

Fig. 7 Data-driven design workflow used to generate new stable compounds. P_{ABX} is the compound probability from the oxidation state analysis, which must be greater than the threshold, t. The structure prediction procedure has a separate threshold, σ. The structure with the highest σ is placed onto a phase diagram constructed using compounds from the MP database and corresponding energies from the AFLOW-ML approach. Density functional theory (DFT) is used to calculate the total energies of competing phases in order to determine the energy above the convex hull, E_{hull}, of a new compound.

includes 90% of the structures in the dataset results in a further three-fold reduction of the search space to <6000 compositions.

If we set a probability threshold of $t = 1$, there are 346 compositions that pass through the filter, equating to 88 distinct sets of three species. In order to demonstrate the rest of our workflow, we take 10 of these sets (Table 2) to the next step, which is to assign them to likely crystal structures (first yellow box in Fig. 7). For this, we adopt the structure substitution algorithm developed by Hautier *et al.*[26] This method also uses a statistical model and relies on a database of known compounds including oxidation state information: a combination of species is substituted onto lattice sites in known structures from the dataset of known materials. Each species substitution is associated with a certain probability, which comes from a model trained on the compounds that already exist in the ICSD. If the overall probability for a given set of substitutions is above a certain threshold, σ, it is added to a list of possible structures. This substitution process is performed on each known crystal structure for each set of species in the MP database. The structure with the highest overall probability for each of the 10 sets of species is taken forward to the next stage of the workflow (second and third green diamonds in Fig. 7). These are listed in Table 2, along with their parent compounds.

Each structure is placed on a phase diagram in order to determine likely competing phases, which requires the total energies to be calculated using DFT. The key quantity of interest is the energy above the convex hull (E_{hull}) that is formed by drawing straight lines between thermodynamically stable phases. It was recently estimated by Sun *et al.* that around half of all known inorganic materials are metastable,[27] so to focus solely on thermodynamically stable compounds would be to potentially overlook kinetically stabilised, useful materials. The likelihood of existence drops off exponentially as E_{hull} increases. The rate of decay depends on the chemistry of the system and we use 100 meV per atom as a guiding principle for the maximum E_{hull}. The set of competing phases on which DFT calculations were performed was determined using a trained machine learning model (AFLOW-ML), in which structures are represented as property-labelled fragments.[28]

Table 2 Energy above the convex hull (E_{hull}) of the proposed compound, along with the chemical formula of the parent compound, found by the structure predictor algorithm for each composition

Species set	Formula	Parent formula	E_{hull} (meV per atom)
Co^{2+} Ru^{3+} Br^-	$CoRu_2Br_8$	$TiAl_2Cl_8$	287
Mn^{2+} Cd^{2+} Br^-	$MnCdBr_4$	$CdCuF_4$	99.5
Mn^{2+} Co^{2+} Br^-	$MnCoBr_4$	$CdCuF_4$	130
Mn^{2+} Ru^{3+} Br^-	$MnRu_2Br_8$	$TiAl_2Br_8$	73.2
Mn^{2+} Zn^{2+} Br^-	**$MnZnBr_4$**	**$GaCuI_4$**	**0**
Sc^{3+} Zn^{2+} F^-	$ScZnF_5$	$MnCdF_5$	48.3
Y^{3+} Co^{2+} I^-	Y_2CoI_8	$TiAl_2Br_8$	181
Y^{3+} Zr^{4+} F^-	**$YZrF_7$**	**$YSnF_7$**	**0**
Zn^{2+} Cd^{2+} Cl^-	$ZnCd_2Cl_6$	$ZnPb_2F_6$	132
Zn^{2+} Co^{2+} Br^-	$ZnCoBr_4$	$CdCuF_4$	40.4

This stage reveals a key advantage of pursuing only those compositions with higher probabilities based on the oxidation state analysis: the parent binary compounds are well defined. Competing binary compounds containing the metals in the same oxidation states as in the ternary (or higher order) compound are more likely to be known and amenable to total energy calculations to determine phase stability. Arbitrary combinations of species can result in stoichiometries that require energies of competitive gas or liquid phases which are subject to larger errors in DFT simulations. Fig. 8 illustrates this point with a comparison between the phase diagrams of three proposed ternary compositions. $MnZnBr_4$ has a probability of 1.0, as both $MnBr_2$ and $ZnBr_2$ are known and these decomposition products do not require a change of oxidation state of either metal. The ternary therefore sits on the tie-line between the two binaries. The proposed compositions of $MnRuBr_6$ and $ScMnI_7$, however, both have probabilities of zero, as in each case one or more species–anion pairs are not known to occur. These compositions sit in an equilibrium triangle as opposed to on a tie-line, and the stabilities of the proposed compounds now depend on the chemical potentials of the anions.

The final E_{hull} values are shown in Table 2. Two of the new compounds, $MnZnBr_4$ and $YZrF_7$, are predicted to be thermodynamically stable with respect to competing phases. Of the remaining eight compounds, four sit within the metastability window of $0 < E_{hull} < 100$, while four are unlikely to form stable compounds. The crystal structures of the two compounds identified as stable are shown in Fig. 9. By comparison with previous work where a similar workflow was employed,[20] this result provisionally indicates that the additional step of considering compound probabilities based on our oxidation state analysis increases the chance of identifying stable compounds.

The main limitation of the procedure outlined here is that it is based on the analysis of known materials with extrapolation to new systems. This assumes that the range of structure types and chemistries found in current materials databases provides a complete basis for materials design. While this is a reasonable starting point, advances in materials synthesis – for example in the area of hybrid organic–inorganic solids – will require adaptations and the development of alternative approaches. We have noted that there are many instances where oxidation states themselves become ill-defined, which are often associated with interesting and important physical behaviour (*e.g.* superconductivity). Before tackling such challenging cases, we have highlighted[19] that a vast amount of "conventional" materials space remains unexplored.

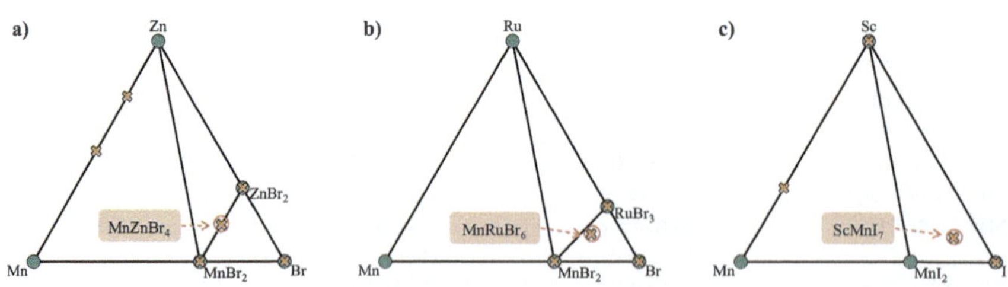

Fig. 8 Ternary phase diagrams of the hypothetical compositions (a) $MnZnBr_4$, (b) $MnRuBr_6$ and (c) $ScMnI_7$. Stable phases (green circles) are connected to form the convex and unstable or proposed phases (orange crosses) sit above the convex hull.

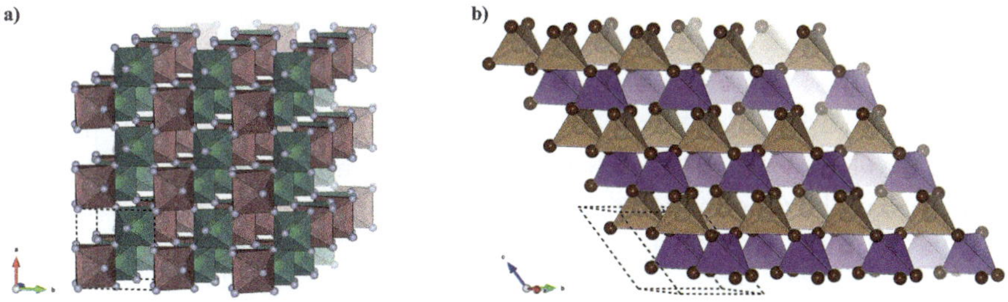

Fig. 9 Two new stable ternary metal halides predicted using this workflow. (a) $YZrF_7$ consists of vertex-sharing irregular polyhedra of YBr_8 (red) and octahedra of $ZnBr_6$ (green). (b) $MnZnBr_4$ consists of vertex-sharing ZnF_4 (orange) and MnF_4 (purple), with both metals in a tetrahedral coordination environment.

III. Conclusion

We have performed a statistical analysis of the occurrence of oxidation states in 16 735 inorganic compounds and shown that qualitative trends in keeping with chemical intuition can be extracted from the data. Many of the highest oxidation states of transition metals are only observed in the presence of the most electronegative anions, O^{2-} and F^-, whilst the absence of these anions is required for many of the lower oxidation states of transition metals. We go on to use the data to construct a model that is applied to inform a high-throughput search for new stable ternary halide materials. The application of the model results in an immediate three-fold reduction in the search space of 54 484 compositions. The search space is reduced to those compositions which are more likely to have known chemically similar compounds as competing phases, such as binary halides, thereby increasing the confidence we have in their calculated stabilities. Our workflow is able to identify two new stable compounds, $YZrF_7$ and $MnZnBr_4$, using modest computing resources.

IV. Methods

A. Dataset

The MP API[22] was used to download the structures of all the compounds that were associated with at least one ICSD entry and had a calculated energy above the hull of <100 meV per atom. An attempt was made to add oxidation states to all the species in each structure using the pymatgen[29] bond_valence module (see the subsection 'Oxidation state assignment' for details). Those compounds for which oxidation states could not be assigned were discarded. Finally, the dataset was limited to compounds that feature at least one metal element and one of the anions of interest, *i.e.* O^{2-}, S^{2-}, Se^{2-}, Te^{2-}, F^-, Cl^-, Br^- or I^-.

B. Oxidation state assignment

In order to assign integer numbers of electrons to atoms, the bond order must be determined. This task is easily carried out for molecules, but not for extended solids. Bond valence (BV) is a quantity similar to bond order that is used instead and, for atoms i and j, is calculated by

$$\mathrm{BV_{ij}} = \exp\left(\frac{R_{\mathrm{ij}}^0 - d_{\mathrm{ij}}}{B}\right) \tag{4}$$

where d is the distance between the atoms and B is a parameter usually fixed to 0.37. R^0 is the single bond length between the two atoms, although in practice it is a function of the coordination number and oxidation state of the approximated cation for a given approximated anion and is fitted to a set of structures. In the general implementation by Brese and O'Keeffe[23] it is calculated as

$$R_{\mathrm{ij}} = r_{\mathrm{i}} + r_{\mathrm{j}} - \frac{r_{\mathrm{i}}r_{\mathrm{j}}\left(\sqrt{c_{\mathrm{i}}} - \sqrt{c_{\mathrm{j}}}\right)^2}{c_{\mathrm{i}}r_{\mathrm{i}} + c_{\mathrm{j}}r_{\mathrm{j}}} \tag{5}$$

where r and c are parameters related to the size and electronegativity of the atoms, respectively.

We use the maximum *a posteriori* (MAP) estimation method to determine oxidation states using the BV approach, as implemented within the pymatgen code[29] with a maximum nearest-neighbour radius of 4 Å.

C. Compound design

Using as the input the metal species for which we have P_{SA} values, we use the SMACT package[19] to generate all the charge-neutral $A_aB_bX_x$ compositions where A and B are d-block metals, X is one of the first four halides, and the stoichiometries a, b and x are integers ≤ 8. For structure prediction, we use the structure substitution algorithm developed by Hautier *et al.*,[26] as implemented in the pymatgen framework[29] with a probability threshold, σ, of 0.00001. The structure with the highest probability that does not contain more than 40 atoms per unit cell is selected as the candidate compound for a given set of species.

D. Total energy calculations

For calculating E_{hull}, first-principles calculations are carried out using Kohn–Sham DFT with a projector-augmented plane wave basis,[30] as implemented in the Vienna *Ab initio* Simulation Package (VASP).[31,32] We use the PBEsol exchange-correlation functional[33] and a k-point grid is generated for each calculation with a density of 120 Å^3 in the reciprocal lattice. The kinetic energy cut-off is set at 600 eV and the force on each atom minimised to below 0.005 eV Å^{-1}.

We note that no Hubbard + U parameters are used in the calculations to correct for the self-interaction error present in the generalised gradient approximation (GGA) for some transition metals.[34,35] The use of GGA + U has been shown the improve the stability estimates of ternary oxides,[36] however, in the absence of any reliable U parameters fitted to metal halides, we use GGA for all calculations for consistency.

V. Data access statement

The SMACT package can be accessed from https://github.com/WMD-group/SMACT. Screening results from these calculations may be reproduced using the Python code available on-line from https://github.com/WMD-group/SMACT/tree/master/examples. Optimised structures are available on-line from https://

github.com/WMD-group/Crystal_structures/tree/master/TM_halides. All other data may be obtained from the authors on request.

Conflicts of interest

There are no conflicts of interest to declare.

Acknowledgements

DWD gratefully acknowledges support from the Engineering and Physical Sciences Research Council (EPSRC) *via* the Centre for Doctoral Training in Sustainable Chemical Technologies (grant no. EP/L016354/1). Calculations were carried out on the Balena HPC cluster at the University of Bath, which is maintained by Bath University Computing Services. AW acknowledges support from the Royal Society and the Leverhulme Trust.

References

1 F. Wöhler, *Unorganische Chemie*, Duncker und Humblot, Berlin, 3rd edn, 1835, p. 4.
2 I. Langmuir, *J. Am. Chem. Soc.*, 1919, **41**, 868–934.
3 W. M. Latimer, *The Oxidation States of the Elements and their Potentials in Aqueous Solutions*, Prentice Hall, 1938.
4 W. P. Jorissen, H. Bassett, A. Damiens, F. Fichter and H. Rémy, *J. Am. Chem. Soc.*, 1941, **63**, 889–897.
5 L. Pauling, *J. Chem. Soc.*, 1948, 1461–1467.
6 L. Pauling, *J. Am. Chem. Soc.*, 1932, **54**, 3570–3582.
7 IUPAC, *Compendium of Chemical Terminology (the"Gold Book")*, Blackwell Scientific Publications, Oxford, 2nd edn, 1997.
8 P. Karen, *Angew. Chem., Int. Ed.*, 2015, **54**, 4716–4726.
9 D. W. Smith, *J. Chem. Educ.*, 2005, **82**, 1202.
10 G. Parkin, *J. Chem. Educ.*, 2006, **83**, 791.
11 H.-P. Loock, *J. Chem. Educ.*, 2011, **88**, 282–283.
12 L. Jiang, S. V. Levchenko and A. M. Rappe, *Phys. Rev. Lett.*, 2012, **108**, 166403.
13 A. Walsh, A. A. Sokol, J. Buckeridge, D. O. Scanlon and C. R. A. Catlow, *J. Phys. Chem. Lett.*, 2017, **8**, 2074–2075.
14 M. Jansen and U. Wedig, *Angew. Chem., Int. Ed.*, 2008, **47**, 10026–10029.
15 B. R. Pamplin, *J. Phys. Chem. Solids*, 1964, **25**, 675–684.
16 C. H. L. Goodman, *J. Phys. Chem. Solids*, 1958, **6**, 305–314.
17 A. Walsh, S.-H. Wei, S. Chen and X. G. Gong, *2009 34th IEEE Photovolt. Spec. Conf.*, 2009, pp. 001875–001878.
18 Z.-H. Cai, P. Narang, H. a. Atwater, S. Chen, C.-G. Duan, Z.-Q. Zhu and J.-H. Chu, *Chem. Mater.*, 2015, **27**, 7757–7764.
19 D. W. Davies, K. T. Butler, A. J. Jackson, A. Morris, J. M. Frost, J. M. Skelton and A. Walsh, *Chem*, 2016, **1**, 617–627.
20 D. W. Davies, K. T. Butler, J. M. Skelton, C. Xie, A. R. Oganov and A. Walsh, *Chem. Sci.*, 2018, **9**, 1022–1030.
21 A. Jain, S. P. Ong, G. Hautier, W. Chen, W. D. Richards, S. Dacek, S. Cholia, D. Gunter, D. Skinner, G. Ceder and K. A. Persson, *APL Mater.*, 2013, **1**, 011002.

22 S. P. Ong, S. Cholia, A. Jain, M. Brafman, D. Gunter, G. Ceder and K. A. Persson, *Comput. Mater. Sci.*, 2015, **97**, 209–215.
23 M. O'Keeffe and N. E. Brese, *J. Am. Chem. Soc.*, 1991, **113**, 3226–3229.
24 R. S. Mulliken, *J. Chem. Phys.*, 1934, **2**, 782.
25 R. G. Pearson, *Inorg. Chem.*, 1988, **27**, 734–740.
26 G. Hautier, C. Fischer, V. Ehrlacher, A. Jain and G. Ceder, *Inorg. Chem.*, 2011, **50**, 656–663.
27 W. Sun, S. T. Dacek, S. P. Ong, G. Hautier, A. Jain, W. D. Richards, A. C. Gamst, K. A. Persson and G. Ceder, *Sci. Adv.*, 2016, **2**, e1600225.
28 O. Isayev, C. Oses, C. Toher, E. Gossett, S. Curtarolo and A. Tropsha, *Nat. Commun.*, 2017, **8**, 15679.
29 S. P. Ong, W. D. Richards, A. Jain, G. Hautier, M. Kocher, S. Cholia, D. Gunter, V. L. Chevrier, K. A. Persson and G. Ceder, *Comput. Mater. Sci.*, 2013, **68**, 314–319.
30 G. Kresse and D. Joubert, *Phys. Rev. B*, 1999, **59**, 1758–1775.
31 G. Kresse and J. Furthmüller, *Comput. Mater. Sci.*, 1996, **6**, 15–50.
32 G. Kresse and J. Furthmüller, *Phys. Rev. B*, 1996, **54**, 11169.
33 J. P. Perdew, A. Ruzsinszky, G. I. Csonka, O. A. Vydrov, G. E. Scuseria, L. A. Constantin, X. Zhou and K. Burke, *Phys. Rev. Lett.*, 2008, **100**, 136406–136414.
34 V. I. Anisimov, J. Zaanen and O. K. Andersen, *Phys. Rev. B*, 1991, **44**, 943–954.
35 V. I. Anisimov, F. Aryasetiawan and A. I. Lichtenstein, *J. Phys.: Condens. Matter*, 1997, **9**, 767–808.
36 G. Hautier, S. P. Ong, A. Jain, C. J. Moore and G. Ceder, *Phys. Rev. B*, 2012, **85**, 155208.

Faraday Discussions

PAPER

Adventures in boron chemistry – the prediction of novel ultra-flexible boron oxide frameworks

Neil L. Allan, *[a] Harvey J. A. Dale, [a] Judy N. Hart [b] and Frederik Claeyssens [c]

Received 1st March 2018, Accepted 16th May 2018
DOI: 10.1039/c8fd00052b

Recent periodic density functional calculations have predicted the existence of ultra-flexible low-energy forms of boron oxides in which rigid boron–oxygen heterocycles are linked by flexible B–O–B bridges. The minima in the energy landscapes of these frameworks are remarkably broad, with widths in excess of those of many hybrid metal–organic frameworks. Enormous changes in cell volume, which can exceed a factor of two, are accompanied by negligible changes in energy. Here we explore the underlying reasons for this behaviour using molecular electronic-structure calculations, periodic density functional theory and template-based geometric simulations. The angular flexibility of the B–O–B bridge depends only upon the geometry of the local B_2O_5 unit, independent of the configuration of neighbouring bridges. Unique cooperativity between the bending and twisting motions of the bridges leads to considerable anisotropy in framework flexibility. Exceptional flexibility is conferred not only by the intrinsic bending flexibility of the bridges but by topological factors, crucially the relaxation of torsional constraints when B_3O_3 rings are present. We test these conclusions by showing how the flexibility of the frameworks can be tuned by decoration with isoelectronic rings. The new nanoporous boron oxides presented in this work are predicted to be potential novel guest–host materials because of their flat energy landscapes. Furthermore, such structures can be generated systematically from silicates by the substitution of $B_2O_5^{4-}$ for SiO_4^{4-}. A borate analogue of β-cristobalite is shown to be isoenergetic with the known B_2O_3-I polymorph. We raise the possibility of new families of frameworks and zeolite analogues.

Introduction

Standard textbooks assert that a solid is distinguished from other states of matter by a fixed volume, a key principle introduced early on in any general science

[a]School of Chemistry, University of Bristol, Cantock's Close, Bristol, BS8 1TS, UK. E-mail: neil.allan@bristol.ac.uk
[b]School of Materials Science and Engineering, UNSW Sydney, NSW 2052, Australia
[c]Materials Science and Engineering Department, Kroto Research Institute, University of Sheffield, Broad Lane, Sheffield, S3 7HQ, UK

course. This reflects the typical form of the energy landscapes of solid-state compounds which have well-defined minima at specific volumes, corresponding to different phases.

Nevertheless, recent years have seen substantial and increasing attention paid to nanoporous hybrid metal–organic frameworks (MOFs),[1] which contain metal centres connected though multidentate organic ligands, forming three-dimensional networks. Numerous crystal structures are adopted by such systems and they have many potential applications as host–guest materials, including gas storage, fluid separation and drug delivery, and as sensors and heterogeneous catalysts.[2–6] The organic ligands can provide framework flexibility, and flexible MOFs exhibit structures which are topologically identical but have very different unit cell volumes and consequent pore sizes.[7–10] For example, calculations[11] on the flexible MOF MIL-53(Cr) ($Cr_4(OH)_4(C_8H_4O_4)$) revealed a structure with a small pore volume that is almost equienergetic with a topologically equivalent form with large pores and a unit cell volume that is larger by a factor of $\approx$1.6. MIL-88 is another example, where due to their rotational freedom, carboxylic acid groups play the role of knee-caps.[12,13] This structural flexibility can allow guest molecules to travel through pores that would otherwise be too small and rigid to allow them to pass, and also gives rise to intriguing mechanical properties such as negative thermal expansion.[14,15]

This framework flexibility is generally less marked in nanoporous materials which are solely inorganic, such as zeolites. The influence of flexibility on properties such as adsorption has only been studied in a few such systems, including silicate-1.[16]

In this paper we examine and predict inorganic boron–oxygen frameworks which show unprecedented flexibility. The minima in the energy landscapes for these three-dimensional crystalline structures are very broad; in some cases, the frameworks can triple their volume with negligible changes in energy, with no change in connectivity and topology. Tripling of volume also takes place in MIL-88, but is accompanied by the adsorption of solvent;[12,13] such ranges of volume are far larger than that[11] for MIL-53(Cr). We probe the underlying atomic mechanisms that permit such extraordinary behaviour, demonstrate how to tune the flexibility of structures in a number of different ways and discuss the wider implications of our results.

Diboron trioxide (B_2O_3) has so far been observed to adopt only a select few crystalline structures and an amorphous phase.[17] The polymorph B_2O_3-I (α-B_2O_3, space group $P3_1$) is obtained at relatively low pressure and comprises interconnected ribbons of trigonal BO_3 units.[18] The application of higher pressure ($\approx$7 GPa) leads to an increase in the coordination number of boron from three to four and the corresponding transformation to orthorhombic B_2O_3-II (β-B_2O_3, space group *Cmc*2), a crystalline network of corner-sharing BO_4 tetrahedra.[19] The vitreous phase of diboron trioxide, B_2O_3-g (γ-B_2O_3), is a disordered covalent framework of trigonal BO_3 units and bridged boroxyl rings (B_3O_6), with the most recent experimental assessments assigning approximately 70–75% of the boron atoms to boroxyl rings.[20,21]

Until recently, the rationale for such limited polymorphism in B_2O_3 was that a significant proportion of these theoretical structures – including B_2O_3-0, proposed on the basis of molecular dynamics simulations to account for the abnormal thermomechanical behaviour of vitreous B_2O_3 – were supposedly thermodynamically inaccessible (>23.8 kJ mol^{-1} less stable than B_2O_3-I).[22,23] However, later *ab initio* calculations by Ferlat *et al.*[17] and Claeyssens *et al.*[24] unveiled the existence of various low-energy B_2O_3 polymorphs. Indeed, the potential existence of numerous low-energy polymorphs of B_2O_3 had been suggested by Ferlat *et al.*[17] as rationalisation for its ease of vitrification. Theoretical studies[25–28] have previously drawn attention to possible nanoporous polymorphs of other oxides, such as ZnO and MgO, but, in contrast to the broadly similar predicted structures for boron oxides, these are all considerably higher in internal energy than the known structures and do not show any exceptional flexibility.

Thus our starting point in this paper is the (small) set of periodic calculations on eight possible structures for B_2O_3 reported in our earlier work.[24] Fig. 1 shows the previously reported energy–volume curves, calculated using periodic density functional theory. Structure A corresponds to B_2O_3-I. Structures B, C and D were built from BO_3 units by interconnecting 2D sheets with B–O–B bridges, as discussed in ref. 24. This generated three nanoporous structures with different

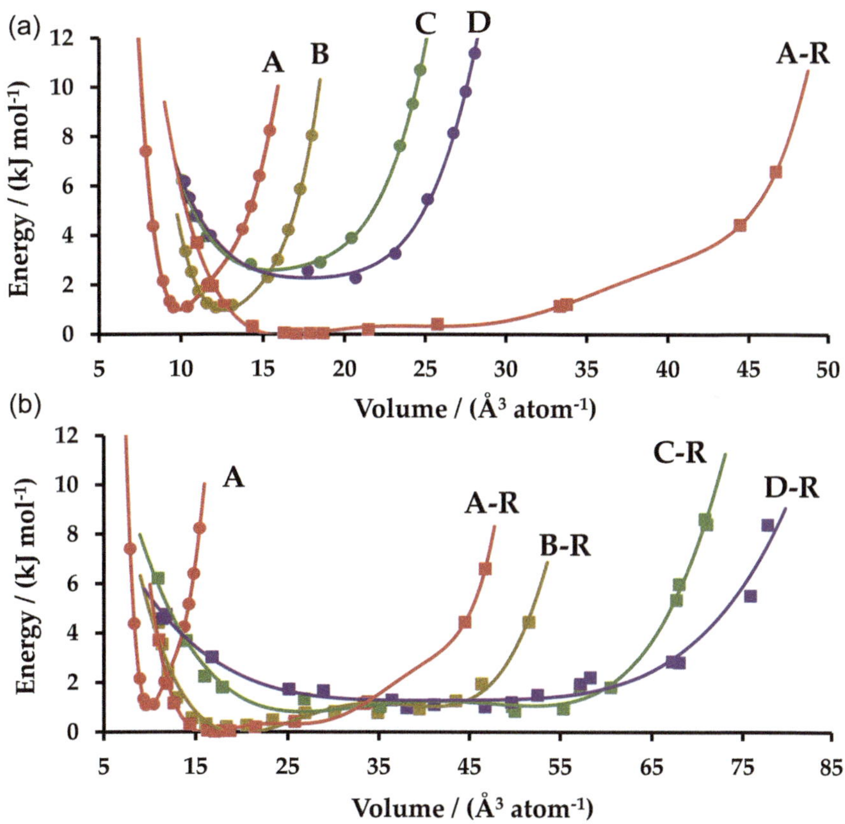

Fig. 1 Energy (per mole of atoms) *vs.* volume curves for (a) the four B_2O_3 structures A–D and (b) their analogues with B_3O_3 rings replacing the BO_3 units, A-R, B-R, C-R and D-R. Structure A is the experimentally known structure B_2O_3-I. All energies are plotted relative to the lowest energy calculated, the minimum for structure A-R. Structure A is included in both plots for comparison. The interconnecting lines are 4th order polynomial fits in order to guide the eye.

numbers of BO_3 units surrounding the nanopores, with the number of such units around the pores increasing from B to D. Structure B is equivalent to that denoted T2 in ref. 17, and structures C and D were constructed from arrangements known to occur in other networks ($B_6O_9(en)_2@(H_2en)Cl_2$, $B_6O_9(en)$ and $(H_2en)_2(Hen)_2B_{16}O_{27}$).[29,30]

The other four structures (A-R, B-R, C-R and D-R) are analogous to A–D and are formed by replacing all the boron atoms in each structure in turn by boroxyl (B_3O_3) rings. This is shown in Fig. 2 for A (B_2O_3-I) and A-R (Fig. 2a and b). We stress that this interchange of three-coordinate building blocks preserves the overall stoichiometry, and the number of building blocks surrounding the nanopores is the same whether the blocks are single boron atoms or B_3O_3 rings. The construction of D-R by interconnecting 2D sheets is shown in Fig. 2c and d.

The calculated unit cell parameters of B_2O_3-I (A) were in good agreement with experiments. All seven other structures examined are comparable in energy to B_2O_3-I; the energies (per mole of atoms) at the minima in the energy–volume plots all lie within 3 kJ mol^{-1} of each other, as shown in Fig. 1.

All seven structures that are as yet unknown in crystalline form (B–D and R-forms) are considerably more flexible than B_2O_3-I. The energy–volume curves for the R-forms (Fig. 1b) are all remarkably flat, varying by less than 5 kJ mol^{-1} over an enormous volume range, which for D-R is as large as 12–75 Å^3 per atom. A

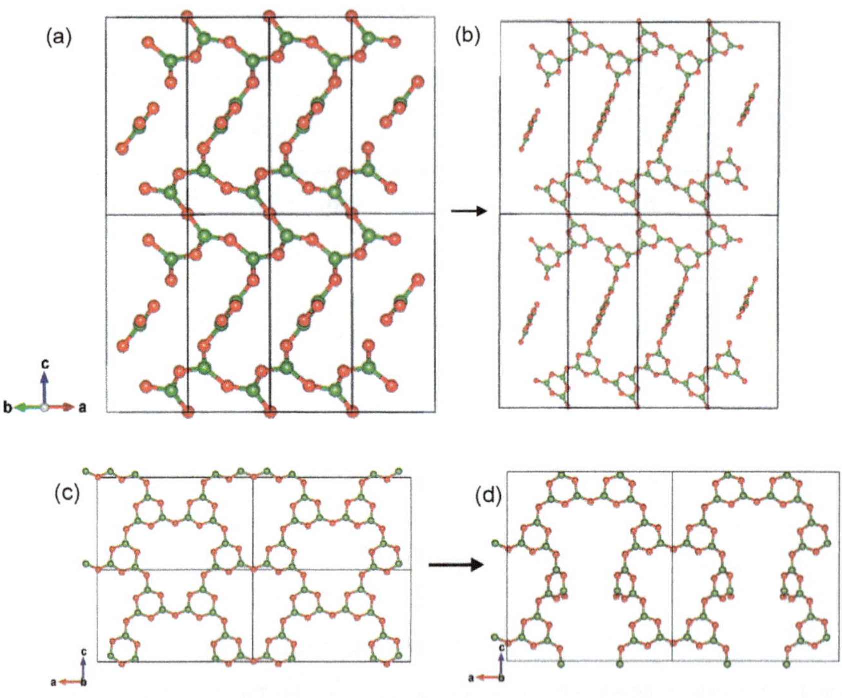

Fig. 2 (a) Structure of B_2O_3-I (A in Fig. 1) and (b) the analogous B_3O_3 ring-based structure A-R. (c) 2D sheet used to generate (d) structure D-R. To construct the 3D structure of D-R shown in (d) from the sheet in (c), one intralayer B–O–B bond per unit cell was replaced by one interlayer B–O–B bond. Boron atoms are green and oxygen atoms are red. Black lines show the unit cell boundaries.

predicted low-energy form of boron monoxide, containing six-membered B_4O_2 rings, also exhibits very similar behaviour.[31] The existence of such flexible structures can be readily invoked to rationalise the difficulty faced by researchers in constructing a theoretical model of vitreous B_2O_3 that is capable of both satisfying the experimental boroxyl content and also replicating the observed density (~1.8 g cm^{-3}).[32]

We can thus predict a large number of virtually equienergetic structures with very broad energy landscapes. The underlying cause of the width of the landscapes must be associated with the electronic structure of the B–O–B bridges. Analogous flexibility is seen in molecular forms; for example, calculations at the MP2/6-311++G(d,p) level show that all the conformations of $(HO)_2B$–O–$B(OH)_2$ for B–O–B angles larger than 120° are within 30 kJ mol^{-1} of each other. Such angular flexibility is integral to the broad energy–volume landscapes observed in the solid state, but by itself it is unable to account for the substantial widening of the landscapes observed for the boroxyl-decorated systems A-R to D-R.

In this paper we more fully elucidate the nature of this behaviour and study the relationship between the intrinsic flexibility of these frameworks and the underlying electronic structures of the units from which they are built. Plane-wave density functional theory calculations are complemented by molecular electronic structure calculations and template-based geometric simulations of the solid state. We subsequently use this information to predict further structures and show how the breadth of the energy landscape may be tuned.

Theoretical methods

Periodic *ab initio* calculations were conducted on the solid-state structures with the CASTEP program, using plane-wave density functional theory, the GGA approximation and the Perdew–Wang 91 (PW91) exchange–correlation functional.[33–35] Core electrons were simulated by default on-the-fly ultrasoft pseudopotentials (USP) and reciprocal space integration was conducted using the Monkhorst–Pack sampling scheme, with a minimum $2 \times 2 \times 2$ k-point grid.[36,37] Variable cell geometry optimisations were executed with the Broyden–Fletcher–Goldfarb–Shanno (BFGS) minimisation algorithm and the following parameters: (i) a precise basis set; (ii) maximum geometry force and stress tolerances of 0.02 eV Å^{-1} and 0.02 GPa, respectively; (iii) an electronic energy tolerance of 2.0×10^{-6} eV; (iv) a geometry energy tolerance of 1.0×10^{-5} eV; and (v) a maximum ionic displacement criterion of 0.001 Å. No symmetry constraints were imposed upon any structure during optimisation. Energy–volume landscapes were obtained by subjecting each optimised structure to a range of isotropic tensile and compressive stresses.

Ab initio calculations on all molecular species were carried out with the GAUSSIAN09 suite of programs[38] using second-order Møller–Plesset perturbation theory (MP2) and a split-valence triple-zeta basis set 6-311++G(d,p), with supplementary d-type polarisation functions on all main-group elements, p-type polarisation functions on hydrogen and diffuse functions on all atoms. Minimum energy structures on the potential energy surface were located using the Berny algorithm and verified by the absence of negative eigenvalues in the diagonalised Hessian matrix. Angular bending potentials were obtained *via* relaxed potential energy surface scans of the B–O–B bridging angle; in each case the HB–OB/BO–BO torsion angle was constrained to a specified value, whilst all other internal coordinates were fully optimised. All calculations were conducted *in vacuo*. Natural bond orbital (NBO) analyses[39] were conducted on the MP2 optimised structures.

Localised geometric simulations were conducted with the GASP program[40] in order to identify the flexibility windows of the B_2O_3-I and B_2O_3-IR frameworks, windows within which the rigid trigonal BO_3 units or 'clusters' remain undistorted from their idealised geometries. Geometric simulations model the bonding within relatively rigid units such as BO_3 by a template which represents the ideal bonding geometry of the cluster. Harmonic constraints connecting each atom to the corresponding vertex of a template impose energetic penalties if there is any deviation from the ideal cluster geometry. During a geometric simulation, atom and template positions are varied by minimising deviations from cluster geometry and the steric overlap of atomic spheres. This cheap geometric technique is quite different from *ab initio* or traditional atomistic potential methods. Energy landscapes are not generated, but it is possible to cheaply investigate framework geometries for large systems that satisfy local steric and bonding geometric constraints. Here, simulations were carried out with an oxygen hard-sphere radius of 1.350 Å and B–O bond lengths of 1.390 Å, in accordance with the bond lengths returned by *ab initio* calculations on the B_2O_3 structure. Default mismatch and minimum move criteria were implemented in all simulations, with geometric relaxation deemed to be successful if the largest disparities between the relaxed and idealised template structures remained lower than 0.001 Å for B–O bond lengths and 0.001° for O–B–O bond angles. The flexibility window is defined as the range of 'framework volumes' over which the BO_3 units that make up the framework can in principle be made geometrically ideal.

Coordination and angular flexibility

In the two experimentally known crystalline B_2O_3 polymorphs – B_2O_3-I and B_2O_3-II – adjacent boron atoms are coupled by an intermediate oxygen atom, forming an interlinked network of B–O–B bridges. The flexibility of these simple frameworks is significantly different, as can be seen by contrasting the energy–volume profiles of B_2O_3-I and B_2O_3-II (Fig. 3). The energy landscape of B_2O_3-I is considerably broader than that of the high-pressure B_2O_3-II polymorph.

In this paper, we quantify the width of the energy–volume landscapes, as measured by the maximum and minimum volumes, $V_5^{\max}$ and $V_5^{\min}$, respectively, at 5 kJ mol^{-1} above the local minimum, by the dimensionless parameter ξ_5:

$$\xi_5 = \frac{V_5^{\max} - V_5^{\min}}{V_5^{\min}} \quad (1)$$

The energy of B_2O_3-I varies by only 5 kJ mol^{-1} over a wide volume range of 8.0–14.6 Å^3 per atom ($\xi_5 = 0.83$), while the B_2O_3-II polymorph is far more rigid, with a corresponding range of only 6.8–8.6 Å^3 per atom ($\xi_5 = 0.26$).

These polymorphs differ in two fundamental respects: the density of B–O–B linkages, which is higher for B_2O_3-II, and the coordination of the boron atoms (3 for B_2O_3-I and 4 for B_2O_3-II). The former confers topological rigidity, whilst the

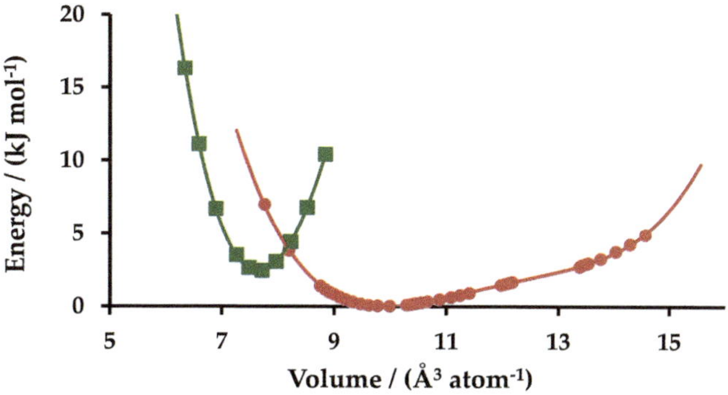

Fig. 3 Energy–volume profiles for B_2O_3-I (●; red) and B_2O_3-II (■; green).

latter dictates the angular flexibility of each individual bridge. Both factors contribute to the relative widths of the energy landscapes for B_2O_3-I and B_2O_3-II.

To examine bridge flexibility, we consider the angular bending potentials of two equivalent molecular systems. To reflect the presence of the trigonal BO_3 clusters in B_2O_3-I, the simple boroxane $H_2B–O–BH_2$ was selected as a molecular analogue. In order to retain the bonding character but with tetrahedral coordination, the corresponding dianion $[H_3B–O–BH_3]^{2-}$ was employed to represent a B–O–B bridge in B_2O_3-II. As in previous studies, two equilibrium structures were identified for $H_2B–O–BH_2$: one has C_{2v} symmetry and an optimised B–O–B bond angle of 123°, and the other has D_{2d} symmetry with an angle of 180° (Fig. 4).[41,42] The two structures are similar in energy, with calculations at the MP2/6-311G++(d,p) level confirming the C_{2v} structure as the global minimum on the potential energy surface. The fully extended planar conformation has D_{2h} symmetry and corresponds to a second-order stationary point on the potential energy surface; this species is 42.9 kJ mol^{-1} less stable than the equilibrium D_{2d} structure and lies 48.4 kJ mol^{-1} above the global minimum.

A single minimum was characterised on the potential energy surface of $[H_3B–O–BH_3]^{2-}$; the corresponding eclipsed structure is of C_{2v} symmetry and also has an optimised B–O–B bond angle of 123°.

The potential energy of the $H_2B–O–BH_2$ molecule as a function of the B–O–B bond and HB–OB torsion angles is shown in Fig. 5. The bending potential in $H_2B–O–BH_2$ is highly dependent upon the torsion angle between the two BH_2 planes. Compressing the B–O–B linkage in the D_{2d} structure (HB–OB = 90°) lowers the

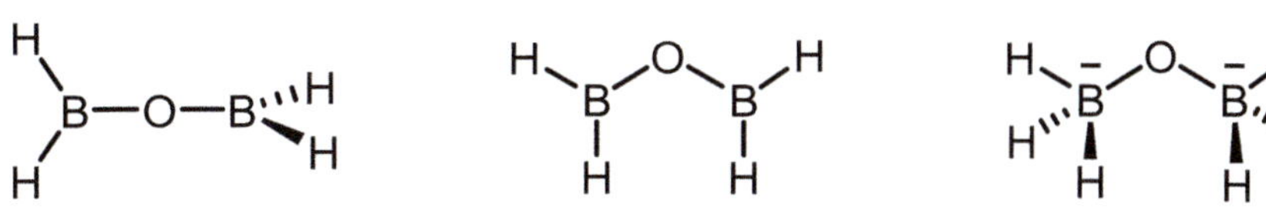

Fig. 4 Equilibrium structures for $H_2B–O–BH_2$ and $[H_3B–O–BH_3]^{2-}$, calculated at the MP2/6-311++G(d,p) level of theory. The C_{2v} structure of $H_2B–O–BH_2$ is the global minimum on the PES, lying 5.5 kJ mol^{-1} below the optimised D_{2d} structure.

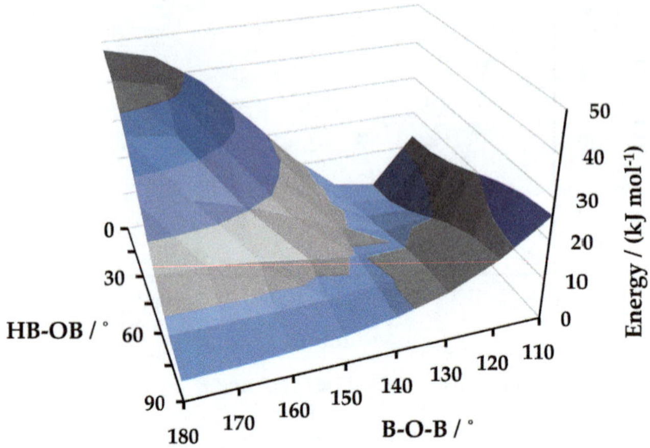

Fig. 5 Potential energy surface of $H_2B{-}O{-}BH_2$, plotted as a function of the B–O–B bond angle (110–180°) and the HB–OB torsion angle (0–90°). The C_{2v} structure corresponds to HB–OB = 0° and the D_{2d} structure to HB–OB = 90°. All other internal coordinates are fully optimised at each point.

symmetry to C_s and the energy increases; in contrast, a reduction in the B–O–B angle from 180° in the C_{2v} structure (HB–OB = 0°) leads to an energy minimum at 123°.

Starting from the global minimum C_{2v} structure, to achieve B–O–B angles greater than 135° without surmounting a significant potential energy barrier, there must be a concurrent increase in the HB–OB torsion angle (Fig. 5) – beyond B–O–B angles of 135°, the structure can be stabilised by abandoning planarity and deforming towards the staggered C_s structure. Thus, cooperation between the bending and twisting motions of the bridge is crucial for its facile compression and expansion; provided there is complete torsional freedom, the B–O–B bond angle can vary between 110° and 180° with a minimal change in energy (<10 kJ mol^{-1}). Without torsional freedom, the B–O–B linkage is far more rigid.

The subtle balance in energy between the two equilibrium structures is a fundamental feature of the B–O–B bridge, and can be readily explained by valence bond theory.[43] A number of plausible resonance forms are readily available for $H_2B{-}O{-}BH_2$ (Fig. 6). A natural bond orbital (NBO) analysis of the MP2 optimised C_{2v} equilibrium structure suggests that the resonance structure that

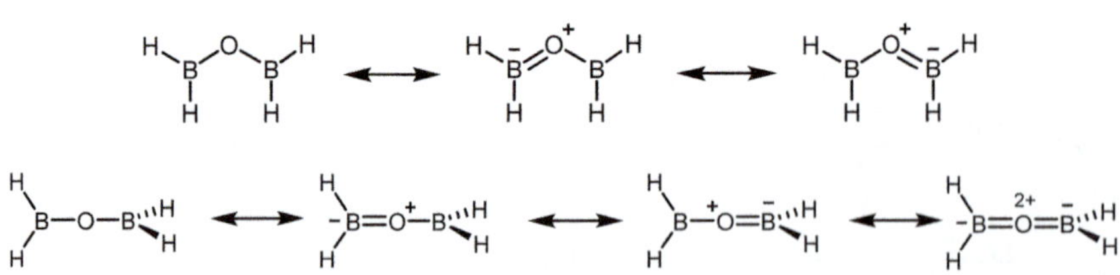

Fig. 6 Resonance forms for the C_{2v} (top) and D_{2d} (bottom) structures of $H_2B{-}O{-}BH_2$. Formally, the π-bonding in the D_{2d} species consists of two π_{B-O} bonding orbitals; in the C_{2v} structure it is best described by a single three-centred bonding orbital τ_{B-O-B}.

Table 1 Key structural parameters for the two equilibrium structures of $H_2B{-}O{-}BH_2$, including the B–O–B bond angles, lengths and Wiberg bonding indices (WBI) of the B–O bonds and the natural charge distribution. Q(B) and Q(O) are natural charges and L_{H-H} denotes the closest through-space H–H separation in each species. r_H is the van der Waals radius of hydrogen

Structure	B–O–B angle/°	B–O length/Å	B–O WBI	$Q(B)/e$	$Q(O)/e$	$L_{H-H}/2r_H$
C_{2v}	123	1.378	0.837	0.801	−0.985	1.03
D_{2d}	180	1.345	0.862	0.801	−0.985	1.71

accounts most comprehensively for the total electron density ($\rho_{Lewis} = 0.997$) incorporates a three-centred bonding orbital τ_{B-O-B}, derived from the normalised linear combination of an oxygen 2p orbital and parallel 2p orbitals from each of the adjoining boron atoms. Each B–O bond is also described by a σ_{B-O} bonding orbital, while the central oxygen atom maintains an occupied valence nonbonding orbital n_o of hybridisation $sp^{3.86}$. The analysis also suggests that the D_{2d} structure of $H_2B{-}O{-}BH_2$ is most accurately represented ($\rho_{Lewis} = 0.995$) by the doubly-bonded resonance form, which resembles the isoelectronic molecule allene (Fig. 6). Each of the two equivalent B–O interactions is best described by two bonding NBOs – σ_{B-O} and π_{B-O} – which are heavily polarised towards the bridging oxygen.

Table 1 summarises the key structural parameters, including the length and Wiberg bonding indices (WBI) of the B–O bonds and the natural charge distribution, of the two equilibrium structures of $H_2B{-}O{-}BH_2$.

The key point is that although conjugation is more extensive in the D_{2d} structure – as shown by the shorter B–O bond length (1.345 Å *vs.* 1.378 Å for the C_{2v} structure) and larger Wiberg bonding index (0.862 *vs.* 0.837) – its stabilising influence is compensated by the hybridisation of the valence nonbonding orbital n_o in the C_{2v} structure.

Central to the flexibility of $H_2B{-}O{-}BH_2$ is the presence of three-coordinate boron. The bending potential of the B–O–B bridge in $[H_3B{-}O{-}BH_3]^{2-}$, where

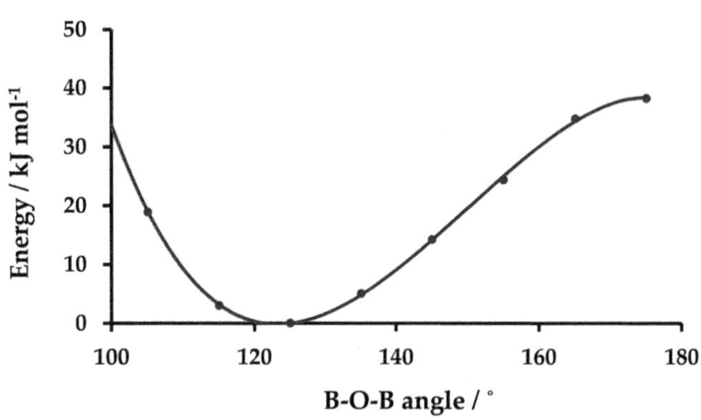

Fig. 7 Bending potential for the C_{2v} structure of $[H_3B{-}O{-}BH_3]^{2-}$, obtained from a relaxed potential energy scan of the B–O–B angle at fixed symmetry (HB–OB = 0°).

boron is four-coordinate, is instead characterised by a relatively deep well for all the HB–OB torsion angles. Fig. 7 shows the bending potential for the eclipsed C_{2v} structure. Increasing the coordination of boron from three to four removes the lone vacant 2p orbital from the system, prevents pπ-donation from the bridging oxygen and stiffens the B–O–B bridge. This rationalises the very different flexibilities of B_2O_3-I and B_2O_3-II shown in Fig. 3.

Coordination and topological rigidity

The trigonal coordination of boron in B_2O_3-I, compared with the tetrahedral coordination in B_2O_3-II, changes not only the B–O–B bending potential but also the topological freedom in the structure, *i.e.* the density of the B–O–B linkages. To decouple these two effects, it is necessary to compare structures with similar bending potentials but containing atoms with different valencies and hence coordinations. To achieve this, we turn our attention to β-cristobalite. This cubic polymorph of silica comprises an interlinked network of rigid SiO_4 tetrahedra connected by Si–O–Si bridges. The silicon atoms are four-coordinate, as in B_2O_3-II, yet the energy–volume landscape is considerably broader. Below the 5 kJ mol^{-1} ceiling defined earlier, the volume occupied per atom varies between 12.6–18.6 $Å^3$ per atom ($\xi_5 = 0.48$). Similar flexibility is present in a range of silicate frameworks.[44,45] As for B_2O_3-I, the origin of its flexibility may be traced to the angular bending potential of a simple molecular analogue, H_3Si–O–SiH_3, which is compared with that of H_2B–O–BH_2 in Fig. 8. Like $[H_3B\text{–}O\text{–}BH_3]^{2-}$, the bending potential is essentially independent of the HSi–OSi torsion angle, but like the C_s species of H_2B–O–BH_2 it is exceptionally flat over the range of 120–180°, with a very shallow minimum at 155°. As a consequence of the flexibility of the Si–O–Si linkage, a wide range of bridging bond angles have been observed experimentally in many different silicate frameworks.[45–47]

Given any arrangement of SiO_4 tetrahedra, it is possible to construct a related structure in which the $SiO_4{}^{4-}$ tetrahedra are exchanged for $B_2O_5{}^{4-}$ units and all the B atoms are three-coordinate. For each tetrahedron, there are three ways in which such a substitution can be realised, since the B–B axis can be aligned along any one of the three tetrahedral C_2 axes.

The DFT energy landscape of β-cristobalite-borate-a (β-CB-a, formula B_2O_3), obtained by making such a substitution and orienting all the B–O–B bridges along the *a*-axis, is superimposed upon that of the silicate precursor in Fig. 8. The energy landscape of β-CB-a is notably broader than its silicate precursor, covering a volume span of 9.7–17.4 $Å^3$ per atom with an energy change < 5 kJ mol^{-1} ($\xi_5 = 0.79$). The change in the form of the energy landscape is more subtle; while for β-cristobalite the minimum energy structure lies towards the high-volume end of the flexible region, for the borate analogue it resides towards the low-volume end. This is consistent with the findings of Wells *et al.*,[40,44] who explored the notion of flexibility windows in zeolites and other covalent frameworks and observed that silicates exhibit an atypical tendency to be maximally extended in their relaxed states. This correlates with the absence of any local minima at small angles in the H_3Si–O–SiH_3 bending potential (Fig. 8).

The replacement of $SiO_4{}^{4-}$ tetrahedra by $B_2O_5{}^{4-}$ units should produce little change in the angular flexibility of the polyhedral bridges – in fact, depending on how twisted the bridges are, such a substitution may rigidify the structure – but

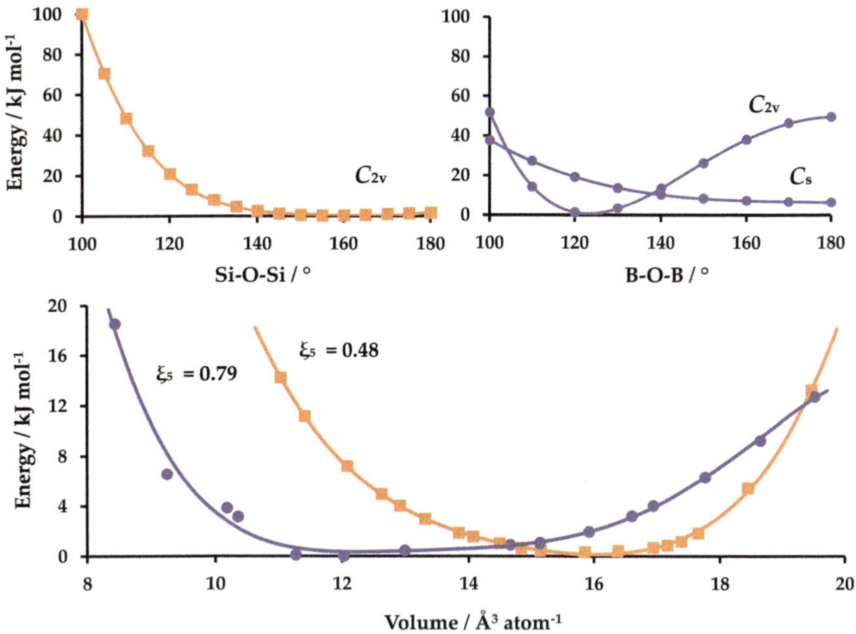

Fig. 8 Upper frames: Molecular bending potentials of $H_3Si-O-SiH_3$ (■; orange) and $H_2B-O-BH_2$ (●; purple); the two profiles for $H_2B-O-BH_2$ correspond to the C_s and C_{2v} structures. Lower frame: Plane-wave DFT energy–volume landscapes of β-cristobalite (■; orange) and β-cristobalite-borate-a (●; purple), in which B_2O_5 units have replaced all the SiO_4 tetrahedra, as described in the text.

nevertheless the energy–volume landscape widens (Fig. 8). As such, we conclude that it is unwise to focus entirely on bending potentials, for it is apparent that topological freedom also dictates the flexibility of covalent frameworks.

A vast array of theoretical B_2O_3 polymorphs may be obtained from silicates in this manner, including new families of zeolites, and we expect the increases in flexibility observed for β-cristobalite will be general. It is also important to stress that the β-cristobalite-borate-a structure is almost isoenergetic with the experimental B_2O_3-I polymorph (higher in energy by less than 0.2 kJ mol^{-1}).

Exploring flexibility in B_2O_3-I and B_2O_3-IR

Although the energy landscape of the crystalline B_2O_3-I polymorph has been reported previously,[24] there has been no examination of the structural transformations observed during the expansion and compression of the framework. Below a ceiling of 5 kJ mol^{-1}, the volume of the B_2O_3-I unit cell is free to vary from 8.0–14.6 Å^3 ($\xi_5 = 0.83$), with the a-lattice parameter varying only in the range of 4.2–4.6 Å, while the c-parameter varies over a much wider range of 7.7–11.8 Å. The B_2O_3-I framework is thus inherently anisotropic in its flexibility, with the shallow region of the energy landscape dominated by the expansion and compression of the framework along the c-axis. During the expansion of B_2O_3-I, a barrierless pressure-induced transition is observed between −1.5 and −1.6 GPa; this transition leads to a substantial anisotropic decompression of the framework along

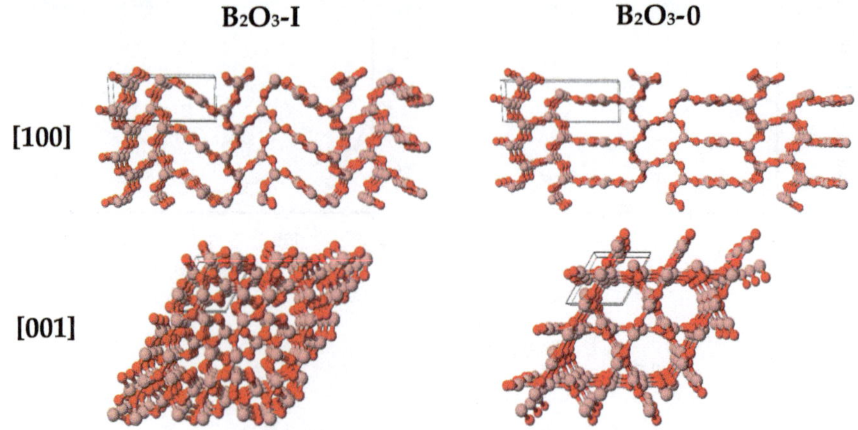

Fig. 9 Optimised structures for the two polymorphs B_2O_3-I and B_2O_3-0, viewed along the [100] and [001] directions. The B_2O_3-I structure is optimised at 0 GPa, the B_2O_3-0 structure at −1.6 GPa. Here and in subsequent figures boron atoms are pink and oxygen atoms are dark red. Black lines show the unit cells.

[001] and the formation of regular hexagonal channels of diameter 4.5 Å, aligned along the same direction. The pleated ribbons of the trigonal BO_3 units undergo a significant reconfiguration to form, as viewed along the [001] direction, a series of interlocking planar sheets (Fig. 9). This expanded structure is very similar to the theoretical polymorph B_2O_3-0, first proposed by Huang and Kieffer on the basis of molecular dynamics simulations.[48]

The B_2O_3-I unit cell contains nine distinct B–O–B bridges, which can be classified into a set of three and a set of six according to their behaviour over the energy landscape. Fig. 10 shows that the six bridges in one group are part of almost planar B_2O_5 units and exhibit B–B vectors with significant components perpendicular to the [001] direction; the three bridges in the other group are part of highly twisted B_2O_5 units and have B–B vectors with significant components parallel to [001], as is also clear from Fig. 10.

During the expansion of the framework, the B_2O_5 units containing the set of six bridges retain their planarity, with the bridge angles expanding from an average of 125.8° at the low-volume end to 145.6° at the high-volume end. The bridges in the group of three, however, undergo a much greater extent of decompression

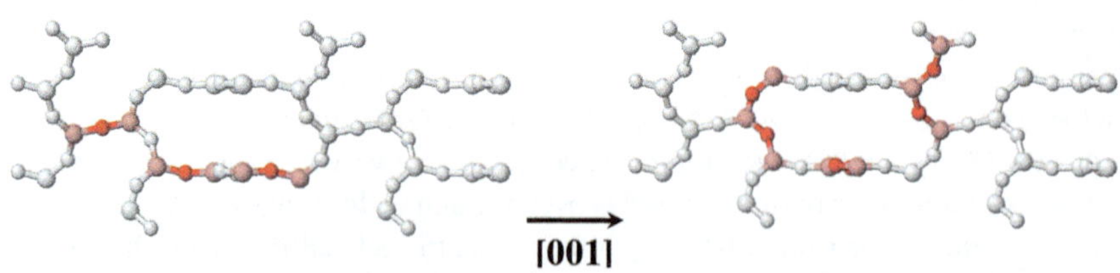

Fig. 10 B_2O_3-I framework, highlighting the two distinct groups of B–O–B bridges. The left figure depicts the three B_2O_5 units that undergo twisting towards the staggered C_s conformation as the framework expands; the figure on the right depicts the six B_2O_5 units that do not undergo any significant twisting during structural expansion.

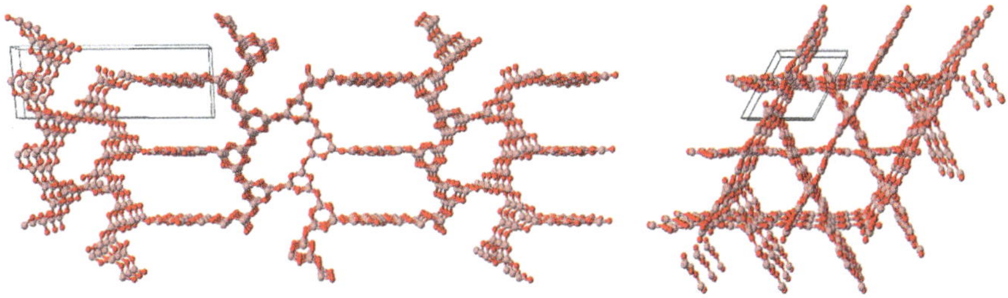

Fig. 11 Optimised structure of the B_2O_3-0R framework, as viewed along the [100] and [001] directions. Black lines show the unit cells.

from 129.5° to 163.5° and are distorted further towards the staggered C_s conformation, with the OB–OB torsion angle increasing over the range of 43.1–58.8°. This suggests that these twisted B_2O_5 units are integral to the anisotropic flexibility of the B_2O_3-I framework, and moreover that the torsional freedom of the bridges in these units should dictate the flatness of the energy–volume landscape.

As emphasised in the introduction, boron atoms and boroxyl rings (B_3O_3) are interchangeable structural units in B_2O_3 frameworks, as their substitution preserves the stoichiometry of the compound, the coordination state of boron and the underlying topology. Fig. 1b shows how the boroxyl-decorated B_2O_3-I framework (A-R in Fig. 1 and ref. 22 or, equivalently, B_2O_3-IR) is characterised by an exceptionally broad energy landscape. For the predefined ceiling of 5 kJ mol^{-1}, the volume of the unit cell is free to vary between 10.6–45.2 Å^3 per atom ($\xi_5 =$ 3.26); this is approximately four times broader than B_2O_3-I ($\xi_5 = 0.83$) and more than one order of magnitude broader than B_2O_3-II ($\xi_5 = 0.26$).

The structural transformations of B_2O_3-IR under tensile and compressive stresses emulate those of its precursor, but the magnitudes of the changes are considerably larger. Over the shallow region (<5 kJ mol^{-1}) of the energy landscape: (i) the a-lattice parameter varies by a relatively small amount between 8.3–9.8 Å, while the c-lattice parameter varies much more, between 7.9–24.6 Å; (ii) the six bridges that are part of planar B_2O_5 units expand from 123.5° to 179.4°; and (iii) the three bridges that are part of twisted B_2O_5 units expand from 121.3° to 179.4°, twisting significantly from an OB–OB torsion angle of 26.1 to 58.1°. As before, a pressure-induced phase transition precipitates the formation of hexagonal channels coincident with the [001] direction (Fig. 11). We denote this low-pressure boroxyl-decorated phase B_2O_3-0R, by analogy[22] with B_2O_3-0. Ferlat *et al.*[17] discounted the B_2O_3-0 phase as insignificant in explaining the ready vitrification of boron oxide, citing both its high energy and the absence of boroxyl rings; we propose that B_2O_3-0R cannot be so dismissed, as our calculations indicate that it is essentially isoenergetic with B_2O_3-I.

Understanding and tuning structure flexibility

Several proposals have been made previously to account for the vastly increased flexibility of B_2O_3-IR, the boroxyl-decorated analogue of the B_2O_3-I framework.[24] One possibility is that when boroxyl rings are incorporated into the framework, each boron atom is now involved in only one flexible B–O–B bridge – with the other two neighbouring oxygens confined to the B_3O_3 ring with fixed O–B–O angles. In contrast, in B_2O_3-I each boron atom is surrounded by flexible bridges in all directions. A further suggestion is that extended π-conjugation in the boroxyl-decorated frameworks could influence the angular flexibility of the bridging linkages.

To test which of these two factors is more important, Fig. 12 compares the bending potentials for X_2B–O–BX_2, where X = H, F and OH. The cooperativity between the B–O–B bridging angle and XB–OB torsion angle present in H_2B–O–BH_2 is damped in both F_2B–O–BF_2 and $(HO)_2B$–O–$B(OH)_2$, due to competing pπ–pπ* donor–acceptor interactions from the peripheral heteroatoms. The angular bending potential of the B–O–B bridge in the boroxyl analogue ($B_3O_3H_2$)–O–($B_3O_3H_2$) (Fig. 15) emulates that of $(HO)_2B$–O–$B(OH)_2$ almost exactly. Similar interactions and angular flexibility are observed in ($B_3O_3H_2$)–O–($C_3N_3H_2$) (Fig. 15) and ($B_3N_3H_5$)–O–($B_3N_3H_5$) (Fig. 16), indicating that provided the heteroatom X has an available lone pair, it can compete effectively with the bridging oxygen for the lone vacant 2p orbital on boron.

Further calculations were conducted on $B(OBH_2)_3$ and the boroxyl analogue $(B_3O_3)(OBH_2)_3$. In $B(OBH_2)_3$, the central boron atom is involved in three flexible bridges, whereas in $(B_3O_3)(OBH_2)_3$ each endocyclic boron is involved in just one flexible bridge. These two molecules thus allow us to examine whether the fixed O–B–O bond angles in the boroxyl system enhance, diminish or do not affect the flexibility of the bridges. In each system the three B–O–B bridges were expanded

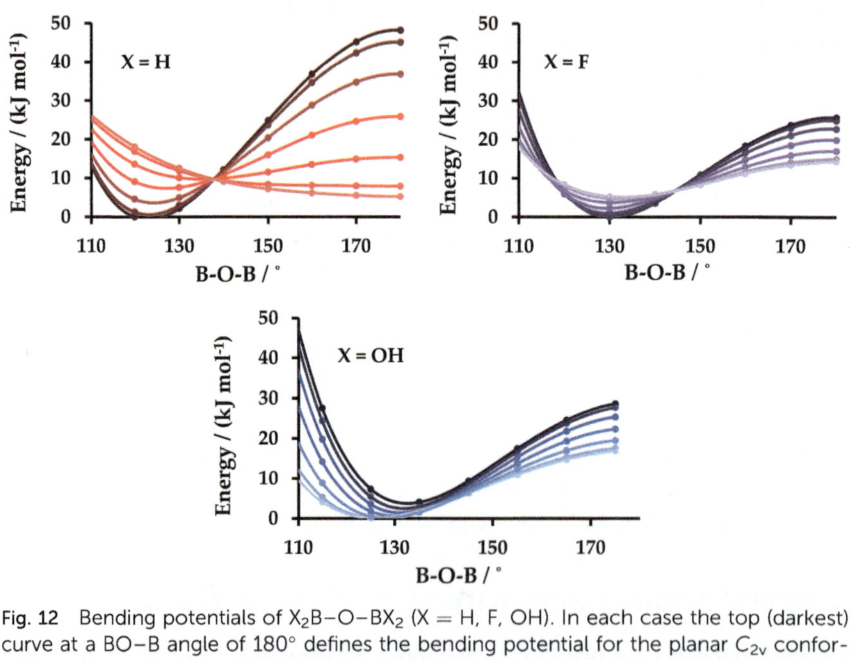

Fig. 12 Bending potentials of X_2B–O–BX_2 (X = H, F, OH). In each case the top (darkest) curve at a BO–B angle of 180° defines the bending potential for the planar C_{2v} conformation (XB–OB = 0°) and the lowest (lightest) curve at 180° is for the staggered C_s conformation (XB–OB = 90°); the intermediate curves correspond to the B–O–B bending potentials at XB–OB torsion angle intervals of 15°.

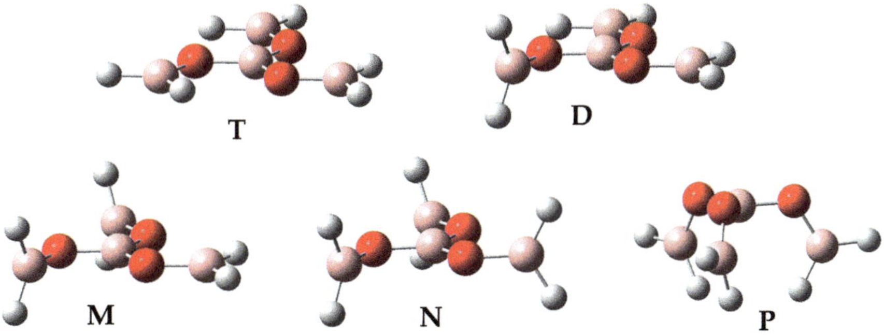

Fig. 13 The five torsional configurations (T, D, M, N and P) studied for $B(OBH_2)_3$ and $(B_3O_3)(OBH_2)_3$. In each case, all three bridges were expanded simultaneously with fixed HB–BO torsion angles. The energy changes are essentially identical in the two molecules, proving that the bending potential of one B–O–B bridge depends only upon the geometry of the local B_2O_5 unit and not upon the configuration of any neighbouring bridges.

concurrently over the range of 120–175°; this was repeated for five torsional configurations – triplanar (T), diplanar (D), monoplanar (M), non-planar (N) and perpendicular (P) (Fig. 13). For all five configurations, the bending potentials of $B(OBH_2)_3$ and of $(B_3O_3)(OBH_2)_3$ exhibit only very minor differences, showing that it is immaterial whether each boron atom is involved in one or three flexible bridges.

Thus, in general, molecular calculations reveal that there is no intrinsic difference between the electronic structures of the bridges in B_2O_3-I and B_2O_3-IR; on the contrary, the angular flexibility of the bridge appears to be governed exclusively by the geometry of the local B_2O_5 unit (*i.e.* O_2B–O–BO_2) and not by any longer range electronic effects.

In the absence of any significant differences in the electronic structures of the bridges, the enhanced flexibility of B_2O_3-IR relative to B_2O_3-I must therefore be geometric or topological in origin. Exchanging boron atoms for boroxyl rings increases the diameters of the framework cavities at high volumes, reduces the number density of B–O–B bridges, and increases the number of bonds between neighbouring bridging oxygens by separating them with a rigid spacer. We suggest that these features increase the torsional freedom of the bridges and their independence within the framework. Across all of the B_2O_3 structures considered here and in ref. 24, there is a very strong inverse correlation, shown in Fig. 14, between the number density of flexible B–O–B bridges, $N_{\text{B–O–B}}$, in the optimised structure and the width of the energy–volume landscape, as measured by the dimensionless parameter ξ_5 (defined in eqn (1)).

Although such an analysis is informative, in reality these frameworks are networks of interlinked bridges. During expansion and compression of the framework, the bridges cannot deform according to their individual preference, but instead are obliged to distort collectively. The manner in which the bridges are connected, and not simply their density, should therefore modulate the flexibility. It is also entirely plausible that the boroxyl rings serve principally as physical spacers between the bridges, offering additional degrees of torsional freedom absent from the B_2O_3-I structure; in frameworks without boroxyl rings,

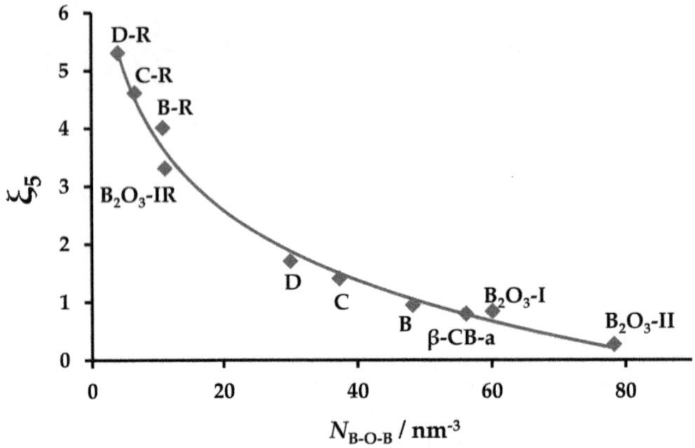

Fig. 14 The width ξ_5 (defined in eqn (1)) of the energy landscape 5 kJ mol^{-1} above the local minimum as a function of the number density of flexible B–O–B bridges (N_{B-O-B}) in each of the polymorphs considered to date. Structures B/B-R, C/C-R and D/D-R were reported in ref. 24. There is a very strong inverse correlation, suggesting that the extreme energy–volume landscapes of the boroxyl-decorated polymorphs may simply be ascribed to their very low density of B–O–B bridging linkages.

the closest pairs of bridging oxygen atoms are separated by two bonds, as opposed to four bonds in exclusive boroxyl systems.

If these topological factors dominate, the frameworks containing boroxyl rings should retain a significant proportion of their flexibility if other six-membered rings replace boroxyl. We therefore consider a range of frameworks obtained by decorating the B_2O_3-IR system with isoelectronic six-membered rings – namely triazine, borazine and benzene – whilst retaining the bridging oxygen atoms.

Exchanging all the boroxyl rings for triazine yields a framework with C–O–C bridges (B_2O_3-IR(T)), whilst replacing alternate boroxyl rings with triazine (B_2O_3-IR(ST)) generates an interlinked network of B–O–C bridges. The corresponding energy landscapes of B_2O_3-IR, B_2O_3-IR(T) and B_2O_3-IR(ST) are depicted in Fig. 15, alongside the planar and staggered bending potentials of their molecular analogues.

The molecular bending potentials show that the angular flexibility of the bridge varies in the order B–O–B > B–O–C > C–O–C, with the well depths increasing more than twofold with each substitution of carbon for boron. This likely reflects the weakening of pπ–pπ* donation from the bridging oxygen to the rings and the increasing dominance of the valence nonbonding orbital on oxygen. At small volumes, the three energy–volume profiles are almost exactly superimposable, suggesting that further compression is prohibited exclusively by the topology of the structure and not at all by the bridge flexibility. It is clear, nevertheless, that the bending potential of the bridge governs the capacity of the framework to undergo expansion, with the width of the energy well decreasing significantly as the triazine content increases. In B_2O_3-IR(ST), over the volume range defined by $\xi_5 = 2.6$, the bridge angles in the six planar B_2O_5 units vary over the range of 122.9–146.6°, and the bridge angles in the three twisted B_2O_5 units vary over the range of 116.2–145.1°; in B_2O_3-IR(T) ($\xi_5 = 1.9$), the analogous ranges, 124.4–137.1° and 115.4–137.5°, respectively, are narrowed at the upper volume

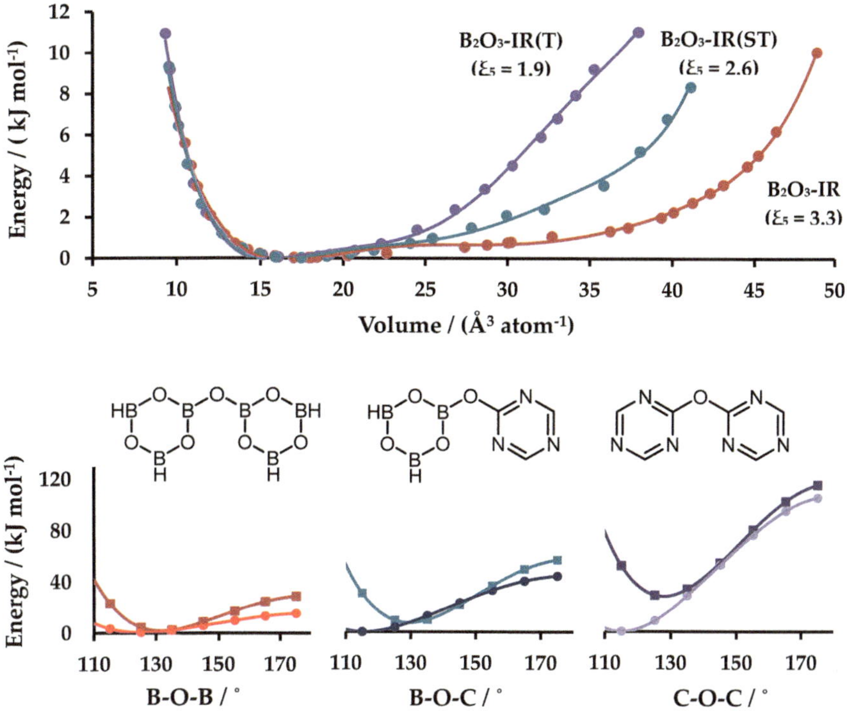

Fig. 15 Bending potentials of $(B_3O_3H_2)–O–(B_3O_3H_2)$, $(B_3O_3H_2)–O–(C_3N_3H_2)$ and $(C_3N_3H_2)–O–(C_3N_3H_2)$, and energy landscapes of the corresponding decorated B_2O_3-IR frameworks. The two bending potentials correspond to the planar C_{2v} conformation (■) and the staggered C_s conformation (●).

limit of the ξ_5 region. The nature of expansion in both B_2O_3-IR(ST) and B_2O_3-IR(T) is the same as in B_2O_3-IR, with the emergence of hexagonal channels along the [001] direction at high volumes. While making the bridges more rigid appears to have a significant impact on the flexibility of the B_2O_3-IR frameworks, it is important to note that the energy landscapes remain considerably broader than that of the B_2O_3-I framework.

Frameworks constructed from borazine and benzene – abbreviated to B_2O_3-IR(Bo) and B_2O_3-IR(Be), respectively – exhibit similar trends in flexibility, as demonstrated in Fig. 16. The presence of clashing *ortho*-hydrogens, however, hinders torsional rotations in the $B_3N_3H_5$ and C_6H_5 units, leading to structures with larger optimised volumes and a reduced capacity for compression. In B_2O_3-IR(Be), the bridges expand from 109.4–138.3° over the ξ_5 region, whereas in B_2O_3-IR(Bo) the bridge angles increase up to 179.3° at the high-volume limit.

The molecular bending potentials of (C_6H_5)–O–(C_6H_5) and $(B_3N_3H_5)$–O–$(B_3N_3H_5)$, shown in Fig. 16, follow the trends in angular flexibility outlined above. However, the structural transformations of the solid-state frameworks B_2O_3-IR(Be) and B_2O_3-IR(Bo) over their energy landscapes differ from those of B_2O_3-IR. As opposed to undergoing an anisotropic extension along the [001] direction, B_2O_3-IR(Be) and B_2O_3-IR(Bo) extend significantly in all directions. In B_2O_3-IR(Bo), the *a*-parameter varies over the range of 6.8–10.2 Å and the *c*-parameter between

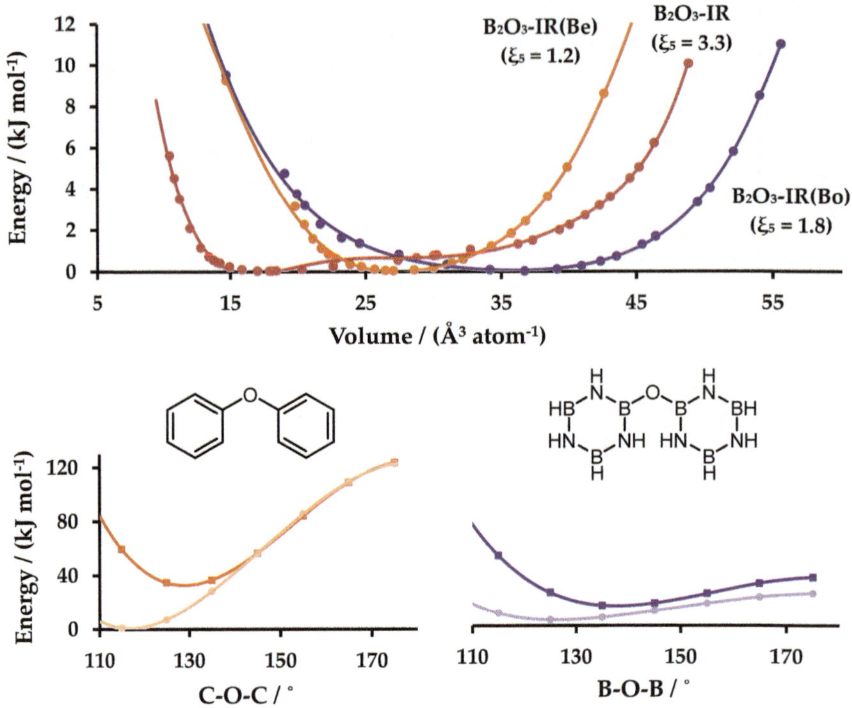

Fig. 16 Bending potentials of (C_6H_5)–O–(C_6H_5) and $(B_3N_3H_5)$–O–$(B_3N_3H_5)$, and energy landscapes of the corresponding decorated B_2O_3-IR frameworks. As in Fig. 15, the energy landscape of B_2O_3-IR is also shown (red curve). The two bending potentials correspond to the planar C_{2v} conformation (■) and the staggered C_s conformation (●).

21.5–25.5 Å, whereas in B_2O_3-IR(Be) the respective ranges are 6.7–9.4 Å and 22.1–23.3 Å.

Both structures appear incapable of accessing lower *c*-lattice parameters, presumably as the $B_3N_3H_5$ and C_6H_5 units are unable to adopt planar conformations at smaller B–O–B angles due to steric repulsions between *ortho*-hydrogens. Despite the significantly rigidified bridges and torsional restrictions, even the energy–volume landscape of B_2O_3-IR(Be) remains wider than that of B_2O_3-I. The density of bridging linkages is seen to be a fundamental feature in dictating solid-state flexibility.

Sources of flexibility and torsional freedom

The molecular species $(B_3O_3H_2)$–O–$(B_3O_3H_2)$ and $(B_3N_3H_5)$–O–$(B_3N_3H_5)$ have strikingly similar angular flexibilities, yet their corresponding solid-state frameworks, B_2O_3-IR and B_2O_3-IR(Bo), behave very differently. B_2O_3-IR(Bo) is far less compressible than B_2O_3-IR because of reduced torsional freedom due to clashing *ortho*-hydrogens. The only significant difference between the two is this torsional freedom of the bridging linkages; herein may lie another clue to the relative flexibilities of B_2O_3-I and B_2O_3-IR. Fig. 17 plots the average OB–OB torsion angle of the set of the three B–O–B bridges contained within twisted B_2O_5 units *versus* the

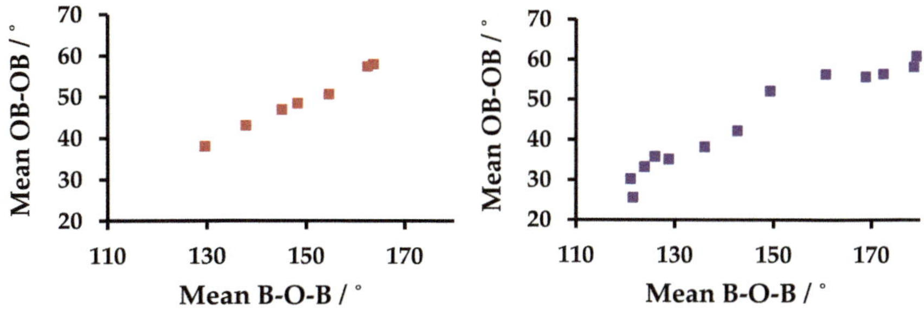

Fig. 17 Plots of the mean OB–OB torsion angle *versus* the mean B–O–B angle for the three B–O–B bridges contained within the twisted B_2O_5 units present in the unit cells of B_2O_3-I (left; red) and B_2O_3-IR (right; purple). The lowest and highest angles correspond to the low- and high-volume thresholds of the energy–volume landscapes (5 kJ mol^{-1}), respectively.

B–O–B angle for B_2O_3-I and B_2O_3-IR, for all points on the energy landscape below the 5 kJ mol^{-1} ceiling. In both frameworks, the upper limit of the torsion angle is 60°, as can be seen in the hexagonal channels along the [001] direction; overall, however, there is clearly greater scope for torsional rotations during compression in B_2O_3-IR, and torsional freedom can help to explain the differences in flexibility between B_2O_3-I and B_2O_3-IR. Flexibility is due to cooperativity between the bending and twisting motions in the bridges during expansion, and this cooperativity requires topological freedom, *i.e.*, a low density of bridges.

The flexibility windows of the B_2O_3-I and B_2O_3-IR frameworks, obtained from template-based geometric simulations, also provide some insight (Fig. 18). The flexibility window defines the range of lattice parameters within which the BO_3 units can in principle remain undistorted from their perfect trigonal geometry, and it therefore indicates the intrinsic flexibility of a particular framework.

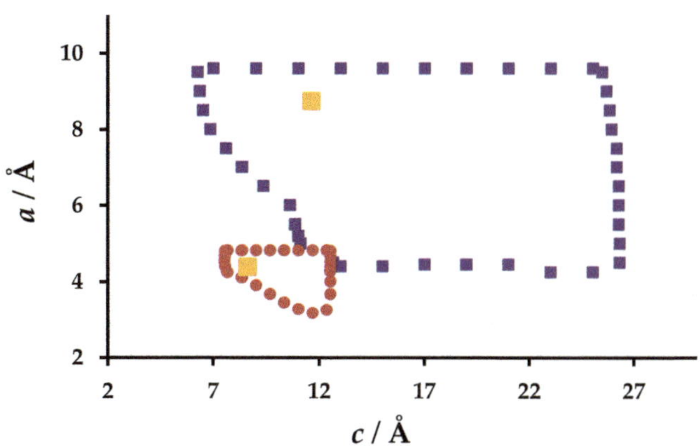

Fig. 18 Flexibility windows (*a*–*c*) for B_2O_3-I (●; red) and B_2O_3-IR (■; purple). The *ab initio* optimised lattice parameters are marked by yellow ■. Within these windows, the trigonal BO_3 units can retain their ideal geometries, so any changes in volume result from distortions of the flexible B–O–B bridges.

Agreement between the geometric simulations and the *ab initio* energy–volume landscapes is remarkably good. The *ab initio* optimised structures of both frameworks reside in the high-density region of their flexibility windows – that is, the B_2O_3-I and B_2O_3-IR structures are highly compressed in their relaxed states – which is in agreement with the presence of global minima at small B–O–B angles on the potential energy surfaces of $H_2B–O–BH_2$, $(OH)_2B–O–B(OH)_2$ and $(B_3O_3H_2)$–$O(B_3O_3H_2)$. As anticipated, the window for B_2O_3-IR is considerably larger than that for B_2O_3-I; in the broadest regions, the *a*-lattice parameter varies over the range of 3.3–4.8 Å in B_2O_3-I and 4.3–9.6 Å in B_2O_3-IR, while the *c*-lattice parameter is free to vary between 7.6–12.5 Å and 6.4–26.4 Å, respectively. This verifies the *ab initio* results, proving that the broad energy landscapes in the boroxyl-decorated frameworks are not artefacts of the DFT calculations.

Two other intriguing features present themselves: (i) for large *a*-parameters, the *c*-axis of the boroxyl-decorated framework can be compressed further than B_2O_3-I; and (ii) decorating the B_2O_3-I framework with boroxyl rings alters the profile of the flexibility window. Across the entire window of B_2O_3-I, increasing the *c*-parameter facilitates greater compression along the *a*-axis, whereas in B_2O_3-IR this interplay between the lattice parameters is quickly exhausted. In B_2O_3-I, the value of the *c*-lattice parameter restricts the height of the window, suggesting that the torsional rotations of the bridges are constrained; in B_2O_3-IR, this only occurs for the smallest values of *c* (<12 Å).

These simple geometric simulations show that the increased flexibility in B_2O_3-IR is not merely a question of electronics, or even the bridge density *per se*; the interdependence of the *a*–*c* lattice parameters is fundamentally different in the two systems, suggesting that the intrinsic geometric constraints present in B_2O_3-I are lifted upon decorating the system with boroxyl rings. From a geometric perspective, it is plausible that the increased cavity diameters in B_2O_3-IR permit larger torsional rotations in the framework, providing the B–O–B bridges with degrees of freedom previously unattainable.

Conclusions and final remarks

In this paper we have explored a wide range of low-energy boron–oxygen frameworks which exhibit unprecedented mechanical flexibility for purely inorganic materials. We anticipate that many of these are already present in the vitreous oxide, given the extensive controversy regarding its density over many years.

We have explored in detail the underlying atomic mechanisms that govern this behaviour and found that the bending potentials of the flexible B–O–B bridges, when considered in isolation, are unable to account for the particularly exceptional flexibility of the boroxyl-decorated frameworks. The flexibility depends on topological and geometric factors, such as the number density of B–O–B bridges. Crucially, the incorporation of boroxyl rings into the frameworks relaxes the torsional constraints on the B–O–B bridges; this allows the frameworks to exploit an intrinsic cooperativity between the bending and twisting motions in the bridges during expansion. The flexibility of the boroxyl-based frameworks can be significantly preserved by decoration with alternative isoelectronic rings, including triazine, borazine and benzene.

We also propose a novel approach to the prediction of further B_2O_3 polymorphs. Given any arrangement of SiO_4 tetrahedra, it is possible to construct a related structure in which the SiO_4^{4-} tetrahedra are exchanged for $B_2O_5^{4-}$ units. Calculations on a borate analogue of β-cristobalite (β-cristobalite-borate-a) constructed in such manner show it to be isoenergetic with the known B_2O_3-I polymorph. An enormous number of boron–oxygen polymorphs can be obtained in this way and we further predict a family of zeolite analogues. We have some evidence to suggest that subtle differences are likely in the energy landscapes of the zeolites and their boron analogues; whilst the minimum energy structures of zeolites tend to lie towards the high-volume end of their flexibility windows, the borate analogues are expected to be highly compressed in their relaxed states. To the best of our knowledge, the templated synthesis of boron oxides similar to that of zeolites has not been attempted so far.

The frameworks that we have examined have access to significantly larger variations in volume than well-known flexible MOFs such as MIL-53(Cr). The facile collapse and expansion of the cavities that pervade the frameworks – in contrast to the vast majority of covalent organic frameworks (COFs), which form a rigid network of fixed nanochannels – opens up the prospect of novel storage capabilities and host–guest chemistry.[49–52] We hope that our work and predicted low-energy structures will encourage further experimental work on these fascinating systems.

Conflicts of interest

There are no conflicts to declare.

Acknowledgements

HJAD acknowledges the award of a Royal Society of Chemistry Undergraduate Research Bursary. We thank Stephen Wells and Asel Sartbaeva for many discussions and their help with the GASP program.

References

1 Themed issue: Metal Organic frameworks, *Chem. Soc. Rev.*, 2009, **38**, 1201–1508.
2 A. U. Czaja, N. Trukhan and U. Müller, *Chem. Soc. Rev.*, 2009, **38**, 1284.
3 P. Horcajada, R. Gref, T. Baati, P. K. Allan, G. Maurin, P. Couvreur, G. Férey, R. E. Morris and C. Serre, *Chem. Rev.*, 2012, **112**, 1232 and references therein.
4 J.-R. Li, J. Sculley and H.-C. Zhou, *Chem. Rev.*, 2012, **112**, 869.
5 H. Furukawa, K. E. Cordova, M. O'Keeffe and O. M. Yaghi, *Science*, 2013, **341**, 974.
6 A. Schneemann, S. Henke, I. Schwedler and R. A. Fischer, *ChemPhysChem*, 2014, **15**, 823.
7 G. Férey and C. Serre, *Chem. Soc. Rev.*, 2009, **38**, 1380.
8 G. Férey, *Dalton Trans.*, 2016, **45**, 4073.
9 A. Schneemann, V. Bon, I. Schwedler, I. Senkovska, S. Kaskel and R. A. Fischer, *Chem. Soc. Rev.*, 2014, **43**, 6062.
10 F. X. Coudert, *Chem. Mater.*, 2015, **27**, 1905.
11 E. Cockayne, *J. Phys. Chem. C*, 2017, **121**, 4312.

12 C. Mellot-Draznieks, C. Serre, S. Surblé, N. Audebrand and G. Férey, *J. Am. Chem. Soc.*, 2005, **127**, 16273.

13 C. Serre, C. Mellot-Draznieks, S. Surblé, N. Audebrand, Y. Flinchuk and G. Férey, *Science*, 2007, **315**, 1828.

14 L. Villaescusa, P. Lightfoot, S. J. Teat and R. E. Morris, *J. Am. Chem. Soc.*, 2001, **123**, 5453.

15 G. D. Barrera, J. A. O. Bruno, T. H. K. Barron and N. L. Allan, *J. Phys.: Condens. Matter*, 2005, **17**, R217.

16 M. Jeffroy, A. H. Fuchs and A. Boutin, *Chem. Commun.*, 2008, 3275.

17 G. Ferlat, A. P. Seitsonen, M. Lazzeri and F. Mauri, *Nat. Mater.*, 2012, **11**, 925.

18 G. E. Gurr, P. W. Montgomery, C. D. Knutson and B. T. Gorres, *Acta Crystallogr.*, 1970, **26**, 906.

19 C. T. Prewitt and R. D. Shannon, *Acta Crystallogr.*, 1968, **24**, 869.

20 G. Ferlat, T. Charpentier, A. P. Seitsonen, A. Takada, M. Lazzeri, L. Cormier, G. Calas and F. Mauri, *Phys. Rev. Lett.*, 2008, **101**, 65504.

21 P. Umari and A. Pasquarello, *Phys. Rev. Lett.*, 2005, **95**, 13401.

22 L. Huang, M. Durandurdu and J. Kieffer, *J. Phys. Chem.*, 2007, **111**, 13712.

23 A. Takada, C. R. A. Catlow and G. D. Price, *Phys. Chem. Glasses*, 2003, **44**, 147.

24 F. Claeyssens, J. N. Hart, N. C. Norman and N. L. Allan, *Adv. Funct. Mater.*, 2013, **23**, 5887.

25 J. Carrasco, F. Illas and S. T. Bromley, *Phys. Rev. Lett.*, 2007, **99**, 235502.

26 M. A. Zwijnenburg and S. T. Bromley, *Phys. Rev. B*, 2011, **83**, 024104.

27 M. A. Zwijnenburg and S. T. Bromley, *J. Mater. Chem.*, 2011, **21**, 15255.

28 W. Sangthong, J. Limtrakul, F. Illas and S. T. Bromley, *J. Mater. Chem.*, 2008, **18**, 5871.

29 M.-S. Wang, G.-C. Guo, W.-T. Chen, G. Xu, W.-W. Zhou, K.-J. Wu and J.-S. Huang, *Angew. Chem., Int. Ed.*, 2007, **46**, 3909.

30 M.-C. Liu, P. Zhou, H.-G. Yao, S.-H. Ji, R.-C. Zhang, M. Ji and Y.-L. An, *Eur. J. Inorg. Chem.*, 2009, **31**, 4622.

31 F. Claeyssens, N. L. Allan, N. C. Norman and C. A. Russell, *Phys. Rev. B*, 2010, **82**, 094119.

32 R. J. Bell and A. Carnevale, *Philos. Mag.*, 1982, **43**, 389.

33 S. J. Clark, M. D. Segall, C. J. Pickard, P. J. Hasnip, M. J. Probert, K. Refson and M. C. Payne, *Z. Kristallogr.*, 2005, **220**, 567.

34 J. P. Perdew, K. Burke and M. Ernzerhof, *Phys. Rev. Lett.*, 1996, **77**, 3865.

35 J. P. Perdew and Y. Wang, *Phys. Rev. B*, 1992, **45**, 13244.

36 D. Vanderbilt, *Phys. Rev. B*, 1990, **41**, 7892.

37 H. J. Monkhorst and J. D. Pack, *Phys. Rev. B*, 1976, **13**, 5188.

38 M. J. Frisch, G. W. Trucks, H. B. Schlegel, G. E. Scuseria, M. A. Robb, J. R. Cheeseman, G. Scalmani, V. Barone, G. A. Petersson, H. Nakatsuji, X. Li, M. Caricato, A. Marenich, J. Bloino, B. G. Janesko, R. Gomperts, B. Mennucci, H. P. Hratchian, J. V. Ortiz, A. F. Izmaylov, J. L. Sonnenberg, D. Williams-Young, F. Ding, F. Lipparini, F. Egidi, J. Goings, B. Peng, A. Petrone, T. Henderson, D. Ranasinghe, V. G. Zakrzewski, J. Gao, N. Rega, G. Zheng, W. Liang, M. Hada, M. Ehara, K. Toyota, R. Fukuda, J. Hasegawa, M. Ishida, T. Nakajima, Y. Honda, O. Kitao, H. Nakai, T. Vreven, K. Throssell, J. A. Montgomery Jr, J. E. Peralta, F. Ogliaro, M. Bearpark, J. J. Heyd, E. Brothers, K. N. Kudin, V. N. Staroverov, T. Keith, R. Kobayashi, J. Normand, K. Raghavachari, A. Rendell, J. C. Burant, S. S. Iyengar,

J. Tomasi, M. Cossi, J. M. Millam, M. Klene, C. Adamo, R. Cammi, J. W. Ochterski, R. L. Martin, K. Morokuma, O. Farkas, J. B. Foresman and D. J. Fox, *Gaussian 09, Revision D.01*, Gaussian, Inc., Wallingford CT, 2016.

39 E. D. Glendening, A. E. Reed, J. E. Carpenter and F. Weinhold, *NBO Version 3.1*.

40 S. A. Wells and A. Sartbaeva, *Mol. Simul.*, 2015, **41**, 1409.

41 F. Gatti, N. Berthe-Gujac, I. Demachy and F. Voltatron, *Chem. Phys. Lett.*, 1995, **232**, 503.

42 R. M. Minyaev and E. A. Lepin, *Mendeleev Commun.*, 1997, 7, 80.

43 F. Wiitkar, S. Kahlal, J. F. Halet, J. Y. Saillard, J. Bauer and P. J. Rogl, *J. Am. Chem. Soc.*, 1994, **116**, 251.

44 S. A. Wells, K. M. Lueng, P. P. Edwards, M. G. Tucker and A. Sartbaeva, *R. Soc. Open Sci.*, 2017, **4**, 170757.

45 A. Sartbaeva, S. A. Wells, M. M. J. Treacy and M. F. Thorpe, *Nat. Mater.*, 2006, **5**, 962.

46 D. S. Wragg, R. E. Morris and A. W. Burton, *Chem. Mater.*, 2008, **20**, 1561.

47 W. H. Baur, *Acta Crystallogr.*, 1980, **36**, 2198.

48 L. Huang and J. Kieffer, *Phys. Rev. B*, 2006, **74**, 224107.

49 A. P. Côté, A. I. Benin, N. W. Ockwig, M. O'Keeffe, A. J. Matzger and O. M. Yaghi, *Science*, 2005, **310**, 1166.

50 S. Y. Wang and W. Wang, *Chem. Soc. Rev.*, 2013, **42**, 548.

51 P. J. Waller, F. Gandara and O. M. Yaghi, *Acc. Chem. Res.*, 2015, **48**, 3053.

52 H. Xu, J. Gao and D. Jiang, *Nat. Chem.*, 2015, **7**, 905.

Faraday Discussions

PAPER

What is the best or most relevant global minimum for nanoclusters? Predicting, comparing and recycling cluster structures with WASP@N†‡

Scott M. Woodley, [*a] Tomas Lazauskas, [a] Malcolm Illingworth,[b] Adam C. Carter [b] and Alexey A. Sokol [a]

Received 7th March 2018, Accepted 13th April 2018
DOI: 10.1039/c8fd00060c

To address the question posed in the title, we have created, and now report details of, an open-access database of cluster structures with a web-assisted interface and toolkit as part of the WASP@N project. The database establishes a map of connectivities within each structure, the information about which is coded and kept as individual labels, called hashkeys, for the nanoclusters. These hashkeys are the basis for structure comparison within the database, and for establishing a map of connectivities between similar structures (topologies). The database is successfully used as a key element in a data-mining study of $(MX)_{12}$ clusters of three binary compounds (LiI, SrO and GaAs) of which the database has no prior knowledge. The structures are assessed on the energy landscapes determined by the corresponding bulk interatomic potentials. Global optimisation, using a Lamarckian genetic algorithm, is used to search for low lying minima on the same energy landscape to confirm that the data-mined structures form a representative sample of the landscapes, with only very few structures missing from the close energy neighbourhood of the respective global minima.

1. Introduction

The application of structure prediction in the field of clusters and nanoparticles has resulted in literally millions of structures being discovered for different compounds, systems with different magnetic ordering, systems containing different dopants, or simply systems of different sizes.[1–4] Crucially, each system can be described as an energy landscape and the initial target or targets are the location of the global minimum (GM) or the locations of low energy local minima

[a]*University College London, Department of Chemistry, 20 Gordon Street, London WC1H 0AJ, UK. E-mail: Scott.Woodley@ucl.ac.uk*

[b]*EPCC, James Clerk Maxwell Building, Peter Guthrie Tait Road, Edinburgh, EH9 3FD, UK*

† Electronic supplementary information (ESI) available. See DOI: 10.1039/c8fd00060c

‡ Free access *via* http://hive.chem.ucl.ac.uk.

(LM).[5] Today when one wants to study a new compound of interest within certain sets of parameters, including stoichiometry, size, environment, *etc.*, a key question springs to mind: is it worth running new simulations that employ one or several contemporary global structure optimisation algorithms? We argue – not necessarily! Thoughtful exploitation of the available data that can be found in the literature presents a viable alternative that turns out to be the most efficient way to discover new structures, materials, and their physics and chemistry.[6–14] Similar considerations, apart from size, can be applied to crystal structures, including molecular, metallic, ionic, covalent, or hybrid organic and inorganic frameworks.[15–18]

Another problem encountered by practically every practitioner of global optimisation for structure prediction is how to ascertain that the newly discovered configuration of a particular compound is not known from competitors' studies, for example, or exists out there under the guise of a different compound of similar stoichiometry, or is not published but is known as a lower ranked local energy minimum (*i.e.* data that has a rank that is beyond a chosen set threshold for publication). The use of slightly different energy functions, unintentional effects of tolerances both in energy definition and local optimisation, or possibly an intentional bias to match measurable properties (for example, infrared data) will all muddle the waters further.

The choice of the best – or most suitable for the investigator's purposes – cost (or fitness) function is uncertain, and could be quite different in different studies even on the same system.

To address these challenges, we have developed a database complemented by a toolkit that includes structure comparison as a key element. Aggregating structures and their properties into one place also enables the sophisticated exploration of structural motifs and particular properties and the discovery of structure–property relationships. Databases are not a new concept in materials modelling,[19–29] even in the field of nanoclusters.[30,31] Crucially, our searchable database generates a map of connections relating different structures. In this article, we describe both the database and the algorithms that generate these mappings, followed by simple showcase examples.

2. Web-assisted structure prediction at the nanoscale (WASP@N)

In the development of the database, our Hive of knowledge, we aimed to arm the scientific community and general public, from professional researchers to school pupils, with a new intelligent tool to search, discover and disseminate structures and properties of new nanoclusters. To allow access and interaction with the Hive, we built a web interface, which we refer to as the WASP toolkit. The mapping between structures and various properties is an essential element, or feature, of the Hive database, which is generated by algorithms that form part of a separate piece of code that we refer to as the Bee software. The Bee software runs on dedicated computing facilities. The WASP interface links the user, the Hive and the Bee software – see Fig. 1. With open access to the Hive, a number of security measures have been employed in order to protect the integrity of the data and the computing facilities from malicious attacks (to complete the analogy, we refer to unwanted visitors to the Hive as hornets).

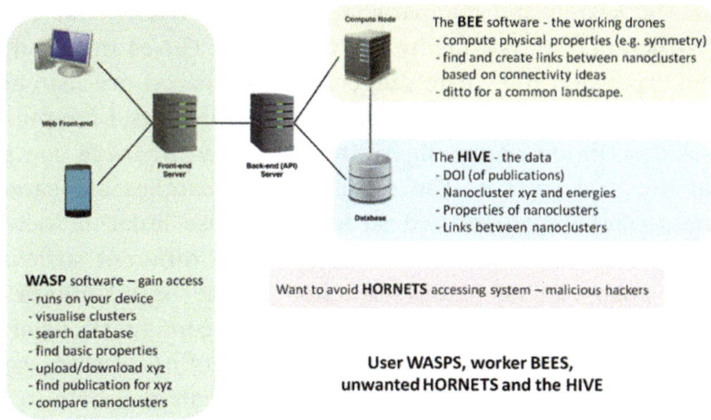

Fig. 1 Schematic of the hardware and software solution for web-assisted structure prediction at the nanoscale.

Datasets within the Hive are organised as follows: (a) published atomic structures, the atomic coordinates of which were originally used to generate a figure (e.g. ball and stick models) or were explicitly given in a table as part of a published paper (or electronic supplementary information) that has a DOI; and (b) atomic structures generated using the Bee software. For the former, the atomic structures are labelled using the DOI of the published article they were taken from, and are uploaded as one or more concatenated *xyz* file(s) using an extended format that contains both the metadata saved on the comment line and the atomic structure, which includes atomic labels: Cartesian coordinates; one additional scalar and one vector record per atom (for example, charges, spin, dipole on atom). Searchable metadata are vital for the use of a database. Values for metadata that can be provided include the definition of energy and software, total charge, energy ranking, total spin, *etc.* For example, the comment line:

"Name=drum; Symmetry=D_{3h}; Definition={FHI-aims, PBE0/PBE, tight}; Energy=210Hartree; Size=6; Atoms=12; Charge=0; Spin=0; Dipole=(0,0,0)"

for the cluster $(ZnO)_6$ indicates that the user refers to the local minimum configuration as a "drum", the atomic coordinates of which have D_{3h} point group symmetry after geometry relaxation using the FHI-aims software with the generalised gradient approximation in density functional theory in the form of the PBE exchange and correlation density functional and the tight basis set, an energy of 210 Ha with the same basis set and the hybrid PBE0 exchange and correlation density functional, a total charge and spin of zero, and no resultant dipole. If not specified upon upload to the Hive, some of these will be calculated along with, for example, stoichiometry, topology, total mass, centre of mass, and principal moments of inertia. Non-searchable metadata like, for example, thumbnail ball and stick images, are generated on-the-fly. The dataset for each DOI string will also contain timestamp metadata (when it was uploaded or last modified) and publication metadata (authors and journal name, volume and page numbers). Generated datasets are given a DOI string by the Bee software that is based on the chosen energy definition, and the atomic configurations result from structural relaxations of all the published datasets.

The essential search and comparison features of WASP enable the user to investigate structural motifs and physical properties. The comparison of clusters can be quite expensive and, therefore, comparison-based pre-searches are performed by the Bee software upon the upload of new datasets, both published and generated. A description of the algorithms employed in these comparisons is provided in the next section. The results of pre-searches are saved as links between (thus establishing) related structures. These links, or new metadata generated by the Bee software, form a map linking different structures in the database. The map can be readily exploited by the user through the WASP interface to ascertain the uniqueness of newly found configurations of clusters of a certain compound and size or to compare clusters of different compounds. Moreover, as we will demonstrate below, this map can also help to reduce the effort needed to explore the energy landscapes of a compound that has yet to be investigated. The computational work and the interaction of the three complementary codes (WASP, Bee, and the Hive) are supported by appropriate hardware solutions – as illustrated in Fig. 1 – and related operating system server software (including task scheduler, *etc.*). In the near future, we plan to expand the solution shown in Fig. 1 to include the exploitation of third party computing platforms.

3. Methodology

Uniqueness and similarity

Being able to quickly recognise similar structures, or measure their similarity, has always been a challenge in materials modelling.[32] Consider comparing the atomic structures of two nanoclusters that are essentially the same but have either small random perturbations (noise) resulting from the applied optimisation tolerances or slight differences because of the different, but similar, density functionals employed. In the comparison procedure, the first task is to correctly align these two configurations: the translation and rotation of each cluster is fixed by positioning the centre of mass at the origin and aligning the principal axes of rotation with the chosen Cartesian axes. Hopefully, upon alignment, a one-to-one match is found for each atom in one configuration with the equivalent atom in the other. If not, then there is a combinatorial problem to solve: which combination of atom pairs minimises the sum of the distances between all pairs (a sum of zero implies a perfect match, with each atom in one configuration positioned exactly on top of the equivalent atom in the other configuration). Minimising this measure of likeness for two dissimilar nanoclusters may also require optimising the relative rotation and translation of the two nanoclusters.

The efficiency of stochastic search algorithms – particle swarm, basin hopping, and genetic or evolutionary algorithms – that are employed to locate local minima (LM) on the energy landscape can be improved if there is a computationally cheap method that provides a measure of how similar two structures are. For example, this could be used to check whether a newly found/generated configuration is unique, whether the starting points are sufficiently spread apart for different random walkers on the energy landscape, or whether the candidate structures in the current population are sufficiently diverse for the evolutionary algorithm (otherwise inbreeding results in the population not evolving, or improving, any further). One may also want to distinguish between enantiomorphic clusters – two clusters that are mirror images of each other. One half of such a pair can easily be lost if the comparison of nanoclusters is simply based on their relative energy of formation (since both enantiomorphic clusters have identical energies).

There are several approaches in the literature designed to measure the similarity between structures,[33–45] which can be classified in two groups: direct one-to-one comparison or an indirect approach that requires the generation of labels, also known as fingerprints or hashkeys, which are then compared.

One-to-one comparison algorithms are typically based around a cost function that measures the degree of similarity between two structures. As introduced above, the cost function will depend on the successful superimposition of the two structures, *i.e.* the translation and rotation of one cluster with respect to the other. Where Dirac delta functions are used to describe the position of an atom, the cost function will also depend on the matching of atomic pairs between the structures. This in itself can pose a formidable task (see for example ref. 33, which employs the Hungarian algorithm).[34–36] This problem is reduced for compounds or alloys if pairs are restricted between like species. Alternatively, where a Gaussian, or a similar function, is centred on each atom, the cost function is typically based on the degree of overlap of atom-centred Gaussians between the two clusters. For compounds and alloys, the overlap of Gaussians can be determined for each species type; there is no explicit need to match pairs of atoms. Goedecker employed a similar scheme, but based on atomic orbitals (see ref. 37). Both types of cost function can also be employed to find out whether, or how well, a smaller cluster matches a fragment of a larger cluster.

In this article, we only compare pairs of clusters that have the same composition, and use only the species type and atomic coordinates as the input. One of the most straightforward and widely used metrics for the comparison of molecular structures is the root-mean-square deviation (RMSD) of the coordinates of equivalent atoms.[38,39] Following a similar idea, the metrics suggested by Ali Sadeghi *et al.*[37] use configurational fingerprints based on eigenvalues of matrices of interatomic distances. The structural fingerprints are then compared by measuring the distances between them, as small fingerprint-based distances correspond to small RMSD distances. The H-FORMS (a hierarchical algorithm for molecular similarity)[46] approach estimates a rigid transformation that aligns structures and computes rotation-invariant descriptors, which are then used to match atoms. Similarly, R. Hundt *et al.* implemented an algorithm in the analysis program KPLOT[40] based on the mapping of atomic patterns constructed using three-atom frame matches. An alternative approach to the problem of structure comparison exploits the properties of the nanoclusters,[41] such as radial distribution functions, vibrational frequencies[42] or principal moments of inertia.

Whichever method is used, when a structure needs to be efficiently compared with vast data for thousands or millions of configurations, the chosen approach needs to be both robust and computationally affordable. The second class of comparison methods – based on comparing unique labels that are generated for every configurationally unique structure – may address this big data challenge.

Within our database, we implemented the approach first adopted in the KLMC software[47] to address the challenge of maintaining the diversity of structures during a genetic algorithm search. The approach relies on the NAUTY software package (No AUTomorphisms, Yes?) written by McKay and Piperno,[48] which can generate canonical labels for graphs and compute automorphisms between them. NAUTY labels graphs canonically by providing a string consisting of three 8-digit hexadecimal numbers depending on the graph, *i.e.* a set of vertices and edges, and, in general, every unique graph will have a unique NAUTY string, also known as a hashkey, or fingerprint. By exploiting the feature of uniqueness, we have incorporated NAUTY in the Bee software in the following way: each cluster is converted to a coloured graph by treating the atoms as vertices and the bonds between them as edges. The number of colours of vertices (atoms) is determined by the number of species in the structure. Thus, $(MgO)_n$ clusters will have two different colours (species), whereas Ti_n clusters will have only one. It is important to note that $(KF)_n$ clusters will also have two different colours, therefore graphs of $(MgO)_n$ and $(KF)_n$ clusters of the same size can be compared explicitly. The edges of the clusters' graphs are generated from the calculated interatomic distances between the atoms (vertices) of a cluster and can be thought of as "bonds" between atoms. The radial cut-off by which the "bonds" are determined depends on the species and is slightly longer than the expected actual bond length. A flowchart of the implemented hashkey generation is given in Fig. 2, where the $(MgO)_5$ GM cluster is used as an example. Here, the $(MgO)_5$ GM cluster (shown as a ball and stick model in Fig. 2a) is transformed into a coloured graph (shown in Fig. 2b). This graph is then processed using the "NAUTY" software package, which in turn generates a unique hashkey for the cluster. An example of a hashkey is shown in Fig. 2d.

Given that the comparison of hashkeys is orders of magnitude faster than comparing atomic structures explicitly, each cluster within the Hive database is labelled with a hashkey. As described above, the hashkeys enable a rapid check of the database for duplicate structures by both the WASP and Bee software and are used in the generation of maps connecting similar structures (the network of links between clusters entered into the database is updated as soon as the atomic coordinates of generated and published LM nanoclusters are uploaded to the Hive) – a feature that is not currently implemented in other structural databases. This feature has proven to be essential when the WASP interface is used to find out whether a newly discovered cluster is already within the Hive. To demonstrate one of the utilities our database provides, we have used the generated hashkeys to identify unique structural motifs for a particular stoichiometry (1 : 1) and size (24

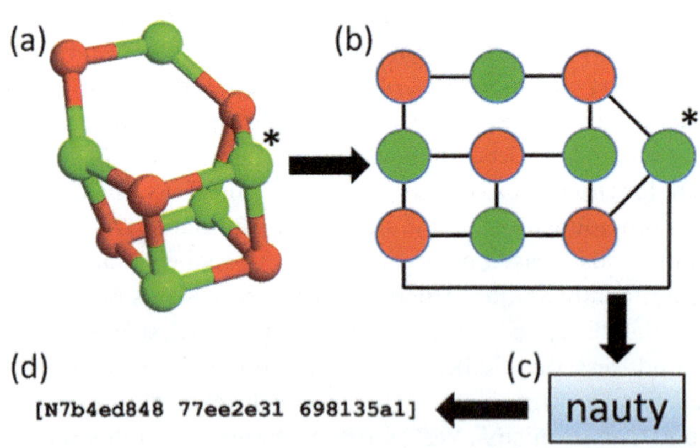

Fig. 2 Flowchart of hashkey (fingerprint) generation illustrated using the $(MgO)_5$ global minimum configuration.

atoms). We then data-mined from this set, rather than a set of LM configurations of one or all compounds in the Hive.

Data normalisation

Published LM cluster structures, which can be uploaded to the database, are, by definition, dependent upon the theory and accuracy of the level of theory employed in the calculation of energy as a measure of stability. Moreover, the measure of fitness may also be based on the deviation from some geometric, physical or chemical observable(s). When LM on a potential energy landscape are targeted, energy calculations at different levels of theory (quantum mechanical (all-electron or pseudopotential), semi-empirical, Hartree–Fock, DFT, tight-binding, semi-classical, or atomistic simulations) yield values that may scatter across a few orders of magnitude. Even if a similar method is chosen, *e.g.* DFT with identical basis sets and, possibly, effective core potentials, employing different exchange and correlation density functionals could still lead to substantially different values. The situation is just as problematic if semi-classical simulations are employed, as there are often many different sets of parameterised interatomic potentials for the same material or compound. One trick commonly used across the field of materials chemistry is to switch from total to binding or cohesion energies, which can be expected to behave better, and do in practice.[49] The scatter in the calculated binding energy values obtained using different approaches is usually, however, still greater than the energy separating low ranking energy minima on the same energy landscape (definition of energy).

In practice, the WASP interface lets users upload their data without any restrictions on how the data were obtained, but encourages the users to provide details of the adopted computational approach as metadata. To support the comparison of individual structures obtained using different energy definitions, we introduced an internal standard attained by a data normalisation routine. In particular, when data are uploaded to the Hive database, they are automatically refined by the Bee software, using the all-electron, full potential electronic structure code FHI-aims[50] with the PBEsol functional,[51–53] the light basis set (which is variationally equivalent to split valence double-zeta Gaussian plus polarisation basis sets but can obtain energies that are much closer to the basis set limit). Further computational parameters are provided in the ESI.† After normalisation, the newly obtained structure is automatically uploaded to the Hive database with a two-way link between the original and normalised configurations, along with similarity links to the whole dataset in the database.

Hence, the user can search for structures that refine to the same LM on our normalised energy landscape (particularly useful for the investigation of nano-clusters of the same compound) or structures of any compound with the same connectivity (structural motif), as explained in the previous section.

Data mining

Starting from a known set of atomic configurations with the target stoichiometry and total number of atoms, the Data Mining (DM) module of the KLMC software package[54] rescales each configuration to obtain an estimate of the expected nearest neighbour interatomic distances for the target compound, and then, using third party software, relaxes the rescaled atomic structures to LM. In the results shown below, we employ GULP[55] as the third party software, *i.e.* a semi-classical level of theory is used for the calculation of energies (and atomic forces). After the rescaling and refinement procedure, KLMC is also employed to analyse the resulting configurations in terms of their energy ranking, uniqueness and geometrical properties.

Global optimisation

A Lamarckian genetic algorithm (GA) approach implemented in the KLMC software package[47] was also used to locate LM on the energy landscape defined by the same set of interatomic potentials (semi-classical level of theory) as those used in the data-mining investigation. We note that the ability of the KLMC GA[47] to locate LM and GM efficiently has been proven for various types of system, and thus it is chosen here as a method for providing reliable data that we can use to assess the results obtained using the data-mining approach. The population of each GA run was set to 200 candidate structures, with the initial random structures generated within a 15 Å × 15 Å × 15 Å cubic simulation box. Default values, as given in ref. 47, were used for the remaining simulation parameters.

4. Results

Isomorphic structures, or structural motifs

As an illustration of how the connectivity maps are employed, we consider the case of a GM nanocluster reported in ref. 56 for $(MgO)_7$ that has the symmetry point group C_{3v}; see Fig. 3a. The topological analysis tool finds that this structure has "7Mg3–7O3" topology, *i.e.* seven Mg and seven O atoms, each with a coordination number of three. When selected using the WASP interface for the Hive, beneath the rotatable ball and stick model of this structure are two lists; one showing the standardised entry for this configuration (as described earlier), and another showing all the "isomorphic structures" found in the Hive based on matching hashkeys (as also described above). A snapshot of the second list is shown in Fig. 4. In our chosen example, the $(MgO)_7$ GM structure currently has eleven isomorphic structures: eleven atomic configurations within the Hive have the same hashkey as our chosen example. The inclusion of a DOI in the entry for a candidate structure in this list indicates that it is a published LM. The remaining five are, therefore, standardised LM (using FHI-aims). As more entries are submitted to the Hive, we would expect many more matches to be found. The

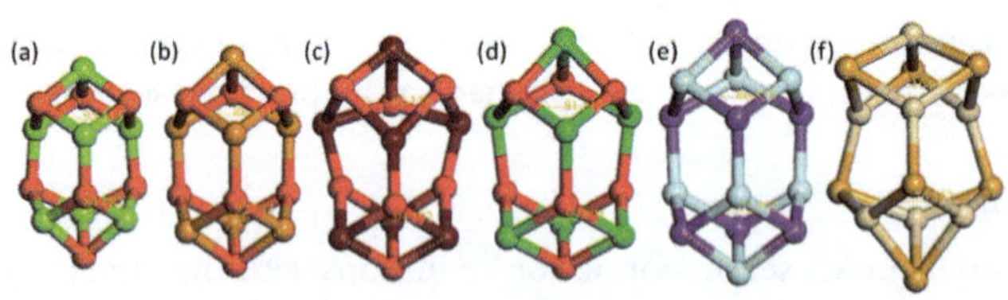

Fig. 3 Six configurations for $(XY)_7$ taken from the Hive database that all have the same hashkey, shown to scale as ball and stick models with a cation shown uppermost, two bond angles measured, and where XY is the compound (a) MgO, (b) CaO, (c) SrO, (d) BaO, (e) KF and (f) CdSe.

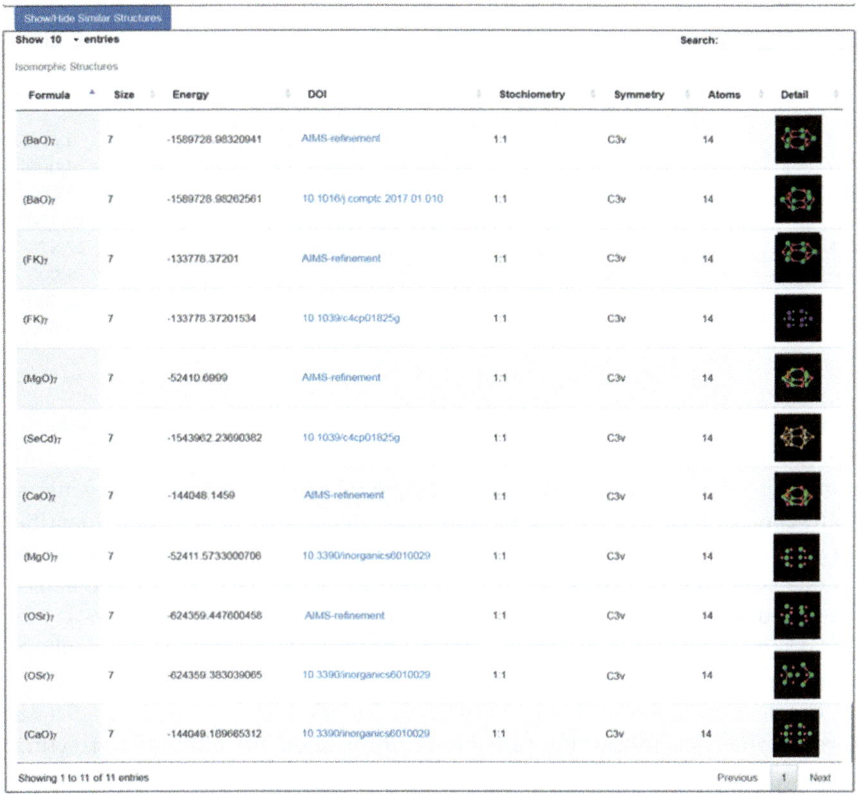

Fig. 4 Snapshot of the lower part of the WASP interface showing 11 results (5 generated by the Bee software) returned from the Hive for structures that have the same hashkey as the GM structure for $(MgO)_7$ as reported in ref. 56.

six published LM show that this structural motif is also reported[54,56,57] to be the GM for $(KF)_7$, $(CaO)_7$, $(SrO)_7$, $(BaO)_7$ and $(CdSe)_7$. There is also another $(MgO)_7$ configuration, which has a different DOI[54] to that of the original chosen structure. Given that there are six different compounds with the same structural motif, we would expect six standardised LM. The two published LM entries for $(MgO)_7$, the same compound, relax to the same standardised LM. To find all the nanoclusters within the Hive that relax to the same standardised LM, the user only needs to click on the thumbnail of the standardised nanocluster. In our example, the missing standardised LM results from the standardised configuration for $(CdSe)_7$ relaxing to a different LM. Therefore, it has a different hashkey as it is a different structure (in fact, it has C_1 point symmetry).

Efficient structure prediction

The Hive contains the LM atomic structures for numerous binary compounds with 1 : 1 stoichiometry and a total charge of 0. We now concentrate on one particular size, clusters composed of 12 cations and 12 anions. To investigate a compound that is missing from the Hive database, one could data-mine structures already in the Hive for a similar compound. The success of this

Table 1 Parameters for the Buckingham potential, $A \exp(-r/\rho) - C/r^6$, applied between ions X and Y

X–Y	A (eV)	ρ (Å^{-1})	C (Å^6 eV)
Li–Li	1153.80	0.13640	0.000
Li–I	8894.70	0.26170	0.000
I–I	5502.50	0.30660	0.000
Sr–O	1952.39	0.33685	19.220
O–O	22 764.00	0.14900	27.879
Ga–Ga	470.18	0.13490	0.000
Ga–As	1544.69	0.42310	0.000
As–As	484.31	0.24810	0.000

approach would rely on the chosen set of initial configurations; the more extensive this set, the greater the probability of finding the target LM. To maximise this probability one could data-mine all the compounds; however, this would generate many copies of each LM. Using the hashkey, which provides a unique identifier for each structural motif, we were able to reduce this initial set to just over 100 unique structural motifs (which we will refer to as the DM-set). If the database contained entries for alkali halides, $(XY)_{12}$, and alkaline earth oxides, $(ZO)_{12}$, for X = Li to Cs, Y = F to I, and Z = Mg to Ba, then potentially there would be a maximum reduction of 96%.

The determination of this reduced set (calculation and comparison of hashkeys) is orders of magnitude faster to perform than the additional structural relaxations (using standard algorithms within an electronic structure code) that would have been necessary if we could not determine equivalent structures. Moreover, data-mining requires the evaluation of far fewer candidate structures than is typically performed in a stochastic approach. It is expected that the number of datasets within the Hive will grow, and that important unique structural motifs may be missed given our search has been performed soon after we have created this database. Stochastic approaches may also miss important LM, and the number of unique motifs is likely to increase much more slowly than the number of entries for clusters of any particular size, charge and stoichiometry.

Using our DM-set of unique LM, we now investigate three different compounds that were not included in the initial dataset taken from the Hive, namely $(LiI)_{12}$,

Table 2 Parameters for the shell model for ions X, where Q and Y are the point-charges of the core and shell, which are connected by a spring with constants k_2 and k_4. The Coulomb contribution to the energy between point-charges of an individual ion X is replaced with the energy associated with the spring, $1/2k_2x^2 + 1/4k_4x^4$, where x is the distance between the core and shell. Note that the strontium cation is treated as a rigid ion and therefore only has one parameter

X	Q (e)	Y (e)	k_2 (eV Å^{-2})	k_4 (eV Å^{-4})
Li	0.295	0.705	15.979	0
I	3.087	−4.087	39.950	0
Sr	2.000			
O	0.869	−2.869	74.920	0
Ga	3.436	−0.436	2418.361	0
As	0.809	−3.809	7.722	300 000

$(SrO)_{12}$ and $(GaAs)_{12}$. As the main focus of this article is the methodology as opposed to the physical and electronic properties of the predicted nanoclusters, we have chosen to present new IP-LM structures, *i.e.* the atomic configurations and ranks of local minima on the energy landscape are defined using interatomic potentials (IP), the parameters of which are given in Tables 1 and 2. For each compound we also perform a search of low energy IP-LM using an evolutionary algorithm; details of both methods are described in the previous section. We note that the potential parameters for LiI were taken from ref. 58. The small spring constant for the lithium cation caused problems during the global optimisation runs; during the relaxations of new candidate structures (particularly the random structures used in the initial population), the initial electric fields were sometimes strong enough that during structural relaxation the shell was stripped away from the cation. It is known that the polarisability of an ion is dependent upon the electric field, which is much stronger for our clusters than that experienced within

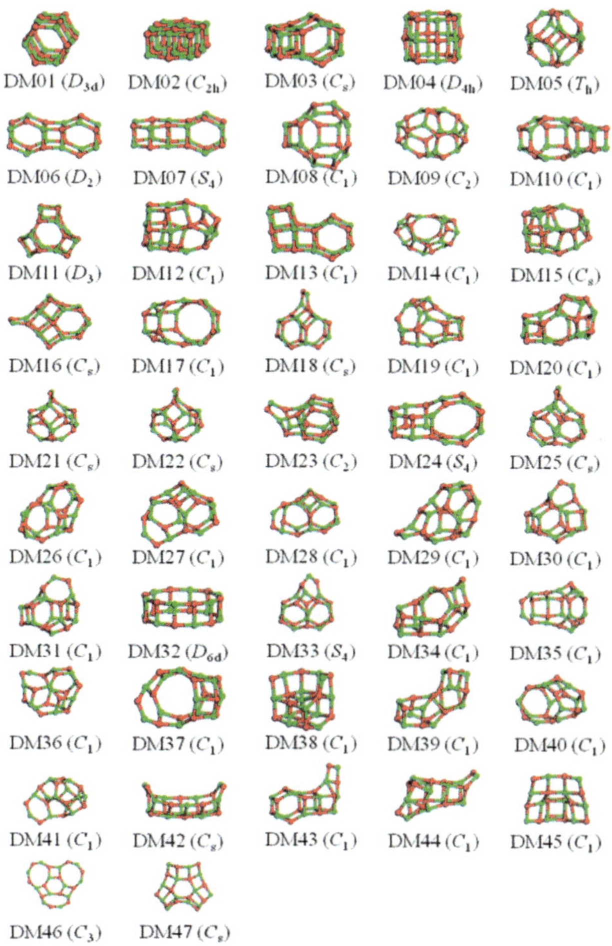

Fig. 5 $(SrO)_{12}$ IP-LM configurations obtained by relaxing the DM-set of unique LM found by data-mining from other 1 : 1 compounds. Green and red atoms represent Sr and O, respectively, and the symmetry point groups of the clusters are given within parentheses.

the bulk. Thus, in our simulations, we doubled the value of the spring constant for lithium cations, which corresponds to an apparent reduction in their coordination number compared to the bulk.

The results from data-mining our DM-set of unique LM are shown in Fig. 5–7. For strontium oxide, lithium iodide and gallium arsenide, 47, 50 and 41 LM structures were generated, respectively, *i.e.* not all the structural motifs of one compound were locally stable for another. Moreover, a different global minimum was found for each compound. Labelled DM01 in Fig. 5, the D_{3d} barrel was found to be the IP-GM for $(SrO)_{12}$, whereas for $(LiI)_{12}$ and $(GaAs)_{12}$ it was ranked fourth and second, respectively. The $2 \times 2 \times 6$ D_{2d} configuration of alternating atoms, labelled DM01 in Fig. 6, was found to be the IP-GM for $(LiI)_{12}$. One can imagine that this cuboid configuration could be cut from the NaCl rock salt structure, and thus it is not surprising that this structural motif was not generated for $(GaAs)_{12}$. The T_h sodalite cage, so named as it is a basic building block of the sodalite bulk structure (given the abbreviation SOD by the zeolite community), was found to be

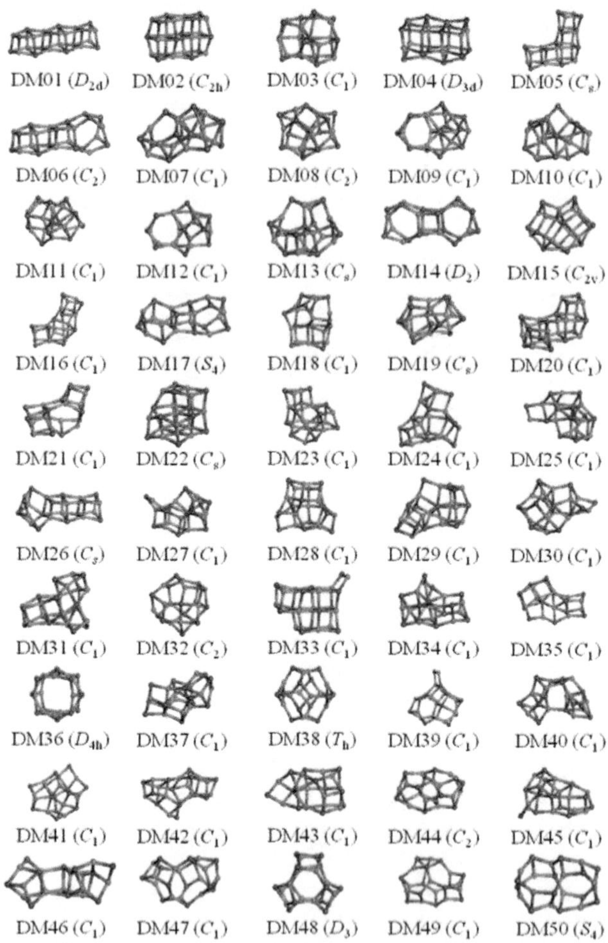

Fig. 6 $(LiI)_{12}$ IP-LM configurations obtained by relaxing the DM-set of unique LM found by data-mining from other 1 : 1 compounds. Pink and red atoms represent Li and I, respectively, and the symmetry point groups of the clusters are given within parentheses.

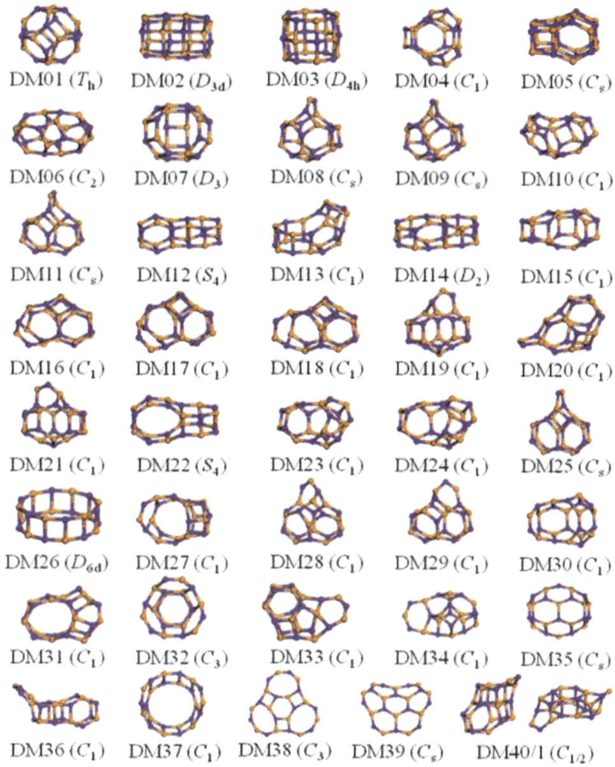

Fig. 7 $(GaAs)_{12}$ IP-LM configurations obtained by relaxing the DM-set of unique LM found by data-mining from other 1 : 1 compounds. Purple and orange atoms represent Ga and As, respectively, and the symmetry point groups of the clusters are given within parentheses.

the IP-GM for $(GaAs)_{12}$. This configuration was ranked fifth and thirty-eighth for $(SrO)_{12}$ and $(LiI)_{12}$, respectively. Comparing the ball and stick models for different compounds but for the same structural motif, one noticeable difference between the LM for lithium iodide and those of the other two compounds is the sharper (more acute) bond angles that directly result from the greater polarisability of the iodide anion. Essentially, the iodide anions are further out from the cluster's centre of mass than the lithium cations.

To check the current success of data-mining the Hive for these three compounds, we also conducted global optimisation on each of the three IP-energy landscapes for low lying LM. We present the results as three densities of LM graphs; see Fig. 8. In the panel insert for each compound it is very clear that the data-mined LM present only a sample of all the possible LM. In terms of ranking, fortunately, the missing LM tend to be mid-range rather than at the more stable end (which, typically, is where there is most interest). Looking more closely at the top ranked LM, we identified which IP-LM structures are missing; these are shown in Fig. 9.

For strontium oxide clusters, the first six missing LM were ranked 6, 7, 8, 9, 13 and 16. The first three of these are basic rock-salt cuts that could have been included in our data-mined set if we had included the structures from ref. 54 (we

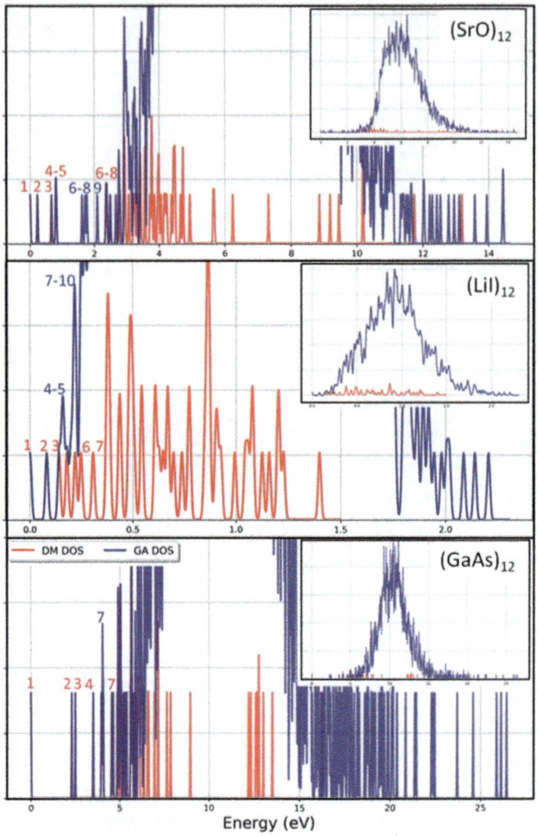

Fig. 8 Density of IP-LM found for $(XY)_{12}$ by data-mining our DM-set of unique structures (red lines) and by employing a genetic algorithm (blue lines) as a function of energy from the GM. The main panels show enlarged regions of the full graphs that are shown in the insets. Red and blue numbers label the ranks of the IP-LM as found by data-mining and the genetic algorithm, respectively.

did not as this paper includes data-mined structures for alkaline oxides, one of which is one of the compounds we chose to investigate). The GA08 cuboid configuration was in fact found as the IP-GM for $(LiI)_{12}$. Generating this LM during the data-mining process was fortuitous given that this structural motif was not included in the DM-set of unique LM. GA09 and GA13 are composed of a $n = 6$ drum (typically the IP-GM for $(XY)_6$) and $2 \times 2 \times m$ cuboids. More interesting is the GA16 configuration, which we have previously seen; it has an unusual distorted planar four-coordinated oxygen anion site.

For lithium iodide clusters, the first six missing LM structures were ranked 3, 4, 5, 7, 8 and 9. Unlike our DM-set, these configurations, which we will refer to as HC, have at least one highly coordinated (greater than 4) anion site and are not one of the possible cuboid cuts from the NaCl rock salt phase. Given the stability of this type of structure, quite a few of the better ranked structures were missed. As already seen, any unstable LM in the DM-set can lead to new structural LM and thus we did not miss all of the HC structures; the enantiomer of GA03 was found (labelled as DM03 in Fig. 6 and ranked equal third).

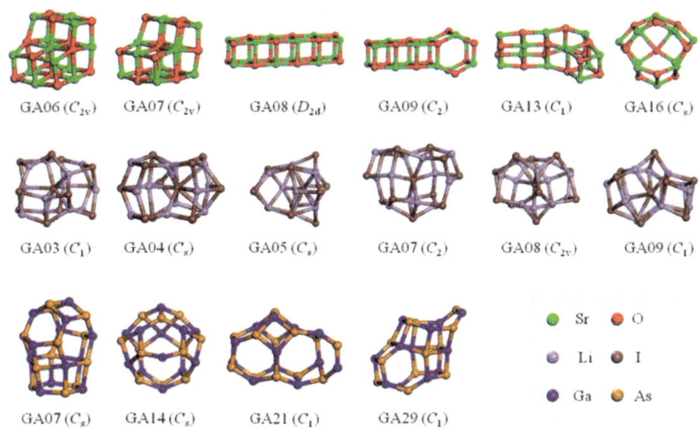

Fig. 9 Ball and stick models of $(XY)_{12}$ IP-LM configurations obtained by the genetic algorithm that were missing from the IP-LM found using the data-mining approach. The colour scheme is shown in the lower right hand panel and is the same as that employed in previous figures. The numbers in the GA** labels indicate the rank found for the nanocluster, where 01 indicates the IP-GM, whereas in the previous labels, DM**, they indicate the rank before the missing IP-LM were found using the GA.

For gallium arsenide clusters, data-mining the DM-set was much more successful in that only four additional IP-LM structures were found in the top thirty; the first four missing LM structures were ranked 7, 14, 21 and 29. Of these, GA07 is the result of merging IP-GM for $n = 6$ (a drum) and $n = 9$ (bubble) across a hexagonal face; GA14 is very similar to the GA16 LM that was missed for $(SrO)_{12}$; GA21 has the same structural motif as DM18, but with all the anions switched for cations, and *vice versa*, *cf.* DM23 and DM24 and also GA06 and GA07 for strontium oxide. We note that the DM and GA runs found different chiral versions of DM23 and DM24.

Finally, we should reiterate that the structures reported above for LiI, SrO and GaAs were obtained on the interatomic potential landscape. These potentials were originally parameterised for bulk compounds, where atoms are typically in higher coordinated environments, and therefore such parameterisations are very limited in scope. For example, arsenide anions are highly polarisable, and more realistic structures should be expected to have more buckled shapes, as seen above in LiI configurations. The latter proved to be easier to optimise due to the relatively low charges on Li and I. Notwithstanding this, the structures obtained here will be uploaded to the Hive and refined using our chosen *ab initio* approach, which will both give the actual findings more credence for future applications, but will also allow the parameters of the interatomic potentials to be refined. The latter is an important element of machine-learning techniques that have been particularly successful in studies of metallic clusters.[59,60]

5. Conclusions

We have presented, for the first time, details of our database of published atomic configurations of nanoclusters. We have described the algorithms employed within this database to establish whether two entries are equivalent LM for a particular compound and whether configurations of different compounds are equivalent when judged using connectivity arguments, and have shown how to exploit these data in order to predict structures for three new compounds. The database provides initial model structures that were traditionally obtained from experiments, configurations that can be employed in structure prediction using a data-mining approach, and a way of checking whether a candidate structure is indeed new. Data-mining the set of configurations for $(XY)_{12}$ structures that have a unique hashkey proved relatively successful in that the top two LM configurations for each of three compounds were found. However, global optimisation techniques are still required for compounds that are chemically distinct enough that their low energy LM structures do not match configurations already in the database, using our connectivity arguments. This will of course change with time, as more data is entered into the database.

Lessons learnt in the creation of the Hive and the associated WASP interface as a toolkit will be of direct use for further work on nucleation and crystallisation processes,[61] crucially the nucleation and growth of small particles on or in solid supports and liquid environments. The LM atomic configurations in the database are also readily usable as secondary building units (SBU) for constructing crystal structures.[6,8,10,62–68] Here, using low energy SBUs that do not resemble cuts from the main phases of the chosen compounds will produce more interesting results.

Conflicts of interest

There are no conflicts to declare.

Acknowledgements

The creation of the Hive, WASP interface, and Bee software are all part of the WASP@N five-year project that is supported by EPSRC (grant numbers EP/K038958, K038559). We are grateful for valuable discussions and help from our colleagues who are also part of this project, namely: Jörg Sassmannshausen (University College London, UK), Martijn A. Zwijnenburg (University College London, UK), Roy L. Johnston (University of Birmingham, UK), Stefan T. Bromley (University of Barcelona, Spain), Joseph J. BelBruno (Dartmouth College, USA), Volker Blum (Duke University, USA) and Julian D. Gale (Curtin University, Australia).

Notes and references

1 M. Chen, *et al.*, Structural and Electronic Property Study of $(ZnO)_n$, $n \leq 168$: Transition from Zinc Oxide Molecular Clusters to Ultrasmall Nanoparticles, *J. Phys. Chem. C*, 2016, **120**(36), 20400–20418.

2 C. R. A. Catlow, *et al.*, Zinc oxide: a case study in contemporary computational solid state chemistry, *J. Comput. Chem.*, 2008, **29**(13), 2234–2249.

3 F. Vines, *et al.*, Size dependent structural and polymorphic transitions in ZnO: from nanocluster to bulk, *Nanoscale*, 2017, **9**(28), 10067–10074.

4 O. Lamiel-Garcia, *et al.*, Predicting size-dependent emergence of crystallinity in nanomaterials: titania nanoclusters *versus* nanocrystals, *Nanoscale*, 2017, **9**(3), 1049–1058.

5 S. M. Woodley, Nanoclusters and Nanoparticles, in *Computational Modeling of Inorganic Nanomaterials*, ed. S. T. Bromley and M. A. Zwijnenburg, CRC Press, Taylor and Francis Group, London, 2016, pp. 3–46.

6 J. Carrasco, F. Illas and S. T. Bromley, Ultralow-density nanocage-based metal-oxide polymorphs, *Phys. Rev. Lett.*, 2007, **99**(23), 235502.

7 S. T. Bromley, A computational study into the viability of new molecular materials polymorphs based on fully-coordinated inorganic nanoclusters, *CrystEngComm*, 2007, **9**(6), 463–466.

8 S. M. Woodley, *et al.*, Construction of nano- and microporous frameworks from octahedral bubble clusters, *Phys. Chem. Chem. Phys.*, 2009, **11**(17), 3176–3185.

9 M. B. Watkins, *et al.*, Bubbles and microporous frameworks of silicon carbide, *Phys. Chem. Chem. Phys.*, 2009, **11**(17), 3186–3200.

10 M. Farrow, *et al.*, From stable ZnO and GaN clusters to novel double bubbles and frameworks, *Inorganics*, 2014, **2**(2), 248–263.

11 Y. Yong, *et al.*, The $Zn_{12}O_{12}$ cluster-assembled nanowires as a highly sensitive and selective gas sensor for NO and NO_2, *Sci. Rep.*, 2017, **7**(1), 17505.

12 Z. Liu, *et al.*, From the ZnO Hollow Cage Clusters to ZnO Nanoporous Phases: A First-Principles Bottom-Up Prediction, *J. Phys. Chem. C*, 2013, **117**(34), 17633–17643.

13 B. Wang, X. Wang and J. Zhao, Atomic Structure of the Magic $(ZnO)_{60}$ Cluster: First-Principles Prediction of a Sodalite Motif for ZnO Nanoclusters, *J. Phys. Chem. C*, 2010, **114**(13), 5741–5744.

14 Y. Yong, *et al.*, Theoretical prediction of low-density nanoporous frameworks of zinc sulfide based on Zn_nS_n (n = 12, 16) nanocaged clusters, *RSC Adv.*, 2014, **4**(70), 37333–37341.

15 J. C. Schön and M. Jansen, Determination, prediction, and understanding of structures, using the energy landscapes of chemical systems – Part III, *Z. Kristallogr.*, 2001, **216**(7), 361–383.

16 C. Mellot-Draznieks, Role of computer simulations in structure prediction and structure determination: from molecular compounds to hybrid frameworks, *J. Mater. Chem.*, 2007, **17**(41), 4348–4358.

17 J. C. Schön and M. Jansen, Determination, prediction, and understanding of structures, using the energy landscapes of chemical systems – Part I, *Z. Kristallogr.*, 2001, **216**(6), 307–325.

18 S. M. Woodley and R. Catlow, Crystal structure prediction from first principles, *Nat. Mater.*, 2008, **7**(12), 937–946.

19 The NOMAD Laboratory, 2018, http://nomad-repository.eu/.

20 ICSD – the Inorganic Crystal Structure Database, 2018, https://www.fiz-karlsruhe.de/en/leistungen/kristallographie/icsd.html.

21 Crystallographic and Crystallochemical Database for Minerals and their Structural Analogues, 2018, http://database.iem.ac.ru/mincryst/.

22 COD: Crystallography Open Database, 2018, http://www.crystallography.net/cod/.

23 Mineralogy Database, 2018, http://webmineral.com/.

24 Crystal Lattice-Structures, 2018,https://homepage.univie.ac.at/michael.leitner/lattice/index.html.

25 American Mineralogist Crystal Structure Database, 2018, http://rruff.geo.arizona.edu/AMS/amcsd.php.

26 CDS National Chemical Database Service, 2018, http://cds.rsc.org/.
27 Database of zeolite structures, 2018, http://www.iza-structure.org/databases/.
28 The Cambridge Structural Database (CSD), 2018, https://www.ccdc.cam.ac.uk/solutions/csd-system/components/csd/.
29 AFLOW: Automatic – Flow for Materials Discovery, 2018, http://aflowlib.org/.
30 The Cambridge Energy Landscape Database, 2018, http://www-wales.ch.cam.ac.uk/CCD.html.
31 Cn Fullerenes, 2018, http://www.nanotube.msu.edu/fullerene/.
32 S. De, *et al.*, Comparing molecules and solids across structural and alchemical space, *Phys. Chem. Chem. Phys.*, 2016, **18**(20), 13754–13769.
33 B. Helmich and M. Sierka, Similarity recognition of molecular structures by optimal atomic matching and rotational superposition, *J. Comput. Chem.*, 2012, **33**(2), 134–140.
34 H. W. Kuhn, The Hungarian method for the assignment problem, *Nav. Res. Logist. Q.*, 1955, **2**(1–2), 83–97.
35 H. W. Kuhn, Variants of the Hungarian method for assignment problems, *Nav. Res. Logist. Q.*, 1956, **3**(4), 253–258.
36 J. Munkres, Algorithms for the Assignment and Transportation Problems, *J. Soc. Ind. Appl. Math.*, 1957, **5**(1), 32–38.
37 A. Sadeghi, *et al.*, Metrics for measuring distances in configuration spaces, *J. Chem. Phys.*, 2013, **139**(18), 184118.
38 A. McLachlan, A mathematical procedure for superimposing atomic coordinates of proteins, *Acta Crystallogr., Sect. A: Cryst. Phys., Diffr., Theor. Gen. Crystallogr.*, 1972, **28**(6), 656–657.
39 A. Wagner and H.-J. Himmel, aRMSD: A Comprehensive Tool for Structural Analysis, *J. Chem. Inf. Model.*, 2017, **57**(3), 428–438.
40 R. Hundt, *et al.*, CCL: an algorithm for the efficient comparison of clusters, *J. Appl. Crystallogr.*, 2013, **46**(3), 587–593.
41 S. Alexander, Structure identification methods for atomistic simulations of crystalline materials, *Modell. Simul. Mater. Sci. Eng.*, 2012, **20**(4), 045021.
42 Y. Dong, *et al.*, Analyzing the properties of clusters: Structural similarity and heat capacity, *Comput. Theor. Chem.*, 2013, **1021**, 16–25.
43 C. X. Su, *et al.*, Construction of crystal structure prototype database: methods and applications, *J. Phys.: Condens. Matter*, 2017, **29**(16), 165901.
44 B. Schaefer and S. Goedecker, Computationally efficient characterization of potential energy surfaces based on fingerprint distances, *J. Chem. Phys.*, 2016, **145**(3), 034101.
45 J. A. Chisholm and S. Motherwell, COMPACK: a program for identifying crystal structure similarity using distances, *J. Appl. Crystallogr.*, 2005, **38**, 228–231.
46 A. Ramirez-Manzanares, *et al.*, A hierarchical algorithm for molecular similarity (H-FORMS), *J. Comput. Chem.*, 2015, **36**(19), 1456–1466.
47 T. Lazauskas, A. A. Sokol and S. M. Woodley, An efficient genetic algorithm for structure prediction at the nanoscale, *Nanoscale*, 2017, **9**(11), 3850–3864.
48 B. D. McKay and A. Piperno, Practical graph isomorphism, II, *J. Symbolic Comput.*, 2014, **60**, 94–112.
49 C. R. A. Catlow, *et al.*, Modelling nano-clusters and nucleation, *Phys. Chem. Chem. Phys.*, 2010, **12**(4), 786–811.
50 V. Blum, *et al.*, *Ab initio* molecular simulations with numeric atom-centered orbitals, *Comput. Phys. Commun.*, 2009, **180**(11), 2175–2196.

51 J. P. Perdew, K. Burke and M. Ernzerhof, Generalized Gradient Approximation made simple, *Phys. Rev. Lett.*, 1996, **77**(18), 3865.

52 J. P. Perdew, K. Burke and M. Ernzerhof, Generalized Gradient Approximation Made Simple, *Phys. Rev. Lett.*, 1996, **77**, 3865; Phys. Rev. Lett., 1997. 78(7): p. 1396.

53 J. P. Perdew, *et al.*, Restoring the density-gradient expansion for exchange in solids and surfaces, *Phys. Rev. Lett.*, 2008, **100**(13), 136406.

54 S. Escher, *et al.*, Synthesis Target Structures for Alkaline Earth Oxide Clusters, *Inorganics*, 2018, **6**(1), 29.

55 J. D. Gale and A. L. Rohl, The General Utility Lattice Program (GULP), *Mol. Simul.*, 2003, **29**(5), 291–341.

56 M. R. Farrow, Y. Chow and S. M. Woodley, Structure prediction of nanoclusters; a direct or a pre-screened search on the DFT energy Landscape?, *Phys. Chem. Chem. Phys.*, 2014, **16**(39), 21119–21134.

57 S. G. E. T. Escher, *et al.*, Structure prediction of $(BaO)_n$ nanoclusters for $n \leq 24$ using an evolutionary algorithm, *Comput. Theor. Chem.*, 2017, **1107**, 74–81.

58 C. R. A. Catlow, M. J. Norgett and T. A. Ross, Ion transport and interatomic potentials in the alkaline-earth-fluoride crystals, *J. Phys. C: Solid State Phys.*, 1977, **10**(10), 1627.

59 C. Zeni, *et al.*, Building machine learning force fields for nanoclusters. 2018, https://arxiv.org/abs/1802.01417.

60 A. P. Bartók, *et al.*, Machine learning unifies the modeling of materials and molecules, *Sci. Adv.*, 2017, **3**(12), e1701816.

61 C. Leitold and C. Dellago, Nucleation and structural growth of cluster crystals, *J. Chem. Phys.*, 2016, **145**(7), 074504.

62 C. Mellot-Draznieks, *et al.*, Computational design and prediction of interesting not-yet-synthesized structures of inorganic materials by using building unit concepts, *Chem.–Eur. J.*, 2002, **8**(18), 4103–4113.

63 Z. Liu, *et al.*, From the ZnO Hollow Cage Clusters to ZnO Nanoporous Phases: A First-Principles Bottom-Up Prediction, *J. Phys. Chem. C*, 2013, **117**(34), 17633–17643.

64 Z. F. Liu, X. Q. Wang and H. J. Zhu, New nanomaterials based on $In_{12}As_{12}$ cages: an *ab initio* bottom-up study, *RSC Adv.*, 2013, **3**(5), 1450–1459.

65 Y. Li, *et al.*, Prediction of open-framework aluminophosphate structures using the automated assembly of secondary building units method with Lowenstein's constraints, *Chem. Mater.*, 2005, **17**(24), 6086–6093.

66 C. M. Draznieks, *et al.*, *De novo* prediction of inorganic structures developed through automated assembly of secondary building units (AASBU method), *Angew. Chem., Int. Ed.*, 2000, **39**(13), 2270–2275.

67 B. K. Teo and H. Zhang, Clusters of clusters – self-organization and self-similarity in the intermediate stages of cluster growth of au ag supraclusters, *Proc. Natl. Acad. Sci. U. S. A.*, 1991, **88**(12), 5067–5071.

68 X. F. F. Duan and L. W. Burggraf, The *closo*-$Si_{12}C_{12}$ molecule from cluster to crystal: A theoretical prediction, *J. Chem. Phys.*, 2016, **144**(11), 114309.

Faraday Discussions

DISCUSSIONS

Applications of crystal structure prediction – inorganic and network structures: general discussion

Virginia Burger, Frederik Claeyssens, Daniel W. Davies, Graeme M. Day, Matthew S. Dyer, Alan Hare, Yi Li, Caroline Mellot-Draznieks, John B. O. Mitchell, Sharmarke Mohamed, Artem R. Oganov, Sarah L. Price, Michael Ruggiero, Matthew R. Ryder, German Sastre, J. Christian Schön, Peter Spackman, Scott M. Woodley and Qiang Zhu

DOI: 10.1039/c8fd90034e

J. Christian Schön opened discussion of the paper by Yi Li: It is not quite clear to me, to what extent you treat symmetry as a deciding criterion as far as the acceptance of a structure candidate is concerned. We had lots of discussions with crystallographers and solid state chemists in the early 1990s at the beginning of crystal structure prediction for inorganic compounds, who essentially claimed that "the higher the symmetry, the better the structure" – an argument which we have always felt to be dangerously flawed.[1] Do you have the impression from your study that for your type of systems, high symmetry is a good indicator of the quality of the structure? After all, there could be many systems with space group *P*1 which are very low in energy, and furthermore, the zeolites with space group *P*1 would be expected to greatly outnumber the ones with higher symmetry. Could you comment on this?

1 J. C. Schön and M. Jansen, *Angew. Chem., Int. Ed.*, 1996, **35**, 1286–1304.

Yi Li replied: Yes, high symmetry is not an indicator of the quality of a zeolite structure. Among the 239 known zeolite framework types approved by the International Zeolite Association, the numbers of zeolites in triclinic, monoclinic, orthorhombic, tetragonal, trigonal/hexagonal, and cubic systems are 2, 40, 83, 36, 53, and 21, respectively. Moreover, in each crystal system, space groups with the highest multiplicities occur much more frequently than the others (such as *C*2/*m* in the monoclinic system, *Cmcm* in the orthorhombic system, *I*4/*mmm* in the tetragonal system, *P*63/*mmc* in the hexagonal system, and *Im*$\bar{3}$*m* in the cubic system). In theory, we cannot rule out the possibility of zeolites with low symmetry, but it will be a good start to predict an unknown zeolite structure from higher symmetry to lower symmetry. Another important fact is that all known zeolite structures exhibit similar lattice energies. Zeolites with low symmetry do not have an obvious advantage in lattice energy over those with high symmetry. To date, there is no report on a zeolite with *P*1 symmetry. On the other hand, our method is not only for high symmetry space groups. In fact, it will have good performance on structures with multiple Wyckoff positions, which may occur in many low symmetry space groups.

J. Christian Schön remarked: In the structures you consider/generate, do you also have T atoms that are not located on special sites?

Yi Li responded: Absolutely yes. Over 70% of the known zeolite structures possess T atoms on general positions. Our approach does not rule out general positions. We simply do not add special constraints on general positions.

J. Christian Schön continued: Conversely to the previous question, do you have symmetry elements (rotation axes, mirror planes), which do not contain any T atoms?

Yi Li answered: Yes, we do have such empty site symmetries. The number of T atoms we put in a unit cell is decided by the expected framework density. Zeolites have relatively lower framework densities than other inorganic materials. In fact, zeolites with all special site symmetries occupied by T atoms are quite rare.

German Sastre asked: Using your algorithm, is there a limit to the degree of complexity of new zeolites that you can find? By complexity I mean the maximum number of symmetry unequivalent tetrahedral sites. Similar approaches tend to stop the search at 6 T sites and so if your approach is able to go further it would be the first to include more complex zeolites in a database. If so, could you plot the number of zeolites found containing x T sites (where x goes from 1 to n, with n being the maximum complexity) for each space group? In other words, are there many complex ($x > 6$) zeolites?

Yi Li answered: We use our home-made program FraGen to generate zeolite models within a given unit cell. FraGen is based on parallel tempering Monte Carlo algorithms, which is a powerful global optimization technique. We do not need to set the maximum number of unequivalent T atoms in our approach. It is decided by the cell volumes and the upper limit in framework density. In our experience, FraGen will find structures with 10 T atoms easily. So, we did generate many structures with >6 T atoms, and the number of the generated structures grew rapidly with the number of unique T atoms. We have tried our methods in two space groups, and the most complex structures we have generated so far contain 14 T atoms.

German Sastre said: Does your algorithm require significantly more computing time for those difficult cases? Could you plot, for a significant space group containing simple (less or equal than 6 T sites) and complex (more than 6 T sites) zeolites, the average CPU time taken for each number of T sites? I mean, take all the zeolites (say n) containing 2 T sites, add all the CPU times and divide by n. And do the same for all the numbers of T sites found in the space group of your choice.

Yi Li answered: The CPU time is determined by not only the complexity of the modeled system but also the number of Monte Carlo steps we set in FraGen. Normally, we use 20 000–30 000 MC steps for each cycle of FraGen, which is already enough for the building of complex structures. Increasing the number of MC steps does not significantly increase the number of plausible structures, since FraGen locates local minima very quickly. In an orthorhombic space group, taking 30 000 MC steps in each cycle, FraGen will find a tetrahedrally coordinated structure with 6 unequivalent T atoms in 2 s, a 10-T structure in 5 s, and a 14-T structure in 7 s.

Scott M. Woodley remarked: Atomic configurations predicted for nanoclusters of a compound (examples are given in my paper) can often resemble a fragment cut from one of its observed bulk phases. Sometimes, as found for $(ZnO)_n$, where n is the number of formula units the cluster is composed of, this is not the case. The global minimum for $n = 12$ actually resembles a secondary building block of the

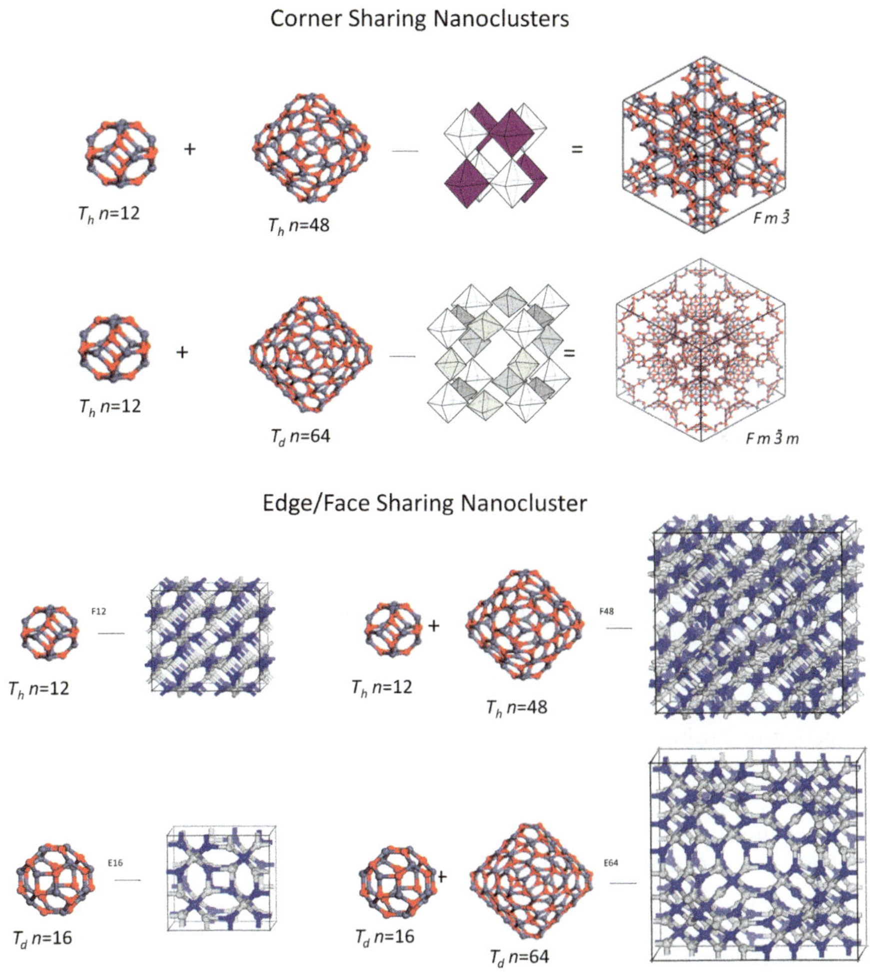

Fig. 1 Framework structures created by using predicted low energy nanoclusters for ZnO as secondary building units.[1,2]

sodalite framework. Thus, we have used the more stable sized global minima to generate new frameworks; one of these is composed of both $n = 12$ and $n = 48$ units (Fig. 1 shows this and other examples). All my atoms are four coordinated, *i.e.* sit on T sites, so switching all the zinc and oxygen atoms with a silicon and placing an oxygen atom between T sites will generate silica frameworks. Using your method, have you already predicted and published the equivalent structures that were shown during the discussions (and in the figure here)? If not, are they, particularly the structure composed of $n = 12$ and $n = 48$ units as it has a relatively low energy for ZnO (or SiC), already known in one of the zeolite databases you refer to? I can supply you with the atomic coordinates. Sometimes this generates a familiar framework, for example, the $n = 12$ example shown in the bottom left hand corner of Fig. 1 generates the sodalite framework (where atoms are located on the so called T sites of the Zeolite).

1 S. M. Woodley, M. B. Watkins, A. A. Sokol, S. A. Shevlin and C. R. A. Catlow, *Phys. Chem. Chem. Phys.*, 2009, **11**, 3176–3185.
2 M. B. Watkins, S. A. Shevlin, A. A. Sokol, B. Slater, C. R. A. Catlow and S. M. Woodley, *Phys. Chem. Chem. Phys.*, 2009, **11**, 3186–3200.

Yi Li responded: We have not performed structure prediction in space group $Fm\bar{3}$ and $Fm\bar{3}m$ yet. So your structure is not yet in our database. If you can provide a CIF file, I can check your structure in other databases, such as those developed by Prof. Deem and Prof. Treacy. If your structure cannot be found in any known database, I think there will be two reasons. The first is that your structure may have a big unit cell, whose dimensions are beyond current approaches. The second reason is that it may contain distortions and have a relatively high lattice energy, whereas current approaches tend to find structures with low energies. Anyway, please provide your structure file.

Caroline Mellot-Draznieks asked: There are many methods for generating thousands/millions of zeolite structures. Your method allows you to focus more efficiently on a single space group and search for structures in that space group. How do you check that the structures generated have not been predicted before, *i.e.* are new? Which is the cost function used to estimate their ranking or feasibility?

Yi Li responded: We have calculated the coordination sequences for all the zeolite structures we predict. By comparing their coordination sequences, we have removed all duplicated structures and ensure that all the structures we generate are unique and new. During model building, we use a simple cost function regarding the coordination numbers, bonding distances, and bonding angles. Then, we perform more sophisticated geometry optimizations on all the models we have built. Finally, the optimized models are evaluated by framework energy and local interatomic distance criteria. The local interatomic distance criteria are a set of structure regulations regarding the relationship between T–O, O–T–O, and T–*O*–T distances, which are obeyed by nearly all known zeolite frameworks. All structures violating any of these criteria are removed from our final result.

Scott M. Woodley asked: How does your method compare to the topological approach that employs tiling theory (as developed by Rob Bell and co-workers,[1] for example)?

1 M. D. Foster, *et al.* Chemically feasible hypothetical crystalline networks, *Nat. Mater.*, 2004, **3**, 234–238.

Yi Li responded: As far as I understand, the tiling theory will produce all four-connected network topologies without considering much geometric information. To my knowledge, the tiling theory method has produced all uni-, bi-, and tri-nodal networks (*i.e.*, networks consisting of one, two, and three unequivalent T atoms). I think the tiling theory will generate too many networks if we consider more T atoms, and most of them are inaccessible in experiment (as shown in the mentioned reference). By considering geometric constraints/restraints, our method can generate much more complicated and chemically feasible zeolite structures (say, structures consisting of ten unequivalent T atoms), but we cannot enumerate all the possibilities because there are too many. I think the aims of these two approaches are different. Tiling theory aims to enumerate all the possible networks in a mathematical sense, whereas ours aims to find feasible synthetic targets/structure solutions as quickly/many as possible, which in most cases are complicated and inaccessible by previous approaches.

Graeme M. Day commented: You mentioned that you use the calculated energy in evaluating structures, but also geometric criteria, and that structures that have low energies are sometimes excluded because they do not fit the structural trends seen in known zeolites. Coming from the field of molecular organic crystal structure prediction, I find this interesting. For organic molecules, if we predict low energy crystal structures that have geometrical features that are different from what we have seen before, we would not exclude them from our predictions, as long as we trust the calculated energies. This is because we want to find structures that are different and might have different properties to known structures. Are those zeolite structures that are excluded based on their geometry of interest in looking for zeolites with potentially new properties?

Yi Li replied: The lattice energies of different known zeolite structures are quite similar. But there are millions of tetrahedral networks exhibiting energies similar to known ones. On the other side, many new zeolites have been synthesized with lattice energies much higher than previously reported ones. This is why we cannot use lattice energy as the main criterion to judge whether a predicted zeolite structure is feasible or not. We find that all known zeolite structures obey a set of geometric rules, the local interatomic distance criteria, which are much more reliable than the energy criterion. Recently reported high-energy zeolites obey these criteria too. The properties of zeolites, such as adsorption, separation, and catalysis, are mainly determined by their porous networks (accessible cavities, channels, and their connectivity and orientation, *etc.*). So the local geometry is not very important to these properties.

Graeme M. Day added: You mentioned that observed zeolite structures can be relatively high in energy compared to the predicted structures. Is this observation dependent on the method used to evaluate their energies? Do the high energy structures that are observed experimentally remain high in energy if they are re-evaluated using a different computational method?

Yi Li responded: Because of their framework complexity, force-field-based calculations are the most used methods for the evaluation of zeolite lattice energies. Many researchers have systematically calculated the lattice energies of known zeolites, and some of them always exhibit much higher energies than the others, no matter what program and potential functions are used. No one has tried systemic periodic DFT calculations on all known zeolites, but I would expect the same trend. One of the most important reasons why high energy zeolites exist in experiment is that there are many other species being introduced into zeolite synthesis. For instance, metal cations and organic amines are often used as key additives in zeolite synthesis. These species do not form zeolite frameworks, but reside in the extraframework voids and co-crystallize with zeolite frameworks. The host–guest interactions may compensate the high lattice energy of zeolite frameworks. Other unconventional synthetic methods, such as high-pressure approaches and ADOR approaches, can also lead to the formation of high-energy zeolites. Unfortunately, we cannot consider these practical synthetic factors among such a large number of hypothetical zeolite structures.

Graeme M. Day said: You found limits for the densities of T atoms (TAD densities) along symmetry axes and in symmetry planes, and use these limits in predicting structures. You demonstrated that the limits on TAD density are related to the distortion of the structure around the T atoms. I would expect that geometrical distortion would lead to high calculated energies. Did you look for a correlation between the TAD densities and the resulting energies of the predicted structures? I wonder if the limits that you apply to the TAD density reflect an energetic limit.

Yi Li responded: Yes. We propose defining the TAD density limit because overly crowed T atoms may lead to distortions in the bonding geometry and violate the required tetrahedral coordination in the most extreme cases, which definitely will increase the framework energy (to an unrealistic level). So, setting an upper limit for the TAD density is equivalent, more or less, to setting an upper limit for the lattice energy. We have not studied the exact correlation between the TAD densities and framework energies, because we believe the correlation is not smooth and is quite specific to different zeolite structures. Our aim is to rule out as many Wyckoff position combinations as possible, so only the highest allowed TAD density matters to us.

Qiang Zhu asked: What is the physics behind the observed empirical relation in the geometries? If you already know the relation, can you impose these geometry constraints when you generate structures for zeolite crystal structure prediction?

Yi Li replied: The local interatomic distance criteria reflect the bonding geometry of zeolite frameworks. The correlation between T–O and O–T–O distances corresponds to the constraints on the shape of a TO_4 tetrahedron. It cannot be far away from an ideal tetrahedron. The correlation between T–O and T–O–T reflects the specific way that TO_4 tetrahedra are linked. In general, we can impose constraints on coordination numbers, bonding distances, and bonding angles for zeolite structure prediction, but we cannot ensure all generated structures will satisfy the local interatomic distance criteria. However, for some specific zeolite structures, such as ABC-6 zeolites, which are constructed by the stacking of 6-rings, we can encode these structural regulations into the structure prediction procedure, which ensures all the structures predicted satisfy the local interatomic distance criteria.

Qiang Zhu continued: If these are purely geometry constraints, could you start the structure generation with building blocks that have these geometry relationships?

Yi Li responded: Yes, our approach is mainly based on geometric constraints/restraints. Currently, we cannot generate structures with building blocks. We notice that some methods are capable of generating structure models with building blocks, such as the AASBU method developed by Caroline Mellot Draznieks *et al.* This is a good suggestion, but it requires a lot of code writing. We may implement this idea in the future in our program, but it will require a lot of code writing.

Matthew S. Dyer opened discussion of the paper by Daniel W. Davies: This could be taken in a positive or negative sense, but one of the two new stable phases proposed in Table 2 and Fig. 9 of the paper has been previously reported. The synthesis of $YZrF_7$ is reported in ref. 1. It is isostructural with $SmZrF_7$, with the crystal structure reported in ref. 2.

This can be seen as a validation of the approach in this paper, as an existing compound outside of the original database has been identified. However, I believe that more care must be taken to perform a wide literature search before proposing a compound as a particularly promising new target.

1 M. Poulain, M. Poulain and J. Lucas, Les fluorozirconates de terres rares $LnZrF_7$, *Mater. Res. Bull.*, 1972, **7**, 319–325.
2 M. Poulain, M. Poulain and J. Lucas, Structure cristalline de $SmZrF_7$. Relations structurales avec le type ReO_3, *J. Solid State Chem.*, 1973, **8**, 132–141.

Daniel W. Davies replied: Thank you for highlighting this important previous work. I would argue that this should be seen as a positive result in the context of this study for the reasons you presented, and would add that for this work only minimal checks were carried out to establish the existence of a compound: the Materials Project database was queried automatically as a screening step and the ICSD was searched manually at the end of the process for any leading compounds. More careful checks should indeed be performed before taking compounds forward to experimental synthesis.

Matthew S. Dyer remarked: When allocating oxidation states on the basis of bond valence sums in a crystal structure, how does your method cope when less common anion oxidation states such as peroxide, superoxide and disulphide anions are present? For example, does it correctly ascribe the +2 oxidation state to Fe in FeS_2?

Daniel W. Davies replied: The approach we used as implemented in Pymatgen[1] uses a maximum *a posteriori* estimation method, so some prior knowledge of the distribution of oxidation states for each particular metal is needed.

For some very uncommon oxidation states, this model would lack the necessary data to assign them correctly. For example, for potassium and chlorine, it only has the required parameters for the +1 and −1 oxidation states, respectively. It would therefore be unable to reconcile the unusual compositions KCl_3, KCl_7 and K_3Cl_5, as characterised under high pressures.[2] For the particular example of FeS_2, the algorithm is able to assign these oxidation states correctly, as it has data on the distribution of the +1 to +4 oxidation state of iron. Furthermore, the benefit of using a structure-based approach can be seen when treating mixed valence compounds, for example magnetite (Fe_3O_4), which contains both Iron(II) and Iron(III).

1 S. P. Ong *et al.*, *Comput. Mater. Sci.*, 2013, **68**, 314–319.
2 W. Zhang *et al.*, *Sci. Rep.*, 2016, 26265.

J. Christian Schön remarked: Concerning the height of the bars in Fig. 3 in your paper, indicating the distribution of oxidation states: did you take into account that these might be actually due to spatial constraints? After all, there are only so many *e.g.* halogen atoms (with oxidation state −1) that you can pack around a cation.

Daniel W. Davies answered: This is would certainly be an explanation for some of the unexplained trends that are seen in Fig. 3–5, and one that we have not explored in the main text. For example, it may go some way to explaining why for higher oxidation states, the likelihood of finding the metal with an anion of moderate electronegativity often goes to zero before the likelihood of finding it with an anion of low electronegativity.

J. Christian Schön added: As a concrete example regarding the previous question, in spite of the fact that F has a higher electronegativity that O, you would end up preferring to use O (in oxidation state −2) when aiming for cations in a high oxidation state.

J. Christian Schön continued: In eqn 3 in your paper, you write down a probability as the product of other probabilities. This only makes sense if the product involves only independent probabilities. I am not sure whether this assumption really holds. If you look at the dataset where you get the probabilities from, this might already contain lots of correlations between certain types of cations/anions (when discovering certain elements, people often talked about "Vergesellschaftung", indicating that in minerals certain elements liked to appear together[1]). Since the formula is very nice and suggestive, it would be great if you could show to what extent the assumption of statistical independence is justified.

1 V. M. Goldschmidt, H. Hauptmann and Cl. Peters, *Naturwissenschaften*, 1933, **20**, 362–365.

Daniel W. Davies replied: This is an important point, as the independence of the individual probabilities is a key assumption of this approach. This is done for simplicity only and I agree that there is likely to be a range of other correlations between species and other trends whose incorporation into the model would improve its predictive power.

For the present study, we are satisfied that sufficient information is captured by the model for it to act as a pre-screening filter, as evidenced by Fig. 6 in the paper. The number of compositions generated using the SMACT code[1] that pass through such a filter decreases rapidly as the threshold is increased. Further work is planned to investigate how many of the compositions that are disregarded are false negatives. This will act as a first step towards establishing which other correlations are the most important to consider and incorporate.

1 D. W. Davies *et al.*, *Chem*, 2016, **1**, 617–627.

Sarah L. Price remarked: This paper was a valuable update for me on inorganic chemistry. I was struck by the statement "it was recently estimated that half of all inorganic compounds are metastable". Whilst it seems reasonable, with our current lack of knowledge about the causes of polymorphism, to assume that there is an exponential decrease in probability with thermodynamic instability, the limit of 100 meV is quite large. Your analysis is more complex than in the organic solid state where we are less often addressing the problem of the relative stability of multi-component systems such as solvates, hydrates, and cocrystals relative to their components or other stoichiometries. When you are comparing your ternary systems with the possible binary phases *etc.*, to what extent does the idea that the structures may be metastable complicate the analysis of whether the predicted structures are feasible?

Daniel W. Davies replied: The recent estimation comes from an analysis carried out by Sun *et al.*[1] in which they explore the "thermodynamic stability limit" for different chemistries. It is certainly true to say that 100 meV above the convex hull is a large limit; for certain chemistries such as chlorides, Sun *et al.* show that the limit is more like 50 meV, and for others such as iodides, it is considerably lower at around 25 meV. 100 meV is a reasonable limit for fluorides, however, so we have used it universally here in our search for mixed halide systems. The concept of metastability does complicate the prediction of new feasible structures significantly. One extension of the workflow presented here that we have successfully used previously[2] is to carry out phonon calculations as a logical next step, to confirm dynamic stability. This comes at a significantly higher computational price than carrying out DFT total energy calculations alone, but does provide useful information about the potential energy surface. Some more recent studies using, for example, energies of the amorphous states of compositions,[3] are beginning to provide a clearer definition to the upper limits of metastability for inorganic systems, but further work is needed to clarify the relationship between metastability and whether or not a compound can be synthesised.

1 Sun *et al.*, *Sci. Adv.*, 2016, **2**, e1600225.
2 D. W. Davies *et al.*, *Chem. Sci.*, 2018, **9**, 1022–1030.
3 M. Aykol *et al.*, *Sci. Adv.*, 2018, **4**, eaaq0148.

German Sastre asked: Can you use your approach with mixed metallic oxides containing molybdenum, vanadium and niobium where the stoichiometry is known? Vanadium and molybdenum can have different oxidation states in the same compound. The challenge is to find, in the unit cell corresponding to a given compound, the oxidation state of each atom.

Daniel W. Davies replied: I believe this can be done using the approach that we used to assign oxidation states,[1] as this employs parameters based on the size and electronegativity of elements to establish bond valence (similar to bond order in molecules).

This is implemented in the Pymatgen code[2] using a maximum *a posteriori* estimation method, so some prior knowledge of the distribution of oxidation states for each particular metal is needed. In your particular example, I would suggest that given the wide range of oxidation states that can be adopted by each transition metal, the limits of this method would be tested. It would be an interesting test case.

1 M. O'Keeffe and N. E. Brese, *J. Am. Chem. Soc.*, 1991, **113**, 3226–3229.
2 S. P. Ong *et al.*, *Comput. Mater. Sci.*, 2013, **68**, 314–319.

Alan Hare enquired: Is the geometrical shape of the YF_8 polyhedron known so precisely that you can calculate the packing density of its ternary halide with ZrF_6? The ternary structure looks potentially to be a 3D space-filler: almost a cube, in fact (or possibly Menger sponge, or something like that). If so, it might be possible to suggest a further solution to Hilbert's 18th problem in mathematics, and perhaps to propose the computation of a supercell on a supercell recursively.

Daniel W. Davies responded: You have raised a very interesting series of points that merit further investigation beyond what has been reported in our work on the role of oxidation states in materials discovery.

John B. O. Mitchell added: I understood Hilbert's 18th problem to have been solved. The anisohedral tiling part was solved by Reinhardt in 1928. As for the sphere packing part, Thomas Hales (1998) produced a computer-assisted enumeration of all the possibilities, which seems to have been accepted as effective proof that cubic and hexagonal close packings can't be bettered.

Artem R. Oganov also responded: You are talking about what was known in mathematics as Kepler's hypothesis and is now known as Kepler's theorem. Let me update you: in 2015, Thomas Hales produced its formal proof, and this is why now this is a theorem.

Alan Hare answered: Yes. I understood this, too. The close packings (of just over 74%, as I recall) can't be bettered by spheres. They can of course be bettered by cubes (which fill space 100%). Octahedra, tetrahedra and icosahedra don't fill space at all (except for their own individual volume). Also, there are (100%) space-fillers beyond these Platonic solids.

An example is the Bilinski rhombic dodecahedron. This comprises two unlike rhombohedra (compare this to Penrose in 2D, with its kites and darts). There may well be other combinations of polyhedra, besides.

When I saw the ternary $YZrF_7$ pairing drawn, I thought this could be another: it looks almost cubic. But I don't know the precise geometry of the YF_8 polyhedron. So I asked the question.

Given molecular orbitals, I suppose it depends if we can ever really be satisfied with either cubes *or* polyhedra.

I recognise that years ago Reinhardt's "jigsaw" tiling solved, and latterly Hales's sphere packing addressed, an important part of the problem; and that close packings had already been seen as maximising the *sphere* packing density. But here in the crystal chemistry we are looking at two unlike *polyhedra* that have come together; and so I wondered if this might suggest a further solution to the part of the problem that concerns anisohedral tiling.

Once again we know that while the cube is the only Platonic solid that fills space, beyond these solids there are numerous space-filling polyhedra; although small unlike pairs, such as we see here perhaps, may be of particular significance physicochemically.

J. Christian Schön asked: In Fig. 8 in your paper, you state a probability of 1.0 for magnesium bromide and zinc bromide. But what about solid solution phases? Are they also captured in your scheme?

Daniel W. Davies responded: The probability of 1.0 is associated with the hypothetical ternary phase $MnZnBr_4$. This comes about due to the existence of stable $ZnBr_2$ and $MnBr_2$ phases (as depicted in Fig. 8), *i.e.*, the necessary species-anion pairs (Mn^{2+}–Br^- and Zn^{2+}–Br^-) both have probabilities of 1.0 according to eqn 2. This scheme cannot strictly be applied to solid solutions in its current form as it is stoichiometry agnostic (eqn 3), however this would form an interesting extension.

Qiang Zhu enquired: You mentioned that you used both first principle CSP and data mining approaches. There are structures from evolutionary structure prediction that belong to (un)known structure types. Could you comment on these two different approaches?

Daniel W. Davies answered: The structure prediction by substitution (data mining) approach[1] finds structures within our thermodynamic limit with very low computational cost; a database of possible parent structures can be searched on a desktop computer within a few minutes. Evolutionary structure prediction explores large areas of the potential landscape at a much higher computational cost, but is not restricted to known structure types. Hence, it was able to find lower energy structures for all four compounds in our previous study.[2]

1 G. Hautier *et al.*, *Inorg. Chem.*, 2011, **50**, 656–663.
2 D. W. Davies *et al.*, *Chem. Sci.*, 2018, **9**, 1022–1030.

Qiang Zhu continued: There are also some structures found by a loose fit that are analogous to known structure types. However, they were not found by the substitution method – do you know why?

Daniel W. Davies answered: Specifically, in the study to which you refer[1] the compound $Cd_5S_4Cl_2$ adopts the same structure type as Li_5BiO_5. The partial inversion in terms of the anion/cation occupancy means that the substitution algorithm does not consider this structure type.

1 D. W. Davies *et al.*, *Chem. Sci.*, 2018, **9**, 1022–1030.

Artem R. Oganov added: Among the four compounds that we have looked at[1] with Daniel W. Davies, data mining with the substitution algorithm failed in all four cases – USPEX found lower-energy structures in all four cases. In 3 cases this was because there is no known prototype. In 1 case a prototype could be found in the database, but the substitution was deemed improbable as it involved a change of charges (through a coupled substitution) and exchange of extremely different atoms.

1 D. W. Davies, K. T. Butler, J. M. Skelton, C. Xie, A. R. Oganovcde and A. Walsh, *Chem. Sci.*, 2018, **9**, 1022–1030.

J. Christian Schön remarked: Based on the many discussions regarding "energy landscapes" and "free energy landscapes" I have had over the past day and a half, I wanted to show a slide to illustrate the central aspects and differences of these entities (Fig. 2). Mathematically, an energy landscape consists of three pieces: the configuration (or state or solution) space of the system, a neighborhood definition (called moveclass in the case of an algorithm), and an energy (or more generally cost) function. In chemistry, the configuration space is the set of all atom arrangements in the system, possibly with some constraints, *e.g.* if we already know what the unit cell of a crystalline structure is supposed to be, and in the case of a general complex optimization problem, we speak of the solution space (in which we search for the optimal solution to our task such as *e.g.* allocating money to different goods we want to invest in).

A neighborhood on the energy landscape of a chemical system is either an algorithmic one or a physical one; in the latter case, the neighborhood is associated with small displacements of individual atoms, or a combination of them

Energy landscapes versus Free Energy landscapes

Energy landscape = Configuration (state) space of the system + neighborhood definition (moveclass) + energy (cost) function
=>
Landscape corresponds to a map from the configuration space to the real numbers

Note: this landscape is independent of observational time scales or temperature; dependence on external fields like pressure is included in the cost function definition

Free Energy landscape = set of locally ergodic regions belonging to the system at temperature T and on observation time scale t_{obs} + connectivity (probability flows / transition rates) + free energy function (sum over states) at temperature T
=>
Landscape corresponds to a map from the set of locally ergodic regions (defined as those regions that are locally ergodic on the time scale t_{obs}, i.e. essentially $t_{equil} << t_{obs} << t_{escape}$) to the real numbers

Note: this landscape strongly depends on time scales and temperature

(Reference discussing landscape concepts: Neelamraju, Oligschleger, Schön, Journal of Chemical Physics 2017)

Fig. 2 Energy landscapes *vs.* free energy landscapes: Basic concepts.[9]

(which can occur on very short time scales during atom-level physical processes). The landscape for this kind of moveclass is the one we are familiar with, and the one we have in mind when we discuss the physics and chemistry of the chemical system. But this is complemented by an algorithmic neighborhood, *i.e.* the general moveclass which defines the way we explore the configuration space and thus tells us which configurations are neighbors to one another for the purpose of the algorithm. For example, in an algorithmic landscape, we can have jumpmoves or large hops such as the exchange of different cations or whole building blocks *etc.*. Typically, such moves would require a much longer time in the natural evolution of the system according to physical laws (Newtonian dynamics, *etc.*). But for the global exploration of the landscape, the moveclass of our algorithm should help us zero in on low-energy states very quickly (*i.e.* make the landscape look very "smooth" with "easy" paths towards low-energy states). Choosing the optimal moveclass for exploring the energy landscape is the "black art" behind every stochastic global optimization algorithm! But this algorithmic landscape is not connected to the stability of the predicted polymorphs – for that we need to look at the first kind of landscape that is defined by the physically relevant neighborhoods! Taken together, an energy landscape is a map from the configuration space to the real numbers, where we have some freedom to choose our neighborhoods (with all that implies for mathematical properties such as differentiability!).

As an aside, we note that *e.g.* basin hopping[1] or evolutionary algorithms (equal to a genetic algorithm operating on local minima)[2] which combine jumpmoves (hops or cross-over moves) with local optimizations actually operate on two different landscapes: for the move part, they allow large jumps from current configurations to some that are far removed (compared to what a physical move would be), but the subsequent local minimization takes place on the physical landscape. As a consequence, it is essentially impossible to derive information about the physical stability from such an exploration – we only gain information about the energy of local minima but nothing about the barriers surrounding them. An important feature is that this landscape is independent of observational time scales or temperature; the dependence on external thermodynamic parameters such as pressure p or electric fields E is introduced *via* additional terms in the potential energy (*e.g.* by adding a $+pV$ term or an $+PE$ term). Another quantity that is independent of observational times would be the density of states for any pre-defined subregion of the configuration space; however, this does not mean that this subregion is physically relevant! From this discussion, we also see that calling an energy–volume or energy–density correlation diagram an "energy landscape" is a misnomer: such correlation plots do not provide any information about the connectivity of these minima, *i.e.* which ones are neighbors on the landscape, or the energy barriers separating them!

What is a free energy landscape now? As is known from statistical mechanics, quantities like the entropy or the free energy cannot be associated with single configurations – instead we need to sum over a set of states; the choice of weight function depends on the type of ensemble we use, and the quantities kept constant, *e.g.* the microcanonical ensemble (E, N, V) with weight one for each state, the canonical ensemble (T, N, V) with the weight being the Boltzmann factor, *etc.*. In experiments, however, our measurements are always time-averages of some quantities, and the applicability of statistical mechanics, and thus our ability to define *e.g.* a free energy, depends on whether the time averages of our observables are (approximately) equal to the ensemble averages of these observables. This property is called ergodicity, and we can only compute a free energy for an ergodic system. For systems with complex (physical) energy landscapes, it often happens that we cannot explore the whole landscape on the time scale of observation, and thus we never achieve global ergodicity. However, one often finds that the system can equilibrate within a subregion R of the full landscape, and "remain" in this region for a long time. On observational time scales between the equilibration time and the escape time of this region, the system behaves as if it were ergodic, and we speak of local ergodicity.[3–5] (Note that you can even have locally ergodic regions that do not contain a local minimum – they are separated by entropic barriers from the rest of the system![6]) In this situation, we can define a local free energy by computing the sum over states restricted to the subregion, *e.g.* via the local density of states we have measured for this subregion. One should note that the equilibration and escape times clearly depend on temperature, *e.g. via* the Arrhenius law for the escape time across energetic barriers. But keep in mind that you can associate a local free energy with a local density of states only if the region of interest is locally ergodic. For a given observational time scale, we can thus determine all the locally ergodic regions and compute their local free energy.

If we now add information about the probability flows between these regions on the observational time scale (*e.g.* measured using transition path sampling[7] or the threshold algorithm[8]), then we have all the pieces together for constructing the free energy landscape of the system as function of the observational time scale: the free energy landscape corresponds to a map from the set of locally ergodic regions to the real numbers where the neighborhood is defined by the probability flows of the system. It is important to keep in mind that this landscape strongly depends on the time scales and temperature. For more details about (free) energy landscape concepts I would refer to the literature.[9]

1 D. J. Wales and J. P. K. Doye, *J. Phys. Chem. A*, 1997, **101**, 5111–5116.
2 D. M. Deaven and K. M. Ho, *Phys. Rev. Lett.*, 1995, **75**, 288.
3 J. C. Schön, in *Proceedings of RIGI-workshop 1998*, ed. J. Schreuer, ETH Zürich, Zürich, 1998, pp. 75–93.
4 J. C. Schön and M. Jansen, *Z. Kristallogr.*, 2001, **216**, 307–325.
5 J. C. Schön and M. Jansen, *Z. Kristallogr.*, 2001, **216**, 361–383.
6 J. C. Schön, M. A. C. Wevers and M. Jansen, *J. Phys.: Condens. Matter*, 2003, **15**, 5479–5486.
7 C. Dellago, P. Bolhuis, F. S. Csajka and D. Chandler, *J. Chem. Phys.*, 1998, **108**, 1964.
8 J. C. Schön, *Ber. Bunsenges. Phys. Chem.*, 1996, **100**, 1388–1391.
9 S. Neelamraju, C. Oligschleger and J. C. Schön, *J. Chem. Phys.*, 2017, **147**, 152713.

J. Christian Schön opened discussion of the paper by Frederik Claeyssens: I am wondering about Fig. 8 in your paper. The energy *vs.* volume curve for $\xi_5 = 0.48$ shows a very unusual shape: when going from the minimum to larger volumes, the slope is much larger than when moving towards smaller volumes – exactly the opposite one finds in most bulk inorganic materials. Could you comment on this?

Frederik Claeyssens responded: You are referring to the energy *vs.* volume curve in Fig. 8 for beta-cristobalite itself. This is in very good agreement with previous theoretical work on this system and the shape of this curve is indeed unusual, as you point out. Nevertheless, it is consistent with the work of Wells and co-workers (*e.g.* ref. 1 and 2), who have developed the concept of flexibility windows in covalent frameworks. They have noted that silicates are atypical in their tendency to be maximally extended in their relaxed (minimum energy) states. The behaviour of the borate analogue (beta-cristobalite-borate-a), also shown in Fig. 8 is different and more common; the minimum energy structure lies towards the low-volume end of the flexible region. Compare also the results for other B_2O_3 polymorphs in Fig. 1 in the paper.

1 S. A. Wells and A. Sartbaeva, *Mol. Simul.*, 2015, **41**, 1409–1421.
2 S. A. Wells *et al.*, *R. Soc. Open Sci.*, 2017, **4**, 170757.

J. Christian Schön commented: Concerning your study, there has been similar work by the group of Catlow in the mid-1990s.[1] How does your work compare to their investigations?

1 A. Takada, C. R. A. Catlow and G. D. Price, *J. Phys.: Condens. Matter*, 1995, 7, 8659–8692.

Frederik Claeyssens responded: We are aware of the work of Catlow and co-workers[1,2] and refer to these in our paper. These authors reported the results of classical molecular dynamics simulations using a conventional harmonic bond-angle potential for the B–*O*–B bond angle. Thus these simulations take no account of the changes in bonding in the B–*O*–B bridges with angle that are responsible for the flexibility which plays such a crucial role in our own findings.

1 A. Takada, C. R. A. Catlow and G. D. Price, *J. Phys.: Condens. Matter*, 1995, 7, 8659–8692.
2 A. Takada, C. R. A. Catlow and G. D. Price, *J. Phys.: Condens. Matter*, 1995, 7, 8693–8722.

Sarah L. Price commented: I was struck by your comments about boron oxide B_2O_3 being a good glass former and wondered how this may link to the plethora of structures being generated as possible polymorphs. The zeolites are polymorphs of SiO_2, the definitive glass former. In the organic/pharmaceutical solid state world, the amorphous state is of great interest. Why won't a given organic molecule crystallise? Can we develop the amorphous form and be confident that it will not crystallise? There is the possibility that the formation of amorphous phases is linked with the generation of a suitably diverse set of hypothetical crystal structures with, for example, a diversity of possible hydrogen bonding motifs, that impedes the formation of a stable crystal structure.[1] The glass forming ability could be linked with having many polymorphs or no crystallisation. Could you comment on whether being able to readily form an amorphous state may make it difficult to experimentally realise your predicted new families of frameworks?

1 M. Habgood *et al.*, *Cryst. Growth Des.*, 2013, **13**, 1771–1779.

Frederik Claeyssens responded: This remains an open question. Zeolites can of course be synthesised even though SiO_2 as you comment is the definitive glass former.

Michael Ruggiero added: This is a follow up to the previous question by Prof. Price – it is important to differentiate between flexibility and stability. It is true that molecular flexibility often increases the number of accessible polymorphs, but once crystallised (or quenched into amorphous solids), care must be taken when referring to the same process, as now the energy barriers to reorganisation must be considered. A highly flexible molecule will inherently have shallow potential curves at some finite displacement, but to completely reorganise might (or might not) require overcoming large energy barriers, limiting the process. This might not be necessarily true in the solution/liquid state, which is why numerous polymorphs might form, each individually stable. This is also why thermodynamically unstable (metastable) polymorphic forms are shown on occasion to never undergo a solid-state phase transformation, since presumably the barriers are simply too large.

Matthew R. Ryder remarked: It would be interesting to have a better idea of the mechanical and thermal stability of these ultra-flexible boron oxide frameworks. Have you done any work on the single-crystal elasticity or thermal expansion?

Frederik Claeyssens answered: As yet, we haven't done so – these remain important studies for the future.

J. Christian Schön remarked: You have rather big empty rings in your structure, reminding me of some unpublished preliminary calculations we did about twenty years ago on B_2O_3. We also found such rings, sometimes even in a three-dimensional "stacking", with B sometimes three- and sometimes fourfold coordinated by O. These rings were big enough to allow other rings to penetrate, in principle, although I am not sure whether this ever happened in our searches. Have you observed such stacking of loops on top of each other?

Frederik Claeyssens replied: We have not so far examined structures in which rings penetrate other rings.

J. Christian Schön continued: Following up on the previous question, since there appear to be rather large "holes" in some of the boron oxide structures, I am wondering whether you can knit them in a loop, or generate some interpenetrating B–O network?

Frederik Claeyssens answered: Further interesting possibilities, and we suspect quite challenging ones experimentally.

Caroline Mellot-Draznieks asked: You show that boron chemistry may lead potentially to new structures directly inspired from zeolite topologies, however with much greater flexibility. When compared to zeolites, have you tried to establish the nature of "universal" descriptors that might indicate whether such structures are attainable experimentally?

Frederik Claeyssens replied: This is an interesting question which we will address further in future work, where we will study in more detail the enthalpies of formation of the new structures derived from zeolite topologies.

Caroline Mellot-Draznieks noted: there is some analogy between the boron-based porous structures you predict and MOFs possessing very large flexibility (breathing, gate opening). Such flexibility in MOFs is highly dependent on temperature, pressure, and guest induced transitions. Are your predictions currently able to incorporate such effects (responsive materials)?

Frederik Claeyssens replied: Pressure effects are readily included in the calculations and indeed we report in our paper a number of results at high pressure (and negative pressure). Guest molecules could be incorporated straightforwardly although calculations on cases where there are many possible sites for their incorporation would be computationally expensive. Temperature effects are more challenging and remain for future work. Expensive *ab initio* molecular dynamics (MD) rather than its much cheaper classical equivalent is required because it is essential to take account of the changes in electronic structure at the boron atoms which are responsible for the crucial angular flexibility of the B–*O*–B bridges. It is worth also nothing that unfortunately it has been usual in classical molecular dynamics of B_2O_3 to use a conventional three-body bond-bending term for the B–*O*–B bond angle which stiffens the angle and so removes this crucial flexibility from such simulations (*e.g.* ref. 1 and 2).

1 A. Takada, C. R. A. Catlow and G. D. Price, *J. Phys.: Condens. Matter*, 1995, 7, 8659–8692.
2 A. Takada, C. R. A. Catlow and G. D. Price, *J. Phys.: Condens. Matter*, 1995, 7, 8693–8722.

Peter Spackman opened discussion of the paper by Scott M. Woodley: About your hashing algorithm for structures: is it actually a hash (*i.e.* a lossy transform) or just a string encoding of the graph of molecular connectivity? If it's a hash, what do you plan to do for potential hash collisions (however unlikely they might be)?

Scott M. Woodley answered: As explained in the main paper, each HASHKEY is a character string for a particular connectivity graph. Yes, actual interatomic distances are lost as we want to be able to determine whether

(a) two configurations are essentially the same and differences (other than connectivity) exist because either different tolerances were used during the optimisation stage or slightly different accuracies with regards to the model employed and

(b) two configurations are the same structural motif even though they are formed of different atom types and therefore have different bond lengths and slightly different bond angles, *e.g.* see Fig. 3 in our main paper, where all configurations shown are considered to have the same structural motif.

We do not consider our solution to be a hash; however, we do intend to implement other approaches to measuring structural similarity between clusters that have been developed by others (as described and cited in the main paper), and allow the user to change the cut-off used in determining the connectivity.

Peter Spackman asked: These sorts of databases would be a useful tool in other areas (*i.e.* for other materials *e.g.* crystals, molecules *etc.*). What sort of standardisation (file formats, structure of submission *etc.*) do you use for the deposition of data?

Scott M. Woodley answered: I also agree that similar databases would be a useful tool in other areas, in fact EPSRC have funded a second project based on producing something similar for surfaces, and further funding is sought for extending the current database so that the environment of the nanoclusters can also be included.

Currently, structural information is uploaded as an xyz file, the atomic coordinates of which are immediately transformed to the reference frame where the cluster's centre of mass is at the origin and the axes of the principal moments of inertia align with the Cartesian axes. The xyz file may contain many configurations, concatenated one after the other, and allows additional information about the cluster (total energy for example) and individual atoms (charge and/or spin, for example). We do not automatically extract this information from the standard output files of materials software, as we have developed our own global optimisation software, KLMC, that automates the process of running third party codes and creating suitably named xyz files.

German Sastre added: I agree with what you say: this database could help to unify chemical knowledge and could help to discover things by analogy. How much human work does it take to upload one entry? Is it to be done by professionals or by the person who published it? How many entries do you have currently?

Scott M. Woodley answered: We have tried to minimise the effort required to upload datasets as ideally we would like the authors of published datasets to upload their dataset(s) into the HIVE. Five of the investigators on the WASP@N project (either listed as a co-author or in the acknowledgments of the main paper) have an extensive track record of publications that report nanocluster structures predicted using global optimisation methods, which we hope will provide the required datasets, or critical mass, for the HIVE to take-off. We have also already uploaded some other datasets after gaining permission from the main author who was not originally part of the WASP@N project. Typically, such authors have perhaps moved away from the field and their xyz files are available (as opposed to only being available as a graphic in their paper).

Anyone can temporarily upload any entry to compare against entries already within the HIVE, whereas permanent entries can only be uploaded by *trusted users* of this project (as this requires level-3 access to the HIVE) who agree to upload their published work for others to use. To apply to become a trusted user requires the applicant to first register online (obtaining automatic level-2 access) and then to send an email to me, Scott M. Woodley, requesting additional access in order to upload their dataset(s) that are already associated with a published journal manuscript. With level-3 access, the user can register the DOI of their manuscript, and then upload one or many xyz files containing one or more atomic structures. To reduce the effort of providing any additional information, there is an initial web form where default values can be provided (for example, the compound and stoichiometry of the set of clusters, and/or the definition of the energy and software used in creating the atomic configurations) for each xyz file. Upon upload, WASP will check whether the uploaded data is consistent with the default values provided and will also warn if any of the uploaded configurations have already been uploaded for the DOI. The uploaded datasets connected to the DOI are only visible once the author uploading these is happy the dataset is correct and makes the *private* DOI *public*. After making the dataset associated with a particular DOI public, the BEE software computes various properties of each configuration, refines each configuration to create another entry, and establishes links between these and other datasets in the HIVE. Importantly, any future changes to published datasets will be documented and these records will form part of the dataset. Originally we had planned that there would be a time limit on how long a DOI can remain private in order to give the owner of the DOI time to upload and check all configurations associated with the DOI (perhaps doing configurations of one cluster size at a time) so that the length of the *record of changes* is minimised; however, as this has not been implemented it requires the author to complete the uploading process by clicking on the "make public" link. During this Faraday Discussion when I checked, there were around 1000 configurations earmarked as public and between 10 and 100 times more that were marked as private. Clearly these need to be made public as soon as possible; moreover, we hope the community will support our efforts and add their datasets so that the HIVE continuously grows with time and becomes progressively more useful.

There is also the question of how we assure the quality of the datasets. Only trusted users have the access level to upload permanent datasets; manually uploaded datasets must come from a publication that has a DOI, simple checks are automatically performed by WASP, and all configurations are refined by the BEE software. Consider, for example, the case of the dataset generated for the main paper, where interatomic potential (IP) local minima (LM) were optimised using the third party code, GULP, in order to minimise the energy of each nanocluster. Before this Faraday Discussion, these were all uploaded and, as a result, each was further refined using an electronic structure approach implemented within FHI-aims. The unique refined configurations are now also available *via* the WASP interface to the HIVE and so one can easily see whether or not the IP LM are reasonable models (assuming the chosen electronic structure approach provides a good representation of these clusters). We leave this analysis as an exercise for anyone who would like to try the WASP interface; after selecting one of our IP nanocluster LM, at the bottom of its webpage there should be two lists, one showing the refined PBEsol configuration the BEE software has generated *via* the call to FHI-aims (which will resemble the IP LM if there is a good match between our IP model and the PBEsol one), and, if the HASHKEY does not

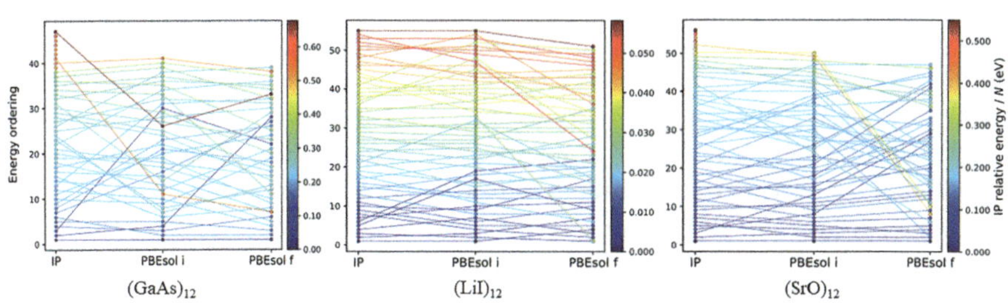

Fig. 3 The energy ranking of the local energy minima reported in the main paper are represented as circles on the left hand side of each of the three panels (one panel per compound), the colour of which indicates the energy difference between the respective local minimum and the global minimum. The circles in the middle and right hand side that are connected to the left hand side are the energies of the left hand side atomic configurations before and after refining the respective configurations using an electronic structure based software, FHI-aims.

change, the PBEsol LM should also be listed as one of the matching structures. The change in ranking when switching between IP and PBEsol can also give an insight into the quality of our interatomic potential parameters in modelling these nanoclusters.

How the rankings of the $(GaAs)_{12}$, $(LiI)_{12}$ and $(SrO)_{12}$ nanoclusters change upon switching between IP and PBEsol is shown graphically in Fig. 3. The first thing to note is that the electronic structure for the original IP atomic configuration for a number of the LM for $(GaAs)_{12}$ and $(SrO)_{12}$ did not readily converge so there is a decrease in the number of LM from IP (far left hand side) to PBEsol energy (middle), and that a number of the IP LM were not stable on the PBEsol landscape, *i.e.* there are fewer PBEsol LM (far right had side) than IP LM. In both cases, the additional IP LM were typically those of the highest energy, greater than 0.5 eV/atom, and not of immediate interest. More importantly, there is good agreement in that the IP global minimum is the lowest PBEsol LM energy found for two of these compounds and is ranked fourth for the third, whereas the new lowest PBEsol configuration was ranked third (of 55 LM) on the IP landscape.

Matthew S. Dyer asked: The unambiguous allocation of edges between nodes on the graphs representing the clusters is clearly important. How is this done (*i.e.* which interatomic distances are chosen to be linked with an edge, and which are not)?

Scott M. Woodley replied: Currently we use a predefined combination of ionic and covalent radii of the elements that the nanocluster is composed of in order to determine the cut-off used to distinguish between atoms that are or are not connected. Typically the cut-off results in edges between nearest neighbours, *i.e.* edges are made between each atom and the atoms within its first coordination shell. In a future version of the WASP interface, we will investigate the possibility of allowing the user to rescale this cut-off and generate a new, perhaps temporary, HASHKEY for the set of configurations already found in a previous search of the HIVE and earmarked for further investigation by the user.

J. Christian Schön enquired: Once you are given the HASHKEY – can you reconstruct the three-dimensional structure from this key?

Scott M. Woodley replied: The HASHKEY was introduced into the WASP@N project as a means to enable a rapid search of the HIVE for matching atomic structures based on connectivity arguments; this can also include the constraint that the nanoclusters must be composed of the same constituent atoms (compound). It was never our intention to reconstruct three-dimensional atomic structures from a HASHKEY, but we have created an alternative SCOTTKEY that is a character string that is formed of element symbols, coordination numbers and the frequency thereof, and thus it is easier to interpret what kind of configuration it represents. We have not attempted to reconstruct three-dimensional atomic structures from a HASHKEY or a SCOTTKEY and currently have no plans to attempt this as, even if it were possible, it would not be a straightforward process and, even for a nanocluster composed of one element, the exact interatomic distances cannot be retrieved without, for example, local optimisation of the original energy function.

J. Christian Schön asked: Concerning "duplicates": if I had performed a study with *e.g.* MgO clusters, and now I would like to find out whether some of the structures have already appeared in the (presumably related) SrO clusters that are in your database, could I find out about this using the tools provided by your database?

Scott M. Woodley answered: Yes, you can use the WASP interface to find whether a structural motif of one compound exists for another compound. If you have published your study on MgO nanoclusters, then we would encourage you to upload these into the HIVE so that the visibility of your work is improved – note that all entries, when displayed, have a hyperlink *via* the DOI of your publication (paper). Searching the HIVE using, for example, the DOI of your paper, you will obtain a list of your MgO nanoclusters (which will look similar to that shown in Fig. 4 of the main paper). Upon selecting any one of your nanoclusters, by clicking on its thumbnail image, will generate a new webpage containing a rotatable ball and stick image of your nanocluster together with details of various properties and, at the bottom of the page, a list of all structures in the HIVE that have the same HASHKEY as your nanocluster. Here, in this list, you would find the equivalent SrO nanocluster, if it is already in the HIVE. Moreover, we have also created the SCOTTKEY for each nanocluster entered into the HIVE, which allows an efficient search based on finding nanoclusters that have a certain structural feature as part of their atomic configuration, *i.e.* one can search for nanoclusters with only three coordinated atoms, or those with at least one six coordinated magnesium or calcium atom, for example.

J. Christian Schön asked: Is the connection to your database possible just *via* the internet, or does one need to get an account on your machine?

Scott M. Woodley answered: WASP@N is a *community* project, sponsored initially by EPSRC for five years, involving partners who either: (a) develop and/or apply global optimisation methods for predicting lowest energy atomic structures of nanoclusters; (b) develop and/or apply materials modelling software; or (c) develop databases. From its conception we agreed that the best way to help our communities, and the general public who may also be interested in what nanoclusters are, is to make the database, or HIVE, and the associated tools developed as part of this project, accessible *via* the internet (the WASP interface). The WASP interface allows four levels of access:

(1) with no registration (a) example nanoclusters can be displayed and (b) an atomic structure can be temporarily uploaded in order to see whether or not it already exists in the HIVE example;

(2) upon registering your name, institution and email address (which is automatically checked) online and agreeing to cite the relevant DOI(s) of any nanoclusters used in your future work and acknowledge use of our website, the HIVE and WASP toolkit is immediately made available;

(3) the registered account can also upload new data into the HIVE;

and (4) has admin rights to determine which accounts have either access level 2 or 3, as well as the ability to prioritise which datasets are processed (see the description about links between data entries in main paper).

Even though I am the PI of this project, I still only use the WASP interface when accessing the HIVE; only the actual developers of the WASP, BEE and HIVE software and the hardware support officers need accounts on the machine hosting the software and database; hence, no you do not need an account on our machine.

J. Christian Schön opened a general discussion of the papers from the meeting: Inspired by the talk of Jonas Nyman, with its history of crystal structure prediction for molecular crystals, I would like to show two slides about the history of energy landscape exploration in chemical systems.

Looking at the history of energy landscapes of chemical systems, we can consider it from the methods point of view and the system type perspective. Let us start with the methods (Fig. 4). There are four different types of tasks for which algorithms and general approaches have to be developed: a) the global optimization, *i.e.* the search for local minima; b) the exploration of the barrier structure, in order to determine generalized (energetic + entropic + kinetic + mixed) barriers and the equilibration, escape and transition rates associated with the barriers; c) the determination of the (local) densities of states needed for the computation of (local) free energies; and d) the low-dimensional representation of energy landscapes to help us visualize the properties of the landscape and the time evolution of the chemical system.

Concerning global optimizations, mathematicians, physicists and computer scientists had already before the 1980s developed Monte Carlo[1] and molecular dynamics[2] simulation algorithms that can be combined with stochastic and deterministic quenches (*i.e.* local minimizations), multiple (random) starting point quenches, genetic algorithms[3] (for discrete configuration spaces that could be represented as a binary string), or systematic exploration methods such as

History of energy landscapes in chemistry: Methods

	before 80's	80's	90's	after 2000
Global optimizations	Gen. Alg. (discr), Many-quench, "MC/MD",EA , Branch+bound	SA, Taboo, Jump algs., Demon alg., NN	GA (min,="EA"), BH, Deluge, MQ, Therm.Cyc., SA-variants	RRT, Ant, PSO, NN (Machine learning)
Barrier explorations	Double-wells	Saddle searches, "NEB"	Lid/Threshold, Transition path sampling, NEB activated MD	Metadynamics
Density of states/minima measurements	Umbrella sampling, LDOS	Thermodyn. Int. Comp. Alchemy, "WHAM"	WHAM, parallel tempering, Lid/ Threshold	WHAM-variants, ParQ
Landscape representations	generically wild, double wells, phase diagrams	Lumped tree graphs (basins, energy barriers), Boltzmannized graphs	Equil. Tree, Prob. flow, (L)tree, (L)DOS, (L)DOM, order param. plots, entr. barr., PCA	Char. Regions, weighted tree graphs, phase space plots, metastable phase diagrams

Fig. 4 History of energy landscapes in chemistry: Methods being employed.

branch-and-bound methods.[4] In the 1980s, *e.g.*, simulated annealing,[5,6] taboo searches,[7] jump/bounce algorithms[8] (equivalent to repeated "quench/simulated annealing followed by a large jump or short high-temperature Monte Carlo simulation" cycles), the Demon algorithm[9] (guided multi-walker stochastic quench procedure), and neural network algorithms[10] (for the parameter landscape of the network) were developed. In the 1990s, various simulated annealing variants were added (for a review see *e.g.* ref. 11) such as jumpmove simulated annealing or basin hopping,[12] shadowing simulated annealing (simulated annealing on minima instead of states), sequential multi-quench[13] and thermal cycling[14] (both similar to the bounce algorithm), parallel tempering[15] or the grand deluge algorithm[16] (where a lid is imposed on top of a system that pushes the random walker down into deep-lying minima). After 2000, biology inspired algorithms such as ant-search (originally invented in the 1990s[17]) or particle-swarm optimization (also already invented in the 1990s[18]) became popular in chemistry, a renaissance of neural networks under the name of machine learning took place, and the rapidly growing random tree algorithm (RRT) in many variants[19,20] appeared on the scene.

Barrier explorations have not seen so many methods, but still quite a few: before 1980, a major focus was on double-well systems and analytical methods supported by simulations. But in the 1980s people started working on studying the landscape by generalizing the double-well methods to multi-minima systems, and various saddle-search procedures such as slowest-slides[21] or eigenvector-following[22] were employed. There followed a sharp increase in the number of algorithms in the 1990s, starting with the Lid algorithm for exhaustive exploration of pockets of discrete landscapes[23] and the analogous threshold algorithm for continuous ones,[24,25] which allowed the determination of saddle/transition regions and probability flows (as often happens in this field, these algorithms were reinvented a couple of times). Molecular dynamics based methods became very successful such as transition path sampling,[26,27] the nudged elastic band,[28,29] or activated molecular dynamics,[30] in order to efficiently cross energy barriers and compute rate constants. In the new millennium, metadynamics[31] was introduced combining molecular dynamics with the taboo and Lid algorithm concepts, as an evolution from the so-called local elevation approach from the 1990s.[32] One should note that the computational effort involved in exploring the barrier structure is typically an order of magnitude larger than the "straightforward" global optimization.

Computation of the density of states and (local) free energies usually requires an even larger computational effort. Before the 1980s, umbrella sampling had been introduced,[33,34] and, of course, the computation of the local density of states around a local minimum in the (quasi-) harmonic approximation was well-established (using empirical/fitted-to-experiment potentials). In the 1980s, direct sampling methods such as the lumped-transition matrix approach[35] were introduced, but also thermodynamic integration/perturbation, *e.g.* ref. 36 (based on early work by Zwanzig[37] and Kirkwood[38]), and its cousin, "computational alchemy",[39] were developed from the umbrella sampling methods, in order to compute the difference in free energy between a known system and the target system. This was supplemented by the first weighted histogram analysis methods,[40] which came into their own in many variants in the 1990s.[41,42] In that time, parallel tempering and subsequently multi-canonical simulations[43] were also introduced that allowed combinations of global optimization tasks with sampling of the density of states. Since 2000, these WHAM-based methods have continued to be developed,[44,45] but also other approaches such as the ParQ-transition-matrix algorithm[46] have been introduced.

Finally, the representation of the energy landscape has evolved during this time. Before the 1980s, we are usually dealing with qualitative sketches of wild-looking landscapes, combining many double-wells, or trying to depict meta-stable phases in phase diagram pictures. In the 1980s, work on discrete combinatorial optimization problems and spin glasses lead to the introduction of (lumped) tree graphs,[47] where all the states that were connected (*via* the move-class) by paths below a given energy lid, to a certain local minimum (always the lowest one in the region), were replaced by a single node at the energy of the lowest minimum in the region. The energy lid was raised step by step, and whenever another previously unconnected region could be reached below the new energy lid, another node was created at that energy level, connecting the two previous nodes. As a consequence, a tree graph was generated, where the leaves were the local minima, and the regular nodes of the graph connected to the leaves (and the other nodes) depicted the energy levels where two lower energy nodes or leaves became connected. (This very powerful description could also include temperature *via* the so-called Boltzmannization procedure,[48] where the transition matrix between the nodes was weighted by Boltzmann-acceptance factors analogous to those in the Monte Carlo algorithm.) In particular for glassy systems, the so-called inherent structures were analyzed[49] (the inherent structures are the local minima that can be reached *via* local, usually gradient-based, minimizations along a MD or MC trajectory). In the 1990s these first tree graphs were expanded to multi-lump trees,[50] which incorporated the local densities of states associated with each node and energy band between two energy levels. In addition, probability flows were being depicted between the nodes and leaves of the tree[51].

Supplementing such landscape visualizations, local densities of states and minima were derived, entropic barriers were depicted as so-called return-probabilities[52] (*i.e.* the probability that a random walker does return to its starting basin), and the so-called principal component analysis (originally developed in the 1960s[53]) was used to classify the minima and to develop a coordinate system on which the set of local minima that had been found was most disperse such that *e.g.* order parameters might be derived. Another common way to depict a "cut" through the landscape was to define an order parameter (or a reaction coordinate for the description of chemical processes and transformations), and then to compute *e.g.* an average energy or a "free energy" for a given value of the order parameter. However, in the latter case, one must wonder to what extent it is justified to assume that on the observational time scales on which one could claim that the system possesses a specific value of the order parameter, it is permitted to treat the subregion of the landscape with a given order parameter value as locally ergodic. (If not, we should not be allowed to define a local free energy!) Tree graphs continued to exert their fascination after 2000, with a variety of modifications adding more features to the tree graph and especially the connections between the nodes.[54]A new concept was the idea of characteristic regions,[55] which puts all the states in a given energy range together into a lumped node, if they "see" the same part of the landscape below them, *i.e.* if we perform *e.g.* twenty stochastic quenches from such a point then, within statistical error, they will always reach *e.g.* minimum one with 50%, minimum 2 with 30% and minimum 3 with 20% probability (we would call such a set of states a multi-minima transition region; one should note that basin regions, *i.e.* where nearly all quenches reach only one local minimum can extend to quite high energies compared to the energy of the first saddle point that can be reached from the minimum). In addition, people have started to plot and study phase space (position+momentum space) trajectories and thus construct connected (locally ergodic) regions of phase space.[56]

Turning to the kind of chemical systems with complex landscapes that have been studied in these years, Fig. 5 we find that before the 1980s, mostly (bio) molecules (key word: Levinthal's paradox in protein folding[57]) and glasses were investigated from a landscape point of view, with the landscape paradigm having been introduced by Goldstein[58] in the late 1960s to gain an understanding of what glassy systems might be like and what their dynamics might be. In the field of crystals, crystallographers (and solid state chemists) introduced a variety of rules (usually based on some intuitive understanding of the energy involved, such as the close-packing rule or the radius–ratio-rule) that allowed them to judge the quality of observed or suggested crystal structures, and generate new candidate structures in a more or less systematic way. Of course, organic molecules had already been "designed" on paper by a complex system of rules for quite some time even before 1980,[59] *e.g.* resulting in the systematic enumeration of possible isomers *etc.*,[60] and estimates of *cis–trans*-barriers (analogous to double-well systems). In the 1980s, protein models (first attempts at "from sequence to structure" predictions) and polymer glasses became a mainstay of landscape studies (for an overview up to *ca.* 2000, see *e.g.* ref. 61), as did structural glasses, spin glasses or solid solutions, especially in intermetallic systems. But in

History of energy landscapes in chemistry: systems

	before 80's	80's	90's	after 2000
Molecules / Biomolecules	cis/trans barriers, exh. enum. isomers, Levinthal paradox: folding	Protein models, polymer glasses, "from sequence to structure"	Small molecule landscapes, struct. pred. for prots./polymers, folding (funnel)	medium/large molecules, molecules on surfaces
Clusters		atomic/nuclear	small/medium struct.prediction + barriers + DOS	large, prob. flows
Glasses / Amorphous compounds	Landscape paradigm, "Coulomb" glass	Struct. glasses, polymers, spin glasses, struct. solution	binary LJ glasses, solid solutions, glass transition	Dynamics
Crystals: elements, salts, intermetallics (alloys), molecules	optimal surfaces shapes (Wulff), defects, vibr. DOS, implicit by chemical rules	struct. compar., surface reconst., alloy models	struct. solution /determ./predic t. optim. lattice occup., verific. alloy PD	free energy landscapes, prediction of (metastable) PD, hierarch. syst.

Fig. 5 History of energy landscapes in chemistry: Types of chemical systems being studied.

addition, atomic (and also nuclear, in physics) clusters became paradigmatic example systems for the development of global optimization algorithms for structure prediction (for an overview up to *ca.* 2000, see *e.g.* ref. 62). Alloy models and simple comparisons of typical structure candidates (*e.g.* comparing sodium metal in the bcc, fcc and sc structure), were becoming of interest in crystal chemistry and solid state physics, as did studies on the energetics of surface reconstruction.

Energy landscape analysis for chemical systems really came into its own in the 1990s, when structure prediction and energy barrier analysis were applied in all fields of chemistry: small molecule landscapes, structure prediction for proteins and polymers, funnel models for protein folding (such models are very intuitive but also quite tricky because of issues of dimensionality – it is not fully clear whether the landscapes of such molecules really show a funnel structure), prediction of small and medium sized clusters including their energy barriers and local densities of states, computational models of the glass transition, and, of course, crystal structure solution (from raw data), determination (prediction but with direct experimental input restricting the allowed region of the landscape), and full unbiased prediction (only based on the energy landscape). By then, this also included optimal defect occupation structures, verification of alloy phase diagrams, and estimates of the stability of various modifications. In the time since 2000, energy landscapes have become ubiquitous all over chemistry, with even (*pace*, my experimental colleagues!) synthetic chemists getting very excited about the richness of chemical modifications present on the energy landscape. Now, we can study the landscape of even large molecules, or small to medium size molecules on surfaces (for an overview, see *e.g.* ref. 63), explore large clusters and the transition from nanocrystals to bulk materials, follow the dynamics of glassy and amorphous systems, and derive true free energy landscapes of crystalline systems, predict stable and metastable phase diagrams, and predict crystal structures of systems that exhibit complex hierarchical structures.

Of course, this two-slide overview must be very far from complete, and similarly, the references given here can only serve as starting points; *e.g.* we have not touched upon the prediction of phase diagrams (for an overview see *e.g.* ref. 64) or molecular crystals (a good starting point for a dive into the history of this field might be the presentation by J. Nyman (DOI: 10.1039/c8fd00033f). Nevertheless, seeing how many researchers from industry are attending this conference demonstrates the importance of the issue of complex energy landscapes of chemical systems and the prediction of their feasible structures in the "real" world outside academia, and I am very confident that we have by now reached a level of understanding of energy landscapes of chemical systems that will allow us to start assisting our experimental colleagues in their difficult task to synthesize the modifications we have been suggesting to them.

1 N. Metropolis, A. W. Rosenbluth, M. N. Rosenbluth and A. H. Teller, *J. Chem. Phys.*, 1953, **21**, 1087.
2 B. J. Alder and T. E. Wainwright, *J. Chem. Phys.*, 1957, 27, 1208.
3 J. H. Holland, *Adaptation in Natural and Artificial Systems*, Michigan Press, Ann Arbor, 1975.
4 E. L. Lawler and D. E. Wood, *Oper. Res.*, 1966, **14**, 555–756.
5 S. Kirkpatrick, C. D. Gelatt Jr. and M. P. Vecchi, *Science*, 1983, **220**, 671–680.
6 V. Czerny, *J. Optim. Theor. Appl.*, 1985, **45**, 41–51.
7 F. Glover, *Interfaces*, 1990, **20**, 74–94.
8 H. Müller-Krumbhaar, *Europhys. Lett.*, 1988, **7**, 479–484.

9 T. Zimmermann and P. Salamon, *Int. J. Comput. Math.*, 1992, **42**, 21–31.
10 J. Hertz, A. Krogh and R. G. Palmer, *Introduction to the theory of neural computations*, Addison-Wesley, Reading, 1991.
11 J. C. Schön and M. Jansen, in *Modern Methods of Crystal Structure Prediction*, ed. A. R. Oganov, Wiley-VCh, Weinheim, 2011, pp. 67–106.
12 D. J. Wales and J. P. K. Doye, *J. Phys. Chem. A*, 1997, **101**, 5111–5116.
13 J. C. Schön and M. Jansen, *Z. Kristallogr.*, 2001, **216**, 307–325.
14 A. Möbius, A. Neklioudov, A. Diaz-Sanchez, K. H. Hoffmann, A. Fachat and M. Schreiber, *Phys. Rev. Lett.*, 1997, **79**, 4297.
15 B. A. Berg and T. Neuhaus, *Phys. Rev. Lett.*, 1992, **68**, 9–12.
16 G. Dueck, *J. Comput. Phys.*, 1993, **104**, 86–92.
17 A. Colorni, M. Dorigo and V. Maniezzo, in *Proceedings of ECAL91 - European Conference on Artificial Life*, Elsevier, Amsterdam, 1991, pp. 124–142.
18 R. Eberhart and J. Kennedy, in *Proceedings of the Sixth International Symposium on Micro machine and Human Science*, IEEE, 1995, pp. 39–43.
19 D. Parsons and J. Canny, in *Proc. Int. Conf. Intel. Sys. Mol. Biol.*, 1994, p. 322.
20 I. Al-Bluwi, T. Simeon and J. Cortes, *Comput. Sci. Rev.*, 2012, **6**, 125–143.
21 A. Banerjee, N. Adams, J. Simmons and R. Shepard, *J. Phys. Chem.*, 1985, **89**, 52–57.
22 D. J. Wales, *J. Chem. Soc., Faraday Trans.*, 1993, **89**, 1305–1313.
23 P. Sibani, J. C. Schön, P. Salamon and J.-O. Andersson, *Europhys. Lett.*, 1993, **22**, 479–485.
24 J. C. Schön, *Ber. Bunsenges. Phys. Chem.*, 1996, **100**, 1388–1391.
25 J. C. Schön, H. Putz and M. Jansen, *J. Phys.: Condens. Matter*, 1996, **8**, 143–156.
26 C. Dellago, P. Bolhuis, F. S. Csaijka and D. Chandler, *J. Chem. Phys.*, 1998, **108**, 1964.
27 P. G. Bolhuis, C. Dellago and D. Chandler, *Faraday Discuss.*, 1998, **110**, 421–436.
28 G. Mills and H. Jonsson, *Phys. Rev. Lett.*, 1994, **72**, 1124.
29 H. Tanaka, *J. Chem. Phys.*, 2000, **113**, 11202.
30 G. T. Barkema and N. Mousseau, *Phys. Rev. Lett.*, 1996, **77**, 4358.
31 A. Laio and M. Parrinello, *Proc. Nat. Acad. Sci. U. S. A.*, 2002, **99**, 12562–12566.
32 T. Huber, A. Torda and W. F. von Gunsteren, *J. Comput.-Aided Mol. Des.*, 1994, **8**, 695–708.
33 G. M. Torrie and J. P. Valleau, *J. Comput. Phys.*, 1977, **23**, 187–199.
34 C. H. Bennett, *J. Comput. Phys.*, 1976, **22**, 245–268.
35 B. Andresen, K. H. Hoffmann, K. Mosegaard, J. Nulton, J. M. Pedersen and P. Salamon, *J. Phys.*, 1988, **49**, 1485–1492.
36 W. L. Jorgensen, *J. Phys. Chem.*, 1983, **87**, 5304–5314.
37 R. Zwanzig, *J. Chem. Phys.*, 1954, **33**, 1420.
38 J. G. Kirkwood, *J. Chem. Phys.*, 1935, **3**, 300.
39 M. Watanabe and W. P. Reinhardt, *Phys. Rev. Lett.*, 1990, **65**, 3301.
40 R. H. Swendsen and A. M. Ferrenberg, *Phys. Rev. Lett.*, 1989, **61**, 2635.
41 S. Kumar, D. Bouzida, R. H. Swendsen, P. A. Kollman and J. M. Rosenberg, *J. Comput. Chem.*, 1992, **13**, 1011–1021.
42 B. Roux, *Comput. Phys. Commun.*, 1995, **91**, 275–282.
43 F. Wang and D. P. Landau, *Phys. Rev. Lett.*, 2001, **86**, 2050.
44 J. Kim, J. E. Straub and T. Keyes, *Phys. Rev. Lett.*, 2006, **97**, 050601.
45 P. Liu and G. A. Voth, *J. Chem. Phys.*, 2007, **126**, 045106.
46 F. Heilmann and K. H. Hoffmann, *Europhys. Lett.*, 2005, **70**, 155–161.
47 K. H. Hoffmann and P. Sibani, *Phys. Rev. A*, 1988, **38**, 4261–4270.
48 P. Sibani, J. C. Schön, P. Salamon and J.-O. Andersson, *Europhys. Lett.*, 1993, **22**, 479–485.
49 F. H. Stillinger and T. A. Weber, *Phys. Rev. A*, 1983, **28**, 2408–2416.
50 J. C. Schön, H. Putz and M. Jansen, *J. Phys.: Condens. Matter*, 1996, **8**, 143–156.
51 M. A. C. Wevers, J. C. Schön and M. Jansen, *J. Phys.: Condens. Matter*, 1999, **11**, 6487–6499.
52 M. A. C. Wevers, J. C. Schön and M. Jansen, *J. Phys.: Condens. Matter*, 1999, **11**, 6487–6499.
53 J. C. Gower, *Biometrika*, 1966, **53**, 325–338.
54 T. Komatsuzaki, K. Hoshino, Y. Matsunaga, G. J. Rylance, R. L. Johnston and D. J. Wales, *J. Chem. Phys.*, 2005, **122**, 084714.
55 M. A. C. Wevers, J. C. Schön and M. Jansen, *J. Phys. A: Math. Gen.*, 2001, **34**, 4041–4052.
56 C. B. Li, A. Shoujiguchi, M Toda and T. Komatsuzaki, *Phys. Rev. Lett.*, 2006, **97**, 028302.
57 C. Levinthal, *J. Chim. Phys. Phys.-Chim. Biol.*, 1968, **65**, 44–45.
58 M. Goldstein, *J. Chem. Phys.*, 1969, **51**, 3728.
59 E. J. Corey, *Pure Appl. Chem.*, 1967, **14**, 19–37.
60 G. Polya, *Acta Mathematica*, 1937, **68**, 145–206.
61 J. C. Schön and M. Jansen, *Z. Kristallogr.*, 2001, **216**, 361–384.
62 D. J. Wales, J. P. K. Doye, M. A. Miller, P. N. Mortenson and T. R. Walsh, *Adv. Chem. Phys.*, 2000, **115**, 1–111.
63 J. C. Schön, C. Oligschleger and J. Cortes, *Z. Naturf. B*, 2016, **71**, 351–374.
64 J. C. Schön and M. Jansen, *Int. J. Mater. Res.*, 2009, **100**, 135–152.

Virginia Burger asked: What do you think is coming next, now that we are approaching 2020?

J. Christian Schön answered: 2020 is a bit too close to go for bold predictions! I would expect future directions in the field of molecular crystals to be strongly influenced by practical (industrial) needs: which information is needed to safeguard intellectual property regarding a system, and to make sure that no unknown polymorph interferes with the applicability of the compound in *e.g.* medicinal applications. Thus, I think estimates of lifetimes/transformation times under various conditions (temperature, humidity, competing medicines that are taken in addition to the one being discussed, *etc.*) will become an important focus, especially since we are talking about shelf-lives of years. (After all, many people keep old medicines in their cabinet and take them out again when a health problem re-appears!) Such estimates will supplement the stability checks performed in the laboratory, and provide information about the mechanism involved (which might be relevant for many other systems, too).

This is not only true for medicine –I recall at a conference five years ago that a company producing detergents and other cleaning agents asked me whether I could help them provide shelf-life estimates. (I did not have the time to take on this problem, but it seemed to be a big issue!) So I would think that structure prediction will branch out to many more chemical systems that have real-life applications.

A much more long-term issue is the one I mentioned earlier: the development of a large tool-box of synthesis modeling techniques to help synthetic chemists to perform target-oriented syntheses in an optimal fashion. For an outline of such a multi-year research program, see ref. 1.

1 J. C. Schön, *Adv. Chem. Phys.*, 2015, **157**, 125–134.

Sharmarke Mohamed opened a discussion of the closing remarks by Artem R. Oganov: It is very promising to see the range of structures and properties that can be predicted using the USPEX code. I was fascinated to see in your slides that multi-parameter optimisations using USPEX can lead to the discovery of novel hard materials and it is also interesting to see that you are letting the code sample all possible binary combinations in the periodic table when searching the range of materials that have a desired property. The composition (*i.e.* assumption of a binary molecular formula of the type MX) appears to be a user-defined parameter in the search algorithm. Can you comment on the feasibility and robustness of USPEX in facilitating the discovery of novel materials with desired properties when the composition of the material (*i.e.* molecular formula) is itself a variable in the parameter space? Clearly the search problem will be an order of magnitude more complex but as you have mentioned in your presentation, assumptions about chemical bonding are not always intuitive or correct so we need to also take into account possible variations in the molecular composition.

Artem R. Oganov responded: Indeed, when you include composition as a search variable, the search problem becomes much harder, but USPEX can deal with it quite efficiently – you can see numerous works that we and users of our code have done using this capability. Since 2010 USPEX has the capability to robustly predict stable compounds, and all you need is to specify the chemical

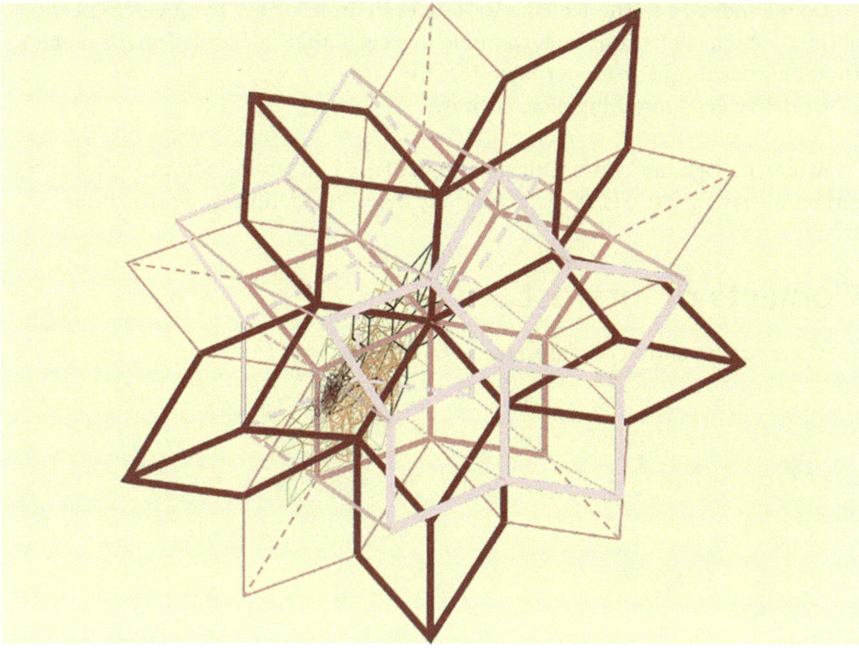

Fig. 6 A stellation of Kepler's rhombic triacontahedron. A rhombic hexecontahedral stellation of the Great Stella class, this non-convex polyhedron has 62 vertices and an internal vertex, with icosahedral symmetry. It comprises 20 golden rhombohedra (all prolate). The stellation is shown enlarged, with self-similarity, by Penrose in 3D. Given that with Shechtman icosahedral symmetry is now regarded as crystalline and that earlier an Al alloy was ascribed the stellar structure, it is reasonable to seek near-icosahedral aluminal species. The fractal nature of the 3D Penrose enlargement and its peculiar characterisation as asymmetric Cantor dust suggests the possibility of a quasi-crystal which may be nowhere dense.

elements. The main limitation is that this works well only for (pseudo)binary and (pseudo)ternary systems (see, *e.g.* an application to a ternary Mg–Si–O system:[1] when the number of compositional search variables becomes too large (>3–4), the number of possible compounds becomes too large).

1 H. Niu, A. R. Oganov, X. Chen, D. Li, *Sci. Rep.*, 2015, **5**, 18347.

J. Christian Schön asked: In your nice presentation, you highlighted "search" and "energy ranking" as the two major features of crystal structure prediction. Let me add "stability" as the third big challenge that we should take a look at in crystal structure prediction research.

Artem R. Oganov responded: Stability is part of what I mentioned: "ranking" is the sorting of all sampled structures by energy, and "search" is finding the structure with lowest possible energy (*i.e.* the most stable structure). One can also discuss mechanical and dynamical stability, as well as chemical stability.

Alan Hare communicated: So, as regards a quasi-crystal, what does happen if we put exactly 63 Al atoms into a crystal predictor together, either with $Cu_{24} + Fe_{13}$, or with $O_{2+10\alpha+50\gamma} + (OH)_{10+20\beta+200\gamma} + (OH_2)_{40\varepsilon+50\eta}$, say?

Do we indeed see the icosahedritic icosahedron; or (as I've been expecting), a Great Stella stellation: the rhombic hexecontahedral stellation of Kepler's rhombic triacontahedron, perhaps (Fig. 6)?

Or do we get something else, entirely?

Artem R. Oganov communicated in reply: 63 Al atoms should give an fcc structure. Trying 24Cu+13Fe is a good idea, and it's definitely worthwhile.

Conflicts of interest

There are no conflicts to declare.

Faraday Discussions

PAPER

Crystal structure prediction: reflections on present status and challenges

Artem R. Oganov [abc]

Received 30th August 2018, Accepted 30th August 2018
DOI: 10.1039/c8fd90033g

Long thought to be impossible, crystal structure prediction (CSP) is a thriving field today, with many important discoveries in fields as diverse as computational materials discovery, drug design, high-pressure chemistry and mineralogy of the Earth's and planetary interiors. However, major challenges remain, warranting more research. In these Concluding Remarks, I try to summarize my personal view of the enormous progress made in the field of CSP and the open questions and challenges that keep this field more exciting than ever.

Introduction

One can say that the first crystal structure predicted by a human was the structure of ice, first drawn by Johannes Kepler in his treatise "On the six-cornered snowflake" in 1611 (Fig. 1) – this was an intuitive model, which was meant to explain the hexagonal symmetry of snowflakes. Kepler's model turned out not to be the correct structure of ice, and corresponds to what we call today the hexagonal close packing – such a structure is adopted by Be, Mg and Cd under normal conditions, and Fe at high pressures. The first correct structure prediction, albeit for a molecule rather than a crystal, was made in 1857 by August Kekulé: in his famous dream by the fireplace he saw a swirling snake biting its tail, and this gave him a model of the cyclic structure of benzene. Then, in 1897 William Barlow created a model of the rocksalt structure – again, this model was just a fruit of intuition, but much later turned out to be correct. I could rate this as the first successful example of crystal structure prediction (CSP), but I won't, because the whole point of CSP is to find the structure based on the laws of physics, rather than human imagination or intuition. By the way, Barlow's result played an important role: W. L. Bragg knew this model, and his determination of the crystal structure of rocksalt can be described as a confirmation of Barlow's model.

[a]Skolkovo Institute of Science and Technology, Skolkovo Innovation Center, 3 Nobel St., Moscow 143026, Russia. E-mail: a.oganov@skoltech.ru
[b]Moscow Institute of Physics and Technology, 9 Institutskiy Lane, Dolgoprudny City, Moscow Region, 141700, Russia
[c]International Center for Materials Discovery, Northwestern Polytechnical University, Xi'an, 710072, China

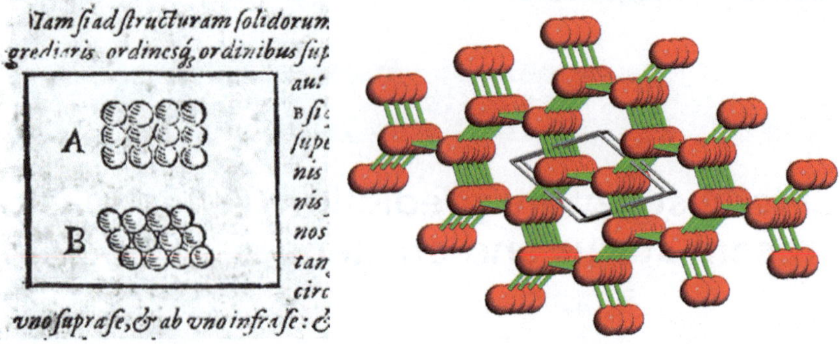

Fig. 1 Crystal structure of ice: (a) structure proposed by Kepler, (b) correct structure. In both drawings, each molecule of H_2O is represented as a sphere.

Since the pioneering works of W. L. Bragg and his father W. H. Bragg,[1,2] where the first crystal structures were experimentally determined using X-ray diffraction, an enormous number of crystal structures were solved, and this knowledge formed much of the foundation of modern solid-state physics, chemistry, biochemistry, and materials science. It all started with the simplest structures (historically, the first structures solved by Braggs were diamond, zincblende and rocksalt), and as time went on, experimental techniques improved, and nowadays it is possible to solve structures containing thousands of atoms in the unit cell. Still, challenges remain – for example, and while structure solution by single-crystal X-ray diffraction is routine, solving structures from powder data is still largely an art.

CSP, on the other hand, has only recently become possible. Compared to the contemporaneous triumphant successes of experimentalists, theorists looked hopeless for a long time. This situation can be understood: predicting a stable crystal structure (*i.e.* the arrangement of atoms with the lowest possible energy) encounters two key problems:

(1) The "Ranking problem", *i.e.* the reliable calculation of relative structural energies. This turned out to be highly non-trivial, because the energy differences between different polymorphs are often very small and correctly ranking the structures by energy has long been (and to some extent still is) a challenge.

(2) The "Search problem": the number of possible arrangements of atoms in space is astronomically large: for a unit cell with N atoms, the number of possible structures is $C \sim \exp(ad)$, where d is the number of degrees of freedom and a is some system-specific constant. If the positions of all N atoms are uncorrelated, then $d = 3N + 3$, resulting in a very high-dimensional problem; roughly, $C \sim 10^N$ (ref. 3). The problem can be greatly simplified if every generated structure is relaxed, *i.e.* brought to a local energy minimum, as this dramatically lowers the number of configurations, but the scaling of the problem is still exponential, *i.e.* it is NP-hard and very challenging.

Both problems have been largely solved, though as I will detail below, the story is far from being over. These successes are making industry more interested, especially in the pharmaceutical world: Lilly, Avant-garde, and XtalPi are examples of companies specializing in CSP for the pharmaceutical industry, and such companies as Novartis actively participate in this research field. CSP is also rapidly revealing its potential for computational materials discovery,[4] and we have sold a number of (expensive) commercial licenses for our USPEX code (http://uspex-team.org) to companies.

Historically, the inorganic and organic crystal structure prediction communities evolved in parallel. In the organic community, the ranking problem always attracted more attention and the focus has been on the development of a sufficiently accurate forcefield; since forcefield calculations are rather cheap, it was thought that even an enumerative search through all local minima was doable. This focus on the ranking problem is partly justified by the fact that organic crystals often have ≤ 1 molecule in the asymmetric unit and the vast majority crystallize in just a dozen space groups. In the inorganic community, the search problem attracted more attention: first, the number of degrees of freedom is typically very large (and equal to $3N + 3$, *i.e.* for a medium-complexity crystal with 30 atoms per cell one has 93 degrees of freedom), making the search problem very acute. Second, it is quite clear that forcefields cannot describe with sufficient accuracy all the types of chemical bonding present in inorganic substances and one must resort to *ab initio* calculations. Third, given the high computational cost of *ab initio* calculations, one is forced to reduce their number, *i.e.* solve the search problem as efficiently (and reliably) as possible.

This historical separation of the two communities is reflected by the fact that traditional blind tests organized by the Cambridge Crystallographic Data Centre (CCDC) were and still are entirely focused on organic crystals and until the 2010s only researchers from the organic CSP community participated in it. One can notice convergence of the two communities: in the 6th blind test,[5] 3 out of 25 participating teams were "inorganic"; a sad note is that they did not score much success (in fact, ours was the only "inorganic" team that scored any successes – getting the right structure in 2 out of 5 cases, which is better than most "organic" teams). It appears that when the conformation of the molecule is known or can be correctly guessed before CSP, the crystal structure can usually be predicted easily. For example, the structures of the newly synthesized polymorphs of resorcinol[6] and coumarin[7] were easily predicted by us and gave a perfect match to experimental X-ray diffraction patterns. The presence of many flexible torsion angles within the molecule, or the possibility of the formation of zwitterions make the problem much harder, both in search and in ranking. In contrast to six organic structure prediction tests, there had been only one such test for inorganic structures,[8] and I think this has to be continued and done regularly too, and "organic" teams will need to be invited (I joking say that I am vengefully looking forward to seeing how they fare on our territory). The gradual convergence of the "organic" and "inorganic" CSP groups is very encouraging.

To sum up, CSP consists of two problems: search and ranking. Ranking is usually done at zero Kelvin, with either forcefields or quantum-mechanical approaches. Forcefields come in different flavors – pair potentials *vs.* many-body potentials, rigid-ion *vs.* shell model potentials, rigid-molecule *vs.* flexible-molecule, point charge *vs.* multipole, parametric *vs.* machine learning forcefields. Quantum-mechanical methods are usually done at the level of density functional theory (DFT) – for molecular crystals, with some dispersion correction needed to correctly describe van der Waals bonding, the most advanced approaches to which seem to be the SCAN meta-GGA functional[9] with the rVV10 van der Waals functional,[10] and various exchange-correlation functionals (*e.g.*, PBE[11] or PBE0 (ref. 12)) with a many-body dispersion potential.[13] It was a big surprise for me when Tkatchenko *et al.*[13] showed that many-body van der Waals effects are essential for molecular crystals, and proposed an elegant way to account for them. Between forcefields and DFT-based calculations are semiempirical methods – these crude quantum-mechanical methods are much cheaper than DFT, but their accuracy has not convinced the CSP community so far. The new study by Sally Price's group (Iuzzolino *et al.*, DOI: 10.1039/c8fd00010g) utilizes a semiempirical tight-binding method for the relatively cheap yet reliable pre-relaxation of crystal structures made of flexible molecules, and perhaps this is a right niche for such methods.

To go beyond zero-Kelvin CSP, it is necessary to compute the free energy, the most difficult part of which is the entropy. The vibrational entropy can be computed quasiharmonically, or at various levels of accounting for anharmonicity. A full account of anharmonicity is possible with molecular dynamics and Monte Carlo methods, but the calculation of the free energy in these methods in not trivial,[14] and for CSP one needs a fast and automatic way. Even that will only take care of the vibrational part of the free energy, and the treatment of other contributions to the entropy (especially configurational entropy, and also magnetic entropy) is a challenge. While the calculation of the free energy at finite temperature (*i.e.* the ranking problem) is much heavier than at zero Kelvin, the search problem should become (much?) simpler as the temperature increases, because shallow energy minima merge into one broader free energy minimum. Indeed, while at 0 K we have an astronomically large number of local minima; near the melting point there are only on the order of one local minima (one or a few crystalline states, and a few or usually one liquid state). I expect the number C of local free energy minima to decrease exponentially with temperature T:

$$C = C_0 \exp\left(\beta \frac{T^* - T}{T^*}\right), \tag{1}$$

where C_0 is the number of local minima at zero Kelvin, β is a constant, and T^* is a characteristic temperature (higher than the melting temperature), at which only one free energy minimum exists.

The search problem, being NP-hard, is bound to have an upper limit of tractable complexity, which today stands at $d < 400$–500, *i.e.* we can deal with non-molecular crystals containing not more than ~150 atoms in the primitive cell. However, for molecular crystals the number of degrees of freedom is much lower than $3N + 3$, *i.e.* if we impose a molecular geometry and search for a minimum-energy packing of prespecified molecules, the problem becomes much simpler than if all atoms were treated independently. Partial or complete freezing of the molecular geometry is needed not only for computational convenience, but also because most organic compounds are metastable, *i.e.* the global minimum will not have the organic molecules we are interested in, but will contain a mixture of such small molecules as H_2O, CH_4, CO_2, NH_3, *etc.* Here, only constrained, as opposed to unconstrained, global optimization is meaningful – *i.e.* searching for the most favorable packing of the molecules of interest.

An interesting approach, nested sampling, was used by Livia Partay and Gabor Csanyi,[15,16] allowing one to compute the partition function Z in a realistic, though long, time:

$$Z = \sum_i e^{-E_i/k_B T}, \tag{2}$$

where E_i are the energies of all the quantum states and k_B is the Boltzmann constant. Taking advantage of the well-known formula for the Helmholtz free energy

$$F = -k_B \ln Z, \tag{3}$$

it is possible to predict phase transitions (they are indicated by a peak of the heat capacity), the stable phases (including even the liquid state) and the free energy as a function of temperature at a given pressure or volume.

Today we develop better algorithms to deal with the NP-hard search problem and make more complex systems tractable – but quantum computers might be able to deal with the search problem efficiently in the future, bringing it to polynomial or even linear complexity. In what follows, I will focus more on different aspects of the search problem, though from time to time I will detour into ranking-related issues and issues beyond both these problems. This is by no means a review, but a personal reflection – hence, there are many references to the works of my laboratory, and my personal thoughts.

Probing the limits of complexity

The "exponential wall", which one encounters in the search problem, is one of the faces of the phenomenon known in mathematics as the "curse of dimensionality". A simple strategy to reduce the curse of dimensionality is to reduce, whenever possible, the dimensionality of the problem.

In the FUSE method (Collins *et al.*, DOI: 10.1039/c8fd00045j) one assembles structures not from atoms, but from atomic pairs or triples – in ionic crystals (for which FUSE is designed) one expects cations to be surrounded by anions, rather than by other cations – hence, the pairs can be cation–anion. This simple trick allows rather complex systems to be studied, and Collins *et al.* (DOI: 10.1039/c8fd00045j) have reported a successful exploration of the challenging quasiternary SrO–TiO_2–Y_2O_3 system (Fig. 2). Another method for generating random structures with alternating cations and anions was proposed by Stevanovic.[17] Such random structures can then, for example, be used as an initial population in an evolutionary search.

For ionic systems, the permissible stoichiometries must satisfy charge balance – this restriction on allowed compositions is also an effective way to reduce the dimensionality of the problem. To use it, one has to know the oxidation states of all the atoms – and Davies *et al.* (DOI: 10.1039/c8fd00032h) have proposed a data mining approach to predict the likely oxidation states of all elements when present together with other elements. Note, however, that the charge balance rule is often violated under pressure: *e.g.*, in the Na–Cl system, such compounds as Na_3Cl, Na_2Cl, $NaCl_3$ and $NaCl_7$ become stable.[18] Even at ambient conditions, in the Ca–Al–O system, a very interesting material with metallic conductivity – the electride $Ca_{12}Al_{14}O_{32}$ – exists,[19] and violates charge balance. Nature, as usual, is smarter than our rules.

There are ways to avoid the curse of dimensionality – for example, the metadynamics method for CSP[20,21] (or its other variant, evolutionary metadynamics[22,23])

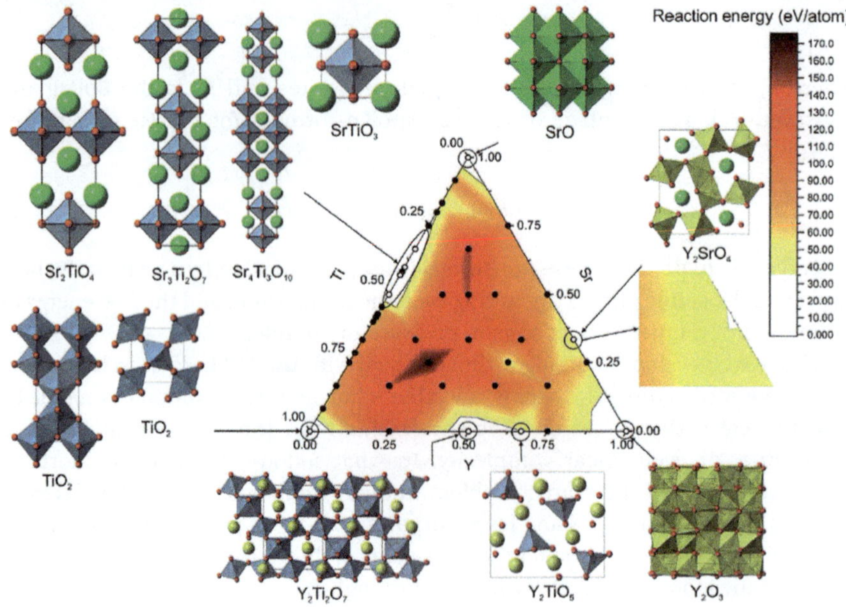

Fig. 2 Results of the FUSE method for the $SrO-TiO_2-Y_2O_3$ system (Collins *et al.*, DOI: 10.1039/c8fd00045j), showing numerous correctly predicted ground states. Reproduced from Collins *et al.*, DOI: 10.1039/c8fd00045j with permission from the Royal Society of Chemistry.

reduces the number of degrees of freedom from $3N + 3$ to a mere 6 – searching for the stable crystal structure in the space of 6 lattice vectors matrix components. In this case, the exponential scaling disappears, and large systems can be dealt with efficiently. For example, a complex structure of the high-pressure phase of $Li_{15}Si_4$ (the starting low-pressure structure has 152 atoms in the unit cell) was predicted at a very modest computational cost using evolutionary metadynamics, and confirmed by powder X-ray diffraction.[24] One of the bonuses of metadynamics is that it can be naturally used at finite temperatures. Piaggi and Parrinello have developed a new version of metadynamics, which was used for simulating *e.g.* the crystallization and solid–solid phase transitions of urea.[25] In their approach, they used a vibrational entropy-like descriptor as an order parameter. Metadynamics is unique in that it has no exponential scaling and allows large systems to be treated. However, there is a price to pay: the success of CSP depends on the initial crystal structure and on the appropriateness of the reduced-dimensionality description (*e.g.*, the 6-dimensional order parameter composed of lattice vectors matrix components).

Even when we have a good algorithm to solve the search problem, structure relaxation and energy evaluations may be too expensive at the *ab initio* level. Here, the greatest promise is given by machine learning forcefields (MLF). Trained on *ab initio* energies, stresses and forces on atoms for a number of configurations, MLF can predict energies, stresses and forces in a new, hitherto unseen, configuration. This works only as an interpolation, *i.e.* the new configurations should be sufficiently similar to the ones used for training the forcefield, and if a very different configuration is found, the errors can be very large, and such a configuration should either be rejected, or calculated at the *ab initio* level and then used for re-training the MLF. In this issue, Deringer *et al.* (DOI: 10.1039/c8fd00034d) showed results on phosphorus, obtained with a combination of random sampling and MLF. Their work shows both the current level of accuracy of such methods (*e.g.*, training a MLF accurate on average to less than 100 meV per atom is cheap) and problems involved (*e.g.*, Hittorf's phosphorus with 84 atoms per cell could not be found in their random sampling search, even after imposing constraints, and errors of the MLF for this metastable phase are unusually large, above 100 meV per atom). Such studies, openly showing the difficulties, are very important for further progress. Recently, we[26] combined a MLF with our evolutionary method USPEX to search for stable and low-energy metastable structures of boron. With a MLF showing a mean average error of 37 meV per atom, we have successfully found α-boron (12 atoms per cell), γ-boron (28 atoms per cell), several versions of tetragonal T52-boron (52 atoms per cell), and even the extremely complex and disordered β-boron (for which we found ∼50 energetically nearly degenerate ordered approximants with 105–108 atoms per cell, see Fig. 3a). In addition to these known phases, we predicted a new low-energy cubic *Im*3 phase (with 54 atoms per cell, see Fig. 3b), which is only 29 meV per atom higher in energy than α-boron and energetically degenerate with the experimentally known T52-boron. The use of a MLF has allowed a speedup of 100–1000 times.

Machine learning approaches, such as neural networks, are traditionally viewed as black boxes that can give the right numbers, but cannot give insight. However, there are ways to obtain insight, too. For example, describing the pair potential by a very flexible inverse-power series with fitted parameters and the many-body potential by a neural network, we[27] obtained an accurate representation of the energies in solid He, Xe and Al, and recovered a Lennard-Jones-like He–He pair potential, while for the Xe–Xe potential there were visible deviations from such shape (probably indicative of significant many-body effects renormalizing the effective pair potential), and for the Al–Al pair potential we obtained a density-dependent oscillating pair potential – which is exactly what one would expect for a metal. In all these pair potentials the minimum corresponded precisely to the equilibrium bond distance in the crystal (or a sum of van der Waals or metallic radii). I think it is a matter of time until more complex insight, currently accessible only to humans – *e.g.*, the derivation of Pauling's rules – becomes derivable from machine learning.

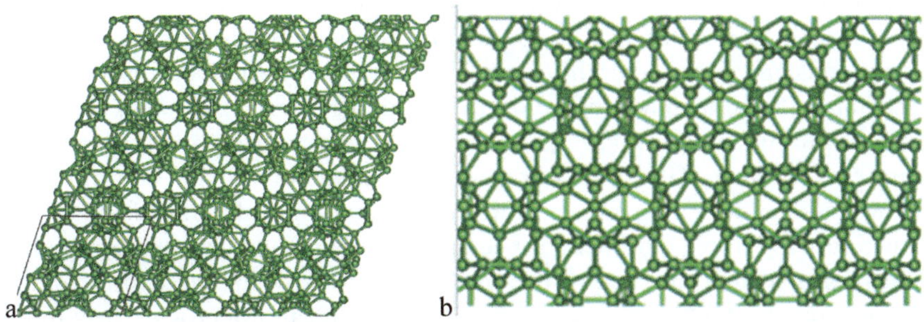

Fig. 3 (a) β-boron approximant with 106 atoms per cell, found in ref. 26. (b) *Im*3 structure predicted in ref. 26.

Though the current progress is impressive, many challenges remain. I don't think anyone can handle cases as complex as the Samson phase, β-Mg_2Al_3: its cubic structure has 1168 atoms in the conventional unit cell[28] and is, I think, too large and complex. This is just one example of a tricky complex metallic alloy, and the worst example would be quasicrystals. For quasicrystals, lacking periodicity, the energy cannot be computed by existing approaches, and one has to use periodic approximants – and, in view of their structural complexity, this should be challenging (though no one has really tried). Another limitation is magnetism: so far, CSP studies have largely avoided magnetic materials, and even when dealing with them, were confined to collinear magnetism. Can any of the existing methods predict the structure of α-Mn with non-collinear magnetism and containing 54 atoms in the unit cell (belonging to four types, one of which is non-magnetic[29])? I doubt it – though with a little bit of method development and at a high computational cost (to deal with a difficult search problem and facing technical issue of having to resolve small energy differences between different magnetic states) this should become possible. Finally, the prediction of protein structures remains an open challenge – in addition to their extreme structural complexity, we are not even sure whether the structure they adopt is thermodynamically controlled (see below).

When there is more than one solution, and searching for useful materials

Finding the stable crystal structure for a given chemical composition, mathematically formulated as global (free) energy minimization, is finally, though with certain important caveats, a solved problem. The solution of this problem is one crystal structure – the global minimum.

There are at least three important problems where the solution is not a single structure, but a set of structures or a set of materials. These problems, which can also be considered as solved, letting us do a lot more, are described below:

(I) Variable-composition structure prediction, searching for all stable compounds (and their crystal structures) formed by given elements. Let me take three simple examples. The Fe–S system has two stable iron sulfides, FeS and FeS_2. The H–O system has two well-known compounds, H_2O and H_2O_2, but the former is stable and the latter is metastable. In the Na–Cl system, only one compound is known, NaCl. Stability or instability of different compounds can be conveniently determined by the convex hull (or Maxwell) construction, an example of which is shown in Fig. 4. This construction is a convenient representation of the free energies of all possible reactions in the system. All stable compounds (*i.e.* those that have a lower free energy than any isochemical assemblage) form a convex figure on the graph, the *y*-axis of which is the free energy of formation (normalized per atom) and *x*-axis of which is the composition. The example of the Mn–B system (Fig. 4) shows that all known stable manganese borides were predicted correctly (and for MnB_4 the theoretical crystal structure was subsequently confirmed by experiment), and a new compound MnB_3 was predicted and then synthesized.[30] Our latest calculations show that this new compound, MnB_3, has very interesting mechanical properties.[31] It is interesting that in such well-studied systems as Mn–B, even at ambient conditions new stable compounds keep being discovered. Three-, four- and more-component

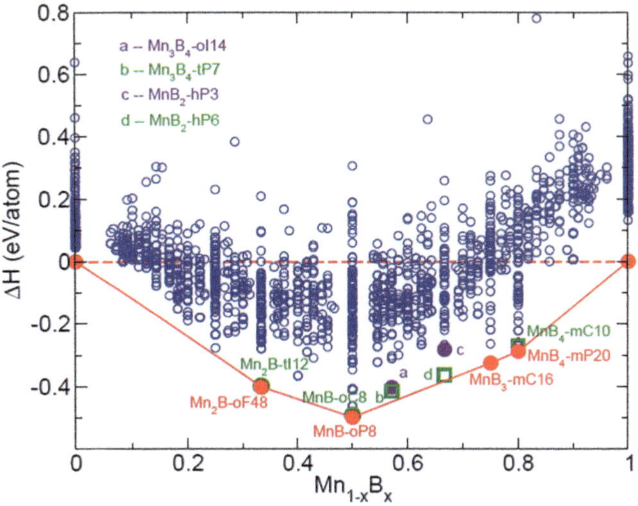

Fig. 4 Convex hull of the Mn–B system.[30] Solid red circles indicate stable phases, which are joined by lines forming the convex hull. Open green squares are experimentally known compounds Mn_3B_4 and MnB_2, which are calculated to be metastable. Reproduced from ref. 30 with permission from the PCCP Owner Societies.

systems can be studied the same way. This formalism can be applied to molecular co-crystals; in this case, one has to consider the free energy of formation normalized per molecule (rather than per atom). As there are many possible stoichiometries, doing CSP for all of them is probably unaffordable. One can limit consideration to the most common cases (*e.g.* AB, AB_2, AB_3, AB_4, and A_2B_3 stoichiometries), but this is risky as many stable compounds can be overlooked. Evolutionary codes USPEX (http://uspex-team.org) or GASP (http://gasp.mse.ufl.edu/) can efficiently sample the entire compositional space in an automatic fashion and without any assumptions about stoichiometries, and locate all stable compounds. The advantage of evolutionary algorithms is that different sampled stoichiometries compete, exchange structural information (greatly speeding up the search), and evolve, producing new stoichiometries. This is much more efficient than sampling all possible compositions independently.

(II) Prediction of materials optimal in two or more properties – *i.e.* multi-objective optimization, or Pareto optimization. Without realizing it, we use Pareto optimization in our own decision making every day. Which car to buy – a perfect and expensive one, or the cheapest but not so good? – here we optimize price and performance. Which school to choose for our children? – here, we optimize the quality of education, cost, and duration of commute. The set of optimal solutions, known as the Pareto front (or first Pareto front), is made of all non-dominated solutions, *i.e.* those solutions that cannot be beaten by any other solution on all target properties at the same time.

Materials scientists also usually want to find materials satisfying several criteria: for example, for many applications one needs materials with the highest possible hardness and fracture toughness. It is generally wise to have stability as one of the optimized properties: nobody wants to predict materials that are so unstable that they cannot be synthesized. For example, we searched[32] for polymorphs of Bi_2Te_3 with the highest thermoelectric figure of merit ZT, and found that maximizing ZT alone leads to crazy structures, which are extremely unstable. Pareto optimization, searching simultaneously for high ZT and low energy, on the other hand, leads to reasonable results: it correctly finds the stable $R\bar{3}m$ polymorph of Bi_2Te_3 (a known good thermoelectric) and locates a number of low-energy polymorphs with very high ZT. This result shows not only the power of Pareto optimization, but also the promise of searching for new thermoelectrics: a 2–3-fold increase of ZT, compared to the current best materials, seems absolutely possible (Fig. 5).

(III) Searching for the highest-performance material(s) among all possible compounds of all elements. With 118 elements in Mendeleev's Periodic Table, we have 7021 binary, 273 937 ternary, and many more quaternary, quinary *etc.* systems, in each of which many stable and metastable compounds are possible. Doing so many CSP runs is impractical. Yet, it is possible to optimize the desired properties and predict the best-performing material(s) among all possible compounds. A coevolutionary method for doing this, called Mendelevian Search, was recently developed.[31] Here, the key is to arrange the chemical space in such a way that neighboring points are chemically similar (*e.g.* Na–Cl and K–Cl systems are similar) and have similar properties: then, the target property has a benign landscape that can be globally optimized. We[31] used this technique to simultaneously optimize, at 0 K, the hardness and stability, and magnetization and stability. In the former case we found diamond and lonsdaleite to be the hardest possible materials, but as these are metastable, a number of other materials also appeared on the Pareto front. In the latter case pure iron in its stable bcc structure was the only material on the first Pareto front, but a number of interesting magnetic materials appeared on higher-order Pareto fronts. Regarding the search

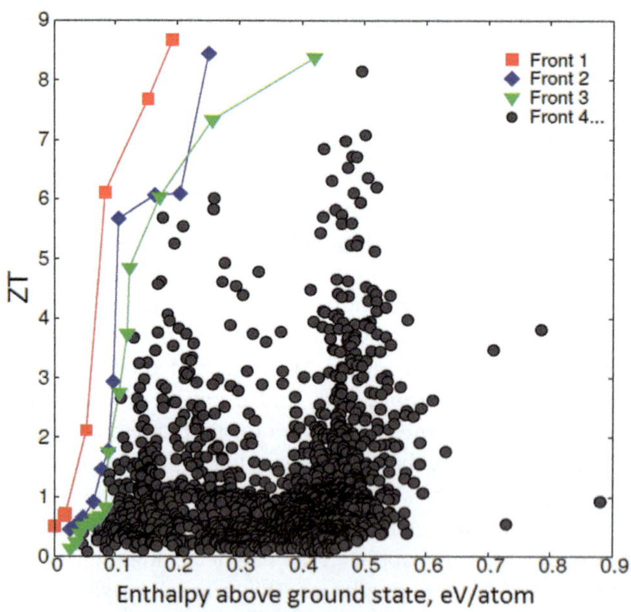

Fig. 5 Pareto optimization of thermoelectric figure of merit ZT and stability. Different Pareto fronts are shown. Reproduced from ref. 32 with permission from Elsevier.

for novel hard materials, two important stories spun off this calculation. In one of these studies,[33] we explored Cr–B, Cr–C, and Cr–N systems, found that the only superhard compound in these systems is CrB_4, and predicted a new stable chromium carbide, Cr_2C, which experimentalists should be able to synthesize. In the other study,[34] we explored the W–B system and predicted a new outstanding and hitherto unknown superhard compound WB_5 to be stable at normal conditions (Fig. 6); subsequently it was synthesized by V. Filonenko and V. Brazhkin and we have filed a patent on this material. This material has extraordinary hardness and fracture toughness, making it likely to replace WC in drilling equipment, machining tools and other applications.

Predicting an array of materials by just one calculation sounds like a dream, and although it is a reality, there are significant limitations. First, there are limitations related to the dimensionality of the problem. Variable-composition CSP can easily be done only for (pseudo)binary systems, already for ternary systems it is a challenge (but still doable, see ref. 35), while for quaternary systems it is probably already out of reach. Likewise, Pareto optimization works well for up to 3–4 simultaneously optimized properties.

Then, whilst some properties can easily be computed, others cannot. In general, physical properties (which characterize a particular energy minimum, and usually are response functions) are easier than chemical properties (which characterize a process of changing state). Biochemical properties, characterizing the complex interaction of a given molecule with receptors, enzymes, ion channels, *etc.*, are the hardest: for example, we do not have a general theory of toxicity, and are unlikely to have one any time soon (because there are so many mechanisms of toxicity). Today, we routinely optimize many physical properties, but I have not seen any direct optimizations of chemical or biochemical properties.

Yet, even for physical properties, we still have much to desire. Some properties are calculable, but at a great computational cost: *e.g.*, the critical temperature T_c of conventional (phonon-mediated) superconductors, or thermal conductivity. In principle, their optimization could be done, but it is just too expensive computationally. There are properties (*e.g.*, the viscosity of solids, or T_c of non-conventional superconductors) which cannot be computed at all at the

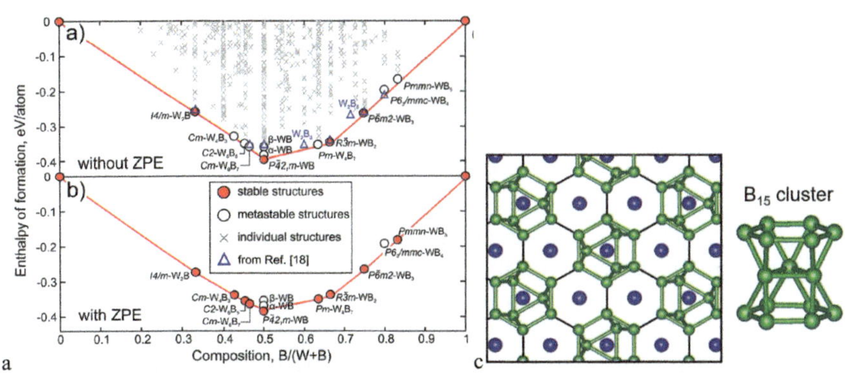

Fig. 6 (a and b) Convex hull of the W–B system with and without account for zero-point energy (ZPE), and (c) crystal structure of superhard WB_5. Interestingly, WB_5 is stabilized by ZPE. Adapted with permission from ref. 34. Copyright (2018) American Chemical Society.

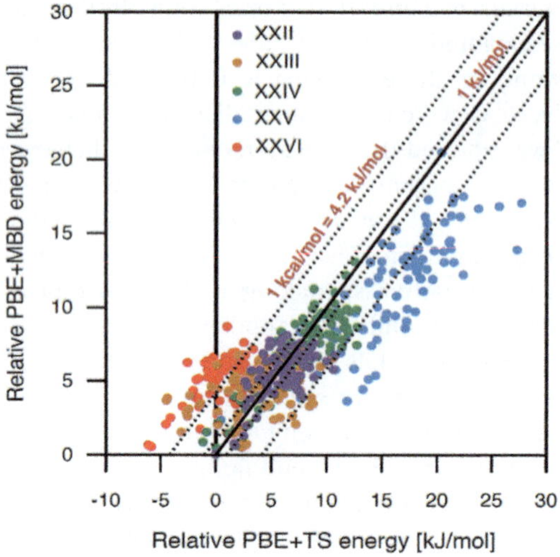

Fig. 7 Comparison of the energies obtained using two approximations for different structures of several blind test cases (PBE + TS and PBE + MBD). Reproduced from Hoja and Tkatchenko, DOI: 10.1039/c8fd00066b with permission from the Royal Society of Chemistry.

moment. In such cases, machine learning models of "difficult" properties may be the best way forward.

Even for the most basic property – the energy – for some systems the accuracy of existing approximations is sorely insufficient: this has been most clearly demonstrated by Tan *et al.* (DOI: 10.1039/c8fd00039e) for ROY (which stands for "red, orange, yellow", the colors of some of its numerous polymorphs – this is the compound with the largest number of coexisting polymorphs, ten). Comparing energies obtained by different modern approximations for different blind test molecules, Hoja and Tkatchenko (DOI: 10.1039/c8fd00066b) found a rather poor correlation, indicating that even for the energy the development of better approximations must continue (Fig. 7).

Discovering new chemical phenomena with CSP

A lot of new phenomena and surprising new compounds have been predicted with CSP (and many have already been experimentally verified). I will mention just a few such discoveries in the field of high-pressure chemistry to illustrate the ability of CSP to discover new knowledge.

In 2009, we published the prediction and experimental synthesis of a transparent insulating high-pressure allotrope of sodium.[36] This was at first surprising: first, because sodium is known as an "alkali metal", but at pressures above ~190 GPa, as it turns out, it is not a metal at all. Second, this was surprising because under pressure one expects closing of the band gap, rather than its opening. However, not long before this discovery, Ashcroft published[37] a model explaining this phenomenon (yes, the explanation appeared before the discovery!), based on the (unexpected to the classical chemist!) importance of core electrons in strongly compressed matter. This example clearly shows a dramatic change of the physics and chemistry of the elements under pressure.

Another "simple" element, helium, long believed to be chemically inert – and for seemingly good reasons of having a closed-shell structure, record-high ionization potential and zero electron affinity – was predicted and experimentally confirmed to form a new stable compound Na_2He (ref. 38). This compound, stable from ~110 GPa to at least 1000 GPa, unlike other pressure-stabilized compounds of helium, is not an inclusion compound, because helium fundamentally changes its properties and makes it insulating. Both transparent sodium and Na_2He are electrides, *i.e.* ionic compounds where Na^+ cores are cations and interstitially localized electron pairs $(2e)^{2-}$ are anions. Another helium compound, Na_2HeO, was predicted to be stable at pressures from just 14 GPa (ref. 38), and then a number of other helium compounds were predicted as well.[39] Thanks to CSP, helium chemistry has suddenly become an active field.

It was also shown that completely new and unexpected stable stoichiometries become stable under pressure. For example, while only NaCl is stable at normal conditions (and we all thought that other stoichiometries are forbidden by the "charge neutrality rule"), nature outsmarted our rules, and at pressures amenable to experiment (starting from 22 GPa) new compounds emerge as stable – Na_3Cl, Na_2Cl, Na_3Cl_2, Na_4Cl_4, $NaCl_3$, and $NaCl_7$ (ref. 18 and 40), see Fig. 8. For two compounds, Na_3Cl and $NaCl_3$, experimental synthesis was attempted and led to their successful synthesis.[18] Most of these newly predicted sodium chlorides are metallic, despite the large electronegativity difference that we would expect to dictate ionic bonding and make Na^+Cl^- the only allowed compound. It remains to

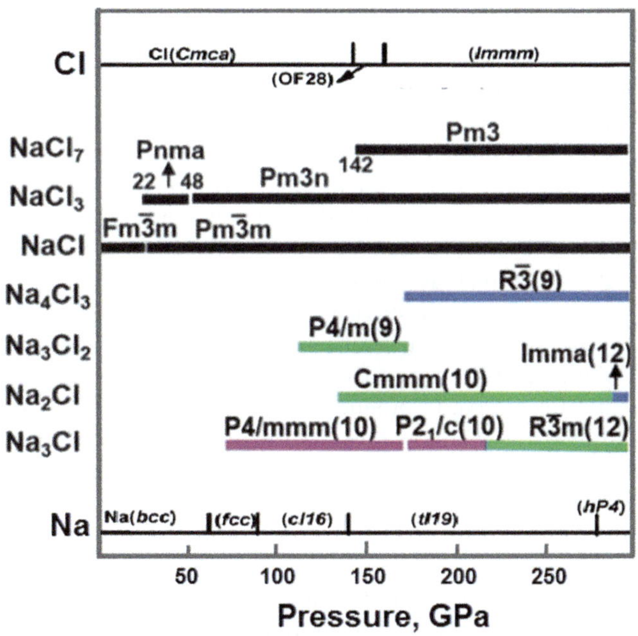

Fig. 8 Pressure-composition phase diagram of the Na–Cl system. Reproduced from ref. 40 with permission from the PCCP Owner Societies.

be explained why such compounds are formed, and whether any rules capable of explaining and predicting their stoichiometries can be formulated.

Such unexpected compounds become stable not only in the Na–Cl system, but probably in all chemical systems under sufficient pressure, and one wonders what role they may play in the interiors of the Earth and other planets, with their typically high pressures. For example, in the Si–O system, at planetary pressures two new stable oxides appear, SiO and SiO_3 (ref. 35), and in the Mg–O system, MgO_2, MgO_3 and Mg_3O_2 become stable.[35,41] Then, while we traditionally thought that ~10% of our planet is comprised by magnesium oxide (Mg,Fe)O – are we really sure that it's not $(Mg,Fe)O_2$? A new compound FeO_2 was recently predicted by CSP and then synthesized by Dave Mao's group,[42] and proposed to play an important role in the structure and evolution of our planet.

Such exotic compounds are bound to have exotic properties. It was predicted[43] that H_2S becomes unstable under pressure and breaks down, forming H_3S, and that compound was predicted to possess extremely high-T_c superconductivity with $T_c \sim 200$ K. One year later, an independent experimental work, confirming this startling prediction and $T_c = 203$ K, appeared.[44] Now there are several predictions of even higher-T_c superconductors, including even room-temperature superconductivity in YH_{10} and LaH_{10} (ref. 45 and 46). LaH_{10} has already been synthesized,[47] and it remains to be seen whether it is indeed a room-temperature superconductor.

Which phases are synthesizable?

So, here we are, a whole community (and a quickly growing one!), predicting numerous new phases, some stable or low-energy metastable (some already known and some not yet known from experiment), and some expected to have very interesting properties. How to synthesize these new phases? And generally, what makes a phase synthesizable? Is the number of phases that can, in principle, be synthesized small or large? Is it even a countable number?

Clearly, some phases are easily obtained, while others (sometimes lower in energy) are very difficult or seemingly even impossible to make. Why? Why is the extremely complex and pretty high in energy Hittorf's phosphorus (see, *e.g.*, Deringer *et al.*, DOI: 10.1039/c8fd00034d in this volume) synthesizable and other, lower-energy structures are seemingly not? Are there any selection rules (ideally based on crystal structure) that would tell us whether the phase is synthesizable at all? These are extremely important questions that have not been answered yet.

Directly related to these questions is the paradox of "disappearing polymorphs",[48] where instead of the easily obtained metastable polymorph (*e.g.* of a pharmaceutical compound), produced for a long time, the stable polymorph suddenly appeared and made synthesis of the metastable polymorph essentially impossible (thus, the old polymorph "disappeared"). This paradox occurs in materials where the ground state is difficult to reach, and was addressed by Neumann and van de Streek (DOI: 10.1039/c8fd00069g). Making the assumption that for rigid molecules the ground state is easily formed in ~100% cases, they estimated that for molecules with the complexity (primarily meaning conformational flexibility) typical of drugs, the ground state has been reached in 55–85% cases (probably because in the solution, from which crystals grew, one has molecules in the "wrong" conformation). Perhaps the percentage of cases where the ground state is reached decreases exponentially with increasing the number of flexible angles. If so, in proteins, the iconic extreme case of conformational flexibility, the ground state is reached in about 0% cases. I do not know (and I think no one really does) whether the conformation of proteins in the living cell, in the solution or in the crystal is really the lowest-energy one. Anfinsen[49] hypothesized that structures of proteins are thermodynamically controlled, but the above considerations make us doubt it.

Synthesizability is related to crystal growth, and it seems to me that we poorly understand the process. It seems natural for crystals to grow by annexing not single atoms or molecules, but entire clusters of atoms or molecules. Using this idea, Anderson *et al.*[50] simulated the growth of zeolite crystals from their secondary building units in excellent agreement with experiment. But even this is a simplified picture of crystal growth. This is illustrated by the mystery of quasicrystal growth: upon hearing that quasicrystals, with structures described by his Penrose tilings, have been discovered, Sir Roger Penrose reportedly exclaimed that he thought this to be impossible because growing a perfect Penrose tiling by just following local rules of assembly is impossible, yet essentially perfect quasicrystals have been grown in experiment (see ref. 51). If we do not understand how quasicrystals grow, we also do not understand the growth of crystals: the mechanism should be similar, as atoms do not know whether they participate in a crystal or quasicrystal structure. The ease of assembling the structure should be an important criterion of its synthesizability. This means that there should exist structure-based selection rules determining the synthesizability of a structure.

Conclusions

In these Concluding Remarks I have covered a number of issues related to CSP and connecting it to other important fields – materials design, drug design, theoretical chemistry and chemical bonding, crystal growth, and planetary sciences. Essentially all fields of science related to the atomic structure of matter benefit from advances in CSP, and CSP can help in resolving many puzzles in these fields. CSP is a new powerful research tool that is already now making great contributions in all these fields, and more is to come.

Conflicts of interest

There are no conflicts to declare.

Acknowledgements

I thank the Russian Science Foundation (grant 16-13-10459) for generously supporting my research.

References

1 W. H. Bragg and W. L. Bragg, The structure of the diamond, *Nature*, 1913, **91**, 557.

2 W. L. Bragg, The structure of some crystals as indicated by their diffraction of X-rays, *Proc. R. Soc. London, Ser. A*, 1913, **89**, 248–277.

3 A. R. Oganov and C. W. Glass, Crystal structure prediction using *ab initio* evolutionary techniques: principles and applications, *J. Chem. Phys.*, 2006, **124**, 244704.

4 *Computational Materials Discovery*, ed. A. R. Oganov, G. Saleh and A. G. Kvashnin, Royal Society of Chemistry, 2018, ISBN: 978-1-78262-961-0.

5 A. M. Reilly, *et al.*, Report on the sixth blind test of organic crystal-structure prediction methods, *Acta Crystallogr., Sect. B: Struct. Sci., Cryst. Eng. Mater.*, 2016, **72**(4), 439–459.

6 Q. Zhu, *et al.*, Resorcinol Crystallization from the Melt: A New Ambient Phase and New "Riddles", *J. Am. Chem. Soc.*, 2016, **138**, 4881–4889.

7 A. G. Shtukenberg, *et al.*, Powder diffraction and crystal structure prediction identify four new coumarin polymorphs, *Chem. Sci.*, 2017, **8**, 4926–4940.

8 A. R. Oganov, J. C. Schön, M. Jansen, S. M. Woodley, W. W. Tipton and R. G. Hennig. First blind test of inorganic crystal structure prediction, in *Modern Methods of Crystal Structure Prediction*, ed. A. R. Oganov, Wiley-VCH, Berlin, 2010, pp. 223–231.

9 J. W. Sun, A. Ruzsinsszky and J. P. Perdew, Strongly constrained and appropriately normed semilocal density functional, *Phys. Rev. Lett.*, 2015, **115**, 036402.

10 H. W. Peng, Z. H. Yang, J. P. Perdew and J. W. Sun, Versatile van der Waals density functional based on a meta-generalized gradient approximation, *Phys. Rev. X*, 2016, **6**, 041005.

11 J. P. Perdew, K. Burke and M. Ernzerhof, Generalized gradient approximation made simple, *Phys. Rev. Lett.*, 1996, **77**, 3865–3868.

12 C. Adamo and V. Barone, Toward reliable density functional methods without adjustable parameters: The PBE0 model, *J. Chem. Phys.*, 1999, **110**, 6158–6170.

13 A. Tkatchenko, R. A. DiStasio, R. Car and M. Scheffler, Accurate and efficient method for many-body van der Waals interactions, *Phys. Rev. Lett.*, 2012, **108**, 236402.

14 M. P. Allen and D. J. Tildesley, *Computer Simulation of Liquids*, Oxford University Press, 2017.

15 L. B. Partay, A. P. Bartok and G. Csanyi, Efficient sampling of atomic configurational space, *J. Phys. Chem. B*, 2010, **114**, 10502–10512.

16 L. B. Partay, On the performance of interatomic potential models of iron: Comparison of the phase diagrams, *Comput. Mater. Sci.*, 2018, **149**, 153–157.

17 V. Stevanovic, Sampling polymorphs of ionic solids using random superlattices, *Phys. Rev. Lett.*, 2016, **116**, 075503.

18 W. W. Zhang, A. R. Oganov, A. F. Goncharov, Q. Zhu, S. E. Boulfelfel, A. O. Lyakhov, M. Somayazulu, V. B. Prakapenka and Z. Konopkova, Unexpected stoichiometries of stable sodium chlorides, *Science*, 2013, **342**, 1502–1505.

19 S. Matsuishi, Y. Toda, M. Miyakawa, K. Hayashi, T. Kamiya, M. Hirano, I. Tanaka and H. Hosono, High-density electron anions in a nanoporous single crystal: $[Ca_{24}Al_{28}O_{64}]^{(4+)}(4e^{-})$, *Science*, 2003, **301**, 626–629.

20 R. Martonak, A. Laio and M. Parrinello, Predicting crystal structures: The Parrinello-Rahman method revisited, *Phys. Rev. Lett.*, 2003, **90**, 075503.

21 R. Martoňák, D. Donadio, A. R. Oganov and M. Parrinello, Crystal structure transformations in SiO_2 from classical and *ab initio* metadynamics, *Nat. Mater.*, 2006, **5**, 623–626.

22 Q. Zhu, A. R. Oganov and A. O. Lyakhov, Evolutionary metadynamics: a novel method to predict crystal structures, *CrystEngComm*, 2012, **14**, 3596–3601.

23 Q. Zhu, A. R. Oganov, A. O. Lyakhov and X. X. Yu, Generalized evolutionary metadynamics for sampling energy landscapes and its applications, *Phys. Rev. B: Condens. Matter Mater. Phys.*, 2015, **92**, 024106.

24 Z. D. Zeng, Q. F. Zeng, N. Liu, A. R. Oganov, Q. S. Zeng, Y. Cui and W. L. Mao, A new phase of $Li_{15}Si_4$ synthesized under pressure, *Adv. Energy Mater.*, 2015, **5**, 1500214.

25 P. Piaggi and M. Parrinello, Predicting polymorphism in molecular crystals using orientational entropy, 2018, arXiv:1806.06006.

26 E. V. Podryabinkin, E. V. Tikhonov, A. V. Shapeev and A. R. Oganov, Accelerating crystal structure prediction by machine-learning interatomic potentials with active learning, 2018, arXiv:1802.07605.

27 P. E. Dolgirev, I. A. Kruglov and A. R. Oganov, Machine learning scheme for fast extraction of interatomic potentials and chemistry, *AIP Adv.*, 2016, **6**, 085318.

28 M. Feuerbacher, *et al.*, The Samson phase, beta-Mg_2Al_3, revisited, *Z. Kristallogr.*, 2007, **222**, 259–288.

29 D. Hobbs, J. Hafner and D. Spisak, Understanding the complex metallic element Mn. I. Crystalline and noncollinear magnetic structure of alpha-Mn, *Phys. Rev. B: Condens. Matter Mater. Phys.*, 2003, **68**, 014407.

30 H. Niu, X. Q. Chen, W. Ren, Q. Zhu, A. R. Oganov, D. Li and Y. Li, Variable-composition structure prediction and experimental verification of MnB_3 and MnB_4, *Phys. Chem. Chem. Phys.*, 2014, **16**, 15866–15873.

31 Z. Allahyari and A. R. Oganov, Coevolutionary search for optimal materials in the space of all possible compounds, 2018, arXiv:1807.00854.

32 M. Nunez-Valdez, Z. Allahyari and A. R. Oganov, Efficient technique for computational design of thermoelectric materials, *Comput. Phys. Commun.*, 2018, **222**, 152–157.

33 A. G. Kvashnin, A. R. Oganov and Z. Allahyari, Computational search for novel hard chromium-based materials, *J. Phys. Chem. Lett.*, 2017, **8**, 755–764.

34 A. G. Kvashnin, H. Zakaryan, C. M. Zhao, Y. F. Duan, Y. A. Kvashnina, C. W. Xie, H. F. Dong and A. R. Oganov, New tungsten borides, their stability and outstanding mechanical properties, *J. Phys. Chem. Lett.*, 2018, **9**, 3470–3477.

35 H. Y. Niu, A. R. Oganov, X. Q. Chen and D. Z. Li, Novel stable compounds in the Mg–Si–O system under exoplanet pressures and their implications in planetary science, *Sci. Rep.*, 2015, **5**, 18347.

36 Y. Ma, M. I. Eremets, A. R. Oganov, Y. Xie, I. Trojan, S. Medvedev, A. O. Lyakhov, M. Valle and V. Prakapenka, Transparent dense sodium, *Nature*, 2009, **458**, 182–185.

37 B. Rousseau and N. W. Ashcroft, Interstitial electronic localization, *Phys. Rev. Lett.*, 2008, **101**, 046407.

38 X. Dong, A. R. Oganov, A. F. Goncharov, E. Stavrou, S. Lobanov, G. Saleh, G. R. Qian, C. Zh Gatti, V. Deringer, R. Dronskowski, X.-F. Zhou, V. Prakapenka, Z. Konopkova, A. I. Popo Boldyrev and H. T. Wang, A stable compound of helium and sodium at high pressure, *Nat. Chem.*, 2017, **9**, 440–445.

39 Z. Liu, J. Botana, A. Hermann, S. Valdez, E. Zurek, D. D. Yan, H. Q. Lin and M. S. Miao, Reactivity of He with ionic compounds under high pressure, *Nat. Commun.*, 2018, **9**, 951.

40 G. Saleh and A. R. Oganov, Alkali subhalides: High-pressure stability and interplay between metallic and ionic bonds, *Phys. Chem. Chem. Phys.*, 2016, **18**, 2840–2849.

41 Q. Zhu, A. R. Oganov and A. O. Lyakhov, Novel stable compounds in the Mg–O system under high pressure, *Phys. Chem. Chem. Phys.*, 2013, **15**, 7696–7700.

42 Q. Hu, D. Y. Kim, W. Yang, L. Yang, Y. Meng, L. Zhang and H. K. Mao, FeO_2 and FeOOH under deep lower-mantle conditions and Earth's oxygen-hydrogen cycles, *Nature*, 2016, **534**, 241–244.

43 D. F. Duan, Y. X. Liu, F. B. Tian, D. Li, X. L. Huang, Z. L. Zhao, H. Y. Yu, B. B. Liu and T. Cui, Pressure-induced metallization of dense $(H_2S)_2H_2$ with high-T_c superconductivity, *Sci. Rep.*, 2014, **4**, 6968.

44 A. P. Drozdov, M. I. Eremets, I. A. Troyan, V. Ksenofontov and S. I. Shylin, Conventional superconductivity at 203 kelvin at high pressures in the sulfur hydride system, *Nature*, 2015, **525**, 73–76.

45 H. Y. Liu, I. I. Naumov, R. Hoffmann, N. W. Ashcroft and R. J. Hemley, Potential high-T_c superconducting lanthanum and yttrium hydrides at high pressure, *Proc. Natl. Acad. Sci. U. S. A.*, 2017, **114**, 6990–6995.

46 F. Peng, Y. Sun, C. J. Pickard, R. J. Needs, Q. Wu and Y. M. Ma, Hydrogen clathrate structures in rare earth hydrides at high pressures: possible route to room-temperature superconductivity, *Phys. Rev. Lett.*, 2017, **119**, 107001.

47 Z. M. Geballe, H. Y. Liu, A. K. Mishra, M. Ahart, M. Somayazulu, Y. Meng, M. Baldini and R. J. Hemley, Synthesis and stability of lanthanum superhydrides, *Angew. Chem.*, 2018, **57**, 688–692.

48 J. D. Dunitz and J. Bernstein, Disappearing polymorphs, *Acc. Chem. Res.*, 1995, **28**, 193–200.

49 C. B. Anfinsen, Principles that govern the folding of protein chains, *Science*, 1973, **181**, 223–230.

50 M. W. Anderson, J. T. Gebbie-Rayet, A. R. Hill, N. Farida, M. P. Attfield, P. Cubillas, V. A. Blatov, D. M. Proserpio, D. Akporiaye, B. Arstad and J. D. Gale, Predicting crystal growth *via* a unified kinetic three-dimensional partition model, *Nature*, 2017, **544**, 456–459.

51 C. Janot, *Quasicrystals: A Primer*, Oxford University Press, 1994.

Poster titles

Accuracy of dispersion corrected DFT calculations for evaluating intermolecular interactions, **S. Tsuzuki,** *National Institute of Advanced Industrial Science and Technology, Japan*

Prediction of topological preferences and energetic properties of metal azolate frameworks, **M. Arhangelskis, A. D. Katsenis, A. J. Morris, T. Friščić,** *McGill University, Canada*

Hypothetical MOF structures from the entire periodic table: AuToGraFS + UFF4MOF, **M. Addicoat,** *Nottingham Trent University, UK*

Use of solid-state DFT for crystal structure prediction: a study of glycine polymorphs, **T. Dierks, T. Korter,** *Syracuse University, USA*

Mapping crystal structure landscapes to different levels of theory by predicting terms in the many-body expansion, **D. McDonagh, C.-K. Skylaris, G. M. Day,** *University of Southampton, UK*

Bridge: An *ab initio* random structure searching approach to the discovery of experimentally accessible zeolite frameworks, **D. McKay, R. F. Moran, R. E. Morris, S. E. Ashbrook,** *University of St. Andrews, UK*

The importance of thermal expansion in the simulation of molecular crystals, **J. McKinley, G. Beran,** *University of California Riverside, USA*

Prediction and characterization of $NaGaS_2$, a high thermal conductivity mid-infrared nonlinear optical material, **D. Hou, Z. Yang,** *Xinjiang Technical Institute of Physics & Chemistry, Chinese Academy of Sciences, China*; and *Charles University, Czech Republic*

Explaining crystallization preferences of two polyphenolic diastereoisomers by crystal structure prediction, **M. Dudek, M. Potrzebowski, G. M. Day,** *University of Southampton, UK*

Nested sampling for exploring potential energy landscapes, **L. Bartok-Partay, R. Baldock, A. Bartok-Partay, N. Bernstein, G. Csanyi,** *University of Reading, UK*

Predicting polymorphism at finite temperature using metadynamics, **P. Piaggi, M. Parrinello,** *Ecole Polytechnique Fédérale de Lausanne, Switzerland*

ChemDASH: Chemically Directed Atom Swap Hopping for crystal structure prediction, **P. Sharp, M. Dyer, G. Darling, M. Rosseinsky,** *University of Liverpool, UK*

Composite and low-cost approaches for molecular crystal structure prediction, **L. LeBlanc, A. Otero-de-la-Roza, E. Johnson,** *Dalhousie University, Canada*

Crystal structure prediction of novel carbon allotropes and zinc oxide polymorphs by random sampling, **S. A. Jobbins, I. A. Baburin, S. Leoni,** *Cardiff University, UK*

Crystal structure prediction: a pharmaceutical perspective upon when and how to use it? **J. van den Ende, U. Schwitter, F. Stowasser, R. Moog, J. Thun, A. Alker, P. Hidber,** *F. Hoffmann-La Roche Ltd,Switzerland*

Thermal behaviour of lead bromide perovskite single crystal, **H. Alsalmah, I. Salaoru, S. Paul,** *De Montfort University, UK*

Reliable and practical computational prediction of molecular crystal polymorphs, **J. Hoja, H.-Y. Ko, M. A. Neumann, R. Car, R. A. DiStasio Jr., A. Tkatchenko,** *University of Luxembourg, Luxembourg*

Evaluating the energetic driving force for co-crystal formation, **C. Taylor, G. M. Day,** *University of Southampton, UK*

Predicting the polymorphs of chlorothalonil: assessment of intermolecular energy contributions, **D. Bowskill, I. Sugden, C. S. Adjiman, C. C. Pantelides,** *Imperial College London, UK*

Free energy calculations of crystalline solids, **S. Konstantinopoulos, I. Sugden, C. Adjiman, C. Pantelides,** *Imperial College London, UK*

Need for crystal structure prediction for organic semiconductors, **T. Hasegawa,** *Department of Applied Physics, The University of Tokyo, Japan*

Applying fast, accurate lattice energies for molecular crystal structure prediction using CrystalExplorer model energies, **P. Spackman, G. M. Day,** *University of Southampton, UK*

Screening for new chiral organic semiconducting molecular materials using crystal structure prediction, **J. Schmidt, J. Nelson, K. E. Jelfs,** *Imperial College London, UK*

Development of a transferable force-field for rapid crystal structure predictions, **S. Nilsson Lill, A. Broo,** *IMED Biotech Unit, AstraZeneca Gothenburg, Sweden*

Relative stabilities from experimental data and thermodynamics, **I. Rietveld, B. Nicolaï, M. Barrio, P. Lloveras, J. Van de Streek, A. Polian, J.-P. Itié, J.-L. Tamarit,** *University of Rouen, France*

Structural determination of the prototype drug LASSBio-1366, **I. Bastos, F. Costa, A. Kümmerle, E. Barreiro, D. Braz, T. Silva, F. Ferreira, R. Barroso,** *Universidade do Estado do Rio de Janeiro, Brazil*

Improved efficiency in the *ab initio* generation of crystal structures, **I. Sugden, C. Adjiman, C. Pantelides,** *Imperial College London, UK*

The Faraday Discussions Poster Prize for the best poster was jointly awarded to Mr Johannes Hoja of the University of Luxembourg, for his poster on the reliable and practical computational prediction of molecular crystal polymorphs, and to Mr Pablo Piaggi of the Ecole Polytechnique Fédérale de Lausanne, Switzerland, for his poster on predicting polymorphism in molecular crystals using orientational entropy.

List of participants

Dr Matthew Addicoat, *Nottingham Trent University, United Kingdom*
Professor Claire S. Adjiman, *Imperial College London, United Kingdom*
Professor Peter Ajibade, *University of KwaZulu-Natal, South Africa*
Mrs Hessa Alsalmah, *De Montfort University, United Kingdom*
Mr Ahmad A. Alwan, *Khalifa University, Abu Dhabi, United Arab Emirates*
Dr Mihails Arhangelskis, *McGill University, Canada*
Dr Livia Bartok-Partay, *University of Reading, United Kingdom*
Professor Gregory J. O. Beran, *University of California, Riverside, U.S.A.*
Mr David Bowskill, *Imperial College London, United Kingdom*
Dr Jan Gerit Brandenburg, *University of Göttingen, Germany*
Dr Doris E. Braun, *Institute of Pharmacy, University of Innsbruck, Austria*
Dr Virginia Burger, *XtalPi, U.S.A.*
Dr Asbjoern Burow, *Avant-garde Materials Simulation Deutschland, Germany*
Professor Eleanor Campbell, *University of Edinburgh, United Kingdom*
Professor Tony Cheetham, *University of Cambridge, United Kingdom*
Dr Frederik Claeyssens, *University of Sheffield, United Kingdom*
Dr Jason Cole, *Cambridge Crystallographic Data Centre (CCDC), United Kingdom*
Dr Christopher Collins, *University of Liverpool, United Kingdom*
Professor Andrew I. Cooper, *University of Liverpool, United Kingdom*
Dr Aurora J. Cruz-Cabeza, *The University of Manchester, United Kingdom*
Dr George R. Darling, *University of Liverpool, United Kingdom*
Mr Daniel W. Davies, *University of Bath, United Kingdom*
Professor Graeme M. Day, *University of Southampton, United Kingdom*
Dr Volker L. Deringer, *University of Cambridge, United Kingdom*
Ms Teresa Dierks, *Syracuse University, U.S.A.*
Mr Victor Do Nascimento, *University of Southampton, United Kingdom*
Dr Marta Dudek, *University of Southampton, United Kingdom*
Dr Matthew S. Dyer, *University of Liverpool, United Kingdom*
Dr Clare Dyer-Smith, *Royal Society of Chemistry, United Kingdom*
Mrs Elsa Gheziel-Neumann, *Avant-garde Materials Simulation Deutschland, Germany*
Professor Hitoshi Goto, *Toyohashi University of Technology, Japan*
Dr Rui Guo, *University College London, United Kingdom*
Dr Alan Hare, *MRSC, without affiliation (retired), United Kingdom*
Professor Tatsuo Hasegawa, *Department of Applied Physics, University of Tokyo, Japan*
Dr Julian Helfferich, *Avant-garde Materials Simulation Deutschland, Germany*
Mr Johannes Hoja, *University of Luxembourg, Luxembourg*
Mr Dianwei Hou, *Charles University, Czech Republic*
Mr Luca Iuzzolino, *University College London, United Kingdom*
Dr Kim E. Jelfs, *Imperial College London, United Kingdom*
Dr Yide Alan Jiang, *XtalPi, U.S.A.*
Dr Samuel A. Jobbins, *Cardiff University, United Kingdom*

Mr Julian Keupp, *Ruhr-Universität Bochum, Germany*
Mr Stefanos Konstantinopoulos, *Imperial College London, United Kingdom*
Miss Johanna Kölbel, *University of Cambridge, United Kingdom*
Mr Luc LeBlanc, *Dalhousie University, Canada*
Mr Paul Lewis, *Royal Society of Chemistry, United Kingdom*
Professor Dr Yi Li, *Jilin University, China*
Miss Qi Li, *University of Cambridge, United Kingdom*
Professor Noa Marom, *Carnegie Mellon University, U.S.A.*
Mr David McDonagh, *University of Southampton, United Kingdom*
Dr David McKay, *University of St Andrews, United Kingdom*
Miss Jessica L. McKinley, *University of California, Riverside, U.S.A.*
Dr Caroline Mellot-Draznieks, *Collège de France, France*
Dr John B. O. Mitchell, *University of St Andrews, United Kingdom*
Dr Sharmarke Mohamed, *Khalifa University, Abu Dhabi, United Arab Emirates*
Dr Sridhar Neelamraju, *National Centre for Biological Sciences, India*
Dr Marcus A. Neumann, *Avant-garde Materials Simulation Deutschland, Germany*
Dr Sten Nilsson Lill, *IMED Biotech Unit, AstraZeneca, Sweden*
Dr Jonas Nyman, *Eli Lilly and Company, U.S.A. and University of Wisconsin - Madison, U.S.A.*
Professor Dr Artem R. Oganov, *Skolkovo Institute of Science and Technology, Russia*
Mr Zhongfu Pang, *University of Liverpool, United Kingdom*
Mr Pablo Piaggi, *Ecole Polytechnique Fédérale de Lausanne, Switzerland*
Professor Sarah L. Price, *University College London, United Kingdom*
Dr Louise Price, *University College London, United Kingdom*
Dr Angeles Pulido, *Cambridge Crystallographic Data Centre (CCDC), United Kingdom*
Dr Susan M. Reutzel-Edens, *Eli Lilly and Company, U.S.A.*
Dr Ivo Rietveld, *Université Paris Descartes, France*
Dr Michael T. Ruggiero, *University of Cambridge, United Kingdom*
Dr Luca Russo, *GlaxoSmithKline, United Kingdom*
Dr Matthew Ryder, *Oak Ridge National Laboratory, U.S.A.*
Dr German Sastre, *Instituto de Tecnologia Quimica UPV-CSIC, Spain*
Ms Julia Schmidt, *Imperial College London, United Kingdom*
Professor Dr J. Christian Schön, *Max Planck Institute for Solid State Research, Germany*
Mrs Ros Searle, *Royal Society of Chemistry, United Kingdom*
Dr Paul Sharp, *University of Liverpool, United Kingdom*
Dr Stephen Shevlin, *Nature Materials, United Kingdom*
Dr Peter Spackman, *University of Southampton, United Kingdom*
Dr Isaac Sugden, *Imperial College London, United Kingdom*
Dr Christopher Taylor, *University of Southampton, United Kingdom*
Ms Fiona Tscherny, *Royal Society of Chemistry, United Kingdom*
Dr Seiji Tsuzuki, *National Institute of Advanced Industrial Science and Technology (AIST), Japan*
Dr Joost van den Ende, *F. Hoffmann-La Roche, Basel, Switzerland*
Dr Susan Weatherby, *Royal Society of Chemistry, United Kingdom*
Dr Alexander Whiteside, *Royal Society of Chemistry, United Kingdom*
Dr Scott M. Woodley, *University College London, United Kingdom*
Mr Grahame R. Woollam, *Novartis, Basel, Switzerland*
Mr Shiyue Yang, *University of Southampton, United Kingdom*
Professor J. Axel Zeitler, *University of Cambridge, United Kingdom*
Professor Qiang Zhu, *University of Nevada, Las Vegas, U.S.A.*

